D0164310

DIFFERENTIAL EQUATIONS

DIFFERENTIAL EQUATIONS

C. Ray Wylie

William R. Kenan Jr., Professor of Mathematics
Furman University

McGraw-Hill Book Company

New York St. Louis San Francisco Auckland Bogotá Düsseldorf
Johannesburg London Madrid Mexico Montreal New Delhi
Panama Paris São Paulo Singapore Sydney Tokyo Toronto

DIFFERENTIAL EQUATIONS

1234567890 DODO 78321098

This book was set in Times Roman.
The editors were Carol Napier and Michael Gardner;
the cover was designed by Joseph Gillians;
the production supervisor was Leroy A. Young.
New drawings were done by Allyn-Mason, Inc.
R. R. Donnelley & Sons Company was printer and binder.

Library of Congress Cataloging in Publication Data

Wylie, Clarence Raymond, date
Differential equations.

Includes bibliographical references and index.
1. Differential equations. 2. Boundary value problems. I. Title.
QA371.W94 515'.35 78-2667
ISBN 0-07-072197-1

CONTENTS

PREFACE

The aim of this book is to provide, simultaneously, an introduction to the theory and techniques of differential equations and an introduction to the use of differential equations in science and engineering. It contains ample material for a two-semester course; and, by a proper selection of topics, it can also be used as a text for a one-semester course with either a moderately theoretical approach or an applied approach. Every effort has been made to keep the presentation detailed and clear, while at the same time maintaining a high level of precision and accuracy. Definitions and theorems are carefully stated; and, except in a few instances where an argument is deemed too difficult for an introductory text, all proofs are carried through in detail. Numerous completely worked examples and carefully drawn figures illustrate all major points. Applications are carefully formulated from the relevant physical principles and the significance of the results is discussed. Every chapter begins with an introductory paragraph which attempts to set the stage for the work to follow and to motivate that work as a natural development from the preceding material.

The book presupposes only a good background in calculus and a modest familiarity with some of the simpler properties of determinants and matrices. An appendix contains a review of much of this material, as well as a few necessary topics, such as the uniform convergence of series and integrals and Taylor's series in several variables, which a student may not have studied in calculus. As far as applications are concerned, the book assumes only a knowledge of such simple physical principles as are ordinarily encountered in a first course in calculus. Anything beyond this is introduced and carefully explained, as needed. A glossary of physical terms, laws, and units is also included in the Appendix.

The book begins with a chapter devoted to the solution of first-order differential equations. Chapter 2 deals in considerable detail with applications of first-order equations in biology, chemistry, physics, and geometry. Chapter 3 develops enough of the theory of linear differential equations to justify the usual solution procedures for linear equations with constant coefficients. Chapter 4 deals with linear differential equations with constant coefficients and with typical applications, particularly the linear oscillator. Chapter 5 deals with simultaneous linear differential equations with constant coefficients. Here matric notation is used extensively, but not exclusively, and only elementary properties of matrices are invoked. Chapters 4 and 5 contain no operational techniques, and the operator D is introduced only as a notational convenience. In contrast, Chapter 6 is devoted entirely to operational methods based on the Laplace transformation. The development here is in more detail than is usual in an introductory text in differential equations. This makes the book especially appropriate for a course taken by engineering students. Chapter 7 deals at a practical level with the analysis of linear mechanical and electrical systems of one or more degrees of freedom, emphasizing the mathematical identity of such systems and the construction of electrical models of mechanical systems. For students not primarily interested in engineering applications, Chapters 6 and 7 can be omitted without interrupting the overall continuity of the book. Chapter 8 is concerned with Fourier series. These will, of course, be used later in the chapter on partial differential equations, but the motivation here is the representation of periodic functions such as might occur as impressed forces or voltages in a physical system. Chapter 9 is devoted to the series solution of linear differential equations with variable coefficients and a fairly detailed discussion of Bessel functions and their elementary properties. In Chapter 10 the discussion of the theory of linear differential equations is resumed from Chapter 3 and extended to include such topics as the adjoint equation, separation theorems, orthogonality properties, and Green's functions. Chapter 11 discusses partial differential equations, beginning with a derivation of the wave equation and the heat equation from physical principles. The method of separation of variables is introduced and used, with relevant material from the earlier work on Fourier series and Bessel functions, to solve a variety of problems in heat flow and vibrations. Chapter 12 deals with the numerical solution of ordinary differential equations by both the Runge-Kutta method and Milne's method. Chapter 13 extends the work in finite differences that was required in Chapter 12 in connection with Milne's method and discusses linear difference equations with constant coefficients. Emphasis here is on the analogy between linear difference equations and linear differential equations. Finally, Chapter 14 is devoted to a discussion of the qualitative theory of differential equations and the use of the phase plane in investigating such descriptive properties as stability and periodicity.

A careful proof of the fundamental existence and uniqueness theorem of Picard should surely be included in any introductory text in differential equations. However, the question of when, or if, it should be taken up in class depends on the background, interests, and objectives of the students. Hence in this book, this theorem is presented in a self-contained appendix that can be introduced when desired, or omitted, at the discretion of the instructor.

The debt of authors to their teachers, students, and colleagues is always too great to be acknowledged adequately. Many persons have assisted me in the writing of this book, often without realizing it, and to all I express my gratitude. I must, however, acknowledge specifically the invaluable assistance of Dr. Louis C. Barrett of Montana State University, who reviewed the manuscript with great care and whose criticisms and suggestions contributed greatly to whatever merit this book may possess, and the devoted labors of my wife, Ellen, and my student, Debbie Burr, who shared with me the task of proofreading the manuscript in all its stages.

C. Ray Wylie

TO THE INSTRUCTOR

This book contains significantly more material than is found in most introductory texts on differential equations, and can easily serve as the basis of a one-year course. This means that for a one-semester course some selection must be made, and the book has been written so that this can be done conveniently with little or no loss of continuity. The following suggestions may be helpful in planning one-semester courses for students with various interests and objectives.

Common core for all options:

Chap. 1: Secs. 1.1 to 1.6
Chap. 2: A sampling of applications
Chap. 3
Chap. 4: Secs. 4.1 to 4.4
Chap. 5: Secs. 5.1 and 5.2
Chap. 12: Secs. 12.1 and 12.2

Additional topics for students majoring in mathematics:

Chap. 9: Secs. 9.1 to 9.3
Chap. 10: Secs. 10.1 to 10.4
Chap. 14
Appendix A

Additional topics for students majoring in chemistry and physics:

Chap. 4: Sec. 4.6
Chap. 8: Secs. 8.1 to 8.3
Chap. 9: Secs. 9.1 to 9.4
Chap. 10: Secs. 10.4 and 10.5
Chap. 11: Secs. 11.1 to 11.5

Additional topics for students majoring in engineering:

Chap. 4: Sec. 4.6
Chap. 6
Chap. 7
Chap. 8
Chap. 11: Secs. 11.1 to 11.5

C. Ray Wylie

CHAPTER
ONE

ORDINARY DIFFERENTIAL EQUATIONS OF THE FIRST ORDER

1.1 INTRODUCTION

An equation involving derivatives (or differentials) of one or more dependent variables $y_1, y_2, \ldots, y_j$ with respect to one or more independent variables $x_1, x_2, \ldots, x_k$ is called a **differential equation.** By a **solution** of such an equation over a region R we mean a set of suitably differentiable functions

$$y_1(x_1, x_2, \ldots, x_k), y_2(x_1, x_2, \ldots, x_k), \ldots, y_j(x_1, x_2, \ldots, x_k)$$

which, when they are substituted into the given equation, reduce it to an identity at all points of R. The study of the existence, nature, and determination of solutions of differential equations is of fundamental importance not only to the pure mathematician but also to anyone engaged in the mathematical analysis of natural phenomena.

In general, mathematicians consider it a triumph if they are able to prove that a given differential equation possesses a solution and if they can deduce a few of the more important properties of that solution. Applied scientists, on the other hand, are usually greatly disappointed if a specific expression for the solution cannot be exhibited. The usual compromise is to find some practical procedure by which the required solution can be approximated with satisfactory accuracy.

Not all differential equations are difficult enough to make this necessary, however, and there are several large and very important classes of equations for which solutions can readily be found. For instance, an equation such as

$$\frac{dy}{dx} = f(x)$$

is really a differential equation, and

$$y = \int f(x)\, dx + c$$

even when the integral cannot be evaluated in terms of elementary functions, is a solution for any constant c. More generally, the equation

$$\frac{d^n y}{dx^n} = g(x)$$

is a differential equation which can be solved by n successive integrations. Except in name, the process of integration is actually an example of a process for solving differential equations.

In this book we shall undertake an elementary discussion of the following four fundamental problems in the field of differential equations:

1. The identification and solution of the major types of differential equations for which exact solutions can be found
2. The development of procedures (usually numerical) for approximating the solutions of differential equations for which exact solutions cannot be found
3. The use of differential equations in the study of problems in such fields as biology, chemistry, engineering, and physics
4. The study of the existence, uniqueness, and properties of the solutions of differential equations, as implied by the equations themselves whether or not they can be solved exactly

The first two of these problems are essentially formal in character and involve for the most part only manipulative techniques from calculus. The third problem is concerned with the applications of differential equations. In addition to a working knowledge of how to solve such equations, it requires familiarity with various laws from elementary chemistry and physics, because it is through the use of such laws that "real-life" problems are rephrased as problems involving differential equations. However, since this book is not addressed primarily to students of applied science, we shall assume only the minimum knowledge of physical principles that one ordinarily acquires in calculus. Anything more than that we shall introduce with appropriate explanation as needed. The last problem is theoretical and leads quickly into areas of current research interest. If explored in depth, it requires much material from linear algebra, advanced calculus, real and complex analysis, and topology. Since we are not assuming these courses as prerequisites for our work, our investigation of the theory of differential equations must necessarily be limited. An adequate summary of the material from postcalculus mathematics which we will need is presented in Appendix B.

In this and the following three chapters, we shall consider differential equations which are next in difficulty after those which can be solved by direct integration. These fall into a number of easily recognized classes for each of which explicit methods of solution exist. They form only a very small part of the set of all differential equations, and yet with a knowledge of them a scientist is equipped to handle a great variety of important applications. To get so much for so little is indeed remarkable.

1.2 FUNDAMENTAL DEFINITIONS

If the derivatives which appear in a differential equation are total derivatives, the equation is called an **ordinary differential equation;** if partial derivatives occur, the equation is called a **partial differential equation.** The **order** of a differential equation is the order of the highest derivative which appears in the equation.

Example 1 The equation $x^2y'' + xy' + (x^2 - 4)y = 0$ is an *ordinary* differential equation of the *second* order connecting the dependent variable y with its first and second derivatives and with the independent variable x.

Example 2 The equation

$$\frac{\partial^4 u}{\partial x^4} + \frac{\partial^4 u}{\partial x^2\, \partial y^2} + \frac{\partial^4 u}{\partial y^4} = 0$$

is a *partial* differential equation of the *fourth* order.

At present we shall be concerned primarily with ordinary differential equations.

The plot of any solution of a differential equation is called an **integral curve** of the equation. If a solution is defined implicitly by an equation of the form $f(x, y) = 0$, we shall call the curve $f(x, y) = 0$ a **solution curve** of the differential equation even though in its entirety it may consist of the plots of several solutions.

Example 3 It is easy to verify that on the interval $-1 < x < 1$ the differential equation $yy' + x = 0$ is satisfied by each of the explicit functions

$$y = \sqrt{1 - x^2} \qquad \text{and} \qquad y = -\sqrt{1 - x^2}$$

and hence by each of the functions defined implicitly by the relation

$$x^2 + y^2 = 1$$

The circle $x^2 + y^2 = 1$ is thus a solution curve whose upper arc is the integral curve defined by the solution $y = \sqrt{1 - x^2}$ and whose lower arc is the integral curve defined by the solution $y = -\sqrt{1 - x^2}$.

An equation which is linear, i.e., of the first degree in the *dependent* variable and its derivatives, is called a **linear differential equation.** From this definition it follows that the most general (ordinary) linear differential equation of order n is of the form

$$p_0(x)y^{(n)} + p_1(x)y^{(n-1)} + \cdots + p_{n-1}(x)y' + p_n(x)y = r(x) \qquad p_0(x) \not\equiv 0 \quad (1)$$

If $r(x) \equiv 0$, Eq. (1) is said to be **homogeneous,** since all terms are alike in containing either the dependent variable or just one of its derivatives as a factor. If $r(x) \not\equiv 0$, Eq. (1) is said to be **nonhomogeneous,** since the term $r(x)$ is unlike the other terms in that it does not contain the dependent variable or any of its derivatives as a factor. A differential equation which is not linear, i.e., cannot be put in the form of Eq. (1), is said to be **nonlinear.** In general, linear equations are much easier to solve than nonlinear ones, and most elementary applications involve linear equations.

Example 4 The equation $y'' + 4xy' + 2y = \cos x$ is a *nonhomogeneous linear* equation of the *second* order. The presence of the terms xy' and $\cos x$ does not alter the fact that the equation is linear because, by definition, linearity is determined solely by the way the *dependent* variable y and its derivatives occur in the equation.

Example 5 The equation $y'' + 4yy' + 2y = \cos x$ is a *nonlinear* equation because of the occurrence of the product of y and one of its derivatives.

Example 6 The equation $y'' + \sin y = 0$ is *nonlinear* because of the presence of $\sin y$, which is a nonlinear function of y.

One particularly important property of linear differential equations is given by the following theorem.

Theorem 1 If y_1 and y_2 are two solutions of a homogeneous, linear differential equation, then for all values of the constants c_1 and c_2 the linear combination $y = c_1 y_1 + c_2 y_2$ is also a solution of the homogeneous equation.

PROOF Our proof is based on two properties of derivatives which we learned in calculus:

1. The derivative of a sum is the sum of the derivatives.
2. The derivative of a constant times a function is the constant times the derivative of the function.

To determine if $y = c_1 y_1 + c_2 y_2$ is a solution of the homogeneous linear equation, we must substitute it into Eq. (1), with $r(x) \equiv 0$, and determine whether the equation is identically satisfied. Thus, using the properties of derivatives we have just recalled, we have successively

$$\begin{aligned}
&p_0(x)[c_1 y_1 + c_2 y_2]^{(n)} + p_1(x)[c_1 y_1 + c_2 y_2]^{(n-1)} + \cdots \\
&\qquad + p_{n-1}(x)[c_1 y_1 + c_2 y_2]' + p_n(x)[c_1 y_1 + c_2 y_2] = \\
&p_0(x)[c_1 y_1^{(n)} + c_2 y_2^{(n)}] + p_1(x)[c_1 y_1^{(n-1)} + c_2 y_2^{(n-1)}] + \cdots \\
&\qquad + p_{n-1}(x)[c_1 y_1' + c_2 y_2'] + p_n(x)[c_1 y_1 + c_2 y_2] = \\
&c_1[p_0(x)y_1^{(n)} + p_1(x)y_1^{(n-1)} + \cdots + p_{n-1}(x)y_1' + p_n(x)y_1] \\
&\qquad + c_2[p_0(x)y_2^{(n)} + p_1(x)y_2^{(n-1)} + \cdots + p_{n-1}(x)y_2' + p_n(x)y_2] = \\
&\qquad\qquad c_1 \cdot 0 + c_2 \cdot 0 = 0
\end{aligned}$$

because, by hypothesis, both y_1 and y_2 are solutions of the homogeneous linear equation and therefore the last two expressions in brackets vanish identically. Thus, no matter what the (constant) values of c_1 and c_2 may be, the expression $y = c_1 y_1 + c_2 y_2$ satisfies the given homogeneous, linear differential equation, and the theorem is proved.

By an immediate extension of the preceding argument, the following corollary can be established.

Corollary 1 If $y_1, y_2, \ldots, y_k$ are k solutions of a homogeneous, linear differential equation, then for all values of the constants $c_1, c_2, \ldots, c_k$ the linear combination

$$y = c_1 y_1 + c_2 y_2 + \cdots + c_k y_k$$

is also a solution of the homogeneous equation.

The significance of Theorem 1 and its corollary is that they assure us that if we have one or more specific solutions of a homogeneous, linear differential equation (no matter how they may have been found), we can immediately generate an infinite family of solutions by forming an arbitrary linear combination of those solutions. This property does not hold for nonlinear differential equations, and this is one of the reasons why, in general, nonlinear equations are much harder to work with than linear equations.†

Example 7 Verify that $y = c_1 e^{-x} + c_2 e^{2x}$ is a solution of the homogeneous linear equation $y'' - y' - 2y = 0$ for all values of the constants c_1 and c_2.

By differentiating y, substituting into the differential equation as indicated, and then collecting terms on c_1 and c_2, we obtain

$$\begin{aligned}(c_1 e^{-x} + 4c_2 e^{2x}) - (-c_1 e^{-x} + 2c_2 e^{2x}) - 2(c_1 e^{-x} + c_2 e^{2x})\\ &= c_1(e^{-x} + e^{-x} - 2e^{-x}) + c_2(4e^{2x} - 2e^{2x} - 2e^{2x})\\ &= c_1 \cdot 0 + c_2 \cdot 0 = 0\end{aligned}$$

regardless of the values of c_1 and c_2. In particular, since $y = c_1 e^{-x} + c_2 e^{2x}$ is a solution for all values of c_1 and c_2, we may first take $c_1 = 1$, $c_2 = 0$, getting $y = e^{-x}$, and then we may take $c_1 = 0$, $c_2 = 1$, getting $y = e^{2x}$. This shows that $y_1 = e^{-x}$ and $y_2 = e^{2x}$ are themselves solutions of the given equation.

Example 8 Show that although $y_1 = e^{-x}$ and $y_2 = e^{2x}$ are solutions of the equation $yy'' - (y')^2 = 0$, the expression $y = c_1 y_1 + c_2 y_2$ is a solution if and only if $c_1 = 0$ or $c_2 = 0$.

To verify that $y_1 = e^{-x}$ is a solution of the given equation, we have

$$(e^{-x})(e^{-x}) - (-e^{-x})^2 = e^{-2x} - e^{-2x} = 0$$

† This property does not hold for nonhomogeneous linear equations either, although this does not mean that these equations are as hard to work with as nonlinear equations.

Similarly, for $y_2 = e^{2x}$, we have

$$(e^{2x})(4e^{2x}) - (2e^{2x})^2 = 4e^{4x} - 4e^{4x} = 0$$

However, for $y = c_1 y_1 + c_2 y_2 = c_1 e^{-x} + c_2 e^{2x}$, we have

$$\begin{aligned}(c_1 e^{-x} + c_2 e^{2x})(c_1 e^{-x} + 4c_2 e^{2x}) - (-c_1 e^{-x} + 2c_2 e^{2x})^2 \\ &= (c_1^2 e^{-2x} + 5c_1 c_2 e^x + 4c_2^2 e^{4x}) - (c_1^2 e^{-2x} - 4c_1 c_2 e^x + 4c_2^2 e^{4x}) \\ &= 9c_1 c_2 e^x\end{aligned}$$

and this is not identically zero for all values of c_1 and c_2. In fact, it is zero if and only if either c_1 or c_2 is zero.

The reason why an arbitrary linear combination of solutions was always a solution for the equation in Example 7 and is not always a solution for the equation in this example is, of course, that the equation $y'' - y' - 2y = 0$ of Example 7 is linear, whereas the equation $yy'' - (y')^2 = 0$ of the present example is nonlinear.

As suggested by Theorem 1 and illustrated by the simple equation

$$\frac{dy}{dx} = 2x$$

and its solution

$$y = x^2 + c$$

differential equations usually possess solutions involving one or more arbitrary constants. A detailed treatment of the question of the maximum number of *essential* arbitrary constants that a solution of a differential equation may contain or even of what is meant by essential constants is quite difficult.† For our purposes, if an expression contains n arbitrary constants, we shall consider them essential if they cannot, through formal rearrangement of the expression, be replaced by any smaller number of constants. For example,

$$a \cos^2 x + b \sin^2 x + c \cos 2x \tag{2}$$

contains three arbitrary constants, a, b, and c. However, since

$$\cos 2x = \cos^2 x - \sin^2 x$$

the expression (2) can be written in the form

$$\begin{aligned}a \cos^2 x + b \sin^2 x + c(\cos^2 x - \sin^2 x) &= (a + c) \cos^2 x + (b - c) \sin^2 x \\ &= d \cos^2 x + e \sin^2 x\end{aligned}$$

where $d = a + c$ and $e = b - c$. On the other hand, since $\cos^2 x$ and $\sin^2 x$ are not proportional, it follows that there is no further rearrangement of the given expression that will permit d and e to be combined into, and replaced by, a single, new arbitrary constant. Hence d and e are essential.

† See, for instance, R. P. Agnew, "Differential Equations," 2d ed., McGraw-Hill, New York, 1960, pp. 103–105.

As our experience in solving differential equations grows, we will see that in general a differential equation of order n possesses solutions containing n essential arbitrary constants, or parameters, but none containing more. However, there are equations such as

$$\left|\frac{dy}{dx}\right| + |y| = 0$$

(which clearly has only the single solution $y = 0$) and

$$\left|\frac{dy}{dx}\right| + 1 = 0$$

(which has no solutions at all) which possess *no* solutions containing *any* arbitrary constants. Moreover, there are also simple differential equations which possess solutions containing more essential parameters than the order of the equation.

To construct such an example, we note first that for all values of the constant c the function $y = cx^2$ is a solution of the differential equation

$$xy' = 2y \tag{3}$$

Then we note that regardless of the value of c, each integral curve passes through the point (0, 0) with slope zero. This suggests that for any two values of c, say c_1 and c_2, the left half of the parabola $y = c_1 x^2$ can be paired with the right half of the parabola $y = c_2 x^2$ to give the graph of a function which for all x is continuous, has a continuous derivative, and satisfies the given differential equation. In other words, for all choices of the *two* essential constants c_1 and c_2, the rule

$$y = \begin{cases} c_1 x^2 & x \le 0 \\ c_2 x^2 & x > 0 \end{cases}$$

defines a function which is continuous and differentiable for all values of x and satisfies the equation $xy' = 2y$ over the entire x axis. Figure 1.1 shows these observations graphically. A still more striking example of this sort appears in Exercise 44 of this section, where a first-order equation with a solution containing infinitely many essential parameters is given.

Almost all applications of differential equations involve equations which possess solutions containing at least one arbitrary constant, and for such equations it is convenient to introduce the following definitions. A solution which contains at least one arbitrary constant is called a **general solution.** A solution obtained from a general solution by assigning particular values to the arbitrary constants which appear in it is called a **particular solution.** For first-order equations, particular solutions arise ordinarily from the requirement that an integral curve pass through a prescribed point (x_0, y_0) in the xy plane, that is, that the solution satisfy the condition that $y = y_0$ when $x = x_0$. For equations of higher order, particular solutions arise, similarly, from the requirement that for some specified value of x, the function y and its first $n - 1$ derivatives take on prescribed values $y_0, y_0', y_0'', \ldots, y_0^{(n-1)}$. The problem of determining solutions that satisfy

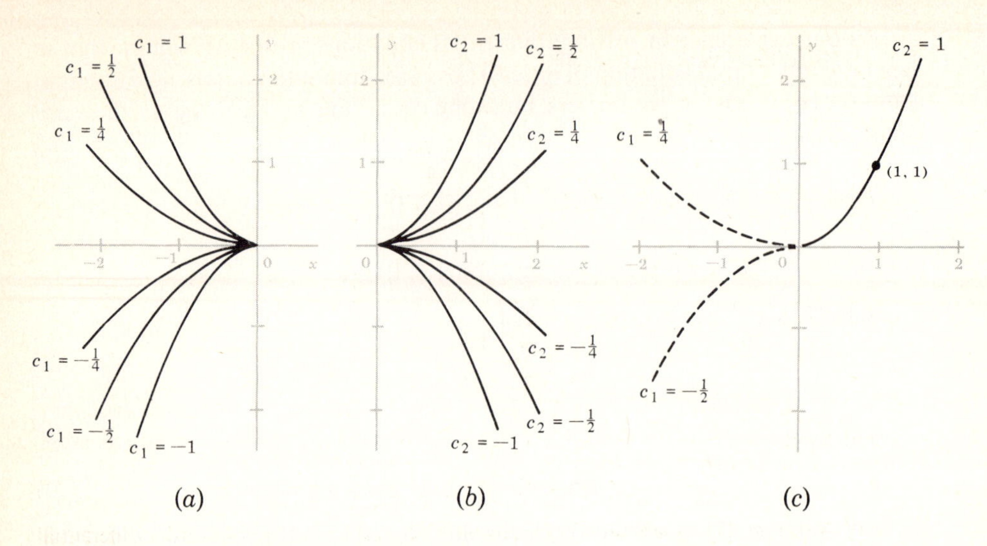

Figure 1.1 Arcs of different parabolas of the family $y = cx^2$ pieced together to give solutions of the differential equation $xy' = 2y$.

conditions of this nature is called an **initial-value problem.** Solutions which cannot be obtained from any general solution by assigning specific values to its arbitrary constants are called **singular solutions.** If a general solution has the property that *every* solution of the differential equation can be obtained from it by assigning suitable values to its arbitrary constants, it is said to be a **complete solution.** A general solution can thus be thought of as a description of some family of particular solutions, and a complete solution can be thought of as a description of the set of all solutions of the given equation.

It is important to note that we speak of *a* general solution and *a* complete solution of a differential equation and not of *the* general solution and *the* complete solution. If an equation has a general solution or a complete solution, it has many such solutions, and these may differ significantly in form. Moreover, in particular problems involving differential equations, the choice of which complete solution to use often has an important bearing on the ease with which the problem can be solved.

Occasionally it is necessary to determine a differential equation of order n which has a given function containing n arbitrary constants as a general solution. This can be done (at least theoretically) by differentiating the given expression n times and then eliminating the arbitrary constants by algebraic manipulation of the original equation and these derived equations.

Example 9 If a and b are arbitrary constants, find a second-order equation which has

$$y = ae^x + b \cos x \tag{4}$$

as a general solution.

By differentiating the given expression, we find

$$y' = ae^x - b \sin x \tag{5}$$

$$y'' = ae^x - b \cos x \tag{6}$$

Then, by adding and subtracting Eqs. (4) and (6), we obtain

$$a = \frac{y + y''}{2e^x} \qquad b = \frac{y - y''}{2 \cos x}$$

Substitution of these into (5) gives

$$y' = \frac{y + y''}{2e^x} e^x - \frac{y - y''}{2 \cos x} \sin x$$

and finally

$$(1 + \tan x)y'' - 2y' + (1 - \tan x)y = 0 \tag{7}$$

Although (7), except for its obvious multiples, is the only second-order differential equation having (4) as a general solution, it is by no means the only equation of which (4) is a general solution. For instance, if (6) is differentiated twice more, we obtain

$$y^{\text{iv}} = ae^x + b \cos x$$

and comparing this with (4), we see that the given function also satisfies the very simple equation

$$y^{\text{iv}} = y \tag{8}$$

Since Eq. (8) is of the fourth order, it presumably possesses general solutions containing four arbitrary constants, and it is easy to verify that

$$y = ae^x + b \cos x + ce^{-x} + d \sin x$$

does in fact satisfy Eq. (8) for all values of a, b, c, and d.

In the rest of this chapter, we shall learn to recognize and solve a number of types of first-order differential equations, including those which are

Separable
Homogeneous
Linear
Exact

Then in subsequent chapters we shall learn to solve linear equations of any order with constant coefficients and certain linear equations of the second order with variable coefficients.

As long as we work only with classes of equations for which we can actually find solutions, the general question of the existence of solutions of differential equations is irrelevant, and discussing it may seem to be belaboring the obvious. However, mathematicians, interested in learning as much as they can about all differential equations, must certainly consider this problem. And so must applied scientists, who today are often confronted with differential equations so complicated that they can be solved, if at all, only through the use of a computer. To ask a computer to solve a differential equation which has no solution is surely an exercise in futility, and the numbers produced by a computer working on such a problem are not only meaningless but actually dangerous if taken seriously.

We shall not interrupt our discussion of methods of solving particular types of differential equations to prove the so-called fundamental existence and uniqueness theorem for differential equations. A detailed proof is given in Appendix A, and every serious student should work through it carefully. It is appropriate, however, to state the theorem here and to indicate its significance.

The fundamental questions are these. Under what conditions does the initial-value problem

$$y' = f(x, y) \qquad y = y_0 \text{ when } x = x_0$$

have a solution? If a solution exists, for what values of x is it defined? And if there is a solution, is it unique? Answers to these questions are contained in the following theorem, originally credited to the French mathematician Émile Picard (1856–1941).

Theorem 2 Consider the differential equation $y' = f(x, y)$ with the initial condition $y(x_0) = y_0$, where both $f(x, y)$ and $f_y(x, y)$ are continuous on the closed rectangle R: $|x - x_0| \le a$, $|y - y_0| \le b$. If $|f(x, y)| \le M$ in R and if h is the smaller of the two numbers $(a, b/M)$, then there exists a unique solution of the initial-value problem $y' = f(x, y)$, $y(x_0) = y_0$, on the interval $x_0 - h < x < x_0 + h$.

By a relatively easy extension of this theorem, similar results can be obtained for differential equations of higher order. The idea here is to rewrite the given nth-order differential equation as a system of simultaneous first-order equations to each of which the reasoning used in the proof of the fundamental theorem can be applied. For example, given the second-order, initial-value problem $y'' = g(x, y, y')$, $y = y_0$, $y' = y'_0$ when $x = x_0$, we put $y' = z$, which implies that $y'' = z'$, yielding the equivalent problem

$$\begin{aligned} y' &= z \\ z' &= g(x, y, z) \end{aligned} \qquad y = y_0,\ z = y'_0 \qquad \text{when} \quad x = x_0$$

These two first-order equations can then be handled very much as the single equation appearing in Picard's theorem.

It is instructive to reconsider Eq. (3) in the light of Theorem 2. For this equation, we have $f(x, y) = 2y/x$, and clearly neither f nor f_y exists when $x = 0$. Hence over an interval containing $x = 0$, neither the existence nor the uniqueness of a solution of Eq. (3) is guaranteed by Theorem 2. Actually, as our earlier discussion pointed out, Eq. (3) does have solutions which are valid for all values of x. However, as Fig. 1.1c illustrates, over any interval which contains $x = 0$, the integral curve which passes through a given point (x_0, y_0), for example (1, 1), is not unique. On the other hand, over any interval which contains x_0 but does not contain $x = 0$, the integral curve which passes through a given point (x_0, y_0) is unique.

Exercises for Section 1.2

Describe each of the following equations, giving its order and telling whether it is ordinary or partial and linear or nonlinear.

1 $y'' + 3y' + 2y = x^4$

2 $y'' + (a + b\cos 2x)y = 0$

3 $y''' + 6y'' + 11y' + 6y = e^x$

4 $y^{\text{iv}} + xy'' + y^2 = 0$

5 $\dfrac{d(xy')}{dx} + xy = 0$

6 $(x + y)\,dy = (x - y)\,dx$

7 $a^2\dfrac{\partial^2 u}{\partial x^2} = \dfrac{\partial^2 u}{\partial t^2}$

8 $\dfrac{\partial^2(x^2\,\partial^2 u/\partial x^2)}{\partial x^2} = \dfrac{\partial^2 u}{\partial t^2}$

9 $\dfrac{\partial^2 u}{\partial x^2} = u\dfrac{\partial u}{\partial t}$

10 $\dfrac{\partial^2 u}{\partial x^2} + \dfrac{\partial^2 u}{\partial y^2} + \dfrac{\partial^2 u}{\partial z^2} = \phi(x, y, z)$

11 In Theorem 1, is it necessary that the coefficients c_1 and c_2 be constants and not functions of x?

Show that not all the constants which appear in the following expressions are essential, and in each case rearrange the expression so that all the constants which remain are essential.

12 Ae^{x+k}

13 $a + \ln bx$

14 $a \ln x^b$

15 $\dfrac{ax + b}{cx + d}$

16 $A\sin(x + b) + C\sin(x + d)$

17 $A[\cos(x + a) + \cos(x - a)]$

18 $a\cosh^2\theta + b\sinh^2\theta + c\cosh 2\theta$

19 $a\sin 3x + b\sin x + c\sin^3 x$

20 $\dfrac{A}{x + 1} + \dfrac{B}{x + 2} + \dfrac{C}{x^2 + 3x + 2}$

***21**† $a(x - 6y - 7) + b(3x + 4y + 5) + c(5x + 3y + 4)$

Verify that each of the following equations has the indicated solution for all values of the constants a and b.

22 $y'' + 4y = 0$ $\qquad y = a\cos 2x + b\sin 2x$

23 $y'' - 4y = 0$ $\qquad y = ae^{2x} + be^{-2x}$

24 $y'' + 3y' + 2y = 12e^{2x}$ $\qquad y = ae^{-x} + be^{-2x} + e^{2x}$

25 $y'' - 6y' + 9y = 0$ $\qquad y = ae^{3x} + bxe^{3x}$

26 $(\cos 2x)y' + (2\sin 2x)y = 2$ $\qquad y = a\cos 2x + \sin 2x$

27 $2xy\,dy = (y^2 - x)\,dx$ $\qquad y^2 = ax - x\ln x$

† Problems deemed significantly longer or more difficult than the others in a given set of exercises are indicated by one, or sometimes two, stars.

28 $y'' + (y')^2 + 1 = 0$ $\qquad y = \ln \cos (x - a) + b$

29 $\dfrac{\partial^2 u}{\partial x^2} = \dfrac{\partial u}{\partial t}$ $\qquad u = ae^{-9t} \cos (3x + b)$

30 $4\dfrac{\partial^2 u}{\partial x^2} = \dfrac{\partial^2 u}{\partial t^2}$ $\qquad u = af(x + 2t) + bg(x - 2t)$

If a and b are arbitrary constants, find a second-order differential equation of which each of the following expressions is a general solution.

31 $y = ae^{-2x} + be^x$ **32** $y = ae^{-2t} + bte^{-2t}$

33 $y = ae^{-t} + be^t$ **34** $y = 2ax + bx^2$

35 $y = a \cosh 2x + b \sinh 2x$ **36** $y = \sin (ax + b)$

37 Find a differential equation which has as a general solution the expression that defines the family of all parabolas which touch the x axis and have their axes vertical.

***38** Find a differential equation which has as a general solution the expression that defines the family of all lines which touch the parabola $2y = x^2$. Verify that the equation of the given parabola defines a function which is a singular solution of the required differential equation.

39 Verify that for all values of the arbitrary constants a and b, both $y_1 = a$ and $y_2 = bx^2$ satisfy each of the differential equations

$$xy'' = y' \qquad \text{and} \qquad 2yy'' = (y')^2$$

but that $y = a + bx^2$ will satisfy only the first of these equations. Explain.

40 Verify that for all values of the arbitrary constants a and b both $y_1 = a$ and $y_2 = b\sqrt{x}$ satisfy each of the differential equations

$$2xy'' + y' = 0 \qquad \text{and} \qquad 8x^3(y'')^2 - yy' = 0$$

but that $y = a + b\sqrt{x}$ will satisfy only the first of these equations. Explain.

41 Verify that for all values of the arbitrary constants a and b both $y_1 = a(x - 1)^2$ and $y_2 = b(x + 1)^2$ satisfy each of the differential equations

$$(x^2 - 1)y'' - 2xy' + 2y = 0 \qquad \text{and} \qquad 2yy'' - (y')^2 = 0$$

but that $y = a(x - 1)^2 + b(x + 1)^2$ will satisfy only the first of these equations. Explain.

42 Verify that for all values of the arbitrary constants c_1 and c_2 the differential equation $xy' = 2y + x$ is satisfied by the function

$$y = \begin{cases} c_1 x^2 - x & x \le 0 \\ c_2 x^2 - x & x > 0 \end{cases}$$

Explain.

***43** Verify that for all values of the arbitrary constants c_1, c_2, and c_3 the differential equation

$$(x^2 - 1)y' = 4xy$$

is satisfied by the function

$$y = \begin{cases} c_1(x^2 - 1)^2 & x < -1 \\ c_2(x^2 - 1)^2 & -1 \le x \le 1 \\ c_3(x^2 - 1)^2 & x > 1 \end{cases}$$

Explain.

***44** Verify that for all values of the arbitrary constants $\{c_n\}$ $(n = \ldots, -2, -1, 0, 1, 2, \ldots)$ the differential equation

$$(1 - \cos x)y' = (\sin x)y$$

is satisfied by the function

$$y = c_n(1 - \cos x) \qquad 2n\pi \le x < 2(n + 1)\pi$$

Explain.

1.3 SEPARABLE FIRST-ORDER EQUATIONS

In many cases a first-order differential equation can be reduced by algebraic manipulations to the form

$$f(x)\,dx = g(y)\,dy \tag{1}$$

Such an equation is said to be **separable** because the variables x and y can be *separated* from each other in such a way that x appears only in the coefficient of dx and y appears only in the coefficient of dy. An equation of this type can be solved at once by integration (see Exercise 25), and we have the general solution

$$\int f(x)\,dx = \int g(y)\,dy + c \tag{2}$$

where c is an arbitrary constant of integration. It must be borne in mind, however, that the integrals which appear in (2) may be impossible to evaluate in terms of elementary functions, and numerical or graphical integration may be required before this solution can be put to practical use.

Other forms which should be recognized as being separable are

$$f(x)G(y)\,dx = F(x)g(y)\,dy \tag{3}$$

$$\frac{dy}{dx} = M(x)N(y) \tag{4}$$

A general solution of Eq. (3) can be found by first dividing by the product $F(x)G(y)$ to separate the variables and then integrating:

$$\int \frac{f(x)}{F(x)}\,dx = \int \frac{g(y)}{G(y)}\,dy + c$$

Similarly, a general solution of Eq. (4) can be found by first multiplying by dx and dividing by $N(y)$ and then integrating:

$$\int \frac{dy}{N(y)} = \int M(x)\,dx + c$$

Clearly, the process of solving a separable equation will often involve division by one or more expressions. In such cases the results are valid where the divisors are not equal to zero but may or may not be meaningful for values of the variables for which the division is undefined. Such values require special consideration and, as we shall see in the next example, may lead us to singular solutions.

Example 1 Solve the differential equation $dx + xy\, dy = y^2\, dx + y\, dy$.

It is not immediately evident that this equation is separable. In any case, however, the best first step in solving an equation of this sort is to collect terms on dx and dy. This gives

$$(1 - y^2)\, dx = y(1 - x)\, dy$$

which is of the form (3). Hence, division by the product $(1 - x)(1 - y^2)$ will separate the variables and reduce the equation to the standard form (1):

$$\frac{dx}{1 - x} = \frac{y\, dy}{1 - y^2}$$

Now, multiplying by -2 and integrating, we obtain the following equation defining y as an implicit function of x:

$$2 \ln |1 - x| = \ln |1 - y^2| + c$$

In this case, as in many problems of this sort, it is possible to write the solution in a more convenient form by first combining the logarithmic terms and then taking antilogarithms:

$$\ln \frac{|1 - x|^2}{|1 - y^2|} = c \qquad \frac{|1 - x|^2}{|1 - y^2|} = e^c = k^2$$

where $k^2 = e^c$ is necessarily positive. Finally, clearing of fractions and eliminating the absolute values, we have

$$(1 - x)^2 = \pm k^2(1 - y^2) \qquad k \neq 0$$

The two possibilities here can, of course, be combined into one by writing

$$(1 - x)^2 = \lambda(1 - y^2)$$

where now λ can take on any real value, positive or negative, except 0. The solution of the differential equation thus leads to the family of conics

$$\frac{(x - 1)^2}{\lambda} + y^2 = 1 \qquad \lambda \neq 0 \tag{5}$$

typical members of which are shown in Fig. 1.2. If $\lambda > 0$, these solution curves are all ellipses; if $\lambda < 0$, they are all hyperbolas.

In most practical problems, a general solution of a differential equation is required to satisfy specific conditions which permit its arbitrary constants to be uniquely determined. For instance, in the present problem we might ask for the particular solution curve which passes through the point $(-\frac{7}{5}, \frac{13}{5})$. Substituting these values of x and y, we then have

$$\frac{(-\frac{7}{5} - 1)^2}{\lambda} + \left(\frac{13}{5}\right)^2 = 1$$

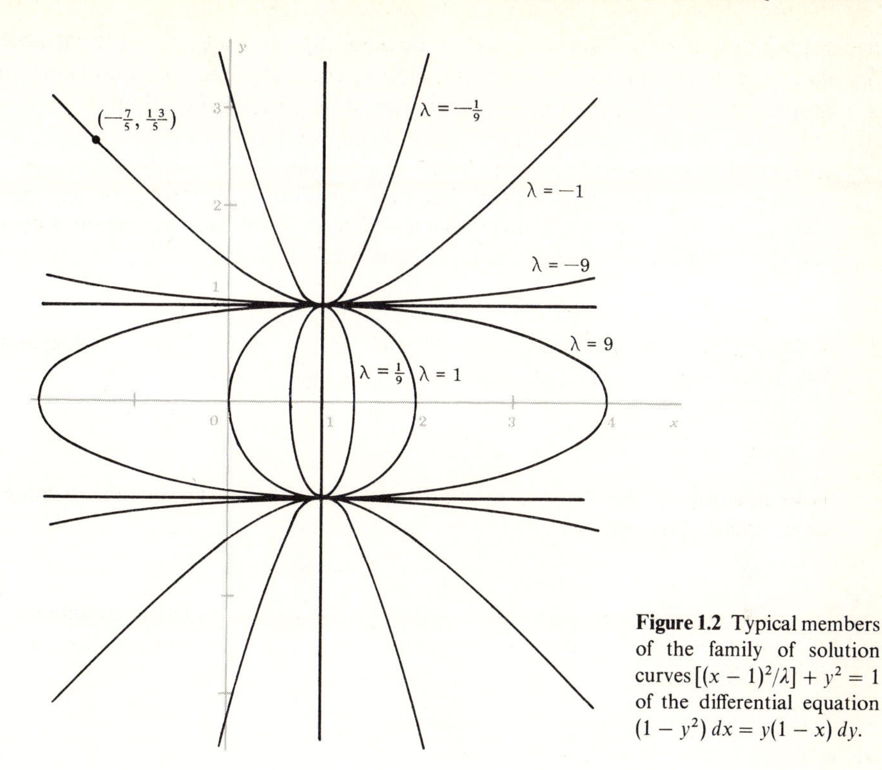

Figure 1.2 Typical members of the family of solution curves $[(x-1)^2/\lambda] + y^2 = 1$ of the differential equation $(1-y^2)\,dx = y(1-x)\,dy$.

from which we find the value $\lambda = -1$ and then the specific solution curve

$$y^2 = 1 + (x-1)^2 \tag{6}$$

Equation (6) defines the unique member of the family of curves (5) which passes through the point $(-\frac{7}{5}, \frac{13}{5})$, and on the interval $(-\infty, 1]$ the function

$$y = \sqrt{1 + (x-1)^2}$$

defines the unique solution which satisfies the condition that $y = \frac{13}{5}$ when $x = -\frac{7}{5}$. However, over any interval which contains both $x = -\frac{7}{5}$ and $x = 1$ as interior points, there are many functions which satisfy the given differential equation and are such that $y = \frac{13}{5}$ when $x = -\frac{7}{5}$. In fact, the upper branch of *any* curve of the family (5) for $x > 1$ can be associated with the upper branch of the curve (6) for $x \leq 1$ to give a function which satisfies the given equation and fulfills the condition that $y = \frac{13}{5}$ when $x = -\frac{7}{5}$. This is, of course, consistent with the fact that according to Theorem 2, Sec. 1.2, the uniqueness of the solution for which $y = \frac{13}{5}$ when $x = -\frac{7}{5}$ can be guaranteed only over an interval around $x = -\frac{7}{5}$ which does not contain $x = 1$, since y' is undefined at $x = 1$.

It should be noted that in separating variables in the given differential equation, it was necessary to divide by $1 - x$ and by $1 - y^2$; hence, the possibility that $x = 1$ and the possibility that $y = \pm 1$ were implicitly ruled out. Therefore, had we desired the particular solution curve which passed through any point with coordinates of the form $(1, y_0)$, $(x_0, 1)$, or $(x_0, -1)$, we could not have found that curve, even if it existed, by starting with the general solution (5) and particularizing the arbitrary constant λ.

Instead, it would have been necessary to return to the differential equation and search for the required solution by some method other than separation of variables. In this case it is obvious that the linear equations $x = 1$, $y = 1$, and $y = -1$ all define solutions of the given differential equation and, moreover, satisfy, respectively, the conditions $(1, y_0)$, $(x_0, 1)$, and $(x_0, -1)$. None of these can be *obtained* from our general solution, although $x = 1$ can be *included* in the first form of it by permitting λ to take on the (previously excluded) value zero. In this case, then, only $y = 1$ and $y = -1$ appear as singular solutions of the given equation.

Exercises for Section 1.3

Find a general solution of each of the following equations.

1 $y' = -2xy$ **2** $(\sin x)\, dy = 2y(\cos x)\, dx$

3 $y' = 3x^2(1 + y^2)$ **4** $x\, dy = 3y\, dx$

5 $2(xy + x)y' = y$ **6** $x\, dy = (y^2 - 3y + 2)\, dx$

7 $(y + x^2 y)\, dy = (xy^2 - x)\, dx$ **8** $y\, dx - x\, dy = x(dy - y\, dx)$

9 $yy' = 2(xy + x)$ **10** $dx + y\, dy = x^2 y\, dy$

11 $ye^{x+y}\, dy = dx$ **12** $x \exp(x^2 + y)$† $dx = y\, dy$

13 $y' = \left(\dfrac{y+1}{x+1}\right)^2$ **14** $y' = \dfrac{2(y^2 + y - 2)}{x^2 + 4x + 3}$

15 $y'' + (y')^2 + 1 = 0$ *Hint:* Observe that $y'' = dy'/dx$.

16 $xy'' = y'$ **17** $yy'' = (y')^2$

Find the particular solution of each of the following equations which satisfies the indicated conditions.

18 $2xy' + y = 0$ $x = 4,\ y = 1$

19 $y' + 2y = 0$ $x = 0,\ y = 100$

20 $2x\, dx - dy = x(x\, dy - 2y\, dx)$ $x = -3,\ y = 1$

21 $dy = x(2y\, dx - x\, dy)$ $x = 1,\ y = 4$

***22** Is there a solution of the equation $x\, dy = 3(y - 1)\, dx$ satisfying the two conditions $y = 3$ when $x = 1$ and $y = 9$ when $x = 2$? Is there a solution of this equation which satisfies the two conditions $y = 3$ when $x = -1$ and $y = 9$ when $x = 2$? Explain.

***23** Find a solution of the equation $(1 - x^2)\, dy + 4xy\, dx = 0$ with the property that $y = 9$ when $x = -2$, $y = 2$ when $x = 0$, and $y = 0$ when $x = 2$.

***24** Show that every solution of the equation $y' = ky$ is of the form $y = Ae^{kx}$. *Hint:* Let y be any solution of the given equation, and consider the derivative of the fraction y/e^{kx}.

***25** A critical student watching the professor integrate the separable equation $f(x)\, dx = g(y)\, dy$ objected that the procedure was incorrect, since one side was integrated with respect to x while the other side was integrated with respect to y. How would you answer the student's objection?

26 Show that there is no loss of generality if the arbitrary constant added when a separable equation is integrated is written in the form $\ln c$ rather than just c. Do you think this would

† The notation $\exp[f(x)]$ is frequently used in place of $e^{f(x)}$, especially when $f(x)$ is a complicated expression.

ever be a convenient thing to do? Is there any loss of generality if the integration constant is written in the form c^2? $\tan c$? $\sin c$? e^c? $\sinh c$? $\cosh c$?

27 Show that the change of dependent variable from y to v defined by the substitution $v = ax + by + c$ will always transform the equation $y' = f(ax + by + c)$ into a separable equation.

Using the substitution described in Exercise 27, find a general solution of each of the following equations.

28 $y' = (x - y)^2$

29 $y' = e^{2x+y-1} - 2$

30 $y' = (x + y - 3)^2 - 2(x + y - 3)$

***31** $y' = (x - y + 1)^2 + x - y$

32 Does the equation

$$y' = \begin{cases} 1 & x \leqq 0 \\ y & x > 0 \end{cases}$$

have a solution which is continuous on $(-\infty, \infty)$? Does it have a solution with a continuous derivative on $(-\infty, \infty)$?

1.4 HOMOGENEOUS FIRST-ORDER EQUATIONS

If all terms in the coefficient functions $M(x, y)$ and $N(x, y)$ in the general, first-order differential equation

$$M(x, y)\,dx = N(x, y)\,dy \tag{1}$$

are of the same total degree in the variables x and y, then either of the substitutions $y = ux$ and $x = vy$ will reduce the equation to one which is separable.

More generally, if $M(x, y)$ and $N(x, y)$ have the property that for all positive values of λ, the substitution of λx for x and λy for y converts them, respectively, into the expressions

$$\lambda^n M(x, y) \qquad \text{and} \qquad \lambda^n N(x, y)$$

then Eq. (1) can always be reduced to a separable equation by either of the substitutions $y = ux$ and $x = vy$.

Functions with the property that the substitutions

$$x \to \lambda x \qquad \text{and} \qquad y \to \lambda y \qquad \lambda > 0$$

merely reproduce the original forms multiplied by λ^n are called **homogeneous functions of degree** n. As a direct extension of this terminology, the differential equation (1) is said to be **homogeneous**† when $M(x, y)$ and $N(x, y)$ are homogeneous functions *of the same degree.*

Example 1 Is the function

$$F(x, y) = x(\ln \sqrt{x^2 + y^2} - \ln y) + ye^{x/y}$$

homogeneous?

† It is unfortunate that the word *homogeneous* should be used here in a sense totally different from its use in describing homogeneous linear equations of all orders (Sec. 1.2). The dual usage is universal, however, and must be accepted.

To decide this question, we replace x by λx and y by λy, getting

$$\begin{aligned} F(\lambda x, \lambda y) &= \lambda x(\ln \sqrt{\lambda^2 x^2 + \lambda^2 y^2} - \ln \lambda y) + \lambda y e^{\lambda x/\lambda y} \\ &= \lambda x[(\ln \sqrt{x^2 + y^2} + \ln \lambda) - (\ln y + \ln \lambda)] + \lambda y e^{x/y} \\ &= \lambda[x(\ln \sqrt{x^2 + y^2} - \ln y) + y e^{x/y}] \\ &= \lambda F(x, y) \end{aligned}$$

The given function is therefore homogeneous of degree 1.

If Eq. (1), assumed now to be homogeneous, is written in the form

$$\frac{dy}{dx} = \frac{M(x, y)}{N(x, y)}$$

it is evident that the fraction on the right is a homogeneous function of degree 0, since the same power of λ will multiply both numerator and denominator when the test substitutions $x \to \lambda x$ and $y \to \lambda y$ are made. But if

$$\frac{M(\lambda x, \lambda y)}{N(\lambda x, \lambda y)} = \frac{M(x, y)}{N(x, y)}$$

is an identity in x and y and $\lambda > 0$, it follows, by assigning to the arbitrary symbol λ the value $1/x$ if x is positive and the value $-1/x$ if x is negative, that

$$\frac{M(x, y)}{N(x, y)} = \frac{M(\lambda x, \lambda y)}{N(\lambda x, \lambda y)} = \begin{cases} \dfrac{M(1, y/x)}{N(1, y/x)} & x > 0 \\ \dfrac{M(-1, -y/x)}{N(-1, -y/x)} & x < 0 \end{cases}$$

In either case, it is clear that the result is a function of the fractional argument y/x. Thus, an alternative standard form for a homogeneous, first-order differential equation is

$$\frac{dy}{dx} = R\left(\frac{y}{x}\right) \tag{2}$$

Although in practice it is not necessary to reduce a homogeneous equation to the form (2) in order to solve it, the theory of the substitution $y = ux$, or $u = y/x$, is most easily developed when the equation is written in this form.

Now if $y = ux$, then $dy/dx = u + x\, du/dx$ (or, equivalently, $dy = u\, dx + x\, du$). Hence, under this substitution Eq. (2) becomes

$$u + x\frac{du}{dx} = R(u)$$

or

$$x\, du = [R(u) - u]\, dx \tag{3}$$

If $R(u) \equiv u$, then Eq. (2) is simply

$$\frac{dy}{dx} = \frac{y}{x}$$

and this is separable at the outset. If $R(u) \not\equiv u$, then we can divide (3) by the product $x[R(u) - u]$, getting

$$\frac{du}{R(u) - u} = \frac{dx}{x}$$

The variables have now been separated, and the equation can be integrated at once. Finally, by replacing u by its value y/x, we obtain an equation defining the original dependent variable y as a function of x.

Example 2 Solve the equation $(x^2 + 3y^2)\,dx - 2xy\,dy = 0$.

By inspection, this equation is homogeneous, since all terms in the coefficient of each differential are of the second degree. Hence we substitute $y = ux$ and $dy = u\,dx + x\,du$, getting

$$(x^2 + 3u^2x^2)\,dx - 2x^2u(u\,dx + x\,du) = 0$$

or, dividing by x^2 and collecting terms,

$$(1 + u^2)\,dx - 2xu\,du = 0$$

Separating variables, we obtain

$$\frac{dx}{x} - \frac{2u\,du}{1 + u^2} = 0$$

and then, by integrating, we find

$$\ln|x| - \ln|1 + u^2| = c$$

This can be written as

$$\ln\left|\frac{x}{1 + u^2}\right| = \ln e^c = \ln k \qquad \text{where } k = e^c > 0$$

Hence $|x/(1 + u^2)| = k$; or, replacing u by y/x and dropping absolute values,

$$\frac{x}{1 + (y/x)^2} = \pm k$$

Finally, clearing of fractions, we have

$$x^3 = K(x^2 + y^2) \qquad K = \pm k$$

From the preceding steps, it appears that K must be different from zero. However, it is easy to verify by direct substitution that the function corresponding to $K = 0$, namely, $x = 0$, is also a solution of the given equation. Hence, in the implicit solution we have just obtained, K is actually unrestricted.

Exercises for Section 1.4

Determine which, if any, of the following functions are homogeneous.

1 $\dfrac{x}{x^2 + y^2}$ **2** $\sin \dfrac{x}{x^2 + y^2}$

3 $x\left[\ln \dfrac{2x^2 + y^2}{x} - \ln (x + y)\right] + y^2 \tan \dfrac{x + 2y}{3x - y}$ **4** $\dfrac{x^2 + y^2 + 1}{xy + 2}$

5 Prove that the substitution $x = vy$ will also transform any homogeneous, first-order differential equation into one which is separable.

6 Under what conditions, if any, do you think that the substitution $x = vy$ would be more convenient than the substitution $y = ux$?

7 Show that the product of a homogeneous function of degree m and a homogeneous function of degree n is a homogeneous function of degree $m + n$.

8 Show that the quotient of a homogeneous function of degree m by a homogeneous function of degree n is a homogeneous function of degree $m - n$.

9 If $f(x, y, c_1) = 0$ and $f(x, y, c_2) = 0$ are two solution curves of a homogeneous, first-order differential equation, and if P_1 and P_2 are, respectively, the points of intersection of these curves and an arbitrary line, $y = mx$, through the origin, prove that the slopes of these two curves at P_1 and P_2 are equal.

Find a general solution of each of the following differential equations.

10 $(x^2 + y^2)\, dx = 2xy\, dy$ **11** $2xy' = y - x$

12 $xy' - y = \sqrt{x^2 - y^2}$ **13** $x^2\, dy = (xy - y^2)\, dx$

Solve each of the following equations and discuss the family of solution curves.

14 $\dfrac{dy}{dx} = \dfrac{2x - y}{x - 2y}$ **15** $\dfrac{dy}{dx} = \dfrac{x - y}{x + 3y}$

16 $\dfrac{dy}{dx} = \dfrac{x + 2y}{2x + y}$ **17** $\dfrac{dy}{dx} = \dfrac{x + y}{x - y}$

Find the particular solution of each of the following equations which satisfies the given conditions.

18 $x^2y\, dx = (x^3 - y^3)\, dy$ $x = 1,\ y = 1$

19 $xy' = y + \sqrt{x^2 + y^2}$ $x = 4,\ y = 3$

20 $(3y^3 - x^3)\, dx = 3xy^2\, dy$ $x = 1,\ y = 2$

21 $(x^4 + y^4)\, dx = 2x^3y\, dy$ $x = 1,\ y = 0$

22 $y' = \dfrac{y}{x} + \sec \dfrac{y}{x}$ $x = 2,\ y = \pi$

23 $(x^3 + y^3)\, dx = 2xy^2\, dy$ $x = 1,\ y = 0$

***24** If $aB \neq bA$, show that by choosing d and D suitably, the equation

$$\frac{dy}{dx} = \frac{ax + by + c}{Ax + By + C}$$

can be reduced to a homogeneous equation in the new variables t and z by the substitutions

$$x = t + d \qquad \text{and} \qquad y = z + D$$

Using the substitutions described in Exercise 24, find a general solution of each of the following equations.

***25** $y' = \dfrac{x - y + 5}{x + y - 1}$ ***26** $y' = \dfrac{2x + 2y + 1}{3x + y - 2}$

***27** Discuss Exercise 24 in the case when $aB = bA$. *Hint:* Recall Exercise 27, Sec. 1.3.

***28** Prove that $b + c = 0$ is a sufficient condition for all solutions of the equation $y' = (ax + by)/(cx + ey)$ to be conics. Prove further that when this is the case, the conics are all ellipses if $c^2 + ae < 0$ and are all hyperbolas if $c^2 + ae > 0$.

****29** Show that $b + c = 0$ is not a necessary condition for the solutions of the equation $y' = (ax + by)/(cx + ey)$ to be conics. *Hint:* Construct a counterexample.

***30** If $M(x, y)\,dx = N(x, y)\,dy$ is a homogeneous equation, prove that if it is expressed in terms of the polar coordinates r and θ by means of the substitutions $x = r\cos\theta$ and $y = r\sin\theta$, it becomes separable.

Solve each of the following equations, using the method described in Exercise 30.

***31** $y' = \dfrac{x + y}{x - y}$ ***32** $y' = \dfrac{x + 2y}{2x - y}$

***33** Give an example of a function which is homogeneous according to our definition but is not homogeneous if the condition $f(\lambda x, \lambda y) = \lambda^n f(x, y)$ is required to hold for all real values of λ.

***34** If $f(x, y)$ is a homogeneous function of degree n, show that

$$x\frac{\partial f}{\partial x} + y\frac{\partial f}{\partial y} = nf$$

What is the generalization of this result to functions of more than two variables? (This result is commonly referred to as **Euler's theorem for homogeneous functions.**)

1.5 EXACT FIRST-ORDER EQUATIONS

Associated with each suitably differentiable function of two variables $f(x, y)$ is an expression called its **total differential,** namely,

$$df = \frac{\partial f}{\partial x}\,dx + \frac{\partial f}{\partial y}\,dy$$

Conversely, if for some $f(x, y)$

$$M(x, y) = \frac{\partial f}{\partial x} \qquad \text{and} \qquad N(x, y) = \frac{\partial f}{\partial y}$$

then the differential equation

$$M(x, y)\,dx + N(x, y)\,dy = 0$$

can be written in the form

$$\frac{\partial f}{\partial x}\,dx + \frac{\partial f}{\partial y}\,dy = 0$$

from which it follows that $f(x, y) = k$ defines a family of solution curves for suitable values of the constant k. A differential equation of this form is said to be **exact** since, as it stands, its left member is an exact differential.

When $M(x, y)$ and $N(x, y)$ are sufficiently simple, it is possible to tell by inspection whether or not there exists a function f with the property that

$$\frac{\partial f}{\partial x} = M(x, y) \qquad \text{and} \qquad \frac{\partial f}{\partial y} = N(x, y)$$

In general, however, this cannot be done, and it is desirable to have a straightforward test to determine when a given first-order equation is exact. Such a criterion is provided by the following theorem.

Theorem 1 If $\partial M/\partial y$ and $\partial N/\partial x$ are continuous in a rectangular region R, then the differential equation $M(x, y)\,dx + N(x, y)\,dy = 0$ is exact if and only if

$$\frac{\partial M}{\partial y} = \frac{\partial N}{\partial x} \qquad \text{in } R$$

PROOF To prove the theorem, let us assume first that the given equation is exact. Under this assumption there exists a function f such that

$$M = \frac{\partial f}{\partial x} \qquad \text{and} \qquad N = \frac{\partial f}{\partial y}$$

Hence,

$$\frac{\partial M}{\partial y} = \frac{\partial^2 f}{\partial y\,\partial x} \qquad \text{and} \qquad \frac{\partial N}{\partial x} = \frac{\partial^2 f}{\partial x\,\partial y}$$

Moreover, $\partial^2 f/(\partial y\,\partial x)$ and $\partial^2 f/(\partial x\,\partial y)$ are continuous since we have just found them to be equal, respectively, to $\partial M/\partial y$ and $\partial N/\partial x$, which are continuous by hypothesis. Therefore it follows, from the familiar properties of partial derivatives, that the order of differentiation is immaterial and

$$\frac{\partial^2 f}{\partial y\,\partial x} = \frac{\partial^2 f}{\partial x\,\partial y}$$

Hence $\partial M/\partial y = \partial N/\partial x$, and the "only if" part of the theorem is established.

To complete the proof, we must now show that if $\partial M/\partial y = \partial N/\partial x$, then there is a function f such that $\partial f/\partial x = M$ and $\partial f/\partial y = N$. To do this, let us first integrate $M(x, y)$ with respect to x, holding y fixed. Introducing the dummy variable t gives us the expression

$$f(x, y) = \int_a^x M(t, y)\,dt + c(y) \qquad (a, y) \text{ in } R \tag{1}$$

in which, since the integration is done with respect to x while y is held constant, the integration "constant" is actually a function of y to be determined. Clearly, $\partial f/\partial x =$

$M(x, y)$, and our proof will be complete if we can determine $c(y)$ so that $\partial f/\partial y = N(x, y)$.

Now, observing that under the hypothesis that $\partial M/\partial y$ is continuous, the operations of integrating with respect to x and differentiating with respect to y can legitimately be interchanged, and recalling our current assumption that $\partial M/\partial y = \partial N/\partial x$, we have, from (1),

$$\begin{aligned}\frac{\partial f}{\partial y} &= \frac{\partial}{\partial y}\int_a^x M(t, y)\,dt + c'(y)\\ &= \int_a^x \frac{\partial M(t, y)}{\partial y}\,dt + c'(y)\\ &= \int_a^x \frac{\partial N(t, y)}{\partial t}\,dt + c'(y)\\ &= N(x, y) - N(a, y) + c'(y)\end{aligned}$$

Thus, $\partial f/\partial y$ will equal $N(x, y)$, as required, if $c(y)$ is determined so that $c'(y) = N(a, y)$, that is, if

$$c(y) = \int_b^y N(a, u)\,du \qquad (a, b) \text{ in } R$$

We have thus shown that if $\partial M/\partial y = \partial N/\partial x$, then

$$f(x, y) = \int_a^x M(t, y)\,dt + \int_b^y N(a, u)\,du \tag{2}$$

is a function such that

$$df = \frac{\partial f}{\partial x}\,dx + \frac{\partial f}{\partial y}\,dy = M(x, y)\,dx + N(x, y)\,dy$$

This establishes the "if" assertion of the theorem, and our proof is complete.

Since the proof of the preceding theorem tells us that when the equation $M(x, y)\,dx + N(x, y)\,dy = 0$ is exact, its left member is, in fact, the total differential of the function f defined by (2), it follows that the solution in this case can be found at once by integration. Thus we have the following corollary.

Corollary 1 If the differential equation $M(x, y)\,dx + N(x, y)\,dy = 0$ is exact in a rectangular region R, then in R

$$\int_a^x M(t, y)\,dt + \int_b^y N(a, u)\,du = k \qquad (a, b) \text{ an arbitrary point in } R$$

defines a family of solution curves of the differential equation for suitable values of the constant k.

Example 1 Show that the equation $(2x + 3y - 2)\,dx + (3x - 4y + 1)\,dy = 0$ is exact, and find a general solution.

Applying the test provided by Theorem 1, we find

$$\frac{\partial M}{\partial y} = \frac{\partial(2x + 3y - 2)}{\partial y} = 3 \qquad \text{and} \qquad \frac{\partial N}{\partial x} = \frac{\partial(3x - 4y + 1)}{\partial x} = 3$$

Since the two partial derivatives are equal, the equation is exact. Its solution can therefore be found by means of Corollary 1, Theorem 1:

$$\int_a^x (2t + 3y - 2)\,dt + \int_b^y (3a - 4u + 1)\,du = k$$

$$(t^2 + 3ty - 2t)\Big|_a^x + (3au - 2u^2 + u)\Big|_b^y = k$$

$$x^2 + 3xy - 2y^2 - 2x + y = k + a^2 + 3ab - 2b^2 - 2a + b = K$$

Occasionally a differential equation which is not exact can be made exact by multiplying it by some simple expression. For example, if the (exact) equation $2xy^3\,dx + 3x^2y^2\,dy = 0$ is simplified by the natural process of dividing out the common factor xy^2, the resulting equation, namely, $2y\,dx + 3x\,dy = 0$, is *not* exact. Conversely, however, the last equation can be restored to its original exact form by multiplying it through by xy^2. This illustrates the general result† that every first-order equation which possesses a family of solutions can be made exact by multiplying it by a suitable factor, called an **integrating factor.** In general, the determination of an integrating factor for a given equation is difficult. However, as the following examples show, in particular cases an integrating factor can often be found by inspection.

Example 2 Show that $1/(x^2 + y^2)$ is an integrating factor for the equation $(x^2 + y^2 - x)\,dx - y\,dy = 0$, and then solve the equation.

The test provided by Theorem 1 shows that in its present form the given equation is not exact. However, if it is multiplied by the indicated factor, it can be rewritten in the form

$$\left(1 - \frac{x}{x^2 + y^2}\right)dx - \frac{y}{x^2 + y^2}\,dy = 0$$

Testing again, we find that the last equation is exact, and we can now use Corollary 1 to find a general solution of it. However, it is simpler to observe that the given equation can also be written

$$dx - \frac{x\,dx + y\,dy}{x^2 + y^2} = 0 \qquad \text{or} \qquad dx - \frac{1}{2}\,d[\ln (x^2 + y^2)] = 0$$

Hence, integrating, we have

$$x - \ln \sqrt{x^2 + y^2} = k$$

† See, for instance, M. Golomb and M. E. Shanks, "Elements of ordinary differential equations," 2d, ed., McGraw-Hill, New York, 1965, pp. 52–53.

Example 3 Find an integrating factor for the equation $y\,dx + (x^2y^3 + x)\,dy = 0$, and solve the equation.

Since this equation can be rewritten in the form

$$(y\,dx + x\,dy) + x^2y^3\,dy = 0$$

and since $y\,dx + x\,dy = d(xy)$, it is natural to multiply the equation by $1/x^2y^2$, getting

$$\frac{d(xy)}{(xy)^2} + y\,dy = 0$$

This equation can now be integrated by inspection, and we have

$$-\frac{1}{xy} + \frac{y^2}{2} = k$$

Example 4 Find an integrating factor for the equation $x\,dy - y\,dx = (4x^2 + y^2)\,dy$, and solve the equation.

In this equation, the terms on the left seem related equally well to

$$d\left(\frac{y}{x}\right) = \frac{x\,dy - y\,dx}{x^2} \qquad \text{or} \qquad d\left(\frac{x}{y}\right) = \frac{y\,dx - x\,dy}{y^2}$$

If we pursue the first suggestion and multiply the equation by $1/x^2$, we obtain

$$d\left(\frac{y}{x}\right) = \left(4 + \frac{y^2}{x^2}\right)dy$$

This equation is still not exact, but it is separable, and division by $4 + y^2/x^2$ gives us

$$\frac{d(y/x)}{4 + (y/x)^2} = dy$$

Integrating this, we have finally

$$\frac{1}{2}\,\text{Tan}^{-1}\frac{y}{2x} = y + k$$

The results of the last three examples suggest the following observations, which are often helpful:

1. If a first-order differential equation contains the combination $x\,dx + y\,dy = \frac{1}{2}\,d(x^2 + y^2)$, try some function of $x^2 + y^2$ as a multiplier to make the equation integrable.
2. If a first-order differential equation contains the combination $x\,dy + y\,dx = d(xy)$, try some function of xy as a multiplier to make the equation integrable.
3. If a first-order differential equation contains the combination $x\,dy - y\,dx$, try $1/x^2$ or $1/y^2$ as a multiplier to make the equation integrable.

Exercises for Section 1.5

Show that the following equations are exact, and integrate each one.

1 $(y^2 - 1)\,dx + (2xy - \sin y)\,dy = 0$

2 $(2xy + x^3)\,dx + (x^2 + y^2)\,dy = 0$

3 $(3x^2 - 6xy)\,dx - (3x^2 + 2y)\,dy = 0$

4 $(x\sqrt{x^2 + y^2} + y)\,dx + (y\sqrt{x^2 + y^2} + x)\,dy = 0$

5 $(2xy^4 + \sin y)\,dx + (4x^2y^3 + x\cos y)\,dy = 0$

Find a general solution of each of the following equations by first multiplying by a suitable factor and then integrating.

6 $y(1 + xy)\,dx + (2y - x)\,dy = 0$ **7** $3(y^4 + 1)\,dx + 4xy^3\,dy = 0$

8 $(xy^2 + y)\,dx + (x - x^2y)\,dy = 0$ **9** $(x^2 + y^2 + 2x)\,dy = 2y\,dx$

10 $x\,dy + 3y\,dx = xy\,dy$

Solve each of the following equations by two methods.

11 $2y\,dx + (3y - 2x)\,dy = 0$ **12** $(x^2 - y^2)\,dy = 2xy\,dx$

13 $x\,dy + y\,dx = \dfrac{dx}{y} - \dfrac{dy}{x}$ **14** $2x \ln y\,dx + \dfrac{1 + x^2}{y}\,dy = 0$

15 $\sqrt{x^2 + y^2}\,dx = x\,dy - y\,dx$

***16** Solve the equation $(xy^2 - y)\,dx + (x^2y - x)\,dy = 0$ first by integrating it as an exact equation and then by multiplying it by $1/x^2y^2$ before integrating it. Reconcile your results.

17 Show that if the equation of Exercise 16 is multiplied by any differentiable function of the product xy, it is still exact.

18 If $\phi(x, y)$ is an integrating factor of the differential equation $M(x, y)\,dx + N(x, y)\,dy = 0$, show that ϕ satisfies the partial differential equation

$$M\frac{\partial \phi}{\partial y} - N\frac{\partial \phi}{\partial x} + \left(\frac{\partial M}{\partial y} - \frac{\partial N}{\partial x}\right)\phi = 0$$

***19** Show that $f(x, y) = k$ is a general solution of the differential equation $M(x, y)\,dx + N(x, y)\,dy = 0$ if and only if

$$M\frac{\partial f}{\partial y} - N\frac{\partial f}{\partial x} \equiv 0$$

***20** Using the result of the preceding exercise, show that if the equation $M(x, y)\,dx + N(x, y)\,dy = 0$ is both homogeneous and exact, its solution is $xM(x, y) + yN(x, y) = k$. *Hint:* Recall the result of Exercise 34, Sec. 1.4.

***21** Show that if $\phi(x, y)$ is an integrating factor leading to the solution $f(x, y) = k$ for the differential equation $M(x, y)\,dx + N(x, y)\,dy = 0$, then $\phi F(f)$ is also an integrating factor, where F is an arbitrary differentiable function.

***22** If the equation $M(x, y)\,dx + N(x, y)\,dy = 0$ is homogeneous, show that $1/(xM + yN)$ is an integrating factor. *Hint:* Observe that

$$\frac{M\,dx + N\,dy}{xM + yN} = \frac{dx}{x} + \frac{(x\,dy - y\,dx)N}{x(xM + yN)}$$

$$= \frac{dx}{x} + \frac{(x\,dy - y\,dx)/x^2}{M/N + y/x}$$

23 Prove the "if" part of Theorem 1 by first integrating $N(x, y)$ with respect to y.

***24** Show that the arbitrary constants a and b which appear in the formula of Corollary 1, Theorem 1, add no generality to the solution. *Hint:* Consider the partial derivatives with respect to a and b of the left-hand side of the formula.

25 Show by an example that the integral formula of Corollary 1 does not necessarily define a solution of the exact equation $M(x, y)\,dx + N(x, y)\,dy = 0$ for *all* values of k.

1.6 LINEAR FIRST-ORDER EQUATIONS

First-order equations which are linear form an important class of differential equations which can always be routinely solved by the use of an integrating factor. By definition, a linear, first-order differential equation cannot contain products, powers, or other nonlinear combinations of y or y'. Hence its most general form is

$$F(x)\frac{dy}{dx} + G(x)y = H(x)$$

If we divide this equation by $F(x)$ and rename the coefficients, it appears in the more usual form

$$\frac{dy}{dx} + P(x)y = Q(x) \tag{1}$$

To determine whether this equation is, or can be made, exact, let us rewrite it in the form

$$[P(x)y - Q(x)]\,dx + dy = 0 \tag{2}$$

Here $M(x, y) = P(x)y - Q(x)$, $N(x, y) = 1$, and the condition for exactness $\partial M/\partial y \equiv \partial N/\partial x$ becomes $P(x) \equiv 0$. This is surely not true in general; and when it is true, Eq. (1) can be solved immediately by integration, and no further investigation is necessary.

Assuming that $P(x) \not\equiv 0$, let us now attempt to find an integrating factor $\phi(x)$ for Eq. (2). Applying the test for exactness to the new equation

$$\phi(x)[P(x)y - Q(x)]\,dx + \phi(x)\,dy = 0 \tag{3}$$

it follows that $\phi(x)$ will be an integrating factor provided

$$\phi(x)P(x) = \phi'(x)$$

This is a simple separable equation, any nontrivial solution of which will serve our purpose. Hence, we can write, in particular,

$$\frac{d\phi(x)}{\phi(x)} = P(x)\,dx$$

$$\ln|\phi(x)| = \int P(x)\,dx$$

$$\phi(x) = e^{\int P(x)\,dx}$$

Thus Eq. (2), or equally well Eq. (1), possesses the integrating factor $\phi(x) = e^{\int P(x)\,dx}$.

When Eq. (1) is multiplied by $e^{\int P(x)\,dx}$, it can be written in the form

$$\frac{d}{dx}\left(ye^{\int P(x)\,dx}\right) = Q(x)e^{\int P(x)\,dx}$$

The left-hand side is now an exact derivative and hence can be integrated at once. Moreover, the right-hand side is a function of x only and therefore can also be integrated, with at most practical difficulties requiring numerical integration. Thus we have, on performing these integrations,

$$ye^{\int P(x)\,dx} = \int Q(x)e^{\int P(x)\,dx}\,dx + c$$

and finally, after dividing by $e^{\int P(x)\,dx}$,

$$y = e^{-\int P(x)\,dx}\int Q(x)e^{\int P(x)\,dx}\,dx + ce^{-\int P(x)\,dx} \tag{4}$$

Equation (4) should *not* be remembered as a formula for the solution of Eq. (1). Instead, a linear first-order equation should be solved by actually carrying out the steps we have described:

1. Compute the integrating factor $e^{\int P(x)\,dx}$.
2. Multiply both sides of the given equation by this factor.
3. Integrate both sides of the resulting equation, taking advantage of the fact that the integral of the left member is *always* just y times the integrating factor.
4. Solve the integrated equation for y.

Example 1 Find the solution of the equation $(1 + x^2)(dy - dx) = 2xy\,dx$ for which $y = 1$ when $x = 0$.

Dividing the given equation by $(1 + x^2)\,dx$ and transposing, we have

$$\frac{dy}{dx} - \frac{2x}{1 + x^2}y = 1 \tag{5}$$

which is a linear first-order equation. In this case $P(x) = -2x/(1 + x^2)$; hence the integrating factor is

$$\exp\left(\int \frac{-2x}{1 + x^2}\,dx\right) = \exp\,[-\ln\,(1 + x^2)] = \exp\,[\ln\,(1 + x^2)^{-1}] = \frac{1}{1 + x^2}\dagger$$

Multiplying Eq. (5) by this factor gives the equation

$$\frac{1}{1 + x^2}\frac{dy}{dx} - \frac{2x}{(1 + x^2)^2}y = \frac{1}{1 + x^2}$$

† Note that $\exp\,(\ln u) \equiv e^{\ln u} = u$, for any expression u for which $\ln u$ is defined.

Integrating this, remembering that the integral of the left member is just y times the integrating factor, we have

$$\frac{y}{1+x^2} = \mathrm{Tan}^{-1}\, x + c \qquad \text{or} \qquad y = (1+x^2)\mathrm{Tan}^{-1}\, x + c(1+x^2)$$

To find the specific solution required, we substitute the given conditions $x = 0$, $y = 1$ into the general solution, getting $1 = 0 + c$. The required solution is, therefore,

$$y = (1+x^2)\,\mathrm{Tan}^{-1}\, x + (1+x^2)$$

Exercises for Section 1.6

Find a general solution of each of the following equations.

1 $(2y + x^2)\, dx = x\, dy$ **2** $y' + 2xy + x = \exp(-x^2)$

3 $y' + y \tan x = \sec x$ **4** $y' + y \cot x = \sin 2x$

5 $x^2\, dy + (2xy - x + 1)\, dx = 0$ **6** $(1 - x^2)y' + xy = 2x$

7 $y' + \dfrac{y}{1-x} = x^2 - x$ **8** $y' = \dfrac{2y}{x+1} + (x+1)^3$

9 $xy' + (1+x)y = e^{-x}$ **10** $xy' + 2(1-x^2)y = 1$

Find the particular solution of each of the following equations which satisfies the indicated conditions.

11 $y' + y = e^x$ $x = 0,\ y = 2$

12 $y' + y = e^{-x}$ $x = 0,\ y = 3$

13 $(x^2 + 1)\, dy = (x^3 - 2xy + x)\, dx$ $x = 1,\ y = 1$

14 $y' + (1 + 2x)y = \exp(-x^2)$ $x = 0,\ y = 3$

15 $(1 + x^2)\, dy = (1 + xy)\, dx$ $x = 1,\ y = 0$

***16** Find a solution of the equation $x\, dy + (x^2 - 3y)\, dx = 0$ which simultaneously meets the condition $x = -1$, $y = 1$ and the condition $x = 1$, $y = -1$.

17 Find a general solution of the equation $y^2\, dx + (3xy - 4y^3)\, dy = 0$. *Hint:* Consider x as the dependent variable.

18 Find a general solution of the equation $y'' + [y'/(x-1)] = x - 1$. *Hint:* Consider y' as the dependent variable.

19 Prove that no extra generality in the final answer results from using

$$\exp\left[\int P(x)\, dx + k\right]$$

instead of just $\exp\left[\int P(x)\, dx\right]$ as an integrating factor of the equation $y' + P(x)y = Q(x)$.

1.7 SPECIAL FIRST-ORDER EQUATIONS

We have now learned how to recognize and solve the four major types of solvable first-order differential equations, namely, those that are separable, homogeneous, exact, or linear. In addition to these, however, there are other, more special classes of first-order equations which can be solved by various ingenious but less general methods. In this section we shall take a brief look at several of these.

Sometimes a differential equation of the second order in y can be regarded as a first-order equation in y'. If this is the case and if the first-order equation provided by this new point of view can be solved, then an expression for y' can be obtained. This, in turn, provides a first-order equation in y from which, if it can be solved, a solution y of the original second-order equation can be found.

Example 1 According to our definition of the order of a differential equation, the equation $y'' + (y')^2 + 1 = 0$ is a second-order equation. However, observing that y'' is just dy'/dx, we see that this equation is also a first-order equation in y', namely,

$$\frac{dy'}{dx} + (y')^2 + 1 = 0$$

This is a separable equation which can easily be solved, and we have

$$\frac{dy'}{1 + (y')^2} + dx = 0$$

$$\operatorname{Tan}^{-1} y' = c_1 - x$$

$$y' = \tan (c_1 - x)$$

Integration now gives y immediately:

$$y = \ln |\cos (c_1 - x)| + c_2$$

Frequently in problems in dynamics, one encounters differential equations of the form $y'' = f(y, y')$. For such equations, the substitution of dy'/dx for y'' is ineffectual since it yields a first-order equation involving not one but *two* dependent variables, y and y'. However, the following important substitution will always reduce such an equation to a first-order equation in which y' appears as the only dependent variable and y plays the role of the independent variable: Beginning with the relation $y'' = dy'/dx$ and using the chain rule, we have

$$y'' = \frac{dy'}{dx} = \frac{dy'}{dy}\frac{dy}{dx} = y'\frac{dy'}{dy}$$

Under this substitution, the original equation becomes

$$y'\frac{dy'}{dy} = f(y, y')$$

and this is a first-order equation in y' which it may be possible to solve. In particular, if f is a function of y only, then the last equation is always separable, and we have

$$y'\, dy' = f(y)\, dy$$

$$\tfrac{1}{2}(y')^2 = \int f(y)\, dy + c_1$$

From this, by solving for y', separating variables, and integrating again, a complete solution for y can always be obtained, although it may involve integrals which cannot be evaluated in terms of elementary functions.

Example 2 Find the solution of the differential equation $y'' = -2y + 2y^3$ for which $y(0) = 0$ and $y'(0) = 1$.

This equation is of the form $y'' = f(y)$. Hence we replace y'' by $y'(dy'/dy)$, getting

$$y' \frac{dy'}{dy} = -2y + 2y^3$$

and, integrating,

$$\tfrac{1}{2}(y')^2 = c_1 - y^2 + \tfrac{1}{2}y^4$$

Now we know that $y = 0$ and $y' = 1$ when $x = 0$. Hence, by substitution we find

$$c_1 = \tfrac{1}{2} \qquad \text{and} \qquad (y')^2 = 1 - 2y^2 + y^4 = (1 - y^2)^2$$

Therefore $y' = \pm(1 - y^2)$. To be consistent with the data of our problem, namely, $y' = 1$ when $y = 0$, we must select the positive sign. Hence we have

$$y' = \frac{dy}{dx} = 1 - y^2$$

$$\frac{dy}{1 - y^2} = dx$$

$$\tanh^{-1} y = x + c_2$$

$$y = \tanh (x + c_2)$$

Finally, since $y = 0$ when $x = 0$, we find that $c_2 = 0$ and

$$y = \tanh x$$

This example worked out more simply than is usually the case because the initial conditions were very carefully chosen. Had the initial data not been chosen to make $c_1 = \tfrac{1}{2}$, the expression for $(y')^2$ would not have been a perfect square and the integral for y could not have been evaluated in terms of elementary functions. In fact, for any value of c_1 except $\tfrac{1}{2}$, the final integral for y requires what are called *elliptic functions* for its evaluation. It is important to note, however, that in any case an explicit integral expression for y could have been obtained.

Another special type of first-order equation which can always be solved is the **Bernoulli equation**.†

$$\frac{dy}{dx} + P(x)y = Q(x)y^n$$

† Named for the Swiss mathematician Jakob Bernoulli (1654–1705), a member of a family which in little more than a century produced eight distinguished mathematicians.

If $n = 0$, this is a nonhomogeneous, linear first-order equation, and if $n = 1$, it is linear and homogeneous. For any other value of n, it can be converted into a linear equation, and then solved by the substitution defined in the following theorem.

Theorem 1 If $n \neq 1$, the change of dependent variable defined by the substitution $z = y^{1-n}$ will always transform the equation $dy/dx + P(x)y = Q(x)y^n$ into a linear equation in z.

PROOF If we divide the given equation by y^n, we obtain

$$y^{-n}\frac{dy}{dx} + P(x)y^{1-n} = Q(x) \tag{1}$$

Also, from the relation $z = y^{1-n}$, we find

$$\frac{dz}{dx} = (1-n)y^{-n}\frac{dy}{dx} \qquad \text{or} \qquad y^{-n}\frac{dy}{dx} = \frac{1}{1-n}\frac{dz}{dx}$$

Hence, substituting into Eq. (1), we have

$$\frac{1}{1-n}\frac{dz}{dx} + P(x)z = Q(x) \tag{2}$$

which is a linear first-order equation in the variable z. After z has been determined from Eq. (2), y can be found by reversing the given substitution; i.e.,

$$y = z^{1/(1-n)}$$

Example 3 Solve the differential equation $3xy' + y + x^2y^4 = 0$.

If we divide the given equation by $3x$ and then transpose the last term, we obtain

$$y' + \frac{1}{3x}y = -\frac{x}{3}y^4$$

which we recognize as a Bernoulli equation with $n = 4$. The substitution $z = y^{1-4} = y^{-3}$ will therefore transform this equation into a linear equation of the form of Eq. (2), namely,

$$\frac{1}{-3}\frac{dz}{dx} + \frac{1}{3x}z = -\frac{x}{3} \qquad \text{or} \qquad \frac{dz}{dx} - \frac{1}{x}z = x$$

The integrating factor for the last equation is $e^{\int(-1/x)\,dx} = e^{-\ln|x|} = e^{\ln(1/|x|)} = 1/|x|$. Multiplying by this factor gives

$$\frac{1}{x}\frac{dz}{dx} - \frac{1}{x^2}z = 1$$

and, integrating, we obtain

$$\frac{z}{x} = x + c \qquad \text{or} \qquad z = x^2 + cx$$

Finally, since $z = y^{-3}$, we have

$$\frac{1}{y^3} = x^2 + cx \qquad \text{or} \qquad y^3 = \frac{1}{x^2 + cx}$$

Another type of first-order differential equation which can always be solved is the **equation of Clairaut**†

$$y = xy' + f(y')$$

whose solutions are described in the following theorem.

Theorem 2 The equation $y = xy' + f(y')$ has $y = mx + f(m)$ as a general solution and the curve defined parametrically by the equations $x = -f'(t)$ and $y = -tf'(t) + f(t)$ as a singular solution.

PROOF If we differentiate the given equation

$$y = xy' + f(y') \tag{3}$$

we obtain

$$y' = xy'' + y' + f'(y')y'' \qquad \text{or} \qquad [x + f'(y')]y'' = 0$$

By inspection, the last equation will be satisfied if $y'' = 0$ or if $x = -f'(y')$. From the first alternative we infer that y' is a constant, say m; and merely substituting this for y' in Eq. (3) gives us the general solution

$$y = mx + f(m) \tag{4}$$

asserted by the theorem. Clearly, Eq. (4) defines a family of straight lines.

From the alternative

$$x = -f'(y') \tag{5}$$

we obtain, by substituting into Eq. (3),

$$y = -y'f'(y') + f(y') \tag{6}$$

By interpreting y' as a parameter, say t, Eqs. (5) and (6) constitute a pair of parametric equations

$$x = -f'(t) \qquad y = -tf'(t) + f(t) \tag{7}$$

defining some curve. This curve is a straight line if and only if $f(t)$ is a linear function (see Exercise 36). Hence, in general, it cannot be one of the lines described by the general solution (4). It must therefore be a singular solution of Eq. (3).

It is interesting to note that the lines defined by the general solution (4) are all tangent to the graph of the singular solution; that is, the singular solution is the envelope of the members of the general solution. To see this, let us determine the

† Named for the French astronomer and mathematician Alexis Claude Clairaut (1713–1765).

equation of the tangent to the singular curve (7) at a general point, say the point where $t = m$, and verify that it is precisely $y = mx + f(m)$.

For the slope of the graph of the singular solution (7), we have

$$\frac{dy}{dx} = \frac{dy/dt}{dx/dt} = \frac{-f'(t) - tf''(t) + f'(t)}{-f''(t)} = t$$

Hence the equation of the tangent at the point where $t = m$ is

$$y - [-mf'(m) + f(m)] = m[x + f'(m)] \qquad \text{or} \qquad y = mx + f(m)$$

as asserted.

Example 4 Solve the differential equation $y = xy' - \frac{1}{4}(y')^2$, and discuss the nature of the solutions.

Using Theorem 2, we obtain the equation of the general solution simply by replacing y' in the given differential equation by an arbitrary constant m, getting $y = mx - m^2/4$. For the singular solution, we have from (7) the parametric equations

$$x = \frac{t}{2} \qquad y = -t\left(-\frac{t}{2}\right) - \frac{t^2}{4} = \frac{t^2}{4}$$

Eliminating the parameter t, we obtain $y = x^2$, which is the equation of a parabola to which all the lines of the general solution are tangent.

This is not the whole story, however. In the proof of Theorem 2, we encountered the equation

$$[x + f'(y')]y'' = 0 \tag{8}$$

and we explored the possibilities that $y'' = 0$ *for all values of* x and that $x + f'(y') = 0$ *for all values of* x. But Eq. (8) is certainly satisfied for all values of x if one factor is zero for some values of x and the other factor is zero for the remaining values of x (!). This leads to the possibility of a continuously differentiable integral curve consisting of a portion of one of the lines of the general solution and an arc of the parabola, connected at the point of tangency of the line, or an integral curve consisting of portions of two of the lines connected by the arc of the parabola between their points of tangency. For instance, if

$$\begin{aligned} y'' &= 0 & -\infty < x &\le x_1 \\ x + f'(y') &= 0 & x_1 < x &< x_2 \\ y'' &= 0 & x_2 \le x &< \infty \end{aligned}$$

the integral curve consists of the left half of the tangent to the parabola at x_1, the arc of the parabola between x_1 and x_2, and the right half of the tangent to the parabola at x_2. Figure 1.3 shows this curve for $x_1 = -\frac{1}{2}$ and $x_2 = 1$. Since x_1 and x_2 can be chosen arbitrarily, it follows that the given equation actually has a family of solutions depending on two arbitrary constants.

Some differential equations for which no elementary solution procedure is available have the property that a *general* solution can be found as soon as one *particular* solution is known. As we shall see in Sec. 3.3, this is always the case for

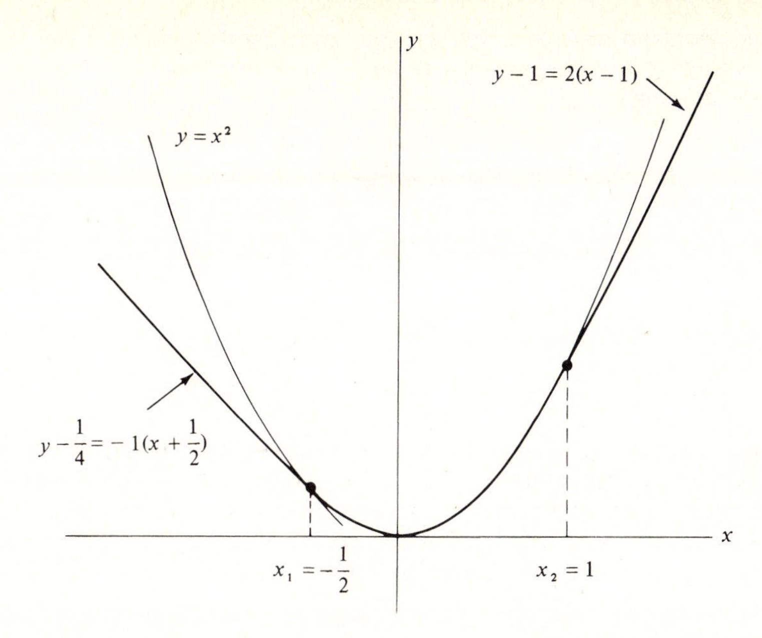

Figure 1.3 A solution of the equation $y = xy' - \frac{1}{4}(y')^2$ defined by three distinct formulas.

homogeneous linear equations of the second order. For first-order equations, it is true for the **Riccati equation**†

$$y' = P(x)y^2 + Q(x)y + R(x)$$

because, as the following theorem assures us, if one particular solution is known, a Riccati equation can always be transformed into another first-order equation which is linear, and hence solvable.

Theorem 3 If $y = y_1$ is a solution of the differential equation $y' = P(x)y^2 + Q(x)y + R(x)$, then the substitution $y = y_1 + 1/z$ transforms this equation into the linear first-order equation

$$z' + [2y_1 P(x) + Q(x)]z = -P(x)$$

PROOF From $y = y_1 + 1/z$ we obtain $y' = y_1' - (1/z^2)z'$. When these expressions for y and y' are substituted into the given equation, the result may be written as

$$y_1' - \frac{1}{z^2}z' = P(x)\left(y_1^2 + \frac{2y_1}{z} + \frac{1}{z^2}\right) + Q(x)\left(y_1 + \frac{1}{z}\right) + R(x)$$

or

$$y_1' - P(x)y_1^2 - Q(x)y_1 - R(x) = \frac{1}{z^2}z' + P(x)\left(\frac{2y_1}{z} + \frac{1}{z^2}\right) + Q(x)\frac{1}{z}$$

† Named for the Italian mathematician J. F. Riccati (1676–1754).

The left-hand side of this equation is equal to zero since, by hypothesis, y_1 is a solution of the given differential equation. Hence, by multiplying by z^2 and rearranging slightly, the last equation becomes

$$z' + [2y_1 P(x) + Q(x)]z = -P(x)$$

which is a linear first-order equation in z, as asserted.

Example 5 Find a general solution of the equation $y' = y^2 + (1 - 2x)y + (x^2 - x + 1)$ given that $y = x$ is a particular solution.

Identifying $P(x) = 1$, $Q(x) = 1 - 2x$, $R(x) = x^2 - x + 1$, and $y_1 = x$, it follows from Theorem 3 that the variable z defined by the substitution $y = x + 1/z$ satisfies the linear first-order equation

$$z' + [2x + (1 - 2x)]z = -1 \qquad \text{or} \qquad z' + z = -1$$

An integrating factor for this equation is $e^{\int dx} = e^x$. Hence, multiplying by this factor and integrating, we have

$$e^x(z' + z) = -e^x$$

$$ze^x = -e^x + c$$

$$z = -1 + ce^{-x}$$

and finally

$$y = x + \frac{1}{z} = x + \frac{1}{ce^{-x} - 1}$$

Exercises for Section 1.7

Find a general solution of each of the following differential equations.

1 $y'' - 2y' = 1$

2 $y'' + y' = e^x$

3 $xy'' = y'$

4 $yy'' = (y')^2$

5 $xy'' + y' = 3x^2 - x$

6 $(\cot x)y'' + y' + 1 = 0$

7 $(1 + x^2)y'' - 2xy' = 2x$

***8** $y'' + 2x(1 + y')^2 = 0$

9 $y' - \dfrac{y}{x} + \dfrac{y^2}{x} = 0$

10 $y^2y' + x^2y^3 = x^2$

11 $dy = (xy^2 + 3xy)\,dx$

12 $xy^2y' - y^3 = x^2$

13 $y\,dy = (x - y^2)\,dx$

14 $y' + y = xy^2$

Find the general solution and the singular solution of each of the following differential equations.

15 $y = xy' - 4(y')^3$

16 $y = xy' + 1/(4y')$

17 $y = xy' - \exp(y')$

18 $y = xy' + \dfrac{1}{1 + y'}$

***19** Construct a Clairaut equation whose singular solution will be $y = x^3$. *Hint:* Let $f(y') = a(y')^n$ and determine a and n appropriately.

***20** Construct a Clairaut equation whose singular solution will be $y = 1/x$.

****21** Construct a Clairaut equation whose singular solution will be $y = x - x^3$.

Solve each of the following initial-value problems.

22 $y'' + 3y^2y' = 0$ $\quad y(1) = 1,\ y'(1) = -1$

23 $y'' = 2yy'$ $\quad y(0) = 1,\ y'(0) = 5$

24 $2yy'' = 1 + (y')^2$ $\quad y(0) = 2,\ y'(0) = -1$

***25** $y'' + (y')^2 + y = 0$ $\quad y(0) = -\frac{1}{2},\ y'(0) = -1$

26 $y' + y/x = (\ln x)y^2$ $\quad y(1) = 1$

***27** $y = xy' - \frac{1}{4}(y')^2$ $\quad y(-2) = 3,\ y(3) = 8$

****28** Show that there is a solution of the Clairaut equation $y = xy' - \frac{1}{4}(y')^2$ satisfying the conditions $y(x_1) = y_1$, $y(x_2) = y_2$ provided $x_1^2 - y_1 > 0$, $x_2^2 - y_2 > 0$, and $4(x_1^2 - y_1)(x_2^2 - y_2) < (y_1 + y_2 - 2x_1x_2)^2$.

Find a general solution of each of the following equations, using the given particular solution.

29 $y' = xy^2 + (1 - 2x)y + x - 1$ $\quad y_1 = 1$

30 $y' = (y - 1)(y + 1/x)$ $\quad y_1 = 1$

31 $y' = (x + y)(x + y - 2)$ $\quad y_1 = 1 - x$

32 $y' = e^{-x}y^2 + y - e^x$ $\quad y_1 = e^x$

33 $y' = x^3(y - x)^2 + y/x$ $\quad y_1 = x$

***34** Show that there are two values of c for which $y = c - x^2$ is a solution of the equation $y' = (x^2 + y + 1)(x^2 + y - \frac{3}{2}) + 1 - 2x$. Find these values and solve this equation using each of the particular solutions determined by these values of c. Are the two general solutions equivalent?

***35** Construct a Riccati equation which can be solved in terms of elementary functions.

36 Show that the values of x and y given in (7) satisfy an equation of the form $ax + by + c = 0$ if and only if $f(t)$ is a linear function.

CHAPTER

TWO

APPLICATIONS OF FIRST-ORDER DIFFERENTIAL EQUATIONS

2.1 INTRODUCTION

Problems in the physical sciences usually involve continuously changing quantities such as distance, velocity, acceleration, or force. On the other hand, many problems in the life sciences deal with aggregates of individuals which clearly are discrete rather than continuous. Since derivatives, and hence differential equations, are meaningful only for quantities that change continuously, one might think that differential equations would arise only in the formulation of physical problems. This is not the case, however, because if the population in a problem in biology is sufficiently large, it can usually be approximated, or modeled, by a continuous system in which rates of change can be expressed as derivatives and the behavior of the system can be described by a differential equation. Whether such an approach is justified depends simply on whether it works: Does it describe past observations with satisfactory accuracy, and does it predict results which can be checked experimentally? If it does, it is useful; if not, it must either be rejected or refined into a more appropriate model. It is a further tribute to the utility of mathematics that this kind of modeling is becoming increasingly effective in the life sciences.

In this chapter we shall illustrate, through examples, how the formulation of problems in both the physical and the biological sciences frequently leads to first-order differential equations of types that we have now learned to solve.

2.2 PROBLEMS FROM PHYSICS

Example 1 According to **Newton's law of cooling,** the rate at which the temperature of a body changes is proportional to the difference between the instantaneous temperature of the body and the temperature of the surrounding medium. If a body whose temperature is initially 100°C is allowed to cool in air whose temperature remains at the constant value 20°C, find the formula which gives the temperature of the body as a function of the time t if it is observed that after 10 min the body has cooled to 40°C.

If we let T denote the instantaneous temperature of the body in degrees Celsius and t denote the time in minutes since the body began to cool, then the rate of cooling, as given by Newton's law, is

$$\frac{dT}{dt} = k(T - 20)$$

This equation can be solved either as a separable equation or as a linear equation. Regarding it as a separable equation, we have

$$\frac{dT}{T - 20} = k\,dt$$

$$\ln(T - 20) = kt + \ln|c|$$

$$\ln\left|\frac{T - 20}{c}\right| = kt$$

$$\frac{T - 20}{c} = e^{kt}$$

$$T = 20 + ce^{kt}$$

Since $T = 100$ when $t = 0$, it follows that the value of the integration constant c is 80, and

$$T = 20 + 80e^{kt}$$

To determine the value of the thermal coefficient k, we use the fact that $T = 40$ when $t = 10$. Hence

$$40 = 20 + 80e^{10k}$$

$$e^{10k} = \tfrac{1}{4}$$

$$10k = -\ln 4 \doteq -1.386$$

$$k \doteq -0.1386 \text{ (min}^{-1}\text{)}$$

The instantaneous temperature of the body is therefore given by the formula

$$T = 20 + 80e^{-0.1386t}$$

Among other things, this example illustrates how problems in differential equations often involve constants of two different sorts, namely, those, like c, which occur as integration constants in the general solution of the equation and those, like k, which occur as initially unknown physical parameters in the differential equation itself. Of course, regardless of their nature, these constants must be determined by making the complete solution of the differential equation satisfy the appropriate number of given conditions.

Example 2 A body of weight w falls from rest under the influence of gravity and a retarding force resulting from air resistance, assumed to be proportional to the velocity of the body. Find the equations expressing the velocity of the body and the distance it has fallen as functions of the time t.

To formulate this problem, we shall use **Newton's second law of motion:**

$$\text{Mass} \times \text{acceleration} = \text{force}$$

We are given that the weight of the body is w. Hence its mass is w/g, where g is the acceleration of gravity. As the dependent variable of the problem we shall take the distance s which the body has fallen from its initial position, the positive direction of s being taken downward. The velocity of the body is then simply $v = ds/dt$. There are only two forces which act on the body. The first is the gravitational attraction, which acts downward, that is, in the positive direction, and which is equal to the weight of the body w. The second is the air resistance, which acts in the upward or negative direction since it clearly opposes the downward motion of the body. From the statement of the problem, this force is equal to $-kv$. If we write the acceleration of the body as dv/dt, Newton's law gives us

$$\frac{w}{g}\frac{dv}{dt} = w - kv \qquad \text{or} \qquad \frac{dv}{dt} + \frac{kg}{w}v = g \tag{1}$$

Treating this as a linear equation (it is also separable), we first compute the integrating factor

$$e^{\int (kg/w)\,dt} = e^{kgt/w}$$

Then multiplying the last version of Eq. (1) by this factor and integrating, we get

$$ve^{kgt/w} = g\int e^{kgt/w}\,dt + c_1 = \frac{w}{k}e^{kgt/w} + c_1$$

or

$$v = \frac{w}{k} + c_1 e^{-kgt/w}$$

Since the body starts to fall from rest, we know that $v = 0$ when $t = 0$. Hence $c_1 = -w/k$ and, finally,

$$v = \frac{w}{k}(1 - e^{-kgt/w})$$

From this we note that as t increases indefinitely, $e^{-kgt/w}$ approaches zero and the velocity approaches the limiting value $v_\infty = w/k$.†

Since v now appears as an explicit function of t, we can find by direct integration the distance s which the body has fallen:

$$s = \int v\,dt = \int \frac{w}{k}(1 - e^{-kgt/w})\,dt = \frac{w}{k}\left(t + \frac{w}{kg}e^{-kgt/w} + c_2\right)$$

† The possibility of t becoming infinite presupposes that the body can fall arbitrarily far without striking the earth, which is impossible. It is interesting, however, that for bodies which experience heavy air resistance, such as packages dropped by parachute, the actual velocity soon becomes almost equal to v_∞.

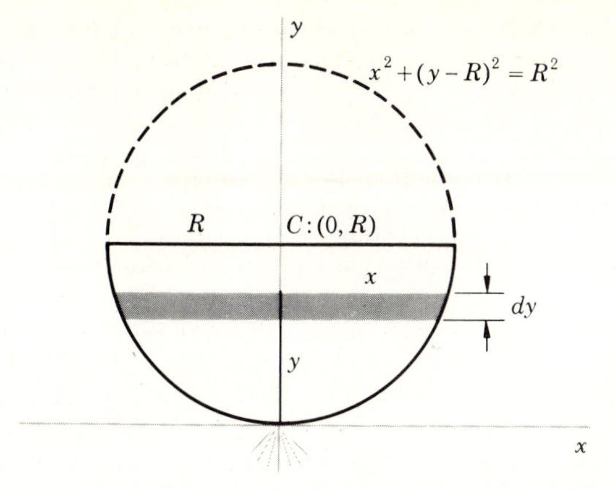

Figure 2.1 A vertical plane section through the center of a hemispherical tank.

From the statement of the problem, we know that $s = 0$ when $t = 0$. Hence, substituting, we find that $c_2 = -w/(kg)$, and therefore

$$s = \frac{w}{k}\left[t + \frac{w}{kg}(e^{-kgt/w} - 1)\right]$$

Example 3 A hemispherical tank of radius R is initially filled with water. At the bottom of the tank, there is a hole of radius r through which the water drains under the influence of gravity. Find the depth of the water in the tank at any time t, and determine how long it will take the tank to drain completely.

Let the origin be chosen at the lowest point of the tank, let y be the instantaneous depth of the water, and let x be the instantaneous radius of the free surface of the water (Fig. 2.1). Then in an infinitesimal interval dt, the water level will fall by the amount dy, and the resultant decrease in the volume of water in the tank will be

$$dV = \pi x^2\, dy$$

This, of course, must equal in magnitude the volume of water that leaves the orifice during the same interval dt. Now by **Torricelli's law**† the velocity with which a liquid issues from an orifice is

$$v = \sqrt{2gh}$$

where g is the acceleration of gravity and h is the instantaneous height, or **head,** of the liquid above the orifice. In the interval dt, then, a stream of water of length $v \times dt = \sqrt{2gy}\, dt$ and of cross-sectional area πr^2‡ will emerge from the outlet. The volume of this amount of water is

$$dV = \text{area} \times \text{length} = \pi r^2\sqrt{2gy}\, dt$$

† Named for the Italian mathematician and physicist Evangelista Torricelli (1608–1647).

‡ This neglects the fact that the stream contracts as it leaves the orifice. How much the cross section of the stream decreases depends in a very complicated way upon both the size and the shape of the orifice and also upon the head. However, in most practical problems, reasonably accurate answers can be obtained by assuming that the cross section of the stream just after it leaves the orifice is 0.6 times the area of the orifice.

Hence, equating the magnitudes of the two expressions for dV, we obtain the differential equation

$$-\pi x^2\, dy = \pi r^2 \sqrt{2gy}\, dt \tag{2}$$

The minus sign indicates that as t increases, the depth y decreases.

Before this equation can be solved, x must be expressed in terms of y. This is easily done through the use of the equation of the circle which describes a maximal vertical cross section of the tank:

$$x^2 + (y - R)^2 = R^2 \qquad \text{or} \qquad x^2 = 2yR - y^2$$

With this, the differential equation (2) can be written

$$\pi(2yR - y^2)\, dy = -\pi r^2 \sqrt{2gy}\, dt$$

This is a simple separable equation which can be solved without difficulty:

$$(2Ry^{1/2} - y^{3/2})\, dy = -r^2\sqrt{2g}\, dt$$

$$\tfrac{4}{3}Ry^{3/2} - \tfrac{2}{5}y^{5/2} = -r^2\sqrt{2g}\, t + c$$

Since $y = R$ when $t = 0$, we find

$$\tfrac{14}{15}R^{5/2} = c$$

and thus

$$\tfrac{4}{3}Ry^{3/2} - \tfrac{2}{5}y^{5/2} = -r^2\sqrt{2g}\, t + \tfrac{14}{15}R^{5/2}$$

To find how long it will take the tank to empty, we must determine the value of t when $y = 0$:

$$0 = -r^2\sqrt{2g}\, t + \tfrac{14}{15}R^{5/2}$$

$$t = \frac{14}{15}\frac{R^{5/2}}{r^2\sqrt{2g}}$$

Exercises for Section 2.2

1 Radium disintegrates at a rate proportional to the amount of radium instantaneously present. If the **half-life** of radium, that is, the time required for one-half of any given amount of radium to disintegrate, is 1,590 years, obtain a formula for the amount of radium present after t years. How long will it be before one-fourth of the original amount has disintegrated? What fraction of the original amount will disintegrate during the first century? During the third century?

2 Work Exercise 1 for the element plutonium, given that the half-life of plutonium is 50 years.

3 It is a fact of common experience that when a rope is wound around a rough cylinder, a small force at one end can resist a much larger force at the other. Quantitatively, it is found that throughout the portion of the rope in contact with the cylinder, the change in tension per unit length is proportional to the tension, the numerical value of the proportionality constant being the coefficient of friction between the rope and the cylinder divided by the radius of the cylinder. If we assume a coefficient of friction of 0.35, how many times must a rope be wound around a post 1 ft in diameter for a person holding one end of the rope to be able to resist a force 200 times greater than the person can exert?

4 According to **Lambert's† law of absorption,** when light passes through a transparent medium, the amount absorbed by any thin layer of the material perpendicular to the direction of the light is proportional to the amount incident on that layer and to the thickness of that layer. In his underwater explorations off Bermuda, Beebe observed that at a depth of 50 ft the intensity of illumination, that is, the amount of light incident on a unit area, was 10 Cd/ft^2 and that at 250 ft it had fallen to 0.2 Cd/ft^2. Find the formula connecting intensity with depth in this case.

5 A rapidly rotating flywheel, after power is cut off, coasts to rest under the influence of a friction torque which is proportional to the instantaneous angular velocity ω. If the angular velocity of the flywheel at the moment when the power is shut off is ω_0, find its instantaneous angular velocity as a function of time. *Hint:* Use **Newton's law in torsional form**

$$\text{Moment of inertia} \times \text{angular acceleration} = \text{torque}$$

to set up the differential equation describing the motion.

6 (*a*) If p lb/in^2 is the atmospheric pressure and ρ lb/in^3 is the density of air at a height of h in above the surface of the earth, show that $(dp/dh) + \rho = 0$.

(*b*) If $p = 14.7$ lb/in^2 at sea level and if it has fallen to half this value at 18,000 ft, find the formula for the pressure at any height, given that p is proportional to ρ (isothermal conditions). What is the predicted height of the atmosphere under this assumption?

(*c*) Work part (*b*), given that p is proportional to $\rho^{1.4}$ (adiabatic conditions).

7 Although water is often assumed to be incompressible, it actually is not. In fact, using pounds and feet as units, the weight of 1 ft^3 of water under pressure p is approximately $w(1 + kp)$, where $w = 64$, $k = 2 \times 10^{-8}$, and p is measured from standard atmospheric pressure as an origin. Using this information, find the pressure at any depth y below the surface of the ocean. At a depth of 6 mi, by what factor does the actual pressure exceed the pressure computed on the assumption that water is incompressible?

8 Work Example 3 if the tank is a right circular cylinder of height h and radius R whose axis is vertical.

9 Work Example 3 if the tank is an inverted right circular cone of height h and radius R.

10 Work Example 3 if the tank is formed by rotating about the y axis the arc of the parabola $y = x^2$ between $x = 0$ and $x = 2$.

11 A cylindrical tank of length l has semicircular end sections of radius R. The tank is placed in an untilted position with its axis horizontal and is initially filled with water. How long will it take the tank to drain through a hole of radius r in the bottom of the tank?

12 What is the shape of a tank which is a surface of revolution if the tank drains so that the water level falls at a constant rate?

13 What is the shape of a perpendicular cross section of a horizontal trough of constant cross section which drains so that the water level falls at a constant rate?

14 A tank having the shape of a right circular cylinder of height h and radius R is filled with water. The tank drains through an orifice whose area is controlled by a float valve in such a way that it is proportional to the instantaneous depth of the water in the tank. Express the depth of the water as a function of time. How long will it take the tank to drain?

† Named for the German physicist Johann Heinrich Lambert (1728–1777).

***15** Water flows into a vertical cylindrical tank of cross-sectional area A ft^2 at the rate of Q ft^3/min. At the same time, the water drains out under the influence of gravity through a hole of area a ft^2 in the base of the tank. If the water is initially h ft deep, find the instantaneous depth as a function of time. What is the limiting depth of the water as time increases indefinitely?

***16** A vertical cylindrical tank of height h and radius R has a narrow crack of width w running vertically from top to bottom. If the tank is initially filled with water and allowed to drain through the crack under the influence of gravity, find the instantaneous depth of the water as a function of time. How long will it take the tank to empty? *Hint:* First imagine the crack to be a series of adjacent orifices, and integrate to find the total efflux from the crack in the infinitesimal time interval dt.

17 A body cools in air of constant temperature 20°C according to Newton's law of cooling. Ten minutes after the body began to cool, its temperature was observed to be 75°C, and 10 min later its temperature was 50°C. What was its temperature when it began to cool?

18 According to **Fourier's law of heat conduction,**† the amount of heat in Btu‡ per unit time flowing through an area is proportional to the area and to the temperature gradient,‡ in degrees per unit length, in the direction perpendicular to the area. Using this law, obtain a formula for the heat lost under steady-state‡ conditions from 1 ft^2 of furnace wall h ft thick if the temperature in the furnace is T_0, the temperature of the air outside the furnace is T_1, and the thermal conductivity (i.e., the proportionality constant in Fourier's law) of the material of the furnace wall is k. What is the temperature distribution through the wall?

***19** Using Fourier's law of heat conduction, obtain a formula for the steady-state heat loss per unit time from a unit length of pipe of radius r_0 carrying steam at temperature T_0 if the pipe is covered with insulation of thickness w, the outer surface of which remains at the constant temperature T_1. What is the temperature distribution through the insulation, i.e., what is the temperature in the insulation as a function of the radius?

***20** A tank and is contents weigh 100 lb. The average heat capacity‡ of the tank and its contents is 0.5 Btu/(lb)(°F). The liquid in the tank is heated by an immersion heater which delivers 100 Btu/min. Heat is lost from the system at a rate proportional to the difference between the temperature of the system, assumed constant throughout at any instant, and the temperature of the surrounding air, the proportionality constant being 2 Btu/(min)(°F). If the air temperature remains constant at 70°F and if the initial temperature of the tank and its contents is 55°F, find the temperature of the tank as a function of t.

21 A stone is dropped from a balloon which is 1,760 ft above the earth and ascending vertically at the rate of 16 ft/s. Neglecting friction, find the equation of motion of the stone referred to an origin fixed in space at the point from which the stone was released. What is the maximum height reached by the stone? When and with what velocity does the stone strike the ground?

22 Using L'Hospital's rule to evaluate the indeterminate forms, verify that the velocity and distance laws derived in Example 2 reduce to the ideal laws $v = gt$ and $s = \frac{1}{2}gt^2$ when the coefficient of air resistance k approaches zero.

***23** An object is projected upward from the surface of the earth with velocity 64 ft/s. Air resistance is proportional to the first power of the velocity (as in Example 2), the proportionality constant being such that if the body were to fall freely, its velocity would approach

† Named for the French mathematician J. B. J. Fourier (1768–1830).
‡ See Appendix B.3 for the definition of technical terms such as this.

the limiting velocity $v_\infty = 256$ ft/s. How high does the body rise? When and with what velocity does it strike the ground?

***24** Work Example 2 given that the retarding force due to air resistance is proportional to the square of the velocity. Verify that the formulas you obtain for v and s reduce to the ideal laws $v = gt$ and $s = \frac{1}{2}gt^2$ when the coefficient of air resistance approaches zero.

***25** A particle of mass m moves along the x axis under the influence of a force which is directed toward the origin and is proportional to the distance of the particle from the origin. If the body starts from rest at the point where $x = x_0$, find the equations which express its velocity and its distance from the origin as functions of time.

***26** Work Exercise 25 if the force, instead of being directed toward the origin, is directed away from the origin.

****27** Work Exercise 25 if the particle moves under the influence of a force which is directed toward the origin and is inversely proportional to the square of the distance of the particle from the origin.

****28** Work Exercise 25 if the particle moves under the influence of a force which is directed away from the origin and is inversely proportional to the square of the distance of the particle from the origin.

****29** A body falls from rest from a height so great that the fact that the force of gravity varies inversely as the square of the distance from the center of the earth cannot be neglected. Find the equations expressing the velocity of fall and the distance fallen as functions of time in the ideal case in which air resistance is neglected.

****30** Under the conditions of Exercise 29, determine the minimum velocity with which a body must be projected upward if it is to leave the earth and never return.

****31** A cylinder of mass m and radius r rolls, without sliding, down an inclined plane of inclination angle α. Neglecting friction, express the distance which the cylinder has rolled down the plane as a function of time. *Hint:* Observe first that since friction is neglected, the principle of the conservation of energy (see p. 136) implies that the sum of the kinetic energy and the potential energy of the cylinder remains constant throughout the motion. Then note that the kinetic energy of the cylinder consists of two parts: that due to the translation of the cylinder and that due to its rotation.

32 A barge is being towed at 16 ft/s when the towline breaks. It continues thereafter in a straight line but slows down at a rate proportional to the square root of its instantaneous velocity. If 2 min after the towline breaks, the velocity of the barge is observed to be 9 ft/s, how far does it move before it comes to rest?

33 When a switch is closed in a circuit containing a resistance R, an inductance L, and a battery which supplies a constant voltage E, the current i builds up at a rate defined by the equation

$$L\frac{di}{dt} + Ri = E$$

Find the current i as a function of time. How long will it take i to reach one-half its final value?

34 When a capacitor of capacitance C is being charged through a resistance R by a battery which supplies a constant voltage E, the instantaneous charge Q on the capacitor satisfies the differential equation

$$R\frac{dQ}{dt} + \frac{Q}{C} = E$$

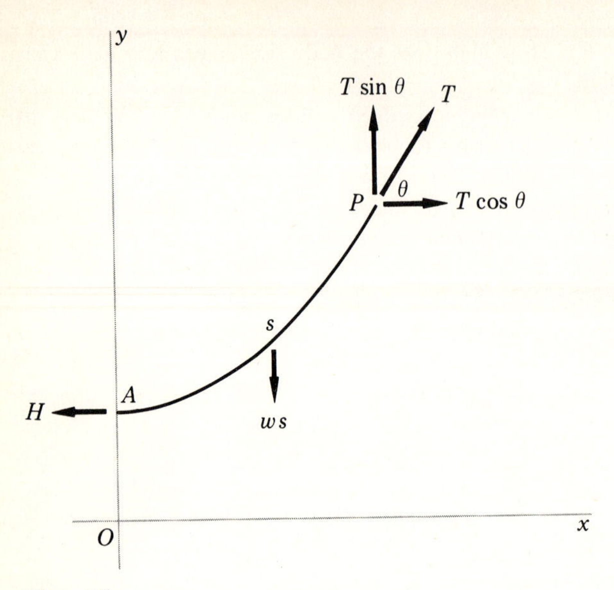

Figure 2.2

Find Q as a function of time if the capacitor is initially uncharged, i.e., if $Q_0 = 0$. How long will it be before the charge on the capacitor is one-half its final value?

35 Determine how an initially charged capacitor will discharge through a resistance; i.e., work Exercise 34 if there is no voltage source in the circuit and the charge on the capacitor is initially some nonzero value Q_0.

***36** Work Exercise 34 if the battery is replaced by a generator which supplies an alternating voltage $E = E_0 \sin \omega t$.

***37** What is the equation of the curve in which a perfectly flexible cable, of uniform weight per unit length w, will hang when it is suspended between two points at the same height? *Hint:* Consider a section of the cable, such as that shown in Fig. 2.2, and note that if H is the horizontal tension in the cable at its lowest point A, if T is the tension in the cable at a general point P, and if s is the length of the cable between A and P, then

$$T \sin \theta = ws \qquad T \cos \theta = H$$

These imply that

$$\tan \theta = y' = \frac{ws}{H} \qquad \text{and hence} \qquad \frac{dy'}{dx} = \frac{w}{H}\frac{ds}{dx}$$

After ds/dx is expressed in terms of y' by means of the familiar formula for the differential of arc length, the integrations required to find y' and then y will be simplified if they are carried out in terms of hyperbolic functions.

38 Find the equation of the curve in which a perfectly flexible cable hangs when, instead of bearing a uniform load per unit length of cable, as in Exercise 37, it bears a uniform horizontal load. (This is approximately the case when the cable is part of a suspension bridge carrying a horizontal roadbed whose weight is much greater than the weight of the cable.)

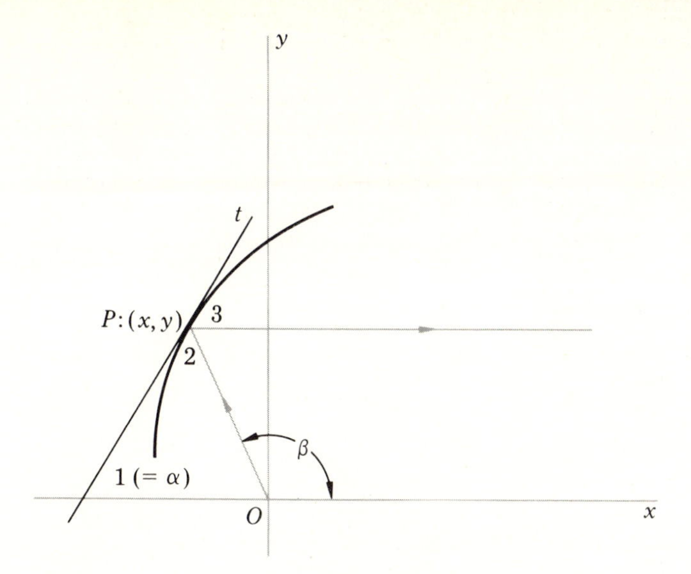

Figure 2.3

***39** Find the equation of a curve with the property that light rays emanating from the origin will all be reflected into rays which are parallel to the x axis. *Hint:* Note in Fig. 2.3 that if t is the tangent to the required curve at a general point P, then the angles 1, 2, and 3 all have the same measure, say α. Then from the exterior-angle theorem $\alpha = \beta - \alpha$, which implies that $\tan \alpha = \tan (\beta - \alpha)$. Expanding $\tan (\beta - \alpha)$ and observing that $\tan \alpha = y'$ and $\tan \beta = y/x$ leads to the differential equation $y = 2xy' + y(y')^2$. This can be solved by first solving for y' and then using an integrating factor, or it can be reduced to a Clairaut equation by multiplying it by y and setting y^2 equal to a new variable u.

***40** A vertical cylindrical tank of radius r is filled with liquid to a depth h. The tank is rotated about its axis with constant angular velocity ω. Find the equation of the curve in which the free surface of the liquid is intersected by a plane through the axis of the cylinder, assuming that the tank is sufficiently deep to prevent liquid from spilling over the edge. *Hint:* The normal force maintaining a particle in equilibrium in the surface of the liquid can be resolved into an upward vertical component w, where w is the weight of the particle, and a centripetal force $wx\omega^2/g$ directed radially inward, where x is the distance of the particle from the axis of rotation.

2.3 PROBLEMS FROM CHEMISTRY

Example 1 When ethyl acetate in dilute aqueous solution is heated in the presence of a small amount of acid, it decomposes according to the equation

$$\underset{\text{(Ethyl acetate)}}{CH_3COOC_2H_5} + \underset{\text{(Water)}}{H_2O} \rightarrow \underset{\text{(Acetic acid)}}{CH_3COOH} + \underset{\text{(Ethyl alcohol)}}{C_2H_5OH}$$

Since this reaction takes place in dilute solution, the quantity of water present is so great that the loss of the small amount which combines with the ethyl acetate produces no appreciable change in the total amount. Hence of the reacting substances only the ethyl acetate suffers a measurable change in concentration. A chemical reaction of this sort, in which the concentration of only one reacting substance changes, is called a **first-order reaction.** It is a law of physical chemistry that the rate at which a substance is being used up, i.e., transformed, in a first-order reaction is proportional to the amount of that substance instantaneously present. If the initial concentration of ethyl acetate is C_0, find the expression for its concentration at any time t.

Let Q be the amount of ethyl acetate instantaneously present in the solution, let V be the (constant) amount of water in which it is dissolved, and let C be the instantaneous concentration of the ethyl acetate. Then $Q = CV$, and, from the given law governing first-order reactions,

$$\frac{dQ}{dt} = -kQ \qquad \text{or} \qquad \frac{d(CV)}{dt} = -k(CV)$$

or finally

$$\frac{dC}{dt} = -kC$$

This is a simple separable equation which can be solved immediately:

$$\frac{dC}{C} = -k\,dt$$

$$\ln C = -kt + \ln A$$

$$\ln \frac{C}{A} = -kt$$

$$C = Ae^{-kt}$$

Since the initial value of the concentration is C_0, it follows that $A = C_0$. Hence

$$C = C_0 e^{-kt}$$

To determine the rate constant k, we would need to know the value of the concentration at some particular time in the course of the experiment. Suppose, for instance, that it was observed that in 30 min the concentration had dropped to four-fifths of its initial value. Then, substituting the values $C = \frac{4}{5}C_0$ and $t = 30$, we would find

$$\tfrac{4}{5}C_0 = C_0 e^{-30k}$$

$$e^{-30k} = \tfrac{4}{5}$$

$$-30k = \ln \tfrac{4}{5} \doteq -0.22314$$

$$k \doteq 0.007438 \ (\text{min}^{-1})$$

Example 2 A tank is initially filled with 100 gal of salt solution containing 1 lb of salt per gallon. Fresh brine containing 2 lb of salt per gallon runs into the tank at the rate of 5 gal/min, and the mixture, assumed to be kept uniform by stirring, runs out at the same rate. Find the amount of salt in the tank at any time t, and determine how long it will take for this amount to reach 150 lb.

Let Q lb be the total amount of salt in the tank at any time t, and let dQ be the increase in this amount during the infinitesimal interval of time dt. Then at any time t, the amount of salt per gallon of solution is $Q/100$ (lb/gal). Now the change dQ in the total amount of salt in the tank is clearly the net gain in the interval dt due to the fresh brine running into, and the mixture running out of, the tank. The rate at which salt enters the tank is

$$5(\text{gal/min}) \times 2(\text{lb/gal}) = 10 \text{ lb/min}$$

Hence in the interval dt the gain in salt from this source is

$$10(\text{lb/min}) \times dt \text{ min} = 10\, dt \text{ lb}$$

Likewise, since the concentration of salt in the mixture as it leaves the tank is the same as the concentration $Q/100$ in the tank itself, the amount of salt leaving the tank in the interval dt is

$$5\left(\frac{\text{gal}}{\text{min}}\right) \times \frac{Q}{100}\left(\frac{\text{lb}}{\text{gal}}\right) \times dt \text{ min} = \frac{Q}{20}\, dt \text{ lb}$$

Therefore, in pounds of salt,

$$dQ = 10\, dt - \frac{Q}{20}\, dt = \left(10 - \frac{Q}{20}\right) dt$$

This equation can be written in the form

$$\frac{dQ}{200 - Q} = \frac{dt}{20} \tag{1}$$

and solved as a separable equation, or it can be written

$$\frac{dQ}{dt} + \frac{Q}{20} = 10 \tag{2}$$

and treated as a linear equation.

Considering it as a linear equation, we must first compute the integrating factor

$$e^{\int P\, dt} = e^{\int dt/20} = e^{t/20}$$

Multiplying Eq. (2) by this factor gives

$$e^{t/20}\left(\frac{dQ}{dt} + \frac{Q}{20}\right) = 10e^{t/20}$$

From this, by integration, we obtain

$$Qe^{t/20} = 200e^{t/20} + c \qquad \text{or} \qquad Q = 200 + ce^{-t/20}$$

Substituting the initial condition $t = 0$, $Q = 100$, we find

$$100 = 200 + c \qquad \text{or} \qquad c = -100$$

Hence
$$Q = 200 - 100e^{-t/20}$$

To find how long it will be before there is 150 lb of salt in the tank, we must find the value of t such that

$$150 = 200 - 100e^{-t/20} \qquad \text{or} \qquad e^{-t/20} = \tfrac{1}{2}$$

From this we have at once

$$-\frac{t}{20} = \ln\frac{1}{2} = -\ln 2 \doteq -0.693 \qquad \text{and} \qquad t \doteq 13.9 \text{ min}$$

Example 3 Under certain conditions it is observed that the rate at which a solid substance dissolves varies directly as the product of the amount of undissolved solid present in the solvent and the difference between the saturation concentration and the instantaneous concentration of the substance. If 40 kg of solute is dumped into a tank containing 120 kg of solvent and at the end of 12 min the concentration is observed to be 1 part in 30, find the amount of solute in solution at any time t if the saturation concentration is 1 part of solute in 3 parts of solvent.

If Q is the amount of the material in solution at time t, then $40 - Q$ is the amount of undissolved material present at that time and $Q/120$ is the corresponding concentration. Hence, according to the given law,

$$\frac{dQ}{dt} = k(40 - Q)\left(\frac{1}{3} - \frac{Q}{120}\right) = \frac{k}{120}(40 - Q)^2$$

This is a simple separable equation, and we have at once

$$\frac{dQ}{(40 - Q)^2} = \frac{k}{120}\,dt \qquad \text{and} \qquad \frac{1}{40 - Q} = \frac{k}{120}t + c$$

Since $Q = 0$ when $t = 0$, we find that $c = \frac{1}{40}$. Also, when $t = 12$, $Q = \frac{1}{30}(120) = 4$. Hence

$$\frac{1}{40 - 4} = \frac{k}{120}12 + \frac{1}{40} \qquad \text{and} \qquad \frac{k}{120} = \frac{1}{4{,}320}$$

Therefore

$$\frac{1}{40 - Q} = \frac{t}{4{,}320} + \frac{1}{40} = \frac{t + 108}{4{,}320}$$

$$40 - Q = \frac{4{,}320}{t + 108}$$

$$Q = 40 - \frac{4{,}320}{t + 108}$$

Exercises for Section 2.3

1 In some chemical reactions where two substances combine to form a third, the amount of each of the reacting substances changes appreciably. A reaction of this sort is called a **second-order reaction,** and in such cases it is observed that the rate at which the resulting compound is being formed is proportional to the product of the untransformed amounts of the two reacting substances. If two substances combine in the ratio 1 : 2, by weight, to form a third substance and if it is observed that 10 min after 10 g of the first substance and 20 g of the second are mixed, the amount of the product that has been formed is 5 g, find an expression for the amount of the product present at any time t. How long will it be before one-half the final amount of the product has been formed?

2 Work Exercise 1 given that 20 g of each substance are mixed.

3 Work Exercise 1 given that 10 g of the first substance and 30 g of the second substance are mixed.

4 Most first-order chemical reactions are reversible; that is, not only is substance A being transformed into substance B, but at the same time substance B is being transformed into substance A. If the rate constant for the reaction $A \to B$ is k_1, if the rate constant for the reaction $B \to A$ is k_2, and if initially the amount of substance A is A_0 and the amount of substance B is zero, find the amount of substance B present at any time t. What is the limiting value of the ratio of the amounts of A and B as the reaction approaches equilibrium? *Hint:* Note that the total rate of change of A consists of the rate at which A is being used up by the reaction $A \to B$ and the rate at which A is being produced by the reaction $B \to A$.

5 Some chemical reactions are **autocatalytic;** that is, the product of the reaction catalyzes its own formation. This means that in an autocatalytic reaction in which a substance A is transformed into a substance B, the rate of formation of B is proportional to the product of the instantaneous amounts of both A and B. If the initial amount of substance A is A_0 and the initial amount of B is B_0, find an expression for the amount of B present at any time t.

6 Work Example 3, given that the amount of solute dumped in the tank is 20 kg instead of 40 kg.

7 Work Example 3, given that the saturation concentration is $\frac{1}{4}$ instead of $\frac{1}{3}$.

***8** Work Example 3 with *concentration* defined as the ratio of solute to solution instead of solute to solvent.

***9** Work Example 3 with *concentration* defined as the ratio of solute to solution instead of solute to solvent, given that the saturation concentration is $\frac{1}{4}$ instead of $\frac{1}{3}$.

10 Work Example 2, given that fresh water rather than brine runs into the tank.

11 Work Example 2, given that the tank is initially filled with pure water.

***12** A tank contains 100 gal of brine in which 50 lb of salt is dissolved. Brine containing 2 lb/gal runs into the tank at the rate of 3 gal/min; and the mixture, assumed to be kept uniform by stirring, runs out at the rate of 2 gal/min. Assuming that the tank is sufficiently large to avoid overflow, find the amount of salt in the tank as a function of the time t. When will the concentration of salt in the tank reach $\frac{3}{2}$ lb/gal?

***13** Work Exercise 12 with the rates of influx and efflux interchanged. When will the tank contain the maximum amount of salt, and what is this amount?

***14** A mothball loses mass by evaporation at a rate proportional to its instantaneous surface area. If half the mass is lost in 100 days, how long will it be before the radius has decreased to one-half its initial value? How long will it be before the mothball disappears completely?

****15** When a volatile substance is placed in a closed container, molecules leave its surface at a rate proportional to the area of the surface and return at a rate proportional to the amount that has evaporated. If a volatile material is spread evenly to a depth h over the bottom of a closed box, find the depth of the material as a function of the time t. Under what conditions, if any, will all the material eventually evaporate?

2.4 PROBLEMS FROM BIOLOGY

Example 1 Consider a colony of microorganisms reproducing through simple cell division under ideal conditions of unlimited food supply and total absence of predators. If it is observed that the population increases by ρ percent each hour, find an expression for the population at any time t.

According to the data of the problem, if N is the number of organisms present at any time t, then 1 h later the number of organisms will have increased by the amount

$\Delta N = (\rho/100)N$. In other words, for any period of length $\Delta t = 1$ h, the corresponding population change ΔN is proportional to the size of the population at the beginning of the hour. From this it seems plausible that a similar relation should hold for the increase in population in any period of time; that is, the increase in any interval should be proportional to the length of the interval Δt as well as the population size N at the beginning of the period. Under this assumption,

$$\Delta N = \left(\frac{r}{100} N\right) \Delta t \qquad \text{or} \qquad \frac{\Delta N}{\Delta t} = \frac{r}{100} N\dagger$$

We cannot logically consider the limit of the ratio $\Delta N/\Delta t$ as $\Delta t \to 0$ because for Δt sufficiently small, the number of new organisms appearing in the interval Δt may be either 0 or 1 but nothing in between. Nonetheless, the last expression *suggests* that we explore the equation

$$\frac{dN}{dt} = \frac{r}{100} N \tag{1}$$

as an approximate description of the behavior of the actual system.

Equation (1) is a very simple separable equation which can be solved immediately:

$$\frac{dN}{N} = \frac{r}{100} dt$$

$$\ln N = \frac{r}{100} t + \ln C$$

$$\ln \frac{N}{C} = \frac{r}{100} t$$

$$N = Ce^{rt/100}$$

When $t = 0$, we know that $N = N_0$. Hence, putting $t = 0$, we find that $C = N_0$ and

$$N = N_0 e^{rt/100} \tag{2}$$

To determine the rate constant r, we must use the given information that when $t = 1$, the population has increased to $N_0 + (\rho/100)N_0$. Thus,

$$N_0 + \frac{\rho}{100} N_0 = N_0 e^{r/100}$$

$$e^{r/100} = 1 + \frac{\rho}{100}$$

$$r = 100 \ln \left(1 + \frac{\rho}{100}\right)$$

† The factor r in this equation is not the same as the factor ρ. In fact, if $r = \rho$, then taking $\Delta t = \frac{1}{2}$ as an illustration, the number of organisms present at the end of $\frac{1}{2}$ h would be $N + (\rho/200)N = N[1 + (\rho/200)]$ and the number present at the end of a second half-hour would be

$$N\left(1 + \frac{\rho}{200}\right) + \left[N\left(1 + \frac{\rho}{200}\right)\right]\frac{\rho}{200} = N\left(1 + \frac{\rho}{200}\right)^2 = N\left(1 + \frac{\rho}{100} + \frac{\rho^2}{40{,}000}\right)$$

which is more than the observed number present after 1 h, namely, $N[1 + (\rho/100)]$.

Table 2.1

Percentage increase in one time period	Number of periods required for population to double
10	7.27
5	14.21
3	23.45
2	35.00
1	69.66
0.5	138.98

With r expressed in terms of the observed growth rate of ρ percent per hour, Eq. (2) now provides a formula for the population size at any time t.

This example is, of course, unrealistic since it neglects such important factors as a limited food supply, the presence of other species competing for the food supply, and the presence of predators. However, it does illustrate the highly important **law of exponential growth,** sometimes called the **compound interest law,** which describes the increase of quantities which grow at a rate proportional to their current size. How dramatic this increase may be is shown in Table 2.1 which gives the time it will take a population to double for various rates of increase. Thus a population growing at only 1 percent per year will double in slightly less than 70 years.

Example 2 In a finite world, no population can become infinite; limiting factors of one kind or another must come into play. One such factor is obviously the food supply, because if this is finite, the amount of food available for each individual decreases as the population increases. This, in turn, through malnutrition or possibly starvation, should tend to make the death rate increase and the birthrate decrease, thereby reducing the growth rate enough to keep the population finite. One way to incorporate such a factor into a population model is to suppose that in the continuous approximation provided by the equation

$$\frac{dN}{dt} = \frac{r}{100}N$$

which we studied in Example 1, the growth factor r (arising from the excess of births over deaths) is not constant but decreases as N increases. One possible formula for the variation of r with N, which biologists have found in good agreement with experiments on fruit flies, is

$$\frac{r}{100} = a - bN \qquad a, b > 0$$

Under this assumption, the differential equation describing the growth of the population becomes

$$\frac{dN}{dt} = (a - bN)N \tag{3}$$

At the outset, we note that both $N = 0$ and $N = a/b$ are solutions of Eq. (3). $N = 0$ is a trivial solution in which we have no interest. $N = a/b$ is a possible solution which asserts that if the size of the population is initially $N_0 = a/b$, it will remain stationary at that value. If $N_0 < a/b$, Eq. (3) implies that $dN/dt > 0$ and the size of the population increases toward a/b as time goes on. Similarly, if $N_0 > a/b$, Eq. (3) implies that $dN/dt < 0$ and the size of the population decreases toward a/b as time goes on.

Assuming now that N is neither 0 nor a/b, we can write Eq. (3) in the form

$$\frac{dN}{(a - bN)N} = dt \tag{3a}$$

To integrate this, we must first use the method of partial fractions, by writing

$$\frac{1}{(a - bN)N} = \frac{A}{N} + \frac{B}{a - bN} = \frac{A(a - bN) + BN}{(a - bN)N} = \frac{Aa + (-Ab + B)N}{(a - bN)N}$$

Then equating coefficients of like terms in the numerators of the first and last fractions, we find

$$1 = aA \qquad \text{and} \qquad 0 = -bA + B$$

$$A = \frac{1}{a} \qquad\qquad B = bA = \frac{b}{a}$$

Thus the differential equation (3a) becomes

$$\frac{1}{a}\left(\frac{1}{N} + \frac{b}{a - bN}\right) dN = dt$$

and, integrating, we have

$$\ln N - \ln |a - bN| = at - \ln |C|$$

$$\ln\left(\frac{|C|N}{|a - bN|}\right) = at$$

$$\frac{|C|N}{|a - bN|} = e^{at}$$

Assuming that N_0 is the number of individuals present when $t = 0$, we find from the last equation that

$$|C| = \frac{|a - bN_0|}{N_0}$$

and

$$\frac{|a - bN_0|N}{|a - bN|N_0} = e^{at} \tag{4}$$

Finally, solving Eq. (4) for N, noting that $a - bN_0$ and $a - bN$ are always of like sign, we obtain

$$N = \frac{aN_0e^{at}}{(a - bN_0) + bN_0e^{at}} = \frac{aN_0}{bN_0 + (a - bN_0)e^{-at}} \tag{5}$$

When $t = 0$, Eq. (5) of course reduces to $N = N_0$. As t becomes infinite, the factor e^{-at} approaches zero and N approaches the limiting value a/b. Thus our model reflects

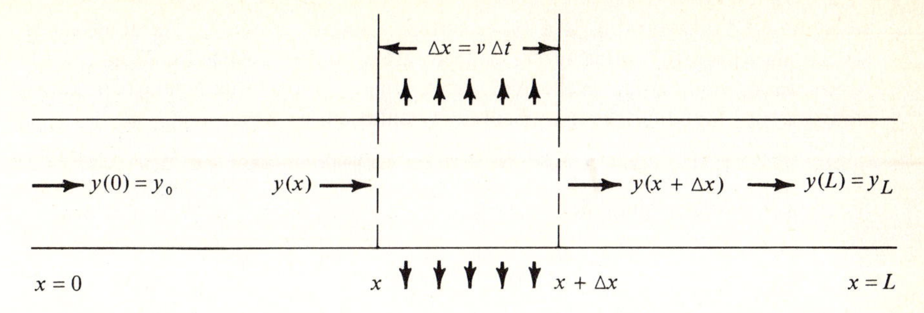

Figure 2.4 Solute diffusing from a tube through which a solution is flowing.

reality at least to the extent that it predicts a finite limit to the growth of the population. To determine the parameters a and b, we would need to know the size of the population at two times other than $t = 0$. Substituting these data, say (t_1, N_1) and (t_2, N_2), into Eq. (5) would give two simultaneous equations in a and b. To solve them would be very difficult, however.

Example 3 As a possible model of a diffusion process in the bloodstream in the human body, consider a solution moving with constant velocity v through a cylindrical tube of length L and radius r. We suppose that as the solution moves through the tube, some of the solute which it contains diffuses through the wall of the tube into an ambient solution of the same solute of lower concentration, while some continues to be transported through the tube. As variables, we let x be a distance coordinate along the tube and $y(x)$ be the concentration of the solute at any point x, assumed uniform over the cross section of the tube. As boundary conditions, we assume that $y(0) = y_0$ and $y(L) = y_L(< y_0)$ are known. We wish to know the concentration $y(x)$ at any point of the tube.

As a principle to use in formulating this problem, we have **Frick's law:**

The time rate at which a solute diffuses through a thin membrane in a direction perpendicular to the membrane is proportional to the area of the membrane and to the difference between the concentrations of the solute on the two sides of the membrane.

We begin by considering conditions in a typical segment of the tube between x and $x + \Delta x$ (Fig. 2.4). The concentration of the solution entering the segment is $y(x)$; the concentration of the solution leaving the segment is $y(x + \Delta x)$. In the time Δt that it takes the solution to move through the segment, an amount of solute equal to

$$\text{Concentration} \times \text{volume} = y(x)\pi r^2\,\Delta x$$

enters the left end of the segment, and the amount $y(x + \Delta x)\pi r^2\,\Delta x$ leaves the right end of the segment. The difference

$$[y(x) - y(x + \Delta x)]\pi r^2\,\Delta x$$

must have left the segment by diffusion through the wall of the tube according to Frick's law. The expression for this amount, as given by Frick's law, is

$$\text{Rate of diffusion} \times \text{time} = k(2\pi r\,\Delta x)[y(x + \theta\,\Delta x) - c]\,\Delta t$$

where $x + \theta\,\Delta x$, $0 < \theta < 1$, is a typical point between x and $x + \Delta x$ at which to assume an "average" value of the concentration, and c, assumed constant, is the concentration of the solute in the fluid surrounding the tube. Equating the two expressions we have found for the loss of solute by diffusion, we have

$$[y(x) - y(x + \Delta x)]\pi r^2\,\Delta x = k(2\pi r\,\Delta x)[y(x + \theta\,\Delta x) - c]\,\Delta t$$

Since $\Delta x = v\,\Delta t$, this simplifies to

$$\frac{y(x) - y(x + \Delta x)}{\Delta x} = \frac{2k}{rv}[y(x + \theta\,\Delta x) - c]$$

and, taking limits (as though the system were continuous),

$$-\frac{dy}{dx} = \frac{2k}{rv}[y(x) - c]$$

By hypothesis, $y > c$; hence $y(x) - c \neq 0$, and we can solve this equation by separating variables:

$$\frac{dy}{y - c} = -\frac{2k}{rv}\,dx$$

$$\ln\,(y - c) = -\frac{2k}{rv}x + \ln B$$

Putting $x = 0$ and $y = y_0$, we find that $\ln B = \ln\,(y_0 - c)$, and

$$\ln\frac{y - c}{y_0 - c} = -\frac{2k}{rv}x \qquad \text{or} \qquad y = c + (y_0 - c)e^{-2kx/(rv)} \tag{6}$$

A (presumably) more convenient form of the answer, involving only the data c, y_0, y_L, and L, can be obtained as follows: If the values $x = L$ and $y = y_L$ are substituted into the first form of Eq. (6), the result is

$$\ln\frac{y_L - c}{y_0 - c} = -\frac{2k}{rv}L \tag{7}$$

Now if the first form of Eq. (6) is divided by Eq. (7), we have

$$\frac{\ln\,[(y - c)/(y_0 - c)]}{\ln\,[(y_L - c)/(y_0 - c)]} = \frac{x}{L} \qquad \text{or} \qquad \ln\frac{y - c}{y_0 - c} = \frac{x}{L}\ln\frac{y_L - c}{y_0 - c} = \ln\left(\frac{y_L - c}{y_0 - c}\right)^{x/L}$$

and finally

$$y = c + (y_0 - c)\left(\frac{y_L - c}{y_0 - c}\right)^{x/L}$$

Exercises to Section 2.4

1 A lake has various streams flowing into it and flowing out of it, the total rates of influx and efflux being equal. For a long time, the streams flowing into the lake were polluted, and pollution in the lake built up to an undesirable level. However, as a result of conservation efforts, the sources of pollution in the streams were eliminated, and now only pure water flows into the lake. If the volume of the lake is V km^3, if the rate of influx and efflux is

r km^3/year, and if the pollutants are always uniformly distributed throughout the lake, obtain formulas for the time it will take for the pollution in the lake to be reduced (*a*) to one-half its level at the time of the clean-up and (*b*) to one-tenth its level at the time of the clean-up. Determine numerical values for these times for Lake Erie and Lake Ontario, given the following data:

	V	r
Lake Erie	460 km^3	175 km^3/yr
Lake Ontario	1,600 km^3	209 km^3/yr

Hint: This problem is very much like Example 2 and Exercise 10, Sec. 2.3.

2 Living tissues, both plant and animal, contain carbon, derived ultimately from the carbon dioxide in the air. Most of this carbon is the stable isotope ^{12}C, but a small, fixed percentage of it is unstable radioactive ^{14}C. It appears that there is little or no segregation of these two forms of carbon in living organisms, and the ratio ^{14}C/^{12}C is essentially constant for all types of tissue. When the tissue dies, the vital processes, of course, end, no more carbon of either form is added to the tissue, and the amount of ^{14}C present at the time of death decreases at a rate proportional to the amount instantaneously present, with a half-life of approximately 5,500 years. If x_0 is the amount of ^{14}C in a given specimen of tissue at the moment of death (determined as a known percentage of the unchanged amount of ^{12}C calculated by chemical analysis of the specimen) and if x is the amount present in the tissue t years after death, express x as a function of t.

3 A piece of charcoal found in the Lascaux Cave in France (the cave with the remarkable paintings of prehistoric animals) contained 14.8 percent of the original amount of ^{14}C. Using the result of Exercise 2, date the occupation of the cave that produced the charcoal.

4 A charred branch of a tree killed by the eruption that formed Crater Lake in Oregon contained 44.5 percent of the original amount of ^{14}C. Date the eruption.

5 For a certain population, both the birthrate and the death rate are constant multiples of the number of individuals instantaneously present. Find the population as a function of time.

***6** In a population in which reproduction is bisexual, rather than asexual, it is probably more realistic to suppose that the birthrate is proportional to the number of pairs of individuals rather than to the number of individuals. Assuming a birthrate of this nature and a death rate proportional to the number of individuals, set up and solve the differential equation governing the population size N, given that the initial size of the population is N_0.

***7** Work Exercise 6 if, as in Example 2, the proportionality factor in the death rate, instead of being a constant, is assumed to vary as a linear function of the population size N.

***8** Discuss the possibility of solving Exercise 6 if the proportionality factors in both the birthrate and the death rate vary as linear functions of the population size N.

In Exercises 5 and 6 we assumed birthrates and death rates that were constant multiples of either the population size or the possible number of pairs of individuals in the population. In Example 2 and in Exercises 7 and 8, we assumed that the proportionality factors in these rates varied as linear functions of the population size N. For a population of rational individuals, i.e., humans, there is still another possibility: It may well be that, *independent of the size of the population*, changing social values and objectives will act to change the birthrate while at the same time medical progress will act to change the death rate. The natural way to take such influences into account is to assume that the proportionality factors in the birthrate and death rate are functions of the time t, say $k_b(t)$ and $k_d(t)$, so that the fundamental differential equation becomes

$$\frac{dN}{dt} = k_b(t)N - k_d(t)N$$

Solve this equation for the choices of $k_b(t)$ and $k_d(t)$ indicated in the following exercises, and determine the limiting behavior of the populations which they predict.

9 $k_b(t) = b_1 - b_2 t, \qquad k_d = d_1 - d_2 t$

***10** $k_b(t) = b_1 e^{-b_2 t}, \qquad k_d(t) = d_1 e^{-d_2 t}$

***11** $k_b(t) = b_1 /(b_2^2 - t^2), \qquad k_d(t) = d_1 /(d_2^2 - t^2)$

***12** Discuss the possibility of solving Exercise 6 if the proportionality factors involved in the birthrate and death rate are nonconstant functions of time.

13 Let V be the volume of the fluids in the human body, and let Q be the concentration of glucose instantaneously present in, and uniformly distributed through, these fluids. Let the tissues of the body absorb glucose at a rate proportional to the instantaneous concentration of glucose in the body fluids. When the glucose level is too low (as it is in many illnesses), glucose solution must be injected into the veins. Suppose that this is done in such a way that A mg of glucose enters the veins per minute but that the accompanying liquid does not appreciably increase the volume of the body fluids. If Q_0 is the initial concentration of the glucose, find an expression for the instantaneous concentration of glucose in the body fluids as a function of time.

14 Consider a fluid-filled cell of volume V and surface area A completely immersed in a fluid in which the concentration of a certain solute is c. Assuming that c remains constant (Why is this a reasonable assumption?), use Frick's law to show that the diffusion of the solute into the cell is described by the differential equation

$$\frac{dy}{dt} = k\frac{A}{V}(c - y)$$

where y is the concentration of the solute in the cell (assumed constant throughout the cell) and k is the permeability coefficient. Solve this equation, given that the initial concentration of the solute in the cell is y_0 ($< c$).

15 In Exercise 14, if the concentration of the solute in the ambient fluid is 0.05, if the initial concentration of the solute in the cell is 0.01, and if after 10 min the concentration is 0.02, how long will it be before the concentration in the cell is 0.03? 0.04?

***16** A diffusion model involves two identical compartments each of volume V which share a common boundary surface of area A. The boundary between the two compartments is a permeable membrane, but the rest of the surface of each compartment is impervious to diffusion. Initially the compartments are filled with solutions of a certain solute of differing concentrations. Let x_0 and x denote, respectively, the initial and instantaneous concentrations of the solute in the more dilute solution, and let y_0 and y denote the initial and instantaneous concentrations in the more concentrated solution. Use Frick's law to show that the diffusion process between the two compartments is described by the equations

$$x + y = x_0 + y_0$$

$$\frac{dx}{dt} = k\frac{A}{V}(x_0 + y_0 - 2x)$$

Solve these equations, and obtain formulas for the concentration of the solute in each compartment as a function of time.

17 If the initial concentrations in the two compartments in Exercise 16 are $x_0 = 0$ and $y_0 = 0.10$ and if after 20 min they are observed to be 0.02 and 0.08, how long will it be before they are 0.04 and 0.06?

18 If the two compartments in Exercise 16 are identical circular cylinders of length 20 cm and radius 4 cm, having one of their circular bases as their common boundary surface, if $x_0 = 0.03$ and $y_0 = 0.12$, and if after 30 min it is observed that $x = 0.05$ and $y = 0.10$, what is the value of the permeability coefficient k?

***19** Work Exercise 16 if the volume of the compartment containing the more dilute solution is 3 times the volume which contains the more concentrated solution.

2.5 PROBLEMS FROM GEOMETRY

Example 1 Let t be the tangent to a curve C at a general point P. Find the equation of the family of curves which have the property that the y axis bisects the segment of the tangent between its point of contact and its intersection with the x axis.

Let P: (u, v) be a general point on the required curve C, let t be the tangent to C at P, and let A and B be, respectively, the intersections of t with the x and y axes (Fig. 2.5). The requirement of the problem is that $|AB| = |BP|$. However, the lengths of the segments AB and BP will be equal if and only if the lengths of their horizontal projections are equal, and so we need consider only these projections. Now, for the case depicted in Fig. 2.5, the length of the horizontal projection of BP is the abscissa of P, namely u. Similarly, the length of the horizontal projection of AB is the negative of the abscissa of A. To find this, we first find the equation of t, namely,

$$y - v = \left.\frac{dy}{dx}\right|_{u,\,v}(x - u) \qquad \text{or} \qquad y - v = v'(x - u)$$

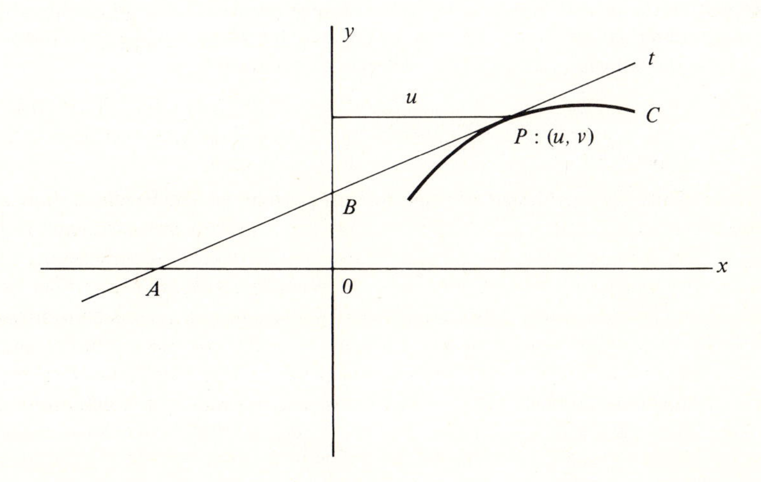

Figure 2.5 A curve and a general tangent.

and then solve for the value of x when $y = 0$, getting $x_A = u - v/v'$. Equating the lengths of the two horizontal projections, we obtain the equation

$$-\left(u - \frac{v}{v'}\right) = u \qquad \text{or} \qquad \frac{v'}{v} = \frac{1}{2u}$$

This simple separable differential equation can be solved at once, and we have

$$\frac{dv}{v} = \frac{du}{2u}$$

$$2 \ln |v| = \ln |u| + \ln |c|$$

$$v^2 = cu$$

or, reverting to the more conventional x, y notation,

$$y^2 = cx$$

Example 2 The curves of a family C are said to be **orthogonal trajectories** of the curves of a family K, and vice versa, if at every intersection of a curve of C with a curve of K the two curves are perpendicular. Using this definition, find the equation of the family of orthogonal trajectories of the curves of the family defined by the equation $y^2 = 4cx$.

To find the equation of the orthogonal trajectories of the curves of the given family, we must first find the expression for the slope of a general curve of this family at a general point. This, of course, is just the process of finding the differential equation satisfied by a given family of functions, which we illustrated in Example 9, Sec. 1.2. In this case, differentiating the equation $y^2 = 4cx$, we have

$$2yy' = 4c \tag{1}$$

Then, by eliminating c, we find $y^2 = 2yy'x$, and from this we obtain the required slope formula

$$y' = \frac{y}{2x}$$

Since the curves $y^2 = 4cx$ and their orthogonal trajectories are to be perpendicular at every intersection, it follows that at every point the slopes of the members of the two families which pass through that point must be negative reciprocals. Thus the required curves must be the solution curves of the differential equation

$$y' = -\frac{2x}{y}\dagger$$

† Note that it would be incorrect to take the slope formula given by Eq. (1), namely

$$y' = \frac{2c}{y}$$

and use its negative reciprocal

$$y' = -\frac{y}{2c}$$

as the differential equation to be satisfied by the required orthogonal trajectories. The curves obtained by integrating this simple separable equation, namely $y = ke^{-x/2c}$, depend on the two constants k and c, and a curve of this family and a curve of the given family $y^2 = 4cx$ will intersect at right angles only if they correspond to the same value of c.

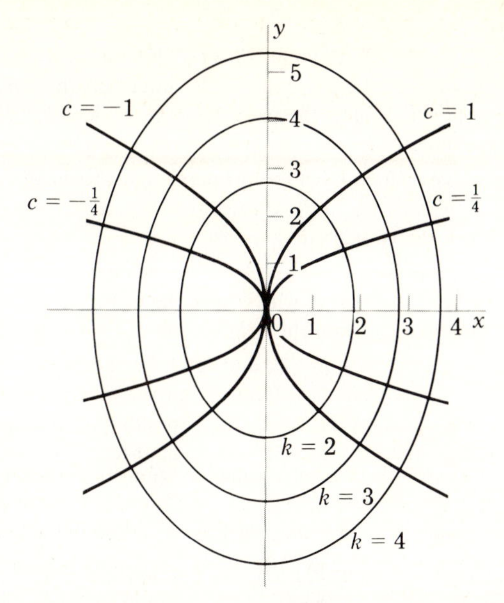

Figure 2.6 The curves $y^2 = 4cx$ and their orthogonal trajectories.

This is a homogeneous equation, but it is simpler to treat it as a separable equation:

$$y\,dy + 2x\,dx = 0$$

$$\tfrac{1}{2}y^2 + x^2 = k^2$$

where the integration constant has been written as a square to emphasize the fact that it cannot be negative since it is equal to the sum of two squares. Typical curves of the two families are shown in Fig. 2.6.

Although the concept of orthogonal trajectories appears to be essentially a geometric one, it actually is intimately related to many important physical problems. For instance, in what are known in physics as two-dimensional field problems, the equipotential lines and the lines of flux are orthogonal trajectories. More specifically, if one knows the family of **isothermal curves,** or curves joining points at the same temperature, in some problem in heat flow, then their orthogonal trajectories are the lines in which the heat flows. Thus in Fig. 2.6, if the ellipses are thought of as isothermal curves, the parabolas are the paths along which the heat flow takes place. Still another interpretation is this: If the curves of one family are thought of as the contour lines on a topographic map, that is, the horizontal projections of the lines of constant elevation on the surface being mapped, then their orthogonal trajectories are the projections of the lines of steepest descent on the surface, that is, the paths in which rain falling on the hill would run off.

Exercises to Section 2.5

Find the orthogonal trajectories of the curves of each of the following families.

1 $x^2 - y^2 = c$ **2** $y^2 = cx^3$

3 $y = (x - c)^2$ **4** $x^2 + 2y^2 = cy$

5 $y^2 = x^2 + cx$ **6** $x^2 + y^2 = cx$

7 Let t be the tangent to a curve C at a general point P, let F be the foot of the perpendicular from P to the x axis, and let T be the point in which t intersects the x axis. Find the equation of the family of curves which have the property that the length TF is equal to the sum of the abscissa and ordinate of P.

8 Find the equation of the family of curves which have the property that the tangent at any point P determines with the coordinate axes a triangle whose area is equal to twice the area of the rectangle determined by P and the coordinate axes.

9 Let n be the normal to a curve C at a general point P, and let N be the intersection of n and the x axis. Find the equation of the family of curves which have the property that the length of the segment PN is equal to the distance from the origin to P.

***10** Let t be the tangent to a curve C at a general point P, and let F be the foot of the perpendicular from P to the x axis. Find the equation of the family of curves which have the property that the distance from F to t is a constant. *Hint:* Recall from analytic geometry that the distance from a point (x_0, y_0) to a line $ax + by + c = 0$ is equal to

$$\left| \frac{ax_0 + by_0 + c}{\sqrt{a^2 + b^2}} \right|$$

***11** Let t be the tangent to a curve C at a general point P. Find the equation of the family of curves which have the property that the distance from the origin to t is equal to the abscissa of P.

***12** Let P be a general point on a curve C which passes through the origin, let F_x and F_y be the feet of the perpendiculars from P to the coordinate axes, and let A_x and A_y be the areas adjacent to the x and y axes, respectively, into which C divides the rectangle OF_xPF_y. Find the equation of the family of curves which have the property that the volume formed when A_x is revolved about the x axis is equal to the volume formed when A_y is revolved about the y axis.

***13** In calculus, the radius of curvature of a curve is defined to be

$$R = \frac{[1 + (y')^2]^{3/2}}{|y''|}$$

Solve this differential equation when R is a constant, and show that the family of solution curves is the set of all circles in the xy plane.

***14** Work Exercise 13 if the radius of curvature varies according to the law $R(x) = \sec x$ and if the slope of each curve is to be zero when $x = 0$.

2.6 MISCELLANEOUS PROBLEMS

Example 1 The spread of technological innovations after they are first introduced is similar in many respects to the population problems we considered in Sec. 2.4. Suppose, for example, that an improved variety of wheat is introduced in a community of N farmers, the first adoption being at $t = 0$. If further adoptions occur only as a result of reports from those who have already switched to the new strain, it is reasonable to assume that the rate of adoption† is jointly proportional to the number x of

† This assumes a continuous model in which derivatives are meaningful. If N is large, such a model should provide an acceptable approximation.

individuals who have adopted the new variety and the number $N - x$ of those who have yet to learn about it. Thus we are led to the separable equation

$$\frac{dx}{dt} = kx(N - x) \qquad \text{or} \qquad \frac{dx}{x(N - x)} = k\,dt$$

as a description of the process by which the innovation spreads. The solution of this equation is immediate, and we have, using familiar partial-fraction techniques,

$$\frac{1}{N}\left(\frac{1}{x} + \frac{1}{N - x}\right) dx = k\,dt$$

$$\ln \frac{x}{N - x} = kNt - \ln c$$

$$\frac{cx}{N - x} = e^{kNt}$$

Since $x = 1$ when $t = 0$, it follows that $c = N - 1$. Hence, solving for x, we finally obtain

$$x = \frac{Ne^{kNt}}{N - 1 + e^{kNt}} = \frac{N}{1 + (N - 1)e^{-kNt}}$$

Example 2 Suppose that at $t = 0$ we put into service a large number $N(0)$ of new items, such as light bulbs or automobiles, having a limited but variable useful life. Suppose, further, that for these items we know a **survival function** $s(t)$ characterized by the following properties:

1. $N(0)s(t)$ is the number of items from the original lot which are still operating at time t.
2. $s(0) = 1$.
3. $s(t)$ is a monotonically decreasing function for $t \geq 0$.
4. $0 \leq s(t) \leq 1$.
5. $s(t) \to 0$ as $t \to \infty$.

As these items wear out, let us suppose that new ones are put into service at times $t = \tau_1, \tau_2, \ldots$, the number of replacements made at $t = \tau_i$ being $r(\tau_i)(\tau_i - \tau_{i-1})$, where $r(t)$ is the so-called **renewal function** for the process. If all replacements are subject to the same survival function as the original items, then the number of the replacements made at $t = \tau_i$ which are still in service at a later time t is

$$r(\tau_i)(\tau_i - \tau_{i-1})s(t - \tau_i)$$

Taking into account both the surviving members of the original population and the surviving members of each replacement batch, the total number of items in service at time t is

$$\begin{aligned} N(t) &= N(0)s(t) + \sum_i r(\tau_i)(\tau_i - \tau_{i-1})s(t - \tau_i) \\ &= N(0)s(t) + \sum_i r(\tau_i)s(t - \tau_i)\,\Delta\tau_i \end{aligned}$$

the summation including all values of τ_i less than or equal to the time t. Under suitable assumptions on $s(t)$ and $r(t)$, the sum in the last expression becomes an integral as each $\Delta\tau_i$ approaches zero, and we have

$$N(t) = N(0)s(t) + \int_0^t r(\tau)s(t-\tau)\,d\tau \tag{1}$$

The important question now is: Given a survival function $s(t)$ (presumably obtained experimentally), what should be the renewal function $r(t)$ in order to maintain a desired operating population $N(t)$? Since the unknown function $r(t)$ occurs as part of an integrand, Eq. (1) is an example of what is called an **integral equation;** and, of course, we have not yet discussed how to solve such equations. However, we can answer the following simpler (and less important) question: If a constant renewal function $r(t) = k$ suffices to maintain the population at its original level, what is the survival function $s(t)$? This involves the following special case of Eq. (1):

$$N(0) = N(0)s(t) + k\int_0^t s(t-\tau)\,d\tau \tag{2}$$

which can be solved as follows: First we transform the integral in (2) by the substitution $u = t - \tau$, so that Eq. (2) becomes

$$N(0) = N(0)s(t) + k\int_t^0 s(u)(-du) = N(0)s(t) + k\int_0^t s(u)\,du$$

Then we differentiate this equation with respect to t, getting

$$0 = N(0)s'(t) + ks(t)$$

This is a simple differential equation which we can solve at once:

$$\frac{ds}{s} = -\frac{k}{N(0)}\,dt$$

$$\ln s = -\frac{k}{N(0)}t + \ln c$$

$$s = c\exp\left[-\frac{k}{N(0)}t\right]$$

Finally, since $s(0) = 1$, it follows that $c = 1$ and

$$s = \exp\left[-\frac{k}{N(0)}t\right]$$

Exercises to Section 2.6

1 If, with a constant renewal function $r(t) = k$, an original population grows according to the law $N(t) = N(0)(1 + at)$, what is the survival function?

***2** The assumptions of Example 1 are somewhat unrealistic because innovations spread not only because of favorable reports from users who have already adopted them but also because of publicity from other sources such as commercial advertisements, trade journals, and government bulletins. Modify the differential equation of Example 1 by introducing a term to take into account the effect of these influences on the $N - x$ individuals who at any given time have not yet heard of the innovation. What is the solution of this equation if $x(0) = 0$?

3 Determine for the equation of Example 1 and the equation of Exercise 2 the value of x for which the rate of adoption is a maximum.

4 Banks often compound interest on savings accounts on a semiannual, quarterly, or even daily basis. If interest at an annual rate of p percent is compounded every Δt years, show that an initial amount P_0 grows to $P_0[1 + (p/100)\,\Delta t]$ in one such (infinitesimal) interest period. Use this result to derive the differential equation satisfied by the amount on deposit P in the limit when interest is compounded continuously. Solve this equation and obtain an expression for P as a function of t.

***5** A weight W lb is to be supported by a column having the shape of a solid of revolution. If the material of the column weighs ρ lb/ft^3 and if the radius of the upper base of the column is to be r_0 ft, determine how the radius of the column should vary if at all cross sections the load per unit area is to be the same.

***6** Work Exercise 5 if the column is hollow, the outer radius of each cross section being h ft more than the inner radius, and the inner radius of the upper base being r_0 ft.

***7** A weight is initially located on the y axis at a distance l from the origin. An inextensible chain of length l is attached to the weight, the free end of the chain being initially at the origin. If the free end of the chain is moved slowly along the x axis, find the equation of the path of the weight as it is dragged across the xy plane by the chain. *Hint:* Note that the direction of the chain will always be tangent to the path of the weight. Hence, if P: (x, y) is the instantaneous position of the weight and if s is the length of the path through which the weight has moved, then $dy/ds = -y/l$. (This important curve is known as the **tractrix.**)

****8** A Coast Guard vessel is pursuing a smuggler in a dense fog. The fog lifts momentarily, and the smuggler is seen d mi away; then the fog descends, and the smuggler can no longer be seen. It is known, however, that in an attempt to escape, the smuggler will set off on a straight course of unknown direction with constant speed v_1. If the velocity of the Coast Guard boat is $v_2(> v_1)$, what course should it follow to be sure of intercepting and capturing the smuggler? *Hint:* Choose the origin at the point where the smuggler was seen, and observe that if the Coast Guard boat spirals the origin in a path such that its distance from the origin is always equal to $v_1 t$, it will necessarily intercept the smuggler. Of course, before the Coast Guard boat can begin its spiral, it must first reach a point where its distance from the origin is $v_1 t_0$, where t_0 is the elapsed time since the smuggler was seen. In implementing these suggestions, it will be convenient to use the polar coordinate equations $x = r(t) \cos \theta(t)$ and $y = r(t) \sin \theta(t)$ to describe the path of the Coast Guard vessel.

****9** One winter morning, it began snowing heavily and continued at a constant rate all day. At noon a snowplow, able to clear c ft^3/h of snow with a blade w ft in width, started plowing. At 1 P.M. it had gone 2 mi, and at 2 P.M. it had gone 1 mi more. When did it start snowing? *Hint:* If t_0 is the number of hours before noon when it began to snow, if t is the time measured in hours after noon, if s ft/h is the constant rate at which the snow is falling, if $y(t)$ is the depth of the snow measured in feet, and if $x(t)$ is the distance the snowplow has traveled measured in feet, then

$$y = s(t_0 + t) \qquad \text{and} \qquad wy\, dx = c\, dt$$

***10** If

$$f(\lambda) = \int_0^\infty \frac{e^{-z} e^{-\lambda/z}}{\sqrt{z}}\, dz$$

show that $f(\lambda)$ satisfies the differential equation

$$f' = -\frac{f}{\sqrt{\lambda}}$$

Solve this equation, given that

$$f(0) = \int_0^\infty \frac{e^{-z}}{\sqrt{z}}\, dz = \sqrt{\pi}$$

Hint: Compute $f'(\lambda)$ by differentiating the defining integral inside the integral sign, i.e., by differentiating the integrand of the defining integral. Then compare this result with the integral obtained when the substitution $u = \lambda/z$ is made in the integral defining $f(\lambda)$.

***11** In Example 2, find the differential equation satisfied by the survival function $s(t)$ if the renewal function is $r(t) = r_0 e^{at}$. *Hint:* Use Leibnitz's rule (see item 10, Appendix B.2) to differentiate Eq. (1), and then apply integration by parts to the integral in the result of this differentiation.

12 Solve the equation in Exercise 11 if $r_0 = 1$, $a = 1/[2N(0)]$, and $N(t) = 1 + t$.

CHAPTER

THREE

THE BASIC THEORY OF LINEAR DIFFERENTIAL EQUATIONS

3.1 INTRODUCTION

In Chaps. 1 and 2, we investigated in some detail the solution and application of various classes of first-order differential equations, many of them nonlinear. The rest of our work, with the exception of the material in Chaps. 12 and 14, will deal exclusively with linear equations, principally those of the second order.

The most general (ordinary) linear differential equation has the form

$$a_0(x)y^{(n)} + a_1(x)y^{(n-1)} + \cdots + a_{n-1}(x)y' + a_n(x)y = A(x) \tag{1}$$

where $a_0, \ldots, a_n$ and A are known functions of x. In our work we shall usually suppose that $a_0, \ldots, a_n$ and A are continuous and that a_0 is different from zero on any interval where we are considering the equation. Under this assumption, we can divide Eq. (1) by $a_0(x)$ and then rename the coefficients, obtaining the standard form

$$y^{(n)} + p_1(x)y^{(n-1)} + \cdots + p_{n-1}(x)y' + p_n(x)y = R(x) \tag{2}$$

Because of the presence of $R(x)$, which is unlike the other terms in that it does not contain the dependent variable y or any of its derivatives, Eq. (2) is said to be **nonhomogeneous.** If $R(x)$ is identically zero, we have the so-called **homogeneous equation**

$$y^{(n)} + p_1(x)y^{(n-1)} + \cdots + p_{n-1}(x)y' + p_n(x)y = 0 \tag{3}$$

As we shall see in Sec. 3.4, to find a complete solution of the nonhomogeneous equation (2), it is first necessary to solve the homogeneous equation (3) obtained from Eq. (2) by deleting the term $R(x)$. For this reason, Eq. (3) is often referred to as the **related homogeneous equation** of Eq. (2).

In general, neither Eq. (2) nor Eq. (3) can be solved in terms of known functions, and the theory associated with such special cases as have been studied at length is difficult. However, there are several general theorems about the solutions of linear differential equations which do not depend upon the form of the solutions or the processes by which they may be found. These results will be of great help to us in constructing complete solutions of specific equations, and we shall investigate them in this chapter. All we will need to know to establish them is that linear differential equations *have* solutions, as guaranteed by the following existence theorem† whose proof we shall leave to more advanced texts.‡

Theorem 1 Let the coefficient functions in the equation $y^{(n)} + p_1(x)y^{(n-1)} + \cdots + p_{n-1}(x)y' + p_n(x)y = R(x)$ be continuous over an interval I, and let x_0 be an arbitrary point in I. Then over I this equation has a unique solution $y(x)$ which at $x = x_0$ satisfies the conditions $y(x_0) = y_0,\ y'(x_0) = y'_0,\ y''(x_0) = y''_0,\ \ldots,\ y^{(n-1)}(x_0) = y_0^{(n-1)}$, where $y_0, y'_0, y''_0, \ldots, y_0^{(n-1)}$ are arbitrary constants.

In the next section we shall develop a criterion which can be applied to n particular solutions of Eq. (3) to tell whether or not a linear combination of these, formed with arbitrary constant coefficients, is a complete solution of Eq. (3). Then in Sec. 3.3 we shall prove that when one particular solution of Eq. (3) is known, the search for additional solutions can be restricted to the solution of a linear equation of one lower order. For the important case $n = 2$, this means that if we can find one particular solution of Eq. (3), then the problem can be reduced to the solution of a linear equation of order 1; and we learned in Sec. 1.6 that this can always be done. Finally, in Sec. 3.4 we shall show that whenever a complete solution of Eq. (3) is known, a complete solution of Eq. (2) can always be constructed.

As a consequence of this remarkable chain of results, it follows that for the very important class of linear differential equations of the second order, *if* we can find just one particular solution (no matter how it is found and no matter what its form may be) of the homogeneous equation

$$y'' + p_1(x)y' + p_2(x)y = 0$$

then a complete solution of both this equation and the nonhomogeneous equation

$$y'' + p_1(x)y' + p_2(x)y = R(x)$$

can always be constructed. Thus our attention will ultimately be focused on finding that first particular solution.

† The existence of solutions of linear differential equations is, of course, implied by Theorem 2, Sec. 1.2, and its generalization to equations of higher order. However, the theorem here quoted, since it applies to linear equations only, is stronger and better suited to our purposes.

‡ See, for instance, E. L. Ince, "Ordinary Differential Equations," Dover, New York, 1944, pp. 73–75.

Exercises for Section 3.1

1 Prove the following corollary of Theorem 1: Let the coefficient functions in Eq. (3) be continuous on an interval I, and let x_0 be a point of I. If $y(x)$ is a solution of Eq. (3) for which $y(x_0) = y'(x_0) = y''(x_0) = \cdots = y^{(n-1)}(x_0) = 0$, then $y(x)$ is zero for all values of x in I.

2 Let
$$y_1(x) = x^4 - x^3 \qquad -2 \le x \le 2$$
and let
$$y_2(x) = \begin{cases} x^3 - x^4 & -2 \le x \le 0 \\ x^4 - x^3 & 0 < x \le 2 \end{cases}$$

Show that both y_1 and y_2 are solutions of the equation $x^2y'' - 6xy' + 12y = 0$ which satisfy the conditions $y(1) = 0$, $y'(1) = 1$. Does the fact that y_1 and y_2 are different contradict Theorem 1? Explain.

3.2 FAMILIES OF SOLUTIONS

In Sec. 1.2 we learned that if we have particular solutions $y_1, y_2, \ldots, y_n$ of an nth-order, homogeneous, linear differential equation
$$y^{(n)} + p_1(x)y^{(n-1)} + \cdots + p_{n-1}(x)y' + p_n(x)y = 0 \tag{1}$$
then for all values of the constants $c_1, c_2, \ldots, c_n$ the linear combination
$$y = c_1y_1 + c_2y_2 + \cdots + c_ny_n \tag{2}$$
is a solution of (1). Since the expression (2) contains n arbitrary constants, it is tempting to hope that it may, in fact, be a complete solution of Eq. (1); but this is not necessarily the case. To see why this is so, suppose that the particular solution y_n is (as, of course, it might well be)
$$y_n = y_1 + y_2 + \cdots + y_{n-1}$$
Then, substituting for y_n in Eq. (2) and collecting terms, we have
$$\begin{aligned} y &= (c_1 + c_n)y_1 + (c_2 + c_n)y_2 + \cdots + (c_{n-1} + c_n)y_{n-1} \\ &= d_1y_1 + d_2y_2 + \cdots + d_{n-1}y_{n-1} \end{aligned} \tag{2a}$$
where $d_i = c_i + c_n$. The last expression, while it is of course a solution of (1), contains only $n - 1$ arbitrary constants. Hence, as our limited experience with linear differential equations suggests, and as Theorem 4 and the subsequent discussion confirms, Eq. (2a) cannot be a complete solution of the nth-order equation (1). Evidently, when we attempt to construct families of solutions from particular solutions, the question of whether or not one or more particular solutions can be expressed in terms of the others must be considered.

Fundamentally, we are dealing here with the important concepts of linear dependence and independence, as customarily developed in courses in algebra.

Definition 1 The n functions $f_1, f_2, \ldots, f_n$ (which need not be solutions of any linear differential equation) are said to be **linearly dependent** over an interval I if and only if there exists a set of n constants $c_1, c_2, \ldots, c_n$, at least one of which is different from zero, such that the equation $c_1f_1(x) + c_2f_2(x) + \cdots + c_nf_n(x) = 0$ holds for all values of x in I.

Definition 2 The n functions $f_1, f_2, \ldots, f_n$ are said to be **linearly independent** over an interval I if and only if they are not linearly dependent over I, i.e., if and only if the only equation of the form $c_1 f_1(x) + c_2 f_2(x) + \cdots + c_n f_n(x) = 0$ which they satisfy identically over I has $c_1 = c_2 = \cdots = c_n = 0$.

Example 1 Show that $\sin x$ and $\cos x$ are linearly independent on the interval $[0, \pi/2]$.

To verify the linear independence of $\sin x$ and $\cos x$ on the interval I: $[0, \pi/2]$, we must, according to Definition 2, show that the only relation of the form $c_1 \sin x + c_2 \cos x = 0$ which they satisfy identically on I has $c_1 = c_2 = 0$. To do this, we note that since the assumed relation must be identically satisfied, that is, for *all* values of x in I, it must hold for any particular values of x in I, say the endpoints $x = 0$ and $x = \pi/2$. Using these values in turn, we obtain from the assumed relation

$$c_1 0 + c_2 1 = 0 \qquad \text{and} \qquad c_1 1 + c_2 0 = 0$$

Hence $c_1 = c_2 = 0$, which proves that $\sin x$ and $\cos x$ are linearly independent on I, as asserted.

It is important to note that the linear independence of $\sin x$ and $\cos x$ on I: $[0, \pi/2]$ does not mean that these functions are unrelated on I. In fact, for all values of x, they are connected by the *nonlinear* identity $\sin^2 x + \cos^2 x = 1$; and, in particular, on I the value of either function is uniquely determined when the other is known.

For two functions f_1 and f_2, linear dependence on an interval I is equivalent to proportionality. In fact, if f_1 and f_2 are linearly dependent, then, by Definition 1, $c_1 f_1 + c_2 f_2 \equiv 0$ where at least one of the constants (c_1, c_2), say c_1, is different from zero. Then, dividing by c_1, we obtain

$$f_1 \equiv -\frac{c_2}{c_1} f_2$$

which shows that f_1 and f_2 are proportional. Conversely, if f_1 and f_2 are proportional, it is evident that they satisfy a relation of the form required by Definition 1.

In any set of linearly dependent functions, at least one (though not necessarily each one) of the functions can be expressed as a linear combination of the others. Suppose, in fact, that the functions $f_1, f_2, \ldots, f_n$ are linearly dependent. Then they satisfy identically an equation of the form

$$c_1 f_1 + c_2 f_2 + \cdots + c_n f_n = 0$$

in which at least one of the c's, say c_i, is different from zero. This being the case, we can divide by c_i and solve for f_i, getting

$$f_i = -\frac{c_1}{c_i} f_1 - \frac{c_2}{c_i} f_2 - \cdots - \frac{c_{i-1}}{c_i} f_{i-1} - \frac{c_{i+1}}{c_i} f_{i+1} - \cdots - \frac{c_n}{c_i} f_n$$

which is an expression for f_i of the asserted form. Of course, since some (though not all) of the c's may be zero, there may be some of the f's that cannot be expressed in terms of the others in this fashion.

A necessary condition for the linear dependence of a set of functions is provided by the following theorem.

Theorem 1 If the functions $f_1, f_2, \ldots, f_n$ are linearly dependent over an interval I and if each function possesses derivatives through the $(n-1)$st, then the determinant

$$\begin{vmatrix} f_1 & f_2 & \cdots & f_n \\ f'_1 & f'_2 & \cdots & f'_n \\ \cdots & \cdots & \cdots & \cdots \\ f_1^{(n-1)} & f_2^{(n-1)} & \cdots & f_n^{(n-1)} \end{vmatrix}$$

is equal to zero at all points of I.

PROOF Because the functions $f_1, f_2, \ldots, f_n$ are linearly dependent over I, there is a set of constants $c_1, c_2, \ldots, c_n$, at least one of which is different from zero, such that

$$c_1 f_1 + c_2 f_2 + \cdots + c_n f_n = 0 \tag{3}$$

at all points of I. By hypothesis, each of the f's possesses derivatives at least through the $(n-1)$st. Hence, by repeated differentiation of Eq. (3), we obtain the additional equations

$$\begin{gathered} c_1 f'_1 + c_2 f'_2 + \cdots + c_n f'_n = 0 \\ \cdots\cdots\cdots\cdots\cdots\cdots\cdots\cdots \\ c_1 f_1^{(n-1)} + c_2 f_2^{(n-1)} + \cdots + c_n f_n^{(n-1)} = 0 \end{gathered} \tag{4}$$

Equation (3) and the $n-1$ equations in (4) constitute a set of n homogeneous linear equations in the n quantities $c_1, c_2, \ldots, c_n$, and this system of equations is satisfied by a set of values at least one of which is different from zero. Now it is a fundamental theorem of algebra (see item 4, Appendix B.1) that a set of n simultaneous, homogeneous linear equations in n unknowns has a solution other than the obvious one in which each unknown is zero,† if and only if the determinant of the coefficients of the system is equal to zero. But the determinant of the coefficients of the system formed by (3) and (4) is just the determinant which appears in the statement of the theorem. Hence, for all values of x in I,

$$\begin{vmatrix} f_1 & f_2 & \cdots & f_n \\ f'_1 & f'_2 & \cdots & f'_n \\ \cdots & \cdots & \cdots & \cdots \\ f_1^{(n-1)} & f_2^{(n-1)} & \cdots & f_n^{(n-1)} \end{vmatrix} = 0$$

as asserted.

By restating Theorem 1 in its logically equivalent contrapositive form, we obtain the following useful result.

Corollary 1 If the determinant

$$\begin{vmatrix} f_1 & f_2 & \cdots & f_n \\ f'_1 & f'_2 & \cdots & f'_n \\ \cdots & \cdots & \cdots & \cdots \\ f_1^{(n-1)} & f_2^{(n-1)} & \cdots & f_n^{(n-1)} \end{vmatrix}$$

is not identically zero over an interval I, then the functions $f_1, f_2, \ldots, f_n$ are linearly independent over I.

† Such a solution is customarily referred to as a **trivial solution.**

In honor of the Polish mathematician Hoene Wronsky (1778–1853), the determinant which appears in Theorem 1 is usually referred to as the **wronskian,** $W(f_1, f_2, \ldots, f_n)$, of the functions $f_1, f_2, \ldots, f_n$. Wronskians, and in particular the second-order wronskian

$$W(f_1, f_2) = \begin{vmatrix} f_1 & f_2 \\ f'_1 & f'_2 \end{vmatrix} = f_1 f'_2 - f'_1 f_2$$

will be very useful to us in our study of linear differential equations.

The converse of Theorem 1 is not true. To show this, it is sufficient to exhibit two functions which are linearly independent over an interval I but whose wronskian vanishes identically on I. One such pair, defined on $(-\infty, \infty)$, is the following:

$$f_1 = x^2 \qquad f_2 = \begin{cases} -x^2 & x \le 0 \\ x^2 & x > 0 \end{cases}$$

Clearly, $f'_1 = 2x$ for all values of x, while

$$f'_2 = \begin{cases} -2x & x \le 0 \\ 2x & x > 0 \end{cases}$$

Therefore for $x \le 0$, the wronskian of f_1 and f_2 is

$$\begin{vmatrix} x^2 & -x^2 \\ 2x & -2x \end{vmatrix}$$

and for $x > 0$, the wronskian is

$$\begin{vmatrix} x^2 & x^2 \\ 2x & 2x \end{vmatrix}$$

and each of these determinants is identically zero. However, f_1 and f_2 are linearly independent over any interval containing the origin. In fact, if $c_1 f_1(x) + c_2 f_2(x) \equiv 0$, then

$$c_1 f_1(1) + c_2 f_2(1) = c_1 + c_2 = 0$$

and

$$c_1 f_1(-1) + c_2 f_2(-1) = c_1 - c_2 = 0$$

Hence $c_1 = c_2 = 0$, and f_1 and f_2 are linearly independent, as asserted.

A correct converse of Theorem 1 for the important case $n = 2$ is the following.

Theorem 2 If the wronskian of f_1 and f_2 vanishes identically over an interval I and if one of these functions is different from zero at all points of I, then the functions are linearly dependent on I.

Proof By hypothesis, $W(f_1, f_2) = f_1 f'_2 - f'_1 f_2 \equiv 0$ on I, and at least one of the functions, say f_1, is different from zero at all points of I. Hence we may divide the wronskian relation by f_1^2, getting

$$\frac{f_1 f'_2 - f'_1 f_2}{f_1^2} \equiv 0$$

But this equation is simply the assertion that the derivative of the fraction f_2/f_1 is identically zero. Therefore the fraction itself must be a constant, say k. In other words, $f_2 = kf_1$, or $kf_1 - f_2 = 0$, which shows that f_1 and f_2 are linearly dependent, as asserted.

If f_1 and f_2, instead of being arbitrary differentiable functions, are known to be solutions of Eq. (1) in the case $n = 2$, then much less than the identical vanishing of their wronskian is required in order to prove that they are linearly dependent. In fact, if their wronskian vanishes at a single point of an interval I of suitable character, then f_1 and f_2 are linearly dependent over I. The following theorem gives the precise result.

Theorem 3 If y_1 and y_2 are two solutions of the differential equation $y'' + p_1(x)y' + p_2(x)y = 0$ on an interval I where $p_1(x)$ and $p_2(x)$ are continuous, and if the wronskian of y_1 and y_2 vanishes anywhere (that is, at even one point) in I, then y_1 and y_2 are linearly dependent over I.

Proof Let I be an interval over which the coefficient functions $p_1(x)$ and $p_2(x)$ in the equation $y'' + p_1(x)y' + p_2(x)y = 0$ are continuous, and let $x = a$ be a point of I at which the wronskian of two solutions, y_1 and y_2, of this equation is equal to zero. Now consider the simultaneous equations

$$\begin{aligned} c_1 y_1(a) + c_2 y_2(a) &= 0 \\ c_1 y_1'(a) + c_2 y_2'(a) &= 0 \end{aligned} \tag{5}$$

By hypothesis, the determinant of the coefficients of this linear system of algebraic equations, namely $W(y_1, y_2)|_{x=a}$, is equal to zero. Hence the system has a nontrivial solution; that is, it is satisfied by a pair of numbers $(\bar{c}_1, \bar{c}_2)$ at least one of which is different from zero. The linear combination

$$y(x) = \bar{c}_1 y_1(x) + \bar{c}_2 y_2(x)$$

formed with these numbers is, of course, a solution of the given differential equation. Moreover, from the fact that $(\bar{c}_1, \bar{c}_2)$ is a solution of (5) it follows that

$$y(a) = y'(a) = 0$$

Thus $y(x)$ must be identically zero since, by the fundamental existence theorem for linear differential equations (Theorem 1, Sec. 3.1), there is only one solution satisfying the conditions $y(a) = y'(a) = 0$, and $y \equiv 0$ is such a solution. Hence over the interval I the solutions y_1 and y_2 satisfy an equation of the form $c_1 y_1(x) + c_2 y_2(x) \equiv 0$ in which at least one of the c's is different from zero. Therefore y_1 and y_2 are linearly dependent on I, as asserted.

Stated contrapositively, Theorem 3 becomes the following result.

Corollary 1 If y_1 and y_2 are two solutions of the differential equation $y'' + p_1(x)y' + p_2(x)y = 0$ which are linearly independent over an interval I in which the coefficient functions $p_1(x)$ and $p_2(x)$ are continuous, then the wronskian of y_1 and y_2 is different from zero at all points of I.

Returning to the problem with which we began this section, it should now be clear that wronskians will be of great help in checking the linear dependence and independence of solutions of Eq. (1). Moreover, wronskians will also be useful in proving that if we have n linearly independent solutions of Eq. (1), then a complete solution of Eq. (1) can always be constructed from them. To prove this, which we shall do in Theorem 4, it will be convenient to use the property of wronskians of solutions of Eq. (1) which is described in the following lemma.

Lemma 1 If the coefficient functions $p_1(x), p_2(x), \ldots, p_n(x)$ in the equation

$$y^{(n)} + p_1(x)y^{(n-1)} + \cdots + p_{n-1}(x)y' + p_n(x)y = 0$$

are continuous over an interval I, then throughout I the wronskian of any n solutions of this equation, $y_1, y_2, \ldots, y_n$, is given by the formula

$$W(y_1, y_2, \ldots, y_n) = ke^{-\int p_1(x)\,dx}$$

where k is a constant depending on the particular solutions involved.

PROOF Partly for reasons of simplicity and convenience, and partly because we are primarily concerned with second-order linear equations, we shall prove this lemma only for the case $n = 2$. Suggestions for constructing a proof when $n = 3$ are given in Exercise 20.

Proceeding, then, with the second-order case, let y_1 and y_2 be any two solutions of the equation

$$y'' + p_1(x)y' + p_2(x)y = 0$$

so that

$$y_1'' + p_1(x)y_1' + p_2(x)y_1 = 0 \tag{6}$$

and

$$y_2'' + p_1(x)y_2' + p_2(x)y_2 = 0 \tag{7}$$

Multiplying Eq. (6) by y_2 and Eq. (7) by y_1 and subtracting the first result from the second, we obtain

$$y_1 y_2'' - y_1'' y_2 + p_1(x)(y_1 y_2' - y_1' y_2) = 0 \tag{8}$$

Now, by definition, $y_1 y_2' - y_1' y_2 = W(y_1, y_2)$, and therefore

$$\frac{dW}{dx} = \frac{d}{dx}(y_1 y_2' - y_1' y_2) = (y_1 y_2'' + y_1' y_2') - (y_1'' y_2 + y_1' y_2')$$

$$= y_1 y_2'' - y_1'' y_2$$

Hence Eq. (8) can be written in the form

$$\frac{dW}{dx} + p_1(x)W = 0$$

This is a simple separable equation whose solution can be written down at once:

$$W = ke^{-\int p_1(x)\,dx} \tag{9}$$

where k is a constant determined by the two particular solutions y_1 and y_2. Thus our proof is complete.

Lemma 1 is due to the Norwegian mathematician Nils Abel (1802–1829), and in his honor Eq. (9) is customarily referred to as **Abel's identity.**

By the hypothesis of Lemma 1, $p_1(x)$ is continuous over I. Hence the exponential function $e^{-\int p_1(x)\,dx}$ is defined and different from zero at all points of I. Therefore, from the formula of Lemma 1 we infer the following useful result.

Corollary 1 If the coefficient functions $p_1(x)$, $p_2(x)$, ..., $p_n(x)$ in the equation

$$y^{(n)} + p_1(x)y^{(n-1)} + \cdots + p_{n-1}(x)y' + p_n(x)y = 0$$

are continuous over an interval I, then throughout I the wronskian of any n solutions of this equation is either always zero or never zero. In the first case, the n solutions are linearly dependent over I; in the second case, they are linearly independent over I.

Example 2 It is easy to verify that $y_1 = e^x$ and $y_2 = x$ are two particular solutions of the equation $(1 - x)y'' + xy' - y = 0$. Writing this in the form

$$y'' + \frac{x}{1-x}y' - \frac{1}{1-x}y = 0$$

it is clear that the coefficient functions

$$p_1(x) = \frac{x}{1-x} \qquad \text{and} \qquad p_2(x) = -\frac{1}{1-x}$$

are continuous over any interval which does not include the point $x = 1$. Let I be such an interval, say an interval to the right of $x = 1$. Then on I

$$W(y_1, y_2) = \begin{vmatrix} e^x & x \\ e^x & 1 \end{vmatrix} = (1 - x)e^x$$

On the other hand, according to Eq. (9), $W(y_1, y_2)$ should be some constant multiple of

$$\begin{aligned} \exp\left[-\int p_1(x)\,dx\right] &= \exp\left[-\int \frac{x}{1-x}\,dx\right] = \exp\left[\int\left(1 + \frac{1}{x-1}\right)dx\right] \\ &= \exp\,[x + \ln\,(x - 1)] \\ &= \exp x \cdot \exp\,[\ln\,(x - 1)] \\ &= (x - 1)e^x \end{aligned}$$

Clearly, the value of $W(y_1, y_2)$, as computed from the solutions y_1 and y_2, satisfies Eq. (9) with $k = -1$.

With Abel's identity available, we can now prove the following fundamental theorem about families of solutions of homogeneous, linear, second-order differential equations.

Theorem 4 Let the coefficient functions $p_1(x)$ and $p_2(x)$ in the equation $y'' + p_1(x)y' + p_2(x)y = 0$ be continuous over an interval I. If y_1 and y_2 are two solutions of this equation which are linearly independent over I, then for any solution y there exist constants c_1 and c_2 such that $y = c_1 y_1 + c_2 y_2$.

PROOF Let y_1 and y_2 be two particular solutions of the equation $y'' + p_1(x)y' + p_2(x)y = 0$ which are linearly independent over an interval I in which $p_1(x)$ and $p_2(x)$ are continuous, and let y_3 be any solution whatsoever of this equation. Then by Lemma 1 and Corollary 1 of Theorem 3,

$$W(y_1, y_2) = y_1 y_2' - y_1' y_2 = k_{12} e^{-\int p_1(x)\,dx} \neq 0$$

where the subscripts on the coefficient k_{12} indicate the two solutions which determine its value. Also, by applying Abel's identity to the pair (y_3, y_1) and the pair (y_3, y_2), we have

$$y_3 y_1' - y_3' y_1 = k_{31} e^{-\int p_1(x)\,dx}$$

$$y_3 y_2' - y_3' y_2 = k_{32} e^{-\int p_1(x)\,dx}$$

These can be regarded as two simultaneous linear equations in the quantities y_3 and y_3'. Multiplying the first by $-y_2$ and the second by y_1 and adding, we obtain

$$(y_1 y_2' - y_1' y_2) y_3 = k_{32} e^{-\int p_1(x)\,dx} y_1 - k_{31} e^{-\int p_1(x)\,dx} y_2$$

As we have already noted, the coefficient of y_3 in this equation is just $W(y_1, y_2) = k_{12} e^{-\int p_1(x)\,dx}$ with $k_{12} \neq 0$. Hence, making this substitution and then solving for y_3, we find

$$y_3 = \frac{k_{32}}{k_{12}} y_1 - \frac{k_{31}}{k_{12}} y_2$$

Thus y_3 has been expressed in the form asserted in the theorem, with $c_1 = k_{32}/k_{12}$ and $c_2 = -k_{31}/k_{12}$. Since y_3 was *any* solution of the given equation, the theorem is proved.

Theorem 4 assures us that for suitable values of c_1 and c_2, *any* solution y_3 of the equation $y'' + p_1(x)y' + p_2(x)y = 0$ can be written in the form $y_3 = c_1 y_1 + c_2 y_2$, provided y_1 and y_2 are linearly independent particular solutions of the equation. Hence it follows that if c_1 and c_2 are arbitrary constants, then $y = c_1 y_1 + c_2 y_2$ is a complete solution of $y'' + p_1(x)y' + p_2(x)y = 0$. Restating this, we have the following corollary of Theorem 4.

Corollary 1 Given the equation $y'' + p_1(x)y' + p_2(x)y = 0$, where $p_1(x)$ and $p_2(x)$ are continuous over an interval I. If y_1 and y_2 are two solutions of this equation which are linearly independent over I, i.e., have nonvanishing wronskian over I, then $y = c_1 y_1 + c_2 y_2$, where c_1 and c_2 are arbitrary constants, is a complete solution of the equation on the interval I.

Two solutions of the equation $y'' + p_1(x)y' + p_2(x)y = 0$ which are linearly independent and from which, therefore, a complete solution can be constructed, are usually referred to as a **fundamental set of solutions** or simply a **basis.**

By using the existence theorem quoted in Sec. 3.1, we can now show that every homogeneous, linear differential equation of order n actually possesses n linearly independent particular solutions and therefore has a complete solution.

To do this, let us define n particular solutions $y_1, y_2, \ldots, y_n$ in the following way. At $x = x_0$ let y_1 satisfy the conditions

$$y_1(x_0) = 1 \qquad y_1'(x_0) = y_1''(x_0) = \cdots = y_1^{(n-1)}(x_0) = 0$$

Let y_2 satisfy the conditions

$$y_2(x_0) = 0 \qquad y_2'(x_0) = 1 \qquad y_2''(x_0) = \cdots = y_2^{(n-1)}(x_0) = 0$$

and in general let $y_j^{(j-1)}(x_0) = 1$ and let all other derivatives of y_j through the $(n-1)$st be zero at x_0. By Theorem 1, Sec. 3.1, solutions satisfying these conditions exist. Then at $x = x_0$ the wronskian of these solutions is

$$W(y_1, y_2, y_3, \ldots, y_n)\big|_{x=x_0} = \begin{vmatrix} 1 & 0 & 0 & \cdots & 0 \\ 0 & 1 & 0 & \cdots & 0 \\ 0 & 0 & 1 & \cdots & 0 \\ \cdots & \cdots & \cdots & \cdots & \cdots \\ 0 & 0 & 0 & \cdots & 1 \end{vmatrix} = 1$$

Thus the value of the wronskian is different from zero at $x = x_0$ and therefore, by Corollary 1 of Lemma 1, is different from zero over any interval I containing x_0 and in which the coefficient functions are continuous. Hence, by Corollary 1 of Theorem 1, the n particular solutions that we have constructed are linearly independent over I and therefore form a fundamental set from which a complete solution of the equation can be constructed.

Exercises for Section 3.2

1 Prove that $\sin x$ and $\cos x$ are linearly independent on $[0, \pi/2]$ by evaluating them for values of x different from the values 0 and $\pi/2$ used in Example 1.

2 Show that the functions 1, x, and x^2 are linearly independent for $-\infty < x < \infty$.

3 Show that the functions $x + 1$, $x + 2$, and $2x + 1$ are linearly dependent for $-\infty < x < \infty$, and find a linear equation which they satisfy.

4 What is the wronskian of the functions e^{mx} and xe^{mx}?

5 What is the wronskian of e^x, e^{-x}, and e^{2x}?

6 Show that $y_1 = \sin 2x$ and $y_2 = \cos 2x$ are solutions of the differential equation $y'' + 4y = 0$, and verify that their wronskian satisfies Abel's identity [Eq. (9)].

7 Show that $y_1 = e^x$, $y_2 = e^{2x}$, and $y_3 = e^{3x}$ are solutions of the differential equation $y''' - 6y'' + 11y' - 6y = 0$, and verify that their wronskian satisfies Abel's identity [Eq. (9)].

Determine whether the functions in the following sets are linearly dependent or linearly independent on the indicated intervals. If the functions are linearly dependent, find a linear equation which they satisfy.

8 $e^x, e^{2x} \qquad -\infty < x < \infty$

9 $x^2, x^2 - 1, x^2 + x + 1 \qquad -\infty < x < \infty$

10 $\cos x \sin 2x, 2 \sin x \cos 2x, 3 \sin 3x \qquad -\infty < x < \infty$

11 $\ln (x - 1), 2 \ln (x + 1), 3 \ln (x^2 - 1) \qquad x > 2$

***12** $x^2 - 1, x^2 + x + 1, x^2 + 3x + 5 \qquad -\infty < x < \infty$

***13** $1/(x+1)$, $1/x$, $1/(x-2)$ $\quad x < -1$

14 Show that there is no point at which two linearly independent solutions of the equation $y'' + p_1(x)y' + p_2(x)y = 0$ are simultaneously zero except possibly at a point where $p_1(x)$ is undefined.

15 Show that there is no point at which two linearly independent solutions of the equation $y'' + p_1(x)y' + p_2(x)y = 0$ simultaneously take on extreme values except possibly at a point where $p_1(x)$ is undefined.

***16** If the wronskian of two functions is different from zero at every point of an interval, show that there is no point of the interval at which either function has a repeated zero.

***17** Given two linearly independent solutions of the equation $y'' + p_1(x)y' + p_2(x)y = 0$, show that between any two consecutive zeros of either solution there is exactly one zero of the other solution. *Hint:* Let y_1 and y_2 be the two solutions, let a and b be two consecutive zeros of y_1, apply Rolle's theorem to the quotient y_1/y_2, and note the contradiction unless $y_2 = 0$ at some point between a and b.

***18** If the quotient of two linearly independent solutions of the equation $y'' + p_1(x)y' + p_2(x)y = 0$ exists at all points of an interval, prove that it either is an increasing function at all points of the interval or is a decreasing function at all points of the interval.

***19** If y_1 and y_2 have a nonvanishing wronskian, show that $y_3 = c_1 y_1 + c_2 y_2$ and $y_4 = k_1 y_1 + k_2 y_2$ have a nonvanishing wronskian if and only if $c_1 k_2 - c_2 k_1 \neq 0$.

***20** Prove Lemma 1 for the case $n = 3$. *Hint:* Consider the three equations which assert that y_1, y_2, and y_3 are solutions of Eq. (1) when $n = 3$ as three simultaneous equations in the quantities $p_1(x)$, $p_2(x)$, and $p_3(x)$. Then use determinants to solve this system for $p_1(x)$, and note that one of the two determinants is $W(y_1, y_2, y_3)$ and the other is $W'(y_1, y_2, y_3)$.

3.3 REDUCTION OF ORDER

If, somehow or other, one knows a particular solution of a homogeneous, nth-order, linear differential equation

$$y^{(n)} + p_1(x)y^{(n-1)} + \cdots + p_{n-1}(x)y' + p_n(x)y = 0 \tag{1}$$

then the search for additional solutions can be narrowed down to a homogeneous linear equation of order $n-1$,† according to the following theorem.

Theorem 1 If y_1 is a solution of a homogeneous, linear differential equation of order n in y, then the substitution $y = \phi(x)y_1$ reduces the given equation to a homogeneous, linear differential equation of order $n-1$ in the dependent variable $\phi'(x)$.

What Theorem 1 tells us is that if y_1 is known to be a solution of Eq. (1), then the product $\phi(x)y_1$ will also be a solution provided that ϕ is a function whose derivative satisfies a related linear equation of order $n-1$ in ϕ'. According to

† This is analogous to the familiar fact that if one root $x = r$ of a polynomial equation of degree n is known, then the remaining roots are solutions of the $(n-1)$st-degree equation obtained when the given equation is divided by the factor $x - r$.

Theorem 1, this related equation is homogeneous and hence has $\phi' = 0$ as one solution. From this we obtain $\phi = c$ and $y_2 = \phi y_1 = cy_1$, but since y_1 and cy_1 are linearly dependent, this gives us no additional information about solutions of Eq. (1). On the other hand, if a nontrivial solution† for ϕ' can be found, then after ϕ is obtained by integration, the product $y_2 = \phi y_1$ becomes a second solution of Eq. (1) which, as we shall see in Theorem 2, is linearly independent of y_1.

Furthermore, when a nontrivial solution ϕ' of the related $(n-1)$st-order equation has been found, Theorem 1 can be applied again; and further solutions of this equation, and hence of Eq. (1), can be found by solving a related equation of order $n-2$. Clearly, this process can be continued as long as we are fortunate enough to be able to find solutions of the successive reduced equations.

Example 1 Given that $y_1 = x$ is a solution of the equation

$$x^3y''' - 3x^2y'' + x(6 - x^2)y' - (6 - x^2)y = 0$$

show that $y = x\phi$ will also be a solution provided that ϕ' satisfies the equation $(\phi')'' - \phi' = 0$.

From $y = x\phi$ we obtain

$$y' = x\phi' + \phi$$

$$y'' = x\phi'' + 2\phi'$$

$$y''' = x\phi''' + 3\phi''$$

Hence, substituting into the given equation, we find that $y = x\phi$ will also be a solution if

$$x^3(x\phi''' + 3\phi'') - 3x^2(x\phi'' + 2\phi') + x(6 - x^2)(x\phi' + \phi) - (6 - x^2)x\phi = 0$$

or, collecting terms,

$$x^4\phi''' - x^4\phi' = 0$$

or, finally,

$$(\phi')'' - \phi' = 0$$

as asserted.

In this particular problem, the reduced equation can be solved by inspection, for it is obvious that both $\phi_1' = e^x$ and $\phi_2' = e^{-x}$ satisfy it. Hence, integrating, we get

$$\phi_1 = e^x \qquad \text{and} \qquad \phi_2 = -e^{-x}$$

Therefore, from the substitution $y = x\phi$, it follows that

$$y_2 = xe^x \qquad \text{and} \qquad y_3 = -xe^{-x}$$

are also solutions of the given equation.

† By a **nontrivial solution** of an equation we mean a solution of the equation which is not identically zero. A **trivial solution,** of course, is one which is identically zero.

We shall not subject ourselves to the purely notational complexities involved in the proof of Theorem 1. The proof when $n = 2$ is fully representative of the general case, and this is the result that we will need in the work ahead of us. Hence (except in the exercises) we shall give only the proof of the following corollaries of Theorem 1.

Corollary 1 If y_1 is a particular solution of the equation

$$y'' + p_1(x)y' + p_2(x)y = 0 \tag{2}$$

then $y_2 = \phi(x)y_1$ is also a solution provided that $\phi'(x)$ is a solution of the linear first-order equation

$$y_1(\phi')' + [2y_1' + p_1(x)y_1]\phi' = 0 \tag{3}$$

PROOF Let y_1 be a solution of the homogeneous linear equation (2), and let us attempt to find a function $\phi(x)$ such that the product $y = \phi(x)y_1$ will also be a solution of (2). Computing the first and second derivatives of $y = \phi(x)y_1$ and substituting into Eq. (2), we obtain

$$\phi''y_1 + 2\phi'y_1' + \phi y_1'' + p_1(x)(\phi'y_1 + \phi y_1') + p_2(x)(\phi y_1) = 0$$

or, collecting terms,

$$y_1\phi'' + [2y_1' + p_1(x)y_1]\phi' + [y_1'' + p_1(x)y_1' + p_2(x)y_1]\phi = 0$$

The last bracketed quantity is identically zero because, by hypothesis, y_1 is a solution of Eq. (2). Hence ϕy_1 will be a solution of (2) if ϕ' satisfies Eq. (3), and the corollary is established.

Equation (3) is a separable first-order equation whose coefficients are known functions of x, since $p_1(x)$ is known from the original equation and y_1 is a known particular solution of that equation. Hence Eq. (3) can be solved explicitly for ϕ':

$$\frac{d\phi'}{\phi'} = -\left[2\frac{y_1'}{y_1} + p_1(x)\right] dx \qquad \phi' \not\equiv 0$$

$$\ln|\phi'| = \ln(1/y_1^2) - \int p_1(x)\,dx + \ln|c_1|$$

$$\phi' = c_1 \frac{1}{y_1^2} e^{-\int p_1(x)\,dx}$$

and, integrating,

$$\phi = c_1 \int \frac{1}{y_1^2} e^{-\int p_1(x)\,dx}\,dx + c_2$$

Finally, taking $c_1 = 1$ and $c_2 = 0$, for convenience, we have as a second solution of Eq. (2)

$$y_2 = \phi y_1 = y_1 \int \frac{1}{y_1^2} e^{-\int p_1(x)\,dx}\,dx$$

Thus we have established a second corollary of Theorem 1.

Corollary 2 If y_1 is a solution of the equation

$$y'' + p_1(x)y' + p_2(x)y = 0$$

then

$$y_2 = y_1 \int \frac{1}{y_1^2} e^{-\int p_1(x)\,dx}\,dx$$

is also a solution.

Example 2 Given that $y_1 = x$ is a solution of the equation $x^2y'' - 3xy' + 3y = 0$, find a second solution.

Rather than using the formula of Corollary 2, let us solve this problem by carrying out the steps used in the proof of Corollary 1. Assuming $y = \phi y_1 = x\phi$, we have

$$y' = x\phi' + \phi \qquad \text{and} \qquad y'' = x\phi'' + 2\phi'$$

and, substituting, we find that if y is to be a solution of the given differential equation, then ϕ must satisfy the equation

$$x^2(x\phi'' + 2\phi') - 3x(x\phi' + \phi) + 3(x\phi) = 0$$

$$x^3\phi'' - x^2\phi' = 0$$

$$x\frac{d\phi'}{dx} = \phi'$$

$$\frac{d\phi'}{\phi'} = \frac{dx}{x}$$

$$\ln|\phi'| = \ln|x| + \ln|c_1|$$

$$\phi' = c_1x$$

$$\phi = c_1\frac{x^2}{2} + c_2$$

Any particular solution ϕ in which $c_1 \neq 0$ will serve our purpose; so, for convenience, let us take $c_1 = 2$ and $c_2 = 0$, getting

$$\phi_1 = x^2 \qquad \text{and} \qquad y_2 = x\phi_1 = x^3$$

In Example 2 it is easy to verify that the wronskian of the two solutions $y_1 = x$ and $y_2 = x^3$, namely,

$$\begin{vmatrix} x & x^3 \\ 1 & 3x^2 \end{vmatrix} = 2x^3$$

is different from zero for all values of x except $x = 0$. Hence y_1 and y_2 are linearly independent, and over any interval not containing $x = 0$ a complete solution of the given equation can be written:

$$y = c_1x + c_2x^3$$

It is now natural to ask if the two solutions y_1 and $y_2 = \phi y_1$ *always* have a nonvanishing wronskian. Clearly, it would be an exercise in futility to apply Corollary 1 and find after all our work that y_1 and $y_2 = \phi y_1$ had a vanishing wronskian; for if this were the case, then y_2 could not serve with y_1 as a basis for constructing a complete solution of the given equation. Fortunately, if we avoid the trivial case $\phi' = 0$, y_1 and $y_2 = \phi y_1$ are always independent, as the next theorem assures us.

Theorem 2 If y_1 is a particular solution of the equation $y'' + p_1(x)y' + p_2(x)y = 0$ and if $\phi = \int (1/y_1^2)e^{-\int p_1(x)\,dx}\,dx$, then over any interval in which $p_1(x)$ and $p_2(x)$ are continuous, y_1 and ϕy_1 have a nonvanishing wronskian, and $y = c_1 y_1 + c_2 \phi y_1$ is a complete solution of the given equation.

PROOF Recalling from the proof of Corollary 2, Theorem 1, that

$$\phi' = \frac{1}{y_1^2}e^{-\int p_1(x)\,dx}$$

we have for the wronskian of y_1 and $y_2 \equiv \phi y_1$

$$W(y_1, y_2) = \begin{vmatrix} y_1 & \phi y_1 \\ y_1' & \phi y_1' + \phi' y_1 \end{vmatrix} = y_1^2\phi' = e^{-\int p_1(x)\,dx}$$

Since $p_1(x)$ is known to be continuous, the last exponential can never vanish, and the theorem is established.

Exercises for Section 3.3

Using the one solution indicated, find a complete solution of each of the following equations.

1 $y'' + y = 0$ $\qquad y_1 = \sin x$

2 $y'' - y' - 2y = 0$ $\qquad y_1 = e^{-x}$

3 $x^2y'' + 4xy' - 4y = 0$ $\qquad y_1 = x$

4 $y'' + 2y' + y = 0$ $\qquad y_1 = e^{-x}$

5 $(1 - 2x)y'' + 2y' + (2x - 3)y = 0$ $\qquad y_1 = e^x$

***6** $(2x - x^2)y'' + 2(x - 1)y' - 2y = 0$ $\qquad y_1 = x - 1$

***7** $y''' - 3y'' + 3y' - y = 0$ $\qquad y_1 = e^x$

***8** $y''' - y'' - y' + y = 0$ $\qquad y_1 = e^x$

9 If y_1 is a solution of the homogeneous equation $y'' + p_1(x)y' + p_2(x)y = 0$, show that the substitution $y = \phi y_1$ will reduce the nonhomogeneous equation $y'' + p_1(x)y' + p_2(x)y = R(x)$ to a linear first-order equation, whose dependent variable is ϕ'.

Using the procedure described in Exercise 9, find a second solution of each of the following equations.

10 $y'' - y = e^{2x}$ $\qquad y_1 = e^x$

11 $y'' - 3y' + 2y = e^x$ $\qquad y_1 = e^x$

12 $y'' - 2y' + y = e^x$ $\qquad y_1 = e^x$

***13** Is the procedure described in Exercise 9 effective if y_1 is a solution of the nonhomogeneous equation rather than the homogeneous equation?

Verify that each of the following equations has the indicated solutions, and in each case construct two different complete solutions using two different bases.

14 $y'' - y = 0$ $\quad y_1 = e^x,\ y_2 = e^{-x}$

15 $y'' - 3y' + 2y = 0$ $\quad y_1 = e^x,\ y_2 = e^{2x}$

16 $y'' + y = 0$ $\quad y_1 = \sin(x + \pi/4),\ y_2 = \sin(x - \pi/4)$

17 $x^2y'' + xy' - y = 0$ $\quad y_1 = x,\ y_2 = 1/x$

18 Explain how Abel's identity can be used to find a second solution of the equation $y'' + p_1(x)y' + p_2(x)y = 0$ when one solution is known. Illustrate this method by applying it to Exercises 1, 2, and 3.

***19** Construct a solution of the equation of Example 2, valid on $(-\infty, \infty)$, which is not contained in the complete solution $y = c_1x + c_2x^3$ obtained in Example 2. Explain.

3.4 VARIATION OF PARAMETERS

So far in this chapter we have considered only the homogeneous linear equation

$$y^{(n)} + p_1(x)y^{(n-1)} + \cdots + p_{n-1}(x)y' + p_n(x)y = 0 \tag{1}$$

If $n \neq 1$, we have as yet no means of finding solutions besides inspection and guessing. However, we know how to form families of solutions from particular solutions, how to use one or more particular solutions to help us find others, and how to determine whether a set of n particular solutions is adequate to serve as a basis for constructing a complete solution.

In this section we shall extend our discussion to the nonhomogeneous linear equation

$$y^{(n)} + p_1(x)y^{(n-1)} + \cdots + p_{n-1}(x)y' + p_n(x)y = R(x) \tag{2}$$

and show how to construct a complete solution of it when a complete solution of (1) is known. The fundamental theorem is the following.

Theorem 1 If Y is a particular solution of the nonhomogeneous linear equation

$$y^{(n)} + p_1(x)y^{(n-1)} + \cdots + p_{n-1}(x)y' + p_n(x)y = R(x)$$

and if $c_1y_1 + c_2y_2 + \cdots + c_ny_n$ is a complete solution of the related homogeneous equation obtained from this by deleting the term $R(x)$, then $y = c_1y_1 + c_2y_2 + \cdots + c_ny_n + Y$ is a complete solution of the nonhomogeneous equation.

PROOF Again, for simplicity and because of our prime concern with equations of the second order, we shall prove this theorem only for the case $n = 2$. The proof for other values of n follows in exactly the same way.

Suppose, then, that Y is a particular solution of the second-order equation

$$y'' + p_1(x)y' + p_2(x)y = R(x) \tag{3}$$

and that $\bar{y}$ is any solution whatsoever of this equation. From these assumptions it follows that

$$\bar{y}'' + p_1(x)\bar{y}' + p_2(x)\bar{y} = R(x)$$

and that

$$Y'' + p_1(x)Y' + p_2(x)Y = R(x)$$

If we subtract these two equations, we obtain

$$\bar{y}'' - Y'' + p_1(x)(\bar{y}' - Y') + p_2(x)(\bar{y} - Y) = 0$$

or

$$(\bar{y} - Y)'' + p_1(x)(\bar{y} - Y)' + p_2(x)(\bar{y} - Y) = 0$$

Thus the quantity $\bar{y} - Y$ satisfies the homogeneous linear equation

$$y'' + p_1(x)y' + p_2(x)y = 0 \tag{4}$$

and hence, by Theorem 4, Sec. 3.2, it must be expressible in the form

$$\bar{y} - Y = c_1 y_1 + c_2 y_2$$

provided that the wronskian $W(y_1, y_2)$ is different from zero, that is, provided that $c_1 y_1 + c_2 y_2$ is a complete solution of Eq. (4), as assumed. Therefore, transposing, we obtain

$$\bar{y} = c_1 y_1 + c_2 y_2 + Y$$

Since $\bar{y}$ was *any* solution of the nonhomogeneous equation (3), the theorem is established (for the case $n = 2$).

The term Y, which can be any solution of Eq. (3) no matter how special, is called a **particular integral** of the nonhomogeneous equation. The expression $c_1 y_1 + c_2 y_2$, which is a complete solution of the homogeneous equation corresponding to Eq. (3), is called the **complementary function** of the nonhomogeneous equation. The steps to be carried out in solving an equation of the form (2) can now be summarized as follows:

1. Delete the term $R(x)$ from the given nonhomogeneous equation, and find n particular solutions of the resulting homogeneous equation which have a nonvanishing wronskian. Then combine these to form the **complementary function** $c_1 y_1 + c_2 y_2 + \cdots + c_n y_n$ of the given equation.
2. Find one particular solution Y of the nonhomogeneous equation itself.
3. Add the **particular integral** Y found in step 2 to the **complementary function** $c_1 y_1 + c_2 y_2 + \cdots + c_n y_n$ found in step 1 to obtain a complete solution of the given equation.

In the next chapter we shall investigate how these theoretical steps can be carried out when $p_1, p_2, \ldots, p_n$ are constants and we have the so-called **linear equation with constant coefficients.**

Various methods are available for finding the particular integral Y required by Theorem 1 for the construction of a complete solution of the nonhomogeneous equation (2). In the next chapter we shall develop a procedure known as the **method of undetermined coefficients,** which is sufficient for most of the usual applications. In this section we shall describe a more general method, known as **variation of parameters,** which, though less convenient in elementary applications, has the advantage that it can be applied to Eq. (2) in all cases, provided a complete solution of the related homogeneous equation (1) is known and the term $R(x)$ in Eq. (2) is integrable. As before, for simplicity we shall give a detailed discussion of the process only for the case $n = 2$.

The fundamental idea behind the process is this. Instead of using two arbitrary *constants* c_1 and c_2 to combine two independent solutions, y_1 and y_2, of the homogeneous equation

$$y'' + p_1(x)y' + p_2(x)y = 0 \tag{5}$$

as we do in constructing the complementary function, we attempt to find two *functions* of x, say u_1 and u_2, such that

$$Y = u_1 y_1 + u_2 y_2$$

will be a solution of the nonhomogeneous equation (3). Having two unknown functions u_1 and u_2, we require two equations for their determination. One of these will be obtained by substituting Y into the given differential equation (3); the other remains at our disposal. As the analysis proceeds, it will become clear what this second condition should be.

From $Y = u_1 y_1 + u_2 y_2$ we have, by differentiation,

$$Y' = (u_1 y_1' + u_1' y_1) + (u_2 y_2' + u_2' y_2) = (u_1 y_1' + u_2 y_2') + (u_1' y_1 + u_2' y_2)$$

Another differentiation will clearly introduce second derivatives of the unknown functions u_1 and u_2, with attendant complications, unless we arrange to eliminate the first-derivative terms u_1' and u_2' from Y'. This can be done if we make

$$u_1' y_1 + u_2' y_2 = 0 \tag{6}$$

which thus becomes the required second condition on u_1 and u_2.

Proceeding now with the simplified expression

$$Y' = u_1 y_1' + u_2 y_2'$$

we find
$$Y'' = (u_1 y_1'' + u_1' y_1') + (u_2 y_2'' + u_2' y_2')$$

Substituting Y, Y', and Y'' into Eq. (3), we obtain

$$(u_1 y_1'' + u_1' y_1' + u_2 y_2'' + u_2' y_2') + p_1(x)(u_1 y_1' + u_2 y_2') + p_2(x)(u_1 y_1 + u_2 y_2) = R(x)$$

or

$$u_1[y_1'' + p_1(x)y_1' + p_2(x)y_1] + u_2[y_2'' + p_1(x)y_2' + p_2(x)y_2] + u_1' y_1' + u_2' y_2' = R(x)$$

The expressions in brackets vanish because, by hypothesis, both y_1 and y_2 are solutions of the homogeneous equation (5). Hence, we find for the other condition on u_1 and u_2

$$u_1' y_1' + u_2' y_2' = R(x) \tag{7}$$

Solving Eqs. (6) and (7) for u_1' and u_2', we obtain

$$u_1' = -\frac{y_2}{y_1 y_2' - y_2 y_1'} R(x) \qquad \text{and} \qquad u_2' = \frac{y_1}{y_1 y_2' - y_2 y_1'} R(x) \tag{8}$$

The functions y_1, y_2, y_1', y_2', and $R(x)$ are all known. Hence, u_1 and u_2 can be found by a single integration. With u_1 and u_2 known, the particular integral

$$Y = u_1 y_1 + u_2 y_2$$

is completely determined.

We should notice, of course, that if $y_1 y_2' - y_2 y_1' = 0$, the solution for u_1' and u_2' cannot be carried out. However, $y_1 y_2' - y_2 y_1'$ is precisely the wronskian of the two solutions y_1 and y_2, and if these are independent, as we suppose them to be, then their wronskian cannot vanish.

Example 1 Find a complete solution of the equation $y'' + y = \sec x$.

By inspection, it is clear that $y_1 = \cos x$ and $y_2 = \sin x$ are two linearly independent particular solutions of the related homogeneous equation $y'' + y = 0$. Hence the complementary function of the given equation is

$$c_1 \cos x + c_2 \sin x$$

Furthermore, from Eqs. (8) we have

$$u_1' = -\frac{\sin x}{\cos x(\cos x) - \sin x(-\sin x)} \sec x = -\tan x$$

$$u_2' = \frac{\cos x}{\cos x(\cos x) - \sin x(-\sin x)} \sec x = 1$$

Therefore, $\quad u_1 = -\int \tan x \, dx = \ln |\cos x| \qquad$ and $\qquad u_2 = \int dx = x$

and thus

$$Y = u_1 y_1 + u_2 y_2 = (\ln |\cos x|) \cos x + x \sin x$$

Finally,

$$\begin{aligned} y &= \text{complementary function} + \text{particular integral} \\ &= c_1 \cos x + c_2 \sin x + (\ln |\cos x|) \cos x + x \sin x \end{aligned}$$

is a complete solution.

Exercises for Section 3.4

Using the indicated solutions of the related homogeneous equation, find a particular integral and then a complete solution of each of the following nonhomogeneous equations.

1 $y'' - 4y' + 3y = e^{-x}$	$y_1 = e^x$, $y_2 = e^{3x}$
2 $y'' - y' - 2y = e^x$	$y_1 = e^{-x}$, $y_2 = e^{2x}$
3 $y'' - y' - 2y = e^{-x}$	$y_1 = e^{-x}$, $y_2 = e^{2x}$
4 $y'' - y' - 2y = x$	$y_1 = e^{-x}$, $y_2 = e^{2x}$
***5** $y'' + y = \sin x$	$y_1 = \sin x$, $y_2 = \cos x$
6 $y'' + 2y' + y = e^{-x}/x$	$y_1 = e^{-x}$, $y_2 = xe^{-x}$
***7** $y'' + y = \tan x$	$y_1 = \sin x$, $y_2 = \cos x$
8 $y'' + 4y' + 4y = e^{-2x}/x^2$	$y_1 = e^{-2x}$, $y_2 = xe^{-2x}$
9 $x^2y'' + xy' - y = x$	$y_1 = x$, $y_2 = 1/x$
10 $x^2y'' + xy' - y = 1/x$	$y_1 = x$, $y_2 = 1/x$
11 $x^2y'' - 2xy' + 2y = x^3e^x$	$y_1 = x$, $y_2 = x^2$
***12** $x^2y'' - 6y = x^3 \ln \lvert x \rvert$	$y_1 = x^3$, $y_2 = 1/x^2$

***13** Given that $y_1 = \sin x$ and $y_2 = \cos x$ are two solutions of the equation $y'' + y = 0$, use the method of variation of parameters to show that $\int_0^x \sin (x - s) f(s)\, ds$ is a particular integral of the nonhomogeneous equation $y'' + y = f(x)$. *Hint:* Introduce the dummy variable s in the integrals which define u_1 and u_2. Then move $y_1(x)$ and $y_2(x)$ into the integrands of the respective integrals and combine the two integrals.

***14** Given that $y_1 = \sinh kx$ and $y_2 = \cosh kx$ are two solutions of the equation $y'' - k^2y = 0$, use the method of variation of parameters to show that $(1/k) \int_0^x \sinh k(x - s) f(s)\, ds$ is a particular integral of the nonhomogeneous equation $y'' - k^2y = f(x)$. *Hint:* Note the hint given in Exercise 13.

***15** Explain how the method of variation of parameters can be used to find a particular integral of a nonhomogeneous, linear, third-order differential equation if *three* linearly independent solutions of the related homogeneous equation are known.

16 Prove Theorem 1 for the case $n = 3$.

CHAPTER

FOUR

LINEAR DIFFERENTIAL EQUATIONS WITH CONSTANT COEFFICIENTS

4.1 INTRODUCTION

In this chapter we shall apply the theoretical results we derived in Chap. 3 to the solution of linear differential equations with constant coefficients, both homogeneous and nonhomogeneous. Then, after having learned how to solve such equations, we shall investigate a number of applications in which equations of this sort are involved.

The most general linear differential equation with constant coefficients has the form

$$a_0 y^{(n)} + a_1 y^{(n-1)} + \cdots + a_{n-1} y' + a_n y = R(x)$$

or, in the important second-order case,

$$ay'' + by' + cy = f(x) \tag{1}$$

In our work we shall always suppose that the coefficients are real numbers.

A second standard form which is also common is based on the so-called **operator notation.** In this, the symbol of differentiation d/dx is replaced by D, so that, by definition,

$$\frac{dy}{dx} = Dy\dagger$$

† Just as the prime notation, y', y'', ..., may in specific instances indicate derivatives with respect to x, t, or any other independent variable, so the operator notation, Dy, D^2y, ..., may also indicate derivatives with respect to an independent variable other than x, depending on the context.

As an immediate extension, the second derivative, which of course is obtained by a repetition of the process of differentiation, is written

$$\frac{d^2y}{dx^2} = D(Dy) = D^2y$$

Similarly,

$$\frac{d^3y}{dx^3} = D(D^2y) = D^3y$$

$$\frac{d^4y}{dx^4} = D(D^3y) = D^4y$$

$$\cdots\cdots\cdots\cdots\cdots\cdots$$

Evidently, positive integral powers of D (which are the only ones we have defined) obey the usual laws of exponents.

If due care is taken to see that variables are not moved across the sign of differentiation by a careless interchange of the order of factors containing variable coefficients,† the operator D can be handled in many respects as though it were a simple algebraic quantity. For instance, after defining $(aD^2 + bD + c)\phi(x)$ to mean $aD^2\phi(x) + bD\phi(x) + c\phi(x)$, we have for the polynomial operator $3D^2 - 10D - 8$ and its factored equivalents, applied to the particular function $\phi(x) = x^2$,

$$(3D^2 - 10D - 8)x^2 = 3(2) - 10(2x) - 8(x^2) = 6 - 20x - 8x^2$$

$$\begin{aligned}(3D + 2)(D - 4)x^2 &= (3D + 2)(2x - 4x^2)\\ &= (6 - 24x) + (4x - 8x^2) = 6 - 20x - 8x^2\end{aligned}$$

$$\begin{aligned}(D - 4)(3D + 2)x^2 &= (D - 4)(6x + 2x^2)\\ &= (6 + 4x) - (24x + 8x^2) = 6 - 20x - 8x^2\end{aligned}$$

These results illustrate how algebraically equivalent forms of an operator yield identical results when applied to the same function.

Using the operator D, we can evidently write Eq. (1) in the alternative standard form

$$(aD^2 + bD + c)y = f(x) \tag{1a}$$

Many writers base the solution of Eq. (1) upon the operational properties of the symbol D. However, we shall postpone all operational methods until the chapter on the Laplace transformation, where operational calculus can be developed easily and efficiently in its proper setting.

Exercises for Section 4.1

1 What is the difference between Dy and yD?

† See Exercise 4.

2 Verify that

$$(D + 1)(D - 2) \cos x = (D - 2)(D + 1) \cos x$$
$$= (D^2 - D - 2) \cos x$$

3 Verify that

$$(D + 1)(D^2 + 2) \sin 3x = (D^2 + 2)(D + 1) \sin 3x$$
$$= (D^3 + D^2 + 2D + 2) \sin 3x$$

4 Is $(D + 1)(D + x)e^x = (D + x)(D + 1)e^x$? Explain.

5 Under what conditions, if any, is

$$[D + r_1(x)][D + r_2(x)]f(x) = [D + r_2(x)][D + r_1(x)]f(x)$$

6 Determine $r(x)$ so that $y = e^x$ will be a solution of the equation $[D + r(x)](D + x)y = 0$.

7 Determine $r(x)$ so that $y = e^x$ will be a solution of the equation $(D + x)[D + r(x)]y = 0$.

***8** Find a complete solution of the equation $(D - 1/x)(D - 1)y = 0$. *Hint:* Note that any solution of the first-order equation $(D - 1)y = 0$ is also a solution of the given equation, since $(D - 1/x)0 = 0$. Solve $(D - 1)y = 0$, and then find a second solution of the given equation by using the method of Sec. 3.3.

***9** Find a complete solution of the equation $(D - 1)(D - 1/x)y = 0$.

4.2 THE HOMOGENEOUS SECOND-ORDER EQUATION WITH CONSTANT COEFFICIENTS

Following the theory of Sec. 3.2, we begin the solution of the second-order equation

$$ay'' + by' + cy = 0 \tag{1a}$$

or

$$(aD^2 + bD + c)y = 0 \tag{1b}$$

by searching for particular solutions which we can ultimately combine into a complete solution. In doing this, it is natural to try

$$y = Ae^{mx}$$

where m is a constant to be determined, because all derivatives of this function are alike except for a numerical coefficient. Substituting into Eq. (1a) and then factoring Ae^{mx} from every term, we have

$$Ae^{mx}(am^2 + bm + c) = 0$$

as the condition to be satisfied if $y = e^{mx}$ is to be a solution. Since e^{mx} can never be zero and since $A = 0$ implies that $y \equiv 0$, which is a trivial solution of no interest to us, it is thus necessary that

$$am^2 + bm + c = 0 \tag{2}$$

This purely algebraic equation is known as the **characteristic** or **auxiliary equation** of either Eq. (1*a*) or Eq. (1*b*) since its roots determine or *characterize* the only possible nontrivial solutions of the assumed form $y = Ae^{mx}$. In practice, it is obtained not by substituting $y = Ae^{mx}$ into the given differential equation and then simplifying, but rather by substituting m^2 for y'', m for y', and 1 for y in the given equation (1*a*) or, still more simply, by equating to zero the operational coefficient of y in Eq. (1*b*) and then letting the symbol D play the role of m:

$$aD^2 + bD + c = 0$$

The characteristic equation (2) is a simple quadratic which will in general be satisfied by two values of m:

$$m = \frac{-b \pm \sqrt{b^2 - 4ac}}{2a}$$

Using these two values, say m_1 and m_2, two particular solutions

$$y_1 = e^{m_1 x} \qquad \text{and} \qquad y_2 = e^{m_2 x}$$

can be constructed. From this pair, according to Theorem 1, Sec. 1.2, an infinite family of solutions

$$y = c_1 y_1 + c_2 y_2 = c_1 e^{m_1 x} + c_2 e^{m_2 x} \tag{3}$$

can be formed. Moreover, by Corollary 1, Theorem 4, Sec. 3.2, if the wronskian of these solutions is different from zero, then (3) is a complete solution of Eq. (1*a*); i.e., it contains all possible solutions of the homogeneous equation. Accordingly, we compute

$$\begin{aligned} W(y_1, y_2) &= y_1 y_2' - y_2 y_1' = e^{m_1 x}(m_2 e^{m_2 x}) - e^{m_2 x}(m_1 e^{m_1 x}) \\ &= (m_2 - m_1)e^{(m_1 + m_2)x} \end{aligned}$$

Since $e^{(m_1+m_2)x}$ can never vanish, it is clear that *a complete solution of Eq. (1a) is always given by (3) except in the special case when $m_1 = m_2$ and the wronskian vanishes identically.*

Example 1 Find a complete solution of the differential equation $y'' + 7y' + 12y = 0$.
The characteristic equation in this case is

$$m^2 + 7m + 12 = 0$$

and its roots are

$$m_1 = -3 \qquad \text{and} \qquad m_2 = -4$$

Since these values are different, a complete solution is

$$y = c_1 e^{-3x} + c_2 e^{-4x}$$

Example 2 Find a complete solution of the equation $y'' + 2y' + 5y = 0$.
The characteristic equation in this case is

$$m^2 + 2m + 5 = 0$$

and its roots are

$$m_1 = -1 + 2i \qquad \text{and} \qquad m_2 = -1 - 2i$$

Since these are distinct, a complete solution is

$$y = c_1 e^{(-1+2i)x} + c_2 e^{(-1-2i)x}$$

Although the last expression is undeniably a complete solution of the given equation, it is unsatisfactory for many practical purposes because it involves complex exponentials, which are awkward to handle and are not tabulated. It is therefore a matter of considerable importance to construct a more convenient complete solution when m_1 and m_2 are conjugate complex numbers.

To do this, let us suppose that

$$m_1 = p + iq \qquad \text{and} \qquad m_2 = p - iq$$

so that a complete solution as first constructed is

$$y = c_1 e^{(p+iq)x} + c_2 e^{(p-iq)x} = c_1 e^{px} e^{iqx} + c_2 e^{px} e^{-iqx}$$

By factoring out e^{px}, this can be written as

$$y = e^{px}(c_1 e^{iqx} + c_2 e^{-iqx})$$

The expression in parentheses can be simplified by using the **Euler formulas** (see item 2, Appendix B.1)

$$e^{i\theta} = \cos\theta + i\sin\theta \qquad \text{and} \qquad e^{-i\theta} = \cos\theta - i\sin\theta$$

taking $\theta = qx$. The result of these substitutions is

$$\begin{aligned} y &= e^{px}[c_1(\cos qx + i\sin qx) + c_2(\cos qx - i\sin qx)] \\ &= e^{px}[(c_1 + c_2)\cos qx + i(c_1 - c_2)\sin qx] \end{aligned}$$

If we now define two new arbitrary constants by the equations

$$A = c_1 + c_2 \qquad \text{and} \qquad B = i(c_1 - c_2)$$

the complete solution can finally be put in the purely real form

$$y = e^{px}(A\cos qx + B\sin qx)$$

Of course, it is not difficult to verify by direct substitution that both $y_1 = e^{px}\cos qx$ and $y_2 = e^{px}\sin qx$ are particular solutions (with nonvanishing wronskian) of the homogeneous equation ($1a$) (see Exercise 26). For a completely satisfactory derivation this should now be done, since we do not yet know that our formal treatment of complex exponentials, as though they obeyed the same laws as real exponentials, is justified.

Example 2 (continued) Applying the preceding reasoning to Example 2, we see that $p = -1$ and $q = 2$. Hence the complete solution can be written

$$y = e^{-x}(A \cos 2x + B \sin 2x)$$

When the characteristic equation has equal roots, the two independent solutions normally arising from the substitution of $y = e^{mx}$ become identical and, as pointed out above, we do not have an adequate basis for constructing a complete solution. To find a second, linearly independent solution in this case, we use the method developed in Sec. 3.3.

By hypothesis, the characteristic equation in this case has equal roots, say $m_1 = m_2 = r$. Hence it must be of the form

$$m^2 - 2rm + r^2 = 0$$

which implies that the differential equation itself is

$$y'' - 2ry' + r^2 y = 0$$

In particular, we observe from this that the coefficient $p_1(x)$ in this case is the number $-2r$. Clearly, $y_1 = e^{m_1 x} = e^{rx}$ is one solution of the differential equation, and, from Corollary 2, Theorem 2, Sec. 3.3, a second linearly independent solution is given by

$$y_1 \int \frac{e^{-\int p_1(x)\,dx}}{y_1^2}\,dx = e^{rx} \int \frac{e^{2rx}}{(e^{rx})^2}\,dx = xe^{rx} = xe^{m_1 x}$$

Thus, *in the exceptional case in which the characteristic equation has equal roots, a complete solution of Eq. (1a) is*

$$y = c_1 e^{m_1 x} + c_2 x e^{m_1 x}$$

Example 3 Find a complete solution of the equation $(D^2 + 6D + 9)y = 0$.

In this case, the characteristic equation $m^2 + 6m + 9 = 0$ is a perfect square, with roots $m_1 = m_2 = -3$. Hence, by our last remark, a complete solution of the given equation is

$$y = c_1 e^{-3x} + c_2 x e^{-3x}$$

The complete process for solving the homogeneous equation (1*a*) in all possible cases is summarized in Table 4.1.

In particular applications, the two arbitrary constants in the complete solution must usually be determined to fit given initial conditions on y and y', or their equivalent. The following examples will clarify the procedure.

Table 4.1

Differential equation: $ay'' + by' + cy = 0$ or $(aD^2 + bD + c)y = 0$
Characteristic equation: $am^2 + bm + c = 0$ or $aD^2 + bD + c = 0$

Nature of the roots of the characteristic equation	Condition on the coefficients of the characteristic equation	Complete solution of the differential equation
Real and unequal $m_1 \neq m_2$	$b^2 - 4ac > 0$	$y = c_1 e^{m_1 x} + c_2 e^{m_2 x}$
Real and equal $m_1 = m_2$	$b^2 - 4ac = 0$	$y = c_1 e^{m_1 x} + c_2 x e^{m_1 x}$
Conjugate complex $m_1 = p + iq$ $m_2 = p - iq$	$b^2 - 4ac < 0$	$y = e^{px}(A \cos qx + B \sin qx)$

Example 4 Find the solution of the equation $y'' - 4y' + 4y = 0$ for which $y = 3$ and $y' = 4$ when $x = 0$.

The characteristic equation of the differential equation is

$$m^2 - 4m + 4 = 0$$

Its roots are $m_1 = m_2 = 2$; hence a complete solution is

$$y = c_1 e^{2x} + c_2 x e^{2x}$$

By differentiating this, we find

$$y' = 2c_1 e^{2x} + c_2(e^{2x} + 2xe^{2x}) = (2c_1 + c_2)e^{2x} + 2c_2 x e^{2x}$$

Substituting the given data into the equations for y and y', respectively, we have

$$3 = c_1 \qquad \text{and} \qquad 4 = 2c_1 + c_2$$

Hence, $c_1 = 3$, $c_2 = -2$, and the required solution is

$$y = 3e^{2x} - 2xe^{2x}$$

Example 5 Find the solution of the equation $(4D^2 + 16D + 17)y = 0$ for which $y = 1$ when $t = 0$ and $y = 0$ when $t = \pi$.

In this case, the statement of the problem makes it clear that the independent variable is not x but t. This does not affect the characteristic equation $4m^2 + 16m + 17 = 0$ or its roots $m = -2 \pm \frac{1}{2}i$. However, the solution which we construct from these roots must be expressed in terms of t rather than x:

$$y = e^{-2t}\left(A \cos \frac{t}{2} + B \sin \frac{t}{2}\right)$$

Substituting the given conditions into this equation, we find

$$1 = A \qquad \text{and} \qquad 0 = e^{-2\pi}B \qquad \text{or} \qquad B = 0$$

Hence, the required solution is $y = e^{-2t} \cos (t/2)$.

Exercises for Section 4.2

Find a complete solution of each of the following equations.

1 $y'' = 0$ **2** $y'' - 3y' + 2y = 0$

3 $y'' - 2y' + y = 0$ **4** $y'' + y' - 2y = 0$

5 $y'' + 5y' + 4y = 0$ **6** $y'' - 5y = 0$

7 $y'' + 5y' = 0$ **8** $(4D^2 + 4D + 1)y = 0$

9 $(9D^2 - 12D + 4)y = 0$ **10** $10y'' + 6y' + y = 0$

11 $y'' + 10y' + 26y = 0$

Find the particular solution of each of the following equations which satisfies the given conditions.

12 $y'' + 3y' - 4y = 0$ $y = 4$, $y' = -2$ when $x = 0$

13 $y'' + 4y = 0$ $y = 2$, $y' = 6$ when $x = 0$

14 $y'' - 4y = 0$ $y = 1$, $y' = -1$ when $x = 0$

15 $25y'' + 20y' + 4y = 0$ $y = y' = 0$ when $x = 0$

16 $(D^2 + 6D + 9)y = 0$ $y = 0$, $y' = 3$ when $x = 0$

17 $(D^2 + 2D + 5)y = 0$ $y = 1$ when $x = 0$, $y = 0$ when $x = \pi$

18 $(D^2 + 2D + 5)y = 0$ $y = 1$ when $x = 0$, $y' = 0$ when $x = \pi$

***19** Show that there is always a unique solution of the equation $y'' + y = 0$ satisfying given conditions of the form $y = y_0$ when $x = x_0$ and $y = y_1$ when $x = x_1$ unless $x_1 = x_0 + n\pi$. What is the situation when $x_1 = x_0 + n\pi$?

20 For what values of λ, if any, are there nontrivial solutions† of the equation $y'' + \lambda^2 y = 0$ which satisfy the conditions $y(0) = 0$ and $y(\pi) = 0$? What are these solutions?

21 For what values of λ, if any, are there nontrivial solutions of the equation $y'' + \lambda^2 y = 0$ which satisfy the conditions $y'(0) = 0$ and $y'(\pi) = 0$? What are these solutions?

22 For what values of λ, if any, are there nontrivial solutions of the equation $y'' + \lambda^2 y = 0$ which satisfy the conditions $y(0) = 0$ and $y'(\pi) = 0$? What are these solutions?

***23** Show that the only values of λ for which there exist nontrivial solutions of the equation $y'' + \lambda^2 y = 0$ satisfying the conditions $y(0) = 0$ and $y(\pi) = y'(\pi)$ are the roots of the equation $\tan \pi\lambda = \lambda$. Show that this equation has infinitely many roots.

24 Work Exercise 20 for the equation $y'' - \lambda^2 y = 0$.

25 Work Exercise 22 for the equation $y'' - \lambda^2 y = 0$.

26 (*a*) Show that if the roots of the characteristic equation of a differential equation are $m = p \pm iq$, then the differential equation itself is $y'' - 2py' + (p^2 + q^2)y = 0$.

(*b*) Verify by direct substitution that $y_1 = e^{px} \cos qx$ and $y_2 = e^{px} \sin qx$ are solutions of the equation

$$y'' - 2py' + (p^2 + q^2)y = 0$$

(*c*) Verify that the particular solutions indicated in part (*b*) have a nonvanishing wronskian.

***27** Show that if $k \neq 0$, both $y = A \cos (kx + B)$ and $y = G \sin (kx + H)$ are complete solutions of the equation $y'' + k^2 y = 0$.

28 Show that if $k \neq 0$, $y = A \cosh kx + B \sinh kx$ is a complete solution of the equation $y'' - k^2 y = 0$.

† See footnote p. 71.

29 Show that $y = e^{px}(A \cosh qx + B \sinh qx)$ is a complete solution of the equation $ay'' + by' + cy = 0$ when the roots of the characteristic equation are $m = p \pm q$ and $q \neq 0$.

***30** If the roots of its characteristic equation are real, show that no nontrivial solution of the equation $ay'' + by' + cy = 0$ can have more than one real zero.

***31** If the characteristic equation of the differential equation $ay'' + by' + cy = 0$ has distinct roots m_1 and m_2, show that

$$y = \frac{e^{m_1 x} - e^{m_2 x}}{m_1 - m_2}$$

is a particular solution of the equation. Determine the limit of this expression as $m_2 \to m_1$, and discuss its relation to the solution of the differential equation when the characteristic equation has equal roots.

***32** (*a*) Show that $(D - a)^2 f(x) = e^{ax} D^2[e^{-ax} f(x)]$.

(*b*) Use the formula of part (*a*) to show that $(D - a)^2[(c_1 + c_2 x)e^{ax}] = 0$.

(*c*) Explain how the formula of part (*b*) can be used to obtain the complete solution of a differential equation whose characteristic equation has equal roots.

***33** (*a*) Show that $D^2(xe^{mx}) = m^2 xe^{mx} + 2me^{mx}$.

(*b*) Using the result of part (*a*), show that if $p(D)$ is a quadratic polynomial in D, then $p(D)(xe^{mx}) = p(m)xe^{mx} + p'(m)e^{mx}$.

(*c*) Explain how the formula of part (*b*) can be used to obtain a second linearly independent solution of a differential equation whose characteristic equation has equal roots. *Hint:* Recall that if a polynomial equation $p(x) = 0$ has a double root $x = r$, then $x = r$ is also a root of the equation $p'(x) = 0$.

34 What meaning, if any, do you think can be assigned to D^0? D^{-1}? D^{-2}?

***35** Show that the change of dependent variable defined by the substitution $y = -z'/zP$ changes the Riccati equation $y' = P(x)y^2 + Q(x)y + R(x)$ into the linear second-order equation $z'' - [Q + (P'/P)]z' + PRz = 0$.

Using the result of Exercise 35, solve each of the following equations.

36 $xy' = x^2y^2 - y + 1$ **37** $x^2y' = x^4y^2 + (3x^2 - 2x)y + 2$

38 $(\cos x)y' = (\cos^2 x)y^2 + (\sin x - 2 \cos x)y + 5$

4.3 THE NONHOMOGENEOUS SECOND-ORDER EQUATION WITH CONSTANT COEFFICIENTS

In the last section we learned how to solve the homogeneous equation $ay'' + by' + cy = 0$, and with this knowledge we can now obtain the complementary function of the nonhomogeneous equation

$$ay'' + by' + cy = f(x) \tag{1}$$

However, we must also have a particular integral, i.e., a particular solution, of Eq. (1) before we can construct its complete solution, namely,

$$y = \text{complementary function} + \text{particular integral}$$

In Sec. 3.4 we developed the method of variation of parameters as a procedure for finding particular integrals of nonhomogeneous equations, and we could, of course, use it here. However, in most elementary problems another process, called the method of **undetermined coefficients,** is simpler. It does not apply to as large a class of nonhomogeneous terms as does the method of variation of parameters, but it has the advantage of involving differentiation rather than integration and is therefore the one we shall use whenever possible. Initially, the method of undetermined coefficients appears to be based on little more than guesswork, but it can easily be formalized into a well-defined procedure applicable to a well-defined and very important class of cases.

To illustrate the method, suppose that we wish to find a particular integral of the equation

$$y'' + 4y' + 3y = 5e^{2x} \tag{2}$$

Since differentiating an exponential of the form e^{kx} merely reproduces that function with, at most, a change in its numerical coefficient, it is natural to "guess" that it may be possible to determine A so that

$$Y = Ae^{2x}$$

will be a solution of (2). To check this, we substitute $Y = Ae^{2x}$ for y in the given equation, getting

$$4Ae^{2x} + 8Ae^{2x} + 3Ae^{2x} = 5e^{2x} \qquad \text{or} \qquad 15Ae^{2x} = 5e^{2x}$$

which will be an identity if and only if $A = \frac{1}{3}$. Thus, the required particular integral is

$$Y = \tfrac{1}{3}e^{2x}$$

Now suppose that instead of $5e^{2x}$, the right-hand side of Eq. (2) is $5 \sin 2x$, so that the equation we have to solve is

$$y'' + 4y' + 3y = 5 \sin 2x \tag{2a}$$

Guided by our previous success in determining a particular solution, we might perhaps be led to try

$$Y = A \sin 2x$$

as a particular integral. Substituting this to check whether or not it can be a solution, we obtain

$$-4A \sin 2x + 8A \cos 2x + 3A \sin 2x = 5 \sin 2x$$

$$-A \sin 2x + 8A \cos 2x = 5 \sin 2x$$

and, since $\sin 2x$ and $\cos 2x$ are linearly independent, this cannot be an identity unless, simultaneously, $A = -5$ and $A = 0$, which is absurd. The difficulty here, of course, is that differentiating $\sin 2x$ introduces the new function $\cos 2x$, which must also be eliminated identically from the equation resulting from the substitution of $Y = A \sin 2x$ for y. Since the one arbitrary constant A cannot satisfy two independent conditions, it is clear that we must arrange to incorporate *two* arbitrary constants in our tentative choice for Y without, at the same time, introducing new terms which will lead to still more conditions. This is easily done by assuming

$$Y = A \sin 2x + B \cos 2x$$

which contains the necessary second parameter yet cannot introduce any further new functions since it already is a linear combination of *all* the independent terms that can be obtained from $\sin 2x$ or from $\cos 2x$ by repeated differentiation. The actual determination of A and B is a simple matter, for substitution into the new equation ($2a$) yields

$$(-4A \sin 2x - 4B \cos 2x) + 4(2A \cos 2x - 2B \sin 2x) + 3(A \sin 2x + B \cos 2x) = 5 \sin 2x$$

$$(-A - 8B) \sin 2x + (8A - B) \cos 2x = 5 \sin 2x$$

and for this to be an identity requires that

$$-A - 8B = 5 \qquad \text{and} \qquad 8A - B = 0$$

from which we find immediately that $A = -\frac{1}{13}$ and $B = -\frac{8}{13}$. Hence, finally,

$$Y = -\frac{\sin 2x + 8 \cos 2x}{13}$$

With these illustrations in mind, we are now in a position to describe more precisely the use of the method of undetermined coefficients for finding particular integrals.

Rule 1 If $f(x)$ is a function for which repeated differentiation yields only a finite number of linearly independent expressions, then, in general, a particular Y for the nonhomogeneous equation $ay'' + by' + cy = f(x)$ can be found by

1. Assuming Y to be an arbitrary linear combination of all the linearly independent terms which arise from $f(x)$ by repeated differentiation
2. Substituting Y into the given differential equation
3. Determining the arbitrary constants in Y so that the equation resulting from the substitution is identically satisfied.

The class of functions $f(x)$ possessing only a finite number of linearly independent derivatives consists of the simple functions

k
x^n $\quad$ n a positive integer
e^{kx}
$\cos kx$
$\sin kx$

and any others obtainable from these by a finite number of additions, subtractions, and multiplications. If $f(x)$ possesses infinitely many independent derivatives, as is the case, for instance, with the simple function $1/x$, it is occasionally convenient to assume for Y an infinite series whose terms are the respective derivatives of $f(x)$, each multiplied by an arbitrary constant. However, the use of the method of undetermined coefficients in such cases involves questions of convergence which never arise when $f(x)$ has only a finite number of independent derivatives.

When $f(x)$ is the sum of several terms, say $f(x) = R_1(x) + R_2(x)$, we can find a particular integral Y of Eq. (1) in either of two slightly different ways. If we wish, we can find Y by applying Rule 1 to $f(x)$ in its entirety. On the other hand, we can also find Y by solving two shorter problems, one involving just $R_1(x)$, the other involving just $R_2(x)$. The basis for this method is provided by the following theorem, whose proof is left as an exercise.

Theorem 1 If Y_1 is a solution of the equation $ay'' + by' + cy = R_1(x)$ and if Y_2 is a solution of the equation $ay'' + by' + cy = R_2(x)$, then $Y = Y_1 + Y_2$ is a solution of the equation

$$ay'' + by' + cy = R_1(x) + R_2(x)$$

Example 1 Find a particular integral for the equation

$$y'' + 3y' + 2y = 10e^{3x} + 4x^2 \tag{3}$$

If we wish, we can find Y by beginning with the expression $Y = Ae^{3x} + Bx^2 + Cx + D$, which means that we are going to handle the various terms in $f(x)$ at the same time. On the other hand, according to Theorem 1, we can also find Y by first finding a particular integral Y_1 for the equation

$$y'' + 3y' + 2y = 10e^{3x} \tag{3a}$$

then finding a particular integral Y_2 for the equation

$$y'' + 3y' + 2y = 4x^2 \tag{3b}$$

and finally taking Y to be the sum $Y_1 + Y_2$.

Using the second method (which means that the expressions we have to substitute are not quite so lengthy and the subsequent collection of terms is not quite so involved), we assume $Y_1 = Ae^{3x}$, substitute into Eq. (3a), and determine A so that the resulting equation will be an identity:

$$9Ae^{3x} + 3(3Ae^{3x}) + 2(Ae^{3x}) = 10e^{3x}$$

$$20Ae^{3x} = 10e^{3x}$$

which implies that $A = \frac{1}{2}$ and $Y_1 = \frac{1}{2}e^{3x}$. Then we assume $Y_2 = Bx^2 + Cx + D$, substitute into Eq. (3b), and determine B, C, and D so that, again, the resulting equation will be an identity:

$$2B + 3(2Bx + C) + 2(Bx^2 + Cx + D) = 4x^2$$

$$2Bx^2 + (6B + 2C)x + (2B + 3C + 2D) = 4x^2$$

$$2B = 4$$

$$6B + 2C = 0$$

$$2B + 3C + 2D = 0$$

Solving these simultaneously, we find at once that

$$B = 2 \qquad C = -6 \qquad D = 7$$

Hence $Y_2 = 2x^2 - 6x + 7$ and, finally,

$$Y = Y_1 + Y_2 = \frac{e^{3x}}{2} + 2x^2 - 6x + 7$$

There is one important exception to the procedure we have just been outlining, which we must now investigate. Suppose, for example, that we wish to find a particular integral for the equation

$$y'' + 5y' + 6y = e^{-3x} \tag{4}$$

Proceeding in the way we have just described, we would start with

$$Y = Ae^{-3x}$$

and substitute into the left member, getting

$$9Ae^{-3x} + 5(-3Ae^{-3x}) + 6(Ae^{-3x}) = e^{-3x}$$
$$0 = e^{-3x} \qquad (!)$$

Thus $Y = Ae^{-3x}$ satisfies the homogeneous equation corresponding to (4), but not (4) itself. Clearly, it is important that we be able to recognize and handle such cases. The source of the difficulty is easily identified, for the characteristic equation of Eq. (4) is

$$m^2 + 5m + 6 = 0$$

and since its roots are $m_1 = -3$ and $m_2 = -2$, the complementary function of Eq. (4) is

$$y = c_1 e^{-3x} + c_2 e^{-2x}$$

Thus, the term on the right-hand side of (4) is proportional to a term in the complementary function; i.e., it is a solution of the related homogeneous equation and hence can yield only zero when it is substituted into the left member.

There are various ways of overcoming this difficulty, several of which are indicated in the exercises. The most natural one for us to try is our powerful general method, variation of parameters. Using the particular solutions $y_1 = e^{-3x}$ and $y_2 = e^{-2x}$ in the formulas (8) of Sec. 3.4, we have

$$u_1' = -\frac{e^{-2x}}{(e^{-3x})(-2e^{-2x}) - (e^{-2x})(-3e^{-3x})} e^{-3x} = -1$$

$$u_2' = \frac{e^{-3x}}{(e^{-3x})(-2e^{-2x}) - (e^{-2x})(-3e^{-3x})} e^{-3x} = e^{-x}$$

Hence, integrating, we get $u_1 = -x$, $u_2 = -e^{-x}$, and

$$Y = u_1 y_1 + u_2 y_2 = (-x)e^{-3x} + (-e^{-x})e^{-2x} = -xe^{-3x} - e^{-3x}$$

Therefore a complete solution of Eq. (4) is

$$\begin{aligned} y &= c_1 e^{-3x} + c_2 e^{-2x} - xe^{-3x} - e^{-3x} \\ &= (c_1 - 1)e^{-3x} + c_2 e^{-2x} - xe^{-3x} \end{aligned}$$

Clearly, the term $-e^{-3x}$ makes no essential contribution to the particular integral Y since it is, in fact, a solution of the related homogeneous equation and therefore can be combined with the corresponding term in the complementary function. It thus appears that had we begun not with the tentative particular integral

$$Ae^{-3x}$$

but rather with

$$Axe^{-3x}$$

the method of undetermined coefficients could have been used to find A so that Axe^{-3x} was a solution of the nonhomogeneous equation (4).

Had we been given the nonhomogeneous equation

$$y'' + 6y' + 9y = e^{-3x} \tag{5}$$

which has $m = 3$ as a double root of the characteristic equation, then both e^{-3x} and xe^{-3x} would have been solutions of the related homogeneous equation and clearly $Y = Axe^{-3x}$ could not have been used as a tentative particular integral. Guided by the preceding discussion, we might in this case be led to try

$$Y = Ax^2e^{-3x}$$

If we did, we would find that by the method of undetermined coefficients A could be determined so that the last expression is a solution of the nonhomogeneous equation (5).

These observations, which can be established in general by the method of variation of parameters (see Exercises 31 and 32), are summarized in the following extension of Rule 1.

Rule 2 In the differential equation $ay'' + by' + cy = f(x)$, let $f(x)$ be a sum $R_1(x) + \cdots + R_n(x)$, and let Y_i be the group of terms normally included in the trial particular integral Y because of $R_i(x)$. If any term in Y_i duplicates a term in the complementary function of the differential equation, then before it is substituted into the equation, Y_i must be multiplied by the lowest positive integral power of x which will eliminate all such duplications. The results of our discussion are summarized in Table 4.2.

Table 4.2

Differential equation: $ay'' + by' + cy = f(x)$ or $(aD^2 + bD + c)y = f(x)$

$f(x)$†		Normal choice for the trial particular integral Y‡
1. α		A
2. αx^n	(n a positive integer)	$A_0x^n + A_1x^{n-1} + \cdots + A_{n-1}x + A_n$
3. αe^{rx}	(r either real or complex)	Ae^{rx}
4. $\alpha \cos kx$§		$A \cos kx + B \sin kx$
5. $\alpha \sin kx$		
6. $\alpha x^n e^{rx} \cos kx$		$(A_0x^n + \cdots + A_{n-1}x + A_n)e^{rx} \cos kx$
7. $\alpha x^n e^{rx} \sin kx$		$+ (B_0x^n + \cdots + B_{n-1}x + B_n)e^{rx} \sin kx$

† When $f(x)$ consists of a sum of several terms, the appropriate choice for Y is the sum of the Y expressions corresponding to these terms individually.

‡ Whenever a term in any of the Ys listed in this column duplicates a term in the complementary function, all terms in that Y expression must be multiplied by the lowest positive integral power of x sufficient to eliminate all such duplications.

§ The hyperbolic functions $\cosh kx$ and $\sinh kx$ can be handled either by expressing them in terms of exponentials or by using formulas entirely analogous to those in lines 4 to 7.

Example 2 Find a complete solution of the equation $y'' + 5y' + 6y = 3e^{-2x} + e^{3x}$.

The roots of the characteristic equation

$$m^2 + 5m + 6 = 0$$

are $m_1 = -2$ and $m_2 = -3$. Hence the complementary function is

$$c_1e^{-2x} + c_2e^{-3x}$$

For the trial solution Y_1 corresponding to the term $3e^{-2x}$, we would normally use Ae^{-2x}. However, e^{-2x} is a part of the complementary function, and thus, following the second footnote to Table 4.2, we must multiply this by x before including it in Y_1. For the term e^{3x} the normal choice for a trial solution, namely, $Y_2 = Be^{3x}$, is satisfactory as it stands, since e^{3x} is not contained in the complementary function. Hence we assume

$$Y = Y_1 + Y_2 = Axe^{-2x} + Be^{3x}$$

Substituting this into the differential equation, we have

$$(4Axe^{-2x} - 4Ae^{-2x} + 9Be^{3x}) + 5(-2Axe^{-2x} + Ae^{-2x} + 3Be^{3x}) + 6(Axe^{-2x} + Be^{3x}) = 3e^{-2x} + e^{3x}$$

or

$$Ae^{-2x} + 30Be^{3x} = 3e^{-2x} + e^{3x}$$

Equating coefficients of like functions, we find $A = 3$ and $B = \frac{1}{30}$. Hence

$$Y = 3xe^{-2x} + \frac{e^{3x}}{30}$$

and a complete solution is

$$y = c_1e^{-2x} + c_2e^{3x} + 3xe^{-2x} + \frac{e^{3x}}{30}$$

Example 3 Find a complete solution of the equation $y'' + y = 3 \sin x$.
The characteristic equation in this case is

$$m^2 + 1 = 0$$

and its roots are $m = \pm i$. Hence the complementary function is

$$c_1 \cos x + c_2 \sin x$$

For the trial particular integral corresponding to the term $3 \sin x$, we would normally use $A \cos x + B \sin x$. However, $\sin x$ occurs as a term in the complementary function. Hence, following the second footnote in Table 4.2, we must multiply $A \cos x + B \sin x$ by x before continuing with the method of undetermined coefficients. Thus we must substitute

$$Y = x(A \cos x + B \sin x)$$

into the given nonhomogeneous equation. This gives us

$$[x(-A \cos x - B \sin x) + 2(-A \sin x + B \cos x)] + x(A \cos x + B \sin x) = 3 \sin x$$

or, collecting terms,

$$-2A \sin x + 2B \cos x = 3 \sin x$$

Hence $A = -\frac{3}{2}$, $B = 0$, $Y = -\frac{3}{2}x \cos x$, and a complete solution is

$$y = c_1 \cos x + c_2 \sin x - \tfrac{3}{2}x \cos x$$

Example 4 Find a complete solution of the equation $y'' - 2y' + y = xe^x - e^x$.
The characteristic equation in this case is

$$m^2 - 2m + 1 = 0$$

and its roots are $m_1 = m_2 = 1$. Hence, the complementary function is

$$c_1 e^x + c_2 x e^x$$

According to line 6 of Table 4.2 (with $n = r = 1$, $k = 0$), we would normally try

$$Y_1 = (A_0 x + A_1)e^x$$

as the particular integral required for the term xe^x. Moreover, since the particular integral for the term $-e^x$, namely, $Y_2 = Ae^x$, is already a part of Y_1, it need not be included separately. However, the terms in Y_1 are both in the complementary function. Hence, following the second footnote in Table 4.2, we must multiply Y_1 by the lowest positive integral power of x which will eliminate all duplication of terms between Y_1 and the complementary function. This means that Y_1 must be multiplied by x^2, since multiplying it by x would still leave the term xe^x common to Y_1 and the complementary function. Thus we continue with the modified trial solution

$$Y = (A_0 x^3 + A_1 x^2)e^x$$

Substituting this into the differential equation, we obtain

$$[A_0 x^3 + (6A_0 + A_1)x^2 + (6A_0 + 4A_1)x + 2A_1]e^x - 2[A_0 x^3 + (3A_0 + A_1)x^2 + 2A_1 x]e^x + (A_0 x^3 + A_1 x^2)e^x = xe^x - e^x$$

or

$$6A_0 xe^x + 2A_1 e^x = xe^x - e^x$$

This will be identically true if and only if $A_0 = \frac{1}{6}$ and $A_1 = -\frac{1}{2}$. Hence

$$Y = \left(\frac{x^3}{6} - \frac{x^2}{2}\right)e^x$$

and so a complete solution is

$$y = c_1 e^x + c_2 x e^x - \frac{x^2 e^x}{2} + \frac{x^3 e^x}{6}$$

Exercises for Section 4.3

Find a complete solution of each of the following equations.

1 $y'' + 4y' + 5y = 2e^x$ **2** $y'' + 4y' + 3y = x - 1$

3 $y'' + y' = x + 2$ **4** $y'' - y = e^x + 2e^{2x}$

5 $y'' + y = \cos x + 3 \sin 2x$ **6** $y'' + 4y' + 13y = \cos 3x - \sin 3x$

7 $y'' + 3y' = \sin x + 2 \cos x$ **8** $y'' + 2y' + 10y = 25x^2 - 3e^{-x}$

9 $(D^2 + 4D + 4)y = xe^{-x}$ **10** $(D^2 + 1)y = e^x \sin x$

11 $y'' + 2y' + y = \cos^2 x$ *Hint:* Recall that $\cos^2 x = (1 + \cos 2x)/2$.

12 $10y'' - 6y' + y = 30 \sin x \cos x$ **13** $y'' - 5y' + 6y = \cosh x$

***14** $y'' - 5y' + 4y = \cosh x$

15 (*a*) Show that $Y = -\cosh x$ is a particular integral of the equation

$$y'' + y' - 2y = e^{-x}$$

(*b*) Determine A so that $Y = A \sinh x$ will be a particular integral of this equation.

Find the solution of each of the following equations which satisfies the given conditions.

16 $y'' + 4y' + 5y = 20e^x$ $y = y' = 0$ when $x = 0$

17 $y'' + 4y' + 4y = 8x - 10$ $y = 2$, $y' = 0$ when $x = 0$

18 $y'' + 4y' + 3y = 4e^{-x}$ $y = 0$, $y' = 2$ when $x = 0$

19 $y'' + 2y' + 5y = 10 \cos x$ $y = 5$, $y' = 6$ when $x = 0$

20 Show that $Y = (\sin \lambda t - \sin kt)/(k^2 - \lambda^2)$ is a particular integral of the equation $y'' + k^2 y = \sin \lambda t$, and investigate the limiting case when $\lambda \to k$.

****21** Construct a solution of the equation $y'' - 2ay' + a^2 y = e^{\lambda t}$ which will approach a solution of the equation $y'' - 2ay' + a^2 y = e^{at}$ as $\lambda \to a$.

22 If y_1 and y_2 are two solutions of the nonhomogeneous equation $y'' + P(x)y' + Q(x)y = R(x)$, determine for what values of c_1 and c_2, if any, $y = c_1 y_1 + c_2 y_2$ is a solution of this equation.

23 If y_1 and y_2 are, respectively, solutions of the equations $y'' + P(x)y' + Q(x)y = R_1(x)$ and $y'' + P(x)y' + Q(x)y = R_2(x)$, show that $y = y_1 + y_2$ is always a solution of the equation $y'' + P(x)y' + Q(x)y = R_1(x) + R_2(x)$.

***24** Using the method of undetermined coefficients, find a particular integral of the equation $y'' - y = 1/x$. For what values of x, if any, is this solution meaningful?

***25** Using the method of undetermined coefficients, find a particular integral of the equation $y'' + y = x^{1/2}$. For what values of x, if any, is this solution meaningful?

26 If y_1 is a solution of the homogeneous equation $y'' + P(x)y' + Q(x)y = 0$, show that the substitution $y = \phi y_1$ will reduce the problem of finding a particular integral of the nonhomogeneous equation $y'' + P(x)y' + Q(x)y = R(x)$ to the solution of a linear, first-order equation in which ϕ' is the dependent variable.

Using the method of Exercise 26, and the given solution y_1, find a particular integral of each of the following equations.

27 (*a*) $y'' - 3y' + 2y = e^x$ $\quad y_1 = e^x$
(*b*) $y'' - 3y' + 2y = e^x$ $\quad y_1 = e^{2x}$

28 $y'' - 2y' + y = xe^x$ $\quad y_1 = e^x$

***29** (*a*) $y'' + y = \cos x$ $\quad y_1 = \sin x$
(*b*) $y'' + y = \cos x$ $\quad y_1 = \cos x$

***30** (*a*) Explain how the result of part (*b*) of Exercise 33, Sec. 4.2, can be used to obtain a particular integral of the equation $ay'' + by' + cy = e^{rx}$ when $m = r$ is a simple root of the characteristic equation.

(*b*) Generalize the results of parts (*a*) and (*b*) of Exercise 33, Sec. 4.2, to show that if $p(D)$ is a quadratic polynomial in D, then $p(D)(x^2e^{mx}) = p(m)x^2e^{mx} + 2p'(m)xe^{mx} + p''(m)e^{mx}$.

(*c*) Explain how the result of part (*b*) can be used to obtain a particular integral of the equation $ay'' + by' + cy = xe^{rx}$ when $m = r$ is a double root of the characteristic equation.

***31** Using the method of variation of parameters, show that if $r_1 \neq r_2$, the coefficient A can always be determined so that $Y = Axe^{r_1x}$ is a solution of the equation

$$y'' - (r_1 + r_2)y' + r_1r_2y = ae^{r_1x}$$

***32** Using the method of variation of parameters, show that A can always be determined so that $Y = Ax^2e^{rx}$ is a solution of the equation $y'' - 2ry' + r^2y = ae^{rx}$.

4.4 EQUATIONS OF HIGHER ORDER

The solution of the general linear differential equation with constant coefficients and order greater than 2,

$$a_0y^{(n)} + a_1y^{(n-1)} + \cdots + a_{n-1}y' + a_ny = f(x) \qquad n > 2,\ a_0 \neq 0 \tag{1}$$

parallels the second-order case in all significant details. For the homogeneous equation

$$a_0y^{(n)} + a_1y^{(n-1)} + \cdots + a_{n-1}y' + a_ny = 0 \tag{2}$$

the substitution

$$y = Ae^{mx}$$

leads, as before, to the characteristic equation

$$a_0m^n + a_1m^{n-1} + \cdots + a_{n-1}m + a_n = 0 \tag{3}$$

which can be obtained in a specific problem simply by replacing each derivative by the corresponding power of m. The degree of this algebraic equation will be the same as the order of the differential equation (2); hence, counting each repeated root the appropriate number of times (i.e., according to its multiplicity), we find that the number of roots $m_1, m_2, \ldots,$ will equal the order of the differential equation. When these roots have been found, a complete solution of the homogeneous equation can be constructed by forming an arbitrary linear combination of the terms that were listed in Table 4.1, Sec. 4.2, as corresponding to each of the various root types. The only extension that is required arises when the characteristic equation (3) has roots of multiplicity greater than 2. In this case, the results of Table 4.1 are to be supplemented by the following rule.

Rule 1 If y_1 is the solution normally corresponding to a root m_1 of the characteristic equation (3), and if this root occurs k (> 2) times, then not only are y_1 and xy_1 solutions (as in the second-order case) but $x^2y_1, x^3y_1, \ldots, x^{k-1}y_1$ are also solutions and must be included in the complementary function.

It is not difficult to verify that the wronskian of the particular solutions formed according to Table 4.1 and Rule 1 is always different from zero, although we shall leave this to the exercises.

For the nonhomogeneous equation (1), it is still true, as we noted in Sec. 3.4, that the complete solution is the sum of the complementary function, obtained by solving the related homogeneous equation (2), and a particular integral. In the important case when $f(x)$ is a function possessing only a finite number of linearly independent derivatives, the particular integral can be found, just as before, by using the tentative choices for Y listed in Table 4.2, Sec. 4.3. Variation of parameters can be extended to those problems which the method of undetermined coefficients cannot handle. An example or two should make these ideas clear.

Example 1 Find a complete solution of the equation $y''' + 3y'' + 3y' + y = 0$.

In this case, the characteristic equation is

$$m^3 + 3m^2 + 3m + 1 \equiv (m + 1)^3 = 0$$

Hence $m = -1$ is a triple root, and not only are e^{-x} and xe^{-x} solutions of the differential equation but so too is x^2e^{-x}. Since these three solutions have a nonvanishing wronskian (see Exercise 24), it follows that a complete solution of the equation is

$$y = c_1e^{-x} + c_2xe^{-x} + c_3x^2e^{-x}$$

Example 2 Find a complete solution of the equation $y''' + 5y'' + 9y' + 5y = 3e^{2x}$.

The characteristic equation in this case is

$$m^3 + 5m^2 + 9m + 5 = 0$$

By inspection $m = -1$ is seen to be a root. Hence c_1e^{-x} must be one term in the complementary function. When the factor corresponding to this root is divided out of the characteristic equation, there remains the quadratic equation

$$m^2 + 4m + 5 = 0$$

Its roots are $m = -2 \pm i$; thus the complementary function must also contain

$$e^{-2x}(c_2 \cos x + c_3 \sin x)$$

The entire complementary function is therefore

$$c_1 e^{-x} + e^{-2x}(c_2 \cos x + c_3 \sin x)$$

For a particular integral we try, as usual, $Y = Ae^{2x}$. Substituting this into the differential equation gives

$$(8Ae^{2x}) + 5(4Ae^{2x}) + 9(2Ae^{2x}) + 5(Ae^{2x}) = 3e^{2x} \qquad \text{or} \qquad 51Ae^{2x} = 3e^{2x}$$

Hence $\qquad A = \frac{1}{17} \qquad Y = \frac{1}{17}e^{2x}$

and therefore $\qquad y = c_1 e^{-x} + e^{-2x}(c_2 \cos x + c_3 \sin x) + \frac{1}{17}e^{2x}$

is the required solution.

Example 3 Find a complete solution of the equation $(D^4 + 8D^2 + 16)y = -\sin x$.

The characteristic equation here is

$$m^4 + 8m^2 + 16 = 0 \qquad \text{or} \qquad (m^2 + 4)^2 = 0$$

The roots of this equation are $m = \pm 2i, \pm 2i$. Hence, the complementary function contains not only the terms

$$\cos 2x \qquad \text{and} \qquad \sin 2x$$

but also these terms multiplied by x, and therefore is

$$c_1 \cos 2x + c_2 \sin 2x + c_3 x \cos 2x + c_4 x \sin 2x$$

To find a particular integral, we try $Y = A \cos x + B \sin x$, which, on substitution into the differential equation, gives

$$(A \cos x + B \sin x) + 8(-A \cos x - B \sin x) + 16(A \cos x + B \sin x) = -\sin x$$

or $\qquad 9A \cos x + 9B \sin x = -\sin x$

This will be an identity if and only if $A = 0$† and $B = -\frac{1}{9}$. Therefore

$$Y = -\frac{\sin x}{9}$$

and the complete solution is

$$y = c_1 \cos 2x + c_2 \sin 2x + c_3 x \cos 2x + c_4 x \sin 2x - \frac{\sin x}{9}$$

Example 4 For what nonzero values of λ, if any, does the equation $y^{\text{iv}} - \lambda^4 y = 0$ have nontrivial solutions which satisfy the four conditions $y(0) = y''(0) = y(l) = y''(l) = 0$? What are these solutions if they exist?

† Since the given differential equation contains only derivatives of even order, we could have foreseen that Y would need to contain only a sine term in order to match the sine term on the right-hand side and that $Y = B \sin x$ would therefore be a satisfactory trial solution. This simplification should be clearly understood, for it can often be applied to the analysis of vibrating mechanical systems with negligible friction or electric circuits with negligible resistance. Of course, it cannot be applied when the nonhomogeneous term duplicates a term in the complementary function. (See Example 3, Sec. 4.3.)

The characteristic equation in this case is $m^4 - \lambda^4 = 0$, and its roots are $m = \pm\lambda$, $\pm i\lambda$. Hence, a complete solution is

$$y = c_1 \cos \lambda x + c_2 \sin \lambda x + c_3 e^{\lambda x} + c_4 e^{-\lambda x}$$

It is more convenient, however, to introduce hyperbolic functions and work with a complete solution of the following form:

$$y = A \cos \lambda x + B \sin \lambda x + C \cosh \lambda x + E \sinh \lambda x$$

Differentiating this twice gives us

$$y'' = \lambda^2(-A \cos \lambda x - B \sin \lambda x + C \cosh \lambda x + E \sinh \lambda x)$$

Hence, substituting the first two of the given conditions, we obtain the relations

$$A + C = 0$$

$$\lambda^2(-A + C) = 0$$

which imply that $A = C = 0$. Using this information and the last two conditions, we have, further,

$$B \sin \lambda l + E \sinh \lambda l = 0$$

$$\lambda^2(-B \sin \lambda l + E \sinh \lambda l) = 0$$

Dividing out λ^2 and then adding these equations, we find that

$$2E \sinh \lambda l = 0$$

Now the hyperbolic sine is zero if and only if its argument is zero. Moreover, by the statement of the problem we are restricted to nonzero values of λ. Hence, $\sinh \lambda l \neq 0$, and therefore $E = 0$, which implies that

$$B \sin \lambda l = 0$$

We have already been forced to the conclusion that $A = C = E = 0$; hence if $B = 0$, the solution would be identically zero, contrary to the requirements of the problem. Thus we must have

$$\sin \lambda l = 0 \qquad \text{or} \qquad \lambda = \frac{n\pi}{l} \qquad n = 1, 2, 3, \ldots$$

For these values of λ, and for these only, there are solutions meeting the requirements of the problem. Clearly, these solutions are all of the form $y = B \sin \lambda x$, or, more specifically, $y_n = B_n \sin (n\pi x/l)$.

Example 5 Using the method of variation of parameters, derive a formula for a particular integral of the equation

$$y''' - 7y' + 6y = f(x) \tag{4}$$

Since the method of variation of parameters is based on the use of particular solutions of the related homogeneous equation, our first step must be to find three independent solutions of the equation $y''' - 7y' + 6y = 0$. The characteristic equation in this case is

$$m^3 - 7m + 6 = 0$$

and by inspection $m = 1$ is one root. Removing the factor $m - 1$ from the left-hand side, we obtain the quadratic equation

$$m^2 + m - 6 = 0$$

whose roots are $m = 2, -3$. Since these values of m are all distinct, the corresponding solutions

$$y_1 = e^x \qquad y_2 = e^{2x} \qquad y_3 = e^{-3x}$$

are linearly independent.

We must now find three functions of x, say, u_1, u_2, u_3, such that

$$Y = u_1 y_1 + u_2 y_2 + u_3 y_3 = e^x u_1 + e^{2x} u_2 + e^{-3x} u_3$$

will be a solution of the given nonhomogeneous equation (4). Differentiating, we have

$$Y' = (e^x u_1 + 2e^{2x} u_2 - 3e^{-3x} u_3) + (e^x u_1' + e^{2x} u_2' + e^{-3x} u_3')$$

To prevent the occurrence of higher derivatives of the unknown functions u_1, u_2, u_3, we set

$$e^x u_1' + e^{2x} u_2' + e^{-3x} u_3' = 0 \tag{5}$$

which thus becomes one of the three conditions required for the determination of u_1, u_2, u_3. Continuing with the simplified expression for Y' and differentiating again, we obtain

$$Y'' = (e^x u_1 + 4e^{2x} u_2 + 9e^{-3x} u_3) + (e^x u_1' + 2e^{2x} u_2' - 3e^{-3x} u_3')$$

We now set

$$e^x u_1' + 2e^{2x} u_2' - 3e^{-3x} u_3' = 0 \tag{6}$$

and this becomes the second of the conditions u_1, u_2, u_3 must satisfy. Differentiating Y'', as now simplified, we obtain

$$Y''' = (e^x u_1 + 8e^{2x} u_2 - 27e^{-3x} u_3) + (e^x u_1' + 4e^{2x} u_2' + 9e^{-3x} u_3')$$

Finally, substituting for Y and its various derivatives in the given equation (4), we obtain the third condition on u_1, u_2, u_3:

$$\begin{aligned}[(e^x u_1 + 8e^{2x} u_2 - 27e^{-3x} u_3) + (e^x u_1' + 4e^{2x} u_2' + 9e^{-3x} u_3')]& \\ -7(e^x u_1 + 2e^{2x} u_2 - 3e^{-3x} u_3) + 6(e^x u_1 + e^{2x} u_2 + e^{3x} u_3) &= f(x)\end{aligned}$$

or, collecting terms,

$$e^x u_1' + 4e^{2x} u_2' + 9e^{-3x} u_3' = f(x) \tag{7}$$

Equations (5), (6), and (7) are three simultaneous linear equations from which u_1', u_2', and u_3' can readily be found. The determinant of the coefficients of this system is

$$\Delta = \begin{vmatrix} e^x & e^{2x} & e^{-3x} \\ e^x & 2e^{2x} & -3e^{-3x} \\ e^x & 4e^{2x} & 9e^{-3x} \end{vmatrix} = 20$$

Hence, using Cramer's rule (see item 3, Appendix B.1), we have

$$u_1' = \frac{\begin{vmatrix} 0 & e^{2x} & e^{-3x} \\ 0 & 2e^{2x} & -3e^{-3x} \\ f(x) & 4e^{2x} & 9e^{-3x} \end{vmatrix}}{\Delta} = -\frac{5e^{-x}f(x)}{20} \quad \text{and} \quad u_1 = -\frac{1}{4}\int e^{-x}f(x)\,dx$$

$$u_2' = \frac{\begin{vmatrix} e^{x} & 0 & e^{-3x} \\ e^{x} & 0 & -3e^{-3x} \\ e^{x} & f(x) & 9e^{-3x} \end{vmatrix}}{\Delta} = \frac{4e^{-2x}f(x)}{20} \quad \text{and} \quad u_2 = \frac{1}{5}\int e^{-2x}f(x)\,dx$$

$$u_3' = \frac{\begin{vmatrix} e^{x} & e^{2x} & 0 \\ e^{x} & 2e^{2x} & 0 \\ e^{x} & 4e^{x} & f(x) \end{vmatrix}}{\Delta} = \frac{e^{3x}f(x)}{20} \quad \text{and} \quad u_3 = \frac{1}{20}\int e^{3x}f(x)\,dx$$

Therefore the required particular integral is

$$Y = e^{x}u_1 + e^{2x}u_2 + e^{-3x}u_3$$

$$= -\frac{e^{x}}{4}\int e^{-x}f(x)\,dx + \frac{e^{2x}}{5}\int e^{-2x}f(x)\,dx + \frac{e^{-3x}}{20}\int e^{3x}f(x)\,dx$$

Exercises for Section 4.4

Find a complete solution of each of the following equations.

1 $(D^3 + 6D^2 + 11D + 6)y = 6x - 7$

2 $(D^4 - 16)y = e^x$

3 $y''' - 2y'' - 3y' + 10y = 40 \cos x$

4 $y^{iv} + 10y'' + 9y = \cos 2x$

5 $(D^4 + 8D^2 - 9)y = 9x^2 + 5 \sin 2x$

6 $(D^3 + D^2 + 3D - 5)y = e^x$

7 $(D^4 + 2D^3 - 3D^2 - 4D + 4)y = e^x$

8 $(D^3 - 7D + 6)y = e^x - 5e^{2x}$

***9** $y^{iv} + 4y = \cos x + \sin 2x$. *Hint:* By adding and subtracting the appropriate term on the left-hand side of the characteristic equation, rewrite it as the difference of two squares.

Find the solution of each of the following equations which satisfies the given conditions.

10 $y''' - 3y' + 2y = 0$ $\quad y = 0,\ y' = 2,\ y'' = 0$ when $x = 0$

***11** $(D^3 + 2D^2 - D - 2)y = 10 \sin x$ $\quad y = 1,\ y' = -2,\ y'' = -1$ when $x = 0$

12 $(D^3 - 2D^2 + D - 2)y = 0$ $\quad y = y' = y'' = 1$ when $x = 0$

13 $y^{iv} + 5y'' + 4y = 0$ $\quad y = y'' = 0$ when $x = 0$; $y = 1,\ y' = 2$ when $x = \pi/2$

****14** Using the method of variation of parameters, find a particular integral of the equation $y''' - 3y'' + 3y' - y = e^x/x$.

Using the method of variation of parameters, obtain a formula for a particular integral of each of the following equations.

***15** $(D^3 - 6D^2 + 11D - 6)y = f(x)$

***16** $y''' - y'' + y' - y = f(x)$

***17** $y''' - y'' - y' + y = f(x)$

****18** $y^{iv} - 5y'' + 4y = f(x)$

****19** For what nonzero values of λ, if any, does the equation $y^{\text{iv}} - \lambda^4 y = 0$ have solutions which satisfy the conditions $y(0) = y''(0) = y(1) = y'(1) = 0$ and are not identically zero? What are these solutions if they exist?

***20** Work Exercise 19 if the given conditions are $y(0) = y''(0) = y''(1) = y'''(1) = 0$.

****21** (*a*) Generalize the results of Exercise 33, Sec. 4.2, and Exercise 30, Sec. 4.3, by showing that if $p(D)$ is a cubic polynomial in D, then

$$p(D)(xe^{mx}) = p(m)xe^{mx} + p'(m)e^{mx}$$

$$p(D)(x^2e^{mx}) = p(m)x^2e^{mx} + 2p'(m)xe^{mx} + p''(m)e^{mx}$$

$$p(D)(x^3e^{mx}) = p(m)x^3e^{mx} + 3p'(m)x^2e^{mx} + 3p''(m)xe^{mx} + p'''(m)e^{mx}$$

(*b*) Discuss the application of these formulas to the solution of the third-order, linear differential equation with constant coefficients.

22 Find three linearly independent solutions of the equation $y''' - 2y'' + y' - 2y = 0$, and verify that their wronskian satisfies Abel's identity.

***23** Prove that the wronskian of the functions e^{m_1x}, e^{m_2x}, and e^{m_3x} is different from zero if and only if m_1, m_2, and m_3 are all different.

***24** Prove that the wronskian of the functions e^{mx}, xe^{mx}, and x^2e^{mx} is always different from zero.

***25** Prove that if a and m are real and if a is different from zero, then the wronskian of the functions e^{mx}, $\cos ax$, and $\sin ax$ is never equal to zero.

***26** Prove that if $\lambda \neq 0$, the wronskian of the functions $\cos \lambda x$, $\sin \lambda x$, $\cosh \lambda x$, and $\sinh \lambda x$ is never equal to zero.

4.5 THE EULER-CAUCHY DIFFERENTIAL EQUATION

So far, this chapter has been devoted to a study of linear differential equations with constant coefficients. However, there is one type of linear equation with variable coefficients which it is appropriate to discuss at this point because by a simple change of independent variable it can always be transformed into a linear equation with constant coefficients. This is the so-called **Euler-Cauchy† equation**

$$a_0x^ny^{(n)} + a_1x^{n-1}y^{(n-1)} + \cdots + a_{n-1}xy' + a_ny = f(x) \tag{1}$$

in which the coefficient of each derivative is a constant multiple of the corresponding power of the independent variable. As we shall soon see, the change of independent variable defined by

$$|x| = e^z \qquad \text{or} \qquad z = \ln |x| \qquad x \neq 0$$

will always convert this equation into a linear equation with constant coefficients. This, in turn, can always be solved by the methods which we have developed in the preceding sections of this chapter. Finally, by replacing z by $\ln |x|$ in the solution

† Leonard Euler (1707–1783) was a great Swiss mathematician. Augustin Louis Cauchy (1789–1857) was a great French mathematician.

of the transformed equation, we obtain the solution of the original differential equation.

Before discussing the solution in the general, nth-order case, let us illustrate the process as applied to a particular second-order differential equation.

Example 1 Find a complete solution of the equation $x^2y'' + xy' + 9y = 0$.

Under the transformation $|x| = e^z$ or $z = \ln |x|$, we have a straightforward application of the chain rule

$$y' = \frac{dy}{dx} = \frac{dy}{dz}\frac{dz}{dx} = \frac{1}{x}\frac{dy}{dz}$$

$$y'' = \frac{d(y')}{dx} = \frac{d}{dx}\left(\frac{1}{x}\frac{dy}{dz}\right) = -\frac{1}{x^2}\frac{dy}{dz} + \frac{1}{x}\frac{d^2y}{dz^2}\frac{dz}{dx} = -\frac{1}{x^2}\frac{dy}{dz} + \frac{1}{x^2}\frac{d^2y}{dz^2}$$

Substituting these into the given differential equation, we have

$$x^2\left(\frac{1}{x^2}\frac{d^2y}{dz^2} - \frac{1}{x^2}\frac{dy}{dz}\right) + x\left(\frac{1}{x}\frac{dy}{dz}\right) + 9y = 0$$

or, simplifying and collecting terms,

$$\frac{d^2y}{dz^2} + 9y = 0$$

The characteristic equation of the last equation is $m^2 + 9 = 0$, and from its roots, $m = \pm 3i$, we obtain immediately the complete solution

$$y = c_1 \cos 3z + c_2 \sin 3z$$

Finally, replacing z by $\ln |x|$, we obtain a complete solution of the original equation:

$$y = c_1 \cos (3 \ln |x|) + c_2 \sin (3 \ln |x|) \qquad x \neq 0$$

By repeated use of the chain rule, derivatives of y with respect to x of any order can be expressed in terms of derivatives of y with respect to z, as we illustrated in Example 1 for $y' = dy/dx$ and $y'' = d^2y/dx^2$. However, the process soon becomes very complicated if we carry it out as we did in Example 1. On the other hand, if we use the operational notation which we introduced in Sec. 4.1, we obtain elegant formulas for the various derivatives which make the transformation of the differential equation a very simple matter.

The clue to this approach comes from an inspection of the formulas for y' and y'' which we obtained in Example 1. If we let the operational symbol D denote the first derivative *with respect to z*, then the formulas of Example 1 can be rewritten

$$y' = \frac{dy}{dx} = \frac{1}{x}Dy \qquad \text{or} \qquad x\frac{dy}{dx} = Dy$$

and

$$y'' = \frac{d^2y}{dx^2} = \frac{1}{x^2}(D^2y - Dy) \qquad \text{or} \qquad x^2\frac{d^2y}{dx^2} = D(D-1)y$$

This suggests the generalization

$$x^n \frac{d^n y}{dx^n} = D(D-1)\cdots(D-n+1)y \tag{2}$$

and it is not difficult to prove that this is correct.

To do this, we use mathematical induction. Having already verified the truth of formula (2) for $n = 1$ and $n = 2$, we assume it true for $n = k$ and attempt to prove it true for $n = k + 1$. Starting with

$$x^k \frac{d^k y}{dx^k} = D(D-1)\cdots(D-k+1)y \tag{3}$$

we obtain, by differentiating both sides with respect to x,

$$\frac{d}{dx}\left(x^k \frac{d^k y}{dx^k}\right) = \frac{d}{dz}[D(D-1)\cdots(D-k+1)y]\frac{dz}{dx}$$

or

$$x^k \frac{d^{k+1} y}{dx^{k+1}} + kx^{k-1}\frac{d^k y}{dx^k} = D[D(D-1)\cdots(D-k+1)y]\frac{1}{x}$$

Now, multiplying through by x, then applying the inductive hypothesis (3) to the second term on the left, and then subtracting that term from both sides, we have

$$\begin{aligned} x^{k+1}\frac{d^{k+1} y}{dx^{k+1}} &= D[D(D-1)\cdots(D-k+1)y] - k[D(D-1)\cdots(D-k+1)]y \\ &= D(D-1)\cdots(D-k+1)(D-k)y \end{aligned}$$

which asserts the correctness of (3) for $n = k + 1$. Thus the induction is complete.

The next example illustrates the application of formula (2) to a nonhomogeneous Euler-Cauchy equation of order greater than 2.

Example 2 Find a complete solution of the equation

$$x^3 \frac{d^3 y}{dx^3} + 4x^2 \frac{d^2 y}{dx^2} - 5x\frac{dy}{dx} - 15y = x^4$$

Using Eq. (2) to transform the various terms on the left-hand side of the given equation, we have

$$D(D-1)(D-2)y + 4D(D-1)y - 5Dy - 15y = (e^z)^4 = e^{4z}$$

From this, by expanding the various operational products and then collecting terms, we find

$$(D^3 + D^2 - 7D - 15)y = e^{4z} \tag{4}$$

The characteristic equation of this equation is

$$m^3 + m^2 - 7m - 15 = (m-3)(m^2 + 4m + 5) = 0$$

From its roots, $m_1 = 3$, $m_2, m_3 = -2 \pm i$, we obtain the complementary function

$$y = c_1 e^{3z} + e^{-2z}(c_2 \cos z + c_3 \sin z)$$

For a particular integral we try $Y = Ae^{4z}$:

$$64Ae^{4z} + 16Ae^{4z} - 7(4Ae^{4z}) - 15(Ae^{4z}) = e^{4z}$$

$$37Ae^{4z} = e^{4z}$$

$$A = \tfrac{1}{37}$$

Therefore $Y = \frac{1}{37}e^{4z}$, and a complete solution of Eq. (4) is

$$y = c_1 e^{3z} + e^{-2z}(c_2 \cos z + c_3 \sin z) + \tfrac{1}{37}e^{4z}$$

Finally, replacing z by $\ln |x|$, we have as a complete solution of the given equation

$$y = c_1 e^{3(\ln |x|)} + e^{-2(\ln |x|)} [c_2 \cos (\ln |x|) + c_3 \sin (\ln |x|)] + \tfrac{1}{37}e^{4(\ln |x|)}$$

$$= c_1 x^3 + \frac{1}{x^2}[c_2 \cos (\ln |x|) + c_3 \sin (\ln |x|)] + \frac{x^4}{37} \qquad x \neq 0$$

Exercises for Section 4.5

Find a complete solution of each of the following equations.

1 $x^2y'' + xy' - y = 0$ **2** $x^2y'' - 6y = 1 + \ln |x|$

3 $x^2y'' - xy' + y = x^5$ **4** $x^3y''' - 3x^2y'' + 7xy' - 8y = 0$

5 $2x^2y'' + 5xy' + y = 3x + 2$ **6** $x^3y''' + 2x^2y'' - xy' + y = 0$

7 $x^4y^{\text{iv}} + 6x^3y''' + 15x^2y'' + 9xy' - 9y = 3 \ln |x| + 1/x^2$

***8** Work Example 2 if the nonhomogeneous term is x^3.

9 Show that the substitution $|Ax + B| = e^z$, or $z = \ln |Ax + B|$, will reduce the equation

$$a(Ax + B)^2 \frac{d^2y}{dx^2} + b(Ax + B)\frac{dy}{dx} + cy = 0$$

to a linear equation with constant coefficients. For $n > 2$, can the equation

$$a_0(Ax + B)^n \frac{d^ny}{dx^n} + a_1(Ax + B)^{n-1}\frac{d^{n-1}y}{dx^{n-1}} + \cdots + a_{n-1}(Ax + B)\frac{dy}{dx} + a_n y = 0$$

be solved in a similar fashion?

10 Using the results of Exercise 9, find a complete solution of the equation

$$(x - 2)^2y'' + 2(x - 2)y' - 6y = 0$$

4.6 APPLICATIONS OF LINEAR DIFFERENTIAL EQUATIONS WITH CONSTANT COEFFICIENTS

Linear differential equations with constant coefficients find their most important application in the study of electric circuits and vibrating mechanical systems. The results of this analysis are so useful to workers in the physical sciences that we shall devote an entire chapter (Chap. 7) to its major features. However, even those

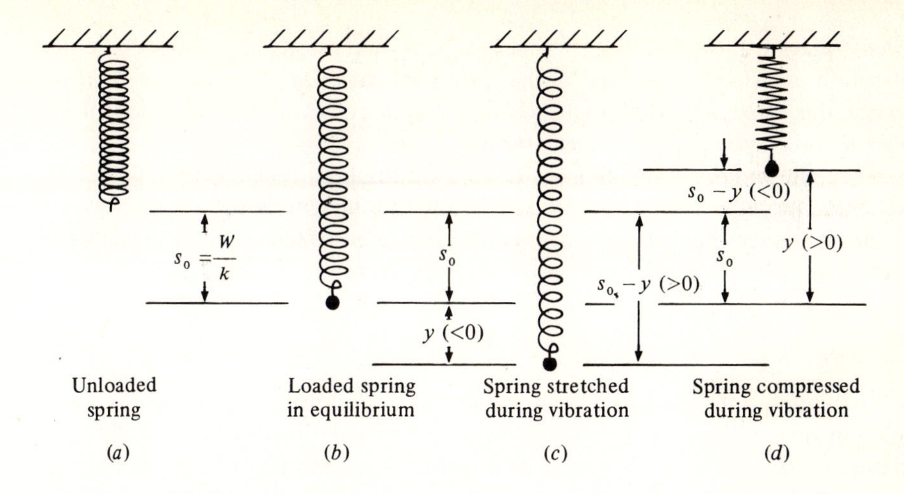

Figure 4.1 A simple mass-spring system.

who are not professionally interested in such applications should have some acquaintance with them, for they provide a remarkable illustration of the power of mathematics to describe important physical phenomena. Moreover, they illustrate dramatically the fact, often unrecognized by those unfamiliar with the applications of mathematics, that "Mathematics is the science of killing many birds with one stone." As we shall see, the mathematical descriptions of the motion of a weight vibrating on a spring and the flow of electric current in a series circuit are identical, and arise directly from the theory of linear differential equations with constant coefficients if we merely identify our abstract variables with force, velocity, displacement, and time, on the one hand, or voltage, current, electric charge, and time on the other. Accordingly, in this section we shall present a sampling of such applications, leaving to Chap. 7 a more detailed discussion for those who want to pursue these matters further.

As a first application, let us consider an object hanging from a spring, as shown in Fig. 4.1*b*. We assume that the object is of mass m and weight w, where w, the force of gravity acting on the mass, is related to m by the formula $w = mg$. As is customary, we shall suppose that the mass is concentrated in a single "point mass" and is not an object of appreciable dimensions. We shall assume that the spring obeys **Hooke's law**:

Force is proportional to displacement.

This means that if the spring is known to be stretched or compressed a distance s (the magnitude of the displacement) from its neutral, or unstretched, length, then the magnitude of the force F exerted on the spring is given by the formula

$$F = ks \tag{1}$$

In this equation the constant k, whose physical dimensions are clearly force per unit length is called the **modulus** of the spring. Because Eq. (1) is a linear relation between the force and the corresponding displacement, a spring which obeys Hooke's law is often called a **linear spring**.

When the weight w hangs from the spring in static (i.e., motionless) equilibrium, the spring is, of course, stretched and the weight is the force F which stretches it. Hence the distance s_0 which the spring is stretched can be found from Eq. (1):

$$w = ks_0 \qquad \text{or} \qquad s_0 = \frac{w}{k}$$

Now suppose that instead of hanging in static equilibrium, the weight is set in motion in such a way that it moves in a purely vertical direction. As a coordinate to describe the motion of the weight, let us use the distance y from the equilibrium position of the weight to its instantaneous position, the positive direction of y being upward (Fig. 4.1c and d). When $y = s_0$, the upward movement of the weight has just restored the length of the spring to its neutral, unstretched length, and the force exerted by the spring is instantaneously zero. In general, at the instant when the displacement of the weight from its equilibrium position is y, the change in the length of the spring is $s_0 - y$ and the force which the spring exerts on the weight is $k(s_0 - y)$. If $s_0 - y < 0$, the spring is compressed and it exerts a force on the weight in the downward or negative direction. If $s_0 - y > 0$, the spring is stretched and it exerts a force on the weight in the upward or positive direction. Hence, in each case the force instantaneously applied to the weight by the spring is given in magnitude and direction by the expression

$$F_{sp} = k(s_0 - y) \tag{2}$$

A realistic analysis of the vertical motion of the weight must take into account not only the force applied by the spring, the so-called **elastic force,** but also the effects of friction arising from the resistance of the air and, to a lesser degree, from internal friction in the spring itself. Friction is a highly complicated phenomenon, but experiments show that under ordinary conditions and for small velocities it can be approximated with satisfactory accuracy by a force (1) whose direction is opposite to the direction of the velocity, and (2) whose magnitude is proportional to the magnitude of the velocity. Under this assumption, the force of friction is given by the formula

$$F_{fr} = -cv = -c\frac{dy}{dt} \tag{3}$$

Friction which obeys this law is said to be **viscous friction,** or **viscous damping.**

Motion under the sole influence of an elastic force (2) and a friction force (3) is said to be **free,** or **unforced.** Often, however, in addition to such forces there are impressed forces which act on a system from external sources. For example, a platoon of soldiers marching in step across a light suspension bridge represents a force, external to the bridge, which will surely cause it to vibrate, perhaps dangerously. Similarly, an unbalanced electric motor may produce severe vibrations in the floor on which it rests. Motion resulting from such external influences is said to be **forced.** The forces impressed on a system are often periodic in nature, and we shall investigate their mathematical description in some detail when we study Fourier series in Chap. 8. In this section the only impressed forces that we shall consider are the simple periodic ones

$$F_{\text{im}} = F_0 \cos \omega t \tag{4a}$$

$$F_{\text{im}} = F_0 \sin \omega t \tag{4b}$$

where F_0 and ω are constants and t denotes time. Of course if $\omega = 0$, Eq. (4*a*) represents a constant force of magnitude F_0. It is interesting that after we have learned how to find the effect of these simple forces and have learned a little about Fourier series, we will be able to determine the effect of any periodic force, however complicated.

Now that we have identified the forces F_{sp}, F_{fr}, F_{im}, and, of course, the force of gravity $F_{\text{gr}} = -w$, which act on the suspended weight, it is an easy matter to set up the differential equation which describes the motion of the weight. To do this, we use **Newton's law:**

$$\textit{Mass} \times \textit{acceleration} = \textit{sum of all forces}$$

We know from elementary physics that the mass of a weight w is

$$\frac{w}{g}$$

The acceleration, of course, is

$$\frac{d^2y}{dt^2}$$

Hence, substituting into Newton's law from Eqs. (2), (3), and (4*a*), say, we obtain the differential equation

$$\frac{w}{g}\frac{d^2y}{dt^2} = F_{\text{sp}} + F_{\text{fr}} + F_{\text{im}} + F_{\text{gr}} = k(s_0 - y) - c\frac{dy}{dt} + F_0 \cos \omega t - w$$

We discovered earlier that $s_0 = w/k$. Hence, when we remove the parentheses, the term ks_0 cancels the last term, $-w$,† leaving

$$\frac{w}{g}\frac{d^2y}{dt^2} + c\frac{dy}{dt} + ky = F_0 \cos \omega t \tag{5}$$

† Because the gravitational force cancels the force represented by the stretch of the spring in its equilibrium position, it is customary in setting up problems involving suspended weights to neglect the force of gravity and to assume that when the weight hangs in static equilibrium, the spring is unstretched.

This is a typical nonhomogeneous, linear, second-order differential equation with constant coefficients whose complete solution we can easily find by the methods we developed in Secs. 4.2 and 4.3. In specific problems, Eq. (5) will be accompanied by initial conditions

$$y(0) = y_0 \qquad \text{and} \qquad \left.\frac{dy}{dt}\right|_{t=0} = y_0'$$

determined by the way in which the system is set in motion. By using these, the constants in any complete solution can be found, and the description of the subsequent motion of the weight will be completely determined.

Before solving Eq. (5), however, we shall undertake a similar derivation of the differential equation which describes the flow of current in a typical series-electric circuit. The result will be an equation mathematically identical with Eq. (5); and this should serve to underscore the fact that although their physical natures are vastly different, the problem of a weight vibrating on a spring and the problem of the flow of current in a series circuit are mathematically the same. In particular, this means that any property we can prove for one system must have its counterpart in the other.

The components which make up a typical electric circuit are

1. Resistors, represented by the symbol -/\/\/\-
2. Inductors, represented by the symbol ൦൦൦൦
3. Capacitors, represented by the symbol ⊣⊢
4. Voltage sources { batteries, represented by the symbol ⊣ ⊢ ; generators, represented by the symbol ◯ }

To say that we have a **series circuit** means that we have some, or all, of these elements connected one after the other in a closed loop, as in Fig. 4.2. The order in which the elements are connected makes no difference in the analysis. Under typical conditions, a current i will flow around the circuit, and our problem is to determine i as a function of time, after the switch is closed, by setting up and solving the appropriate differential equation.

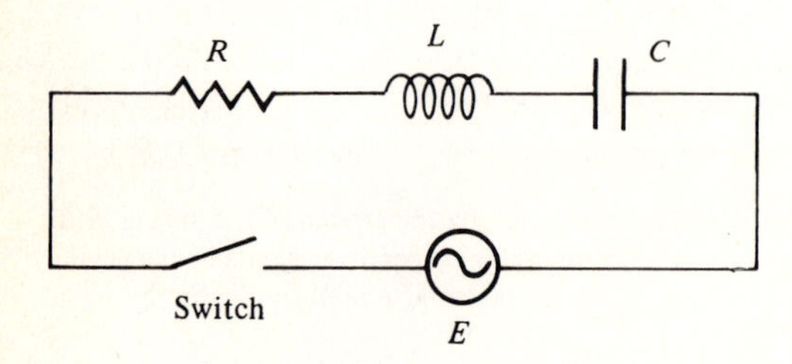

Figure 4.2 A simple series circuit.

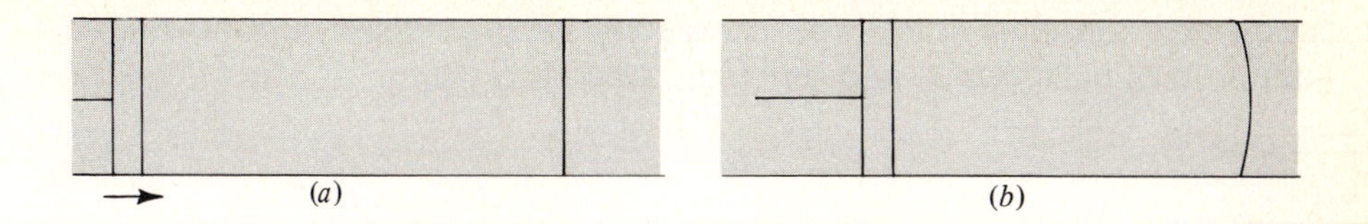

Figure 4.3 A hydraulic analog of a capacitor.

Almost everyone is familiar with **Ohm's law:**

$$E = iR \tag{6}$$

which asserts that if a voltage drop or potential difference E exists across a resistor of resistance R, then a current $i = E/R$ will flow through the resistor; or, conversely, if a current i is known to be flowing through a resistor, then a potential difference equal to iR must exist across that resistor.

Less familiar, perhaps, are the corresponding properties of inductors and capacitors. Inductors differ from resistors in that when a current flows through an inductor, there is no voltage drop *unless the current is changing.* When this is the case, the voltage drop across the inductor is proportional to the rate of change of the current, and is therefore given by the formula

$$E = L\frac{di}{dt} \tag{7}$$

where L is a constant called the **inductance.** The constant L measures the inductive properties of an inductor in the same sense that the resistance R measures the resistive properties of a resistor.

A capacitor is a little harder to appreciate because actually no current can flow through it at all! However, electric charges, whose flow constitutes the observed electric current, can be stored in a capacitor; and as the total stored charge Q increases and decreases, current appears to flow into and out of the capacitor. It may be helpful, in trying to understand this, to think of a capacitor as being somewhat like an elastic membrane stretched across a tube filled with water (Fig. 4.3*a*). Clearly, as long as the membrane is unbroken, no water can flow through it. However, if a piston in the tube applies force to the water, tending to make it move through the tube, the membrane, being elastic, can yield up to a point (Fig. 4.3*b*), and a current of water will appear to flow. Of course, if the piston exerts too great a force, the membrane will be ruptured† and water will obviously be able to flow where before it could not. The magnitude of the charge Q which is

† More than one capacitor has been "burnt out" in laboratory experiments by students who applied too great an electric force, i.e., voltage, across the terminals of the capacitor!

stored on a capacitor which experiences a voltage drop E across its terminals is proportional to E, that is, is given by the formula

$$Q = EC \tag{8a}$$

where C is a constant, called the **capacitance.** Conversely, if a capacitor of capacitance C carries a charge Q, then the potential difference across its terminals is given by the formula

$$E = \frac{Q}{C} \tag{8b}$$

Since the total charge instantaneously present on a capacitor is the result of the flow of charges into the capacitor up to that instant, it follows that the charge Q and the current i are connected by the important relation

$$Q = \int_{-\infty}^{t} i\,dt \tag{9a}$$

or, differentiating,

$$\frac{dQ}{dt} = i \tag{9b}$$

The units in terms of which the quantities E, i, Q, R, L, and C are measured are shown in Table 4.3.

With the basic electrical properties of the various circuit elements now identified, it is a simple matter to set up the differential equation whose solutions

Table 4.3

Quantity	Unit	Abbreviation	Source of name
E (voltage)	Volt	V	Named for the Italian physicist Alessandro Volta (1745–1827)
i (current)	Ampere	A	Named for the French physicist André Marie Ampere (1775–1836)
Q (charge)	Coulomb	C	Named for the French physicist Charles Augustin Coulomb (1736–1806)
R (resistance)	Ohm	Ω	Named for the German physicist Georg Simon Ohm (1787–1854)
L (inductance)	Henry	H	Named for the American physicist Joseph Henry (1797–1878)
C (capacitance)	Farad	F	Named for the English physicist Michael Faraday (1791–1867)

describe the flow of current in a series circuit. To do this, we use **Kirchhoff's† second law:**

> *The algebraic sum of the potential differences around any closed loop in an electric network is zero,* or *the voltage impressed on a closed loop is equal to the sum of the voltage drops in the rest of the loop.*

In this section we shall consider only impressed voltages of the form

$$E = E_0 \cos \omega t \tag{10a}$$

$$E = E_0 \sin \omega t \tag{10b}$$

where E_0 and ω are constants and t denotes time. These correspond to the periodic voltages supplied by ordinary generators. However, we note that if $\omega = 0$, then Eq. (10a) describes a constant voltage of magnitude E_0 such as an ordinary battery would supply.

Assuming an impressed voltage of the form (10a) Kirchhoff's second law gives us the relation

$$\begin{aligned} E_0 \cos \omega t = {} & \text{voltage drop across the inductor} \\ & + \text{voltage drop across the resistor} \\ & + \text{voltage drop across the capacitor} \end{aligned}$$

or, substituting from Eqs. (6), (7), and (8b),

$$E_0 \cos \omega t = L\frac{di}{dt} + Ri + \frac{Q}{C} \tag{11a}$$

Finally, if we use Eq. (9b), noting that $di/dt = d^2Q/dt^2$, we obtain the differential equation

$$L\frac{d^2Q}{dt^2} + R\frac{dQ}{dt} + \frac{Q}{C} = E_0 \cos \omega t \tag{11b}$$

Equation (11b), like Eq. (5), is a typical nonhomogeneous, linear, second-order differential equation with constant coefficients whose solution presents no difficulty. In particular problems it will ordinarily be accompanied by initial conditions

$$Q(0) = Q_0 \qquad \text{(the charge on the capacitor when } t = 0\text{)}$$

$$\left.\frac{dQ}{dt}\right|_{t=0} = \left. i \right|_{t=0} = i_0 \qquad \text{(the current in the circuit when } t = 0\text{)}$$

† Gustav Robert Kirchhoff (1824–1887) was a German physicist. **Kirchhoff's first law,** which we shall use in Chap. 7 when we discuss another type of electrical problem, asserts that *the algebraic sum of the currents flowing toward any point in an electrical network is zero.*

Table 4.4

Mechanical quantity		Electrical quantity	
Displacement	y	Charge	Q
Velocity	$v = \dfrac{dy}{dt}$	Current	$i = \dfrac{dQ}{dt}$
Impressed force	F	Impressed voltage	E
Mass	$m = \dfrac{w}{g}$	Inductance	L
Coefficient of friction	c	Resistance	R
Spring modulus	k	Reciprocal of capacitance (The reciprocal of the capacitance is usually called the **elastance.**)	$\dfrac{1}{C}$
Time	t	Time	t

The solution of Eq. (11*b*) will, of course, give us Q as a function of time. On the other hand, i rather than Q is the physical variable of greatest interest. However, since $i = dQ/dt$, it is a simple matter to find i by differentiation once Q has been determined.

Equations (5) and (11*b*) obviously have the same mathematical form. A more careful examination of their structure reveals an interesting correspondence between their various components, as shown in Table 4.4. The relations noted in Table 4.4 are often referred to, collectively, as **electromechanical analogies.** They have proved useful in enabling engineers to investigate complicated mechanical systems by setting up analogous electric circuits which in general are easier to study experimentally.

Let us begin our investigation of Eq. (5) by considering the simplest possible motion of a spring-suspended weight, namely, free motion with no damping. In this (ideal) case, the coefficient of friction c is zero, and there is no impressed force, that is, $F_0 = 0$. Equation (5) then becomes simply

$$\frac{w}{g}\frac{d^2y}{dt^2} + ky = 0 \qquad \text{or} \qquad \frac{d^2y}{dt^2} + \frac{kg}{w}y = 0$$

The characteristic equation of this differential equation is

$$m^2 + \frac{kg}{w} = 0$$

and from its roots, $m = \pm i\sqrt{kg/w}$, we obtain at once the complete solution

$$y = A\cos\sqrt{\frac{kg}{w}}\,t + B\sin\sqrt{\frac{kg}{w}}\,t \tag{12}$$

Regardless of the values of A and B, that is, regardless of how the system is set in motion, Eq. (12) describes periodic motion with period

$$2\pi\sqrt{\frac{w}{kg}}$$

or frequency

$$\frac{1}{2\pi}\sqrt{\frac{kg}{w}}$$

If w is measured in kilograms, k in kilograms per centimeter, and g in centimeters per second per second, the period is given in seconds and the frequency in cycles per second.† Whether there is damping in the system or not, the quantity $[1/(2\pi)]\sqrt{kg/w}$ is called the **natural frequency** of the system.

By similar reasoning, or simply by using the analogies listed in Table 4.4 to transform Eq. (12), we find from Eq. (11*b*) that in a series circuit containing no resistance ($R = 0$) and subject to no impressed voltage ($E_0 = 0$), the charge on the capacitor varies according to the equation

$$Q = A \cos \frac{1}{\sqrt{LC}} t + B \sin \frac{1}{\sqrt{LC}} t \tag{13}$$

From this, by differentiating with respect to time, we find that the current in the circuit varies according to the equation

$$i = \frac{dQ}{dt} = \frac{1}{\sqrt{LC}}\left(-A \sin \frac{1}{\sqrt{LC}} t + B \cos \frac{1}{\sqrt{LC}} t\right) \tag{14}$$

Regardless of the values of A and B, both Eq. (13) and Eq. (14) describe periodic behavior with period

$$2\pi\sqrt{LC}$$

or frequency

$$\frac{1}{2\pi\sqrt{LC}}$$

If L is measured in henries and C is measured in farads, the period is given in seconds and the frequency in cycles per second (or hertz).

Example 1 A weight of 7 kg is suspended from a spring of modulus $\frac{36}{35}$ kg/cm. At $t = 0$, while the weight is hanging in static equilibrium, it is suddenly given an initial velocity of 48 cm/s in the downward, or negative, direction. Taking the acceleration of gravity g to be 980 cm/s^2, find the vertical displacement of the weight as a function of t. What are the period and frequency of the subsequent motion? Through what amplitude does

† The unit *cycles per second* is customarily given the name hertz (abbreviated Hz) in honor of the German physicist Heinrich Hertz (1857–1894).

the weight move? At what times does the weight reach its extreme displacements above and below its equilibrium position?

The differential equation to be solved in this problem is

$$\frac{7}{980}\frac{d^2y}{dt^2} + \frac{36}{35}y = 0 \qquad \text{or} \qquad \frac{d^2y}{dt^2} + 144y = 0$$

and a complete solution is

$$y = A\cos 12t + B\sin 12t \tag{15}$$

The period of the motion is therefore

$$\frac{2\pi}{12} = \frac{\pi}{6}\ \text{s}$$

and the frequency is

$$\frac{6}{\pi}\ \text{Hz}$$

To determine A and B, we use the fact that $y(0) = 0$ (since the weight starts to move from its equilibrium position) and $v(0) = -48$ (since this is the velocity imparted to the weight at $t = 0$). Substituting the initial displacement condition into Eq. (15), we find $A = 0$. Substituting the initial velocity condition into the velocity equation

$$v = 12B\cos 12t$$

we find that $B = -4$. Thus the motion of the weight for $t \geq 0$ is described by the equation

$$y = -4\sin 12t \tag{16}$$

From Eq. (16) it is clear that the weight alternately rises and falls to a maximum of 4 cm above and below its equilibrium position. In other words, the amplitude of its motion is 4 cm. The extrema ($y = \pm 4$) occur when $\sin 12t = \pm 1$, that is, when

$$12t = \frac{\pi}{2} + n\pi \qquad \text{or} \qquad t = \left(\frac{\pi}{24} + \frac{n\pi}{12}\right)\ \text{s} \qquad n = 0, 1, 2, \ldots$$

Odd values of n correspond to maxima, even values to minima.

Example 2 A series circuit consists of an inductor for which $L = 0.02$ H and a capacitor for which $C = 8 \times 10^{-6}$ F. At $t = 0$, with the capacitor bearing a charge of 16×10^{-4} C, a switch in the circuit is closed, allowing the capacitor to discharge through the (now) closed circuit. Discuss the subsequent behavior of the circuit.

The differential equation to be solved in this problem is

$$0.02\frac{d^2Q}{dt^2} + \frac{1}{8 \times 10^{-6}}Q = 0 \qquad \text{or} \qquad \frac{d^2Q}{dt^2} + 6.25 \times 10^6 Q = 0$$

and a complete solution is

$$Q = A\cos 2{,}500t + B\sin 2{,}500t \tag{17}$$

and, by differentiation,

$$i = -2{,}500A\sin 2{,}500t + 2{,}500B\cos 2{,}500t \tag{18}$$

Both the charge on the capacitor and the current in the circuit thus vary periodically with period

$$\frac{2\pi}{2{,}500} = \frac{\pi}{1{,}250}\ \text{s}$$

and frequency

$$\frac{1{,}250}{\pi}\ \text{Hz}$$

To determine A and B, we use the given fact that $Q(0) = 16 \times 10^{-4}$ and the inferred fact that $i(0) = 0$ (since no current could have been flowing when the switch was closed at $t = 0$). Substituting the initial condition of charge into Eq. (17), we find that

$$A = 16 \times 10^{-4}$$

Substituting the initial current condition into Eq. (18), we find that

$$B = 0$$

Hence, for $t \geq 0$, the charge on the capacitor and the current flowing in the circuit are given by the respective equations

$$Q = 0.0016 \cos 2{,}500t\ \text{(C)} \qquad i = -4 \sin 2{,}500t\ \text{(A)}$$

By Eq. (7), the instantaneous voltage drop across the inductor is

$$L\frac{di}{dt} = (0.02)(-10{,}000 \cos 2{,}500t) = -200 \cos 2{,}500t$$

By Eq. (8*b*), the voltage drop across the capacitor is

$$\frac{Q}{C} = \frac{1}{8 \times 10^{-6}}(0.0016 \cos 2{,}500t) = 200 \cos 2{,}500t$$

The fact that the sum of these two voltage drops is equal to zero confirms the first form of Kirchhoff's second law.

When friction is not negligible, the differential equation describing the free, damped motion of a suspended weight is

$$\frac{w}{g}\frac{d^2y}{dt^2} + c\frac{dy}{dt} + ky = 0 \qquad \text{or} \qquad \frac{d^2y}{dt^2} + \frac{cg}{w}\frac{dy}{dt} + \frac{kg}{w}y = 0 \tag{19}$$

The corresponding characteristic equation is

$$m^2 + \frac{cg}{w}m + \frac{kg}{w} = 0$$

and its roots are

$$m_1, m_2 = -\frac{cg}{2w} \pm \frac{1}{2}\sqrt{\left(\frac{cg}{w}\right)^2 - 4\frac{kg}{w}} = -\frac{cg}{2w} \pm \frac{g}{2w}\sqrt{c^2 - 4\frac{kw}{g}}$$

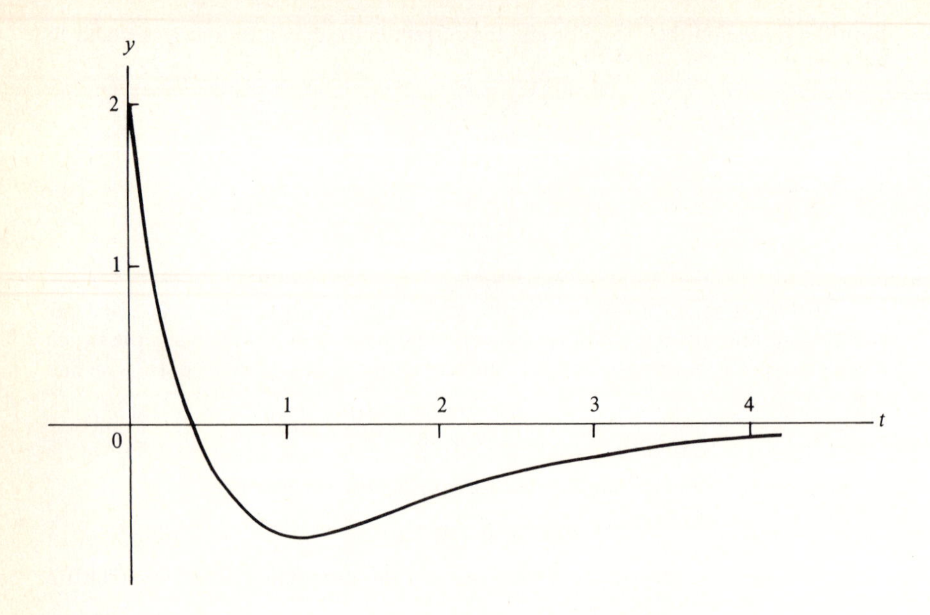

Figure 4.4 A typical displacement-time plot for overdamped motion.

Since g, w, and k are all positive and c is nonnegative, and since the radical, when real, is certainly less than c, it follows that the roots m_1 and m_2 are always negative if they are real and have real parts which are negative or zero if they are complex. As we discovered in Sec. 4.2, there are now three cases to consider:

$$c^2 - 4\frac{kw}{g}\begin{cases} > 0 \\ = 0 \\ < 0 \end{cases}$$

If $c^2 - 4kw/g > 0$, that is, if $c > 2\sqrt{kw/g}$, there is a relatively large amount of friction, and the system is said to be **overdamped.** In this case, the roots m_1 and m_2 of the characteristic equation are real and unequal, and a complete solution of Eq. (19) is given by

$$y = c_1 e^{m_1 t} + c_2 e^{m_2 t} \tag{20}$$

Clearly, the motion described by this equation is not periodic; and in fact, since m_1 and m_2 are negative, the displacement y approaches zero as time increases indefinitely. A typical plot of y vs. t in this case is shown in Fig. 4.4.

If $c^2 - 4kw/g = 0$, we have the borderline case where the roots of the characteristic equation are real and equal, i.e.,

$$m_1 = m_2 = -\frac{cg}{2w}$$

When this occurs, the motion is said to be **critically damped,** and the value of the coefficient of friction which produces it, namely,

$$c_c = 2\sqrt{\frac{kw}{g}}$$

is called the **critical damping.** In this case, a complete solution of Eq. (19) is

$$y = c_1 e^{m_1 t} + c_2 t e^{m_1 t} \tag{21}$$

Again, it is clear that the motion described by this equation is not periodic either, and again the displacement y approaches zero as $t \to \infty$. Plots of the motion of a critically damped system are essentially the same as those of an overdamped system (Fig. 4.4).

If $c^2 - 4kw/g < 0$, the system is said to be **underdamped.** The roots of the characteristic equation are now the complex numbers

$$m_1, m_2 = -\frac{cg}{2w} \pm i\frac{g}{2w}\sqrt{4\frac{kw}{g} - c^2} = -p \pm iq$$

where $\qquad p = \dfrac{cg}{2w} \qquad$ and $\qquad q = \dfrac{g}{2w}\sqrt{4\dfrac{kw}{g} - c^2}$

A complete solution of Eq. (19) is then

$$y = e^{-pt}(c_1 \cos qt + c_2 \sin qt) \tag{22}$$

In this case also, the motion is not periodic, and the displacement y approaches zero as $t \to \infty$. However, there is one important difference between the motion of an underdamped system and the motion of a system which is critically damped or overdamped. In the latter cases (as suggested by Fig. 4.4), there is at most one value of t for which $y = 0$; that is, in the course of the motion, the weight passes through its equilibrium position (where $y = 0$) at most once. On the other hand, in the underdamped case, there are infinitely many equally spaced values of t for which $y = 0$. To see this, we note from Eq. (22) that $y = 0$ if and only if $c_1 \cos qt + c_2 \sin qt = 0$, which implies that

$$\frac{\sin qt}{\cos qt} = \tan qt = -\frac{c_1}{c_2} \qquad \text{or} \qquad qt = \text{Tan}^{-1}\left(-\frac{c_1}{c_2}\right) + n\pi$$

or, finally,

$$t = \frac{1}{q}\text{Tan}^{-1}\left(-\frac{c_1}{c_2}\right) + \frac{n\pi}{q} \qquad \begin{array}{ll} n = 0, 1, 2, \ldots & \text{if } -c_1/c_2 \geq 0 \\ n = 1, 2, \ldots & \text{if } -c_1/c_2 < 0 \end{array}$$

Thus, as time goes on, the weight vibrates with decreasing amplitude in such a way that it passes through its equilibrium position at regular intervals of π/q. Motion of this sort is said to be a **damped oscillation** and has the general appearance shown in Fig. 4.5.

With only a change of symbols, the preceding analysis can be repeated exactly

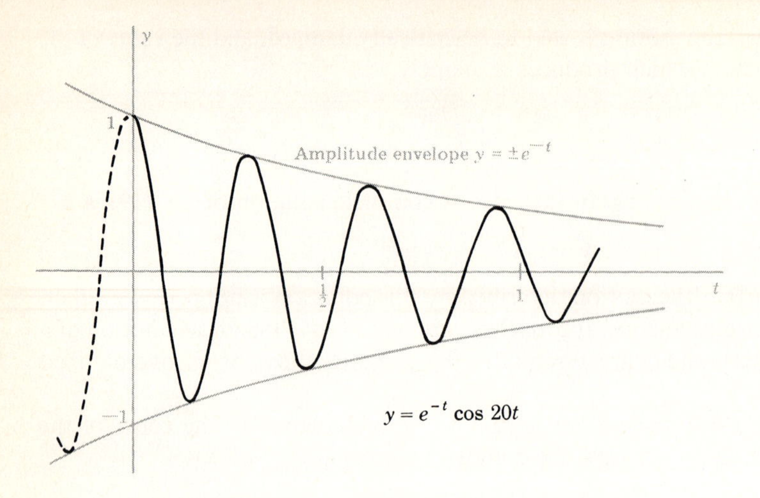

Figure 4.5 A typical displacement-time plot for an underdamped system.

for a series circuit with no impressed voltage but with appreciable resistance. The differential equation to be solved is

$$L\frac{d^2Q}{dt^2} + R\frac{dQ}{dt} + \frac{1}{C}Q = 0$$

and there are three cases to consider as the discriminant of the characteristic equation $R^2 - 4L/C$ is greater than, equal to, or less than zero. Using the terminology of the mechanical case, an electric circuit is said to be **overdamped** if $R^2 - 4L/C > 0$, **critically damped** if $R^2 - 4L/C = 0$, and **underdamped** if $R^2 - 4L/C < 0$. In the critically damped case, the exact value of the resistance which produces it is called the **critical resistance,**

$$R_c = 2\sqrt{\frac{L}{C}}$$

If the actual resistance is greater than or equal to the critical resistance, then the discharge of the capacitor and the flow of current are nonoscillatory, and a plot of either as a function of time resembles the plot in Fig. 4.4. If the actual resistance is less than the critical resistance, the discharge of the capacitor and the flow of current are damped oscillations (Fig. 4.5) and both Q and i are zero at infinitely many equally spaced values of t.

Example 3 A weight of 49 g is suspended from a spring of modulus $\frac{5}{2}$ g/cm. The coefficient of friction in the system is estimated to be $\frac{1}{10}$ g/(cm/s). At $t = 0$, the weight is pulled down 6 cm from its equilibrium position and released from that point with an upward velocity of 20 cm/s. Find the subsequent displacement of the weight as a function of time. When does the weight pass through its equilibrium position? Take $g = 980$ cm/s^2.

The differential equation to be solved is

$$\frac{49}{980}\frac{d^2y}{dt^2} + \frac{1}{10}\frac{dy}{dt} + \frac{5}{2}y = 0 \qquad \text{or} \qquad \frac{d^2y}{dt^2} + 2\frac{dy}{dt} + 50y = 0$$

The characteristic equation of this equation is $m^2 + 2m + 50 = 0$, and its roots are $m_1, m_2 = -1 \pm 7i$. Hence

$$y = e^{-t}(c_1 \cos 7t + c_2 \sin 7t)$$

and, differentiating,

$$v = -e^{-t}(c_1 \cos 7t + c_2 \sin 7t) + e^{-t}(-7c_1 \sin 7t + 7c_2 \cos 7t)$$

Substituting the data $y = -6$, $t = 0$ into the equation for y, we find

$$c_1 = -6$$

Substituting the data $v = 20$, $t = 0$ into the velocity equation, we find

$$20 = -c_1 + 7c_2 \qquad \text{or} \qquad c_2 = 2$$

The displacement of the weight is thus a damped oscillation described by the equation

$$y = e^{-t}(-6 \cos 7t + 2 \sin 7t) \qquad t \geq 0$$

The weight passes through its equilibrium position when $y = 0$, that is, when

$$-6 \cos 7t + 2 \sin 7t = 0$$

$$\tan 7t = 3$$

$$t = \frac{1}{7}\operatorname{Tan}^{-1} 3 + \frac{n\pi}{7} \doteq 0.178 + \frac{n\pi}{7} \qquad \text{s}$$

Example 4 A series circuit contains an inductor for which $L = 1$ H, a resistor for which $R = 1{,}000\ \Omega$, and a capacitor for which $C = 6.25 \times 10^{-6}$ F. At $t = 0$, with the capacitor bearing a charge of 1.5×10^{-3} C, a switch is closed and the capacitor discharges through the (now) closed circuit. Find Q and i as functions of t. When is the absolute value of the current a maximum? What is the extreme value of the current?

The differential equation to be solved is

$$\frac{d^2Q}{dt^2} + 1{,}000\frac{dQ}{dt} + \frac{Q}{6.25 \times 10^{-6}} = 0$$

The characteristic equation of this equation is

$$m^2 + 1{,}000m + 160{,}000 = 0$$

and its roots are $m_1 = -200$ and $m_2 = -800$. Hence

$$Q = c_1 e^{-200t} + c_2 e^{-800t}$$

and, differentiating,

$$i = -200c_1 e^{-200t} - 800c_2 e^{-800t}$$

Substituting the data $Q_0 = 1.5 \times 10^{-3}$, $t = 0$ into the equation for Q, we find

$$1.5 \times 10^{-3} = c_1 + c_2$$

Substituting the data $i_0 = 0$, $t = 0$ into the equation for i, we find

$$0 = -200c_1 - 800c_2$$

Solving these two equations simultaneously, we obtain

$$c_1 = 2 \times 10^{-3} \qquad c_2 = -5 \times 10^{-4}$$

Hence $$Q = 2 \times 10^{-3}e^{-200t} - 5 \times 10^{-4}e^{-800t}$$

and $$i = -0.4e^{-200t} + 0.4e^{-800t}$$

The extreme value of the current will occur when $di/dt = 0$. Hence we must solve the equation

$$\frac{di}{dt} = 80e^{-200t} - 320e^{-800t} = 0$$

From this, by multiplying by e^{200t} and dividing by 320, we obtain

$$e^{-600t} = \tfrac{1}{4}$$

Hence $e^{-200t} = 1/\sqrt[3]{4}$, $-200t = -\frac{1}{3}\ln 4$, and $t \doteq 0.00231$ s. To find the extreme value of the current, it is convenient to note that for the critical value of t we have

$$e^{-200t} = \frac{1}{\sqrt[3]{4}} \qquad \text{and therefore} \qquad e^{-800t} = \frac{1}{4\sqrt[3]{4}}$$

Hence $$i = -0.4\left(\frac{1}{\sqrt[3]{4}}\right) + 0.4\left(\frac{1}{4\sqrt[3]{4}}\right) = -\frac{3}{40}\sqrt[3]{16} \doteq -0.189 \text{ A}$$

Having discussed in detail the free motion of a weight suspended from a spring, we now return to Eq. (5):

$$\frac{w}{g}\frac{d^2y}{dt^2} + c\frac{dy}{dt} + ky = F_0 \cos \omega t \tag{5}$$

and discuss the forced motion of such a system. As we know, the complete solution of an equation of the form (5) consists of the sum of the complementary function and a particular integral. The complementary function will be a complete solution of the related homogeneous equation

$$\frac{w}{g}\frac{d^2y}{dt^2} + c\frac{dy}{dt} + ky = 0 \tag{19}$$

and will be of one or the other of the following forms:

$$y_c = c_1e^{m_1t} + c_2e^{m_2t} \qquad c^2 > 4kw/g \tag{20}$$

$$y_c = c_1e^{m_1t} + c_2te^{m_1t} \qquad c^2 = 4kw/g \tag{21}$$

$$y_c = e^{-pt}(c_1 \cos qt + c_2 \sin qt) \qquad c^2 < 4kw/g \tag{22}$$

In general, the particular integral, which is a solution of the nonhomogeneous equation (5), is of the form

$$Y = A \cos \omega t + B \sin \omega t \tag{23}$$

However, if $c = 0$ and $\omega = \omega_n = \sqrt{kg/w}$, then the nonhomogeneous term in Eq. (5) duplicates a term in the complementary function, and Eq. (23) is no longer valid. The physical significance of this special case we shall consider shortly.

The constants A and B will be determined by substituting Y into Eq. (5) and choosing them so that the resulting equation is identically satisfied. Subsequently, the constants c_1 and c_2 in the complete solution $Y =$ *complementary function* + *particular integral* will be determined by making the complete solution satisfy the appropriate initial conditions.

The complementary function, being the solution of the related homogeneous equation (19), is just a description of the free motion that the weight would experience *if it were not being forced.* As we have already noted, if $c \neq 0$, that is, if there is any friction in the system, then the free motion in each of the three cases (20), (21), and (22) dies away to zero as time goes on. Hence it eventually disappears as a component of the motion of the weight, and for that reason it is called the **transient.** On the other hand, the particular integral (23) is not attenuated but continues indefinitely its regular periodic behavior. Hence, although it does vary with time, it is called the **steady state.** With these definitions, the forced motion of the weight, which we first described as

$$y = \text{complementary function} + \text{particular integral}$$

can now be described in more vivid physical terms as

$$y = \text{transient} + \text{steady state}$$

In Chap. 7 we shall make a detailed study of the steady-state behavior associated with Eq. (5). However, there are certain descriptive features of it which it is appropriate to note at this point. First, it is obvious that *regardless of the values of A and B, the frequency of the steady-state motion is the same as the frequency of the impressed force which produces it,* namely $\omega/(2\pi)$.

To draw other conclusions, we will need to know the values of A and B. Hence let us substitute for Y from Eq. (23) into Eq. (5) and determine A and B so that the result is an identity:

$$\frac{w}{g}(-\omega^2 A \cos \omega t - \omega^2 B \sin \omega t) + c(-\omega A \sin \omega t + \omega B \cos \omega t)$$
$$+ k(A \cos \omega t + B \sin \omega t) = F_0 \cos \omega t$$

or, collecting terms,

$$\left[\left(k - \frac{\omega^2 w}{g}\right)A + \omega c B\right] \cos \omega t + \left[-\omega c A + \left(k - \frac{\omega^2 w}{g}\right)B\right] \sin \omega t = F_0 \cos \omega t$$

Hence, equating coefficients of like trigonometric functions,

$$\left(k - \frac{\omega^2 w}{g}\right)A + \omega c B = F_0$$

$$-\omega c A + \left(k - \frac{\omega^2 w}{g}\right)B = 0$$

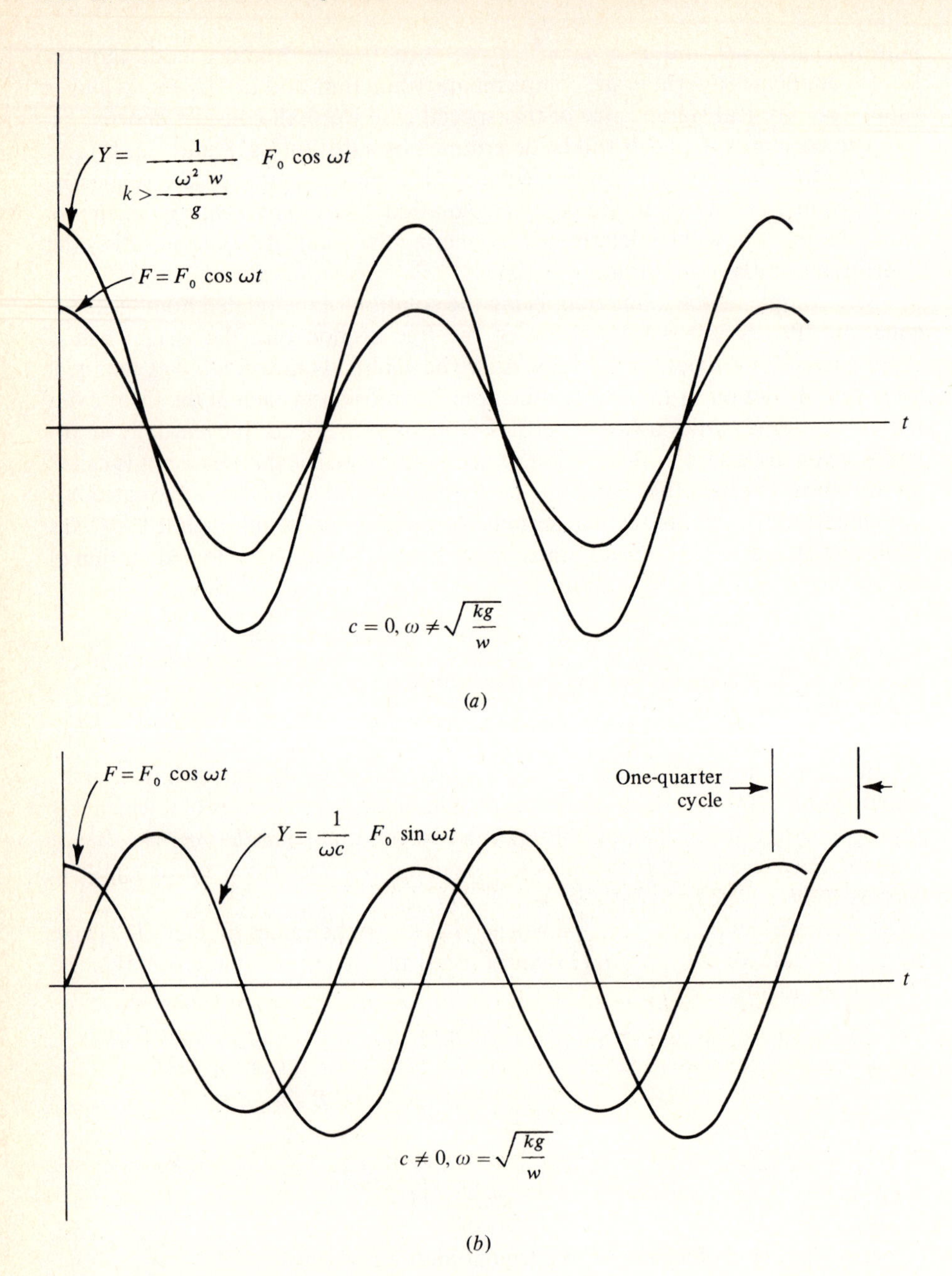

Figure 4.6 Displacement-time plots for two special cases of forced motion.

and, solving simultaneously,

$$A = \frac{k - \omega^2 w/g}{(k - \omega^2 w/g)^2 + (\omega c)^2} F_0 \qquad B = \frac{\omega c}{(k - \omega^2 w/g)^2 + (\omega c)^2} F_0 \tag{24}$$

From these formulas it is clear that in general neither A nor B is zero. However, there are two special cases in which they may be. If $c = 0$ but $k - \omega^2 w/g \neq 0$, then $B = 0$ and

$$Y = \frac{F_0}{k - \omega^2 w/g} \cos \omega t$$

In this case, both the impressed force and the steady-state response are multiples of $\cos \omega t$. If $k > \omega^2 w/g$, they are in phase and reach corresponding maxima and minima at the same times (Fig. 4.6*a*). If $k < \omega^2 w/g$, they are 180° out of phase and one reaches its maxima as the other reaches its minima, and vice versa. On the other hand, if $k - \omega^2 w/g = 0$ but $c \neq 0$, that is, if ω is equal to the natural frequency of the undamped system, namely, $\sqrt{kg/w}$, but some friction is present, then $A = 0$ and

$$Y = \frac{F_0}{\omega c} \sin \omega t$$

In this case, the impressed force and the forced response are out of phase, and the forced motion reaches its extreme values a quarter of a cycle after the impressed force reaches its extrema (Fig. 4.6*b*). In general, with neither $c = 0$ nor $\omega^2 = kg/w$, neither A nor B is zero, and the forced motion lags the impressed force by a fractional part of a cycle that depends on the parameters of the system.

It is important to note that if c is very close to zero and if simultaneously ω^2 is very close to kg/w, then A and B are both very large, even if F_0 itself is small. This is the condition of **approximate resonance.** In it, the weight is being forced to vibrate at a frequency very close to its natural frequency, and with little or no friction to dissipate the energy being fed into the system by the impressed force, it responds with vibrations of large and possibly disastrous amplitudes.

If $c = 0$ and $\omega^2 = kg/w$, formulas (24) are meaningless. We could have foreseen this, of course, because in this case, terms in the particular integral are already in the complementary function; and in place of (23) we should have used

$$Y = t(A \cos \omega t + B \sin \omega t) \tag{25}$$

Substituting this into Eq. (5), equating the coefficients of like terms, and then solving for A and B, we find in this case that

$$A = 0 \qquad B = \frac{gF_0}{2\omega w} \qquad \text{and} \qquad Y = \frac{gF_0}{2\omega w} t \sin \omega t$$

Because of the factor t, the steady-state response in this case becomes larger and larger without limit as time goes on, and the system must surely destroy itself. This behavior is known as **pure resonance.** Of course, the preceding formulas and the properties they imply can all be restated as corresponding results for the series-electric circuit.

Example 5 If the system discussed in Example 3 is acted upon by the force $F = 4 \cos 6t$ g, find the subsequent displacement of the weight as a function of time.

The differential equation to be solved in this problem is

$$\frac{49}{980}\frac{d^2y}{dt^2} + \frac{1}{10}\frac{dy}{dt} + \frac{5}{2}y = 4 \cos 6t \qquad \text{or} \qquad \frac{d^2y}{dt^2} + 2\frac{dy}{dt} + 50y = 80 \cos 6t$$

Since we already have the complementary function, or transient, namely,

$$y_c = e^{-t}(c_1 \cos 7t + c_2 \sin 7t)$$

we need only determine a particular integral to have a complete solution. Assuming $Y = A \cos 6t + B \sin 6t$, substituting, equating the coefficients of like functions, and solving for A and B give us

$$(-36A \cos 6t - 36B \sin 6t) + 2(-6A \sin 6t + 6B \cos 6t)$$

$$+ 50(A \cos 6t + B \sin 6t) = 80 \cos 6t$$

$$(14A + 12B) \cos 6t + (-12A + 14B) \sin 6t = 80 \cos 6t$$

$$14A + 12B = 80$$

$$-12A + 14B = 0$$

$$A = \tfrac{56}{17} \qquad B = \tfrac{48}{17}$$

Therefore $y = y_c + Y = e^{-t}(c_1 \cos 7t + c_2 \sin 7t) + \dfrac{56 \cos 6t + 48 \sin 6t}{17}$

Substituting the given initial data $y = -6$, $t = 0$ (from Example 3), we find

$$-6 = c_1 + \tfrac{56}{17} \qquad c_1 = -\tfrac{158}{17}$$

Before we can determine c_2 by substituting the initial velocity data, we must find v by differentiating y:

$$v = -e^{-t}(c_1 \cos 7t + c_2 \sin 7t) + e^{-t}(-7c_1 \sin 7t + 7c_2 \cos 7t)$$

$$+ \frac{-336 \sin 6t + 288 \cos 6t}{17}$$

Substituting into this equation the values $v = 20$, $t = 0$, we obtain

$$20 = -c_1 + 7c_2 + \tfrac{288}{17} \qquad \text{and} \qquad c_2 = -\tfrac{106}{119}$$

The final solution is thus

$$y = e^{-t}\left(-\frac{158}{17} \cos 7t - \frac{106}{119} \sin 7t\right) + \frac{56 \cos 6t + 48 \sin 6t}{17}$$

Example 6 A series circuit contains the components $L = 1$ H, $R = 1{,}000\ \Omega$, and $C = 4 \times 10^{-6}$ F. At $t = 0$, while the circuit is completely passive (that is, while $Q = i = 0$), a battery supplying a constant voltage of $E = 24$ V is suddenly switched into the circuit. Find the charge on the capacitor and the resultant current as functions of time.

The differential equation to be solved in this problem is

$$\frac{d^2Q}{dt^2} + 1{,}000\frac{dQ}{dt} + \frac{Q}{4 \times 10^{-6}} = 24$$

The characteristic equation is $m^2 + 1{,}000m + 250{,}000 = 0$ and its roots are $m_1 = m_2 = -500$. Hence the complementary function is

$$Q_c = c_1 e^{-500t} + c_2 te^{-500t}$$

To find a particular integral, we substitute $Q = A$ into the nonhomogeneous differential equation, getting

$$A = 9.6 \times 10^{-5}$$

A complete solution is therefore

$$Q = c_1 e^{-500t} + c_2 te^{-500t} + 9.6 \times 10^{-5}$$

and, differentiating,

$$\frac{dQ}{dt} = i = -500c_1 e^{-500t} + c_2(e^{-500t} - 500te^{-500t})$$

Substituting $Q = 0$, $t = 0$, we find

$$c_1 = -9.6 \times 10^{-5}$$

Substituting $i = 0$, $t = 0$, we find

$$0 = -500c_1 + c_2 \qquad c_2 = -4.8 \times 10^{-2}$$

Hence $\quad Q = -9.6 \times 10^{-5}e^{-500t} - 4.8 \times 10^{-2}te^{-500t} + 9.6 \times 10^{-5} \quad \text{C}$

$$i = 24te^{-500t} \quad \text{A}$$

There are numerous mechanical problems besides those involving spring-suspended weights which lead to linear differential equations with constant coefficients, and we shall conclude this section with several such examples.

Example 7 A perfectly flexible cable of length $2L$ cm, weighing w g/cm, hangs over a frictionless peg of negligible diameter. At $t = 0$, the cable is released from rest in a position in which the portion of the cable on one side is a cm longer than that on the other. Find the equation of motion of the cable as it slips over the peg.

At any time t, let y be the distance that the short end of the cable has risen from its initial position. Then at any instant, the cable on the long side is $(a + 2y)$ cm longer than the cable on the short side (Fig. 4.7). The unbalanced weight of this much cable, namely, $(a + 2y)w$ g, is the force which acts to make the cable move. Since the weight of the entire cable is $2Lw$ g, we then have from Newton's law

$$\frac{2Lw}{g}\frac{d^2y}{dt^2} = (a + 2y)w \qquad \text{or} \qquad \frac{d^2y}{dt^2} - \frac{g}{L}y = \frac{ag}{2L}$$

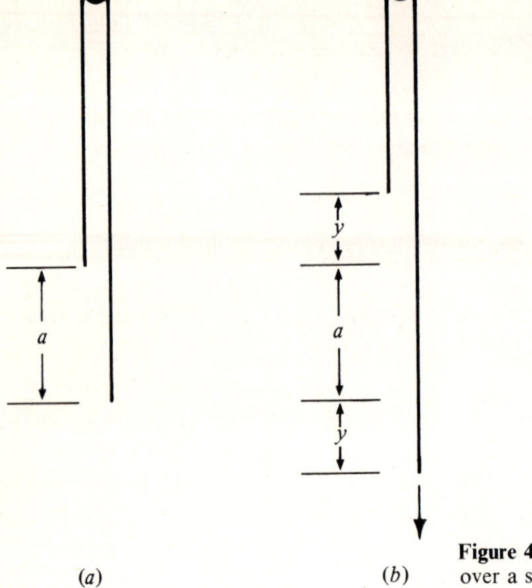

Figure 4.7 A flexible cable slipping over a smooth peg.

Using hyperbolic functions, we can write the complementary function of the last equation in the form

$$y_c = c_1 \cosh \sqrt{\frac{g}{L}}\,t + c_2 \sinh \sqrt{\frac{g}{L}}\,t$$

By inspection, a particular integral is $Y = -a/2$. Hence a complete solution for y is

$$y = c_1 \cosh \sqrt{\frac{g}{L}}\,t + c_2 \sinh \sqrt{\frac{g}{L}}\,t - \frac{a}{2}$$

and, differentiating,

$$v = \sqrt{\frac{g}{L}}\left(c_1 \sinh \sqrt{\frac{g}{L}}\,t + c_2 \cosh \sqrt{\frac{g}{L}}\,t\right)$$

Under the conditions of the problem, the cable starts to move from rest in a position in which $y = 0$. Therefore, substituting $y = 0$, $t = 0$ into the equation for y, and $v = 0$, $t = 0$ into the equation for v, we obtain $c_1 = a/2$ and $c_2 = 0$. Hence

$$y = \frac{a}{2}\left(\cosh \sqrt{\frac{g}{L}}\,t - 1\right)$$

This is valid until the short end of the cable reaches the peg, that is, until $y = L - a/2$.

Newton's law is not always the most convenient way to set up the differential equation that describes the behavior of a mechanical system. Sometimes this is most easily done by using the **principle of the conservation of energy:**

If no energy is lost through friction or other irreversible conversions of energy, then in a mechanical system the sum of the instantaneous kinetic and potential energies of the system must remain constant.

To use this principle, it is, of course, necessary to have formulas for the various energy forms that may be involved. The following are the ones most commonly encountered:

1. The kinetic energy (KE) of a mass m moving with velocity v is given by the formula $\text{KE} = \frac{1}{2}mv^2$.
2. The kinetic energy of a body of moment of inertia I rotating with angular velocity ω is given by the formula $\text{KE} = \frac{1}{2}I\omega^2$.
3. The potential energy (PE) stored in a spring of modulus k when it is stretched or compressed from an initial length s_0 to a final length s is given by the formula $\text{PE} = \frac{1}{2}k(s^2 - s_0^2)$.
4. The change in the potential energy of a weight w when it is moved from an initial height h_0 in the earth's gravitational field to a final height h is given by the formula $\text{PE} = w(h - h_0)$ provided that h and h_0 are small in comparison with the radius of the earth.

Example 8 A pulley of moment of inertia I and radius R turns on frictionless bearings, as shown in Fig. 4.8. A weight w hangs from an inextensible cord of negligible weight

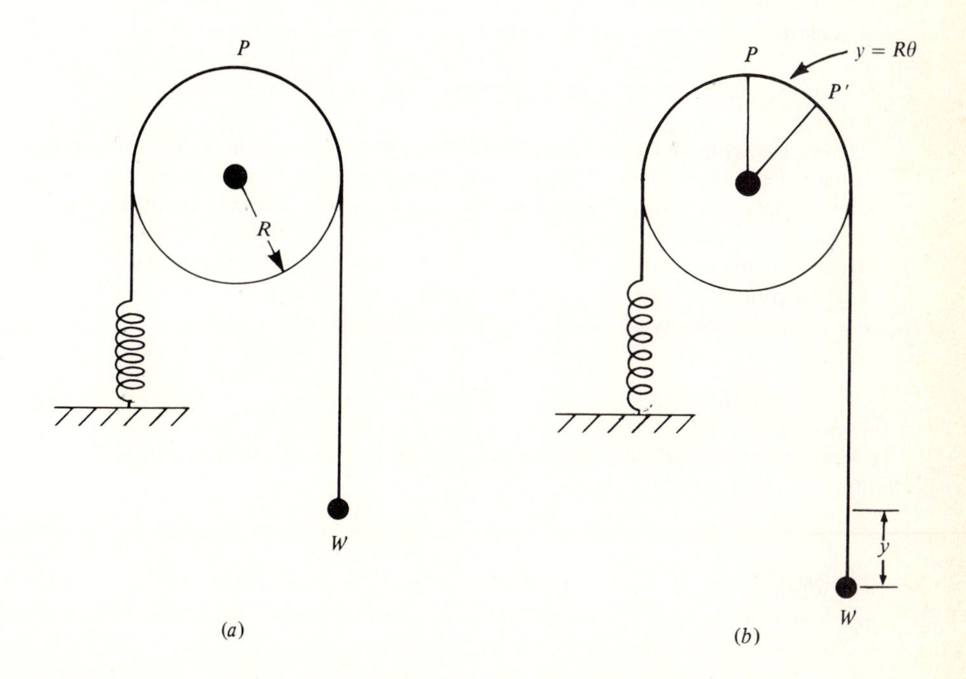

Figure 4.8 An unusual spring-suspended weight in equilibrium and after vertical displacement.

which passes over the pulley to a spring of modulus k. Constraints which need not be specified prevent the weight from swinging and permit it to move only in the vertical direction. Friction between the pulley and the cord prevents any slippage and results in negligible energy loss, but all other frictional effects are to be neglected. With what frequency will the weight vibrate if it is pulled down slightly from its equilibrium position and released?

At the outset, we should note that less is asked for in this problem than in any of the earlier examples in this section. We are not asked to find an expression for the displacement of the weight under specific initial conditions. We are asked only what its frequency would be *if* it began to move. This is typical of many physical systems in which vibratory motion is possible but undesirable. In such cases, it is important to know the frequency at which vibration *can* take place in order to avoid or modify external influences that might be in resonance with the natural frequency of the system. For simple linear systems in which (as is usually the case) energy loss due to friction is neglected, the underlying differential equation is eventually reducible to the form

$$y'' + \omega^2 y = 0$$

Since the complete solution of this equation is

$$y = c_1 \cos \omega t + c_2 \sin \omega t$$

and since both $\cos \omega t$ and $\sin \omega t$ describe periodic behavior of frequency

$$\omega \text{ rad/unit time} \qquad \text{or} \qquad \frac{\omega}{2\pi} \text{ cycles/unit time}$$

it is clear that the frequency ω can be read just as well from the differential equation itself as from any of its solutions, general or particular. The important part of such a frequency calculation, then, is the formulation of the differential equation and not its solution.

As a coordinate to describe the instantaneous position of the weight, let us take y to be the distance the weight has moved from its equilibrium position, the positive direction of y being downward. Since the cord is inextensible and cannot slip relative to the circumference of the pulley, it follows that when the weight has dropped a distance y, the pulley must have turned through an angle θ such that the arc of this angle is equal to y (Fig. 4.8b), that is, an angle such that $R\theta = y$ or $\theta = y/R$.

Now the velocity of the weight is dy/dt. Hence its kinetic energy is

$$\frac{1}{2}\frac{w}{g}\left(\frac{dy}{dt}\right)^2$$

Likewise, the angular velocity of the pulley is $d\theta/dt$. Hence its kinetic energy is

$$\frac{1}{2}I\left(\frac{d\theta}{dt}\right)^2 = \frac{1}{2}I\left(\frac{1}{R}\frac{dy}{dt}\right)^2$$

When the weight has moved through the vertical distance y, the spring, perforce, has stretched the same distance from its equilibrium length, say s_0. Hence its new length is $s_0 + y$, and the potential energy stored in it is

$$\tfrac{1}{2}k[(s_0 + y)^2 - s_0^2] = \tfrac{1}{2}k(2s_0 y + y^2)$$

Finally, the potential energy of the weight is

$$-wy + E_0 \qquad (E_0 = \text{PE when } y = 0)$$

since the weight has moved *down* a distance y if y is *positive*, and vice versa.

By the principle of the conservation of energy, the sum of these four energies must remain constant during any motion of the weight. Hence

$$\frac{1}{2}\frac{w}{g}\left(\frac{dy}{dt}\right)^2 + \frac{1}{2}\frac{I}{R^2}\left(\frac{dy}{dt}\right)^2 + \frac{1}{2}k(2s_0 y + y^2) - wy + E_0 = \text{constant}$$

If we differentiate this with respect to time, we obtain

$$\left(\frac{w}{g} + \frac{I}{R^2}\right)\frac{dy}{dt}\frac{d^2y}{dt^2} + \frac{k}{2}(2s_0 + 2y)\frac{dy}{dt} - w\frac{dy}{dt} = 0$$

Now $ks_0 = w$, since in equilibrium the spring force ks_0 must equal the weight w that produces the elongation s_0. Moreover, under any conditions of motion, the velocity dy/dt cannot be identically zero. Hence we may divide by dy/dt and simplify, getting

$$\left(\frac{w}{g} + \frac{I}{R^2}\right)\frac{d^2y}{dt^2} + ky = 0 \qquad \text{or} \qquad \frac{d^2y}{dt^2} + \frac{kR^2g}{wR^2 + Ig}y = 0$$

From the form of the last equation, it is clear that

$$\omega^2 = \frac{kR^2g}{wR^2 + Ig}$$

Therefore the required frequency is

$$\frac{1}{2\pi}\sqrt{\frac{kR^2g}{wR^2 + Ig}} \qquad \text{cycles/unit time}$$

Another important field in which linear differential equations often arise is the bending of beams. Consider a beam which in its undeflected position extends in the direction of the positive x axis. When the beam is bent, it is obvious that the fibers near the concave surface of the beam are compressed whereas those near the convex surface are stretched. Somewhere between these regions of compression and tension there must, from considerations of continuity, be a surface whose fibers are neither compressed nor stretched. This is known as the **neutral surface** of the beam, and the curve of any particular fiber in this surface is known as the **elastic curve,** or **deflection curve,** of the beam. The line in which the neutral surface is cut by any plane cross section of the beam is known as the **neutral axis** of that cross section (Fig. 4.9).

The loads which cause a beam to bend may be of two sorts: they may be concentrated at one or more points along the beam, or they may be continuously distributed with a density $w(x)$ known as the **load per unit length.** In either case, we have two important related quantities. One is the **shear** $V(x)$ at any point along the beam, which is defined to be the algebraic sum of all the transverse forces†

† **A transverse force** is one whose direction is perpendicular to the length of the beam.

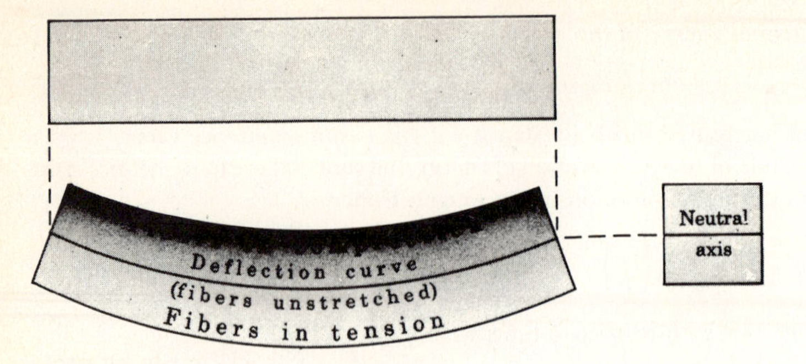

Figure 4.9 A beam before and after bending.

which act on the beam on the positive side of the point in question (Fig. 4.10). The other is the **moment** $M(x)$, which is defined as the total moment produced at a general point along the beam by all the forces, transverse or not, which act on the beam on one side or the other of the point in question. We shall consider the load per unit length and the shear to be positive if they act in the direction of the negative y axis (the direction in which loads usually act on a beam). The moment we shall take to be positive if it acts to bend the beam so that it is concave toward the positive y axis. With these conventions of sign (which are not universally adopted), it is shown in the study of the strength of materials that the deflection curve of the beam satisfies the second-order differential equation

$$EIy'' = M \tag{26}$$

where E is the modulus of elasticity of the material of the beam, and I, which may be a function of x, is the moment of inertia of the cross-sectional area of the beam

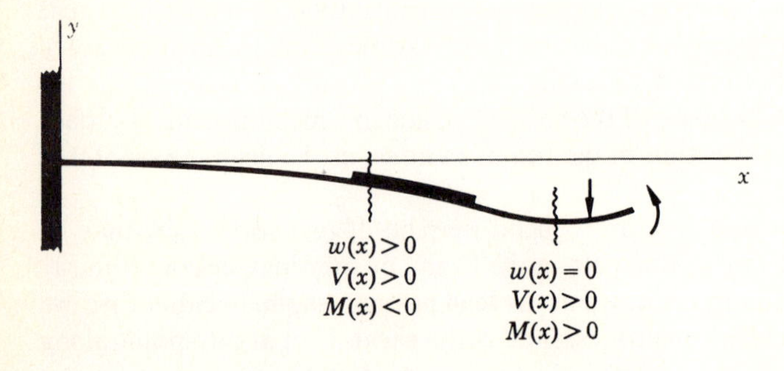

Figure 4.10 The conventions for the signs of the moment, shear, and load per unit length at a general point of a beam.

about the neutral axis of the cross section. If the beam bears only transverse loads, it can be shown further that we have the two additional relations:

$$\frac{dM}{dx} = \frac{d(EIy'')}{dx} = V \tag{27}$$

$$\frac{d^2M}{dx^2} = \frac{dV}{dx} = \frac{d^2(EIy'')}{dx^2} = -w \tag{28}$$

In most elementary applications, the moment M is an explicit function of x; hence Eq. (26) can be solved and the deflection $y(x)$ determined simply by performing two integrations. However, in problems in which the load has a component in the direction of the length of the beam, M depends on y, and Eq. (26) can be solved only through the use of techniques from the field of differential equations.

Example 9 A cantilever beam of uniform cross section and length L bears a concentrated transverse load P at its free end. A tensile force F also acts at the free end in the direction of the undeflected beam. Find the equation of the deflection curve of the beam and the deflection of the free end relative to the fixed end.

By choosing coordinates as shown in Fig. 4.11, it is clear that the moment arm of the load P about a general point X is the abscissa x of that point. Moreover, P acts to bend the beam so that it is concave downward; therefore the moment it produces is negative. Hence

$$M_P = -Px$$

It is also clear from Fig. 4.11 that the moment arm of the force F about the point X is the ordinate y at that point. Hence since F tends to bend the beam so that it is concave upward, its moment is positive, and we have

$$M_F = Fy$$

Therefore, from Eq. (26),

$$EIy'' = Fy - Px \qquad \text{or} \qquad y'' - \frac{F}{EI}y = -\frac{P}{EI}x$$

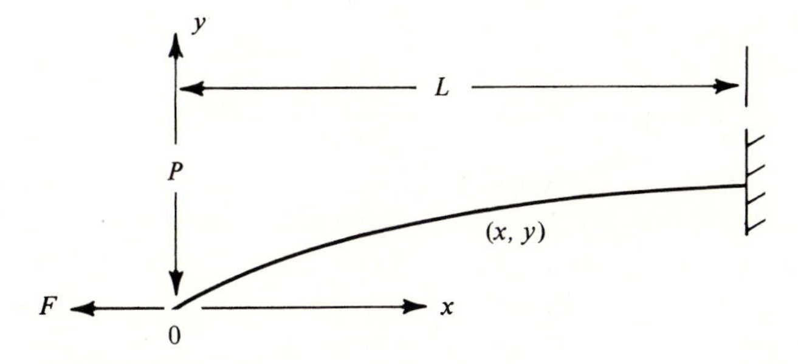

Figure 4.11 An end-loaded cantilever beam.

It is now an easy matter to solve this differential equation, and we have for its complete solution

$$y = c_1 \cosh \sqrt{\frac{F}{EI}}\,x + c_2 \sinh \sqrt{\frac{F}{EI}}\,x + \frac{P}{F}x \tag{29}$$

From our choice of coordinate system, it follows that $y = 0$ when $x = 0$. Hence, substituting these values into Eq. (29), we find $c_1 = 0$. Likewise, since the beam is built in rigidly at $x = L$, it follows that the slope of the deflection curve is zero at $x = L$. Hence, differentiating the expression for y (recalling that we now know that $c_1 = 0$) and substituting the values $y' = 0$, $x = L$, we find

$$y' = \sqrt{\frac{F}{EI}}\,c_2 \cosh \sqrt{\frac{F}{EI}}\,x + \frac{P}{F} \qquad \text{and} \qquad c_2 = -\frac{\dfrac{P}{F}\sqrt{\dfrac{EI}{F}}}{\cosh \sqrt{\dfrac{F}{EI}}\,L}$$

Thus

$$y = \frac{P}{F}\left[x - \sqrt{\frac{EI}{F}}\,\frac{\sinh \sqrt{F/EI}\,x}{\cosh \sqrt{F/EI}\,L}\right]$$

The deflection of the free end relative to the fixed end is simply the value of y at $x = L$, namely

$$\frac{P}{F}\left(L - \sqrt{\frac{EI}{F}} \tanh \sqrt{\frac{F}{EI}}\,L\right)$$

Exercises for Section 4.6

For each of the following mechanical systems, find the displacement y as a function of time if there are no impressed forces. Take $g = 32$ ft/s^2, $g = 384$ in/s^2, or $g = 980$ cm/s^2, as appropriate.

1 $w = 40$ lb, $k = 20$ lb/ft, $c = 0$; $y_0 = 2$ ft, $v_0 = -8$ ft/s

2 $w = 15$ g, $k = 3$ g/cm, $c = 0$; $y_0 = -1$ cm, $v_0 = 7$ cm/s

3 $w = 16$ lb, $k = 4.5$ lb/ft, $c = 5$ lb/(ft/s); $y_0 = 2$ ft, $v_0 = 0$

4 $w = 35$ kg, $k = 7$ kg/cm, $c = 1$ kg/(cm/s); $y_0 = 0$, $v_0 = 1$ cm/s

5 $w = 96$ lb, $k = 25$ lb/in, $c = 0.6\,c_c$; $y_0 = 0$, $v_0 = 16$ in/s

6 Find the steady-state motion of the weight in Exercise 1 if a force $F = 5 \cos 3t$ lb is applied to the weight.

7 Find the steady-state motion of the weight in Exercise 3 if a force $F = 2 \sin t$ lb is applied to the weight.

8 Find the steady-state motion of the weight in Exercise 5 if a force $F = \sin 10t$ is applied to the weight.

9 A weight w hangs from a spring of modulus k. Assuming that friction is negligible, derive the differential equation describing the motion of the weight by using the principle of the conservation of energy.

10 At $t = 0$, a weight w is suddenly attached to the end of a hanging spring of modulus k. Assuming that friction is negligible, find the subsequent displacement of the weight as a function of time.

Find the charge on the capacitor and the current in each of the following series circuits as functions of time.

11 $L = 0.1$ H, $R = 0$, $C = 10^{-5}$ F; $Q_0 = 10^{-3}$ C, $i_0 = 0$

12 $L = 1$ H, $R = 800\ \Omega$, $C = 4 \times 10^{-6}$ F; $Q_0 = 3 \times 10^{-4}$ C, $i_0 = 0$

13 What is the steady-state current in the circuit in Exercise 11 if a voltage $E = 100 \cos 500t$ V is inserted in the circuit?

****14** What is the steady-state current in the circuit in Exercise 12 if a voltage $E = 100 \sin 100t$ V is inserted in the circuit?

***15** A circular cylinder of radius r and height h, made of material of weight density ρ, floats in water in such a way that its axis is always vertical. Neglecting all forces except gravity and the buoyant force of the water, as given by the principle of Archimedes, determine the period with which the cylinder will vibrate in the vertical direction if it is depressed slightly from its equilibrium position and released.

***16** A cylinder weighing 50 lb floats in water with its axis vertical. When depressed slightly and released, it vibrates with period 2 s. Neglecting all frictional effects, find the diameter of the cylinder.

***17** A straight, hollow tube rotates with constant angular velocity ω about a vertical axis which is perpendicular to the tube at its midpoint. A pellet of mass m slides without friction in the interior of the tube. Find the equation describing the radial motion of the pellet until it emerges from the tube, assuming that it starts from rest at a radial distance a from the midpoint of the tube.

***18** A straight, hollow tube rotates with constant angular velocity ω about a horizontal axis which is perpendicular to the tube at its midpoint. Show that if the initial conditions are suitably chosen, a pellet sliding without friction in the tube will never be ejected but will execute simple harmonic motion within the tube.

***19** A tensile force F is applied at the free end of a uniform cantilever beam of length L. If the force makes an angle θ with the equilibrium direction of the beam, find the equation of the deflection curve of the beam.

****20** A uniform shaft of length L rotates about its axis with constant angular velocity ω. The ends of the shaft are held in bearings which are free to swing out of line, as shown in Fig. 4.12, if the shaft deflects from its neutral position. Show that there are infinitely many critical speeds at which the shaft can rotate in a deflected position, and find these speeds and the associated deflection curves. *Hint:* As a consequence of its rotation, the shaft experiences a load per unit length in the radial direction equal to

$$w(x) = -\frac{\rho A \omega^2}{g} y$$

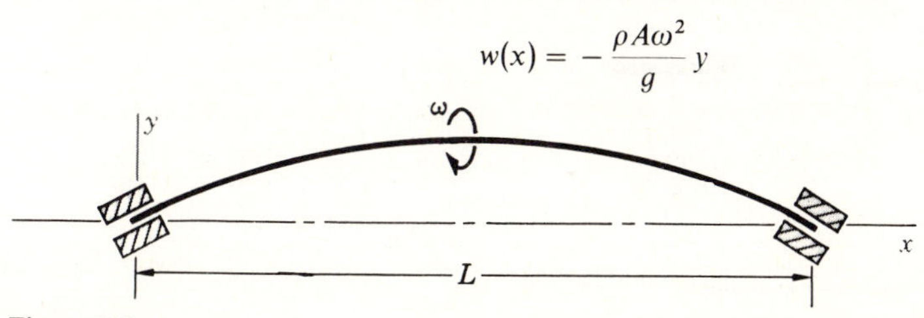

Figure 4.12

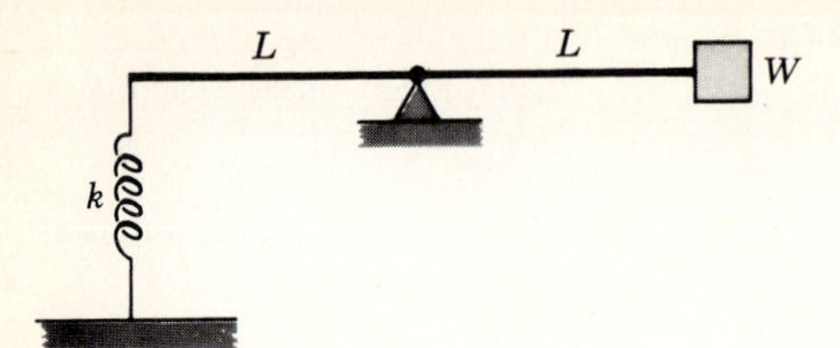

Figure 4.13

where A is the cross-sectional area of the shaft and ρ is the density of the material of the shaft. Substitute this into Eq. (28), solve the resulting differential equation, and then impose the conditions that at $x = 0$ and at $x = L$ both the deflection of the shaft and the moment are zero. It will be convenient to use hyperbolic rather than exponential functions in taking account of the real roots of the characteristic equation.

****21** Work Exercise 20 if the bearings are fixed in position and cannot swing out of line.

***22** Under the assumption of very small motions (so that the end of the spring and the weight W may be considered to move in a purely vertical direction) and neglecting friction, determine the natural frequency of the system shown in Fig. 4.13 if the bar is of uniform cross section, absolutely rigid, and of weight w. *Hint:* Use the energy method to obtain the differential equation of the system, recalling that the moment of inertia of a uniform bar of length l and weight w about its midpoint is $\frac{1}{12}wl^2/g$.

***23** Work Exercise 22 for the system shown in Fig. 4.14. *Hint:* The moment of inertia of a uniform bar of length L and weight w about one end is $\frac{1}{3}wl^2/g$.

***24** A perfectly flexible cable of length L and weight per unit length w lies in a straight line on a frictionless tabletop, with a units of the cable hanging over the edge. At $t = 0$, the cable is released and begins to slide off the edge of the table. Assuming that the height of the table is greater than L, determine the motion of the cable until it leaves the tabletop.

***25** Neglecting friction and assuming angular displacements θ so small that θ is a satisfactory approximation to $\sin\theta$ and $\theta^2/2$ is a satisfactory approximation to $1 - \cos\theta$, find the natural frequency of the system shown in Fig. 4.15 if the bar is absolutely rigid but weightless.

***26** Work Exercise 25 for the system shown in Fig. 4.16. In what significant respect does this system differ from that shown in Fig. 4.15?

27 A pendulum, consisting of a mass m at the end of an inextensible cord of length l and negligible mass, swings between maximum angular displacements of $\pm\alpha$. If θ is the instan-

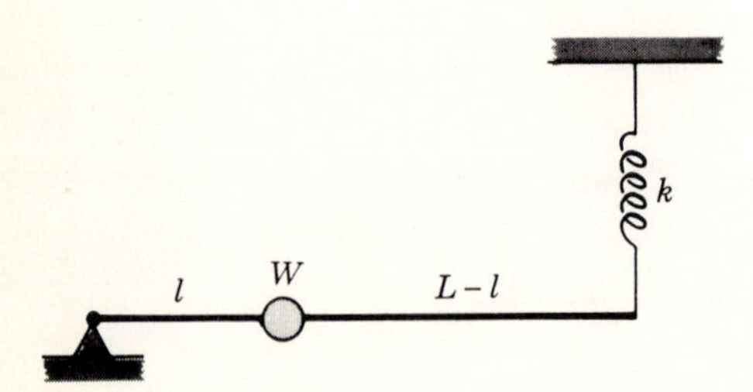

Figure 4.14

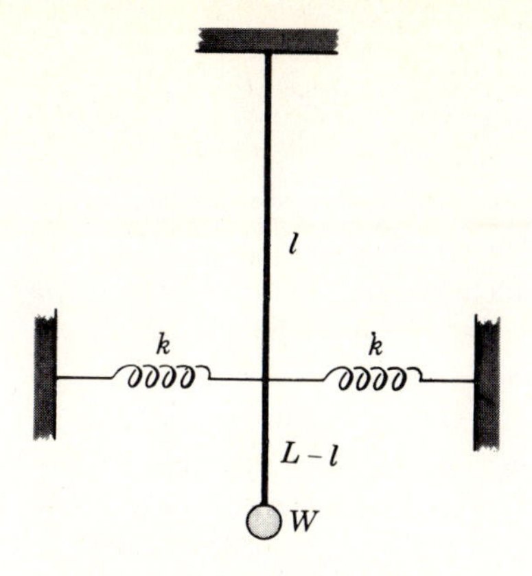

Figure 4.15

taneous angular displacement of the pendulum from the vertical, use the energy method to show that

$$(\dot{\theta})^2 = 2\frac{g}{l}(\cos\theta - \cos\alpha)\dagger$$

From this, determine the period of the pendulum if it swings through an angle small enough to make θ a satisfactory approximation to $\sin\theta$.

***28** (*a*) In the expression for $\dot{\theta}$ from Exercise 27, use the half-angle formula $\cos u = 1 - 2\sin^2(u/2)$ to show that

$$(\dot{\theta})^2 = \frac{4g}{l}\left(\sin^2\frac{\alpha}{2} - \sin^2\frac{\theta}{2}\right)$$

Integrate this differential equation, assuming that $\theta = 0$ when $t = 0$.

(*b*) In the integral obtained in part (*a*), change the variable of integration from θ to ϕ by the substitution

$$\sin\frac{\theta}{2} = \sin\frac{\alpha}{2}\sin\phi$$

† In applied mathematics and physics, derivatives *with respect to time* are frequently indicated by dots placed above the symbols rather than by primes.

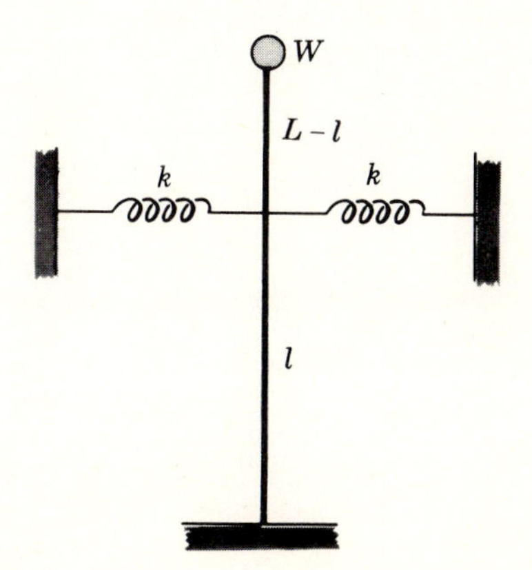

Figure 4.16

and show that

$$t = \sqrt{\frac{l}{g}} \int_0^{\phi} \frac{d\phi}{\sqrt{1 - k^2 \sin^2 \phi}} \qquad \text{where } k^2 = \sin^2 \frac{\alpha}{2}$$

This integral is known as an **elliptic integral of the first kind** and is commonly denoted $F(\phi,k)$. The function $F(\phi,k)$ is tabulated in most elementary handbooks.

29 Using the results of Exercises 27 and 28, together with tables of $F(\phi,k)$, compare the true period of a pendulum swinging through an angle of 90° on each side of the vertical with the period computed under the simplifying assumption that $\sin \theta \doteq \theta$.

CHAPTER

FIVE

SIMULTANEOUS LINEAR DIFFERENTIAL EQUATIONS

5.1 INTRODUCTION

In many problems in applied mathematics there is not one but several dependent variables, each a function of a single independent variable, usually time. The formulation of such problems in mathematical terms frequently leads to a system of simultaneous differential equations, with as many equations as there are dependent variables.

By a **solution** of a set of simultaneous differential equations involving the dependent variables $x_1, x_2, \ldots, x_n$ and the independent variable t, we mean a set of suitably differentiable functions $x_1(t), x_2(t), \ldots, x_n(t)$ which when substituted into the given equations, reduce each equation to an identity at all points of some interval $a \le t \le b$.

Simultaneous differential equations, as they arise in applied problems or are studied for their own sake in pure mathematics, are often nonlinear and exceedingly difficult to solve, even with the aid of a computer. In many important cases, however, they not only are linear but also have constant coefficients and are relatively easy to solve. There are various methods of solving such systems. In one, which bears a strong resemblance to the solution of a system of simultaneous algebraic equations, the system is reduced by successive elimination of the unknowns until a single differential equation in a single unknown remains. This is solved and then, working backward, solutions for the other variables are found, one by one, until the problem is completed. A second method considers the system to be a single matric equation and generalizes the concepts of complementary function and particular integral. Finally, the use of the Laplace transformation provides a straightforward operational procedure for solving systems of linear differential equations with constant coefficients which is probably preferable in most applications to either of the other methods.

In this chapter, we shall attempt through examples to present the first two methods, leaving the third to Chap. 6, where we shall discuss the Laplace transformation and its applications in detail. For the most part, our work will be formal, and we will leave to more advanced texts the theoretical justification of the techniques we present. In every instance, however, the correctness of our formal procedures can be checked by verifying that the solutions to which they lead do satisfy each of the equations in the given system.

5.2 THE REDUCTION OF A SYSTEM TO A SINGLE EQUATION

Consider the following system of equations:

$$\frac{dx}{dt} + 2x + \frac{dy}{dt} + 6y = 2e^t$$

$$2\frac{dx}{dt} + 3x + 3\frac{dy}{dt} + 8y = -1 \tag{1}$$

We intend to eliminate x and its derivatives from this system and obtain a differential equation in y from which y can be found as a function of t by the methods of the last chapter. Then with y determined, we should be able to return to either of the equations (1) and find x.

However, when operations of any kind are performed on one or more equations of a system, to obtain a new system, questions arise concerning the equivalence of the two systems, and it is important that we realize what this means. When we say that two systems are **equivalent,** we mean that they have the same solution set, that is, that every solution of one system is a solution of the other, and conversely. Thus if we can transform a given system into an *equivalent* one which can be solved, we know that its solutions are precisely those of the original system. On the other hand, if the derived system is not equivalent to the original one, its solutions may not include all those of the original system or may include additional, extraneous ones. For example, we learned in elementary algebra that squaring both sides of an equation may introduce extraneous roots, that is, may lead to a system which is not equivalent to the original one. Similarly, differentiation of one or more of the equations in a system of differential equations will raise the order of those equations and thus introduce extraneous solutions.

The conclusion to be drawn from this discussion is not that we are limited to operations that will always lead to an equivalent system. But as we manipulate a given system, we must be aware of the possibility of extraneous solutions and consistently check our purported answers to determine if they are indeed solutions of the original problem. With these observations in mind, we now return to the solution of system (1).

If we first eliminate dx/dt by subtracting the second equation from twice the first, we obtain

$$x - \frac{dy}{dt} + 4y = 4e^t + 1 \tag{2}$$

If we next eliminate x by subtracting 3 times the first equation from twice the second, we obtain

$$\frac{dx}{dt} + 3\frac{dy}{dt} - 2y = -6e^t - 2 \tag{3}$$

Finally, if we differentiate Eq. (2) and subtract the result from Eq. (3), all occurrences of x and its derivatives will be eliminated, and we will have an equation in y alone:

$$\frac{d^2y}{dt^2} - \frac{dy}{dt} - 2y = -10e^t - 2$$

It is now a simple matter to solve this equation by the methods of Chap. 4, and we find without difficulty

$$y = c_1e^{-t} + c_2e^{2t} + 5e^t + 1 \tag{4}$$

In view of our earlier discussion, it should be clear that at this stage there is no assurance that y, as given by (4), is a member of the solution pair $[x(t), y(t)]$ that we are seeking. Nonetheless, we shall continue with this expression for y and investigate several possibilities for obtaining x.

By far the simplest is to use Eq. (2), which gives x directly in terms of y and its first derivative. Thus

$$\begin{aligned} x &= \frac{dy}{dt} - 4y + 4e^t + 1 \\ &= (-c_1e^{-t} + 2c_2e^{2t} + 5e^t) \\ &\qquad - 4(c_1e^{-t} + c_2e^{2t} + 5e^t + 1) + 4e^t + 1 \\ &= -5c_1e^{-t} - 2c_2e^{2t} - 11e^t - 3 \end{aligned} \tag{5}$$

On the other hand, we might (less efficiently) substitute y and dy/dt into either of the given equations (1) and solve the resulting first-order differential equation for x; or we might substitute y into Eq. (3) and then find x by a single integration. Pursuing the latter possibility, we have

$$\begin{aligned} \frac{dx}{dt} &= -3\frac{dy}{dt} + 2y - 6e^t - 2 \\ &= -3(-c_1e^{-t} + 2c_2e^{2t} + 5e^t) \\ &\qquad + 2(c_1e^{-t} + c_2e^{2t} + 5e^t + 1) - 6e^t - 2 \\ &= 5c_1e^{-t} - 4c_2e^{2t} - 11e^t \end{aligned}$$

and, integrating,

$$x = -5c_1 e^{-t} - 2c_2 e^{2t} - 11e^t + c_3 \tag{6}$$

This appears to be a more general solution for x than the one given by Eq. (5) since it contains c_3 as a new *arbitrary* integration constant, whereas Eq. (5) contains only the *specific* additive constant -3. Which is the correct expression for x, Eq. (5) or Eq. (6)?

The need to resolve this question becomes even more pressing if we attempt to find x by using one of the original equations. Specifically, substituting y and dy/dt into the first of the given equations (1), we have

$$\begin{aligned}\frac{dx}{dt} + 2x &= -\frac{dy}{dt} - 6y + 2e^t \\ &= -(-c_1 e^{-t} + 2c_2 e^{2t} + 5e^t) \\ &\quad - 6(c_1 e^{-t} + c_2 e^{2t} + 5e^t + 1) + 2e^t \\ &= -5c_1 e^{-t} - 8c_2 e^{2t} - 33e^t - 6\end{aligned}$$

This is a nonhomogeneous, linear differential equation whose characteristic equation is $m + 2 = 0$ and whose complementary function is therefore

$$c_4 e^{-2t}$$

For a particular integral we substitute, as usual,

$$X = Ae^{-t} + Be^{2t} + Ce^t + E$$

getting

$$(-Ae^{-t} + 2Be^{2t} + Ce^t) + 2(Ae^{-t} + Be^{2t} + Ce^t + E) = -5c_1 e^{-t} - 8c_2 e^{2t} - 33e^t - 6$$

whence, collecting terms and then equating the coefficients of like functions, we find

$$A = -5c_1 \qquad B = -2c_2 \qquad C = -11 \qquad E = -3$$

and finally

$$x = -5c_1 e^{-t} - 2c_2 e^{2t} - 11e^t + c_4 e^{-2t} - 3 \tag{7}$$

This expression agrees with neither of the previous expressions for x, and although it contains the same additive constant as the solution given by Eq. (5), it also contains a variable term $c_4 e^{-2t}$ contained in neither Eq. (5) nor Eq. (6). Now, our question must be: Which is the correct expression for x, Eq. (5), Eq. (6), Eq. (7), or perhaps none of these?

To settle this matter, let us do precisely what we would do at this stage in a simple problem involving simultaneous algebraic equations: let us check our answers by substituting them into the given equations. To check the solution apparently given by Eq. (4) and Eq. (7), we need substitute only into the second of the original equations, since (7) was derived directly from the first equation and so, perforce, (4) and (7) must satisfy it. Thus, substituting as indicated, we have

$$\begin{aligned}
&2(5c_1e^{-t} - 4c_2e^{2t} - 11e^t - 2c_4e^{-2t}) \\
&+ 3(-5c_1e^{-t} - 2c_2e^{2t} - 11e^t + c_4e^{-2t} - 3) \\
&+ 3(-c_1e^{-t} + 2c_2e^{2t} + 5e^t) \\
&+ 8(c_1e^{-t} + c_2e^{2t} + 5e^t + 1) = -1
\end{aligned}$$

or

$$-c_4e^{-2t} - 1 = -1$$

This will be an identity if and only if $c_4 = 0$, which means that the expression for x in (7) reduces to the one given by (5). Similarly, if the tentative solution given by (4) and (6) is checked in the second equation of the original system, the result will be an identity if and only if $c_3 = -3$. Thus a solution (actually a complete solution) of the problem is given by the pair of expressions (4) and (5).

The general situation illustrated by the preceding work can be summarized as follows: if after one of the unknowns has been determined, the other is found by integration or the solution of a differential equation, extraneous constants (such as c_3 and c_4 in the preceding discussion) will usually be introduced. The values of these constants or their relation to other constants already present must be determined by substituting the tentative solutions for the two dependent variables into the original equations and making sure that they are identically satisfied.

In general, the steps in the reduction of a system of differential equations to a single equation are not so obvious as they were in the example we have just worked. For this reason, it is frequently convenient to rewrite the given equations in the D notation. Then, if we regard the operational coefficients of the variables as ordinary algebraic coefficients, the method of elimination will usually be apparent. Still more systematically, determinants can be used to obtain the equation satisfied by any one of the unknowns, very much as in the case of linear algebraic equations.

Suppose, for definiteness, that we have the second-order system

$$(a_{11}D^2 + b_{11}D + c_{11})x + (a_{12}D^2 + b_{12}D + c_{12})y = \phi_1(t)$$
$$(a_{21}D^2 + b_{21}D + c_{21})x + (a_{22}D^2 + b_{22}D + c_{22})y = \phi_2(t)$$

or, more compactly,

$$P_{11}(D)x + P_{12}(D)y = \phi_1(t)$$
$$P_{21}(D)x + P_{22}(D)y = \phi_2(t)$$

where the Ps denote the polynomial operators which act on x and y. If these were, as indeed they appear to be, two algebraic equations in x and y, we could eliminate x at once by subtracting $P_{21}(D)$ times the first equation from $P_{11}(D)$ times the second equation, getting

$$[P_{11}(D)P_{22}(D) - P_{12}(D)P_{21}(D)]y = P_{11}(D)\phi_2(t) - P_{21}(D)\phi_1(t) \tag{8}$$

Moreover, this procedure is clearly justified even though the system consists of differential equations rather than algebraic equations. For "multiplying" the first equation by

$$P_{21}(D) \equiv a_{21} D^2 + b_{21} D + c_{21}$$

is simply a way of performing in one step the operations of adding a_{21} times the second derivative of the equation and b_{21} times the first derivative of the equation to c_{21} times the equation itself; and these steps are individually well defined and completely correct. Similarly, multiplying the second equation by

$$P_{11}(D) \equiv a_{11} D^2 + b_{11} D + c_{11}$$

merely furnishes in one step the sum of a_{11} times the second derivative of the equation, b_{11} times the first derivative of the equation, and c_{11} times the equation itself. Finally, the subtraction of the two equations obtained by the multiplications we have just described eliminates x and each of its derivatives because these operations produce in each equation exactly the same combination of x and its various derivatives. Similarly, of course, y can be eliminated from the system by subtracting $P_{12}(D)$ times the second equation from $P_{22}(D)$ times the first equation, leaving a differential equation which x must satisfy.

The preceding observations can easily be formulated in determinant notation. In fact, the (operational) coefficient of y in Eq. (8) is simply the determinant of the (operational) coefficients of the unknowns in the original system, namely,

$$\begin{vmatrix} P_{11}(D) & P_{12}(D) \\ P_{21}(D) & P_{22}(D) \end{vmatrix}$$

expanded as though D were an algebraic quantity. Furthermore, the right-hand side of (8) can be identified as the expanded form of the determinant

$$\begin{vmatrix} P_{11}(D) & \phi_1(t) \\ P_{21}(D) & \phi_2(t) \end{vmatrix}$$

Thus, Eq. (8) can be written in the form

$$\begin{vmatrix} P_{11}(D) & P_{12}(D) \\ P_{21}(D) & P_{22}(D) \end{vmatrix} y = \begin{vmatrix} P_{11}(D) & \phi_1(t) \\ P_{21}(D) & \phi_2(t) \end{vmatrix} \tag{9}$$

which is precisely what Cramer's rule (item 3, Appendix B.1) would yield if applied to the given system as though it were purely algebraic. In just the same way, the result of eliminating y from the original system, namely,

$$[P_{11}(D)P_{22}(D) - P_{12}(D)P_{21}(D)]x = P_{22}(D)\phi_1(t) - P_{12}(D)\phi_2(t)$$

can be written

$$\begin{vmatrix} P_{11}(D) & P_{12}(D) \\ P_{21}(D) & P_{22}(D) \end{vmatrix} x = \begin{vmatrix} \phi_1(t) & P_{12}(D) \\ \phi_2(t) & P_{22}(D) \end{vmatrix} \tag{10}$$

provided we keep in mind that in the determinant on the right the operators

$P_{12}(D)$ and $P_{22}(D)$ must operate on $\phi_2(t)$ and $\phi_1(t)$, respectively, and hence the diagonal products must be interpreted to mean

$$P_{22}(D)\phi_1(t) \qquad \text{and} \qquad P_{12}(D)\phi_2(t)$$

and not
$$\phi_1(t)P_{22}(D) \qquad \text{and} \qquad \phi_2(t)P_{12}(D)$$

The use of Cramer's rule to obtain the differential equations satisfied by the individual dependent variables is in no way restricted to the case of two equations in two unknowns. Exactly the same procedure can be applied to systems of any number n of equations, regardless of the degree of the polynomial operators which appear as the coefficients of the unknowns.† Moreover, as Eqs. (9) and (10) illustrate, the polynomial operators appearing in the left members of the equations which result when the original system is "solved" for the various unknowns are identical. Hence the characteristic equations of these differential equations are identical, and therefore, except for the presence of different arbitrary constants, the complementary functions in the solutions for the various unknowns are all the same. The constants in these complementary functions are not all independent, however, and relations will always exist between them, serving to reduce their number to the figure stipulated by the following theorem.‡

Theorem 1 If the determinant of the operational coefficients of the unknowns in a system of n linear differential equations with constant coefficients is not identically zero, then the total number of independent arbitrary constants in any complete solution of the system is equal to the degree of the determinant of the operational coefficients, regarded as a polynomial in D. In particular cases in which the determinant of the operational coefficients is identically zero, the system may have no solution or it may have solutions containing any number of independent constants.

The necessary relations between the constants appearing initially in the solutions for the unknowns can always be found by substituting these solutions into the n equations in the original system and then equating to zero the net coefficients of the linearly independent terms that occur in each of these equations.§

† The determinants appearing on the right in the individual differential equations obtained by the formal application of Cramer's rule will always consist of $n - 1$ columns of operators and one column consisting of functions of t. These determinants are to be interpreted in the following way: Expand each determinant in terms of the elements in the column which contains the functions of t. Then in this expansion regard the formal cofactor of each function of t as an operator acting on that function.

‡ For a proof of this result see, for instance, E. L. Ince, "Ordinary Differential Equations," Dover, New York, 1944, pp. 144–150.

§ In most problems, the necessary relations between the constants can be found by substituting the tentative solutions for the variables into all but one of the given differential equations (though not necessarily into each set of $n - 1$ equations). However, there are systems for which natural solution procedures require substitution into *all* the given equations to obtain the necessary relations between the arbitrary constants. Examples of this sort will be found in Exercises 14, 28, and 29.

Example 1 Find a complete solution of the system

$$(2D^2 + 3D - 9)x + (D^2 + 7D - 14)y = 4$$
$$(D + 1)x + (D + 2)y = -8e^{2t} \tag{11}$$

From the preceding discussion we know that if $x(t)$ and $y(t)$ form a solution of (11), then $x(t)$ satisfies the equation

$$\begin{vmatrix} 2D^2 + 3D - 9 & D^2 + 7D - 14 \\ D + 1 & D + 2 \end{vmatrix} x = \begin{vmatrix} 4 & D^2 + 7D - 14 \\ -8e^{2t} & D + 2 \end{vmatrix}$$

or, expanding the determinants and operating, as required, on the known functions 4 and $-8e^{2t}$,†

$$(D^3 - D^2 + 4D - 4)x = 8 + 32e^{2t}$$

The roots of the characteristic equation of this differential equation are $\pm 2i$, 1. Hence the complementary function is

$$c_1 \cos 2t + c_2 \sin 2t + c_3 e^t$$

It is easy to see that

$$X = -2 + 4e^{2t}$$

is a particular integral, and therefore

$$x = c_1 \cos 2t + c_2 \sin 2t + c_3 e^t - 2 + 4e^{2t} \tag{12}$$

The solution for y can now be found by substituting the last expression into either of the original equations and solving the resulting differential equation for y. However, it is usually a little simpler to use Cramer's rule again. Doing this, we find that y must satisfy the equation

$$\begin{vmatrix} 2D^2 + 3D - 9 & D^2 + 7D - 14 \\ D + 1 & D + 2 \end{vmatrix} y = \begin{vmatrix} 2D^2 + 3D - 9 & 4 \\ D + 1 & -8e^{2t} \end{vmatrix}$$

or

$$(D^3 - D^2 + 4D - 4)y = -4 - 40e^{2t}$$

The solution of this equation presents no difficulty, and we find at once that

$$y = k_1 \cos 2t + k_2 \sin 2t + k_3 e^t + 1 - 5e^{2t} \tag{13}$$

However, Eqs. (12) and (13) do not yet constitute a solution of the given system. Collectively, they contain six independent arbitrary constants, whereas, according to Theorem 1, a complete solution of (11) can be found containing only three arbitrary constants. To reduce the number of constants from the six $(c_1, c_2, c_3, k_1, k_2, k_3)$ that presently appear in Eqs. (12) and (13) to three, as stipulated by Theorem 1, we must

† In carrying out these expansions, it must be borne in mind that the operational elements in the determinant on the right operate on the algebraic elements 4 and $-8e^{2t}$, whereas in the determinant on the left these elements all operate on the variable x and not on one another. This is the reason why in expanding the determinant on the right, we have the reduction $D(4) = 0$, whereas in expanding the determinant on the left, we have only formal multiplications such as $2D^2 2 = 4D^2$ and $3D2 = 6D$.

now substitute the expressions for x and y into at least one of the original equations (11). Beginning with the second, because it is a little simpler, we have

$$(D + 1)(c_1 \cos 2t + c_2 \sin 2t + c_3 e^t - 2 + 4e^{2t})$$
$$+ (D + 2)(k_1 \cos 2t + k_2 \sin 2t + k_3 e^t + 1 - 5e^{2t}) = -8e^{2t}$$

or, performing the indicated differentiations and collecting terms,

$$(c_1 + 2c_2 + 2k_1 + 2k_2) \cos 2t + (-2c_1 + c_2 - 2k_1 + 2k_2) \sin 2t$$
$$+ (2c_3 + 3k_3)e^t - 8e^{2t} = -8e^{2t}$$

As it stands, with all six constants completely arbitrary, this equation is not identically satisfied. It will be an identity if and only if

$$c_1 + 2c_2 + 2k_1 + 2k_2 = 0$$
$$-2c_1 + c_2 - 2k_1 + 2k_2 = 0$$
$$2c_3 + 3k_3 = 0$$

From these we find (among many equivalent possibilities)

$$k_1 = \frac{-3c_1 - c_2}{4} \qquad k_2 = \frac{c_1 - 3c_2}{4} \qquad k_3 = -\frac{2}{3}c_3$$

Since these three relations among the six constants are sufficient to reduce the number of constants to three, there is, by Theorem 1, no theoretical need to substitute x and y into the first of the given equations, though, of course, that would provide a practical check on the accuracy of our work. A complete solution of our problem is thus given by the pair of functions

$$x = c_1 \cos 2t + c_2 \sin 2t + c_3 e^t - 2 + 4e^{2t}$$
$$y = -\tfrac{1}{4}(3c_1 + c_2) \cos 2t + \tfrac{1}{4}(c_1 - 3c_2) \sin 2t - \tfrac{2}{3}c_3 e^t + 1 - 5e^{2t}$$

Example 2 Solve the system of equations

$$Dx + (D - 1)y + (D + 2)z = 2e^t$$
$$(D - 1)x + Dy + (D - 2)z = ae^t$$
$$(D + 1)x + (D - 2)y + (D + 6)z = e^t$$

From the preceding discussion we expect that the differential equation satisfied by z is

$$\begin{vmatrix} D & D-1 & D+2 \\ D-1 & D & D-2 \\ D+1 & D-2 & D+6 \end{vmatrix} z = \begin{vmatrix} D & D-1 & 2e^t \\ D-1 & D & ae^t \\ D+1 & D-2 & e^t \end{vmatrix}$$

However, expanding the determinants and operating, as required,† on the known functions $2e^t$, ae^t, and e^t, we obtain

$$0z = (a - 3)e^t$$

† See the first footnote on p. 153.

Clearly, unless $a = 3$, this equation, and hence the system itself, has no solution. On the other hand, if $a = 3$, this equation is satisfied by any function z. In fact, if $a = 3$, it is easy to verify that the third equation in the given system is equal to twice the first equation minus the second. Hence, when $a = 3$, the last equation is dependent upon the first two and is automatically satisfied by any functions $x(t)$, $y(t)$, $z(t)$ which satisfy them. Thus, considering only the first two equations, we can write

$$\begin{aligned} Dx + (D-1)y &= 2e^t - (D+2)z \\ (D-1)x + \quad Dy &= 3e^t - (D-2)z \end{aligned}$$

and for every differentiable function z this system can be solved for x and y. Specifically,

$$\begin{vmatrix} D & D-1 \\ D-1 & D \end{vmatrix} x = \begin{vmatrix} 2e^t - (D+2)z & D-1 \\ 3e^t - (D-2)z & D \end{vmatrix}$$

or

$$(2D-1)x = 2e^t - (5D-2)z \tag{14}$$

and

$$\begin{vmatrix} D & D-1 \\ D-1 & D \end{vmatrix} y = \begin{vmatrix} D & 2e^t - (D+2)z \\ D-1 & 3e^t - (D-2)z \end{vmatrix}$$

or

$$(2D-1)y = 3e^t + (3D-2)z \tag{15}$$

From Eqs. (14) and (15), x and y can be found in terms of z. Moreover, since z is subject only to the restriction that it be differentiable, it may contain any number of arbitrary constants, and hence, when $a = 3$, but not otherwise, the solution of the original system may also contain any number of independent arbitrary constants, as asserted by Theorem 1.

Exercises for Section 5.2

With the understanding that $D \equiv d/dt$, find a complete solution of each of the following systems of equations.

1 $(2D-1)x + (D-2)y = 0$
$(3D-2)x + (2D-3)y = 0$

2 $(D+5)x + (D+4)y = 3e^{-t}$
$(D+2)x + (D+1)y = 3$

3 $(D+5)x + (D+3)y = 6e^{-t}$
$(D+2)x + (D+1)y = 3$

4 $(5D+10)x + (D+3)y = 0$
$(D+2)x + (3D-5)y = 0$

5 $(6D-2)x + (D+3)y = 0$
$(4D-6)x + (3D+2)y = 7e^t$

6 $(6D+8)x - (D+2)y = 2$
$(7D+1)x - (D+1)y = -2$

7 $(2D+5)x + (3D+1)y = 0$
$(D+4)x + (3D+2)y = 0$

8 $(D+1)x + (7D-5)y = 0$
$(D+2)x + (10D-7)y = 0$

9 $(D-1)x + (D-2)y = 0$
$(D-5)x + (2D-7)y = e^{-t}$

10 $D^2x + Dy = 0$
$Dx + (D-3)y = 5e^{-t}$

11 $(5D^2-8)x - 3Dy = 0$
$(5D-4)x - (4D-5)y = 0$

12 $(9D^2+8)x + (3D^2+4)y = 4$
$(2D^2+1)x + (D^2+2)y = -1$

13 $(D-1)x - y = 0$
$-2x + (D-1)y - z = 0$
$-2y + (D-1)z = 6e^{2t}$

***14** $3(D-1)x - (D+1)y = 0$
$(D-2)y - 3(D-1)z = 0$
$2(D-2)x + (D+1)z = 0$

***15** $(D+1)x + (D+5)y + (2D+5)z = 15e^t$
$(2D+1)x + (D+2)y + (3D+1)z = 10e^t$
$(D+3)x + (3D+4)y + (4D+6)z = 21e^t$

16 $(D+1)x + (D+3)y + (2D+3)z = e^t$
$(2D+1)x + (D+2)y + (3D+1)z = 0$
$(D+3)x + (3D+11)y + (4D+13)z = 0$

17 $(3D^2+3D+2)x + (D^2+2D+3)y = 0$
$(2D^2 - D - 2)x + (D^2 + D + 1)y = 8$

Find the solution of each of the following systems which satisfies the indicated conditions.

18 $(2D+1)x + (D+2)y = 0$ $(D-1)x + (D+3)y = 5$ (a) $x_0 = 0$, $y_0 = 7$ (b) $x_0 = 1$, $y'_0 = -3$

***19** $(2D^2+3D-9)x + (D^2+7D-14)y = 4$ $(D+1)x + (D+2)y = 0$ (a) $x_0 = -2$, $y_0 = 0$, $y'_0 = -6$ (b) $x_0 = 3$, $y_0 = -3$, $x'_0 = 7$

20 $(D+4)x - (3D+2)y = 0$ $(D-1)x + (2D+3)y = 5e^t$ (a) $x_0 = 2$, $y_0 = 1$ (b) $x_0 = 2$, $y'_0 = 3$

***21** Find a system of differential equations having

$$x = c_1 e^t + c_2 e^{2t} \qquad y = c_1 e^t + 3c_2 e^{2t}$$

as a complete solution.

***22** Find a system of differential equations having

$$x = c_1 e^{-t} + c_2 e^t + c_3 e^{2t} \qquad y = c_1 e^{-t} - c_2 e^t + 2c_3 e^{2t}$$

as a complete solution.

23 If (x_1, y_1) and (x_2, y_2) are two solutions of the system

$$P_{11}(D)x + P_{12}(D)y = 0$$
$$P_{21}(D)x + P_{22}(D)y = 0$$

prove that $(c_1x_1 + c_2x_2, c_1y_1 + c_2y_2)$ is also a solution of the system for all values of the constants c_1 and c_2.

***24** In Exercise 17, determine (operational) multiples of the two equations which, when added, will yield an equation expressing y directly in terms of x and its various derivatives. Can this be done for the system of equations in Exercise 19? Do you think that this can be done in general?

***25** A system consists of two tanks each containing V gal of brine. The brine in the first tank initially contains s_1 lb of salt per gallon; the brine in the second tank initially contains s_2 lb of salt per gallon. Fresh brine containing s lb of salt per gallon flows into the first tank at the rate of a gal/min, and the mixture, kept uniform by stirring, runs into the second tank at the same rate. From the second tank, the mixture, kept uniform by stirring, runs out at the same rate. Find the amounts of salt in each tank as functions of time. Under what conditions, if any, will the amount of salt in the second tank reach a relative maximum or minimum value?

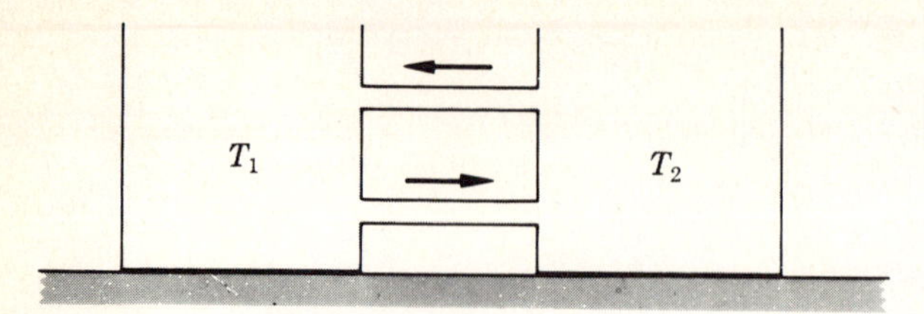

Figure 5.1

*26 Two tanks are connected as shown in Fig. 5.1. The first tank contains 100 gal of pure water; the second contains 100 gal of brine containing 2 lb of salt per gallon. Liquid circulates through the tanks at a constant rate of 5 gal/min. If the brine in each tank is kept uniform by stirring, find the amount of salt in each tank as a function of time.

**27 Three tanks are connected as shown in Fig. 5.2. The first tank contains 100 gal of pure water; the second contains 100 gal of brine containing 1 lb of salt per gallon; the third contains 100 gal of brine containing 2 lb of salt per gallon. Liquid circulates through the tanks at a constant rate of 5 gal/min. If the brine in each tank is kept uniform by stirring, find the amount of salt in each tank as a function of time.

28 If the system

$$(2D^2 - D - 1)x + (2D^2 + 4D - 6)y = 0$$

$$(D^2 + 2D - 3)x + (D^2 + 7D - 8)y = 0$$

is solved by determining both x and y from the differential equations obtained by using Cramer's rule, show that x and y must be substituted into *each* of the given equations to obtain the necessary relations between the arbitrary constants.

29 Solve the system

$$(2D^2 - D - 1)x + (D - 1)y = 0$$

$$(D^2 - 1)x + (D - 1)y = 0$$

*30 If each of the operational coefficients in the system

$$P_{11}(D)x + P_{12}(D)y = 0$$

$$P_{21}(D)x + P_{22}(D)y = 0$$

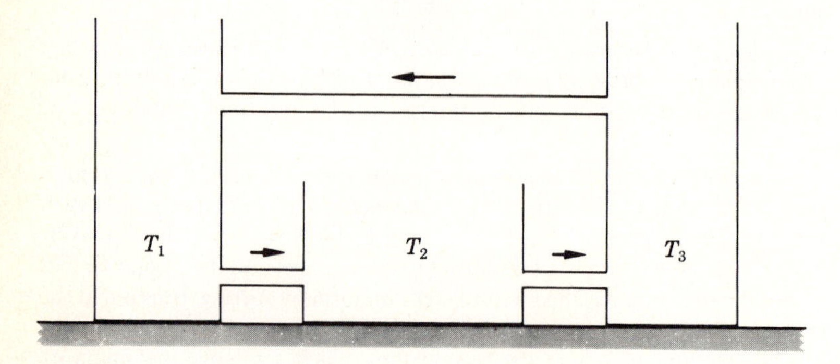

Figure 5.2

contains $D - a$ as a simple factor and if the determinant of the operational coefficients contains $D - a$ only as a double factor, show that neither x nor y contains a term of the form te^{at} even though the characteristic equation of the differential equations for both x and y has a as a double root. *Hint:* Factor $D - a$ from each operational coefficient, set $(D - a)x = u$ and $(D - a)y = v$, and note that neither u nor v can contain a term of the form e^{at}.

5.3 COMPLEMENTARY FUNCTIONS AND PARTICULAR INTEGRALS FOR SYSTEMS OF LINEAR EQUATIONS

The concepts of *characteristic equation, complementary function,* and *particular integral* can be extended from a single, linear differential equation to a system of such equations. The extension is not entirely obvious, however, and is most effectively carried out by using the notation and some of the simple ideas of matrix algebra.

To illustrate this generalization in its simplest form, let us consider the following system of equations:

$$\begin{aligned} (D+1)x + (D+2)y + (D+3)z &= -e^{\alpha t} \\ (D+2)x + (D+3)y + (2D+3)z &= e^{\alpha t} \\ (4D+6)x + (5D+4)y + (20D-12)z &= 7e^{\alpha t} \end{aligned} \tag{1}$$

Using the definitions of the product of two matrices and the equality of two matrices, we can write this system in the form

$$\begin{bmatrix} D+1 & D+2 & D+3 \\ D+2 & D+3 & 2D+3 \\ 4D+6 & 5D+4 & 20D-12 \end{bmatrix} \begin{bmatrix} x \\ y \\ z \end{bmatrix} = \begin{bmatrix} -e^{\alpha t} \\ e^{\alpha t} \\ 7e^{\alpha t} \end{bmatrix} \tag{1a}$$

As in the case of a single differential equation, we first make this system homogeneous by deleting the nonhomogeneous terms in (1) or, equivalently, by replacing the column vector

$$\begin{bmatrix} -e^{\alpha t} \\ e^{\alpha t} \\ 7e^{\alpha t} \end{bmatrix}$$

in (1*a*) by the null vector

$$\begin{bmatrix} 0 \\ 0 \\ 0 \end{bmatrix}$$

getting

$$\begin{aligned} (D+1)x + (D+2)y + (D+3)z &= 0 \\ (D+2)x + (D+3)y + (2D+3)z &= 0 \\ (4D+6)x + (5D+4)y + (20D-12)z &= 0 \end{aligned} \tag{2}$$

or, in matric notation,

$$\begin{bmatrix} D+1 & D+2 & D+3 \\ D+2 & D+3 & 2D+3 \\ 4D+6 & 5D+4 & 20D-12 \end{bmatrix} \begin{bmatrix} x \\ y \\ z \end{bmatrix} = \begin{bmatrix} 0 \\ 0 \\ 0 \end{bmatrix} \tag{2a}$$

Guided by our experience in solving single linear equations, where we began by assuming a solution of the form Ae^{mt} for the dependent variable (see Sec. 4.2), we now replace the arbitrary constant A by an arbitrary constant vector and attempt to find solutions of the form

$$\begin{bmatrix} x \\ y \\ z \end{bmatrix} = \begin{bmatrix} a \\ b \\ c \end{bmatrix} e^{mt}$$

that is, solutions of the form

$$x = ae^{mt} \qquad y = be^{mt} \qquad z = ce^{mt} \tag{3}$$

Substituting these into the equations (2) and dividing out the common factor e^{mt} lead to the set of homogeneous, linear algebraic equations

$$\begin{aligned} (m+1)a + (m+2)b + (m+3)c &= 0 \\ (m+2)a + (m+3)b + (2m+3)c &= 0 \\ (4m+6)a + (5m+4)b + (20m-12)c &= 0 \end{aligned} \tag{4}$$

or, in matric notation,

$$\begin{bmatrix} m+1 & m+2 & m+3 \\ m+2 & m+3 & 2m+3 \\ 4m+6 & 5m+4 & 20m-12 \end{bmatrix} \begin{bmatrix} a \\ b \\ c \end{bmatrix} = \begin{bmatrix} 0 \\ 0 \\ 0 \end{bmatrix} \tag{4a}$$

Naturally, we are interested in solutions other than the obvious trivial solution $a = b = c = 0$. Hence it follows (item 4, Appendix B.1) that the determinant of the coefficients of a, b, and c in (4) must be zero or, equivalently, that the coefficient matrix in (4a) must be singular. Thus we must have

$$\begin{vmatrix} m+1 & m+2 & m+3 \\ m+2 & m+3 & 2m+3 \\ 4m+6 & 5m+4 & 20m-12 \end{vmatrix} = 0 \tag{5}$$

or, expanding and collecting terms,

$$-(m-1)(m-2)(m-3) = 0$$

This equation, which defines all the values of m for which nontrivial solutions of (4), and hence of (2), can exist, is the **characteristic equation** of the system. It is, of course, nothing but the determinant of the operational coefficients of the system equated to zero, with D replaced by m; and in its expanded form it is precisely the characteristic equation of the differential equations obtained for the individual variables by the elimination process described in Sec. 5.2.

We must now find the (nontrivial) solutions of (4) which correspond to the respective roots $m = 1, 2, 3$ of the characteristic equation. When these have been found, the constant vector appearing in (3) will be determined for each possible value of m, and three particular solutions of the system (2), or (2*a*), will be available.

When $m = 1$, Eqs. (4) become

$$\begin{aligned} 2a + 3b + 4c &= 0 \\ 3a + 4b + 5c &= 0 \\ 10a + 9b + 8c &= 0 \end{aligned}$$

and it is easy to verify (item 4, Appendix B.1) that for all values of k_1, they are satisfied by

$$a = -k_1 \qquad b = 2k_1 \qquad c = -k_1$$

Hence, we have as one particular solution of the system (2),

$$x = -e^t \qquad y = 2e^t \qquad z = -e^t$$

or, in matric notation,

$$\begin{bmatrix} x \\ y \\ z \end{bmatrix} = \begin{bmatrix} -1 \\ 2 \\ -1 \end{bmatrix} e^t$$

Similarly, when $m = 2$, Eqs. (4) become

$$\begin{aligned} 3a + 4b + 5c &= 0 \\ 4a + 5b + 7c &= 0 \\ 14a + 14b + 28c &= 0 \end{aligned}$$

and it is easy to verify that for all values of k_2 they are satisfied by

$$a = 3k_2 \qquad b = -k_2 \qquad c = -k_2$$

Thus, we have as another particular solution of (2)

$$x = 3e^{2t} \qquad y = -e^{2t} \qquad z = -e^{2t}$$

or

$$\begin{bmatrix} x \\ y \\ z \end{bmatrix} = \begin{bmatrix} 3 \\ -1 \\ -1 \end{bmatrix} e^{2t}$$

Finally, when $m = 3$, Eqs. (4) become

$$\begin{aligned} 4a + 5b + 6c &= 0 \\ 5a + 6b + 9c &= 0 \\ 18a + 19b + 48c &= 0 \end{aligned}$$

and from these we find

$$a = 9k_3 \qquad b = -6k_3 \qquad c = -k_3$$

Thus as a third particular solution of the homogeneous system (2), we may take

$$x = 9e^{3t} \qquad y = -6e^{3t} \qquad z = -e^{3t}$$

or

$$\begin{bmatrix} x \\ y \\ z \end{bmatrix} = \begin{bmatrix} 9 \\ -6 \\ -1 \end{bmatrix} e^{3t}$$

Since the equations of the homogeneous system (2) are all linear, linear combinations of solutions will also be solutions (see Exercise 23, Sec. 5.2) just as in the case of a single, homogeneous linear equation. Hence combining arbitrary constant multiples of the three particular solutions we have just found, we have the general solution

$$\begin{bmatrix} x \\ y \\ z \end{bmatrix} = k_1 \begin{bmatrix} -1 \\ 2 \\ -1 \end{bmatrix} e^{t} + k_2 \begin{bmatrix} 3 \\ -1 \\ -1 \end{bmatrix} e^{2t} + k_3 \begin{bmatrix} 9 \\ -6 \\ -1 \end{bmatrix} e^{3t} \tag{6}$$

or, equivalently,

$$\begin{aligned} x &= -k_1 e^{t} + 3k_2 e^{2t} + 9k_3 e^{3t} \\ y &= 2k_1 e^{t} - k_2 e^{2t} - 6k_3 e^{3t} \\ z &= -k_1 e^{t} - k_2 e^{2t} - k_3 e^{3t} \end{aligned} \tag{6a}$$

We note that in either form, this solution contains the three independent constants stipulated by Theorem 1, Sec. 5.2, and is, in fact, a complete solution of the homogeneous system (2).

By analogy with a single, linear differential equation, a complete solution of the homogeneous system obtained from a nonhomogeneous system by deleting the nonhomogeneous terms in the various equations is called the **complementary function** of the nonhomogeneous system. Thus either (6) or (6*a*) is the complementary function of the nonhomogeneous system (1).

To complete our problem, we now need to find a **particular integral** of the nonhomogeneous system (1), that is, some specific solution of the system (1). After a particular integral has been determined, a complete solution of the nonhomogeneous system can be found, just as in the case of a single linear equation, by forming the sum of the complementary function and the particular integral.

In the usual case where the roots of the characteristic equation are all distinct and there is no duplication between the complementary function and the nonhomogeneous terms in the various equations of the given system, we assume for the individual dependent variables tentative solutions exactly as described in Table 4.2., Sec. 4.3. When the complementary function has repeated roots or there is duplication between the complementary function and the nonhomogeneous terms in the system, certain modifications in the procedures described in Table 4.2 are required, and these we shall leave until the next section.

In the present problem, following Table 4.2, we assume

$$\begin{bmatrix} X \\ Y \\ Z \end{bmatrix} = \begin{bmatrix} A \\ B \\ C \end{bmatrix} e^{\alpha t} \tag{7}$$

or

$$X = Ae^{\alpha t} \qquad Y = Be^{\alpha t} \qquad Z = Ce^{\alpha t} \tag{7a}$$

where A, B, C are constants to be determined. Substituting this trial solution in (1), collecting terms, and dividing out $e^{\alpha t}$, we find that A, B, and C must satisfy the equations

$$\begin{aligned} (\alpha + 1)A + (\alpha + 2)B + (\alpha + 3)C &= -1 \\ (\alpha + 2)A + (\alpha + 3)B + (2\alpha + 3)C &= 1 \\ (4\alpha + 6)A + (5\alpha + 4)B + (20\alpha - 12)C &= 7 \end{aligned} \tag{8}$$

Now a system of n nonhomogeneous linear equations in n unknowns admits of a unique solution if the determinant of the coefficients is different from zero (item 3, Appendix B.1). For the system (8) the determinant of the coefficients is precisely (5) with the unknown parameter m replaced by the known value α. Hence, as long as α is different from 1, 2, or 3, that is, any one of the values of m for which the determinant (5) is equal to zero, the system (8) can be solved for A, B, and C and a particular integral of the type (7) can be found. In other words, just as for a single differential equation, a particular integral of the expected form (7) can always be found as long as the nonhomogeneous exponential term is not a part of the complementary function. If $\alpha = 1, 2, 3$, that is, if $e^{\alpha t}$ is one of the terms e^t, e^{2t}, e^{3t} in the complementary function of the system (1), a particular integral of the form (7) will usually not exist, though it may in particular problems (see Exercise 11). In the next section, we shall examine the procedure for finding a particular integral corresponding to an exponential $e^{\alpha t}$ which is a part of the complementary function. Somewhat surprisingly, the tentative choice

$$X = Ate^{\alpha t} \qquad Y = Bte^{\alpha t} \qquad Z = Cte^{\alpha t} \qquad \text{or} \qquad \begin{bmatrix} X \\ Y \\ Z \end{bmatrix} = \begin{bmatrix} A \\ B \\ C \end{bmatrix} te^{\alpha t}$$

which we might be led to try by our experience with a single differential equation in Sec. 4.3 is not adequate (see Exercise 15).

To complete the present illustration, let us suppose, specifically, that $\alpha = -1$. Then Eqs. (8) become

$$\begin{aligned} B + 2C &= -1 \\ A + 2B + C &= 1 \\ 2A - B - 32C &= 7 \end{aligned}$$

from which we find at once that $A = 3$, $B = -1$, and $C = 0$. Therefore

$$X = 3e^{-t} \qquad Y = -e^{-t} \qquad Z = 0$$

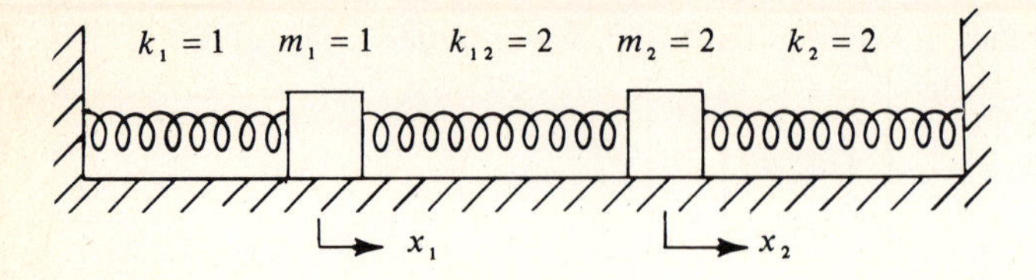

Figure 5.3 A simple two-mass system.

The complete solution of the original system (1), with $\alpha = -1$, is then

$$\begin{bmatrix} x \\ y \\ z \end{bmatrix} = k_1 \begin{bmatrix} -1 \\ 2 \\ -1 \end{bmatrix} e^t + k_2 \begin{bmatrix} 3 \\ -1 \\ -1 \end{bmatrix} e^{2t} + k_3 \begin{bmatrix} 9 \\ -6 \\ -1 \end{bmatrix} e^{3t} + \begin{bmatrix} 3 \\ -1 \\ 0 \end{bmatrix} e^{-t}$$

or

$$\begin{aligned} x &= -k_1 e^t + 3k_2 e^{2t} + 9k_3 e^{3t} + 3e^{-t} \\ y &= 2k_1 e^t - k_2 e^{2t} - 6k_3 e^{3t} - e^{-t} \\ z &= -k_1 e^t - k_2 e^{2t} - k_3 e^{3t} \end{aligned}$$

Example 1 Two masses, $m_1 = 1$ and $m_2 = 2$, are connected by springs of moduli $k_1 = 1$, $k_{12} = 2$, and $k_2 = 2$, as shown in Fig. 5.3. Neglecting all frictional effects and assuming that each spring is unstretched when the system is in its equilibrium position, determine the frequencies of the free vibrations of the system and discuss the motion of the system at each of these frequencies. If the system starts to move from rest in a position in which m_1 is displaced 1 unit to the left and m_2 is displaced 2 units to the right, find the subsequent displacements of m_1 and m_2 as functions of time.

As coordinates to describe the instantaneous positions of the masses,† let us take their displacements, x_1 and x_2, from their equilibrium positions, the positive direction being to the right. At a general time when the displacements of the masses are x_1 and x_2, the lengths of the springs have changed by the amounts

Left spring: x_1
Middle spring: $x_2 - x_1$
Right spring: $-x_2$

and, by Hooke's law, the forces represented by these changes are

Left spring: $1 \cdot x_1$
Middle spring: $2(x_2 - x_1)$
Right spring: $2(-x_2)$

† As usual (in spite of our figure), we assume that m_1 and m_2 are point masses and not objects of appreciable dimensions.

If $x_1 > 0$, the left spring is stretched and therefore pulls on m_1 with a force in the negative direction. If $x_1 < 0$, the left spring is compressed and therefore pushes on m_1 with a force in the positive direction. Hence in each case the force which the left spring applies to m_1 is correctly given by the formula

$$F_1 = -x_1$$

If $x_2 > x_1$, the middle spring is stretched and pulls back on both m_1 and m_2; that is, it applies a force in the positive direction to m_1 and a force in the negative direction to m_2. If $x_2 < x_1$, the middle spring is compressed and therefore pushes outward on both m_1 and m_2; that is, it applies a force in the negative direction to m_1 and a force in the positive direction to m_2. Thus in each case, the force which the middle spring applies to m_1 is correctly given by the formula

$$F_2 = 2(x_2 - x_1)$$

and the force which the middle spring applies to m_2 is given by the formula

$$F_3 = -2(x_2 - x_1)$$

Finally, if $x_2 > 0$, the right spring is compressed and pushes on m_2 with a force in the negative direction. If $x_2 < 0$, the right spring is stretched and pulls on m_2 with a force in the positive direction. Hence in each case, the force which the right spring applies to m_2 is given by the formula

$$F_4 = -2x_2$$

Since only the left spring and the middle spring apply forces to m_1, Newton's law applied to m_1 gives us the differential equation

$$1 \cdot \frac{d^2x_1}{dt^2} = F_1 + F_2 = -x_1 + 2(x_2 - x_1) \qquad \text{or} \qquad (D^2 + 3)x_1 - 2x_2 = 0$$

Likewise, since only the middle spring and the right spring apply forces to m_2, Newton's law applied to m_2 gives us the equation

$$2 \cdot \frac{d^2x_2}{dt^2} = F_3 + F_4 = -2(x_2 - x_1) - 2x_2 \qquad \text{or} \qquad -2x_1 + (2D^2 + 4)x_2 = 0$$

Our task, then, is to solve the system of differential equations

$$\begin{aligned} (D^2 + 3)x_1 - \qquad\quad 2x_2 &= 0 \\ -2x_1 + (2D^2 + 4)x_2 &= 0 \end{aligned} \tag{9}$$

The form of Eqs. (9) suggests a modification of the basic technique of assuming a solution of the form $x_1 = Ae^{mt}$, $x_2 = Be^{mt}$ which we used in the preceding discussion. Since the system (9) involves only derivatives of even order, it is clear that after either of the substitutions

$$\begin{aligned} x_1 &= A \cos \omega t \\ x_2 &= B \cos \omega t \end{aligned} \qquad \text{that is} \qquad \begin{bmatrix} x_1 \\ x_2 \end{bmatrix} = \begin{bmatrix} A \\ B \end{bmatrix} \cos \omega t \tag{10}$$

or

$$\begin{aligned} x_1 &= A \sin \omega t \\ x_2 &= B \sin \omega t \end{aligned} \qquad \text{that is} \qquad \begin{bmatrix} x_1 \\ x_2 \end{bmatrix} = \begin{bmatrix} A \\ B \end{bmatrix} \sin \omega t \tag{11}$$

the trigonometric functions can be divided out, leaving a pair of algebraic equations from which A and B can be found. If, specifically, we make the substitution (10), we obtain

$$\begin{aligned} (-\omega^2 + 3)A \cos \omega t - \qquad\qquad 2B \cos \omega t &= 0 \\ -2A \cos \omega t + \qquad (-2\omega^2 + 4)B \cos \omega t &= 0 \end{aligned}$$

or, dividing out $\cos \omega t$,

$$\begin{aligned} (-\omega^2 + 3)A - \qquad\qquad 2B &= 0 \\ -2A + (-2\omega^2 + 4)B &= 0 \end{aligned} \tag{12}$$

These two equations will have a solution other than the obvious trivial solution $A = B = 0$ if and only if the determinant of the coefficients is equal to zero. Thus we are led to the equation

$$\begin{vmatrix} -\omega^2 + 3 & -2 \\ -2 & -2\omega^2 + 4 \end{vmatrix} = 2\omega^4 - 10\omega^2 + 8 = 2(\omega^2 - 1)(\omega^2 - 4) = 0$$

The values $\omega_1 = \pm 1$ and $\omega_2 = \pm 2$ are thus the only values of ω for which there exist nonzero values of A and B satisfying Eqs. (12). Hence, from (10) it follows that the only frequencies at which the given system can vibrate are $\omega_1 = 1$ and $\omega_2 = 2$. In other words, these are the natural frequencies of the system.

If $\omega = 1$, Eqs. (12) become

$$\begin{aligned} 2A - 2B &= 0 \\ -2A + 2B &= 0 \end{aligned}$$

Hence we may take $A = B = 1$, and one particular solution of (9) is

$$\begin{aligned} x_1 &= \cos t \\ x_2 &= \cos t \end{aligned} \qquad \text{or} \qquad \begin{bmatrix} x_1 \\ x_2 \end{bmatrix} = \begin{bmatrix} 1 \\ 1 \end{bmatrix} \cos t \tag{13}$$

If $\omega = 2$, Eqs. (12) become

$$\begin{aligned} -A - 2B &= 0 \\ -2A - 4B &= 0 \end{aligned}$$

Hence we may take $A = 2$, $B = -1$, and a second particular solution of (9) is

$$\begin{aligned} x_1 &= 2 \cos 2t \\ x_2 &= -\cos 2t \end{aligned} \qquad \text{or} \qquad \begin{bmatrix} x_1 \\ x_2 \end{bmatrix} = \begin{bmatrix} 2 \\ -1 \end{bmatrix} \cos 2t \tag{14}$$

The first of these solutions describes periodic motion with frequency $\omega_1 = 1$ rad/unit time or $1/(2\pi)$ cycles/unit time in which the masses move in phase, that is, in the same direction, through equal amplitudes. The second solution describes periodic motion with frequency $\omega_2 = 2$ rad/unit time or $1/\pi$ cycles/unit time in which the masses move in opposite directions, the amplitude of m_1 being twice the amplitude of m_2.

Had we begun with the assumption $x_1 = A \sin \omega t$, $x_2 = B \sin \omega t$, the result of substituting into Eqs. (9) and simplifying would also have been Eqs. (12). Then, by exactly the same steps, we would have obtained the additional particular solutions

$$\begin{bmatrix} x_1 \\ x_2 \end{bmatrix} = \begin{bmatrix} 1 \\ 1 \end{bmatrix} \sin t \qquad \text{and} \qquad \begin{bmatrix} x_1 \\ x_2 \end{bmatrix} = \begin{bmatrix} 2 \\ -1 \end{bmatrix} \sin 2t \tag{15}$$

These, too, describe periodic motions with the same characteristics we noted for the particular solutions described, respectively, by (13) and (14).

Finally, by forming an arbitrary linear combination of the particular solutions given by (13), (14), and (15), we obtain the complete solution

$$\begin{bmatrix} x_1 \\ x_2 \end{bmatrix} = k_1 \begin{bmatrix} 1 \\ 1 \end{bmatrix} \cos t + k_2 \begin{bmatrix} 2 \\ -1 \end{bmatrix} \cos 2t + k_3 \begin{bmatrix} 1 \\ 1 \end{bmatrix} \sin t + k_4 \begin{bmatrix} 2 \\ -1 \end{bmatrix} \sin 2t$$

or

$$x_1 = k_1 \cos t + 2k_2 \cos 2t + k_3 \sin t + 2k_4 \sin 2t$$

$$x_2 = k_1 \cos t - k_2 \cos 2t + k_3 \sin t - k_4 \sin 2t\dagger$$

We are told that the system starts to move from rest in a position in which $x_1 = -1$ and $x_2 = 2$. Substituting these initial displacements and $t = 0$ into the equations for x_1 and x_2, we obtain the equations

$$-1 = k_1 + 2k_2$$

$$2 = k_1 - k_2$$

from which we find that $k_1 = 1$ and $k_2 = -1$. To determine k_3 and k_4, we must substitute the initial velocity conditions into the velocity equations. Hence, differentiating the expressions for x_1 and x_2 and then substituting the values $v_1 = v_2 = 0$ and $t = 0$, we obtain

$$0 = k_3 + 4k_4$$

$$0 = k_3 - 2k_4$$

From these we conclude that $k_3 = k_4 = 0$, and therefore

$$\begin{aligned} x_1 &= \cos t - 2 \cos 2t \\ x_2 &= \cos t + \cos 2t \end{aligned} \qquad \text{or} \qquad \begin{bmatrix} x_1 \\ x_2 \end{bmatrix} = \begin{bmatrix} 1 \\ 1 \end{bmatrix} \cos t + \begin{bmatrix} -2 \\ 1 \end{bmatrix} \cos 2t$$

Thus the actual motion of the system is a combination, or superposition, of motion at two frequencies.

Exercises for Section 5.3

Find a complete solution of each of the following systems.

1 $(D + 2)x + (D + 4)y = 1$
$(D + 1)x + (D + 5)y = 2$

2 $(2D + 1)x + (D + 2)y = 0$
$(D + 3)x + (D + 6)y = -3e^t$

3 $(D + 1)x + (4D - 2)y = t - 1$
$(D + 2)x + (5D - 2)y = 2t - 1$

4 $(D + 5)x + (D + 7)y = 4e^{2t}$
$(2D + 1)x + (3D + 1)y = 0$

5 $(2D + 1)x + (D + 2)y = 6e^t$
$(D + 2)x + (D + 4)y = 4e^{-t}$

6 $(2D + 1)x + (D - 1)y = -3 \cos t$
$(D + 2)x + (D + 3)y = 5 \sin t$

7 $(2D + 1)x + (D + 2)y = 8e^{-t}$
$(D^2 + D + 9)x + (D^2 - 2D + 12)y = 6$

8 $(2D + 1)x + (D^2 + 6D + 1)y = 0$
$(D + 2)x + (D^2 + 2D + 5)y = 6e^{2t}$

† Clearly, the roots $\omega = -1$ and $\omega = -2$ lead to solutions which duplicate those we have now constructed, and therefore they need not be considered.

9 $(2D^2 + 5)x + (D^2 + 3)y = -8 \sin 3t$
$(D^2 + 7)x + (D^2 + 5)y = 8 \sin 3t$

Hint: Noting that only derivatives of even order appear in the two equations, assume first $x = a \cos mt$, $y = b \cos mt$ and then $x = c \sin mt$, $y = d \sin mt$, where m is a parameter to be determined. But note also the footnote on p. 107.

10 $(2D + 11)x + (D + 3)y + (D - 2)z = 14e^t$
$(D - 2)x + (D - 1)y + Dz = -2e^t$
$(D + 1)x + (D - 3)y + (2D - 4)z = 4e^t$

11 Verify that the system (1) discussed in this section has no solution of the form (3) if $\alpha = 1$. If the terms on the right-hand sides of the equations in (1) are $9e^t$, $12e^t$, and $27e^t$, respectively, show that a solution of the form (3) exists and find it.

12 Show that if the equation $(aD^2 + bD + c)y = 0$ is written as a pair of simultaneous differential equations by means of the substitution $z = Dy$, the characteristic equation of the system formed by these two equations is the same as the characteristic equation of the original equation.

13 Show that if the equation $(aD^3 + bD^2 + cD + d)y = 0$ is written as a system of three equations by means of the substitutions $z = Dy$ and $w = Dz$, the characteristic equation of this system is the same as the characteristic equation of the original differential equation.

14 Verify that the characteristic equation of the system

$$(2D + 1)x + (D + 1)y = 0$$

$$(D - 4)x + (D - 3)y = 0$$

has the repeated root $m = 1$. Show, further, that although a_1 and b_1 can be related so that $x = a_1 e^t$ and $y = b_1 e^t$ will constitute a particular solution of the system, it is impossible to determine a_2 and b_2 so that $x = a_2 te^t$ and $y = b_2 te^t$ will form a particular solution of the system. Show, however, that a complete solution can be found by considering simultaneously the two tentative particular solutions and assuming $x = a_1 e^t + a_2 te^t$ and $y = b_1 e^t + b_2 te^t$.

15 Verify that in the system of equations

$$(2D + 1)x + (D + 1)y = e^t$$

$$(D - 7)x + (D - 5)y = 0$$

the nonhomogeneous term e^t duplicates a term in the complementary function of the system. Verify, further, that it is impossible to determine A and B so that $X = Ate^t$ and $Y = Bte^t$ will form a particular integral of the system but that A, B, C, and E can be determined so that $X = Ate^t + Ce^t$ and $Y = Bte^t + Ee^t$ will constitute a particular integral of the system.

16 Work Example 1 if $m_1 = 8$, $m_2 = 10$, $k_1 = 18$, $k_{12} = 10$, and $k_2 = 5$.

17 Work Example 1 if $m_1 = 2$, $m_2 = 3$, $k_1 = 8$, $k_{12} = 6$, and $k_2 = 12$, given that initially m_1 and m_2 are each displaced 1 unit to the right and set in motion from those positions with velocities 3 and -2, respectively.

18 Two particles, each of weight w, are attached to a perfectly flexible, elastic string of negligible weight, stretched under tension T as shown in Fig. 5.4. The particles vibrate in a direction perpendicular to the length of the string through amplitudes so small that (a) the tension in the string remains constant, and (b) the angles shown in Fig. 5.4 are so small that their sines can with satisfactory accuracy be approximated by their tangents.

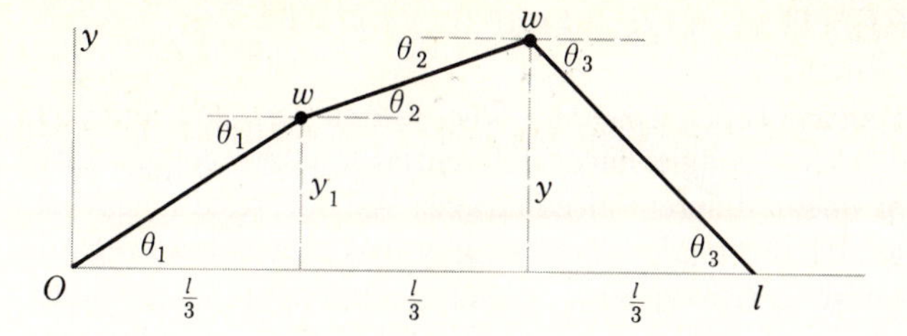

Figure 5.4

Neglecting all forces but the elastic forces supplied by the string, set up the differential equations describing the behavior of the system and find the natural frequencies of the system and the ratios of the amplitudes of the two particles at each frequency.

***19** Work Exercise 18 for the system of three particles shown in Fig. 5.5.

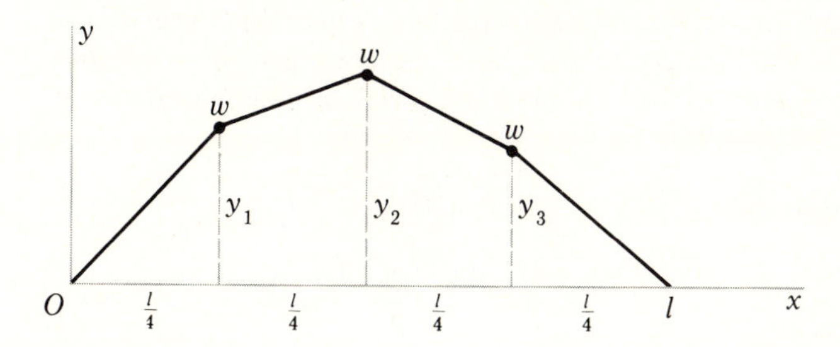

Figure 5.5

***20** When the switch is closed in the network shown in Fig. 5.6, the current and the charge on the capacitor in the closed loop are zero, and the capacitor in the open loop bears a charge Q_0. Find the current in each loop as a function of time after the switch is closed. *Hint:* Let $Q_1(t)$ and $Q_2(t)$ be the instantaneous charges on the two capacitors, so that dQ_1/dt and dQ_2/dt are the currents i_1 and i_2 indicated in Fig. 5.6. Then note that the current flowing through the resistor in the branch common to the two loops is $i_1 - i_2$. Finally, apply Kirchhoff's second law to each loop to obtain the differential equations which describe the behavior of the system.

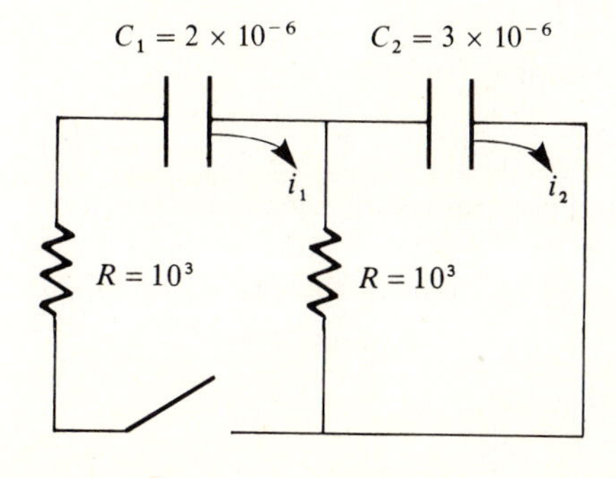

Figure 5.6 A simple two-loop series network.

5.4 MATRIC DIFFERENTIAL EQUATIONS

In Sec. 5.3 we saw that the ideas of *complementary function* and *particular integral* which we developed in Sec. 4.3 for a single, linear differential equation could easily be extended to systems of linear differential equations. As our restatement of results in matric form suggested, this analogy is especially striking when we regard a system of linear differential equations with constant coefficients as a single matric equation, in the same way that in algebra we regard a system of linear algebraic equations as a single matric equation. Moreover, the procedure for handling systems of equations when the characteristic equation has repeated or complex roots or when a term on the right-hand side duplicates a term in the complementary function is best described in the language of matrices. Hence we shall conclude this chapter with a brief discussion of matric differential equations. For our purposes it will be sufficient to know the meaning of matric addition, multiplication, and equality; the definition of the derivative of a matrix; a few of the elementary properties of systems of linear algebraic equations; and what is meant by the real part, the imaginary part, and the conjugate of a complex number. Appendix B.1 should provide an adequate review of this material.

Let the system we are given be

$$\begin{aligned} p_{11}(D)x_1 + p_{12}(D)x_2 + \cdots + p_{1n}(D)x_n &= f_1(t) \\ p_{21}(D)x_1 + p_{22}(D)x_2 + \cdots + p_{2n}(D)x_n &= f_2(t) \\ \cdots\cdots\cdots\cdots\cdots\cdots\cdots\cdots\cdots\cdots\cdots \\ p_{n1}(D)x_1 + p_{n2}(D)x_2 + \cdots + p_{nn}(D)x_n &= f_n(t) \end{aligned} \tag{1}$$

where the p_{ij}'s are polynomials in the operator D with coefficients which may be functions of t, but which in our work we shall suppose to be real constants. If we define the matrices

$$\mathbf{P}(D) = \begin{bmatrix} p_{11}(D) & p_{12}(D) & \cdots & p_{1n}(D) \\ p_{21}(D) & p_{22}(D) & \cdots & p_{2n}(D) \\ \cdots & \cdots & \cdots & \cdots \\ p_{n1}(D) & p_{n2}(D) & \cdots & p_{nn}(D) \end{bmatrix} \qquad \mathbf{X} = \begin{bmatrix} x_1 \\ x_2 \\ \vdots \\ x_n \end{bmatrix} \qquad \mathbf{F}(t) = \begin{bmatrix} f_1(t) \\ f_2(t) \\ \vdots \\ f_n(t) \end{bmatrix}$$

the system (1) can be written in the compact form

$$\mathbf{P}(D)\mathbf{X} = \mathbf{F}(t) \tag{2}$$

The associated homogeneous equation is, of course,

$$\mathbf{P}(D)\mathbf{X} = \mathbf{0} \tag{3}$$

where the symbol $\mathbf{0}$ denotes an n-dimensional column vector each component of which is the scalar 0.

The first step in finding the complementary function of Eq. (2) is to assume that solutions of Eq. (3) exist in the form

$$\mathbf{X} = \mathbf{A}e^{mt}$$

where the scalar m and the column vector of constants $\mathbf{A}$ have yet to be determined.† Since

$$D^r(e^{mt}) = m^r e^{mt}$$

it follows that the result of applying any polynomial operator $p(D)$ to e^{mt} is simply to multiply e^{mt} by $p(m)$. Hence if we substitute the vector $\mathbf{X} = \mathbf{A}e^{mt}$ into the homogeneous equation (3), we obtain just

$$\mathbf{P}(m)\mathbf{A}e^{mt} = \mathbf{0}$$

or, dividing out the scalar factor e^{mt},

$$\mathbf{P}(m)\mathbf{A} = \mathbf{0}\ddagger \tag{4}$$

Now, Eq. (4) will have a nontrivial solution if and only if

$$|\mathbf{P}(m)| = 0 \tag{5}$$

and for each root m_j of this equation there will be a solution vector $\mathbf{A}_j$ of (4) determined to within an arbitrary scalar factor k_j. If the characteristic equation (5) is a polynomial of degree N and if its roots $\{m_j\}$ are all distinct, a complete solution of Eq. (3) [and the complementary function of Eq. (2)] is then

$$\mathbf{X} = k_1 \mathbf{A}_1 e^{m_1 t} + k_2 \mathbf{A}_2 e^{m_2 t} + \cdots + k_N \mathbf{A}_N e^{m_N t}\S$$

As in the case of a single scalar differential equation, if the set of roots $\{m_j\}$ includes one or more pairs of conjugate complex roots, it is desirable to reduce the corresponding complex exponential solution to a purely real form. To see how this can be accomplished, let $p \pm iq$ be a pair of conjugate complex roots of Eq. (5), and let $\mathbf{A}$ be a particular solution vector of (4) corresponding to the root $m = p + iq$; that is, let

$$\mathbf{P}(m)\mathbf{A} \equiv \mathbf{P}(p + iq)\mathbf{A} = \mathbf{0}$$

Then, since all the coefficients in (4) are real, it follows by taking conjugates throughout the last equation that

$$\mathbf{P}(\bar{m})\bar{\mathbf{A}} \equiv \mathbf{P}(p - iq)\bar{\mathbf{A}} = \mathbf{0}$$

Thus $\bar{\mathbf{A}}$ is a solution vector corresponding to the conjugate root $\bar{m} = p - iq$, and therefore we have the two particular solutions of Eq. (3),

$$\mathbf{A}e^{(p+iq)t} \qquad \text{and} \qquad \bar{\mathbf{A}}e^{(p-iq)t}$$

† The expressions $x = ae^{mt}$, $y = be^{mt}$, $z = ce^{mt}$ in Eq. (3), Sec. 5.3, are, of course, just the scalar form of this assumption for the special case $n = 3$.

‡ This is just the matric equivalent of the algebraic system in Eq. (4), Sec. 5.3, which we obtained in our scalar treatment of the specific system of differential equations considered in that section.

§ Equation (6), Sec. 5.3, is a special case of this, with $N = 3$ and

$$\mathbf{A}_1 = \begin{bmatrix} -1 \\ 2 \\ -1 \end{bmatrix} \qquad \mathbf{A}_2 = \begin{bmatrix} 3 \\ -1 \\ -1 \end{bmatrix} \qquad \mathbf{A}_3 = \begin{bmatrix} 9 \\ -6 \\ -1 \end{bmatrix}$$

By combining these as follows and applying the Euler formulas, just as we did in handling the case of complex roots in Sec. 4.2, we obtain the two independent, real solutions:

$$\begin{aligned}\frac{\mathbf{A}e^{(p+iq)t}+\bar{\mathbf{A}}e^{(p-iq)t}}{2} &= \frac{e^{pt}}{2}[\mathbf{A}(\cos qt + i\sin qt) + \bar{\mathbf{A}}(\cos qt - i\sin qt)] \\ &= e^{pt}\left(\frac{\mathbf{A}+\bar{\mathbf{A}}}{2}\cos qt - \frac{\mathbf{A}-\bar{\mathbf{A}}}{2i}\sin qt\right) \\ &= e^{pt}[\mathscr{R}(\mathbf{A})\cos qt - \mathscr{I}(\mathbf{A})\sin qt] \qquad (6a)\end{aligned}$$

$$\begin{aligned}\frac{\mathbf{A}e^{(p+iq)t}-\bar{\mathbf{A}}e^{(p-iq)t}}{2i} &= \frac{e^{pt}}{2i}[\mathbf{A}(\cos qt + i\sin qt) - \bar{\mathbf{A}}(\cos qt - i\sin qt)] \\ &= e^{pt}\left[\frac{\mathbf{A}-\bar{\mathbf{A}}}{2i}\cos qt + \frac{\mathbf{A}+\bar{\mathbf{A}}}{2}\sin qt\right] \\ &= e^{pt}[\mathscr{I}(\mathbf{A})\cos qt + \mathscr{R}(\mathbf{A})\sin qt] \qquad (6b)\end{aligned}$$

where $\mathscr{R}(\mathbf{A})$ and $\mathscr{I}(\mathbf{A})$ denote the column vectors whose components are, respectively, the real parts of the components of $\mathbf{A}$ and the imaginary parts of the components of $\mathbf{A}$. In many cases, this method of solving (3) is simpler than the alternative process of substituting the expressions

$$x_j = e^{pt}(a_j \cos qt + b_j \sin qt)$$

into the original differential equations, collecting terms, and equating the resulting coefficients to zero.

If $|\mathbf{P}(m)| = 0$ has a double root, say $m = r$, we proceed very much as in the case of a single differential equation. If $\mathbf{A}$ is a solution of the equation $\mathbf{P}(r)\mathbf{A} = \mathbf{0}$, then, of course,

$$\mathbf{A}e^{rt}$$

is one solution of (3). In exceptional cases (see Exercise 15) it may be that Eq. (4) will have two linearly independent solution vectors, $\mathbf{A}_1$ and $\mathbf{A}_2$, corresponding to the repeated root $m = r$. When this is the case, there are then two linearly independent solutions of Eq. (3), namely, $\mathbf{A}_1 e^{rt}$ and $\mathbf{A}_2 e^{rt}$, and we need look no further. In general, this will not happen, and so we must seek a second solution of another form. However, as a second independent solution we must try not $\mathbf{B}te^{rt}$, as strict analogy with the scalar case would suggest, but rather

$$\mathbf{B}_1 te^{rt} + \mathbf{B}_2 e^{rt} \qquad (7)$$

To see why $\mathbf{X} = \mathbf{B}te^{rt}$ by itself will not serve as a second independent solution of Eq. (3), we note first that

$$D(te^{mt}) = mte^{mt} + e^{mt}$$

$$D^2(te^{mt}) = m^2te^{mt} + 2me^{mt}$$

$$D^3(te^{mt}) = m^3te^{mt} + 3m^2e^{mt}$$

and, by an obvious induction,

$$D^k(te^{mt}) = m^k te^{mt} + km^{k-1}e^{mt}$$

Next we observe that the coefficient of e^{mt} in the second term is just the derivative with respect to m of the coefficient of te^{mt} in the first term. Hence, if a polynomial operator $p(D)$ is applied to the function te^{mt}, the result is

$$p(D)te^{mt} = p(m)te^{mt} + p'(m)e^{mt}$$

Furthermore, since the derivative of a matrix $\mathbf{P}$ is the matrix $\mathbf{P}'$ whose elements are the derivatives of the corresponding elements of $\mathbf{P}$, it follows that if the vector $\mathbf{B}te^{rt}$ is substituted into the equation $\mathbf{P}(D)\mathbf{X} = \mathbf{0}$, the result is

$$\mathbf{P}(r)\mathbf{B}te^{rt} + \mathbf{P}'(r)\mathbf{B}e^{rt} = \mathbf{0}$$

For this equation to hold, $\mathbf{B}$ must be a vector such that simultaneously

$$\mathbf{P}(r)\mathbf{B} = \mathbf{0} \qquad \text{and} \qquad \mathbf{P}'(r)\mathbf{B} = \mathbf{0}$$

Now since r is a root of the characteristic equation (5), we know that there is a nontrivial vector $\mathbf{B}$ such that $\mathbf{P}(r)\mathbf{B} = \mathbf{0}$. In fact, this is just the vector $\mathbf{A}$ in our first solution $\mathbf{A}e^{rt}$. In general, however, $|\mathbf{P}'(r)|$ will not be zero, and therefore neither $\mathbf{A}$ nor any other nontrivial vector can satisfy the equation $\mathbf{P}'(r)\mathbf{X} = \mathbf{0}$. Hence, in general, no solution of the form $\mathbf{B}te^{rt}$ exists for Eq. (3).

On the other hand, when the characteristic equation (5) has a double root r for which there is a single solution vector $\mathbf{A}$ of Eq. (4) and a corresponding solution of Eq. (3) of the form $\mathbf{X} = \mathbf{A}e^{rt}$, a second linearly independent solution of (3) can be found by assuming it has the form

$$\mathbf{X} = \mathbf{B}_1 te^{rt} + \mathbf{B}_2 e^{rt} \tag{8}$$

The inclusion of the term $B_2 e^{rt}$ is unnecessary in the case of a single (scalar) differential equation because, regardless of the value of the scalar coefficient B_2, the term $B_2 e^{rt}$ can always be combined with the corresponding term in the complentary function. For a matric equation, however, the vector coefficient $\mathbf{B}_2$ in the term $\mathbf{B}_2 e^{rt}$ will usually not be a multiple of the vector $\mathbf{A}$ in the first solution $\mathbf{A}e^{rt}$, and hence the two terms cannot be combined into a single term with an arbitrary scalar coefficient.

When the tentative solution $\mathbf{X} = \mathbf{B}_1 te^{rt} + \mathbf{B}_2 e^{rt}$ is substituted into the left-hand member of the equation $\mathbf{P}(D)\mathbf{X} = \mathbf{0}$, the result is

$$\begin{aligned}\mathbf{P}(D)[\mathbf{B}_1 te^{rt} + \mathbf{B}_2 e^{rt}] &= [\mathbf{P}(r)\mathbf{B}_1 te^{rt} + \mathbf{P}'(r)\mathbf{B}_1 e^{rt}] + \mathbf{P}(r)\mathbf{B}_2 e^{rt} \\ &= \mathbf{P}(r)\mathbf{B}_1 te^{rt} + [\mathbf{P}'(r)\mathbf{B}_1 + \mathbf{P}(r)\mathbf{B}_2]e^{rt}\end{aligned}$$

and for this to equal zero it is necessary that

$$\mathbf{P}(r)\mathbf{B}_1 = \mathbf{0} \qquad \text{and} \qquad \mathbf{P}(r)\mathbf{B}_2 + \mathbf{P}'(r)\mathbf{B}_1 = \mathbf{0}$$

The equation $\mathbf{P}(r)\mathbf{B}_1 = \mathbf{0}$ is, of course, satisfied by the vector $\mathbf{A}$ that appears in our first solution $\mathbf{A}e^{rt}$. Hence $\mathbf{B}_2$ is to be found from the nonhomogeneous equation

$$\mathbf{P}(r)\mathbf{B}_2 = -\mathbf{P}'(r)\mathbf{B}_1 = -\mathbf{P}'(r)\mathbf{A} = \mathbf{C}, \text{ say.} \tag{9}$$

At first glance, this appears impossible, since the determinant $|\mathbf{P}(r)|$ is zero because r is a root of the characteristic equation (5). However, as item 7, Appendix B.1, assures us, there are nonhomogeneous systems of linear algebraic equations which have solutions even though the determinant of the coefficients is equal to zero. Fortunately (although we cannot prove the fact here), the system (9) is always one of these exceptional cases, and a matrix $\mathbf{B}_2$ can be found so that (8) is a second solution of Eq. (3) corresponding to the double root r.

Similar observations hold for roots of (5) of higher multiplicity. Thus, for a k-fold root r, the appropriate trial solutions are not $\mathbf{A}e^{rt}$, $\mathbf{B}te^{rt}$, $\mathbf{C}t^2e^{rt}, \ldots, \mathbf{K}t^{k-1}e^{rt}$, but

$$\mathbf{A}e^{rt}$$

$$\mathbf{B}_1 te^{rt} + \mathbf{B}_2 e^{rt}$$

$$\mathbf{C}_1 t^2e^{rt} + \mathbf{C}_2 te^{rt} + \mathbf{C}_3 e^{rt}$$

$$\cdots\cdots\cdots\cdots\cdots\cdots\cdots\cdots\cdots\cdots\cdots\cdots$$

$$\mathbf{K}_1 t^{k-1}e^{rt} + \mathbf{K}_2 t^{k-2}e^{rt} + \cdots + \mathbf{K}_{k-1} te^{rt} + \mathbf{K}_k e^{rt}$$

In this case, to within arbitrary scalar factors, the matrices $\mathbf{A}$, $\mathbf{B}_1$, $\mathbf{C}_1, \ldots, \mathbf{K}_1$ are identical (see Exercises 17 to 19).

Example 1 Find a complete solution of the system

$$(D^2 + D + 8)x_1 + (D^2 + 6D + 3)x_2 = 0$$

$$(D + 1)x_1 + \qquad (D^2 + 1)x_2 = 0$$

In this case, the characteristic equation (5) is

$$\begin{vmatrix} m^2 + m + 8 & m^2 + 6m + 3 \\ m + 1 & m^2 + 1 \end{vmatrix} = m^4 + 2m^2 - 8m + 5 = 0$$

with roots 1, 1, $-1 \pm 2i$. For the root $-1 + 2i$, Eq. (4) becomes

$$\begin{bmatrix} (-1+2i)^2 + (-1+2i) + 8 & (-1+2i)^2 + 6(-1+2i) + 3 \\ (-1+2i) + 1 & (-1+2i)^2 + 1 \end{bmatrix} \begin{bmatrix} a_1 \\ a_2 \end{bmatrix} = \begin{bmatrix} 0 \\ 0 \end{bmatrix}$$

or

$$\begin{bmatrix} 4 - 2i & -6 + 8i \\ 2i & -2 - 4i \end{bmatrix} \begin{bmatrix} a_1 \\ a_2 \end{bmatrix} = \begin{bmatrix} 0 \\ 0 \end{bmatrix}$$

This is equivalent to the two scalar equations

$$(2 - i)a_1 + (-3 + 4i)a_2 = 0$$

$$ia_1 - \quad (1 + 2i)a_2 = 0$$

Since $m = -1 + 2i$ is a root of the characteristic equation (5), these two equations are dependent and values of a_1 and a_2 satisfying both can be found by observing that the second equation is satisfied if $a_1 = 1 + 2i$ and $a_2 = i$, so that

$$\mathbf{A} \equiv \begin{bmatrix} a_1 \\ a_2 \end{bmatrix} = \begin{bmatrix} 1 + 2i \\ i \end{bmatrix}$$

Hence, $\mathscr{R}(\mathbf{A}) = \begin{bmatrix} 1 \\ 0 \end{bmatrix}$ and $\mathscr{I}(\mathbf{A}) = \begin{bmatrix} 2 \\ 1 \end{bmatrix}$

and thus from (6*a*) and (6*b*) we have the two particular solutions

$$\mathbf{X}_1 = e^{-t}\left(\begin{bmatrix} 1 \\ 0 \end{bmatrix} \cos 2t - \begin{bmatrix} 2 \\ 1 \end{bmatrix} \sin 2t\right)$$

$$\mathbf{X}_2 = e^{-t}\left(\begin{bmatrix} 2 \\ 1 \end{bmatrix} \cos 2t + \begin{bmatrix} 1 \\ 0 \end{bmatrix} \sin 2t\right)$$

For the repeated root $m = 1$, we have one solution of the form $\mathbf{B}e^t$, where, from (4),

$$\mathbf{P}(1)\mathbf{B} \equiv \begin{bmatrix} 10 & 10 \\ 2 & 2 \end{bmatrix} \begin{bmatrix} b \\ \beta \end{bmatrix} = \begin{bmatrix} 0 \\ 0 \end{bmatrix}$$

so that we can take

$$\mathbf{B} = \begin{bmatrix} b_1 \\ \beta_1 \end{bmatrix} = \begin{bmatrix} 1 \\ -1 \end{bmatrix} \qquad \text{and} \qquad \mathbf{X}_3 = \begin{bmatrix} 1 \\ -1 \end{bmatrix} e^t$$

As a second solution corresponding to $m = 1$, we try, from (7),

$$\mathbf{B}_1 te^t + \mathbf{B}_2 e^t$$

or, since $\mathbf{B}_1 = \mathbf{B}$, as we observed above,

$$\begin{bmatrix} 1 \\ -1 \end{bmatrix} te^t + \begin{bmatrix} b_2 \\ \beta_2 \end{bmatrix} e^t$$

In this case

$$\mathbf{P}'(D) = \begin{bmatrix} 2D + 1 & 2D + 6 \\ 1 & 2D \end{bmatrix} \qquad \text{and} \qquad \mathbf{P}'(1) = \begin{bmatrix} 3 & 8 \\ 1 & 2 \end{bmatrix}$$

Hence Eq. (9) becomes $\mathbf{P}(1)\mathbf{B}_2 = -\mathbf{P}'(1)\mathbf{B}_1 = -\mathbf{P}'(1)\mathbf{B}$, or

$$\begin{bmatrix} 10 & 10 \\ 2 & 2 \end{bmatrix} \begin{bmatrix} b_2 \\ \beta_2 \end{bmatrix} = -\begin{bmatrix} 3 & 8 \\ 1 & 2 \end{bmatrix} \begin{bmatrix} 1 \\ -1 \end{bmatrix}$$

This is equivalent to the nonhomogeneous scalar system

$$10b_2 + 10\beta_2 = 5$$

$$2b_2 + 2\beta_2 = 1$$

The determinant of the coefficients of this system, namely, $|\mathbf{P}(1)|$, is obviously zero; but, as we observed above in our general discussion, the system does have a nontrivial solution. The equations, of course, are dependent, and from either one we conclude that

$$b_2 = \lambda \qquad \beta_2 = (1 - 2\lambda)/2 \qquad \lambda \text{ arbitrary}$$

The solutions associated with the double root $m = 1$ are therefore

$$\mathbf{X}_3 = \mathbf{B}e^t = \begin{bmatrix} 1 \\ -1 \end{bmatrix} e^t \qquad \text{and} \qquad \mathbf{X}_4 = \mathbf{B}_1 te^t + \mathbf{B}_2 e^t = \begin{bmatrix} 1 \\ -1 \end{bmatrix} te^t + \begin{bmatrix} \lambda \\ (1 - 2\lambda)/2 \end{bmatrix} e^t$$

$$= \begin{bmatrix} 1 \\ -1 \end{bmatrix} te^t + \lambda \begin{bmatrix} 1 \\ -1 \end{bmatrix} e^t + \begin{bmatrix} 0 \\ \frac{1}{2} \end{bmatrix} e^t$$

The term

$$\lambda\begin{bmatrix}1\\-1\end{bmatrix}e^{t}$$

is proportional to the solution $\mathbf{X}_3$ and adds no generality to the final result. A complete solution of the problem is thus

$$\mathbf{X}=\begin{bmatrix}x_1\\x_2\end{bmatrix}=k_1e^{-t}\left(\begin{bmatrix}1\\0\end{bmatrix}\cos 2t-\begin{bmatrix}2\\1\end{bmatrix}\sin 2t\right)+k_2e^{-t}\left(\begin{bmatrix}2\\1\end{bmatrix}\cos 2t+\begin{bmatrix}1\\0\end{bmatrix}\sin 2t\right)$$
$$+k_3\begin{bmatrix}1\\-1\end{bmatrix}e^{t}+k_4\left(\begin{bmatrix}0\\\frac{1}{2}\end{bmatrix}e^{t}+\begin{bmatrix}1\\-1\end{bmatrix}te^{t}\right)$$

or, in scalar form,

$$x_1=e^{-t}[(k_1+2k_2)\cos 2t-(2k_1-k_2)\sin 2t]+k_3e^{t}+k_4te^{t}$$
$$x_2=e^{-t}(k_2\cos 2t-k_1\sin 2t)-(k_3-\tfrac{1}{2}k_4)e^{t}-k_4te^{t}$$

To find a particular integral for the nonhomogeneous system (2) when the vector function $\mathbf{F}(t)$ has only a finite number of linearly independent derivatives, we proceed very much as in the case of a single scalar differential equation. At the outset, it is convenient to identify the linearly independent functions $\phi_1(t)$, $\phi_2(t)$, $\ldots$, $\phi_j(t)$ which appear in the components of $\mathbf{F}(t)$ and then express $\mathbf{F}(t)$ in the form

$$\mathbf{F}(t)=\mathbf{K}_1\phi_1(t)+\mathbf{K}_2\phi_2(t)+\cdots+\mathbf{K}_j\phi_j(t)$$

where the **K**s are appropriate constant column vectors. Then for such terms as do not duplicate vectors already in the complementary function, particular integrals can be constructed as described in Table 4.2, Sec. 4.3, provided that the arbitrary scalar constants appearing in the entries in the table are replaced by arbitrary constant vectors. The trial solutions are then substituted into the nonhomogeneous system, and the arbitrary components of the coefficient vectors are determined to make the resulting equations identically true. For terms in the expanded form of $\mathbf{F}(t)$ which duplicate vectors in the complementary function, the results of Table 4.2 are still valid with one additional provision: not only must the usual choice for a trial particular integral be multiplied by the lowest positive integral power of the independent variable which will eliminate the duplication, but the products of the normal choice and all lower nonnegative integral powers of the independent variable must also be included in the actual choice. An example should clarify the details of the procedure.

Example 2 Find a particular integral for the system of equations in Example 1 if the terms $2e^{t}$ and $2e^{t}+e^{-2t}$ appear on the right-hand sides of the respective equations.

At the outset, it is convenient to group like terms on the right-hand side of the equivalent matric equation by writing

$$\mathbf{P}(D)\mathbf{X}=\mathbf{F}_1+\mathbf{F}_2$$

where $\quad \mathbf{F}_1 = \begin{bmatrix} 0 \\ e^{-2t} \end{bmatrix} = \begin{bmatrix} 0 \\ 1 \end{bmatrix} e^{-2t} \quad$ and $\quad \mathbf{F}_2 = \begin{bmatrix} 2e^t \\ 2e^t \end{bmatrix} = \begin{bmatrix} 2 \\ 2 \end{bmatrix} e^t$

Since $\mathbf{F}_1$ does not duplicate any vector in the complementary function, i.e., the complete solution found in Example 1, we assume as a trial particular integral simply

$$\mathbf{X}_p = \mathbf{A}e^{-2t}$$

where
$$\mathbf{A} = \begin{bmatrix} a \\ \alpha \end{bmatrix}$$

is a constant vector to be determined. Then, substituting, we have

$$\mathbf{P}(D)\mathbf{X}_p \equiv \mathbf{P}(D)\mathbf{A}e^{-2t} = \mathbf{P}(-2)\mathbf{A}e^{-2t} = \mathbf{F}_1 \equiv \begin{bmatrix} 0 \\ 1 \end{bmatrix} e^{-2t}$$

or, dividing out e^{-2t},

$$\mathbf{P}(-2)\mathbf{A} \equiv \begin{bmatrix} 10 & -5 \\ -1 & 5 \end{bmatrix} \begin{bmatrix} a \\ \alpha \end{bmatrix} = \begin{bmatrix} 0 \\ 1 \end{bmatrix}$$

This is equivalent to the scalar system

$$10a - 5\alpha = 0$$

$$-a + 5\alpha = 1$$

from which it follows that $a = \frac{1}{9}$, $\alpha = \frac{2}{9}$, $\mathbf{A} = \frac{1}{9}\left[\begin{smallmatrix} 1 \\ 2 \end{smallmatrix}\right]$. Thus $\frac{1}{9}\left[\begin{smallmatrix} 1 \\ 2 \end{smallmatrix}\right]e^{-2t}$ is one term in the particular integral we are seeking.

To find the terms in the particular integral arising from $\mathbf{F}_2$, we note that since both e^t and te^t occur as terms in the complementary function, the normal choice for a trial particular integral, namely, $\mathbf{B}_1 e^t$, must be modified by multiplying it by t^2 *and including the terms* $\mathbf{B}_2 te^t$ *and* $\mathbf{B}_3 e^t$, where

$$\mathbf{B}_1 = \begin{bmatrix} b_1 \\ \beta_1 \end{bmatrix} \qquad \mathbf{B}_2 = \begin{bmatrix} b_2 \\ \beta_2 \end{bmatrix} \qquad \text{and} \qquad \mathbf{B}_3 = \begin{bmatrix} b_3 \\ \beta_3 \end{bmatrix}$$

are constant matrices to be determined. Then, substituting

$$\mathbf{X}_p = \mathbf{B}_1 t^2 e^t + \mathbf{B}_2 te^t + \mathbf{B}_3 e^t$$

into the equation $\mathbf{P}(D)\mathbf{X} = \mathbf{F}_2$ and using the results of Exercise 17, we obtain

$$\begin{aligned} \mathbf{P}(D)(\mathbf{B}_1 t^2 e^t + \mathbf{B}_2 te^t + \mathbf{B}_3 e^t) &= \mathbf{P}(1)\mathbf{B}_1 t^2 e^t + 2\mathbf{P}'(1)\mathbf{B}_1 te^t + \mathbf{P}''(1)\mathbf{B}_1 e^t \\ &\quad + \mathbf{P}(1)\mathbf{B}_2 te^t + \mathbf{P}'(1)\mathbf{B}_2 e^t + \mathbf{P}(1)\mathbf{B}_3 e^t \\ &= \mathbf{F}_2 = \begin{bmatrix} 2 \\ 2 \end{bmatrix} e^t \end{aligned}$$

Hence, equating the coefficients of like terms on the two sides of this equation, we find that

$$\begin{aligned} \mathbf{P}(1)\mathbf{B}_1 &= \mathbf{0} \\ \mathbf{P}(1)\mathbf{B}_2 + 2\mathbf{P}'(1)\mathbf{B}_1 &= \mathbf{0} \\ \mathbf{P}(1)\mathbf{B}_3 + \mathbf{P}'(1)\mathbf{B}_2 + \mathbf{P}''(1)\mathbf{B}_1 &= \begin{bmatrix} 2 \\ 2 \end{bmatrix} \end{aligned} \tag{10}$$

The first of these equations is simply

$$\begin{bmatrix} 10 & 10 \\ 2 & 2 \end{bmatrix} \begin{bmatrix} b_1 \\ \beta_1 \end{bmatrix} = \begin{bmatrix} 0 \\ 0 \end{bmatrix}$$

which implies that

$$\mathbf{B}_1 \equiv \begin{bmatrix} b_1 \\ \beta_1 \end{bmatrix} = \begin{bmatrix} \lambda \\ -\lambda \end{bmatrix} \qquad \lambda \text{ arbitrary}$$

The second of the equations in (10) now becomes

$$\begin{bmatrix} 10 & 10 \\ 2 & 2 \end{bmatrix} \begin{bmatrix} b_2 \\ \beta_2 \end{bmatrix} + 2 \begin{bmatrix} 3 & 8 \\ 1 & 2 \end{bmatrix} \begin{bmatrix} \lambda \\ -\lambda \end{bmatrix} = \begin{bmatrix} 0 \\ 0 \end{bmatrix}$$

or

$$\begin{bmatrix} 5 & 5 \\ 1 & 1 \end{bmatrix} \begin{bmatrix} b_2 \\ \beta_2 \end{bmatrix} = \begin{bmatrix} 5\lambda \\ \lambda \end{bmatrix}$$

from which, without loss of generality (see Exercise 21) we conclude that

$$\mathbf{B}_2 \equiv \begin{bmatrix} b_2 \\ \beta_2 \end{bmatrix} = \begin{bmatrix} \lambda \\ 0 \end{bmatrix}$$

The third equation in the set (10) now becomes

$$\begin{bmatrix} 10 & 10 \\ 2 & 2 \end{bmatrix} \begin{bmatrix} b_3 \\ \beta_3 \end{bmatrix} + \begin{bmatrix} 3 & 8 \\ 1 & 2 \end{bmatrix} \begin{bmatrix} \lambda \\ 0 \end{bmatrix} + \begin{bmatrix} 2 & 2 \\ 0 & 2 \end{bmatrix} \begin{bmatrix} \lambda \\ -\lambda \end{bmatrix} = \begin{bmatrix} 2 \\ 2 \end{bmatrix}$$

or

$$\begin{bmatrix} 10 & 10 \\ 2 & 2 \end{bmatrix} \begin{bmatrix} b_3 \\ \beta_3 \end{bmatrix} = \begin{bmatrix} 2 \\ 2 \end{bmatrix} - \begin{bmatrix} 3\lambda \\ \lambda \end{bmatrix} - \begin{bmatrix} 0 \\ -2\lambda \end{bmatrix} = \begin{bmatrix} 2 - 3\lambda \\ 2 + \lambda \end{bmatrix}$$

which is equivalent to the scalar system

$$10b_3 + 10\beta_3 = 2 - 3\lambda$$

$$2b_3 + 2\beta_3 = 2 + \lambda$$

Since the left member of the first equation is 5 times the left member of the second equation, it follows that the ratio of the right members must also be 5. Hence we must have

$$2 - 3\lambda = 5(2 + \lambda)$$

which implies that $\lambda = -1$. For this value of λ we may, without loss of generality (see Exercise 21), take $b_3 = \frac{1}{2}$ and $\beta_3 = 0$. Thus

$$\mathbf{B}_3 = \begin{bmatrix} \frac{1}{2} \\ 0 \end{bmatrix} \qquad \mathbf{B}_2 = \begin{bmatrix} -1 \\ 0 \end{bmatrix} \qquad \text{and} \qquad \mathbf{B}_1 = \begin{bmatrix} -1 \\ 1 \end{bmatrix}$$

Finally, putting our results together, we have the entire particular integral

$$\frac{1}{9} \begin{bmatrix} 1 \\ 2 \end{bmatrix} e^{-2t} + \begin{bmatrix} -1 \\ 1 \end{bmatrix} t^2 e^t + \begin{bmatrix} -1 \\ 0 \end{bmatrix} t e^t + \begin{bmatrix} \frac{1}{2} \\ 0 \end{bmatrix} e^t$$

Up to this point, we have said nothing about the theoretical aspects of the existence and uniqueness of solutions of systems of linear differential equations. It would take us too far afield to embark on a detailed investigation of such matters, but it is appropriate to quote the fundamental theorem as we conclude this chapter.

In matric notation, the systems we have studied have all been of the form $\mathbf{P}(D)\mathbf{X} = \mathbf{F}(t)$ where the elements $p_{ij}(D)$ of the operator matrix $\mathbf{P}(D)$ could be polynomials of any degree. The theory of systems of linear differential equations, however, is based on the following more specific standard form:

$$\begin{aligned} Dy_1 &= a_{11}(t)y_1 + a_{12}(t)y_2 + \cdots + a_{1n}(t)y_n + F_1(t) \\ Dy_2 &= a_{21}(t)y_1 + a_{22}(t)y_2 + \cdots + a_{2n}(t)y_n + F_2(t) \\ &\cdots\cdots\cdots\cdots\cdots\cdots\cdots\cdots\cdots\cdots \\ Dy_n &= a_{n1}(t)y_1 + a_{n2}(t)y_2 + \cdots + a_{nn}(t)y_n + F_n(t) \end{aligned} \tag{11}$$

in which each equation is of the first order. Actually, this involves no restriction since any system of linear differential equations can be reduced to the standard form (11) provided the determinant of the operational coefficients is not a constant.† For example, under the substitutions

$$\begin{aligned} x_1 &= y_1 \qquad Dx_1 = Dy_1 = y_2 \qquad D^2x_1 = Dy_2 \\ x_2 &= y_3 \qquad Dx_2 = Dy_3 = y_4 \qquad D^2x_2 = Dy_4 \end{aligned} \tag{12}$$

the differential equations of Example 1 become

$$\begin{aligned} (Dy_2 + y_2 + 8y_1) + (Dy_4 + 6y_4 + 3y_3) &= 0 \\ (y_2 + y_1) + (Dy_4 + y_3) &= 0 \end{aligned} \tag{13}$$

The last equation gives us

$$Dy_4 = -y_1 - y_2 - y_3 \tag{14}$$

and if we substitute this into the first of the equations in (13) and solve for Dy_2, we find

$$Dy_2 = -7y_1 - 2y_3 - 6y_4 \tag{15}$$

Thus from Eqs. (12), (13), (14), and (15) we see that the original system is equivalent to

$$\begin{aligned} Dy_1 &= \qquad\quad y_2 \\ Dy_2 &= -7y_1 \qquad - 2y_3 - 6y_4 \\ Dy_3 &= \qquad\qquad\qquad\qquad y_4 \\ Dy_4 &= -\ y_1 - y_2 - y_3 \end{aligned} \tag{16}$$

which is a system in standard form.

† If the determinant of the operational coefficients is a nonzero constant, the unknowns can be expressed uniquely in terms of the functions F_i and their derivatives by purely algebraic operations (see Exercise 2, Sec. 5.2, for instance). If the constant value of the determinant is zero, then, in general, the system has no solution (see Example 2, Sec. 5.2, for instance).

The fundamental existence and uniqueness theorem is now the following.

Theorem 1 Let the coefficient functions $a_{ij}(t)$ and the nonhomogeneous terms $F_i(t)$ in the system of equations

$$\begin{aligned}
Dy_1 &= a_{11}(t)y_1 + a_{12}(t)y_2 + \cdots + a_{1n}(t)y_n + F_1(t) \\
Dy_2 &= a_{21}(t)y_1 + a_{22}(t)y_2 + \cdots + a_{2n}(t)y_n + F_2(t) \\
&\cdots\cdots\cdots\cdots\cdots\cdots\cdots\cdots\cdots \\
Dy_n &= a_{n1}(t)y_1 + a_{n2}(t)y_2 + \cdots + a_{nn}(t)y_n + F_n(t)
\end{aligned}$$

be continuous over an interval I and let t_0 be an arbitrary point in I. Then over I this system has a unique solution

$$\mathbf{Y} = \begin{bmatrix} y_1 \\ y_2 \\ \vdots \\ y_n \end{bmatrix}$$

which at $t = t_0$ satisfies the condition

$$\mathbf{Y}(t_0) = \begin{bmatrix} y_1(t_0) \\ y_2(t_0) \\ \vdots \\ y_n(t_0) \end{bmatrix} = \begin{bmatrix} c_1 \\ c_2 \\ \vdots \\ c_n \end{bmatrix}$$

where $c_1, c_2, \ldots, c_n$ are arbitrary constants.

Exercises for Section 5.4

Find a complete solution of each of the following systems.

1 $(D+1)x + (2D-6)y = 0$
$(2D+1)x + (3D-11)y = 0$

2 $Dx + (D-5)y = 0$
$(D+1)x + (2D-8)y = 0$

3 $(D-4)x - 3y = e^{-t}$
$-5x + (D-6)y = 0$

4 $(D-4)x - 3y = 0$
$-5x + (D-6)y = e^t$

5 $(3D+1)x + (D+7)y = e^{-t}$
$(2D+1)x + (D+5)y = e^{-t}$

6 $(D+5)x + (2D+1)y = e^{-t} + e^t$
$(D+7)x + (3D+1)y = 0$

7 $(D+5)x + (D+7)y = 2e^t$
$(2D+1)x + (3D+1)y = e^t$

8 $(D+2)x + (D+3)y = -4$
$(2D-6)x + (3D-4)y = 2$

9 $(D+1)x + (D+2)y = -e^t$
$(3D+1)x + (4D+7)y = -7e^t$

***10** $(D+1)x + (D+2)y = -t+1$
$(5D+1)x + (6D+3)y = -2t+1$

11 $(2D+1)x + (D+2)y = e^{-t}$
$(3D-7)x + (3D+1)y = 0$

***12** $(2D+1)x + (D+2)y = \sin t$
$(3D+1)x + (3D+5)y = \cos t$

13 $(D-4)x - 6y - 6z = 0$
$-x + (D-3)y - 2z = 0$
$x + 4y + (D+3)z = 0$

14 $(D-3)x + y + z = 0$
$x + (D-3)y + z = 0$
$x + y + (D-3)z = 0$

****15** $(D-1)x - z = 0$
$(D-1)x + 2(D-1)y - 5z = 0$
$(D-1)y + (D-3)z = 0$

16 $(6D^2+6)x - 3y = 0$
$-3x + (4D^2+6)y - 3z = 0$
$-3y + (4D^2+4)z = 0$

***17** Show that

$$D^r(t^2e^{mt}) = m^rt^2e^{mt} + 2rm^{r-1}te^{mt} + r(r-1)m^{r-2}e^{mt}$$

Hence show that

$$p(D)t^2e^{mt} = p(m)t^2e^{mt} + 2p'(m)te^{mt} + p''(m)e^{mt}$$

and
$$\mathbf{P}(D)t^2e^{mt} = \mathbf{P}(m)t^2e^{mt} + 2\mathbf{P}'(m)te^{mt} + \mathbf{P}''(m)e^{mt}$$

where $p(D)$ is a polynomial in the operator D and $\mathbf{P}(D)$ is a matrix whose elements are polynomials in D.

***18** Using the results of Exercise 17, show that if $m = m_1$ is a triple root of the characteristic equation $|\mathbf{P}(m)| = 0$, then $\mathbf{X}_1 = \mathbf{A}e^{m_1t}$, $\mathbf{X}_2 = \mathbf{B}_1te^{m_1t} + \mathbf{B}_2e^{m_1t}$, and $\mathbf{X}_3 = \mathbf{C}_1t^2e^{m_1t} + \mathbf{C}_2te^{m_1t} + \mathbf{C}_3e^{m_1t}$ are three independent solutions of the system $\mathbf{P}(D)\mathbf{X} = \mathbf{0}$ provided the matric coefficients $\mathbf{A}$, $\mathbf{B}_1$, $\mathbf{B}_2$, $\mathbf{C}_1$, $\mathbf{C}_2$, $\mathbf{C}_3$ satisfy the equations

$$\mathbf{P}(m_1)\mathbf{A} = \mathbf{0} \qquad \mathbf{P}(m_1)\mathbf{B}_1 = \mathbf{0} \qquad \mathbf{P}(m_1)\mathbf{B}_2 + \mathbf{P}'(m_1)\mathbf{B}_1 = \mathbf{0}$$

$$\mathbf{P}(m_1)\mathbf{C}_1 = \mathbf{0} \qquad \mathbf{P}(m_1)\mathbf{C}_2 + 2\mathbf{P}'(m_1)\mathbf{C}_1 = \mathbf{0}$$

$$\mathbf{P}(m_1)\mathbf{C}_3 + \mathbf{P}'(m_1)\mathbf{C}_2 + \mathbf{P}''(m_1)\mathbf{C}_1 = \mathbf{0}$$

***19** Generalize the results of Exercise 17 to the function t^3e^{mt}.

***20** Generalize the results of Exercise 18 to the solution of the equation $\mathbf{P}(D)\mathbf{X} = \mathbf{0}$ arising from a fourfold root of the characteristic equation.

***21** (*a*) In Example 2, verify that no generality is lost in taking $b_2 = \lambda$, $\beta_2 = 0$ by showing that if the general solution $b_2 = g + \lambda$, $\beta_2 = -g$ is used, the term arising from the arbitrary constant g can be absorbed in the term X_4 in the complementary function.

(*b*) In Example 2, verify that no generality is lost in taking $b_3 = \frac{1}{2}$ and $\beta_3 = 0$.

22 Verify that the characteristic equation of the system (16) is the same as the characteristic equation obtained for this system in Example 1.

23 Express the system of equations in Example 1, Sec. 5.2, in the standard form (11) and verify that the characteristic equation of the system is the same in either representation.

***24** If **A**, **B**, and **C** are constant column matrices and $\mathbf{P}(D)$ is a matrix whose elements are polynomials in the operator D, show that

(*a*) $\mathbf{P}(D)\mathbf{A} = \mathbf{P}(0)\mathbf{A}$ (*b*) $\mathbf{P}(D)\mathbf{B}t = \mathbf{P}(0)\mathbf{B}t + \mathbf{P}'(0)\mathbf{B}$

(*c*) $\mathbf{P}(D)\mathbf{C}t^2 = \mathbf{P}(0)\mathbf{C}t^2 + 2\mathbf{P}'(0)\mathbf{C}t + \mathbf{P}''(0)\mathbf{C}$

***25** Verify, for a system of two linear differential equations in two unknowns for which the characteristic equation has a double root m_1, that the equation

$$\mathbf{P}(m_1)\mathbf{B}_2 + \mathbf{P}'(m_1)\mathbf{B}_1 = \mathbf{0}$$

is always solvable for $\mathbf{B}_2$ when $\mathbf{B}_1$ is determined so that $\mathbf{P}(m_1)\mathbf{B}_1 = \mathbf{0}$.

CHAPTER

SIX

THE LAPLACE TRANSFORMATION

6.1 INTRODUCTION

In Sec. 3.1 we introduced the symbol D to represent the operation of taking the derivative of a function with respect to its independent variable, whatever that variable might be. Since then we have made extensive use of the symbol D as a notational convenience, but we have made no attempt to assign operational properties to it, though this can be done.

To explore the matter briefly, consider the equation

$$(D - a)y = f(t) \tag{1}$$

A naive student, misled by the algebraic appearance of Eq. (1), might "divide" by $D - a$ and claim that $y = f(t)/(D - a)$ was the required answer; but this is just meaningless formalism. On the other hand, a more experienced student might do the same thing and then inquire if there was any way in which $1/(D - a)$ could be interpreted as a meaningful operator in its own right. Pursuing this thought, the student might return to Eq. (1) and solve it by the methods of Sec. 1.6, getting

$$y = e^{at} \int e^{-at} f(t)\, dt + c_1 e^{at}$$

and then assert that $1/(D - a)$ is an operator which gives this result when applied to a function $f(t)$:

$$\frac{1}{D - a} f(t) = e^{at} \int e^{-at} f(t)\, dt + c_1 e^{at} \tag{2}$$

Initially, this operational interpretation of $1/(D - a)$ merely provides an alternative way of solving certain types of linear first-order equations, but perhaps it can be extended. Proceeding in a purely formal way, let us try to apply formula (2) to the equation

$$(D^2 - 3D + 2)y = e^t \tag{3}$$

"Dividing" by $(D^2 - 3D + 2)$, then using partial fractions, and finally applying formula (2), we have

$$y = \frac{e^t}{D^2 - 3D + 2} = \frac{e^t}{(D-1)(D-2)} = \left(\frac{1}{D-2} - \frac{1}{D-1}\right)e^t$$

$$= \left(e^{2t} \int e^{-2t}e^t\,dt + c_1 e^{2t}\right) - \left(e^t \int e^{-t}e^t\,dt + c_2 e^t\right)$$

$$= [e^{2t}(-e^{-t}) + c_1 e^{2t}] - [e^t(t) + c_2 e^t]$$

$$= c_1 e^{2t} - (c_2 + 1)e^t - te^t$$

Surprisingly, perhaps, our formal manipulations have led to the correct answer, for $c_1 e^{2t} - (c_2 + 1)e^t$ is the complementary function of Eq. (3), and it is easy to verify that $-te^t$ is a particular integral. Moreover, the particular integral $-te^t$ emerged without any special treatment, even though the nonhomogeneous term in Eq. (3) duplicated a term in the complementary function!

Although the interpretation of D as an operator goes back to Leibnitz (1646–1716), the English engineer Oliver Heaviside (1850–1925) was the first to make effective and extensive use of it. With his *operational calculus*, Heaviside solved a great variety of difficult and important physical problems that classical methods had been unable to handle. Because his work was formal and not rigorous, it was scorned by purists; but because it "worked," it was widely accepted by engineers and applied mathematicians.† As the great power of Heaviside's operational calculus became more and more apparent, mathematicians who had previously scoffed at it became interested in trying to justify it. These efforts continued with varying degrees of success for several decades until finally it was recognized that an integral transform, originally constructed by Laplace (1749–1827) almost a century before, not only furnished an adequate theoretical foundation for Heaviside's work but, in fact, provided a more systematic alternative to the methods themselves.

In this chapter we shall study the modern form of operational calculus based on the Laplace transform and investigate its application to the solution of linear differential equations with constant coefficients. As we shall see, the Laplace transform is especially effective in handling physical problems, both mechanical and electrical, involving initial conditions and discontinuous forcing functions.

† Reproached by someone because of the lack of a logical justification for his methods, Heaviside is said to have replied, "Shall I refuse to eat my dinner because I do not understand the processes of digestion?"

Exercises for Section 6.1

Using Eq. (2), find a complete solution of each of the following equations.

1 $(D^2 - 5D + 6)y = e^t$
2 $(D^2 + D - 2)y = e^t$
3 $(D^2 + 4D + 3)y = t$
4 $(D^2 - D - 6)y = 0$
5 $(D^2 - 2D + 1)y = e^{2t}$
6 $(D^2 - 4D + 4)y = e^{2t}$
7 $(D^3 - 6D^2 + 11D - 6)y = 6$
8 $(D^3 + 2D^2 - D - 2)y = e^t$

6.2 THEORETICAL PRELIMINARIES

The Laplace transformation is an operation, denoted by the symbol $\mathscr{L}$, which associates with each function $f(t)$, satisfying suitable conditions for $t \geqq 0$, a unique function $\phi(s)$, called the **Laplace transform** of $f(t)$, according to the rule

$$\mathscr{L}\{f(t)\} = \phi(s) = \int_0^\infty f(t)e^{-st}\,dt\dagger \tag{1}$$

Even when $f(t)$ is continuous (which it need not be), the integral which appears in the Laplace transform of $f(t)$ is improper and must be interpreted in the usual way, namely,

$$\int_0^\infty f(t)e^{-st}\,dt = \lim_{b\to\infty} \int_0^b f(t)e^{-st}\,dt$$

According to Eq. (1), the Laplace transform of $f(t)$ exists if and only if the integral in (1) converges for at least some values of s. There are many functions for which this is not the case; that is, there are many functions which do not have Laplace transforms. For instance, the function $f(t) = e^{t^2}$ grows so rapidly as t becomes infinite that there is no value of the parameter s for which the decreasing factor e^{-st} is able to keep the product $e^{-st}e^{t^2}$ bounded as $t \to \infty$. Hence the integral corresponding to $\mathscr{L}\{e^{t^2}\}$ in (1) does not converge. Likewise, because of its unbounded behavior in the neighborhood of the origin, the function $f(t) = 1/t$ has no Laplace transform.

To state conditions on $f(t)$ which are sufficient to ensure that $\mathscr{L}\{f(t)\}$ exists, we first define a **piecewise-continuous function** and a **function of exponential order.**

Definition 1 A function f is said to be **piecewise continuous** on a finite interval I: $a \leq t \leq b$ if and only if I can be subdivided into a finite number of subintervals such that

1. f is continuous throughout the interior of each subinterval.
2. In each subinterval, f approaches a finite limit as t approaches either endpoint through the interior of the subinterval.

† The variable of integration t is, of course, a dummy variable and can be replaced at pleasure by any other symbol. From time to time, we shall find it convenient to do this in our work.

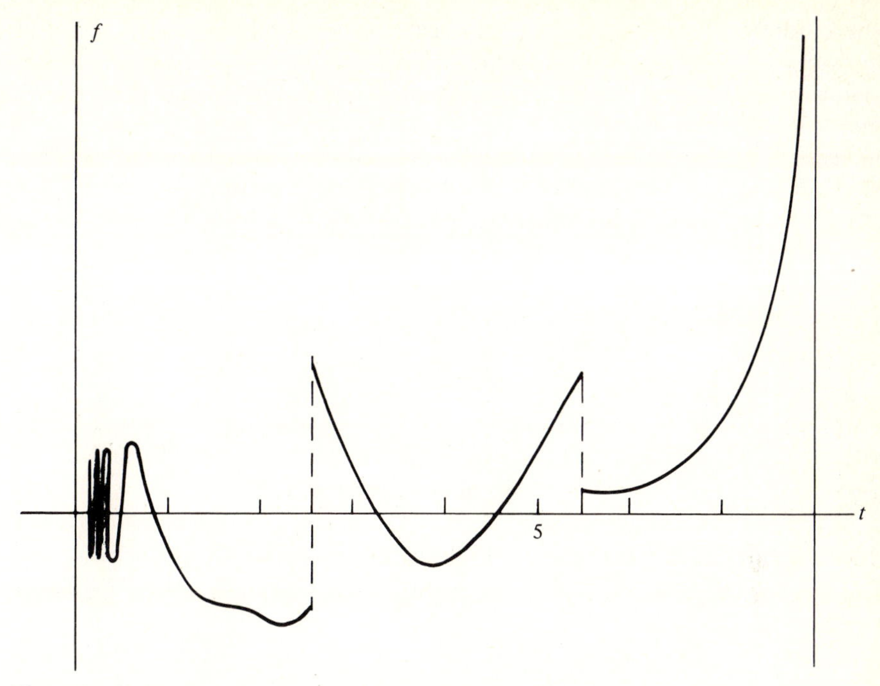

Figure 6.1 The graph of a function which is piecewise continuous for $1 \le t \le 7$.

The function f may or may not be defined at the endpoints of the various subintervals. A piecewise-continuous function is also called a **sectionally continuous function**.

Of course, if f is continuous on a closed interval I, it is also piecewise continuous on I. In fact, at every interior point of an interval where a function is continuous, its right- and left-hand limits not only exist but are equal. Figure 6.1 shows a function which is piecewise continuous on the interval $1 \le t \le 7$ but not on the interval $0 \le t \le 7$ or the interval $1 \le t \le 8$.

Definition 2 A function is said to be **piecewise continuous** on an infinite interval $[a, \infty)$ if and only if it is piecewise continuous on every finite interval of the form $[a, b]$, where $b > a$.

Definition 3 A function $f(t)$ is said to be of **exponential order** if there exist numbers α, M, and T such that

$$e^{-\alpha t}|f(t)| < M$$

for all $t > T$ at which $f(t)$ is defined.

From the limits on the integral in (1) it is evident that with the possible exception of a countable set of points, the domain of any function to which the

Laplace transformation can be applied must include all positive values of t; and throughout this chapter we shall make this assumption. Now for $t > 0$, $e^{-\alpha t}$ is a *monotonically decreasing* function of α; that is, $\alpha_1 > \alpha$ implies that $e^{-\alpha_1 t} < e^{-\alpha t}$. Hence it is clear that if

$$e^{-\alpha t}|f(t)| < M \qquad \text{for all } t > T$$

then for all $\alpha_1 > \alpha$, it is also true that

$$e^{-\alpha_1 t}|f(t)| < M \qquad \text{for all } t > T$$

Thus the α required by Definition 3 is not unique. The greatest lower bound α_0 of the set of all α's which can be used in Definition 3 is often called the **abscissa of convergence** of $f(t)$.

The abscissa of convergence α_0 of a function $f(t)$ may or may not itself be one of the α's which will serve in Definition 3. For instance, if $f(t) = t$, then for every positive α and no others, the product $e^{-\alpha t}|f(t)| = |t|e^{-\alpha t}$ remains bounded, and in fact approaches zero, as t becomes infinite. Since the greatest lower bound of the set of all positive numbers is the number zero, it follows that in this case $\alpha_0 = 0$. However, for α_0 itself, the product $|t|e^{-\alpha_0 t} = |t|$ increases beyond all bounds as $t \to \infty$. Thus for the function $f(t) = t$, the abscissa of convergence, namely $\alpha_0 = 0$, is not one of the α's that can be used in Definition 3.

On the other hand, if $f(t) = e^{2t}$, then for every α greater than or *equal* to 2, $e^{-\alpha t}|f(t)| = e^{-\alpha t}e^{2t} = e^{-(\alpha-2)t}$ is bounded as $t \to \infty$. Since the greatest lower bound of all numbers equal to or greater than 2 is 2, it is clear that in this case the abscissa of convergence α_0 is 2 and moreover is a value of α which will serve in Definition 3.

Since $e^{-\alpha t}|f(t)| < M$ implies only that $|f(t)| < Me^{\alpha t}$, it follows that if a function is of exponential order, its absolute value need not remain bounded as $t \to \infty$; but it will not increase more rapidly than some constant multiple of a simple exponential function of t. As the particular function $f(t) = \sin e^{t^2}$ shows, *the derivative of a function of exponential order is not necessarily of exponential order*. On the other hand, it is not difficult to prove that *the integral of a function of exponential order is also of exponential order.*

The existence of the Laplace transform of any function of exponential order which is piecewise continuous on $[0, \infty)$ follows as a corollary of the following fundamental theorem.

Theorem 1 If $f(t)$ is piecewise continuous on $[0, \infty)$ and of exponential order, then for any value of s which is greater than the abscissa of convergence of $f(t)$, the integral $\int_0^\infty f(t)e^{-st}\,dt$ converges absolutely.

PROOF To establish this theorem, we must show that under the hypotheses of the theorem, $\int_0^\infty |f(t)e^{-st}|\,dt$ converges; that is, we must show that

$$\lim_{b\to\infty}\int_0^b |f(t)e^{-st}|\,dt = \lim_{b\to\infty}\int_0^b |f(t)|e^{-st}\,dt \tag{2}$$

exists. To do this, we first prove that there is an exponential function whose value is greater than or equal to the corresponding value of $|f(t)|$ over the entire range of integration $t \geq 0$. Now, by hypothesis, $f(t)$ is of exponential order and therefore has an abscissa of convergence α_0. Hence, there exist numbers M_1 and T such that for all $t > T$ and any α greater than, but bounded from, α_0, that is, any α such that $\alpha > \alpha_1 > \alpha_0$, we have

$$|f(t)| < M_1 e^{\alpha t}$$

We now seek a bounding function of the same form, namely $|f(t)| < Me^{\alpha t}$, for $|f(t)|$ on the rest of the range of integration, that is, the interval $0 \leq t \leq T$. To do this, we note that by hypothesis, $f(t)$ is piecewise continuous over any finite interval. Hence it is bounded over any such interval, and in particular there exists a positive number M_2 such that

$$|f(t)| < M_2 = (M_2 e^{-\alpha t})e^{\alpha t} \qquad \text{for } 0 \leq t \leq T$$

Now if $\alpha \geq 0$, then $0 < e^{-\alpha t} \leq 1$ for $t \geq 0$, and we have $M_2 e^{-\alpha t} \leq M_2$. On the other hand, if $\alpha < 0$, then for $0 \leq t \leq T$ we have $e^{-\alpha t} \leq e^{-\alpha T}$, and hence $M_2 e^{-\alpha t} \leq M_2 e^{-\alpha T}$. Thus if we let M be the largest of the three numbers M_1, M_2, and $M_2 e^{-\alpha T}$, it is clear that

$$|f(t)| < Me^{\alpha t} \qquad \text{for } all\ t \geq 0$$

Hence, returning to the second integral in (2) and replacing $|f(t)|$ by $Me^{\alpha t}$, we have

$$I \equiv \int_0^b |f(t)|e^{-st}\,dt \leq \int_0^b Me^{\alpha t}e^{-st}\,dt = \frac{Me^{-(s-\alpha)t}}{-(s-\alpha)}\bigg|_0^b = \frac{M}{s-\alpha}(1 - e^{-(s-\alpha)b})$$

Now if $s > \alpha$, the exponential in the last expression decreases monotonically and hence the expression itself increases monotonically and approaches $M/(s - \alpha)$ as b becomes infinite. Therefore

$$I \leq \frac{M}{s-\alpha} \qquad s > \alpha > \alpha_0$$

Since the integrand of I is everywhere nonnegative, it is clear that I is a monotonically increasing function of b. Hence, being bounded above, as we have just shown, it must approach a limit as b becomes infinite. Since $s > \alpha > \alpha_0$ is equivalent to the condition $s > \alpha_0$, the theorem is established.

In calculus we learned that an infinite series converges if it converges absolutely, and the same thing is true for infinite integrals. Hence, since we have just shown that $\int_0^\infty |f(t)|e^{-st}\,dt$ converges, it follows that $\int_0^\infty f(t)e^{-st}\,dt$ also converges, and we have the following corollary.

Corollary 1 Sufficient (but not necessary) conditions for a function to have a Laplace transform are that the function be piecewise continuous on $[0, \infty)$ and of exponential order.

Since the absolute value of an integral is always equal to or less than the integral of the absolute value, it follows from the proof of Theorem 1 that

$$\left|\int_0^b f(t)e^{-st}\,dt\right| \leq \int_0^b |f(t)|e^{-st}\,dt \leq \frac{M}{s-\alpha}$$

Hence, letting $b \to \infty$, we have the following useful result.

Corollary 2 If $f(t)$ is piecewise continuous on $[0, \infty)$ and of exponential order with abscissa of convergence α_0, then for all values of s and α such that $s > \alpha > \alpha_0$

$$|\mathscr{L}\{f(t)\}| \le \frac{M}{s - \alpha} \qquad \text{where } M \text{ is independent of } s$$

Finally, from Corollary 2 we draw the following interesting conclusions.

Corollary 3 If $f(t)$ is piecewise continuous on $[0, \infty)$ and of exponential order, then $\mathscr{L}\{f(t)\}$ approaches zero as s becomes infinite.

Corollary 4 If $f(t)$ is piecewise continuous on $[0, \infty)$ and of exponential order, then $s\mathscr{L}\{f(t)\}$ is bounded as s becomes infinite.

Corollaries 3 and 4 make it clear that not all functions of s are Laplace transforms—or at least not Laplace transforms of functions of the "respectable" type that are piecewise continuous on $[0, \infty)$ and of exponential order. For instance, $\phi(s) = s/(s - 1)$ does not approach zero as s becomes infinite; hence it is not the Laplace transform of any "respectable" function. Also, although $\phi(s) = 1/\sqrt{s}$ does approach zero as s becomes infinite, it too is not the transform of any "respectable" function, since $s\phi(s) = \sqrt{s}$ is not bounded as s becomes infinite.

As will soon become apparent, the derivation of the fundamental properties of the Laplace transformation will involve manipulation of the definitive integral (1). In particular, it will occasionally be necessary for us to differentiate or integrate a Laplace transform $\phi(s)$ with respect to s by differentiating or integrating with respect to s inside the integral sign in formula (1). The convergence or even the absolute convergence of the Laplace transform integral is not enough to justify these operations. What is required is the stronger type of convergence known as **uniform convergence**.

Definition 4 The improper integral $\int_a^\infty F(s, t)\, dt$ is said to **converge uniformly** over a given set S of s values if and only if given any $\varepsilon > 0$, there exists a number B, depending on ε but not on s, such that

$$\left| \int_b^\infty F(s, t)\, dt \right| < \varepsilon \qquad \text{for } b > B \text{ and all } s \text{ in the set } S$$

To help us appreciate the concept of uniform convergence as a refinement of the idea of ordinary convergence, suppose that the integral

$$\int_a^\infty F(s, t)\, dt \tag{3}$$

converges for some particular value of s, say $s = s_0$. This means that

$$\lim_{b\to\infty} \int_a^b F(s_0, t)\, dt$$

exists. This, in turn, means that given any $\varepsilon > 0$, there exists a number B, depending on ε, such that

$$\left| \int_a^b F(s_0, t)\, dt - \int_a^\infty F(s_0, t)\, dt \right| < \varepsilon \qquad \text{for all } b > B$$

If we combine the two integrals, this becomes the equivalent statement

$$\left| \int_b^\infty F(s_0, t)\, dt \right| < \varepsilon \qquad \text{for all } b > B \tag{4}$$

If the integral (3) converges for other values of s, then for each of these values there must exist a corresponding value of B to serve in the counterpart of condition (4). In general, these Bs will all be different, and no one will suffice for all the s's in the set S under consideration. If it should happen, however, that there is a single B that can be used for *all* the s values in S,† then for these values the convergence is uniform, i.e., defined *uniformly*, or equally well, by the same B independent of the value of s.

The important property that the integral defining the Laplace transform converges uniformly is guaranteed by the following theorem.

Theorem 2 If $f(t)$ is piecewise continuous on $[0, \infty)$ and of exponential order with abscissa of convergence α_0, then for any number $s_0 > \alpha_0$

$$\mathscr{L}\{f(t)\} = \int_0^\infty f(t)e^{-st}\, dt$$

converges uniformly for all values of s such that $s \geq s_0$.

PROOF To prove this theorem, we must show that given any $\varepsilon > 0$, there exists a number B, depending on ε but not on s, such that

$$\left| \int_b^\infty f(t)e^{-st}\, dt \right| < \varepsilon \qquad \text{for all } b > B \text{ and all } s \geq s_0 > \alpha_0$$

Now

$$\left| \int_b^\infty f(t)e^{-st}\, dt \right| \leq \int_b^\infty |f(t)|e^{-st}\, dt$$

and we know that for $s > \alpha_0$ the integral on the right approaches zero as b becomes infinite, since this is implied by the fact that

$$\int_0^\infty |f(t)|e^{-st}\, dt$$

† This will be the case if and only if the set of Bs is bounded. In this case, the natural choice for the single B that will serve for each s in S is the least upper bound of the set of Bs.

is convergent for $s > \alpha_0$ (Theorem 1). In other words, given any $\varepsilon > 0$ and any $s_0 > \alpha_0$, there exists a number B such that

$$\int_b^\infty |f(t)|e^{-s_0 t}\,dt < \varepsilon \qquad \text{for all } b > B$$

Now if $s \geq s_0$, then for all $t \geq 0$, $e^{-st} \leq e^{-s_0 t}$. Hence

$$\left|\int_b^\infty f(t)e^{-st}\,dt\right| \leq \int_b^\infty |f(t)|e^{-st}\,dt \leq \int_b^\infty |f(t)|e^{-s_0 t}\,dt$$

and so for any $s \geq s_0$ the integral on the left is less than ε for all values of b greater than the particular B which suffices for the integral on the right. This value of B is clearly independent of s (since it arises from the *specific* value $s = s_0$), and so the proof of the theorem is complete.

In succeeding sections we shall find that many relatively complicated operations upon $f(t)$, such as differentiation and integration, for instance, can be replaced by simple algebraic operations such as multiplication or division by s upon the transform of $f(t)$. This is analogous to the way in which such operations as multiplication and division of numbers are replaced by the simpler processes of addition and subtraction when we work not with the numbers themselves but with their logarithms. Our primary purpose in this chapter is to develop rules of transformation and tables of transforms which can be used, like tables of logarithms, to facilitate the manipulation of functions and by means of which we can recover the proper function from its transform at the end of a problem.

Exercises for Section 6.2

1 Which of the following functions are of exponential order?
(*a*) t^n (*b*) $\tan t$ (*c*) e^{t^2} (*d*) $\cosh t$ (*e*) $1/t$ (*f*) t^2e^{3t}

2 Show by an example that it is possible for the abscissa of convergence of a function to be negative.

3 What is the abscissa of convergence α_0 of each of the following functions?
(*a*) $\cos kt$ (*b*) $\sin kt$ (*c*) t^2 (*d*) $\cosh kt$ (*e*) $\sinh kt$ (*f*) $\ln(1 + t)$

4 For which of the functions in Exercise 3 is α_0 a value of α which will serve in Definition 3?

***5** Show that each of the following integrals converges uniformly over the indicated set of s values:

(*a*) $\displaystyle\int_0^\infty \frac{\sin st}{1 + t^2}\,dt$ all real values of s

(*b*) $\displaystyle\int_1^\infty \frac{1}{s^4 + t^4}\,dt$ all real values of s

(*c*) $\displaystyle\int_0^\infty e^{-st^2}\,dt$ all values of $s > s_0 > 0$

(*d*) $\displaystyle\int_0^\infty \frac{\sin t}{1 + st}\,dt$ all values of $s > s_0 > 0$

Hint: Consider the alternating series which results when the integration is performed over the successive subintervals $(0, \pi)$, $(\pi, 2\pi)$, ...

***6** Prove that if $f(t)$ is piecewise continuous on $[0, \infty)$ and of exponential order, then $\int_0^t f(t)\,dt$ is also piecewise continuous on $[0, \infty)$ and of exponential order. Show further that if α_0 and α_1 are, respectively, the abscissas of convergence of $f(t)$ and $\int_0^t f(t)\,dt$ and if $\alpha_0 \geq 0$, then $\alpha_1 \leq \alpha_0$. Is it necessarily true that $\alpha_1 \leq \alpha_0$ if $\alpha_0 < 0$?

***7** In the proof of Theorem 1, why must α be bounded from α_0? That is, why cannot it simply be assumed that $\alpha > \alpha_0$? *Hint:* Consider the function $f(t) = t$ as a counterexample.

***8** Prove that a function $f(t)$ is of exponential order if and only if s can be chosen so that $\lim_{t\to\infty} e^{-st}f(t) = 0$. If $f(t)$ is of exponential order, show that its abscissa of convergence α_0 is the greatest lower bound of all values of s such that

$$\lim_{t\to\infty} e^{-st}f(t) = 0$$

6.3 THE GENERAL METHOD

The utility of the Laplace transformation is based primarily upon the following three theorems.

Theorem 1 $\mathscr{L}\{c_1 f_1(t) \pm c_2 f_2(t)\} = c_1 \mathscr{L}\{f_1(t)\} \pm c_2 \mathscr{L}\{f_2(t)\}$ provided that both $\mathscr{L}\{f_1(t)\}$ and $\mathscr{L}\{f_2(t)\}$ exist.

PROOF To prove this, we have by definition

$$\begin{aligned}\mathscr{L}\{c_1 f_1(t) \pm c_2 f_2(t)\} &= \int_0^\infty [c_1 f_1(t) \pm c_2 f_2(t)]e^{-st}\,dt \\ &= c_1 \int_0^\infty f_1(t)e^{-st}\,dt \pm c_2 \int_0^\infty f_2(t)e^{-st}\,dt \\ &= c_1 \mathscr{L}\{f_1(t)\} \pm c_2 \mathscr{L}\{f_2(t)\}\end{aligned}$$

as asserted. The extension of Theorem 1 to linear combinations of more than two functions is immediate.

The property ascribed to the Laplace transformation by Theorem 1 is, of course, the property of linearity. In other words, *the Laplace transformation is a linear operator*. (See item 12, Appendix B.2.)

Theorem 2 If $f(t)$ is a continuous function of exponential order on $[0, \infty)$ whose derivative is also of exponential order and at least piecewise continuous on $[0, \infty)$, and if $f(t)$ approaches the limit $f(0^+)$ as t approaches zero from the right, then the Laplace transform of $f'(t)$ is given by the formula

$$\mathscr{L}\{f'(t)\} = s\mathscr{L}\{f(t)\} - f(0^+)$$

provided s is greater than the abscissa of convergence of $f(t)$.

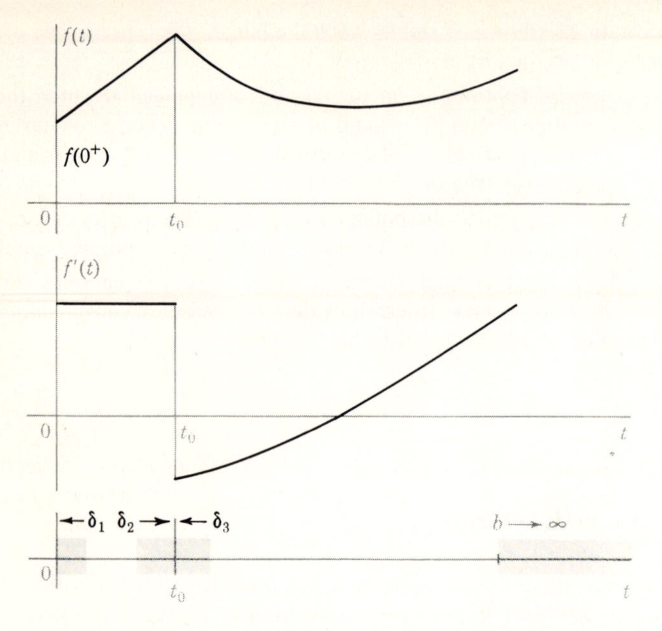

Figure 6.2 A continuous function whose derivative has a point of discontinuity.

Proof To prove this, let us suppose for definiteness that there is a single point, say $t = t_0$, where though $f(t)$ is continuous, its derivative has a finite jump, as suggested by Fig. 6.2. Then, by definition,

$$\mathscr{L}\{f'(t)\} = \int_0^\infty f'(t)e^{-st}\,dt$$

$$= \lim_{\substack{\delta_1,\,\delta_2,\,\delta_3 \to 0 \\ b \to \infty}} \left[\int_{\delta_1}^{t_0-\delta_2} f'(t)e^{-st}\,dt + \int_{t_0+\delta_3}^{b} f'(t)e^{-st}\,dt\right]$$

If we use integration by parts on these integrals, choosing

$$u = e^{-st} \qquad dv = f'(t)\,dt$$

$$du = -se^{-st}\,dt \qquad v = f(t)$$

we have

$$\mathscr{L}\{f'(t)\} = \lim_{\substack{\delta_1,\,\delta_2,\,\delta_3 \to 0 \\ b \to \infty}} \left[e^{-st}f(t)\Big|_{\delta_1}^{t_0-\delta_2} + s\int_{\delta_1}^{t_0-\delta_2} f(t)e^{-st}\,dt \right.$$

$$\left. + \, e^{-st}f(t)\Big|_{t_0+\delta_3}^{b} + s\int_{t_0+\delta_3}^{b} f(t)e^{-st}\,dt \right]$$

In the limit the two integrals which remain combine to give precisely

$$s\int_0^\infty f(t)e^{-st}\,dt = s\mathscr{L}\{f(t)\}$$

Similarly, the first evaluated portion yields

$$e^{-st_0}f(t_{0-}) - f(0^+)$$

and the second yields simply

$$0 - e^{-st_0}f(t_{0+})$$

because, since $f(t)$ is of exponential order, s can be chosen sufficiently large, i.e., greater than the abscissa of convergence of $f(t)$, for the contribution from the upper limit to be zero. Now $f(t)$ was assumed to be continuous. Hence at t_0 (as at all other points) its right- and left-hand limits must be equal. Therefore, the terms

$$e^{-st_0}f(t_{0-}) \qquad \text{and} \qquad -e^{-st_0}f(t_{0+})$$

cancel, leaving finally

$$\mathscr{L}\{f'(t)\} = s\mathscr{L}\{f(t)\} - f(0^+)$$

as asserted. The preceding proof is readily extended to functions whose derivatives have more than one finite jump. The extension of the theorem to the relatively unimportant case in which $f(t)$ itself is permitted to have finite jumps is indicated in Exercise 3.

Corollary 1 If both $f(t)$ and $f'(t)$ are continuous functions of exponential order on $[0, \infty)$ and if $f''(t)$ is also of exponential order and at least piecewise continuous on $[0, \infty)$, then

$$\mathscr{L}\{f''(t)\} = s^2\mathscr{L}\{f(t)\} - sf(0^+) - f'(0^+)$$

where $f(0^+)$ and $f'(0^+)$ are, respectively, the values that $f(t)$ and $f'(t)$ approach as t approaches zero from the right.

PROOF This result follows immediately by applying Theorem 2 twice to $f''(t)$:

$$\begin{aligned}\mathscr{L}\{f''(t)\} = \mathscr{L}\{[f'(t)]'\} &= s\mathscr{L}\{f'(t)\} - f'(0^+)\\ &= s[s\mathscr{L}\{f(t)\} - f(0^+)] - f'(0^+)\\ &= s^2\mathscr{L}\{f(t)\} - sf(0^+) - f'(0^+)\end{aligned}$$

as asserted. The extension of this result to derivatives of higher order is indicated in Exercise 1.

Since we are primarily concerned in this book with *differential* equations, it is natural that we should be interested in a result like Theorem 2 which expresses the Laplace transform of the derivative of a function in terms of the transform of the function itself. On the other hand, there is no obvious reason why we should be concerned about the Laplace transform of the integral of a function. However, if we recall our discussion of the series-electrical circuit in Sec. 4.6, we find a reason. Through the use of Kirchhoff's second law, we obtained initially

$$L\frac{di}{dt} + iR + \frac{Q}{C} = E$$

as the equation describing the behavior of the general series circuit. There were two dependent variables, i and Q, in this equation; and one had to be eliminated before we could proceed. To do this, we could use either of two relations between i and Q, namely,

$$i = \frac{dQ}{dt} \qquad \text{or} \qquad Q = \int i\,dt$$

We chose the first because it led to a *differential* equation that we knew how to solve, whereas the second led to the *integrodifferential* equation

$$L\frac{di}{dt} + iR + \frac{1}{C}\int i\,dt = E \tag{1}$$

which we did not know how to solve. Once formulas for the Laplace transforms of both derivatives and integrals are available, we will be able to attack Eq. (1) directly and find i immediately, without first solving for Q and then obtaining i by differentiation. Theorem 3 gives us the formula we need to solve equations like (1).

Theorem 3 If $f(t)$ is piecewise continuous on $[0, \infty)$ and of exponential order, then the Laplace transform of $\int_a^t f(t)\,dt$ is given by the formula

$$\mathscr{L}\left\{\int_a^t f(t)\,dt\right\} = \frac{1}{s}\mathscr{L}\{f(t)\} + \frac{1}{s}\int_a^0 f(t)\,dt$$

Proof To prove this theorem, we have by definition

$$\mathscr{L}\left\{\int_a^t f(t)\,dt\right\} = \int_0^\infty \left[\int_a^t f(x)\,dx\right] e^{-st}\,dt$$

where the dummy variable x has been introduced in the inner integral for convenience. If we integrate the last integral by parts, with

$$u = \int_a^t f(x)\,dx \qquad dv = e^{-st}\,dt$$

$$du = f(t)\,dt \qquad v = \frac{e^{-st}}{-s}$$

we have
$$\mathscr{L}\left\{\int_a^t f(t)\,dt\right\} = \left[\frac{e^{-st}}{-s}\int_a^t f(x)\,dx\right]_0^\infty + \frac{1}{s}\int_0^\infty f(t)e^{-st}\,dt$$

Since $f(t)$ is of exponential order, so too is its integral (Exercise 6, Sec. 6.2). Hence s can be chosen sufficiently large for the integrated portion to vanish at the upper limit, leaving

$$\mathscr{L}\left\{\int_a^t f(t)\,dt\right\} = \frac{1}{s}\int_a^0 f(x)\,dx + \frac{1}{s}\mathscr{L}\{f(t)\}$$

as asserted.

The extension of this result to repeated integrals of $f(t)$ is indicated in Exercise 2.

Although we need many more formulas before the Laplace transformation can be applied effectively to specific problems, Theorems 1 to 3 allow us to outline all the essential steps in the usual application of this method to the solution of differential and integrodifferential equations with constant coefficients. Suppose, for instance, that we are given the equation

$$ay'' + by' + cy = f(t) \qquad a, b, c \text{ constants}$$

If we take the Laplace transform of both sides, we have by Theorem 1 (assuming that all terms are transformable)

$$a\mathscr{L}\{y''\} + b\mathscr{L}\{y'\} + c\mathscr{L}\{y\} = \mathscr{L}\{f(t)\}$$

Now applying Theorem 2 and its corollary, we have

$$a(s^2\mathscr{L}\{y\} - sy_0 - y_0') + b(s\mathscr{L}\{y\} - y_0) + c\mathscr{L}\{y\} = \mathscr{L}\{f(t)\}$$

where y_0 and y_0' are the given initial values of y and y'. Collecting terms on $\mathscr{L}\{y\}$ and then solving for $\mathscr{L}\{y\}$, we obtain finally

$$\mathscr{L}\{y\} = \frac{\mathscr{L}\{f(t)\} + (as + b)y_0 + ay_0'}{as^2 + bs + c}$$

Now $f(t)$ is a given function of t; hence its Laplace transform (if it exists) is a perfectly definite function of s [although, except for Rule (1), Sec. 6.2, we have as yet no formulas for finding it]. Moreover, y_0 and y_0' are definite numbers known from the data of the problem. Hence the transform of y is a completely determined function of s. Thus if we had available a sufficiently extensive table of transforms, we could find in it the function $y(t)$ having the right-hand side of the last equation for its transform, *and this function would be the formal solution to our problem, initial conditions and all.*† The formal solution could then be substituted into the differential equation to verify that it was indeed the genuine solution.

This brief discussion illustrates the two great advantages of the Laplace transformation in solving linear constant-coefficient differential equations: it reduces the problem to one in algebra, and it takes care of initial conditions without the necessity of first constructing a complete solution and then specializing the arbitrary constants it contains. Clearly, our immediate task is to implement this process by establishing an adequate table of transforms.

† This, of course, assumes the "obvious" theorem that the function having a given function of s as its Laplace transform is unique or, in other words, that if two functions have the same transform, they are identical. This is strictly true if the functions are continuous. If discontinuities are permitted, the most we can say is that two functions with the same transform cannot differ over any interval of positive length, although they may differ at various isolated points. A detailed discussion of this result (Lerch's theorem) would take us too far afield, but we shall assume it and use it repeatedly throughout this chapter. (For a proof of Lerch's theorem, see D. V. Widder, "The Laplace Transform," Princeton University Press, Princeton, N.J., 1941, pp. 59–63.)

Exercises for Section 6.3

1 Show that under the appropriate conditions

$$\mathscr{L}\{f'''\} = s^3\mathscr{L}\{f\} - s^2 f_0 - sf'_0 - f''_0$$

What is $\mathscr{L}\{f^{(n)}\}$?

2 Show that

$$\mathscr{L}\left\{\int_a^t \int_a^t f(r)\,dr\,dt\right\} = \frac{1}{s^2}\mathscr{L}\{f\} + \frac{1}{s^2}\int_a^0 f(t)\,dt + \frac{1}{s}\int_a^0 \int_a^t f(r)\,dr\,dt$$

3 If $f(t)$ satisfies all the conditions of Theorem 2 except that it has an upward jump of magnitude J_0 at $t = t_0$, show that

$$\mathscr{L}\{f'(t)\} = s\mathscr{L}\{f\} - f_0 - J_0 e^{-st_0}$$

4 Is the proof of Theorem 3 valid if $f(t)$ has a finite jump at $t = t_0$?

5 Devise a proof of Theorem 3 based upon the use of Theorem 2.

6 Show that

$$\mathscr{L}\{f(at)\} = \frac{1}{a}\mathscr{L}\{f(t)\}\bigg|_{s \to s/a}$$

7 (*a*) Given $\mathscr{L}\{\cos t\} = s/(s^2 + 1)$, use the result of Exercise 6 to determine $\mathscr{L}\{\cos bt\}$.
(*b*) Given $\mathscr{L}\{\sin t\} = 1/(s^2 + 1)$, use the result of Exercise 6 to determine $\mathscr{L}\{\sin bt\}$.

8 (*a*) Given $\mathscr{L}\{\sin t\} = 1/(s^2 + 1)$, use Theorem 2 to obtain $\mathscr{L}\{\cos t\}$.
(*b*) Given $\mathscr{L}\{\cos t\} = s/(s^2 + 1)$, use Theorem 2 to obtain $\mathscr{L}\{\sin t\}$.
(*c*) Use Theorem 3 to obtain $\mathscr{L}\{\sin t\}$ from $\mathscr{L}\{\cos t\}$.
(*d*) Use Theorem 3 to obtain $\mathscr{L}\{\cos t\}$ from $\mathscr{L}\{\sin t\}$.

9 Given that $y = y_0$ when $t = 0$, find the Laplace transform of y if $a_0(dy/dt) + a_1 y + a_2 \int_0^t y\,dt = f(t)$.

10 Explain how the Laplace transformation can be used to solve a system of simultaneous, linear differential equations with constant coefficients. In particular, given that $y = y_0$ and $z = z_0$ when $t = 0$, obtain formulas for the Laplace transforms of y and z if

$$a_1\frac{dy}{dt} + b_1 y + c_1\frac{dz}{dt} + d_1 z = f_1(t)$$

$$a_2\frac{dy}{dt} + b_2 y + c_2\frac{dz}{dt} + d_2 z = f_2(t)$$

***11** Let $T\{f(t)\}$ be a general integral transform

$$T\{f(t)\} = \int_a^b f(t)K(s, t)\,dt$$

where $K(s, t)$ is the so-called **kernel** of the transformation. Obtain conditions on $K(s, t)$ so that $T\{f'\}$ and $T\{f''\}$ contain no terms involving the evaluation of f or any of its derivatives. Find at least one kernel satisfying these conditions.

***12** If $f(t)$ is continuous and $f'(t)$ is at least piecewise continuous on $[0, \infty)$, if both $f(t)$ and $f'(t)$ are of exponential order, and if $f(0^+) = 0$, show that as s becomes infinite, $\mathscr{L}\{f(t)\}$ tends to zero at least as rapidly as c/s^2 where c is a constant independent of s. Can this result be generalized?

6.4 THE TRANSFORMS OF SPECIAL FUNCTIONS

Among all the functions whose transforms we might now think of tabulating, the most important are the simple ones

$$e^{-at} \qquad \cos bt \qquad \sin bt \qquad t^n$$

and the **unit step function,**

$$u(t) = \begin{cases} 0 & t < 0 \\ 1 & t > 0 \end{cases}$$

shown in Fig. 6.3. Once we know the transforms of these functions, nearly all the formulas we shall need can be obtained through the use of a few additional theorems which we shall establish in the next section. The specific results are the following.

Formula 1 $\quad \mathscr{L}\{e^{-at}\} = \dfrac{1}{s+a}$

Formula 2 $\quad \mathscr{L}\{\cos bt\} = \dfrac{s}{s^2+b^2}$

Formula 3 $\quad \mathscr{L}\{\sin bt\} = \dfrac{b}{s^2+b^2}$

Formula 4 $\quad \mathscr{L}\{t^n\} = \begin{cases} \dfrac{\Gamma(n+1)}{s^{n+1}} & n > -1 \\ \dfrac{n!}{s^{n+1}} & n \text{ a positive integer} \end{cases}$

Formula 5 $\quad \mathscr{L}\{u(t)\} = \dfrac{1}{s}$

To prove Formula 1, we have simply

$$\mathscr{L}\{e^{-at}\} = \int_0^\infty e^{-at}e^{-st}\,dt = \left.\frac{e^{-(s+a)t}}{-(s+a)}\right|_0^\infty = \frac{1}{s+a} \qquad \text{if } s+a>0$$

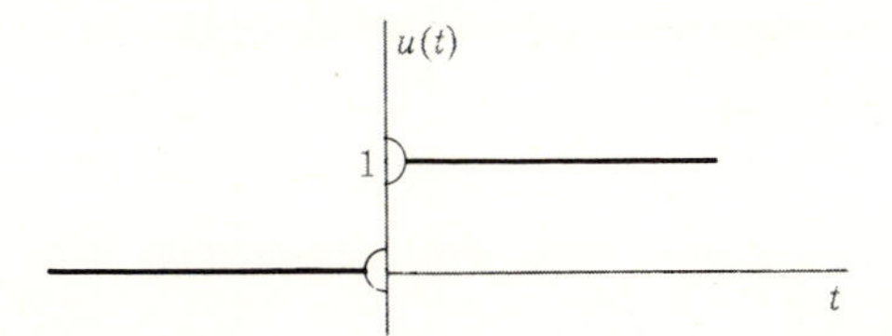

Figure 6.3 The unit step function $u(t)$.

To prove Formula 2, we have

$$\mathscr{L}\{\cos bt\} = \int_0^\infty \cos bt\, e^{-st}\, dt = \frac{e^{-st}}{s^2 + b^2}(-s \cos bt + b \sin bt)\bigg|_0^\infty$$

$$= \frac{s}{s^2 + b^2} \qquad \text{if } s > 0$$

To prove Formula 3, we have

$$\mathscr{L}\{\sin bt\} = \int_0^\infty \sin bt\, e^{-st}\, dt = \frac{e^{-st}}{s^2 + b^2}(-s \sin bt - b \cos bt)\bigg|_0^\infty$$

$$= \frac{b}{s^2 + b^2} \qquad \text{if } s > 0$$

Before we can prove Formula 4, it will be necessary for us to investigate briefly the so-called **gamma function** or **generalized factorial function,** defined by the equation

$$\Gamma(x) = \int_0^\infty e^{-t} t^{x-1}\, dt \tag{1}$$

This improper integral can be shown to be convergent for all $x > 0$.†

To determine the simple properties of the gamma function and its relation to the familiar factorial function

$$n! = n(n-1) \cdots 3 \cdot 2 \cdot 1$$

defined in elementary algebra for positive integral values of n, let us apply integration by parts to the definitive integral (1), taking

$$u = e^{-t} \qquad dv = t^{x-1}\, dt$$

$$du = -e^{-t}\, dt \qquad v = \frac{t^x}{x}$$

Then

$$\Gamma(x) = \frac{t^x e^{-t}}{x}\bigg|_0^\infty + \frac{1}{x}\int_0^\infty e^{-t} t^x\, dt$$

Under the restriction $x > 0$, the integrated portion vanishes at both limits. By comparison with (1), it is clear that the integral which remains is simply $\Gamma(x + 1)$. Thus we have established the important recurrence relation

$$\Gamma(x) = \frac{\Gamma(x+1)}{x} \qquad x > 0 \tag{2}$$

or

$$x\Gamma(x) = \Gamma(x+1) \tag{2a}$$

† See, for instance, P. Franklin, "A Treatise on Advanced Calculus," John Wiley & Sons, New York, 1947, p. 559.

Moreover, we have specifically

$$\Gamma(1) = \int_0^\infty e^{-t}\, dt = -e^{-t}\Big|_0^\infty = 1$$

Therefore, using (2a),

$$\Gamma(2) = 1 \cdot \Gamma(1) = 1$$

$$\Gamma(3) = 2 \cdot \Gamma(2) = 2 \cdot 1 = 2!$$

$$\Gamma(4) = 3 \cdot \Gamma(3) = 3 \cdot 2! = 3!$$

and in general

$$\Gamma(n + 1) = n! \qquad n = 1, 2, 3, \ldots \tag{3}$$

The connection between the gamma function and ordinary factorials is now clear. However, the gamma function constitutes an essential extension of the idea of a factorial, since its argument x is not restricted to positive integral values but can vary continuously.

From (2) and the fact that $\Gamma(1) = 1$, it is evident that $\Gamma(x)$ becomes infinite as x approaches zero. It is thus clear that $\Gamma(x)$ cannot be defined for $x = 0, -1, -2, \ldots$ in a way consistent with Eq. (2); hence we shall leave it undefined for these values of x. For all other values of x, however, $\Gamma(x)$ is well defined, the use of the recurrence formula (2a) effectively removing the restriction that x be positive, which the integral definition (1) requires. By methods which need not concern us here, tables of $\Gamma(x)$ have been constructed and can be found, usually as tables of $\log \Gamma(x)$, in most elementary handbooks. Because of the recurrence formula which the gamma function satisfies, these tables ordinarily cover only a unit interval on x, usually the interval $1 \le x \le 2$. A plot of $\Gamma(x)$ is shown in Fig. 6.4.

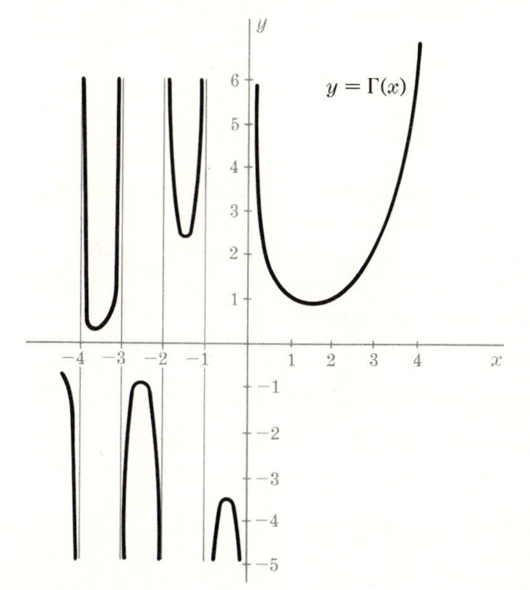

Figure 6.4 The function $y = \Gamma(x)$.

Example 1 What is the value of $I = \int_0^\infty \sqrt{z}\, e^{-z^3}\, dz$?

This integral is typical of many which can be reduced to the standard form of the gamma function by a suitable substitution. In this case a comparison of the given integral with (1) suggests that we should let

$$z = t^{1/3} \qquad dz = \tfrac{1}{3}t^{-2/3}\, dt$$

getting
$$I = \int_0^\infty \sqrt{t^{1/3}}\, e^{-t}\left(\frac{1}{3}t^{-2/3}\, dt\right) = \frac{1}{3}\int_0^\infty e^{-t}t^{1/2-1}\, dt = \frac{1}{3}\Gamma\left(\frac{1}{2}\right)$$

Since $\Gamma(\frac{1}{2})$ cannot be found in the usual table, which lists $\Gamma(x)$ only for $1 \le x \le 2$, it is necessary to use the recurrence relation (2) to bring the argument of the gamma function into this interval:

$$I = \frac{1}{3}\Gamma\left(\frac{1}{2}\right) = \frac{1}{3}\frac{\Gamma(\frac{3}{2})}{\frac{1}{2}} \doteq \frac{2}{3}(0.86623) \doteq 0.59082\dagger$$

Returning now to Formula 4, we have

$$\mathscr{L}\{t^n\} = \int_0^\infty t^n e^{-st}\, dt$$

In an attempt to reduce this integral to the standard form of the gamma function, let us make the substitution

$$t = \frac{z}{s} \qquad dt = \frac{dz}{s}$$

Then
$$\mathscr{L}\{t^n\} = \int_0^\infty \left(\frac{z}{s}\right)^n e^{-z}\frac{dz}{s} = \frac{1}{s^{n+1}}\int_0^\infty e^{-z}z^n\, dz = \frac{\Gamma(n+1)}{s^{n+1}}$$

Since $\Gamma(n+1) = n!$ when n is a positive integer, this establishes the second part of Formula 4 also.

It is interesting to note that if n is negative,

$$s\mathscr{L}\{t^n\} = \frac{\Gamma(n+1)}{s^n}$$

is not bounded as $s \to \infty$. Hence, according to Corollary 4, Theorem 1, Sec. 6.2, this function of s is not the Laplace transform of a piecewise continuous function of exponential order. This, of course, is not surprising since when n is negative, t^n, though of exponential order (with abscissa of convergence $\alpha_0 = 0$), is not bounded in the neighborhood of the origin and so is not piecewise continuous. It can be shown, however, that the improper integral defining $\mathscr{L}\{t^n\}$ exists for $n > -1$ although it does not exist for $n \le -1$. The first version of Formula 4 must therefore be qualified by the restriction $n > -1$.

Formula 5 can be obtained immediately by taking $n = 0$ in Formula 4.

† Actually, the value of $\Gamma(\frac{1}{2})$ is known exactly and in fact is equal to $\sqrt{\pi}$ (Exercise 14). Hence, in this example $I = \sqrt{\pi}/3$.

Example 2 What is the Laplace transform of $\sinh bt$?

Since $\sinh bt = (e^{bt} - e^{-bt})/2$, we have

$$\mathscr{L}\{\sinh bt\} = \mathscr{L}\left\{\frac{e^{bt} - e^{-bt}}{2}\right\} = \frac{1}{2}\left(\frac{1}{s-b} - \frac{1}{s+b}\right) = \frac{b}{s^2 - b^2}$$

The analogy with Formula 3 for the transform of $\sin bt$ is apparent.

Example 3 If

$$\mathscr{L}\{y\} = \frac{s+1}{s^2 + s - 6}$$

what is $y(t)$?

None of our formulas yields a transform resembling this one. However, using the method of partial fractions, we can write

$$\frac{s+1}{s^2+s-6} = \frac{s+1}{(s-2)(s+3)} = \frac{A}{s-2} + \frac{B}{s+3} = \frac{A(s+3) + B(s-2)}{(s-2)(s+3)}$$

For this to be an identity, we must have

$$s + 1 = A(s+3) + B(s-2)$$

Setting $s = 2$ and $s = -3$ in turn, we find from this that $A = \frac{3}{5}$, $B = \frac{2}{5}$. Hence

$$\mathscr{L}\{y(t)\} = \frac{1}{5}\left(\frac{3}{s-2} + \frac{2}{s+3}\right)$$

Formula 1 can now be applied (in reverse) to the individual terms, and we find

$$y(t) = \tfrac{1}{5}(3e^{2t} + 2e^{-3t})$$

Example 4 Find the particular solution of the differential equation $y'' - 3y' + 2y = 12e^{-2t}$ for which $y = 2$ and $y' = 6$ when $t = 0$.

Taking the Laplace transformation of each side of the given equation, using Theorem 2 and its corollary from Sec. 6.3 and Formula 1 of this section, we have

$$\mathscr{L}\{y''\} - 3\mathscr{L}\{y'\} + 2\mathscr{L}\{y\} = \mathscr{L}\{12e^{-2t}\}$$

$$(s^2\mathscr{L}\{y\} - 2s - 6) - 3(s\mathscr{L}\{y\} - 2) + 2\mathscr{L}\{y\} = \frac{12}{s+2}$$

$$(s^2 - 3s + 2)\mathscr{L}\{y\} = 2s + \frac{12}{s+2} = \frac{2s^2 + 4s + 12}{s+2}$$

and finally

$$\mathscr{L}\{y\} = \frac{2s^2 + 4s + 12}{(s-1)(s-2)(s+2)}$$

By an easy application of the method of partial fractions, we can now express $\mathscr{L}\{y\}$ in the form

$$\mathscr{L}\{y\} = -\frac{6}{s-1} + \frac{7}{s-2} + \frac{1}{s+2}$$

Hence, applying Formula 1 in reverse, we have

$$y = -6e^t + 7e^{2t} + e^{-2t}$$

It is, of course, an easy matter to verify that y satisfies the given initial value problem.

Example 5 Solve for $y(t)$ from the simultaneous equations

$$y' + 2y + 6\int_0^t z\,dt = -2u(t)$$

$$y' + z' + z = 0$$

if $y_0 = -5$ and $z_0 = 6$.

We begin by taking the Laplace transformation of each equation, term by term, using Theorems 2 and 3, Sec. 6.3 and Formulas 1 and 5:

$$(s\mathscr{L}\{y\} + 5) + 2\mathscr{L}\{y\} + \frac{6}{s}\mathscr{L}\{z\} = -\frac{2}{s}$$

$$(s\mathscr{L}\{y\} + 5) + (s\mathscr{L}\{z\} - 6) + \mathscr{L}\{z\} = 0$$

Obvious simplifications then lead to the following pair of linear algebraic equations in the transforms of the unknown functions $y(t)$ and $z(t)$:

$$(s^2 + 2s)\mathscr{L}\{y\} + 6\mathscr{L}\{z\} = -2 - 5s$$

$$s\mathscr{L}\{y\} + (s+1)\mathscr{L}\{z\} = 1$$

Since it is $y(t)$ that we are asked to find, we solve these simultaneous equations for $\mathscr{L}\{y\}$, getting

$$\mathscr{L}\{y\} = \frac{\begin{vmatrix} -2-5s & 6 \\ 1 & s+1 \end{vmatrix}}{\begin{vmatrix} s^2+2s & 6 \\ s & s+1 \end{vmatrix}} = \frac{-5s^2 - 7s - 8}{s^3 + 3s^2 - 4s}$$

Applying the method of partial fractions to this expression, we have

$$\mathscr{L}\{y\} = \frac{-5s^2 - 7s - 8}{s^3 + 3s^2 - 4s} = \frac{2}{s} - \frac{4}{s-1} - \frac{3}{s+4}$$

Finally, determining the inverse of each of these terms, we find

$$y(t) = 2u(t) - 4e^t - 3e^{-4t}$$

Exercises for Section 6.4

1 Plot each of the following functions:
(*a*) $u(t-2)$ (*b*) $u(t^2)$ (*c*) $u(t^2-1)$ (*d*) $u(t^3 - 6t^2 + 11t - 6)$

2 Plot each of the following functions:
(*a*) $u(t-2) - u(t-1)$ (*b*) $u(t) + u(t-1) + u(t-2) + u(t-3) + \cdots$
(*c*) $u(\sin t)$ (*d*) $2u(\sin \pi t) - 1$

3 What is $\mathscr{L}\{\cosh kt\}$?

4 What is $\mathscr{L}\{\cos(at+b)\}$? *Hint:* First express $\cos(at+b)$ as the difference of two terms.

5 What is $\mathscr{L}\{\cos^2 bt\}$? *Hint:* First express $\cos^2 bt$ as a function of $2bt$.

6 What is $\mathscr{L}\{(t+1)^2\}$?

7 Find the inverse Laplace transform of each of the following functions:

$$(a)\ \frac{1}{s+3} \qquad (b)\ \frac{1}{s^4} \qquad (c)\ \frac{1}{s^2+9} \qquad (d)\ \frac{2s+3}{s^2+9} \qquad (e)\ \frac{s+3}{(s+1)(s-3)}$$

8 Find the solution of each of the following differential equations which satisfies the given conditions:

(*a*) $y'' + 4y' - 5y = 0$ $\qquad y_0 = 1,\ y_0' = 0$
(*b*) $y'' - 4y = 0$ $\qquad y_0 = -1,\ y_0' = 1$
(*c*) $4y'' + y = 0$ $\qquad y_0 = 2,\ y_0' = 1$
(*d*) $y'' + 4y = u(t)$ $\qquad y_0 = y_0' = 0$
(*e*) $y'' + 3y' + 2y = e^t$ $\qquad y_0 = 1,\ y_0' = 0$
(*f*) $y''' + 6y'' + 11y' + 6y = 0$ $\qquad y_0 = 2,\ y_0' = 1,\ y_0'' = -1$
(*g*) $y''' - y'' + 4y' - 4y = t$ $\qquad y_0 = y_0' = 0,\ y_0'' = 1$

9 Solve for $z(t)$ in Example 5.

10 Find the solution of the following system of equations:

$$y' + y + 2z' + 3z = e^{-t}$$

$$3y' - y + 4z' + z = 0 \qquad y_0 = -1,\ z_0 = 0$$

11 Find the solution of the following system of equations:

$$(D+1)y + (2D+3)z = 0$$

$$(D-4)y + (3D-8)z = \sin t \qquad y_0 = 2,\ z_0 = -1$$

12 Evaluate each of the following integrals:

$$(a)\ \int_0^\infty \frac{e^{-x}}{\sqrt{x}}\,dx \qquad (b)\ \int_0^\infty \exp(-\sqrt{x})\,dx \qquad (c)\ \int_0^\infty (x+1)^2 e^{-x^3}\,dx$$

***13** Evaluate each of the following integrals:

(*a*) $\displaystyle\int_0^\infty \frac{x^c}{c^x}\,dx$ $\qquad$ *Hint:* Recall that $c^x = \exp(x \ln c)$.

(*b*) $\displaystyle\int_0^1 \frac{dx}{\sqrt{\ln(1/x)}}$ $\qquad$ *Hint:* Let $\ln \dfrac{1}{x} = z$.

(*c*) $\displaystyle\int_0^1 x^m \left(\ln \frac{1}{x}\right)^n dx$

***14** Show that $\Gamma(\frac{1}{2}) = \sqrt{\pi}$. *Hint:* First show that

$$\Gamma(\tfrac{1}{2}) = 2\int_0^\infty e^{-x^2}\,dx = 2\int_0^\infty e^{-y^2}\,dy$$

Then multiply these integrals and evaluate the resulting double integral by changing to polar coordinates.

***15** A particle of mass m moves along the x axis under the influence of a force which varies inversely as the distance from the origin. If the particle begins to move from rest at the point $x = a$, find the time it takes to reach the origin. *Hint:* After the equation of motion is set up, recall the discussion that preceded Example 2, Sec. 1.7.

6.5 FURTHER GENERAL THEOREMS

We are now in a position to derive a number of theorems that will be of considerable use in the application of the Laplace transformation to practical problems. We begin with a result which allows us to infer the behavior of a function $f(t)$ for small positive values of t from the behavior of $\mathscr{L}\{f(t)\}$ for large positive values of s.

Theorem 1 If $f(t)$ is continuous on $(0, \infty)$, if $\lim_{t\to 0^+} f(t)$ exists, if $f'(t)$ is at least piecewise continuous on $[0, \infty)$ and if both $f(t)$ and $f'(t)$ are of exponential order, then

$$\lim_{s\to\infty} s\mathscr{L}\{f(t)\} = \lim_{t\to 0^+} f(t) = f(0^+)$$

PROOF From Theorem 2, Sec. 6.3, we have

$$\mathscr{L}\{f'(t)\} = s\mathscr{L}\{f(t)\} - f(0^+)$$

Hence, taking the limit of each side, we have

$$\lim_{s\to\infty} \mathscr{L}\{f'(t)\} = \lim_{s\to\infty} s\mathscr{L}\{f(t)\} - f(0^+) \tag{1}$$

By hypothesis, $f'(t)$ is piecewise continuous on $[0, \infty)$ and of exponential order. Hence by Corollary 3, Theorem 1, Sec. 6.2, its Laplace transform must approach zero as s becomes infinite. Therefore, from (1),

$$\lim_{s\to\infty} s\mathscr{L}\{f(t)\} = f(0^+)$$

as asserted.

Just as the symbol f^{-1} is commonly used to denote the inverse of a function f, so it is customary to use the symbol $\mathscr{L}^{-1}$ to denote the operation which is the inverse of the Laplace transformation. In this notation, $\mathscr{L}^{-1}\{\phi(s)\}$ then denotes the **inverse Laplace transform** of a given function $\phi(s)$, that is, the function of t (if it exists) which has $\phi(s)$ as its Laplace transform. The next two theorems contain results which are frequently useful in determining inverse Laplace transforms.

When the Laplace transform of an unknown function $f(t)$ contains the factor s,† it is often convenient to find $f(t)$ by means of the following theorem.

Theorem 2 If $f(t)$ is piecewise continuous on $[0, \infty)$ and of exponential order, if $\mathscr{L}\{f(t)\} = s\phi(s)$, and if the inverse of the factor $\phi(s)$ is continuous for $t > 0$ and has a piecewise continuous derivative of exponential order on $[0, \infty)$, then

$$f(t) = \frac{d}{dt}\mathscr{L}^{-1}\{\phi(s)\}$$

† This can always be arranged, of course, by multiplying and dividing the transform by s; that is, $\Phi(s) = s[\Phi(s)/s] = s\phi(s)$.

PROOF To prove this theorem, let $F(t) \equiv \mathscr{L}^{-1}\{\phi(s)\}$ be the function which has the factor $\phi(s)$ as its Laplace transform. From the hypotheses on $F(t)$, it follows from Theorem 2, Sec. 6.3, that

$$\mathscr{L}\{F'(t)\} = s\mathscr{L}\{F(t)\} - F(0^+) = s\phi(s) - F(0^+) \tag{2}$$

However, by Theorem 1,

$$F(0^+) = \lim_{s\to\infty} s\mathscr{L}\{F(t)\} = \lim_{s\to\infty} s\phi(s) = \lim_{s\to\infty} \mathscr{L}\{f(t)\} = 0$$

the last step following from Corollary 3, Theorem 1, Sec. 6.2, since $f(t)$, though unknown, is assumed to be piecewise continuous and of exponential order. Hence, from (2),

$$\mathscr{L}\{f(t)\} = \mathscr{L}\{F'(t)\}$$

since each is equal to $s\phi(s)$. Therefore†

$$f(t) = \frac{dF(t)}{dt} = \frac{d}{dt}\mathscr{L}^{-1}\{\phi(s)\}$$

as asserted.

As a working rule, Theorem 2 can be restated in the following way:

If a Laplace transform contains the factor s, the inverse of that transform can be found by suppressing the factor s, determining the inverse of the remaining portion of the transform, and finally differentiating that inverse with respect to t.

Example 1 What is $\mathscr{L}^{-1}\{s/(s^2+4)\}$?

By Formula 2, Sec. 6.4, we see immediately that the required inverse is $f(t) = \cos 2t$. However, it is interesting that we can also obtain this result by suppressing the factor s, finding the inverse $F(t)$ of the remaining portion of the transform, namely,

$$\frac{1}{s^2+4}$$

and then differentiating this inverse according to Theorem 2:

$$f(t) = \frac{d}{dt}\mathscr{L}^{-1}\left\{\frac{1}{s^2+4}\right\} = \frac{d}{dt}\left(\frac{\sin 2t}{2}\right) = \cos 2t$$

as before. The usual applications of this theorem are, of course, not of this trivial character.

When the Laplace transform of an unknown function $f(t)$ contains the factor $1/s$,‡ it is often convenient to find $f(t)$ by means of the following theorem.

† See footnote, p. 195.

‡ This can always be arranged, of course, by multiplying and dividing the transform by s; that is, $\Phi(s) \equiv s\Phi(s)/s = \phi(s)/s$.

Theorem 3 If $f(t)$ is piecewise continuous on $[0, \infty)$ and of exponential order, if $\mathscr{L}\{f(t)\} = \phi(s)/s$, and if $\phi(s)$ possesses an inverse, then

$$f(t) = \int_0^t \mathscr{L}^{-1}\{\phi(s)\}\, dt$$

PROOF To prove this theorem, let $F(t) \equiv \mathscr{L}^{-1}\{\phi(s)\}$ be the function which has the factor $\phi(s)$ for its transform. Then by Theorem 3, Sec. 6.3,

$$\mathscr{L}\left\{\int_0^t F(t)\, dt\right\} = \frac{1}{s}\mathscr{L}\{F(t)\} + \frac{1}{s}\int_0^0 F(t)\, dt = \frac{1}{s}\mathscr{L}\{F(t)\} = \frac{\phi(s)}{s}$$

Thus both $f(t)$ and $\int_0^t F(t)\, dt \equiv \int_0^t \mathscr{L}^{-1}\{\phi(s)\}\, dt$ have $\phi(s)/s$ for their Laplace transform and so must be equal, as asserted.

As a working rule, Theorem 3 can be restated in the following way:

If a Laplace transform contains the factor $1/s$, the inverse of that transform can be found by suppressing the factor $1/s$, determining the inverse of the remaining portion of the transform, and finally integrating that inverse with respect to t from 0 to t.

Example 2 What is

$$\mathscr{L}^{-1}\left\{\frac{1}{s(s^2+4)}\right\}$$

Here, using the last theorem, we first suppress the factor $1/s$, getting

$$\phi(s) = \frac{1}{s^2+4}$$

By Formula 3, Sec. 6.4, the inverse of this is $F(t) = \frac{1}{2}\sin 2t$. Finally, we obtain $f(t)$ by integrating $F(t)$ from 0 to t:

$$f(t) = \int_0^t \frac{\sin 2t}{2}\, dt = -\frac{\cos 2t}{4}\bigg|_0^t = \frac{1-\cos 2t}{4}$$

One of the most useful properties of the Laplace transformation is contained in the so-called **first shifting theorem.**

Theorem 4 If $f(t)$ is piecewise continuous on $[0, \infty)$ and of exponential order, then $\mathscr{L}\{e^{-at}f(t)\} = \mathscr{L}\{f(t)\}_{s \to s+a}$.

PROOF By definition,

$$\mathscr{L}\{e^{-at}f(t)\} = \int_0^\infty e^{-at}f(t)e^{-st}\, dt = \int_0^\infty f(t)e^{-(s+a)t}\, dt$$

and the last integral is in structure exactly the Laplace transform of $f(t)$ itself, except that $s + a$ takes the place of s.

In words, Theorem 4 says that *the transform of e^{-at} times a function of t is equal to the transform of the function itself, with s replaced by $s + a$.* Conversely, as a tool for finding inverses, this theorem states that if we reverse the substitution $s \to s + a$, that is, if we replace $s + a$ by s or s by $s - a$ in the transform of a function, then the inverse of the modified transform $\phi(s - a)$ must be multiplied by e^{-at} to obtain the inverse of the original transform $\phi(s)$. This procedure is summarized in the following result.

Corollary 1 $\mathscr{L}^{-1}\{\phi(s)\} = e^{-at}\mathscr{L}^{-1}\{\phi(s - a)\}$.

By means of Theorem 4 we can easily establish the following important formulas.

Formula 1 $$\mathscr{L}\{e^{-at} \cos bt\} = \frac{s + a}{(s + a)^2 + b^2}$$

Formula 2 $$\mathscr{L}\{e^{-at} \sin bt\} = \frac{b}{(s + a)^2 + b^2}$$

Formula 3 $$\mathscr{L}\{e^{-at}t^n\} = \begin{cases} \dfrac{\Gamma(n + 1)}{(s + a)^{n+1}} & n > -1 \\[2ex] \dfrac{n!}{(s + a)^{n+1}} & n \text{ a positive integer} \end{cases}$$

Example 3 If

$$\mathscr{L}\{y\} = \frac{2s + 5}{s^2 + 4s + 13}$$

what is y?

A comparison of $\mathscr{L}\{y\}$ with Formulas 1 and 2 suggests that we write

$$\mathscr{L}\{y\} = \frac{2(s + 2) + 1}{(s + 2)^2 + 3^2} = 2\frac{s + 2}{(s + 2)^2 + 3^2} + \frac{1}{3}\frac{3}{(s + 2)^2 + 3^2}$$

Hence, by Formulas 1 and 2,

$$y = 2e^{-2t} \cos 3t + \tfrac{1}{3}e^{-2t} \sin 3t$$

Example 4 What is the solution of the differential equation

$$y'' + 2y' + y = te^{-t}$$

for which $y_0 = 1$ and $y'_0 = -2$?

Transforming both sides of the given equation, we have

$$(s^2\mathscr{L}\{y\} - s + 2) + 2(s\mathscr{L}\{y\} - 1) + \mathscr{L}\{y\} = \frac{1}{(s + 1)^2}$$

$$(s^2 + 2s + 1)\mathscr{L}\{y\} = \frac{1}{(s + 1)^2} + s$$

$$\mathscr{L}\{y\} = \frac{1}{(s + 1)^4} + \frac{s}{(s + 1)^2}$$

By Formula 3, the inverse of the first fraction in $\mathscr{L}\{y\}$ is

$$\frac{t^3e^{-t}}{3!}$$

To find the inverse of the second fraction, we can write it in the form

$$\frac{s+1-1}{(s+1)^2} = \frac{1}{s+1} - \frac{1}{(s+1)^2}$$

and take the inverse of each term; or we can suppress the factor s, take the inverse of what remains, and differentiate this result, according to Theorem 2. By either method we obtain immediately $e^{-t} - te^{-t}$. Hence

$$y = \frac{t^3e^{-t}}{3!} + e^{-t} - te^{-t}$$

In this example the characteristic equation of the differential equation has repeated roots, and moreover the term on the right is a part of the complementary function; yet neither of these features requires any special treatment in the operational solution of the problem. This is another of the many advantages of the Laplace transform method of solving linear differential equations with constant coefficients.

Example 5 Find the particular solution of the system

$$(2D+1)x + (D+3)y = 0$$

$$(D-6)x + (D-1)y = 0$$

for which $x = -1$ and $y = 2$ when $t = 0$.

Taking the Laplace transformation of each equation and then collecting terms, we have

$$\begin{aligned} 2(s\mathscr{L}\{x\}+1) + \mathscr{L}\{x\} + (s\mathscr{L}\{y\}-2) + 3\mathscr{L}\{y\} &= 0 \\ (s\mathscr{L}\{x\}+1) - 6\mathscr{L}\{x\} + (s\mathscr{L}\{y\}-2) - \mathscr{L}\{y\} &= 0 \\ (2s+1)\mathscr{L}\{x\} + (s+3)\mathscr{L}\{y\} &= 0 \\ (s-6)\mathscr{L}\{x\} + (s-1)\mathscr{L}\{y\} &= 1 \end{aligned} \tag{3}$$

When these equations are solved for $\mathscr{L}\{x\}$, we obtain

$$\mathscr{L}\{x\} = \frac{\begin{vmatrix} 0 & s+3 \\ 1 & s-1 \end{vmatrix}}{\begin{vmatrix} 2s+1 & s+3 \\ s-6 & s-1 \end{vmatrix}} = -\frac{s+3}{s^2+2s+17} = -\frac{s+1}{(s+1)^2+4^2} - \frac{1}{2}\frac{4}{(s+1)^2+4^2}$$

Hence, using Formulas 1 and 2, with $a = 1$ and $b = 4$, we have

$$x = -e^{-t}\cos 4t - \tfrac{1}{2}e^{-t}\sin 4t = -\tfrac{1}{2}e^{-t}(2\cos 4t + \sin 4t)$$

Similarly, solving Eqs. (3) for $\mathscr{L}\{y\}$, we obtain

$$\mathscr{L}\{y\} = \frac{\begin{vmatrix} 2s+1 & 0 \\ s-6 & 1 \end{vmatrix}}{\begin{vmatrix} 2s+1 & s+3 \\ s-6 & s-1 \end{vmatrix}} = \frac{2s+1}{s^2+2s+17} = \frac{2(s+1)}{(s+1)^2+4^2} - \frac{1}{4}\frac{4}{(s+1)^2+4^2}$$

and $$y = 2e^{-t}\cos 4t - \tfrac{1}{4}e^{-t}\sin 4t = \tfrac{1}{4}e^{-t}(8\cos 4t - \sin 4t)$$

This example is another illustration of how the use of the Laplace transformation eliminates the need to obtain a complete solution before a required particular solution can be found.

We have already made repeated use of Theorems 2 and 3, Sec. 6.3, on the transforms of derivatives and integrals. On the other hand, it is sometimes convenient or necessary to consider the derivatives and integrals of transforms. The basis for this is contained in the next two theorems.

Theorem 5 If $f(t)$ is piecewise continuous on $[0, \infty)$ and of exponential order, and if $\mathscr{L}\{f(t)\} = \phi(s)$, then $\mathscr{L}\{tf(t)\} = -\phi'(s)$.

PROOF By definition, we have

$$\mathscr{L}\{f(t)\} = \int_0^\infty f(t)e^{-st}\,dt = \phi(s)$$

and differentiating this with respect to s, we obtain

$$\frac{d}{ds}\int_0^\infty f(t)e^{-st}\,dt = \phi'(s) \tag{4}$$

What we would like to do here is perform the indicated differentiation *inside* the integral by differentiating the integrand partially with respect to s, getting

$$\int_0^\infty -tf(t)e^{-st}\,dt = \phi'(s) \tag{5}$$

However, first integrating with respect to t and then differentiating with respect to s, as indicated in (4), is not always equivalent to first differentiating with respect to s and then integrating with respect to t, as in (5). To justify this interchange of the order of these operations, we note first that the integral in (5) is just the Laplace transform of $-tf(t)$. Moreover, since $f(t)$ is assumed to be piecewise continuous and of exponential order, so too is the product $-tf(t)$. Therefore, by Theorem 2, Sec. 6.2, the integral in (5) converges *uniformly* and thus by item 5, Appendix B.2, is equal to the derivative of the integral in (4). Hence Eq. (5) is correct, and we conclude from it that

$$\mathscr{L}\{tf(t)\} = -\phi'(s)$$

as asserted.

By taking inverses in the assertion of Theorem 5 and then solving for $f(t)$, we obtain the following useful result.

Corollary 1 If $\mathscr{L}\{f(t)\} = \phi(s)$, then

$$f(t) \equiv \mathscr{L}^{-1}\{\phi(s)\} = -\frac{1}{t}\mathscr{L}^{-1}\{\phi'(s)\}$$

Corollary 1 is often helpful when the inverse of a transform cannot conveniently be found but the inverse of the derivative of the transform is known. The extension of Theorem 5 and its corollary to repeated differentiation of transforms is immediate, as our next example illustrates.

Example 6 What is $\mathscr{L}\{t^2 \sin 2t\}$?

By a repeated application of Theorem 5, we have

$$\mathscr{L}\{t^2 \sin 2t\} = (-1)^2 \frac{d^2 \mathscr{L}\{\sin 2t\}}{ds^2} = \frac{d^2}{ds^2}\left(\frac{2}{s^2+4}\right) = \frac{12s^2 - 16}{(s^2+4)^3}$$

Example 7 What is y if $\mathscr{L}\{y\} = \ln [(s+1)/(s-1)]$?

Using Corollary 1 of Theorem 5, we have immediately

$$y = -\frac{1}{t}\mathscr{L}^{-1}\left\{\frac{d}{ds}\left(\ln \frac{s+1}{s-1}\right)\right\} = -\frac{1}{t}\mathscr{L}^{-1}\left\{\frac{1}{s+1} - \frac{1}{s-1}\right\}$$

$$= \frac{e^{-t} - e^t}{-t} = \frac{2 \sinh t}{t}$$

Theorem 6 If $f(t)$ is piecewise continuous on $[0, \infty)$ and of exponential order, if $\mathscr{L}\{f(t)\} = \phi(s)$, and if $f(t)/t$ has a limit as t approaches zero from the right, then

$$\mathscr{L}\left\{\frac{f(t)}{t}\right\} = \int_s^\infty \phi(s)\, ds$$

Proof By definition,

$$\phi(s) = \mathscr{L}\{f(t)\} = \int_0^\infty f(t)e^{-st}\, dt$$

Hence, integrating from s to ∞, we obtain (introducing the dummy variable σ)

$$\int_s^\infty \phi(\sigma)\, d\sigma = \int_s^\infty \left[\int_0^\infty f(t)e^{-\sigma t}\, dt\right] d\sigma$$

Now, by Theorem 2, Sec. 6.2, $\int_0^\infty f(t)e^{-st}\, dt$ converges uniformly. Hence, by item 9, Appendix B.2, the order of integration in the repeated integral can be reversed, giving

$$\int_s^\infty \phi(s) = \int_0^\infty \int_s^\infty f(t)e^{-\sigma t}\, d\sigma\, dt = \int_0^\infty f(t)\left[\frac{e^{-\sigma t}}{-t}\right]_s^\infty dt$$

$$= \int_0^\infty \frac{f(t)}{t} e^{-st}\, dt$$

Since our hypotheses ensure that $f(t)/t$ is piecewise continuous and of exponential order on $[0, \infty)$, it follows that the last integral converges. Hence

$$\int_s^\infty \phi(s)\, ds = \mathscr{L}\left\{\frac{f(t)}{t}\right\}$$

as asserted.

By taking inverses in the assertion of Theorem 6 and then solving for $f(t)$, we obtain the following useful result.

Corollary 1 If $\mathscr{L}\{f(t)\} = \phi(s)$, then

$$f(t) \equiv \mathscr{L}^{-1}\{\phi(s)\} = t\mathscr{L}^{-1}\left\{\int_s^\infty \phi(s)\, ds\right\}$$

Corollary 1 is often useful in finding inverses when the integral of a transform is simpler to work with than the transform itself. The extension of Theorem 6 and its corollary to repeated integration of transforms is immediate.

Example 8 What is $\mathscr{L}\{(\sin kt)/t\}$?

By Theorem 6, we have

$$\mathscr{L}\left\{\frac{\sin kt}{t}\right\} = \int_s^\infty \mathscr{L}\{\sin kt\}\, ds = \int_s^\infty \frac{k}{s^2 + k^2}\, ds = \operatorname{Tan}^{-1}\frac{s}{k}\bigg|_s^\infty$$

$$= \frac{\pi}{2} - \operatorname{Tan}^{-1}\frac{s}{k} = \operatorname{Cot}^{-1}\frac{s}{k}$$

Example 9 What is y if $\mathscr{L}\{y\} = s/(s^2 - 1)^2$?

Using Corollary 1, Theorem 6, we have immediately

$$y = t\mathscr{L}^{-1}\left\{\int_s^\infty \frac{s}{(s^2 - 1)^2}\, ds\right\} = t\mathscr{L}^{-1}\left\{\frac{-1}{2(s^2 - 1)}\bigg|_s^\infty\right\}$$

$$= t\mathscr{L}^{-1}\left\{\frac{1}{2(s^2 - 1)}\right\}$$

$$= t\mathscr{L}^{-1}\left\{\frac{1}{4}\left(\frac{1}{s - 1} - \frac{1}{s + 1}\right)\right\}$$

$$= \frac{t}{4}(e^t - e^{-t}) = \frac{t \sinh t}{2}$$

Example 10 Find the solution of the differential equation $y'' + 4y = \cos 2t$ for which $y = 1$ and $y' = -1$ when $t = 0$.

Taking the Laplace transformation of each side of the given equation, we have

$$(s^2\mathscr{L}\{y\} - s + 1) + 4\mathscr{L}\{y\} = \frac{s}{s^2 + 4}$$

and, solving for $\mathscr{L}\{y\}$,

$$\mathscr{L}\{y\} = \frac{s - 1}{s^2 + 4} + \frac{s}{(s^2 + 4)^2} = \frac{s}{s^2 + 4} - \frac{1}{2}\frac{2}{s^2 + 4} + \frac{s}{(s^2 + 4)^2}$$

By inspection, the inverses of the first two terms on the right-hand side are

$$\cos 2t - \tfrac{1}{2}\sin 2t$$

To find the inverse of the third term, it is convenient to integrate this term from s to ∞, then find the inverse of the integrated result, and finally multiply this inverse by t, according to Corollary 1, Theorem 6. Doing this, we have first

$$\int_s^\infty \frac{s}{(s^2 + 4)^2}\, ds = -\frac{1}{2}\frac{1}{s^2 + 4}\bigg|_s^\infty = \frac{1}{2}\frac{1}{s^2 + 4} = \frac{1}{4}\frac{2}{s^2 + 4}$$

The inverse of this integrated result is

$$\tfrac{1}{4} \sin 2t$$

Hence, multiplying this by t and adding it to the sum of the inverses of the first two terms in $\mathscr{L}\{y\}$, we have the final answer:

$$y = \cos 2t - \frac{1}{2} \sin 2t + \frac{t}{4} \sin 2t$$

Again we note that even though the term on the right-hand side of the given equation duplicated a term in the complementary function, this required no special treatment when the equation was solved by means of the Laplace transformation.

Exercises for Section 6.5

Find the Laplace transform of each of the following functions.

1 te^{2t}

2 $t \cos 2t$

3 $t^2 \cos 3t$

4 $te^{-3t} \sin 2t$

5 $t \int_0^t e^{-3t} \sin 2t \, dt$

6 $e^{-3t} \int_0^t t \sin 2t \, dt$

7 $\int_0^t te^{-3t} \sin 2t \, dt$

8 $\dfrac{e^{2t} - 1}{t}$

9 $\dfrac{1 - \cos 3t}{t}$

10 $\dfrac{e^{-3t} \sin 2t}{t}$

Find the inverse of each of the following transforms.

11 $\dfrac{1}{(s + 2)^4}$

12 $\dfrac{s}{(s + 2)^4}$

13 $\dfrac{s + 1}{s^2 + 4s + 4}$

14 $\dfrac{1}{s(s^2 + 4)}$

15 $\dfrac{1}{s(s + 2)^2}$

16 $\dfrac{1}{s^2(s + 1)}$

17 $\dfrac{1}{(s + 1)(s^2 + 2s + 5)}$

18 $\ln \dfrac{s + a}{s + b}$

19 $\ln \dfrac{s^2 - 1}{s^2}$

20 $\ln \dfrac{s^2 + 1}{s(s + 1)}$

21 $\dfrac{2}{(s^2 + 4)^2}$ *Hint:* Multiply and divide the transform by s.

22 $\dfrac{s + 2}{(s^2 + 4s + 3)^2}$

Find the solution of each of the following differential equations which satisfies the indicated conditions.

23 $y'' + 4y' + 5y = 0 \qquad y_0 = 1,\ y'_0 = -2$

24 $y'' + 4y' + 5y = e^{-2t} \qquad y_0 = y'_0 = 0$

25 $y'' + 2y' + y = te^{-t} \qquad y_0 = 0,\ y'_0 = 1$

26 $y'' + 4y' + 3y = e^{-t} \qquad y_0 = y'_0 = 1$

27 $y''' + y'' - y' - y = e^{-t} \qquad y_0 = y'_0 = 0,\ y''_0 = 2$

28 $y^{\text{iv}} + 2y'' + y = 0 \qquad y_0 = y'_0 = y'''_0 = 0,\ y''_0 = 1$

29 $2x' + 5x + 3y' + y = 0$
$x' + 4x + 3y' + 2y = 0 \qquad x_0 = 1,\ y_0 = 2$

30 $6x' + 8x - y' - 2y = e^{2t}$
$7x' + x - y' - y = 0 \qquad x_0 = y_0 = 0$

31 Find the value of $f(0^+)$ if $\mathscr{L}\{f(t)\}$ is

(a) $\dfrac{s^2+1}{s^3+6s^2+11s+6}$ (b) $\dfrac{s+3}{2s^3-3s^2-2s}$

(c) $\dfrac{s^2+s+1}{s^3-s^2+2}$ (d) $\dfrac{s+2}{s^3+3s^2+4s+2}$

***32** Show that under appropriate conditions

$$\lim_{s\to\infty} s[s\mathscr{L}\{f(t)\} - f(0^+)] = f'(0^+)$$

and that

$$\lim_{s\to\infty} s[s^2\mathscr{L}\{f(t)\} - sf(0^+) - f'(0^+)] = f''(0^+)$$

What conditions beyond those of Theorem 1 are necessary for the validity of these results? Can the value of $f^{(n)}(0^+)$ be obtained by an extension of these formulas?

***33** What is $\int_0^\infty (e^{-t} - e^{-2t})/t\,dt$? *Hint:* Use item 3, Appendix B.2, to justify the observation that

$$\int_0^\infty \frac{e^{-t}-e^{-2t}}{t}\,dt = \int_0^\infty \left(\frac{e^{-t}-e^{-2t}}{t}\lim_{s\to 0} e^{-st}\right)dt = \lim_{s\to 0}\mathscr{L}\left\{\frac{e^{-t}-e^{-2t}}{t}\right\}$$

***34** What is

$$\int_0^\infty e^{-t}\frac{1-\cos t}{t}\,dt$$

***35** What is

$$\int_0^\infty \frac{e^{-at}\cos bt - e^{-pt}\cos qt}{t}\,dt$$

***36** Use the Laplace transform to solve the variable-coefficient, linear differential equation $ty'' + 2(t-1)y' + (t-2)y = 0$. *Hint:* Use Theorem 5 in transforming the equation. Then note that the result is a linear, first-order differential equation in $\mathscr{L}\{y\}$ which can be solved by the methods of Sec. 1.6.

***37** Using the hint of Exercise 36, solve the equation

$$ty'' + 2(2t-1)y' + 4(t-1)y = 0$$

***38** Solve the equation $ty'' - 2y' + ty = 0$.

***39** Find the solution of the equation $ty'' + 2(2t - 1)y' - 4y = 0$ for which $y_0 = 0$ and $y'_0 = 8$.

***40** Discuss the following problem as a possible application of the Laplace transformation. The behavior of a certain system is governed by a linear, second-order differential equation with constant, though unknown, coefficients. The response of the system to a specific test disturbance, a unit step function, say, can be recorded. Is it possible, using numerical integration, to calculate the Laplace transform of such a response for several values of s and thus obtain a set of linear algebraic equations from which the coefficients in the differential equation can be obtained?

6.6 THE TRANSFORMS OF DELAYED EXCITATIONS

In some problems, a system which becomes active at $t = 0$ because of some initial disturbance is subsequently acted upon by another disturbance beginning at a later time, say $t = a$. The analytical representation of such functions and the nature of their Laplace transforms are, therefore, matters of some importance. To illustrate, suppose that we wish an expression describing the function whose graph is shown in Fig. 6.5*a*, the curve being congruent to the right half of the parabola $y = t^2$ shown in Fig. 6.5*b*. It is not enough to recall the translation formula from analytic geometry and write $f(t) = (t - a)^2$, because this equation, even with the usual qualification that $f(t) \equiv 0$ for $t < 0$, defines the curve shown in Fig. 6.5*c* and not the required graph. However, if we take the unit step function and translate it a units to the right by writing $u(t - a)$, we obtain the function shown in Fig. 6.5*d*. Since this vanishes for $t < a$ and is equal to 1 for $t > a$, the product $(t - a)^2 u(t - a)$ will be identically zero for $t < a$ and will be identically equal to $(t - a)^2$ for $t > a$ and hence will define precisely the arc we want. More generally, the expression

$$f(t - a)u(t - a)$$

represents the function obtained by translating $f(t)$ a units to the right and cutting it off, i.e., making it vanish identically to the left of a.

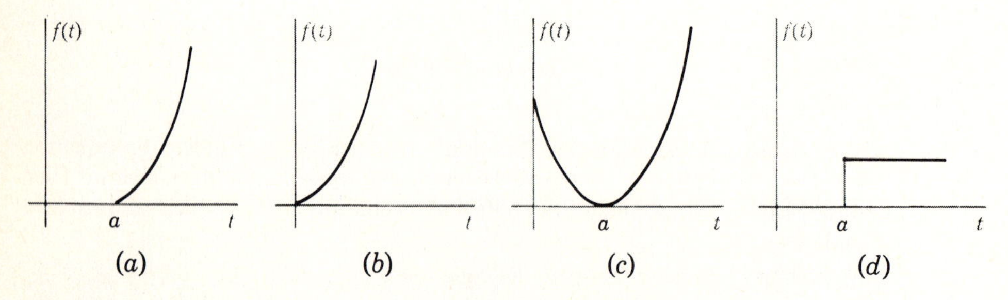

Figure 6.5 Plot describing the graph of a function which has been translated and cut off.

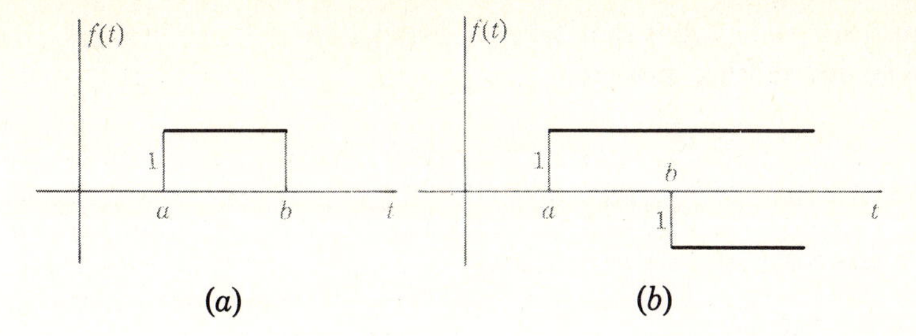

Figure 6.6 The construction of a rectangular pulse, or filter function, from two step functions.

Example 1 What is the equation of the function whose graph is shown in Fig. 6.6?

Clearly we can regard this function as the sum of the two translated step functions shown in Fig. 6.6(*b*). Hence its equation is

$$u(t-a)-u(t-b)$$

Although the function shown in Fig. 6.6(*a*) is not ordinarily given a name, it could appropriately be referred to as a **filter function;** for when any other function is multiplied by this filter function, it is annihilated completely, i.e., reduced identically to zero, outside the "passband" $a<t<b$, and reproduced without any change whatsoever for values of t within the passband.

Example 2 What is the equation of the function whose graph is shown in Fig. 6.7?

To obtain the segment of this function between 1 and 2, we must multiply the expression $2(t-1)$ by a factor which will be zero to the left of 1, unity between 1 and 2, and zero to the right of 2. By Example 1, such a function is $u(t-1)-u(t-2)$. Hence

$$2(t-1)[u(t-1)-u(t-2)]$$

defines the segment of the given function between 1 and 2 and vanishes elsewhere. Similarly

$$(-t+4)[u(t-2)-u(t-4)]$$

defines the segment of the given function between 2 and 4 and vanishes elsewhere. The complete representation of the function is therefore

$$2(t-1)[u(t-1)-u(t-2)]+(-t+4)[u(t-2)-u(t-4)]$$
$$=2(t-1)u(t-1)-3(t-2)u(t-2)+(t-4)u(t-4)$$

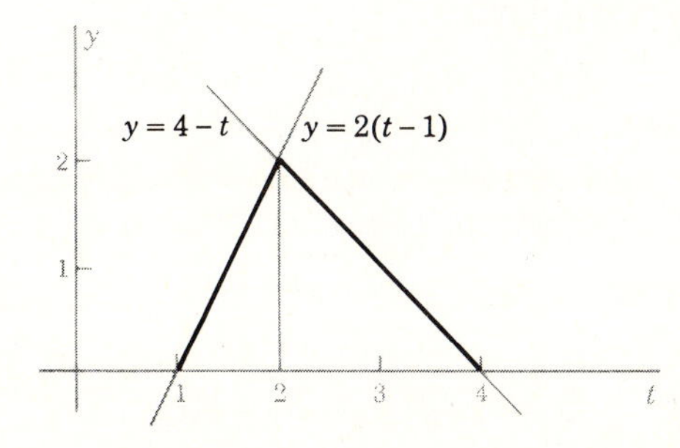

Figure 6.7 A graph consisting of straight-line segments.

The transforms of functions that have been translated and cut off are given by the so-called **second shifting theorem.**

Theorem 1 $\mathscr{L}\{f(t-a)u(t-a)\} = e^{-as}\mathscr{L}\{f(t)\}$ for $a \geq 0$.

PROOF To prove this, we have by definition

$$\mathscr{L}\{f(t-a)u(t-a)\} = \int_0^\infty f(t-a)u(t-a)e^{-st}\,dt = \int_a^\infty f(t-a)e^{-st}\,dt$$

the last step following since the integration effectively starts not at $t = 0$ but at $t = a$ because $f(t-a)u(t-a)$ vanishes identically to the left of $t = a$. Now let $t - a = T$ and $dt = dT$. Then the last integral becomes

$$\int_0^\infty f(T)e^{-s(T+a)}\,dT = e^{-as}\int_0^\infty f(T)e^{-sT}\,dT = e^{-as}\mathscr{L}\{f(t)\}$$

as asserted.

Before Theorem 1 can be applied, the function being transformed must be expressed in terms of the binomial argument which appears in the unit step function. This will not often be the case initially, so it will frequently be necessary to alter the form of the function, as originally given, before it can be conveniently transformed. In many cases, this can be done by inspection. On the other hand, we can always proceed in the following systematic way. Suppose we wish to transform

$$f(t)u(t-a)$$

As it stands, this cannot be handled by Theorem 1; so we rewrite it in the form

$$f[(t-a)+a]u(t-a) \equiv F(t-a)u(t-a)$$

where, by definition, $F(t-a) = f[(t-a)+a] = f(t)$, or

$$F(t) = f(t+a)$$

Theorem 1 can now be applied, and we have

$$\mathscr{L}\{f(t)u(t-a)\} = \mathscr{L}\{F(t-a)u(t-a)\} = e^{-as}\mathscr{L}\{F(t)\} = e^{-as}\mathscr{L}\{f(t+a)\}$$

Thus we have established the following useful result.

Corollary 1 $\mathscr{L}\{f(t)u(t-a)\} = e^{-as}\mathscr{L}\{f(t+a)\}$ for $a > 0$.

As a tool for finding inverses, it is convenient to restate Theorem 1 in the following form.

Corollary 2 If $\mathscr{L}^{-1}\{\phi(s)\} = f(t)$, then $\mathscr{L}^{-1}\{e^{-as}\phi(s)\} = f(t-a)u(t-a)$.

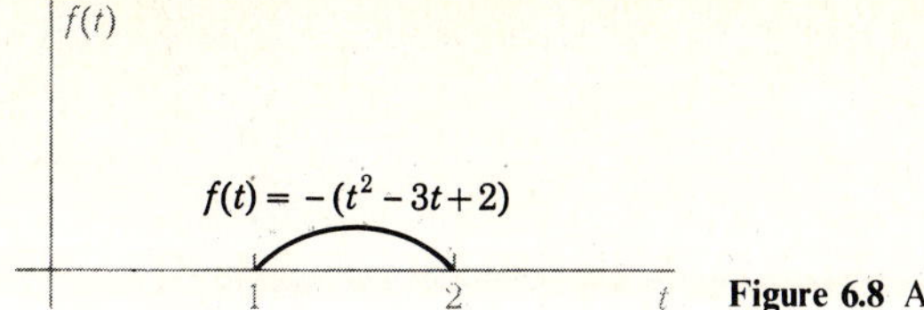

Figure 6.8 A parabolic pulse.

In words, Corollary 2 states that *suppressing the factor* e^{-as} *in a transform requires that the inverse of what remains be translated a units to the right and cut off to the left of the point* $t = a$.

Example 3 What is the transform of the function whose graph is shown in Fig. 6.8?

The equation of this function is obviously

$$\begin{aligned} g(t) &= -(t^2 - 3t + 2)[u(t-1) - u(t-2)] \\ &= -f(t)u(t-1) + f(t)u(t-2) \end{aligned}$$

where $f(t) = t^2 - 3t + 2$. However, the form of $f(t)$ is such that Theorem 1 cannot be applied directly to either term in the expression for $g(t)$. Hence we use Corollary 1, observing that

$$f(t+1) = (t+1)^2 - 3(t+1) + 2 = t^2 - t$$

and
$$f(t+2) = (t+2)^2 - 3(t+2) + 2 = t^2 + t$$

The required transform is therefore

$$-e^{-s}\mathscr{L}\{t^2 - t\} + e^{-2s}\mathscr{L}\{t^2 + t\} = -e^{-s}\left(\frac{2}{s^3} - \frac{1}{s^2}\right) + e^{-2s}\left(\frac{2}{s^3} + \frac{1}{s^2}\right)$$

Example 4 Find the solution of the equation $y' + 3y + 2\int_0^t y\,dt = f(t)$ for which $y_0 = 1$ if $f(t)$ is the function whose graph is shown in Fig. 6.9.

In this case, $f(t) = 2u(t-1) - 2u(t-2)$, and thus the differential equation can be written

$$y' + 3y + 2\int_0^t y\,dt = 2u(t-1) - 2u(t-2)$$

Taking transforms, we have

$$(s\mathscr{L}\{y\} - 1) + 3\mathscr{L}\{y\} + \frac{2}{s}\mathscr{L}\{y\} = \frac{2e^{-s}}{s} - \frac{2e^{-2s}}{s}$$

or
$$(s^2 + 3s + 2)\mathscr{L}\{y\} = 2e^{-s} - 2e^{-2s} + s$$

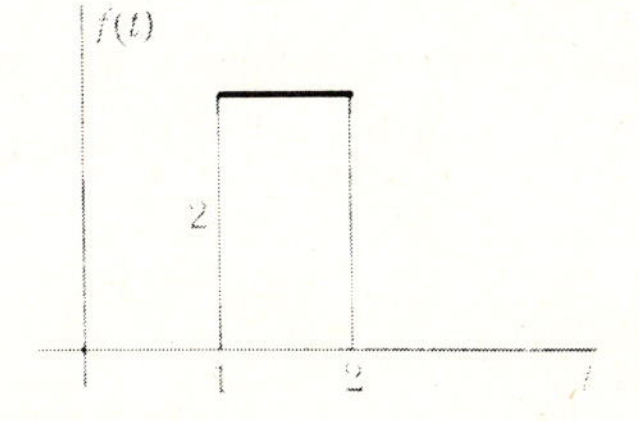

Figure 6.9 A rectangular pulse.

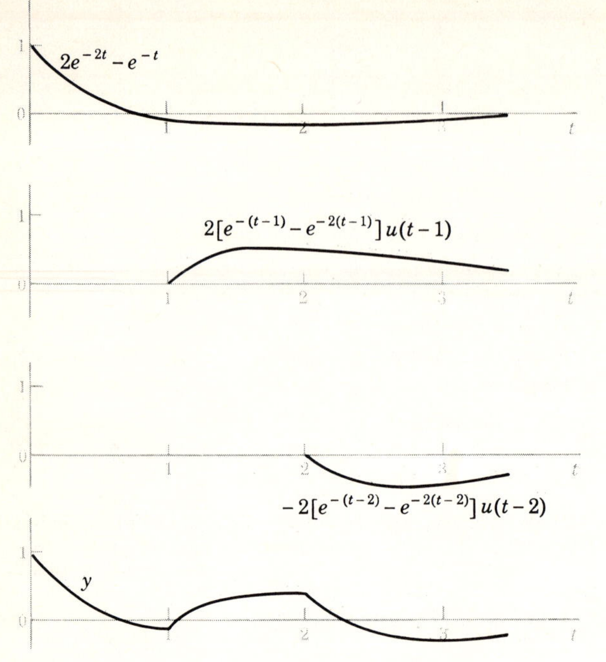

Figure 6.10 The solution of Example 4.

and
$$\mathscr{L}\{y\} = \frac{s}{(s+1)(s+2)} + \frac{2e^{-s}}{(s+1)(s+2)} - \frac{2e^{-2s}}{(s+1)(s+2)}$$

The first term in $\mathscr{L}\{y\}$ can be written

$$\frac{2}{s+2} - \frac{1}{s+1}$$

Hence its inverse is $2e^{-2t} - e^{-t}$. If the exponential factors are suppressed in the second and third terms in $\mathscr{L}\{y\}$, the algebraic portion which remains can be written

$$2\left(\frac{1}{s+1} - \frac{1}{s+2}\right)$$

and the inverse of this is $2e^{-t} - 2e^{-2t}$. However, because the factors e^{-s} and e^{-2s} were neglected, it is necessary to take the last expression, translate it 1 unit to the right and cut it off to the left of $t = 1$, and also translate it 2 units to the right and cut it off to the left of $t = 2$ in order to obtain the inverses of the original terms. This gives for y

$$y = (2e^{-2t} - e^{-t}) + 2(e^{-(t-1)} - e^{-2(t-1)})u(t-1) - 2(e^{-(t-2)} - e^{-2(t-2)})u(t-2)$$

Plots of these three terms and of their sum, that is, y itself, are shown in Fig. 6.10.

Exercises for Section 6.6

Find the Laplace transform of each of the following functions.

1 $u(t-a)$ **2** $\cos(t-1)u(t-1)$

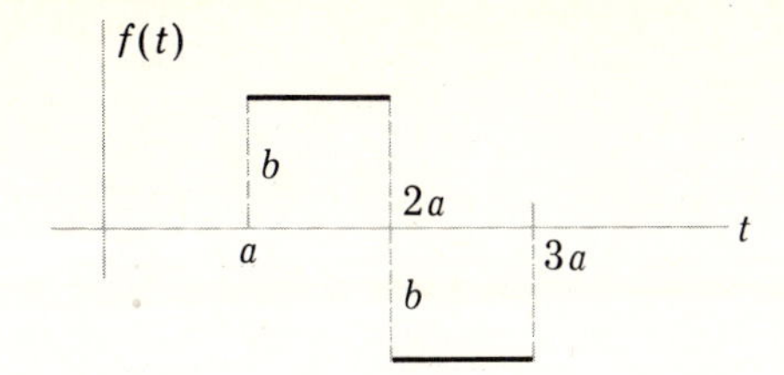

Figure 6.11

3 $u(1 - e^{-t})$

4 $u(t^3 - 1)$

5 $t^2u(t - 2)$

6 $(t^2 - 1)u(t - 1)$

7 $\cos t\, u(t - 1)$

8 $e^{2t}u(t - 2)$

9 $f(t) = \begin{cases} \sin t & 0 < t < \pi \\ 0 & \pi < t \end{cases}$

10 $f(t) = \begin{cases} t & 0 < t < 2 \\ 2 & 2 < t \end{cases}$

11 The function graphed in Fig. 6.11.

Find the solution of each of the following equations which satisfies the indicated conditions.

12 $y'' + 3y' + 2y = u(t - 1)$ $\quad y_0 = 0,\ y'_0 = 1$

13 $y'' + 2y' + y = e^{-t} - e^{-(t-1)}u(t - 1)$ $\quad y_0 = y'_0 = 0$

14 $y'' + 2y' + y = e^{-t} - e^{-t}u(t - 1)$ $\quad y_0 = y'_0 = 0$

15 $y' + 2y + 2\int_0^t y\, dt = u(t - 2)$ $\quad y_0 = 1$

16 $x' + 5x + y' + 3y = u(t) - u(t - 1)$
$x' + 2x + y' + y = 0$ $\quad x_0 = 1,\ y_0 = -2$

6.7 THE HEAVISIDE EXPANSION THEOREMS

The frequent use we have had to make of partial fractions indicates clearly the importance of this technique in operational calculus. It is therefore highly desirable to have the procedure systematized as much as possible. The following theorems, usually associated with the name of Heaviside, are very useful in this connection.

Theorem 1 If $f(t) = \mathscr{L}^{-1}\{p(s)/q(s)\}$, where $p(s)$ and $q(s)$ are polynomials and the degree of $q(s)$ is greater than the degree of $p(s)$, then the term in $f(t)$ corresponding to an unrepeated linear factor $s - a$ of $q(s)$ is equally well

$$\frac{p(a)}{q'(a)}e^{at} \quad \text{or} \quad \frac{p(a)}{Q(a)}e^{at}$$

where $Q(s)$ is the product of all the factors of $q(s)$ except $s - a$.

PROOF In the familiar partial-fraction decomposition of $p(s)/q(s)$, an unrepeated linear factor $s - a$ of $q(s)$ gives rise to a single fraction of the form $A/(s - a)$. Hence, if we denote by $h(s)$ the sum of all the fractions corresponding to the other factors of $q(s)$, we can write

$$\frac{p(s)}{q(s)} = \frac{A}{s-a} + h(s)$$

where, since $s - a$ is an *unrepeated* factor of $q(s)$, the term $h(s)$ does not contain $s - a$ as a factor of its denominator and hence remains finite as s approaches a. Multiplying this identity by $s - a$ then gives

$$\frac{(s-a)p(s)}{q(s)} = \frac{p(s)}{q(s)/(s-a)} = A + (s-a)h(s)$$

If we now let s approach a, the second term on the right vanishes, and we have

$$A = \lim_{s \to a} \frac{p(s)}{q(s)/(s-a)}$$

The limit of the numerator here is evidently $p(a)$. The denominator appears as an indeterminate of the form 0/0. However, if we evaluate it as usual according to L'Hospital's rule, by differentiating numerator and denominator with respect to s and then letting s approach a, we obtain just $q'(a)$. Hence

$$A = \frac{p(a)}{q'(a)}$$

On the other hand, we could have eliminated the indeterminacy before passing to the limit simply by canceling $s - a$ into $q(s)$, which by hypothesis contains this factor. Doing this, we obtain the equivalent form of A:

$$A = \frac{p(a)}{Q(a)}$$

Finally, taking inverses, we see that the fraction $A/(s - a)$ gives rise to the term

$$Ae^{at} = \frac{p(a)}{q'(a)} e^{at} = \frac{p(a)}{Q(a)} e^{at}$$

in the inverse $f(t)$, as asserted.

If $q(s)$ contains only unrepeated linear factors, then by applying Theorem 1 to each factor in turn, we obtain the following useful result.

Corollary 1 If $f(t) = \mathscr{L}^{-1}\{p(s)/q(s)\}$, where the degree of $q(s)$ is greater than the degree of $p(s)$, and if $q(s)$ is completely factorable into unrepeated, real linear factors

$$(s - a_1), (s - a_2), \ldots, (s - a_n)$$

then
$$f(t) = \sum_{i=1}^{n} \frac{p(a_i)}{q'(a_i)} e^{a_i t} = \sum_{i=1}^{n} \frac{p(a_i)}{Q_i(a_i)} e^{a_i t}$$

where $Q_i(s)$ is the product of all the factors of $q(s)$ except the factor $s - a_i$.

Theorem 2 If $f(t) = \mathscr{L}^{-1}\{p(s)/q(s)\}$, where $p(s)$ and $q(s)$ are polynomials and the degree of $q(s)$ is greater than the degree of $p(s)$, then the terms in $f(t)$ corresponding to a repeated linear factor $(s-a)^r$ in $q(s)$ are

$$\left[\frac{\phi^{(r-1)}(a)}{(r-1)!} + \frac{\phi^{(r-2)}(a)}{(r-2)!}\frac{t}{1!} + \cdots + \frac{\phi'(a)}{1!}\frac{t^{r-2}}{(r-2)!} + \phi(a)\frac{t^{r-1}}{(r-1)!}\right]e^{at}$$

where $\phi(s)$ is the quotient of $p(s)$ and all the factors of $q(s)$ except $(s-a)^r$.

Proof From the familiar theory of partial fractions, we recall that a repeated linear factor $(s-a)^r$ of $q(s)$ gives rise to the component fractions

$$\frac{A_1}{s-a} + \frac{A_2}{(s-a)^2} + \cdots + \frac{A_{r-1}}{(s-a)^{r-1}} + \frac{A_r}{(s-a)^r}$$

If, as before, we let $h(s)$ denote the sum of the fractions corresponding to all the other factors of $q(s)$, we have

$$\frac{p(s)}{q(s)} \equiv \frac{\phi(s)}{(s-a)^r} = \frac{A_1}{s-a} + \frac{A_2}{(s-a)^2} + \cdots + \frac{A_{r-1}}{(s-a)^{r-1}} + \frac{A_r}{(s-a)^r} + h(s)$$

Multiplying this identity by $(s-a)^r$ gives

$$\phi(s) = A_1(s-a)^{r-1} + A_2(s-a)^{r-2} + \cdots + A_{r-1}(s-a) + A_r + (s-a)^r h(s)$$

If we put $s = a$ in this identity, we obtain

$$\phi(a) = A_r$$

If we now differentiate $\phi(s)$, we have

$$\phi'(s) = A_1(r-1)(s-a)^{r-2} + A_2(r-2)(s-a)^{r-3} + \cdots$$
$$+ A_{r-1} + [r(s-a)^{r-1}h(s) + (s-a)^r h'(s)]$$

Again setting $s = a$, we find this time

$$\phi'(a) = A_{r-1}$$

Continuing this process of differentiation and evaluation, and noting that the first $r-1$ derivatives of the product $(s-a)^r h(s)$ will all vanish when $s = a$, we obtain successively

$$\phi''(a) = 2!\,A_{r-2}$$

$$\phi'''(a) = 3!\,A_{r-3}$$

$$\cdots\cdots\cdots\cdots$$

$$\phi^{(r-1)}(a) = (r-1)!\,A_1$$

or

$$A_{r-k} = \frac{\phi^{(k)}(a)}{k!} \qquad k = 0, 1, 2, \ldots, r-1$$

The terms in the expansion of $p(s)/q(s)$ which correspond to the factor $(s-a)^r$ are therefore

$$\frac{\phi^{(r-1)}(a)}{(r-1)!}\frac{1}{s-a} + \frac{\phi^{(r-2)}(a)}{(r-2)!}\frac{1}{(s-a)^2} + \cdots + \frac{\phi'(a)}{1!}\frac{1}{(s-a)^{r-1}} + \phi(a)\frac{1}{(s-a)^r}$$

When we recall that

$$\mathscr{L}^{-1}\left\{\frac{1}{(s-a)^n}\right\} = \frac{t^{n-1}e^{at}}{(n-1)!}$$

it is evident that the terms in $f(t)$ which arise from these fractions are

$$\frac{\phi^{(r-1)}(a)}{(r-1)!}e^{at} + \frac{\phi^{(r-2)}(a)}{(r-2)!}\frac{te^{at}}{1!} + \cdots + \frac{\phi'(a)}{1!}\frac{t^{r-2}e^{at}}{(r-2)!} + \phi(a)\frac{t^{r-1}e^{at}}{(r-1)!}$$

Finally, if we factor out e^{at} from this expression, we have precisely the assertion of the theorem.

Theorem 3 If $f(t) = \mathscr{L}^{-1}\{p(s)/q(s)\}$, where $p(s)$ and $q(s)$ are polynomials and the degree of $q(s)$ is greater than the degree of $p(s)$, then the term in $f(t)$ which corresponds to an unrepeated, irreducible† quadratic factor $(s+a)^2 + b^2$ of $q(s)$ is

$$\frac{e^{-at}}{b}(\phi_i \cos bt + \phi_r \sin bt)$$

where ϕ_r and ϕ_i are, respectively, the real and the imaginary parts of $\phi(-a+ib)$ and where $\phi(s)$ is the quotient of $p(s)$ and all the factors of $q(s)$ except the factor $(s+a)^2 + b^2$.

PROOF From the familiar theory of partial fractions, we recall that an unrepeated, irreducible quadratic factor $(s+a)^2 + b^2$ of $q(s)$ gives rise to a single fraction of the form

$$\frac{As + B}{(s+a)^2 + b^2}$$

in the partial-fraction expansion of $p(s)/q(s)$. If again we let $h(s)$ denote the sum of the fractions corresponding to all the other factors of $q(s)$, we can write

$$\frac{p(s)}{q(s)} \equiv \frac{\phi(s)}{(s+a)^2 + b^2} = \frac{As + B}{(s+a)^2 + b^2} + h(s)$$

Multiplying this identity by $(s+a)^2 + b^2$, we obtain

$$\phi(s) = As + B + [(s+a)^2 + b^2]h(s)$$

Now put $s = -a + ib$. This value, of course, makes $(s+a)^2 + b^2$ vanish; hence the last product drops out, leaving

$$\phi(-a+ib) = (-a+ib)A + B$$

or, reducing $\phi(-a+ib)$ to its standard complex form $\phi_r + i\phi_i$,

$$\phi_r + i\phi_i = (-aA + B) + ibA$$

Equating real and imaginary terms, respectively, in this equality, we find

$$\phi_r = -aA + B \qquad \text{and} \qquad \phi_i = bA$$

† See footnote p. 225.

or, solving for A and B,

$$A = \frac{\phi_i}{b} \qquad B = \frac{b\phi_r + a\phi_i}{b}$$

Thus the partial fraction which corresponds to the quadratic factor $(s + a)^2 + b^2$ is

$$\frac{As + B}{(s + a)^2 + b^2} = \frac{1}{b}\frac{\phi_i s + (b\phi_r + a\phi_i)}{(s + a)^2 + b^2} = \frac{1}{b}\left[\frac{(s + a)\phi_i}{(s + a)^2 + b^2} + \frac{b\phi_r}{(s + a)^2 + b^2}\right]$$

By Formulas 1 and 2, Sec. 6.5, the inverse of the last expression is

$$\frac{1}{b}(\phi_i e^{-at} \cos bt + \phi_r e^{-at} \sin bt)$$

Finally, factoring out e^{-at}, we have the assertion of the theorem.

There is a fourth theorem dealing with repeated, irreducible quadratic factors but because of its complexity and limited usefulness, we shall not develop it here. Fortunately, many of the simpler transforms involving repeated quadratic factors can be handled by other means, e.g., the convolution theorem of Sec. 6.8.

Example 1 If $\mathscr{L}\{f(t)\} = (s^2 + 2)/s(s + 1)(s + 2)$, what is $f(t)$?

The roots of the denominator in this case are $s = 0, -1, -2$. Hence we must compute the values of

$$p(s) = s^2 + 2 \qquad \text{and} \qquad q'(s) = 3s^2 + 6s + 2$$

for these values of s. The results are

$$p(0) = 2 \qquad p(-1) = 3 \qquad p(-2) = 6$$

$$q'(0) = 2 \qquad q'(-1) = -1 \qquad q'(-2) = 2$$

From the corollary of Theorem 1 we now have at once

$$f(t) = \frac{2}{2}e^{0t} + \frac{3}{-1}e^{-t} + \frac{6}{2}e^{-2t} = 1 - 3e^{-t} + 3e^{-2t}$$

Equally well, of course, we could have obtained the coefficients in the inverse by suppressing, in turn, each of the factors of the denominator and evaluating the rest of the fraction at the root corresponding to the suppressed factor. In particular, we note that

$$Q_1(0) = (s + 1)(s + 2)\Big|_{s=0} = 2 = q'(0)$$

$$Q_2(-1) = s(s + 2)\Big|_{s=-1} = -1 = q'(-1)$$

$$Q_3(-2) = s(s + 1)\Big|_{s=-2} = 2 = q'(-2)$$

Example 2 What is the solution of the differential equation

$$y''' + 4y'' + 14y' + 20y = -2e^{-2t}$$

for which $y = y' = 0$ and $y'' = 1$ when $t = 0$?

Taking the Laplace transformation of each side of the given equation, we have

$$(s^3\mathscr{L}\{y\} - 1) + 4s^2\mathscr{L}\{y\} + 14s\mathscr{L}\{y\} + 20\mathscr{L}\{y\} = -\frac{2}{s+2}$$

and, solving for $\mathscr{L}\{y\}$,

$$\mathscr{L}\{y\} = \frac{1}{s^3 + 4s^2 + 14s + 20}\left(1 - \frac{2}{s+2}\right) = \frac{s}{(s^3 + 4s^2 + 14s + 20)(s+2)}$$

Before we can proceed, it is, of course, necessary that the denominator of $\mathscr{L}\{y\}$ be completely factored into irreducible real factors, and in serious applications this is often the most difficult part of the work. Here, however, it is easy to verify that $s = -2$ is a zero of the cubic factor. With the corresponding factor $s + 2$ recognized, the cubic factor becomes

$$(s+2)(s^2 + 2s + 10)$$

and the entire transform can be written in the fully factored form

$$\mathscr{L}\{y\} = \frac{s}{(s+2)^2(s^2 + 2s + 10)}$$

Considering first the repeated linear factor, we identify

$$\phi(s) = \frac{s}{s^2 + 2s + 10} \qquad \text{and} \qquad \phi'(s) = \frac{-s^2 + 10}{(s^2 + 2s + 10)^2}$$

Evaluating these for the root $s = -2$, we obtain

$$\phi(-2) = -\tfrac{1}{5} \qquad \text{and} \qquad \phi'(-2) = \tfrac{3}{50}$$

Hence, by Theorem 2, the terms in y corresponding to $(s+2)^2$ are

$$\left(\frac{3}{50} - \frac{t}{5}\right)e^{-2t} = \frac{(3 - 10t)e^{-2t}}{50}$$

For the quadratic factor $s^2 + 2s + 10 \equiv (s+1)^2 + 3^2$, we have

$$\phi(s) = \frac{s}{(s+2)^2}$$

Hence $\quad \phi(-a + ib) = \phi(-1 + 3i) = \dfrac{-1 + 3i}{[(-1 + 3i) + 2]^2}$

$$= \frac{-1 + 3i}{(1 + 3i)^2} = \frac{-1 + 3i}{-8 + 6i} = \frac{13 - 9i}{50}$$

and thus $\phi_r = \frac{13}{50}$, $\phi_i = -\frac{9}{50}$. The term in y corresponding to the factor $s^2 + 2s + 10$ is therefore

$$\frac{1}{3}\,\frac{e^{-t}(-9\cos 3t + 13\sin 3t)}{50}$$

Adding the two partial inverses, we have finally

$$y = \frac{(3 - 10t)e^{-2t}}{50} + \frac{e^{-t}(-9\cos 3t + 13\sin 3t)}{150}$$

Exercises for Section 6.7

Find the functions which have the following transforms.

1 $\dfrac{s^2 - s + 3}{s^3 + 6s^2 + 11s + 6}$ **2** $\dfrac{s + 2}{(s + 1)(s^2 + 4)}$

3 $\dfrac{s}{(s + 2)^2(s^2 + 1)}$ **4** $\dfrac{s}{(s + 1)(s + 2)^3}$

5 $\dfrac{s}{s^4 - 2s^2 + 1}$ **6** $\dfrac{s + 1}{(s^2 + 1)(s^2 + 4s + 13)}$

7 $\dfrac{s + 2}{s^4 + 4s^3 + 4s^2 - 4s - 5}$ **8** $\dfrac{s^2 + 2}{(s^2 + 4s + 5)(s^2 + 6s + 10)}$

9 $\dfrac{1}{(s^2 - s - 6)^4}$ **10** $\dfrac{s + 2}{s^4 - 16s^2 + 100}$

Solve the following differential equations.

11 $y''' - 2y'' - y' + 2y = u(t - 2)$ $\quad y_0 = y_0' = 0,\ y_0'' = 1$

12 $y''' + 3y'' + 3y' + y = \cosh t$ $\quad y_0 = y_0' = y_0'' = 0$

13 $y^{\text{iv}} + 2y''' + 2y'' + 2y' + y = e^{-t}$ $\quad y_0 = y_0' = y_0'' = y_0''' = 0$

Solve each of the following systems of equations.

14 $x'' + 2x' + \int_0^t y\,dt = 0$

$4x'' - x' + y = e^{-t} \qquad x_0 = 0,\ x_0' = 1$

***15** $(D^2 + D + 1)x + (D - 1)y = u(t)$

$(D^2 + 2D + 3)x + (3D^2 + 4D - 3)y = 0 \qquad x_0 = x_0' = y_0 = y_0' = 0$

***16** $y' - 3z = 5$

$y + z' - w = 3 - 2t \qquad y_0 = 1,\ z_0 = 0,\ w_0 = -1$

$z + w' = -1$

17 In the proof of Theorem 3, verify that if the identity

$$\phi(s) = As + B + [(s + a)^2 + b^2]h(s)$$

is evaluated for $s = -a - ib$, instead of for $s = -a + ib$, the same inverse is obtained.

18 Find the inverse of $s/(s + 1)(s^2 + 2s + 5)$ by factoring the irreducible quadratic factor $s^2 + 2s + 5$ into the unrepeated linear factors $s + 1 + 2i$ and $s + 1 - 2i$† and then applying Theorem 1 to these factors as well as to the real factor $s + 1$. Does your answer agree with the result obtained by using Theorem 3 to handle the quadratic factor? Do you think that this alternative procedure could be used to handle irreducible quadratic factors in general?

19 Using the procedure suggested in Exercise 18, find the inverse of $s/[(s^2 + 4)^2]$. Does your answer agree with the result obtained by using Corollary 1, Theorem 6, Sec. 6.5?

***20** If $q(s)$ is a polynomial of degree n containing only unrepeated, real linear factors, show that the sum of the numerators of the fractions in the partial-fraction decomposition of $p(s)/q(s)$ is equal to the coefficient of s^{n-1} in $p(s)$. Is there a comparable result if $q(s)$ contains only unrepeated, real linear factors and unrepeated quadratic factors?

† There is no contradiction in factoring an expression previously described as irreducible because *irreducible* means, technically, "having no *real* factors."

6.8 THE CONVOLUTION INTEGRAL

We conclude this chapter by establishing a result concerning the product of transforms which is of considerable theoretical as well as practical interest.

Theorem 1 $\mathscr{L}\{f(t)\}\mathscr{L}\{g(t)\} = \mathscr{L}\left\{\int_0^t f(t-\lambda)g(\lambda)\,d\lambda\right\}$

$$= \mathscr{L}\left\{\int_0^t f(\lambda)g(t-\lambda)\,d\lambda\right\}$$

PROOF Working with the term on the right in the first equality, we have by definition

$$\mathscr{L}\left\{\int_0^t f(t-\lambda)g(\lambda)\,d\lambda\right\} = \int_0^\infty \left[\int_0^t f(t-\lambda)g(\lambda)\,d\lambda\right]e^{-st}\,dt \tag{1}$$

Now
$$u(t-\lambda) = \begin{cases} 1 & \lambda < t \\ 0 & \lambda > t \end{cases}$$

and thus
$$f(t-\lambda)g(\lambda)u(t-\lambda) = \begin{cases} f(t-\lambda)g(\lambda) & \lambda < t \\ 0 & \lambda > t \end{cases}$$

Since this product vanishes for all values of λ greater than the upper limit t, the inner integration in (1) can be extended to infinity if the factor $u(t-\lambda)$ is inserted in the integrand. Hence,

$$\mathscr{L}\left\{\int_0^t f(t-\lambda)g(\lambda)\,d\lambda\right\} = \int_0^\infty \left[\int_0^\infty f(t-\lambda)g(\lambda)u(t-\lambda)\,d\lambda\right]e^{-st}\,dt \tag{2}$$

Now our usual assumptions about the functions we transform are sufficient to permit the order of integration in (2) to be interchanged:

$$\mathscr{L}\left\{\int_0^t f(t-\lambda)g(\lambda)\,d\lambda\right\} = \int_0^\infty \left[\int_0^\infty f(t-\lambda)g(\lambda)u(t-\lambda)e^{-st}\,dt\right]d\lambda$$

$$= \int_0^\infty g(\lambda)\left[\int_0^\infty f(t-\lambda)u(t-\lambda)e^{-st}\,dt\right]d\lambda \tag{3}$$

Because of the presence of $u(t-\lambda)$, the integrand of the inner integral is identically zero for all $t < \lambda$. Hence, the inner integration effectively starts not at $t = 0$ but at $t = \lambda$. Therefore

$$\mathscr{L}\left\{\int_0^t f(t-\lambda)g(\lambda)\,d\lambda\right\} = \int_0^\infty g(\lambda)\left[\int_\lambda^\infty f(t-\lambda)e^{-st}\,dt\right]d\lambda \tag{4}$$

Now, in the inner integral on the right of (4), let $t - \lambda = \tau$ and $dt = d\tau$. Then

$$\mathscr{L}\left\{\int_0^t f(t-\lambda)g(\lambda)\,d\lambda\right\} = \int_0^\infty g(\lambda)\left[\int_0^\infty f(\tau)e^{-s(\tau+\lambda)}\,d\tau\right]d\lambda$$

$$= \int_0^\infty g(\lambda)e^{-s\lambda}\left[\int_0^\infty f(\tau)e^{-s\tau}\,d\tau\right]d\lambda$$

$$= \left[\int_0^\infty f(\tau)e^{-s\tau}\,d\tau\right]\left[\int_0^\infty g(\lambda)e^{-s\lambda}\,d\lambda\right]$$

$$= \mathscr{L}\{f(t)\}\mathscr{L}\{g(t)\}$$

as asserted. The second equality asserted by the theorem follows in the same way, or may be obtained from the first by the substitution of $t - \lambda = \mu$.

The **convolution**, or **Faltung,**† integral

$$\int_0^t f(t - \lambda)g(\lambda)\, d\lambda$$

is frequently denoted simply by $f(t) * g(t)$. In this notation Theorem 1 becomes

$$\mathscr{L}\{f(t)\}\mathscr{L}\{g(t)\} = \mathscr{L}\{f(t) * g(t)\}$$

Example 1 Find the solution of the differential equation

$$y'' + 4y' + 13y = \tfrac{1}{3}e^{-2t} \sin 3t$$

for which $y = 1$ and $y' = -2$ when $t = 0$.

Taking the Laplace transformation of each side of the given equation, we have

$$(s^2\mathscr{L}\{y\} - s + 2) + 4(s\mathscr{L}\{y\} - 1) + 13\mathscr{L}\{y\} = \frac{1}{(s+2)^2 + 3^2}$$

and, solving for $\mathscr{L}\{y\}$,

$$\mathscr{L}\{y\} = \frac{s+2}{(s+2)^2 + 3^2} + \frac{1}{[(s+2)^2 + 3^2]^2}$$

The inverse of the first term is $e^{-2t} \cos 3t$. To find the inverse of the second term, it is convenient to begin by using the corollary of the first shifting theorem [Theorem 4, Sec. 6.5] to obtain

$$\mathscr{L}^{-1}\left\{\frac{1}{[(s+2)^2 + 3^2]^2}\right\} = e^{-2t}\mathscr{L}^{-1}\left\{\frac{1}{(s^2 + 3^2)^2}\right\} \tag{5}$$

Now
$$\frac{1}{(s^2 + 3^2)^2} = \mathscr{L}\left\{\frac{\sin 3t}{3}\right\}\mathscr{L}\left\{\frac{\sin 3t}{3}\right\}$$

Hence, by the convolution theorem,

$$\begin{aligned}
\mathscr{L}^{-1}\left\{\frac{1}{(s^2 + 3^2)^2}\right\} &= \frac{1}{9}\int_0^t \sin 3(t - \lambda) \sin 3\lambda\, d\lambda \\
&= \frac{1}{9}\int_0^t \frac{\cos(6\lambda - 3t) - \cos 3t}{2}\, d\lambda \\
&= \frac{1}{18}\left[\frac{\sin(6\lambda - 3t)}{6} - \lambda \cos 3t\right]_0^t \\
&= \frac{1}{18}\left(\frac{\sin 3t}{3} - t \cos 3t\right)
\end{aligned}$$

Therefore, from (5),

$$\mathscr{L}^{-1}\left\{\frac{1}{[(s+2)^2 + 3^2]^2}\right\} = \frac{e^{-2t}(\sin 3t - 3t \cos 3t)}{54}$$

† German for *folding*.

and, combining the inverses of the two terms in $\mathscr{L}\{y\}$,

$$y = e^{-2t}\cos 3t + \frac{e^{-2t}(\sin 3t - 3t\cos 3t)}{54}$$

Incidentally, this example illustrates how in certain cases the convolution theorem can be used in place of a fourth Heaviside theorem to handle repeated quadratic factors in the denominator of a transform.

Example 2 Find a particular integral of the differential equation

$$y'' + 2ay' + (a^2 + b^2)y = f(t)$$

Taking the Laplace transform of the given equation and assuming $y_0 = y'_0 = 0$, since we desire only a *particular* solution, we find

$$\mathscr{L}\{y\} = \frac{1}{(s+a)^2 + b^2}\mathscr{L}\{f(t)\}$$

Now $$\frac{1}{(s+a)^2 + b^2} = \mathscr{L}\left\{\frac{e^{-at}\sin bt}{b}\right\}$$

Hence $$\mathscr{L}\{y\} = \mathscr{L}\{f(t)\}\mathscr{L}\left\{\frac{e^{-at}\sin bt}{b}\right\}$$

and thus, by the convolution theorem,

$$y = \frac{1}{b}\int_0^t f(t-\lambda)e^{-a\lambda}\sin b\lambda\, d\lambda$$

or, equally well,

$$y = \frac{1}{b}\int_0^t f(\lambda)e^{-a(t-\lambda)}\sin b(t-\lambda)\, d\lambda = \frac{e^{-at}}{b}\int_0^t f(\lambda)e^{a\lambda}\sin b(t-\lambda)\, d\lambda$$

It is interesting to compare this procedure with the method of variation of parameters (Sec. 3.4) for the determination of particular integrals of linear differential equations. The two give identical results in the case of constant-coefficient differential equations.

An especially important application of the convolution theorem makes it possible to determine the response of a system to a general excitation if its response to a unit step function is known. To develop this idea, we shall need the concepts of **transfer function** and **indicial admittance.**

Any physical system capable of responding to an excitation can be thought of as a device by which an input function is transformed into an output function. If we assume that all initial conditions are zero at the moment when a single excitation, or **input,** $f(t)$ begins to act, then by setting up the differential equations describing the system, taking Laplace transforms (assuming that the equations are linear and have constant coefficients), and solving for the transform of the **output** $y(t)$, we obtain a relation of the form

$$\mathscr{L}\{y(t)\} = \frac{\mathscr{L}\{f(t)\}}{Z(s)} \tag{6}$$

where $Z(s)$ is a function of s whose coefficients depend solely on the parameters of the system. Moreover, in the usual applications to linear systems, $Z(s)$ will be just the quotient of two polynomials in s. For both mechanical systems and electrical systems, the function

$$\frac{1}{Z(s)} = \frac{\mathscr{L}\{y(t)\}}{\mathscr{L}\{f(t)\}} = \frac{\mathscr{L}\{\text{output}\}}{\mathscr{L}\{\text{input}\}}$$

is an important quantity, usually called the **transfer function.**

If a unit step function is applied to a system with transfer function $1/Z(s)$, then from (6) we have

$$\mathscr{L}\{y(t)\} = \frac{\mathscr{L}\{u(t)\}}{Z(s)} = \frac{1}{sZ(s)}$$

The response in this particular case is called the **indicial admittance** $A(t)$; that is,

$$\mathscr{L}\{A(t)\} = \frac{1}{sZ(s)} \tag{7}$$

Using (7), we can now rewrite (6) in the form

$$\mathscr{L}\{y(t)\} = \frac{\mathscr{L}\{f(t)\}}{Z(s)} = \frac{s\mathscr{L}\{f(t)\}}{sZ(s)} = s\mathscr{L}\{A(t)\}\mathscr{L}\{f(t)\}$$

Hence, by the convolution theorem,

$$\mathscr{L}\{y(t)\} = s\mathscr{L}\left\{\int_0^t A(t-\lambda)f(\lambda)\,d\lambda\right\} = s\mathscr{L}\left\{\int_0^t A(\lambda)f(t-\lambda)\,d\lambda\right\}$$

Because of the factor s, it follows from Theorem 2, Sec. 6.5, that

$$y(t) = \frac{d}{dt}\left[\int_0^t A(t-\lambda)f(\lambda)\,d\lambda\right] = \frac{d}{dt}\left[\int_0^t A(\lambda)f(t-\lambda)\,d\lambda\right]$$

Therefore, performing the indicated differentiations,† we have equivalently,

$$y(t) = \int_0^t A'(t-\lambda)f(\lambda)\,d\lambda + A(0)f(t) \tag{8}$$

and

$$y(t) = \int_0^t A(\lambda)f'(t-\lambda)\,d\lambda + A(t)f(0) \tag{9}$$

Since $A(t)$ is by definition the response of a system which is initially passive, that is, has all initial conditions equal to zero, it follows that $A(0) = 0$. Hence Eq. (8) becomes simply

$$y(t) = \int_0^t A'(t-\lambda)f(\lambda)\,d\lambda \tag{10}$$

† See item 10, Appendix B.2.

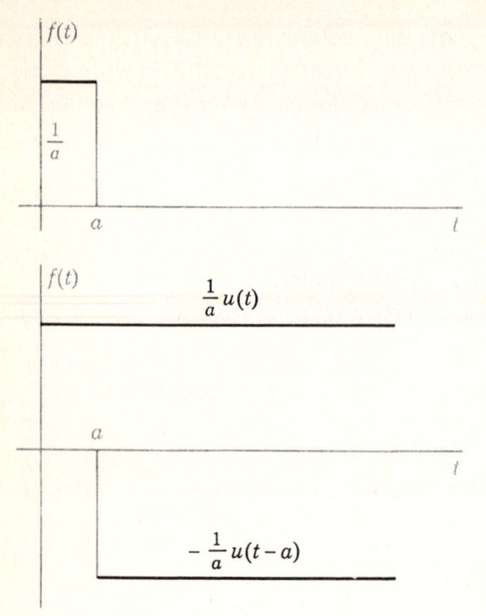

Figure 6.12 Plot suggesting the nature of a unit impulse.

Finally, by making the change of variable $\tau = t - \lambda$ in the integrals in (9) and (10), we obtain the related expressions

$$y(t) = \int_0^t A'(\tau)f(t - \tau)\, d\tau \tag{11}$$

$$y(t) = \int_0^t A(t - \tau)f'(\tau)\, d\tau + A(t)f(0) \tag{12}$$

Equations (9) to (12) all serve to express the response of a system to a general driving function $f(t)$ in terms of the experimentally accessible response $A(t)$ to a unit step function. They are often referred to, collectively, as **Duhamel's formulas,** after the French mathematician J. M. C. Duhamel (1797–1872).

To obtain added insight into Eqs. (10) and (11), it is convenient to introduce the concept of a unit impulse. To do this, suppose that we have the function shown in Fig. 6.12. This consists of a suddenly applied excitation of constant magnitude acting for a certain interval of time and then suddenly ceasing, the product of duration and magnitude being unity. If a is very small, the period of application is correspondingly small, but the magnitude of the excitation is very great. It is sometimes convenient to pursue this idea to the limit and imagine an input function of ever-increasing magnitude acting for an ever shorter interval of time, the product of intensity and duration remaining unity as $a \to 0$. The resulting "function" is usually referred to as the **unit impulse** $I(t)$ or the **δ function** $\delta(t)$.†

† More specifically, $\delta(t)$ is often called the **Dirac δ function,** after the British theoretical physicist P. A. M. Dirac (1902–).

In somewhat different terms, the δ function $\delta(t - t_0)$ is often described by the following purported definition:

$$\delta(t - t_0) = \begin{cases} 0 & t \neq t_0 \\ \infty & t = t_0 \end{cases} \qquad \int_{-\infty}^{\infty} \delta(t - t_0)\, dt = 1 \tag{13}$$

Taken literally, this is nonsense, for the area under a curve which coincides with the t axis at every point but one, if it means anything, must surely be zero and not unity, as (13) suggests. However, if (13) is considered to be merely suggestive of the limiting process by which we first described the unit impulse, then whatever its shortcomings as a definition, it is at least as useful as certain other reasonably respectable concepts in applied mathematics, specifically, the concept of a nonzero load concentrated at a single point on a beam.

The unit impulse is only the first of an infinite sequence of so-called **singularity functions.** As a direct generalization of the unit impulse, we have the **unit doublet** (Fig. 6.13), defined (loosely) as

$$D(t) = \lim_{a \to 0} \frac{u(t) - 2u(t - a) + u(t - 2a)}{a^2}$$

the unit triplet, defined similarly as

$$T(t) = \lim_{a \to 0} \frac{u(t) - 3u(t - a) + 3u(t - 2a) - u(t - 3a)}{a^3}$$

and so on, indefinitely. Some of the properties of these "functions" will be found among the exercises at the end of this section.

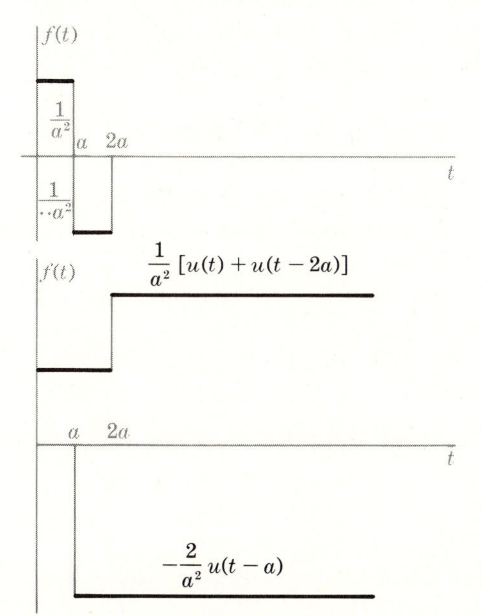

Figure 6.13 Plot suggesting the nature of a unit doublet.

Whatever its theoretical limitations, the δ function is a useful device which, at worst, can be used formally provided the answers to which it leads are subsequently checked experimentally or by independent analysis. It is interesting and important that in many applications the use of the δ function can be rigorously justified by arguments based on what is known as the **Stieltjes integral,**† a generalization of the familiar Riemann integral. More generally, all the singularity functions are examples of mathematical objects known as **generalized functions,** or **distributions,** which are studied in the recently developed **theory of distributions.**‡

To determine the Laplace transform of a unit impulse, we return to the prelimiting approximation

$$\frac{u(t) - u(t-a)}{a}$$

shown in Fig. 6.13. Transforming this expression, we have, for all $a > 0$,

$$\frac{1}{a}\left(\frac{1}{s} - \frac{e^{-as}}{s}\right) = \frac{1 - e^{-as}}{as}$$

As $a \to 0$, this transform assumes the indeterminate form 0/0, but evaluating it in the usual way by L'Hospital's rule, we obtain immediately the limiting value 1. In the same way, we can show that the transforms of the unit doublet and the unit triplet are, respectively, s and s^2, and the transforms of the other singularity functions follow exactly the same pattern. Since these transforms do not approach zero as s becomes infinite, we know from Corollary 3 of Theorem 1, Sec. 6.2, that they are not the transforms of piecewise continuous functions of exponential order. In fact, although the singularity functions are all of exponential order, they are limiting forms involving unbounded behavior in the neighborhood of the origin and hence are not piecewise continuous.

We are now in a position to resume our attempt to give a physical interpretation to Eq. (10). For convenience, let us denote by $h(t)$ the response of the system under discussion when the input is a unit impulse. We have already seen [Eq. (6)] that

$$\mathscr{L}\{y(t)\} = \frac{\mathscr{L}\{f(t)\}}{Z(s)}$$

Hence, if $f(t)$ is a unit impulse, so that $\mathscr{L}\{f(t)\} = 1$ and $y(t) = h(t)$, we have

$$\mathscr{L}\{h(t)\} = \frac{1}{Z(s)} = s\frac{1}{sZ(s)} = s\mathscr{L}\{A(t)\}$$

† Named for the Dutch mathematician T. J. Stieltjes (1856–1894).

‡ An introductory account of the theory of distributions can be found in Athanasios Papoulis, "The Fourier Integral and Its Applications," McGraw-Hill, New York, 1962, pp. 269–282.

Thus, from Theorem 2, Sec. 6.5, it follows that

$$h(t) = \frac{dA(t)}{dt} = A'(t)$$

or, in words, *the response of a system to a unit impulse is the derivative of the response of the system to a unit step function.* Thus Eqs. (10) and (11) express the response of a system to a general driving function $f(t)$ in terms of its response, $h(t) = A'(t)$, to a unit impulse function. In many physical systems, it is possible to approximate a unit impulse input with satisfactory accuracy and obtain $h(t)$ experimentally.

It is interesting that among other things, the unit impulse and the unit doublet provide a hypothetical mechanism for the instantaneous introduction of initial displacements and velocities into a system. To verify this, consider a system described by the equation

$$ay'' + by' + cy = 0 \qquad y(0) = y_0,\ y'(0) = y'_0$$

Taking Laplace transforms, we have

$$a(s^2\mathscr{L}\{y\} - y_0 s - y'_0) + b(s\mathscr{L}\{y\} - y_0) + c\mathscr{L}\{y\} = 0$$

and
$$\mathscr{L}\{y\} = \frac{ay_0 s + (ay'_0 + by_0)}{as^2 + bs + c} \tag{14}$$

Now consider the same system to be completely passive but acted upon by an impulse of magnitude $ay'_0 + by_0$ and a doublet of magnitude ay_0. The differential equation describing this system is

$$ay'' + by' + cy = ay_0 D(t) + (ay'_0 + by_0)I(t)$$

Again taking Laplace transforms, we have this time

$$as^2\mathscr{L}\{y\} + bs\mathscr{L}\{y\} + \mathscr{L}\{y\} = ay_0 s + (ay'_0 + by_0)$$

and
$$\mathscr{L}\{y\} = \frac{ay_0 s + (ay'_0 + by_0)}{as^2 + bs + c} \tag{15}$$

Clearly, the Laplace transforms (14) and (15), and hence the corresponding responses, are identical.

Example 3 Find $A(t)$ and $h(t)$ for the system described by the equation $y'' + 5y' + 4y = 0$, and verify that $h(t) = A'(t)$.

To find $A(t)$, we must solve the initial-value problem

$$y'' + 5y' + 4y = u(t) \qquad y_0 = y'_0 = 0$$

Taking Laplace transforms and solving for $\mathscr{L}\{y\}$, we obtain at once

$$\mathscr{L}\{y\} = \frac{1}{s(s^2 + 5s + 4)} = \frac{1}{s(s + 1)(s + 4)}$$

Hence, by Corollary 1, Theorem 1, Sec. 6.7, we have

$$y \equiv A(t) = \tfrac{1}{4}u(t) - \tfrac{1}{3}e^{-t} + \tfrac{1}{12}e^{-4t}$$

Similarly, to find $h(t)$, we must solve the initial-value problem

$$y'' + 5y' + 4y = I(t) \qquad y_0 = y_0' = 0$$

Again, taking Laplace transforms throughout the equation, we find

$$\mathscr{L}\{y\} = \frac{1}{(s+1)(s+4)}$$

and

$$y \equiv h(t) = \tfrac{1}{3}e^{-t} - \tfrac{1}{3}e^{-4t}$$

For all $t > 0$, $h(t)$ is the derivative of $A(t)$ since $u'(t) = 0$.

Exercises for Section 6.8

Find the Laplace transform of each of the following functions.

1 $\int_0^t \sin \lambda \cos (t - \lambda)\, d\lambda$ **2** $\int_0^t \sin (t - \lambda) \cos \lambda\, d\lambda$

3 $\int_0^t \lambda \cos (t - \lambda)\, d\lambda$ **4** $\int_0^t \lambda^2 e^{-\lambda} \sin (t - \lambda)\, d\lambda$

5 $e^{-2t} * \sin t$ **6** $t^2 * \cos 3t$

Find the inverse of each of the following transforms.

7 $\dfrac{1}{(s^2 + 4)^2}$ **8** $\dfrac{s}{(s^2 + 9)^3}$

9 $\dfrac{s}{s + 2}$ **10** $\dfrac{s^2}{s^2 + 4}$

***11** $\dfrac{s^2 + 4s + 4}{(s^2 + 4s + 13)^2}$ ***12** $\dfrac{s^4 + 2s^2 - s}{(s + 1)(s^2 + 1)^2}$

13 Using the convolution theorem, find a particular integral of the equation $y'' + 2ay' + a^2 y = f(t)$.

14 Using the convolution theorem, find a particular integral of the equation $y'' + (a + b)y' + aby = f(t)$.

***15** Show that

$$\mathscr{L}^{-1}\left\{\frac{1}{\sqrt{s}\,(s - 1)}\right\} = \frac{2e^t}{\sqrt{\pi}} \int_0^{\sqrt{t}} e^{-v^2}\, dv$$

Hint: Use the convolution theorem, and then in the resulting integral let $\sqrt{\lambda} = v$.

***16** Using the results of Exercise 15, find the inverse of $1/(s\sqrt{s + 1})$.

17 Verify that the Laplace transform of the unit doublet is s and that the transform of the unit triplet is s^2.

18 What is $\mathscr{L}\{I(t - t_0)\}$?

19 What is $\mathscr{L}\{D(t - t_0)\}$?

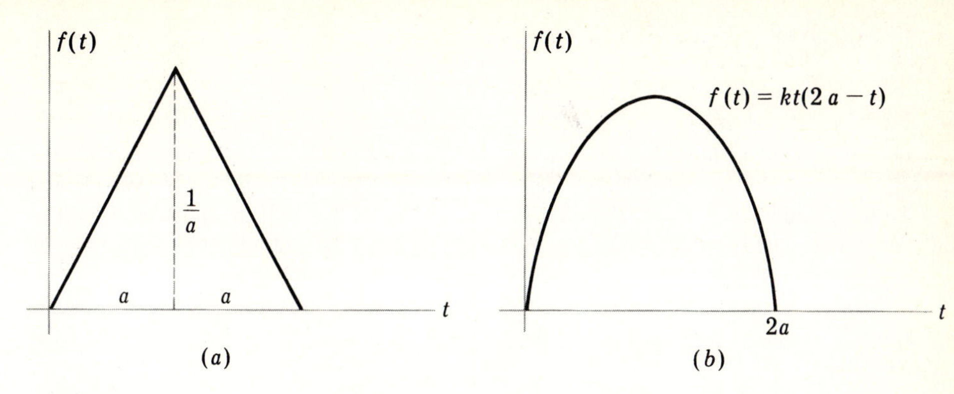

Figure 6.14

***20** Find the transform of the triangular pulse shown in Fig. 6.14*a*. What is the limit of the transform of this pulse as $a \to 0$?

***21** Determine k so that the parabolic pulse shown in Fig. 6.14*b* will have unit area. What is the transform of this unit pulse? What is the limit of the transform of this pulse as $a \to 0$?

***22** Give an argument making plausible the formula $\int_{-\infty}^{\infty} f(t)\delta(t - t_0)\,dt = f(t_0)$. *Hint:* Apply the law of the mean for integrals to the given integral after $\delta(t - t_0)$ has been replaced by its prelimiting representation.

***23** Give an argument making plausible the formula

$$\int_{-\infty}^{\infty} f(t)D(t - t_0)\,dt = -f'(t_0)$$

24 Show that $f(t) * [g(t) \pm h(t)] = f(t) * g(t) \pm f(t) * h(t)$.

25 Show that $f(t) * [g(t) * h(t)] = \int_0^t \int_0^\lambda f(t - \lambda)g(\lambda - \mu)h(\mu)\,d\mu\,d\lambda$.

***26** Show that $f(t) * [g(t) * h(t)] = [f(t) * g(t)] * h(t)$.

27 Show that $\mathscr{L}\{f(t)\}\mathscr{L}\{g(t)\}\mathscr{L}\{h(t)\} = \mathscr{L}\{f(t) * g(t) * h(t)\}$.

28 Show that $1 * 1 = t$ and that $1 * 1 * 1 = t^2/2$. What is the generalization of these results to n "factors"?

***29** What is (*a*) $\delta(t - a) * f(t)$, (*b*) $u(t - a) * f(t)$, (*c*) $t^m * t^n$ where m and n are nonnegative integers?

***30** If $f(0) = g(0) = 0$, show that $f'(t) * g(t) = f(t) * g'(t)$ and that

$$[f(t) * g(t)]' = \frac{f'(t) * g(t) + f(t) * g'(t)}{2}$$

***31** If $f(t)$ and $g(t)$ are given functions, is it possible to solve the equation $f(t) * x(t) = g(t)$ for $x(t)$?

32 Consider the integral equation $x(t) + \int_0^t f(t - \lambda)x(\lambda)\,d\lambda = g(t)$† in which $f(t)$ and $g(t)$ are known functions and $x(t)$ is an unknown function to be determined. By taking Laplace transforms throughout the equation, obtain a formula for $\mathscr{L}\{x(t)\}$.

† Equations of this form are known as **Volterra integral equations** in honor of the Italian mathematician Vito Volterra (1860–1940).

****33** If $x(t)$ is the solution of the integral equation

$$x(t) + \int_0^t (t - \lambda)x(\lambda)\, d\lambda = \sin t$$

and if $\phi(t)$ is a function defined by the relation $\phi''(t) = x(t)$, show that $\phi''(t) + \phi(t) = \sin t - \phi'(0)t - \phi(0)$. Hence show that the given integral equation is equivalent to the initial-value problem $\phi''(t) + \phi(t) = \sin t$, $\phi'(0) = \phi(0) = 0$. *Hint:* Integrate the equation $\phi''(t) = x(t)$ twice from 0 to t using appropriate dummy variables of integration, and then reverse the order of integration in the resulting repeated integral.

***34** Using the procedure suggested in Exercise 32, solve the integral equation $x(t) + \int_0^t (t - \lambda)x(\lambda)\, d\lambda = \sin t$ and check your answer by solving the associated initial-value problem

$$\phi''(t) + \phi(t) = \sin t \qquad \phi'(0) = \phi(0) = 0$$

Solve each of the following integral equations.

***35** $x(t) + \int_0^t (t - \lambda)x(\lambda)\, d\lambda = \sin 2t$

***36** $x(t) + \int_0^t e^{(t-\lambda)}x(\lambda)\, d\lambda = \cos t$

***37** $x(t) + \int_0^t \sin (t - \lambda)x(\lambda)\, d\lambda = t$

***38** $x(t) + \int_0^t e^{(\lambda - t)}x(\lambda)\, d\lambda = te^{-t}$

****39** Find $A(t)$ and $h(t)$ for the equation $y'' + 3y' + 2y = 0$. Verify that $h(t) = A'(t)$ and then verify Eqs. (10) and (12) when this equation is "driven" by the function $f(t) = e^t$.

****40** Find $A(t)$ and $h(t)$ for the system shown in Fig. 6.15 if the input is applied to m_1 and the output is the response, i.e., the displacement, of m_2. Verify that $h(t) = A'(t)$. What is the response of m_2 to an arbitrary force applied to m_1 when the system is in its equilibrium position? What is the response of m_1 to an arbitrary force applied to m_2?

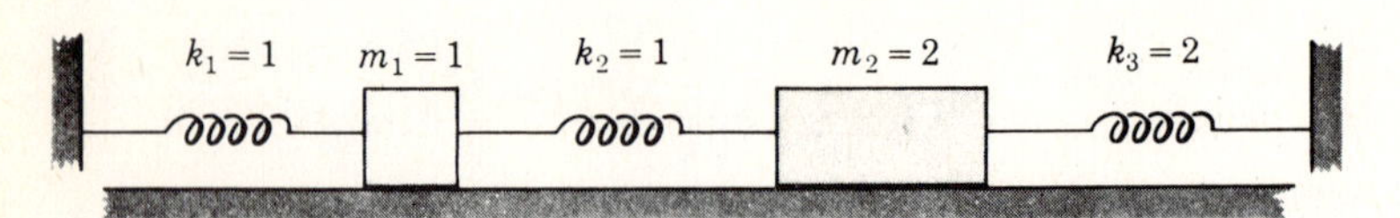

Figure 6.15

CHAPTER

SEVEN

MECHANICAL SYSTEMS AND ELECTRICAL CIRCUITS

7.1 INTRODUCTION

In Sec. 4.6 we discussed at some length the formulation of the differential equations describing the behavior of spring-suspended weights and series-electrical circuits. We also made the important observation that the two differential equations were mathematically identical, and we interpreted various mathematical properties of their solutions in physical terms. This led us to such important physical concepts as free and forced motion, underdamped and overdamped systems, critical damping, natural frequency, resonance, and transient and steady-state behavior.

In this chapter we shall extend the work of Sec. 4.6 in several directions. First, we shall expand the analogy between mechanical and electrical systems by discussing both torsionally vibrating mechanical systems and parallel-electrical circuits and noting that they, too, are governed by differential equations which are mathematically identical to those we encountered in Sec. 4.6. Then we shall investigate in more detail the various concepts we introduced in Sec. 4.6. Finally, we shall indicate how simultaneous differential equations can be used to study mechanical systems involving more than one mass and electrical circuits involving more than one loop.

7.2 TORSIONAL SYSTEMS AND PARALLEL CIRCUITS

To extend the range of physical phenomena to which our limited knowledge of differential equations may fruitfully be applied, let us consider two more simple yet important systems: a torsional mechanical system (Fig. 7.1*a*) and a parallel-electrical circuit (Fig. 7.1*b*).

Figure 7.1*a* shows a prototype of a simple mechanical system capable of vibrating torsionally. A disk of moment of inertia I is connected to an elastic shaft which, if twisted, exerts a restoring torque on the disk. We assume as a realistic approximation that the shaft obeys Hooke's law in torsional form; that is, if one end of the shaft is twisted through an angle of θ rad with respect to the other end, then the elastic restoring torque is equal to $-k\theta$, where k is the **modulus** of the shaft, thought of as a torsional spring. Realistically (though we may sometimes neglect it), there will be friction in the system; and as a reasonable assumption, we assume that it is proportional to the first power of the angular velocity and opposite in sign. In many important applications to rotating machine parts, an impressed torque may act on a disk, forcing it to vibrate torsionally. These are often of a periodic nature, and so we shall consider the possibility of a disturbing torque of either of the forms

$$T = T_0 \cos \omega t \tag{1a}$$

$$T = T_0 \sin \omega t \tag{1b}$$

where T_0 and ω are constants. Of course, if $\omega = 0$, Eq. (1*a*) reduces to a constant torque $T = T_0$.

The analysis of the system shown in Fig. 7.1*a* is based on **Newton's law in torsional form:**

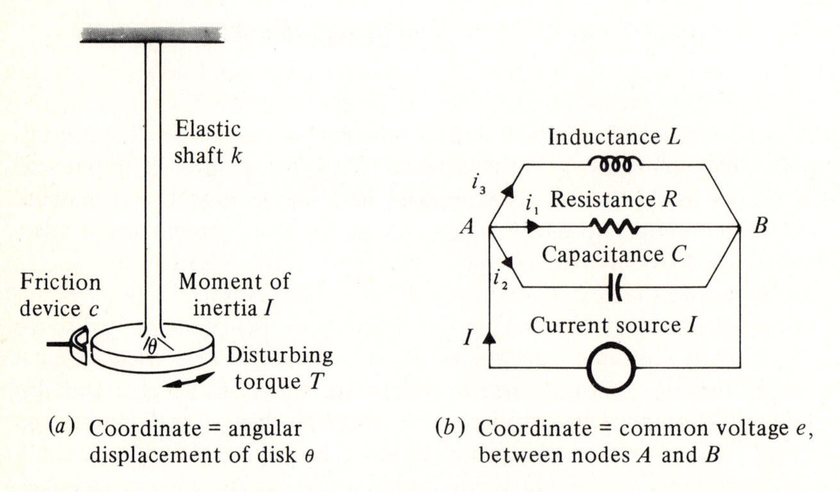

Figure 7.1 Two simple systems of one degree of freedom. (*a*) Torsional mechanical; (*b*) parallel electrical.

Moment of inertia × angular acceleration = sum of all torques

As a coordinate to describe the motion of the disk, we choose the angle θ through which the disk has rotated from its equilibrium position. Thus the angular velocity of the disk is $d\theta/dt$, and its angular acceleration is $d^2\theta/dt^2$. Hence, by assuming for definiteness that $T = T_0 \cos \omega t$, Newton's law gives us

$$I\frac{d^2\theta}{dt^2} = \text{elastic torque} + \text{frictional torque} + \text{impressed torque}$$

$$= -k\theta - c\frac{d\theta}{dt} + T_0 \cos \omega t$$

or

$$I\frac{d^2\theta}{dt^2} + c\frac{d\theta}{dt} + k\theta = T_0 \cos \omega t \tag{2}$$

Clearly, this is a nonhomogeneous, linear, constant-coefficient, second-order differential equation which we now know how to solve. Of course, in particular applications we would expect to be given the initial values of the angular displacement $\theta(0)$ and the angular velocity $d\theta/dt\,|_{t=0}$.

As its name suggests, a parallel-electrical circuit consists of some or all of the typical circuit elements—resistors, capacitors, inductors, and current (rather than voltage) sources connected not in series but in parallel branches having common endpoints, or **nodes.** To establish the differential equation describing the behavior of a typical parallel circuit, such as the one shown in Fig. 7.1*b*, we must use **Kirchhoff's first law:**

> *The algebraic sum of the currents flowing toward any point in an electrical network must be zero;* or *the sum of the currents flowing toward any point in an electrical network must equal the sum of the currents flowing away from that point.*

To describe the behavior of the system, we choose the voltage, or potential difference e, which exists between the nodes A and B in Fig. 7.1*b*. To implement Kirchhoff's first law, we must now express the current flowing through each of the circuit elements in terms of the (common) potential difference e across that element. For the current through the resistor, we have from Ohm's law

$$i_1 = \frac{e}{R}$$

For the capacitor, we have the fundamental law [Eq. (8*a*), Sec. 4.6] $Q = eC$. Hence, differentiating this with respect to time and recalling that $dQ/dt = i$, we obtain

$$i_2 = C\frac{de}{dt}$$

as the current (apparently) flowing through the capacitor. For the inductor, we have the fundamental law [Eq. (7), Sec. 4.6] $e = L(di/dt)$. Hence, integrating this with respect to time, we have

$$i_3 = \frac{1}{L}\int^t e\,dt$$

as the current flowing through the inductor. Assuming, for definiteness, that the impressed current provided by the current source is

$$I = I_0 \cos \omega t$$

we thus have from Kirchhoff's first law

$$\begin{aligned} I_0 \cos \omega t &= \text{current through the resistor} \\ &\quad + \text{current through the capacitor} \\ &\quad + \text{current through the inductor} \end{aligned}$$

or

$$C\frac{de}{dt} + \frac{1}{R}e + \frac{1}{L}\int^t e\,dt = I_0 \cos \omega t \tag{3}$$

This integrodifferential equation can easily be converted into a second-order differential equation by letting $\int^t e\,dt$ be a new variable, say u, so that

$$e = \frac{du}{dt} \qquad \text{and} \qquad \frac{de}{dt} = \frac{d^2u}{dt^2}$$

This stratagem is, of course, analogous to the substitution $Q = \int^t i\,dt$ by which we converted the integrodifferential equation (11*a*) into the differential equation (11*b*) in Sec. 4.6. However, since we now have the Laplace transformation as a solution technique and are able to express the Laplace transform of the integral of a function in terms of the transform of the function, there is no need to alter the form of Eq. (3) in order to solve it.

Various new analogies between mechanical and electrical circuits are now apparent. In particular, if we compare Eq. (3) when expressed in terms of u, with the equation governing the motion of a spring-suspended mass, namely,

$$\frac{w}{g}\frac{d^2y}{dt^2} + c\frac{dy}{dt} + ky = F_0 \cos \omega t$$

we have the following interesting correspondence:

Mechanical quantity	*Electrical quantity*
Mass w/g	Capacitance C
Friction c	Conductance $1/R$
Spring modulus k	Susceptance $1/L$
Impressed force F	Impressed current I
Displacement y	$\int^t e\,dt$
Velocity v	Voltage e

This set of analogies provides an alternative to the use of series circuits for engineers studying mechanical systems by constructing equivalent electrical circuits.

We shall not explore either the torsional system or the parallel circuit any further, since the mathematical identity of these systems with the translational-mechanical system and the series-electrical circuit implies that any results derived for the latter systems can immediately be restated for the former. Instead, in the following sections we shall investigate in greater detail the properties of the translational-mechanical and series-electrical systems.

Exercises for Section 7.2

1 If there is no friction in the system shown in Fig. 7.1*a*, what will be the frequency of its free vibrations?

2 What is the value of the critical damping for the system shown in Fig. 7.1*a*?

3 If there is no resistance in the system shown in Fig. 7.1*b*, what will be the frequency of its free vibrations?

4 For what values of R will the free behavior of the system shown in Fig. 7.1*b* be oscillatory? nonoscillatory?

7.3 THE TRANSLATIONAL-MECHANICAL SYSTEM

In Sec. 4.6 we found that the motion of a spring-suspended weight was described by the differential equation

$$\frac{w}{g}\frac{d^2y}{dt^2} + c\frac{dy}{dt} + ky = F_0 \cos \omega t \tag{1}$$

Applying the techniques of Secs. 4.2 and 4.3, we discovered further that the free, or transient, motion of the weight, that is, the motion described by the complementary function or the motion in the absence of any impressed force, was of one of three types, according to whether the amount of friction in the system was greater than, equal to, or less than the **critical damping**

$$c_c = 2\sqrt{\frac{kw}{g}} \tag{2}$$

For reference, these results are summarized in Table 7.1.

In the overdamped and critically damped cases, a plot of the displacement of the weight as a function of time resembles one or the other of the graphs shown in Fig. 7.2. In particular, in each of these cases, the weight can pass through its equilibrium position at most once in the course of its entire motion. In the underdamped case, the motion of the weight is a damped oscillation, and a plot of the displacement as a function of time resembles the graph shown in Fig. 7.3.

Clearly, because of the decreasing amplitudes of the oscillations in the underdamped case, the motion is not periodic. However, the weight does pass through

Table 7.1

Case	Condition	Solution	Roots
Overdamped	$c^2 - \frac{4kw}{g} > 0$ that is, $c > c_c$	$y = c_1 e^{m_1 t} + c_2 e^{m_2 t}$	$m_1, m_2 = -\frac{cg}{2w} \pm \frac{g}{2w}\sqrt{c^2 - \frac{4kw}{g}}$
Critically damped	$c^2 - \frac{4kw}{g} = 0$ that is, $c = c_c$	$y = c_1 e^{m_1 t} + c_2 t e^{m_1 t}$	$m_1 = m_2 = -\frac{cg}{2w}$
Underdamped	$c^2 - \frac{4kw}{g} < 0$ that is, $c < c_c$	$y = e^{-pt}(c_1 \cos qt + c_2 \sin qt)$	$m_1, m_2 = -p \pm iq$ where $p = \frac{cg}{2w}$ and $q = \frac{g}{2w}\sqrt{\frac{4kw}{g} - c^2}$

its equilibrium position an infinite number of times at regular intervals of π/q. To verify this (which we did in Sec. 4.6 by another line of reasoning), it is convenient to express the solution

$$y = e^{-pt}(c_1 \cos qt + c_2 \sin qt) \tag{3}$$

in a form that involves a single trigonometric function. To do this, we first multiply and divide the expression on the right in (3) by $\sqrt{c_1^2 + c_2^2} = G$, say, getting

$$y = Ge^{-pt}\left(\frac{c_1}{\sqrt{c_1^2 + c_2^2}} \cos qt + \frac{c_2}{\sqrt{c_1^2 + c_2^2}} \sin qt\right) \tag{4}$$

Next we define an angle H by the triangle shown in Fig. 7.4 and interpret the coefficients of $\cos qt$ and $\sin qt$ to be $\cos H$ and $\sin H$, respectively, getting

$$y = Ge^{-pt}(\cos H \cos qt + \sin H \sin qt) \tag{5}$$

Finally, we recognize that the expression in parentheses in (5) is just the expanded form of $\cos(qt - H)$, and so we have

$$y = Ge^{-pt} \cos(qt - H) \tag{6}$$

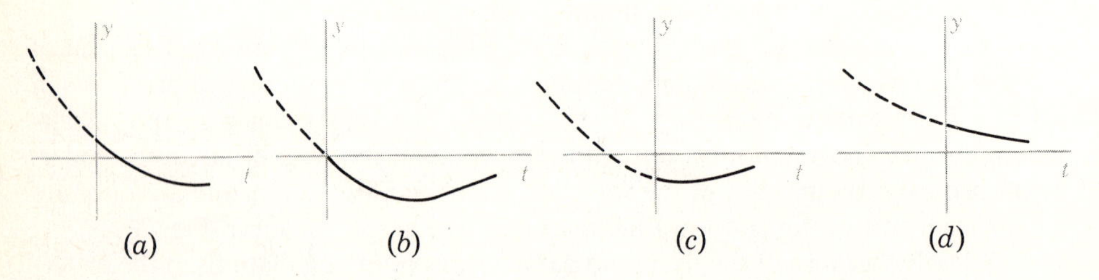

Figure 7.2 Displacement-time plots for free, overdamped, and critically damped motion.

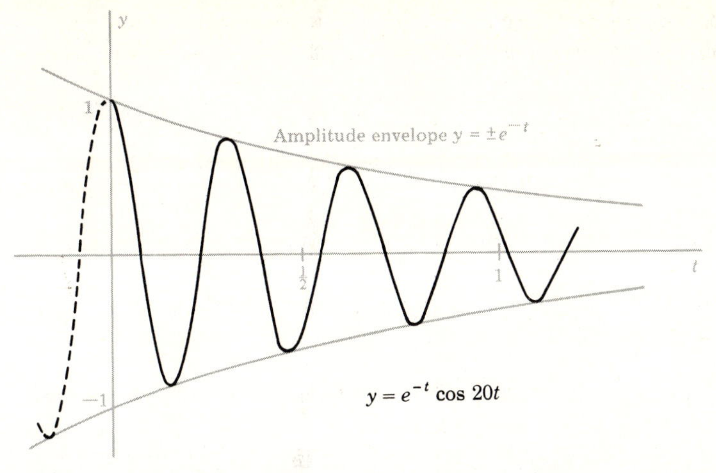

Figure 7.3 A typical displacement-time plot for an underdamped system.

From the description of the motion provided by Eq. (6), it is clear that $y = 0$, and the weight passes through its equilibrium position whenever

$$\cos(qt - H) = 0 \qquad t \geq 0$$

i.e., when $qt - H = \pi/2 + n\pi$ or

$$t = \frac{1}{q}\left(H + \frac{\pi}{2}\right) + \frac{n\pi}{q} \qquad n = 0, 1, 2, \ldots$$

Hence, we can speak of the **pseudoperiod** $2\pi/q$ and of the **pseudofrequency** or **frequency with damping** ω_d, defined by

$$\frac{\omega_d}{2\pi} = \frac{q}{2\pi} = \frac{1}{2\pi}\frac{g}{2w}\sqrt{\frac{4kw}{g} - c^2} = \frac{1}{2\pi}\sqrt{\frac{kg}{w} - \frac{c^2g^2}{4w^2}} \qquad \text{cycles/unit time} \qquad (7)$$

If $c = 0$, that is, if there is no damping in the system, the motion is strictly periodic and its frequency, which we shall call the **undamped natural frequency** ω_n, is, from (7), given by

$$\frac{\omega_n}{2\pi} = \frac{1}{2\pi}\sqrt{\frac{kg}{w}} \qquad \text{cycles/unit time} \qquad (8)$$

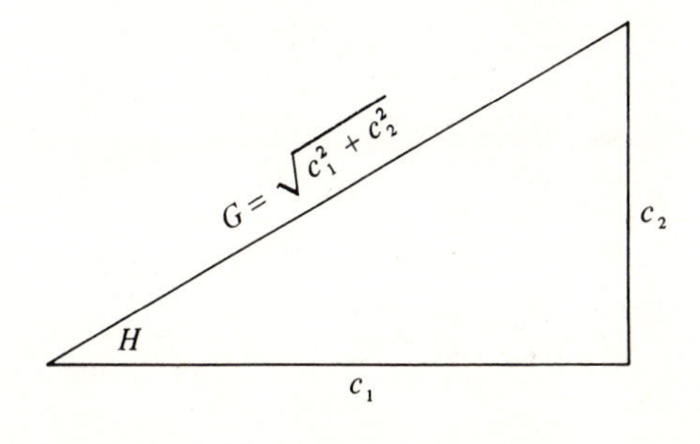

Figure 7.4 The triangle used in converting $c_1 \cos qt + c_2 \sin qt$ into a single trigonometric function.

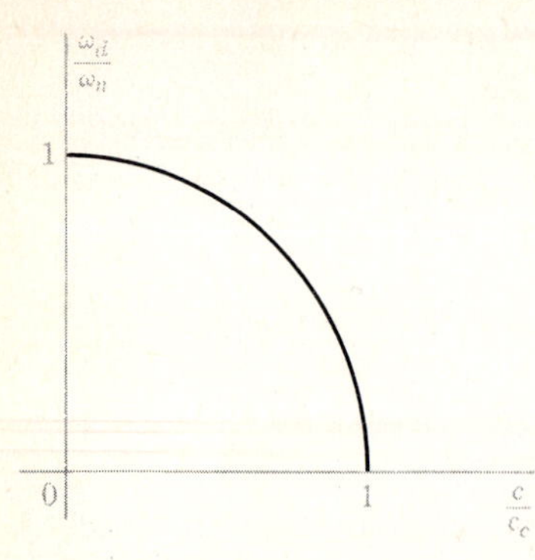

Figure 7.5 The effect of friction on frequency in an underdamped system.

Clearly, the frequency when damping is present is always less than the undamped natural frequency. The ratio of the two frequencies is

$$\frac{\omega_d}{\omega_n} = \frac{\sqrt{kg/w - c^2g^2/4w^2}}{\sqrt{kg/w}}$$

$$= \sqrt{1 - \frac{c^2g}{4kw}} = \sqrt{1 - \frac{c^2}{c_c^2}}$$

since, from Eq. (2), $c_c^2 = 4kw/g$. Figure 7.5 shows a plot of ω_d/ω_n vs. c/c_c. Evidently, if the actual damping is only a small fraction of the critical damping, as it often is, $\omega_d/\omega_n \doteq 1$ and the effect of friction on the frequency of the motion is very small. This explains why friction is usually neglected in natural-frequency calculations.

Still using Eq. (6), it is clear that the extreme values of y occur when

$$\frac{dy}{dt} = G[-pe^{-pt} \cos (qt - H) - qe^{-pt} \sin (qt - H)] = 0$$

i.e., when $\tan (qt - H) = -p/q$ or, finally, when

$$t = \frac{H}{q} - \frac{1}{q} \operatorname{Tan}^{-1} \frac{p}{q} + \frac{n\pi}{q} = T + \frac{n\pi}{q}$$

where T denotes the constant $H/q - (1/q) \operatorname{Tan}^{-1} (p/q)$.

The ratio of successive extreme displacements on the same side of the equilibrium position is a quantity of considerable importance. Its value is, from (6),

$$\frac{y_n}{y_{n+2}} = \frac{y(T + n\pi/q)}{y[T + (n + 2)\pi/q]}$$

$$= \frac{G \exp [-p(T + n\pi/q)] \cos [q(T + n\pi/q) - H]}{G \exp \{-p[T + (n + 2)\pi/q]\} \cos \{q[T + (n + 2)\pi/q] - H\}}$$

$$= e^{2\pi p/q} \frac{\cos (qT + n\pi - H)}{\cos (qT + n\pi - H + 2\pi)}$$

$$= e^{2\pi p/q} \tag{9}$$

Since this result depends only on the parameters of the system and not on n, we have thus established the following remarkable result.

Theorem 1 The ratio of successive maximum (or minimum) displacements remains constant throughout the entire free motion of an underdamped system.

If we take the natural logarithm of the expression in (9), we have

$$\ln \frac{y_n}{y_{n+2}} = \frac{2\pi p}{q} \tag{10}$$

This quantity, known as the **logarithmic decrement** δ, is a convenient measure, in **nepers per cycle,**† of the rate at which the motion dies away. Substituting for p and q from the third line in Table 7.1 into Eq. (10), we find

$$\delta = \frac{2\pi p}{q} = 2\pi \frac{cg/(2w)}{[g/(2w)]\sqrt{4kw/g - c^2}} = 2\pi \frac{c}{\sqrt{c_c^2 - c^2}}$$

Solved for c/c_c, this becomes

$$\frac{c}{c_c} = \frac{\delta}{\sqrt{\delta^2 + 4\pi^2}} \tag{11}$$

Since y_n and y_{n+2} are quantities which are relatively easy to measure, δ can easily be computed. Then from Eq. (11) the fraction of critical damping present in a given system can be found at once.

Now that we have investigated the free motion of the translational mechanical system in the overdamped, critically damped, and underdamped cases, it remains for us to consider the forced motion. To do this we must, of course, find a particular integral of Eq. (1):

$$\frac{w}{g}\frac{d^2y}{dt^2} + c\frac{dy}{dt} + ky = F_0 \cos \omega t \tag{1}$$

Assuming, as usual,

$$Y = A \cos \omega t + B \sin \omega t$$

and substituting into (1), collecting terms, and equating the coefficients of $\cos \omega t$ and $\sin \omega t$ on each side of the equation, we obtain the two conditions

$$\left(k - \omega^2 \frac{w}{g}\right)A + \omega c B = F_0$$

$$-\omega c A + \left(k - \omega^2 \frac{w}{g}\right)B = 0$$

† Equivalently, though less conventionally, the rate of attenuation could be expressed in **decibels per cycle** by means of the definition: Decibels = 20 $\ln(y_n/y_{n+2})$.

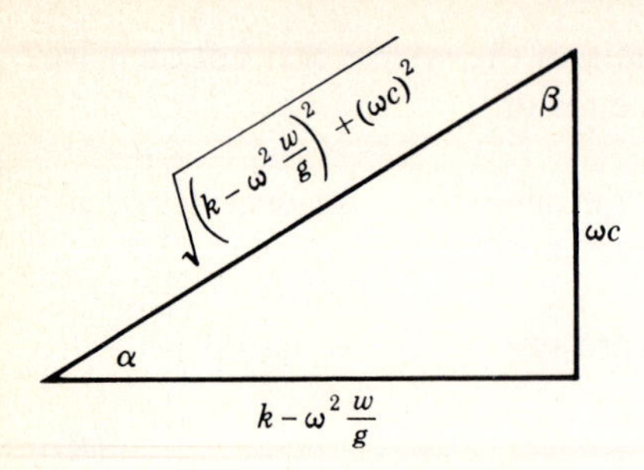

Figure 7.6 The triangle defining the phase angles appearing in Eqs. (12*a*) and (12*b*).

from which we find immediately

$$A = \frac{k - \omega^2(w/g)}{[k - \omega^2(w/g)]^2 + (\omega c)^2} F_0$$

$$B = \frac{\omega c}{[k - \omega^2(w/g)]^2 + (\omega c)^2} F_0$$

Hence

$$\begin{aligned} Y &= \frac{[k - \omega^2(w/g)] \cos \omega t + \omega c \sin \omega t}{[k - \omega^2(w/g)]^2 + (\omega c)^2} F_0 \\ &= \frac{F_0}{\sqrt{[k - \omega^2(w/g)]^2 + (\omega c)^2}} \\ &\quad \times \left\{ \frac{k - \omega^2(w/g)}{\sqrt{[k - \omega^2(w/g)]^2 + (\omega c)^2}} \cos \omega t + \frac{\omega c}{\sqrt{[k - \omega^2(w/g)]^2 + (\omega c)^2}} \sin \omega t \right\} \end{aligned}$$

Referring to the triangle shown in Fig. 7.6, it is evident that Y can be written in either of the equivalent forms

$$\begin{aligned} Y &= \frac{F_0}{\sqrt{[k - \omega^2(w/g)]^2 + (\omega c)^2}} (\cos \omega t \cos \alpha + \sin \omega t \sin \alpha) \\ &= \frac{F_0}{\sqrt{[k - \omega^2(w/g)]^2 + (\omega c)^2}} \cos (\omega t - \alpha) \end{aligned} \tag{12a}$$

or

$$\begin{aligned} Y &= \frac{F_0}{\sqrt{[k - \omega^2(w/g)]^2 + (\omega c)^2}} (\cos \omega t \sin \beta + \sin \omega t \cos \beta) \\ &= \frac{F_0}{\sqrt{[k - \omega^2(w/g)]^2 + (\omega c)^2}} \sin (\omega t + \beta) \end{aligned} \tag{12b}$$

The first of these equations is the more convenient because it involves the same function (the cosine) as the excitation term in the differential equation. Hence, the phase relation between the response of the system and the disturbing force can easily be inferred. Accordingly, we shall continue with the first expression for Y.

If we divide the numerator and denominator by k and rearrange slightly, we obtain

$$Y = \frac{F_0/k}{\sqrt{[1 - \omega^2 w/(kg)]^2 + (\omega c/k)^2}} \cos(\omega t - \alpha)$$

$$= \frac{F_0/k}{\sqrt{[1 - \omega^2/(kg/w)]^2 + [(\omega/\sqrt{kg/w})(2c/\sqrt{4kw/g})]^2}} \cos(\omega t - \alpha)$$

$$= \frac{\delta_{st}}{\sqrt{(1 - \omega^2/\omega_n^2)^2 + [2(\omega/\omega_n)(c/c_c)]^2}} \cos(\omega t - \alpha)$$

where $\delta_{st} = F_0/k$ is the **static deflection** which a *constant* force of magnitude F_0 would produce in a spring of modulus k and, as before, $\omega_n^2 = kg/w$ and $c_c^2 = 4kw/g$.

The quantity

$$M = \frac{1}{\sqrt{(1 - \omega^2/\omega_n^2)^2 + [2(\omega/\omega_n)(c/c_c)]^2}} \tag{13}$$

is called the **magnification ratio.** It is the factor by which the static deflection produced in a spring of modulus k by a steady force F_0 must be multiplied in order to give the amplitude of the vibrations which result when the same force acts dynamically with frequency ω. Curves of the magnification ratio M plotted against the **frequency ratio** ω/ω_n for various values of the **damping ratio** c/c_c are shown in Fig. 7.7. An inspection of Fig. 7.7 reveals the following interesting facts:

1. $M = 1$, regardless of the amount of damping, if $\omega/\omega_n = 0$.
2. If $0 < c/c_c < 1/\sqrt{2}$, M rises to a maximum as ω/ω_n increases from 0, the peak value of M occurring in all cases before the impressed frequency reaches the undamped natural frequency ω_n.
3. The smaller the amount of friction, the larger the maximum of M, until for conditions of undamped resonance, namely, $c/c_c = 0$ and $\omega/\omega_n = 1$, infinite magnification, i.e., a response of infinite amplitude, occurs.
4. If $c/c_c \geq 1/\sqrt{2}$, the magnification ratio decreases steadily as ω/ω_n increases from 0.
5. For all values of c/c_c, M approaches zero as the impressed frequency is raised indefinitely above the undamped natural frequency of the system.

The angle

$$\alpha = \tan^{-1} \frac{\omega c}{k - \omega^2(w/g)} \qquad 0 \leq \alpha \leq \pi$$

which appears in Eq. (12*a*) and is shown in Fig. 7.6, is known as the **phase angle** or **angle of lag** of the response. Like the magnification ratio, it can easily be expressed in terms of the dimensionless parameters ω/ω_n and c/c_c. To do this, we need only

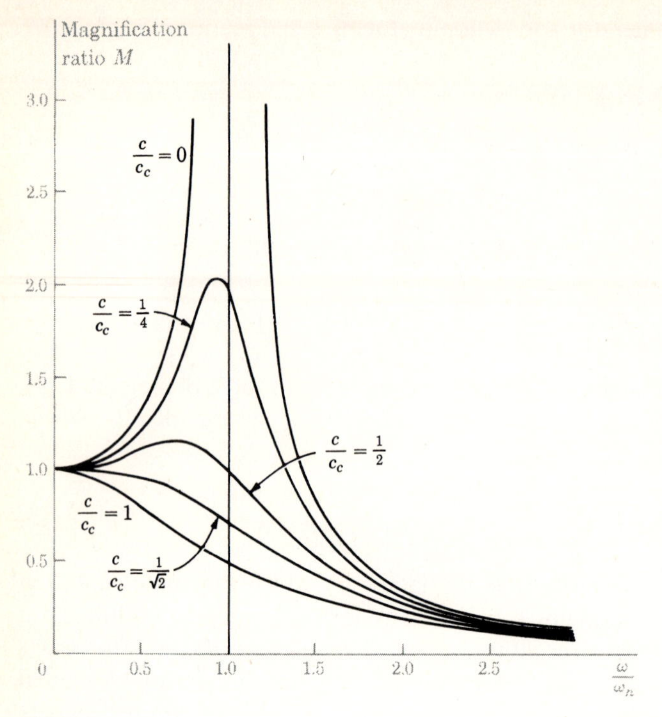

Figure 7.7 Curves of the magnification ratio M as a function of the impressed frequency ratio ω/ω_n for various amounts of damping.

divide the numerator and denominator of the right-hand side of the last expression by k and rearrange slightly:

$$\alpha = \tan^{-1} \frac{\omega c/k}{1 - \omega^2[w/(kg)]}$$

$$= \tan^{-1} \frac{(\omega/\sqrt{kg/w})(2c/\sqrt{4kw/g})}{1 - \omega^2/(kg/w)}$$

$$= \tan^{-1} \frac{2(\omega/\omega_n)(c/c_c)}{1 - (\omega/\omega_n)^2} \tag{14}$$

It is important to note that α is *not* to be read from the principal-value branch of the arctangent relation, for it is evident from Fig. 7.6 that $\sin \alpha$ is always positive, whereas $\cos \alpha$ can be either positive or negative. Hence, α must be an angle between 0 and π and not an angle in the principal-value range $(-\pi/2, \pi/2)$. Plots of α vs. the frequency ratio ω/ω_n for various values of the damping ratio c/c_c are shown in Fig. 7.8.

The physical significance of α is shown in Fig. 7.9. The displacement Y reaches its maxima α/ω units of time *after* or *later than* the driving force reaches its corresponding peak values. When the frequency of the disturbing force is well

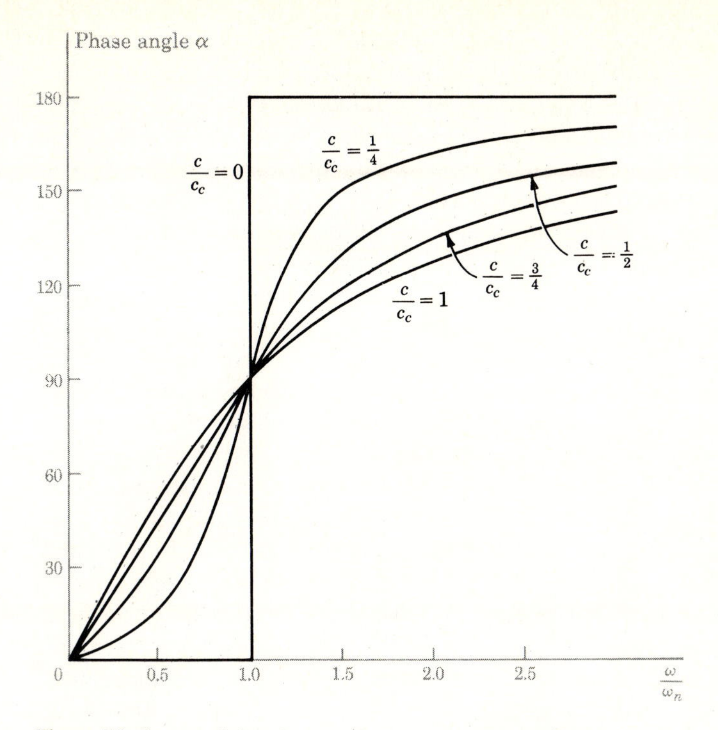

Figure 7.8 Curves of the phase angle α as a function of the impressed frequency ratio ω/ω_n for various amounts of damping.

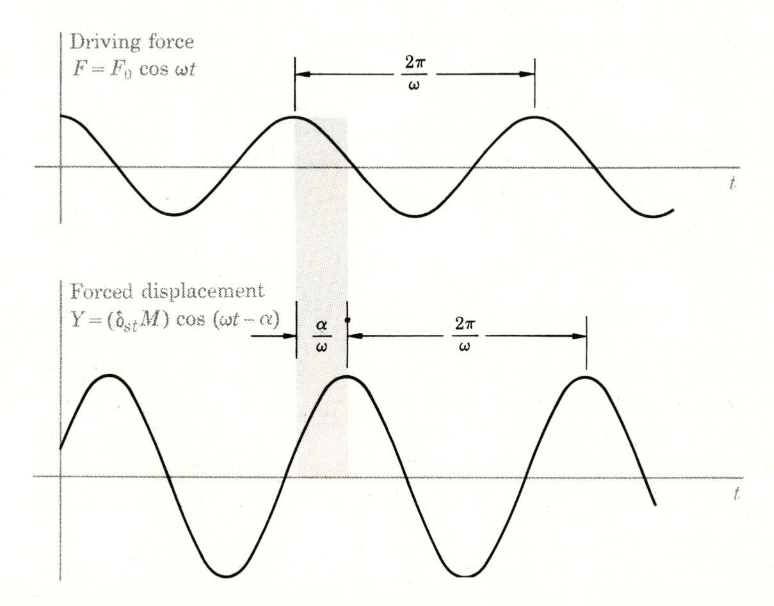

Figure 7.9 The significance of the phase angle as a measure of the time by which the response lags the excitation in a mechanical system.

below the undamped natural frequency of the system, α is small and the forced vibrations lag only slightly behind the driving force. When the impressed frequency is equal to the natural frequency, $\alpha = \pi/2$ and the response of the system lags the excitation by one-fourth cycle. As ω increases indefinitely, the lag of the response approaches half a cycle; or, in other words, the response becomes 180° out of phase with respect to the driving force.

The results of our detailed study of the vibrating weight can now be summarized. The complete motion of the system consists of two parts. The first is described by the complementary function of the underlying differential equation and may be either oscillatory or nonoscillatory, according as the amount of friction in the system is less than or more than the critical damping figure for the system. In any case, however, this part of the solution contains factors which decay exponentially, and it soon becomes vanishingly small. For this reason it is known as the **transient.** The general expression for the transient contains two arbitrary constants, which, after the complete solution has been constructed, must be determined to fit the initial conditions of displacement and velocity. The second part of the solution is described by the particular integral. In the highly important case in which the system is acted upon by a pure harmonic disturbing force (we considered only $F = F_0 \cos \omega t$, but without exception all our conclusions are equally valid for $F = F_0 \sin \omega t$), this term represents a harmonic displacement of the same frequency as the excitation but lagging behind the latter. The amplitude of this displacement is a determinate multiple of the steady deflection which would be produced in the system by a constant force of the same magnitude as the actual, alternating force. This factor of magnification, like the amount of lag, depends only on two dimensionless parameters, ω/ω_n, which is the ratio of the impressed frequency to the undamped natural frequency of the system, and c/c_c, which is the ratio of the actual amount of damping to the critical damping of the system. The motion described by the particular integral does not decay as time goes on but continues its periodic behavior indefinitely. For this reason, although it is obviously not independent of time, it is known as the **steady state.**

Example 1 An object weighing 50 lb is suspended from a spring of modulus 20 lb/in. When the system is vibrating freely, it is observed that in consecutive cycles the maximum displacement decreases by 40 percent. If a force equal to $10 \cos \omega t$ acts upon the system, find the amplitude and phase lag of the resultant steady-state motion if (*a*) $\omega = 6$, (*b*) $\omega = 12$, and (*c*) $\omega = 18$ rad/s.

The first step here is to determine the amount of damping present in the system. From the given data it is clear that

$$y_{n+2} = 0.60 y_n$$

and thus that

$$\delta = \ln \frac{y_n}{y_{n+2}} = \ln \frac{1}{0.60} = 0.511$$

Hence, by Eq. (11),

$$\frac{c}{c_c} = \frac{\delta}{\sqrt{\delta^2 + 4\pi^2}} = \frac{0.511}{\sqrt{(0.511)^2 + 4\pi^2}} = 0.081$$

Next we must compute the undamped natural frequency of the system. Using Eq. (8), we have

$$\omega_n = \sqrt{\frac{kg}{w}} = \sqrt{\frac{20 \times 384}{50}} = 12.4 \text{ rad/s}$$

Knowing c/c_c and ω_n, we can now use Eq. (13) to compute the magnification ratio and Eq. (14) to compute the phase shift for $\omega = 6$, 12, and 18. Direct substitution gives the values:

ω	M	α
6	1.30	0.10
12	5.94	1.19
18	0.88	2.93

Finally, it is clear that a 10-lb force, acting statically, will stretch a spring of modulus 20 lb/in a distance

$$\delta_{st} = \tfrac{10}{20} = 0.5 \text{ in}$$

Hence, multiplying this static deflection by the appropriate values of the magnification ratio, we find for the amplitude A of the steady-state motion the values

ω	6	12	18
A	0.65	2.97	0.44

Using these values and the corresponding values of α, we can now write the equations describing the steady-state motion in the three given cases:

ω	Y
6	$0.65 \cos (6t - 0.10)$
12	$2.97 \cos (12t - 1.19)$
18	$0.44 \cos (18t - 2.93)$

The amplitude corresponding to the impressed frequency $\omega = 12$ is much larger than either of the others because this frequency very nearly coincides with the natural frequency of the system, $\omega_n = 12.4$.

Example 2 A system containing a negligible amount of damping is disturbed from its equilibrium position by the sudden application at $t = 0$ of a force equal to $F_0 \sin \omega t$. Discuss the subsequent motion of the system if ω is close to the natural frequency ω_n.

The differential equation to be solved here is

$$\frac{w}{g}\frac{d^2y}{dt^2} + ky = F_0 \sin \omega t$$

The complementary function is, clearly,

$$A \cos \sqrt{\frac{kg}{w}}\,t + B \sin \sqrt{\frac{kg}{w}}\,t = A \cos \omega_n t + B \sin \omega_n t$$

and it is easy to verify that a particular integral is

$$Y = \frac{F_0}{k - \omega^2(w/g)} \sin \omega t = \frac{(F_0/k)(kg/w)}{(kg/w) - \omega^2} \sin \omega t = \frac{\omega_n^2 \delta_{st}}{\omega_n^2 - \omega^2} \sin \omega t$$

Hence a complete solution can be written

$$y = A \cos \omega_n t + B \sin \omega_n t + \frac{\omega_n^2 \delta_{st}}{\omega_n^2 - \omega^2} \sin \omega t$$

Since $y = 0$ when $t = 0$, we must have $A = 0$, leaving

$$y = B \sin \omega_n t + \frac{\omega_n^2 \delta_{st}}{\omega_n^2 - \omega^2} \sin \omega t \tag{15}$$

and
$$v = \frac{dy}{dt} = \omega_n B \cos \omega_n t + \frac{\omega_n^2 \delta_{st}}{\omega_n^2 - \omega^2}\, \omega \cos \omega t$$

Substituting $v = 0$ and $t = 0$ in the last equation, we obtain

$$0 = \omega_n B + \frac{\omega \omega_n^2 \delta_{st}}{\omega_n^2 - \omega^2} \qquad \text{or} \qquad B = \frac{\omega \omega_n \delta_{st}}{\omega^2 - \omega_n^2}$$

Hence, substituting into (15), we find for the required solution

$$y = \frac{\omega_n \delta_{st}}{\omega^2 - \omega_n^2} (\omega \sin \omega_n t - \omega_n \sin \omega t) \tag{16}$$

If the impressed frequency ω is very close to the natural frequency ω_n, we may, for descriptive purposes, substitute ω_n for ω in the first term in the expression in parentheses (although of course we cannot do this in the denominator of the coefficient fraction). This gives us

$$y \doteq \frac{\omega_n^2 \delta_{st}}{\omega^2 - \omega_n^2} (\sin \omega_n t - \sin \omega t)$$

If we now convert the difference of the sine terms into a product, we get

$$y \doteq -\omega_n^2 \delta_{st} \frac{2 \cos [(\omega + \omega_n)/2]t \sin [(\omega - \omega_n)/2]t}{(\omega + \omega_n)(\omega - \omega_n)}$$

If we denote the small quantity $\omega - \omega_n$ by 2ε and note that $\omega + \omega_n$ is approximately equal to both 2ω and $2\omega_n$, the last expression can be written in the form

$$y \doteq -\omega_n \delta_{st} \frac{\sin \varepsilon t}{2\varepsilon} \cos \omega t \tag{17}$$

Since ε is a small quantity, the period $2\pi/\varepsilon$ of the factor $\sin \varepsilon t$ is large. Hence, the form of the last expression shows that y can be regarded as essentially a periodic function, $\cos \omega t$, of frequency ω, with slowly varying amplitude

$$\omega_n \delta_{st} \frac{\sin \varepsilon t}{2\varepsilon}$$

Figure 7.10*a* shows the general nature of this behavior when ω is nearly but not quite equal to ω_n, and Fig. 7.10*b* depicts this behavior in the limiting case when $\omega = \omega_n$ and conditions of **pure resonance** exist.

This is one of the simplest illustrations of the phenomenon of **beats,** which occurs whenever an impressed frequency is close to a natural frequency of a system or whenever two slightly different frequencies are impressed upon a system regardless of what its natural frequencies may be. A waveform of variable amplitude, like that shown in Fig. 7.10*a*, is said to be **amplitude-modulated,** and the lighter curves to which the actual wave periodically rises and falls are called its **envelope.**

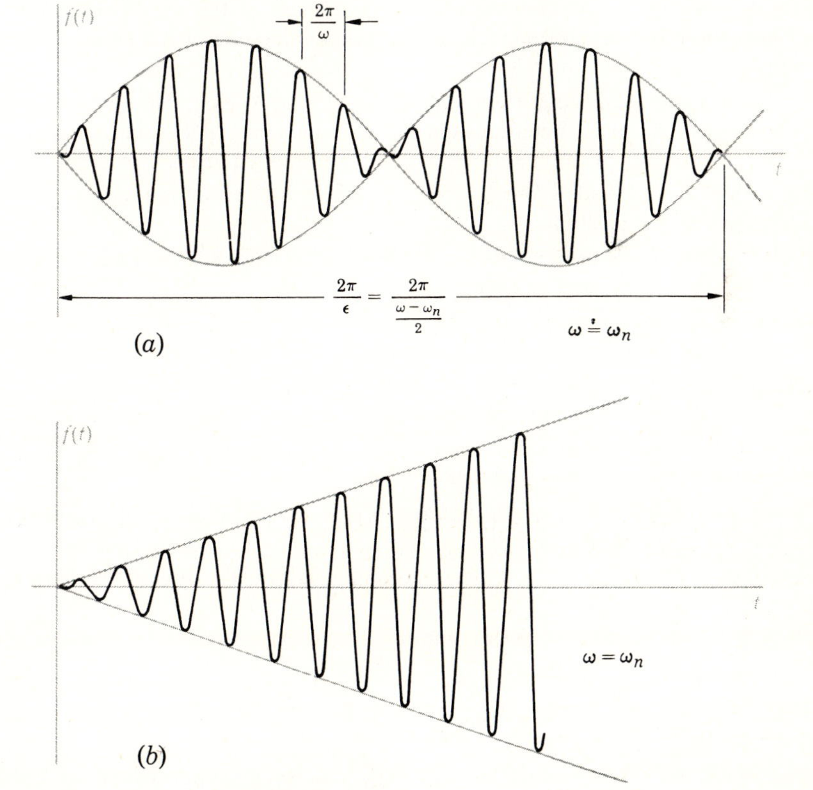

Figure 7.10 The phenomenon of beats.

Exercises for Section 7.3

1 If friction is neglected, show that the natural frequency of a system consisting of a mass on an elastic suspension is approximately equal to $3.13/\sqrt{\delta_{st}}$ Hz, where δ_{st} is the deflection, in inches, produced in the suspension when the mass hangs in static equilibrium.

2 A motor of unknown weight is set on a felt mounting pad of unknown spring constant. What is the natural frequency of the system if the motor is observed to compress the pad $\frac{1}{16}$ in?

***3** Show that the logarithmic decrement δ can also be computed by the formula

$$\delta = \frac{1}{k} \ln \frac{y_n}{y_{n+2k}} \qquad k = 1, 2, 3, \ldots$$

4 Prove that the logarithmic decrement δ is equal to the natural logarithm of the ratio of *any* nonzero displacement to the displacement one full cycle later.

5 For a given value of c/c_c, determine the minimum number of cycles required to produce a reduction of at least 50 percent in the maxima of a damped oscillation.

6 Investigate the motion of a weight hanging on a spring when the disturbing force is equal to $F_0 \sin \omega t$ instead of $F_0 \cos \omega t$. In particular, show that Eqs. (13) and (14) for the magnification ratio and phase shift, respectively, are still the same.

7 A weight of 54 lb hangs from a spring of modulus 36 lb/in. During the free motion of the system it is observed that the maximum displacement of the weight decreases to one-tenth of its value in five complete cycles of the motion. Find the equation describing the steady-state motion produced by a force equal to $6 \sin 15t$.

8 An object weighing 245 g hangs from a spring of modulus 25 g/cm. The damping in the system is 60 percent of critical. Determine the motion of the weight if it is pulled downward 1 cm from its equilibrium position and released with an upward velocity of 2 cm/s.

9 Work Exercise 8 if a constant force of 50 g is suddenly applied to the system when it is at rest in its equilibrium position.

10 Find the steady-state motion of the weight in Exercise 8 if a force equal to (*a*) $50 \cos 10t$ g and (*b*) $50 \cos t$ g acts on it.

11 A weight of 96 lb hangs from a spring of modulus 25 lb/in. The damping in the system is 60 percent of critical. Determine the motion of the weight if it is pulled downward 1 in from its equilibrium position and released with an upward velocity of 2 in/s.

12 Work Exercise 11 if a constant force of 50 lb is suddenly applied to the weight when it is at rest in its equilibrium position.

13 Work Exercise 8 if the system is critically damped.

14 Determine the motion of the weight in Exercise 8 if a force equal to $50e^{-t}$ is suddenly applied while the system is in equilibrium.

15 In the critically damped case, show that the common value of the roots of the characteristic equation is $-\omega_n$.

16 Show that Eq. (1) can be written in the form

$$\frac{d^2y}{dt^2} + 2\frac{c}{c_c}\omega_n\frac{dy}{dt} + \omega_n^2 y = \delta_{st}\omega_n^2 \cos \omega t$$

***17** If c/c_c is small, show that $\delta \doteq (y_n - y_{n+2})/y_n = \Delta y_n/y_n$ is a good approximation to the logarithmic decrement.

***18** Show that the energy dissipated during the nth cycle of a damped oscillation is equal to $(k/2)(y_n^2 - y_{n+2}^2)$. Hence, using the result of Exercise 17, show that when c/c_c is small, the energy loss during the nth cycle is approximately $ky_n^2\delta$. *Hint:* Since the force of friction is $c(dy/dt)$, the energy dissipated in one cycle is $\int c(dy/dt)\,dy$ taken over one cycle.

19 Show that the maxima of the curves of the magnification ratio vs. the frequency ratio occur when

$$\frac{\omega}{\omega_n} = \sqrt{1 - 2\left(\frac{c}{c_c}\right)^2}$$

****20** If y_0 and v_0 are, respectively, the initial displacement and the initial velocity with which an overdamped system begins its motion, show that

$$\frac{w}{g}v_0^2 + cv_0 y_0 + ky_0^2 > 0$$

is the condition that the complementary function have a real zero.

***21** Show that the extreme displacements during the free motion of an underdamped system do not occur midway between the zeros of the displacement but precede the midpoints by the constant amount

$$\frac{\text{Sin}^{-1}(c/c_c)}{\omega_n\sqrt{1-(c/c_c)^2}}$$

****22** A uniform bar of length l and weight w rests on two parallel rollers which rotate about fixed axes, as shown in Fig. 7.11. Friction between the bar and each roller is assumed to be "**dry**," or **coulomb**, i.e., proportional to the normal force between the bar and the roller, the proportionality constant being the so-called **coefficient of friction** μ. When the bar, which always remains in a line perpendicular to the axes of the rollers, is displaced slightly from a symmetrical position, it executes small oscillations in the horizontal direction. Determine the period of these oscillations and show how the value of μ can thus be found experimentally.

****23** A particle of weight w moves along the x axis under the influence of a force equal to $-kx$. Friction in the system is assumed to be dry rather than viscous; i.e., it is proportional to the normal force between the particle and the surface on which it moves and does not depend on the velocity. Show that the motion of the particle is governed by the differential equations

$$\frac{w}{g}\frac{d^2x}{dt^2} + kx = \begin{cases} \mu w & \text{particle moving to left} \\ -\mu w & \text{particle moving to right} \end{cases}$$

If the particle starts from rest at the point $x = x_0$, find x as a function of t. What is the decrease in amplitude per cycle? When will the particle come to rest?

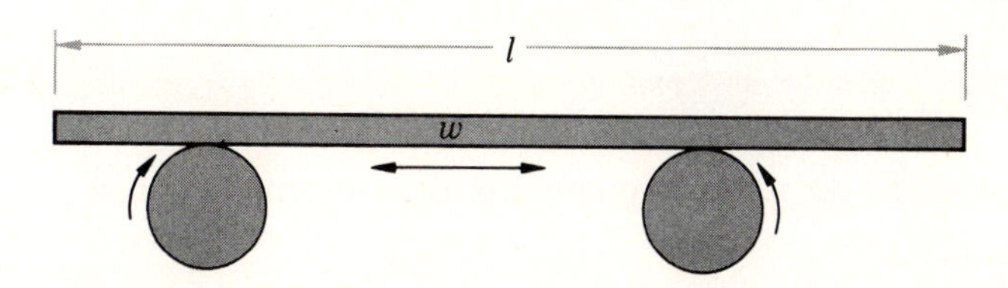

Figure 7.11

*24 An object weighing 48 lb hangs from a spring of modulus 50 lb/in. In 10 cycles of the motion it is observed that the maximum displacement decreases by 50 percent. Determine the steady-state motion of the system if it is acted upon simultaneously by forces equal to $F_0 \cos 15t$ and $F_0 \cos 16t$. Do you think that these two forces will produce the phenomenon of beats? Why?

**25 A spring-suspended weight is acted upon by two forces

$$F_1 \sin \omega_1 t \qquad \text{and} \qquad F_2 \sin \omega_2 t$$

Friction, though present in the system, is so small that it can be neglected in determining the forced motion. Discuss the steady-state behavior of the system if ω_1 and ω_2 are nearly equal but neither is close to the natural frequency of the system. In particular, show that the response consists of a term of frequency $(\omega_1 + \omega_2)/2$ whose amplitude is modulated by a factor of frequency $(\omega_1 - \omega_2)/2$, and determine the limits between which the amplitude varies. *Hint:* Note first that the assumption of at least a little friction in the system implies that only the particular integrals contribute to the steady-state motion. (Why?) Then, after the particular integrals have been determined, note that the expression $K_1 \sin \omega_1 t + K_2 \sin \omega_2 t$ can be written

$$\frac{K_1 + K_2}{2}(\sin \omega_1 t + \sin \omega_2 t) + \frac{K_1 - K_2}{2}(\sin \omega_1 t - \sin \omega_2 t)$$

*26 In Example 2, show that if the substitution $\omega = \omega_n + 2\varepsilon$ is made in Eq. (16), the result (without any approximations) is

$$y = -\frac{\omega_n^2 \delta_{st}}{2(\omega_n + \varepsilon)\varepsilon} \sin \varepsilon t \cos (\omega_n + \varepsilon)t + \frac{\omega_n \delta_{st}}{2(\omega_n + \varepsilon)} \sin \omega_n t$$

Discuss the extent to which the approximation provided by Eq. (17) is consistent with the last equation as a description of the motion when ε is close to zero.

7.4 THE SERIES-ELECTRICAL CIRCUIT

All the results of the last section can, after a suitable change in terminology, be applied to any of the other systems we have considered. However, the concepts central in one field are not always of equal importance in related fields, and it seems desirable to illustrate the minor differences in the application of our general theory to various classes of systems by considering the series-electrical circuit in some detail.

For the simple series circuit with an alternating impressed voltage, we derived (among several equivalent forms) the equation

$$L\frac{d^2Q}{dt^2} + R\frac{dQ}{dt} + \frac{1}{C}Q = E_0 \cos \omega t \tag{1}$$

and on comparing this with the differential equation of the vibrating weight,

$$\frac{w}{g}\frac{d^2y}{dt^2} + c\frac{dy}{dt} + ky = F_0 \cos \omega t$$

we noted the correspondences:

$$\text{Mass } \frac{w}{g} \leftrightarrow \text{inductance } L$$

$$\text{Friction } c \leftrightarrow \text{resistance } R$$

$$\text{Spring modulus } k \leftrightarrow \text{elastance } \frac{1}{C}$$

$$\text{Impressed force } F \leftrightarrow \text{impressed voltage } E$$

$$\text{Displacement } y \leftrightarrow \text{charge } Q$$

$$\text{Velocity } v \leftrightarrow \text{current } i$$

Extending this correspondence to the derived results by making the appropriate substitutions, we infer from the undamped natural frequency of the mechanical system

$$\omega_n = \sqrt{\frac{kg}{w}}$$

that the electric circuit has a natural frequency

$$\Omega_n = \sqrt{\frac{1}{LC}}$$

when no resistance is present. Furthermore, the concept of critical damping

$$c_c = 2\sqrt{\frac{kw}{g}}$$

leads to the concept of critical resistance

$$R_c = 2\sqrt{\frac{L}{C}}$$

which determines whether the free behavior of the electrical system will be oscillatory or nonoscillatory.

The notion of magnification ratio can also be extended to the electrical case, but it is not customary to do so because the extension would relate to Q (the analog of the displacement y), whereas in most electrical problems it is not Q but i which is the variable of interest. To see how a related concept arises in the electrical case, let us convert the particular integral Y given by Eq. (12*b*), Sec. 7.3, into its electrical equivalent. By direct substitution (using Fig. 7.6 to obtain the phase angle β) the result is found to be

$$Q = \frac{E_0 \sin(\omega t + \beta)}{\sqrt{(1/C - \omega^2 L)^2 + (\omega R)^2}} \qquad \beta = \text{Tan}^{-1} \frac{1/C - \omega^2 L}{\omega R}$$

To obtain the current i, we differentiate this, getting

$$\frac{dQ}{dt} = i = \frac{E_0 \omega \cos(\omega t + \beta)}{\sqrt{(1/C - \omega^2 L)^2 + (\omega R)^2}}$$

or, dividing numerator and denominator by ω in the expressions for both i and β and then introducing a new phase angle $\delta = -\beta$,

$$i = \frac{E_0 \cos(\omega t - \delta)}{\sqrt{R^2 + [\omega L - 1/(\omega C)]^2}} \tag{2}$$

where
$$\delta = -\beta = \operatorname{Tan}^{-1} \frac{\omega L - 1/(\omega C)}{R} \tag{3}$$

From Eq. (2) we infer that the steady-state current produced by an alternating voltage is of the same frequency as the voltage but differs from it in phase by

$$\frac{\delta}{\omega} \text{ units of time} \qquad \text{or} \qquad \frac{\delta/\omega}{2\pi/\omega} = \frac{\delta}{2\pi} \text{ cycles}$$

Moreover, from Eq. (3) it is clear that the numerator of $\tan \delta$ (which is proportional to $\sin \delta$) can be either positive or negative, whereas the denominator of $\tan \delta$ (which is proportional to $\cos \delta$) is always positive. Hence δ must be an angle between $-\pi/2$ and $\pi/2$, and so the principal-value designation in Eq. (3) is appropriate. If δ is positive, the steady-state current *lags* the voltage; if δ is negative, the steady-state current *leads* the voltage.

Furthermore, from Eq. (2) we see that the amplitude of the steady-state current is obtained by dividing the amplitude of the impressed voltage E_0 by the expression

$$\sqrt{R^2 + [\omega L - 1/(\omega C)]^2} \tag{4}$$

By analogy with Ohm's law, $I = E/R$, the quantity (4) thus appears as a generalized resistance, although it is actually called the **impedance** of the circuit. While not the analog of the magnification ratio, the impedance is clearly a similar concept. Since impedance is defined as

$$\frac{\text{Voltage}}{\text{Current}}$$

the mechanical quantity corresponding to this is the ratio

$$\frac{\text{Force}}{\text{Velocity}}$$

This is called the **mechanical impedance** by some writers and in certain mechanical problems has proved a useful notion.†

† See, for instance, T. von Kármán and M. A. Biot, "Mathematical Methods in Engineering," McGraw-Hill, New York, 1940, pp. 370–78.

There is another approach to the problem of determining the steady-state current produced by a harmonic voltage that is well worth investigating. Suppose that given *either* $E = E_0 \cos \omega t$ *or* $E_0 \sin \omega t$, we write the basic differential equation (1) in the form

$$L\frac{d^2Q}{dt^2} + R\frac{dQ}{dt} + \frac{1}{C}Q = E_0 e^{j\omega t} = E_0 (\cos \omega t + j \sin \omega t)\dagger \tag{5}$$

This includes both possibilities for the voltage, and if the real and the imaginary terms retain their identity throughout the analysis, then the real part of the particular integral corresponding to $E_0 e^{j\omega t}$ will be the particular integral for $E_0 \cos \omega t$ and the imaginary part will be the particular integral for $E_0 \sin \omega t$.

To see that this is actually the case, we must first find a particular integral of Eq. (5). As usual, we do this by assuming

$$Q = Ae^{j\omega t}$$

and substituting into the differential equation. This gives

$$L(-\omega^2 Ae^{j\omega t}) + R(j\omega Ae^{j\omega t}) + \frac{1}{C}(Ae^{j\omega t}) = E_0 e^{j\omega t}$$

which will be an identity if and only if

$$A = \frac{E_0}{-\omega^2 L + j\omega R + 1/C}$$

Hence
$$Q = \frac{E_0}{j\omega R - \omega^2 L + 1/C} e^{j\omega t}$$

From this, by differentiation, we find that

$$\frac{dQ}{dt} = i = \frac{j\omega E_0}{j\omega R - \omega^2 L + 1/C} e^{j\omega t} = \frac{E_0}{R + j(\omega L - 1/\omega C)} e^{j\omega t}$$

To find the real and imaginary parts of this expression, it is convenient to use the fact that any complex number $a + jb$ can be written in the form $a + jb = re^{j\delta}$, where the magnitude r and the angle δ of the complex number are related to the components a and b as shown in Fig. 7.12. Applied to the denominator of the second expression for i, this gives

$$R + j\left(\omega L - \frac{1}{\omega C}\right) = \sqrt{R^2 + \left(\omega L - \frac{1}{\omega C}\right)^2} e^{j\delta}$$

where
$$\delta = \text{Tan}^{-1} \frac{\omega L - 1/(\omega C)}{R}$$

† To avoid confusing $i = \sqrt{-1}$ with i = current, we shall throughout the rest of this chapter follow the practice, standard in electrical engineering, of writing $\sqrt{-1} = j$.

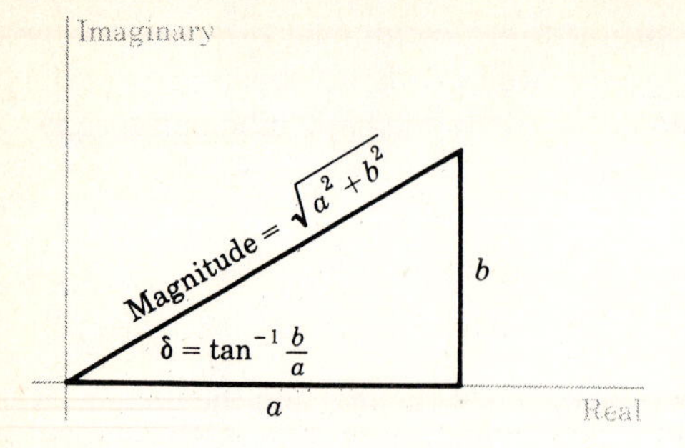

Figure 7.12 The relations between the magnitude, angle, and components of a general complex number $a + jb$.

Hence we can rewrite i in the form

$$i = \frac{E_0}{\sqrt{R^2 + [\omega L - 1/(\omega C)]^2}e^{j\delta}} e^{j\omega t}$$

$$= \frac{E_0}{\sqrt{R^2 + [\omega L - 1/(\omega C)]^2}} e^{j(\omega t - \delta)}$$

$$= E_0 \frac{\cos(\omega t - \delta) + j \sin(\omega t - \delta)}{\sqrt{R^2 + [\omega L - 1/(\omega C)]^2}}$$

Comparing this with Eqs. (2) and (3) makes it clear that the real part here is exactly the particular integral corresponding to $E_0 \cos \omega t$, as we derived it directly. Similarly, had we taken the trouble to work it out explicitly, we would have found for the particular integral corresponding to $E_0 \sin \omega t$ precisely the imaginary part of the last expression. Since it is much easier to find the particular integral corresponding to an exponential term than it is to find the particular integral corresponding to a cosine or sine term, the advantage of using $E_0 e^{j\omega t}$ in place of $E_0 \cos \omega t$ or $E_0 \sin \omega t$ is obvious.

The expression

$$R + j\left(\omega L - \frac{1}{\omega C}\right) \qquad \text{or} \qquad j\omega L + R + \frac{1}{j\omega C}$$

is called the **complex impedance** Z. Its magnitude is the quantity (4) which we referred to simply as the impedance. Its angle δ is the **phase shift.** The real part of Z is clearly a resistance. The imaginary part of Z is called the **reactance.** The reciprocal of Z is called the **admittance.** The real part of the admittance is called the **conductance,** and the imaginary part is called the **susceptance.**

The most striking property of the complex impedance is that when any electric elements are connected in series or in parallel, the corresponding impedances combine just as simple resistances do. Thus the steady-state current through a series of Z's (Fig. 7.13) can be found by dividing the impressed voltage by the single impedance

$$Z = Z_1 + Z_2 + \cdots + Z_n$$

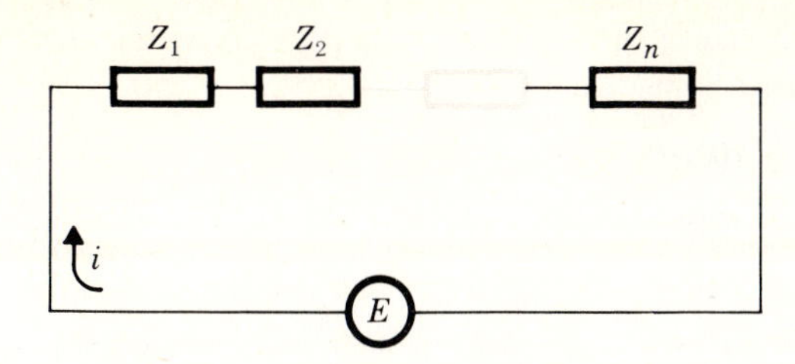

Figure 7.13 Impedances connected in series.

Similarly, the current through a set of elements connected in parallel (Fig. 7.14) can be found by dividing the impressed voltage by the single impedance Z defined by the relation

$$\frac{1}{Z} = \frac{1}{Z_1} + \frac{1}{Z_2} + \cdots + \frac{1}{Z_n}$$

This makes it unnecessary to use differential equations in determining the *steady-state* behavior of an electric network (or of a mechanical system, if the concept of mechanical impedance is used).

Example 1 A series circuit contains the elements $L = 1$ H, $R = 1{,}000\ \Omega$, and $C = 6.25 \times 10^{-6}$ F. If an alternating voltage $E = 100 \cos 200t$ is switched into the circuit, find the resultant steady-state current. What is the amplitude of the resultant current? By what fraction of a cycle does it lag or lead the voltage which produces it?

The behavior of the given circuit is described by the differential equation

$$\frac{d^2Q}{dt^2} + 1{,}000\frac{dQ}{dt} + \frac{Q}{6.25 \times 10^{-6}} = 100 \cos 200t \tag{6}$$

However, since we are asked only for the steady-state current, there is no need to obtain the complementary function of this equation, for that gives only the transient current. To find the steady-state current, it will be convenient to use the impedance concept, since that leads most directly to the amplitude and phase relations that we are

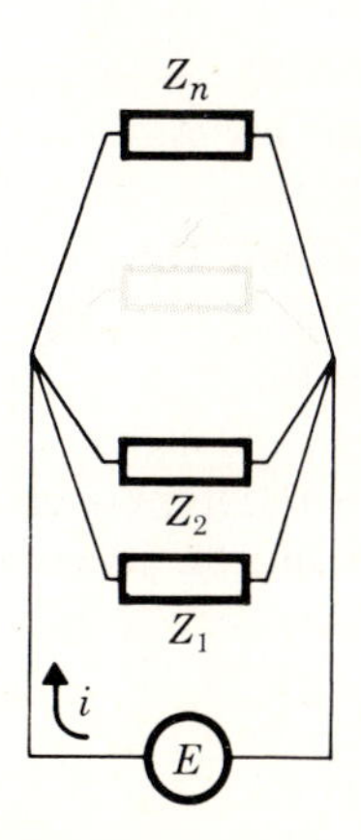

Figure 7.14 Impedances connected in parallel.

asked to determine. Accordingly, we first imagine that the actual voltage $E = 100 \cos 200t$ in Eq. (6) is replaced by the complex voltage

$$100e^{j200t}$$

Next, we compute the complex impedance

$$Z = R + j\left(\omega L - \frac{1}{\omega C}\right) = 1{,}000 + j\left(200 - \frac{160{,}000}{200}\right) = 1{,}000 - 600j$$

and convert it to exponential form, getting

$$Z = \sqrt{1{,}000^2 + 600^2}\, e^{j\delta}$$

$$= 200\sqrt{34}\, e^{j\delta} \qquad \text{where } \delta = \text{Tan}^{-1}\left(-\frac{600}{1{,}000}\right) = -0.29 \text{ rad}$$

Then we divide the complex voltage by the complex impedance to get the complex current:

$$\frac{E_0 e^{j\omega t}}{Z} = \frac{100e^{j200t}}{200\sqrt{34}\, e^{-0.29j}} = \frac{\sqrt{34}}{68} e^{j(200t+0.29)}$$

$$\doteq 0.086[\cos(200t + 0.29) + j \sin(200t + 0.29)]$$

The real part of the last expression is the required steady-state current:

$$i_{ss} \doteq 0.086 \cos(200t + 0.29)$$

The amplitude of the steady-state current is 0.086 A. The phase angle of 0.29 rad represents $0.29/(2\pi) = 0.046$ cycle, and because it is positive here, i.e., negative in the standard form (2), the current leads the voltage by this amount.

Exercises for Section 7.4

1 Work Example 1 if instead of the voltage $E = 100 \cos 200t$ V, a constant voltage $E = 24$ V is switched into the circuit when $t = 0$.

2 In Exercise 1 find the potential difference across each element as a function of time.

3 An open series circuit contains the elements $L = 0.01$ H, $R = 250\ \Omega$, $C = 10^{-6}$ F. At $t = 0$, with the capacitor charged to the value $Q_0 = 10^{-5}$ C, the circuit is closed. Find the resultant current as a function of time.

4 Work Exercise 3, given that the circuit elements are $L = 6.4 \times 10^{-3}$ H, $R = 1.6 \times 10^2\ \Omega$, $C = 10^{-6}$ F.

5 Work Exercise 3, given that the circuit elements are $L = 0.01$ H, $R = 120\ \Omega$, $C = 10^{-6}$ F.

6 A series circuit in which $Q_0 = i_0 = 0$ contains the elements $L = 0.15$ H, $R = 800\ \Omega$, $C = 4 \times 10^{-6}$ F. If a constant voltage $E = 25$ V is suddenly switched into the circuit, find the resultant current as a function of time.

7 Work Exercise 6, given that the circuit elements are $L = 0.16$ H, $R = 800\ \Omega$, $C = 10^{-6}$ F.

8 For the series circuit, what is the analog of the static deflection δ_{st}?

9 Show that Eq. (1) can be written in the form

$$\frac{d^2Q}{dt^2} + 2\frac{R}{R_c}\Omega_n\frac{dQ}{dt} + \Omega_n^2 Q = Q_{st}\Omega_n^2 \cos \omega t$$

where Q_{st} is the quantity identified in Exercise 8 as the analog of the static deflection δ_{st}.

10 Find the particular integral of Eq. (1) corresponding to a driving voltage $E_0 \sin \omega t$, and verify that it leads to a current equal to the imaginary part of the current produced by the complex voltage $E_0 e^{j\omega t}$.

11 (*a*) Prove that if a set of elements with impedance Z_1 is connected in series with a set of elements with impedance Z_2, the impedance of the resultant combination is $Z_1 + Z_2$.

(*b*) Prove that if a set of elements with impedance Z_1 is connected in parallel with a set of elements with impedance Z_2, the impedance Z of the resultant combination is given by

$$\frac{1}{Z} = \frac{1}{Z_1} + \frac{1}{Z_2}$$

12 For what value of ω is the impedance $\sqrt{R^2 + [\omega L - 1/(\omega C)]^2}$ a minimum? How does this compare with the corresponding property of the magnification ratio? Explain.

***13** If the frequency of the voltage $E_0 \cos \omega t$ impressed on a series circuit is the same as the (natural) frequency of the circuit in the absence of any resistance, show that the amplitudes of the steady-state potential differences across the inductor and the capacitor are each equal to $E_0 R_c/(2R)$.

***14** Instead of using the ratio R/R_c as a dimensionless parameter in circuit analysis, it is customary to use the so-called **quality factor Q** (not to be confused with the charge Q) defined to be $R_c/(2R)$. Express the impedance and the phase angle for a simple series circuit in terms of the resistance R, the frequency ratio ω/Ω_n, and the quality factor Q.

***15** A constant voltage is suddenly switched into a nonoscillatory series circuit in which $Q_0 = i_0 = 0$. Show that the potential difference across the capacitor can never overshoot its final value.

7.5 SYSTEMS WITH SEVERAL DEGREES OF FREEDOM

A system which can be described by one coordinate, i.e., by one physical datum such as a displacement, an angle, a current, or a voltage, is called a **system of one degree of freedom.** A system requiring more than one coordinate (or dependent variable) for its complete description is called a **system of several degrees of freedom.** As the work of Sec. 4.6 and the preceding work in this chapter illustrate, a single differential equation suffices for the mathematical description of a system of one degree of freedom. On the other hand, the analysis of a system consisting of several spring-connected masses or several interconnected series circuits involves a set of simultaneous differential equations, as many equations as there are degrees of freedom. In general, regardless of the number of degrees of freedom, these equations arise from the application of the laws of Newton or Kirchhoff to the component parts of the total system, essentially as in the case of a system with a single degree of freedom. The details of such applications can best be made clear through examples.†

Example 1 While the system shown in Fig. 7.15 is at rest in its equilibrium position, a force $F = 90 \sin 25t$ g is suddenly applied to the upper weight. Assuming that friction is negligible, find the subsequent displacement of each weight as a function of time.

As usual, we suppose that the weights are guided, by constraints that need not be specified, so that they can move only in the vertical direction. The instantaneous displacements of the masses from their equilibrium positions we shall use as coordinates to describe the system, with displacements above the equilibrium positions being considered positive. Since friction is assumed to be negligible, the only forces acting on the weights besides the attraction of gravity are those transmitted to them by the attached springs. Moreover, as suggested by the derivation of Eq. (5), Sec. 4.6, and confirmed for systems with two degrees of freedom by Exercise 20 at the end of this section, the force of gravity can be neglected provided that we also neglect the initial elongations of the springs and assume that each is unstretched when the system is in equilibrium.

When the displacements of the weights w_1 and w_2 are y_1 and y_2, respectively, the upper spring is changed in length by the amount y_1 and the lower spring is changed in length by the amount $y_1 - y_2$. Because of these changes in length, the springs exert forces equal to

$$45y_1 \qquad \text{and} \qquad 60(y_1 - y_2)$$

respectively. The masses of the weights are

$$m_1 = \frac{w_1}{g} = \frac{147}{980} = \frac{3}{20} \qquad \text{and} \qquad m_2 = \frac{w_2}{g} = \frac{196}{980} = \frac{1}{5}$$

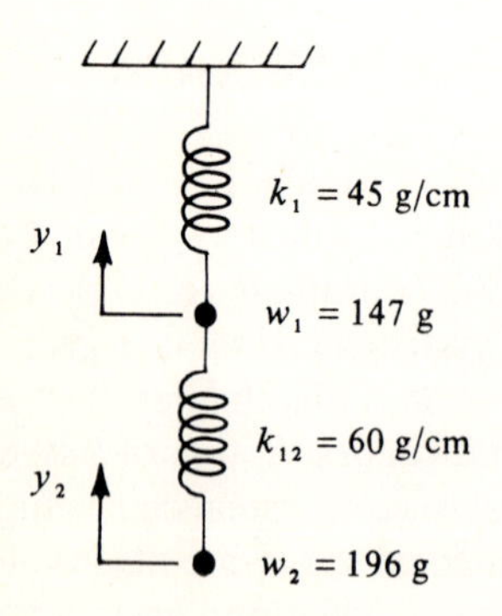

Figure 7.15 A simple mass-spring system.

† One such illustration is provided by Example 1, Sec. 5.3.

Hence, applying Newton's law to each weight and taking due account of the direction of the forces applied to each mass by the attached springs, we have

$$\frac{3}{20}\frac{d^2y_1}{dt^2} = -45y_1 - 60(y_1 - y_2) + 90 \sin 25t$$

$$\frac{1}{5}\frac{d^2y_2}{dt^2} = 60(y_1 - y_2)$$

or

$$(D^2 + 700)y_1 - 400y_2 = 600 \sin 25t$$

$$-300y_1 + (D^2 + 300)y_2 = 0$$

Taking the Laplace transformation of each equation, remembering that by hypothesis all initial displacements and velocities are zero, we have

$$(s^2 + 700)\mathscr{L}\{y_1\} - 400\mathscr{L}\{y_2\} = \frac{15{,}000}{s^2 + 625}$$

$$-300\mathscr{L}\{y_1\} + (s^2 + 300)\mathscr{L}\{y_2\} = 0$$

Hence

$$\mathscr{L}\{y_1\} = \frac{\begin{vmatrix} \dfrac{15{,}000}{s^2 + 625} & -400 \\ 0 & s^2 + 300 \end{vmatrix}}{\begin{vmatrix} s^2 + 700 & -400 \\ -300 & s^2 + 300 \end{vmatrix}} = \frac{15{,}000(s^2 + 300)}{(s^2 + 100)(s^2 + 900)(s^2 + 625)}$$

$$\mathscr{L}\{y_2\} = \frac{\begin{vmatrix} s^2 + 700 & \dfrac{15{,}000}{s^2 + 625} \\ -300 & 0 \end{vmatrix}}{\begin{vmatrix} s^2 + 700 & -400 \\ -300 & s^2 + 300 \end{vmatrix}} = \frac{4{,}500{,}000}{(s^2 + 100)(s^2 + 900)(s^2 + 625)}$$

Applying the third Heaviside expansion theorem (Theorem 3, Sec. 6.7) to $\mathscr{L}\{y_1\}$, we have:
For the factor $s^2 + 100$:

$$\phi(s) = \frac{15{,}000(s^2 + 300)}{(s^2 + 900)(s^2 + 625)}$$

$$\phi(a + jb) = \phi(0 + j10) = \frac{15{,}000(200)}{(800)(525)} = \frac{50}{7}$$

Therefore $\phi_r = \frac{50}{7}$, $\phi_j = 0$, and the corresponding inverse is

$$\tfrac{1}{10}(\tfrac{50}{7} \sin 10t) = \tfrac{5}{7} \sin 10t$$

For the factor $s^2 + 900$:

$$\phi(s) = \frac{15{,}000(s^2 + 300)}{(s^2 + 100)(s^2 + 625)}$$

$$\phi(a + jb) = \phi(0 + j30) = \frac{15{,}000(-600)}{(-800)(-275)} = -\frac{450}{11}$$

Therefore $\phi_r = -\frac{450}{11}$, $\phi_j = 0$, and the corresponding inverse is

$$\tfrac{1}{30}(-\tfrac{450}{11} \sin 30t) = -\tfrac{15}{11} \sin 30t$$

For the factor $s^2 + 625$:

$$\phi(s) = \frac{15{,}000(s^2 + 300)}{(s^2 + 100)(s^2 + 900)}$$

$$\phi(a + jb) = \phi(0 + j25) = \frac{15{,}000(-325)}{(-525)(275)} = \frac{2{,}600}{77}$$

Therefore $\phi_r = 2{,}600/77$, $\phi_j = 0$, and the corresponding inverse is

$$\tfrac{1}{25}(\tfrac{2{,}600}{77} \sin 25t) = \tfrac{104}{77} \sin 25t$$

Thus, combining the three inverses,

$$y_1 = \tfrac{5}{7} \sin 10t - \tfrac{15}{11} \sin 30t + \tfrac{104}{77} \sin 25t$$

In exactly the same way, we find that the inverse of $\mathscr{L}\{y_2\}$ is

$$y_2 = \tfrac{15}{14} \sin 10t + \tfrac{15}{22} \sin 30t - \tfrac{96}{77} \sin 25t$$

It is worth noting that the terms involving $\sin 10t$ and $\sin 30t$ are the components of the complementary function of the system; and if there were even the slightest amount of friction, they would each contain a negative exponential factor and would therefore decay to zero. Thus if friction were not neglected, the steady-state motion of the weights would be approximately described by the terms

$$y_1 = \tfrac{104}{77} \sin 25t \qquad y_2 = -\tfrac{96}{77} \sin 25t$$

The true steady-state would differ from these expressions because of slightly different amplitudes and the presence of phase angles in each term, the changes depending on the amount of friction.

Example 2 A uniform bar 4 ft long and weighing 16 lb/ft is supported as shown in Fig. 7.16, on springs of moduli 24 and 15 lb/in, respectively. If the springs are guided

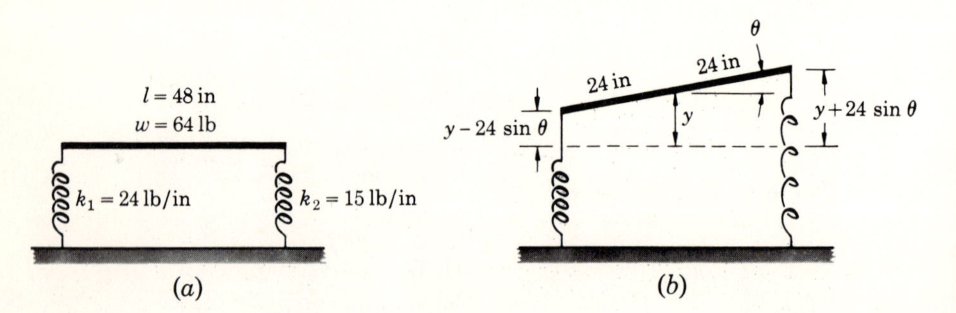

Figure 7.16 A spring-mounted bar.

so that only vertical displacement of the center of the bar is possible and if friction is neglected, find the (natural) frequencies at which the system would *begin* to vibrate if disturbed slightly from its equilibrium position.

This is a system with two degrees of freedom, because two coordinates suffice to specify it completely, for instance, the vertical displacement y of the center of gravity of the bar (with y positive upward), and the angle of rotation θ of the bar about its center of gravity (with θ positive in the counterclockwise direction). Under the requirements of the problem, we are to find the frequencies at which vibrations of arbitrarily small amplitudes can occur. In particular, this means that the bar turns through angles so small that the horizontal displacements of its ends can be neglected and each end can be assumed to move in a purely vertical line.

From Fig. 7.16*b*, it is clear that the instantaneous deflections of the left and right springs are, respectively,

$$y - 24 \sin \theta \qquad \text{and} \qquad y + 24 \sin \theta$$

or, if we make the usual small-angle approximation $\sin \theta = \theta$,

$$y - 24\theta \qquad \text{and} \qquad y + 24\theta$$

Hence the forces which the springs apply to the ends of the bar are

$$-24(y - 24\theta) \qquad \text{and} \qquad -15(y + 24\theta)$$

Newton's law applied to the translation of the center of gravity of the bar therefore gives the equation†

$$\frac{64}{384}\frac{d^2y}{dt^2} = -24(y - 24\theta) - 15(y + 24\theta)$$

or

$$\frac{d^2y}{dt^2} + 234y - 1{,}296\theta = 0 \tag{1}$$

Recalling our assumption that θ is so small that the ends of the bar move in a purely vertical direction, it is clear that the moment arm of each spring force about the center of gravity of the bar is 24 in. Hence, computing the torques applied to the ends of the bar by the spring forces and then applying Newton's law in torsional form to the rotation of the bar about its center of gravity, we have (using the fact that the moment of inertia of a uniform bar of length l about its midpoint is $ml^2/12$)

$$\frac{64}{384}\frac{(48)^2}{12}\frac{d^2\theta}{dt^2} = 24[24(y - 24\theta)] - 24[15(y + 24\theta)]$$

or

$$\frac{d^2\theta}{dt^2} - \frac{27}{4}y + 702\theta = 0 \tag{2}$$

† As we pointed out in the derivation of Eq. (5), Sec. 4.6, we can neglect the gravitational force on the bar and the portion of the elastic forces due to the compression of the springs in their equilibrium position. An equivalent point of view is to imagine that the motion takes place not in a vertical plane but on a frictionless horizontal plane, so that gravitational effects are irrelevant.

To find the solutions of the simultaneous, homogeneous differential equations (1) and (2), we can set up the characteristic equation and proceed as we did in Chap. 5;† or we may take advantage of the fact that because friction was neglected, neither of these equations contains any first-derivative terms. To capitalize on this, we assume solutions of the special form‡

$$y = A \cos \omega t \qquad \text{and} \qquad \theta = B \cos \omega t$$

and substitute them into the differential equations, dividing out the common factor $\cos \omega t$ as we go. The result is the pair of algebraic equations

$$(-\omega^2 + 234)A - 1{,}296B = 0$$

$$-\tfrac{27}{4}A + (-\omega^2 + 702)B = 0$$

In order that there should be nontrivial values of A and B satisfying these equations, it is necessary and sufficient that the determinant of the coefficients should vanish. This gives the condition

$$\begin{vmatrix} -\omega^2 + 234 & -1{,}296 \\ -\frac{27}{4} & -\omega^2 + 702 \end{vmatrix} = 0 \qquad \text{or} \qquad \omega^4 - 936\omega^2 + 155{,}520 = 0$$

Solving this equation, we find

$$\omega_1^2 = 216 \qquad \omega_2^2 = 720$$

or

$$\omega_1 = 6\sqrt{6} \qquad \omega_2 = 12\sqrt{5} \text{ rad/s}$$

For these frequencies, and for these only, there are nontrivial solutions of the two simultaneous differential equations. In other words, for all other frequencies the only solutions are $y = \theta = 0$. Thus ω_1 and ω_2 are the natural frequencies of the system, under the assumption of small-amplitude vibrations. Converted from radians per second to cycles per second, they are

$$f_1 = \frac{6\sqrt{6}}{2\pi} \doteq 2.34 \text{ Hz} \qquad \text{and} \qquad f_2 = \frac{12\sqrt{5}}{2\pi} \doteq 4.27 \text{ Hz}$$

This problem is typical of many in the field of mechanical vibrations, in that only the natural frequencies were required. Accordingly there was no need to construct a complete solution or to determine arbitrary constants to fit prescribed initial conditions.

Example 3 In the circuit shown in Fig. 7.17, find the current in each loop as a function of time, given that all charges and currents are zero when the switch is closed at $t = 0$.

We take as variables the currents i_1 and i_2 in the respective loops, noting that the current in the common branch is therefore $i_1 - i_2$. Applying Kirchhoff's second law to each loop, we obtain the equations

$$0.5\frac{di_1}{dt} + 200(i_1 - i_2) = 50$$

$$300i_2 + 200(i_2 - i_1) + \frac{1}{50 \times 10^{-6}}\int^t i_2\,dt = 0$$

† Of course, we could also use the Laplace transformation.

‡ The assumptions $y = A \sin \omega t$ and $\theta = B \sin \omega t$ would have worked just as well.

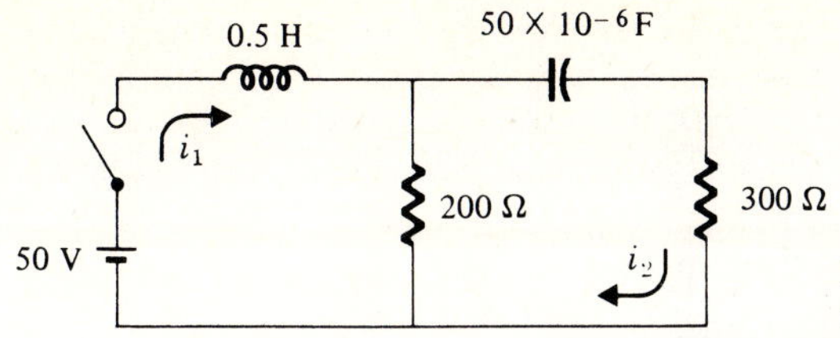

Figure 7.17 A simple two-loop electric circuit.

or

$$\frac{di_1}{dt} + 400i_1 - 400i_2 = 100$$

$$-2i_1 + 5i_2 + 200\int^t i_2\,dt = 0$$

Taking the Laplace transform of each equation, recalling that all initial currents and charges are zero, we obtain

$$s\mathscr{L}\{i_1\} + 400\mathscr{L}\{i_1\} - 400\mathscr{L}\{i_2\} = \frac{100}{s}$$

$$-2\mathscr{L}\{i_1\} + 5\mathscr{L}\{i_2\} + 200\frac{1}{s}\mathscr{L}\{i_2\} = 0$$

or

$$(s^2 + 400s)\mathscr{L}\{i_1\} - 400s\mathscr{L}\{i_2\} = 100$$

$$-2s\mathscr{L}\{i_1\} + (5s + 200)\mathscr{L}\{i_2\} = 0$$

Hence

$$\mathscr{L}\{i_1\} = \frac{\begin{vmatrix} 100 & -400s \\ 0 & 5s + 200 \end{vmatrix}}{\begin{vmatrix} s^2 + 400s & -400s \\ -2s & 5s + 200 \end{vmatrix}} = \frac{500s + 20{,}000}{5s^3 + 1{,}400s^2 + 80{,}000s} = \frac{100s + 4{,}000}{s(s + 80)(s + 200)}$$

$$\mathscr{L}\{i_2\} = \frac{\begin{vmatrix} s^2 + 400s & 100 \\ -2s & 0 \end{vmatrix}}{\begin{vmatrix} s^2 + 400s & -400s \\ -2s & 5s + 200 \end{vmatrix}} = \frac{200s}{5s^3 + 1{,}400s^2 + 80{,}000s} = \frac{40}{(s + 80)(s + 200)}$$

Finally, using the corollary of the first Heaviside expansion theorem (Theorem 1, Sec. 6.7), we have from $\mathscr{L}\{i_1\}$ and $\mathscr{L}\{i_2\}$

$$i_1 = \frac{4{,}000}{(80)(200)} + \frac{100(-80) + 4{,}000}{(-80)(120)}e^{-80t} + \frac{100(-200) + 4{,}000}{(-200)(-120)}e^{-200t}$$

$$= \frac{1}{4} + \frac{5}{12}e^{-80t} - \frac{2}{3}e^{-200t}$$

$$i_2 = \frac{40}{120}e^{-80t} + \frac{40}{-120}e^{-200t} = \frac{1}{3}e^{-80t} - \frac{1}{3}e^{-200t}$$

Exercises for Section 7.5

1 Assuming friction to be negligible, find the natural frequencies of the system shown in Fig. 7.18*a*. What are the relative amplitudes with which the masses vibrate at each of the

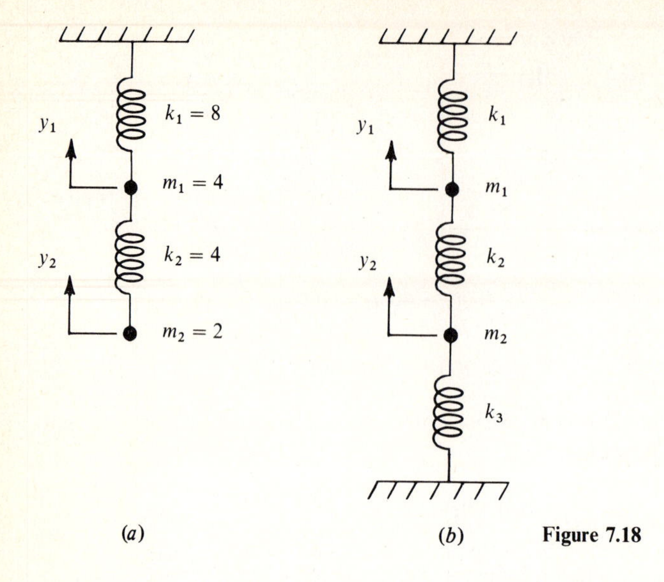

Figure 7.18

natural frequencies? What are the equations of motion of the respective masses if while the system is in equilibrium, m_2 is given an upward velocity of 2?

2 In Exercise 1, find the displacement of each mass as a function of t if a force $F = 40 \sin 3t$ is suddenly applied to m_1 while the system is in equilibrium.

3 In Exercise 1, find the displacement of each mass as a function of t if a force $F = 40 \sin 3t$ is suddenly applied to m_2 while the system is in equilibrium.

4 In Exercise 1, find the particular integral associated with a force $F = F_0 \sin \omega t$ acting on m_1, and discuss the corresponding steady-state motion as a function of ω.

5 Work Exercise 4 if the force $F = F_0 \sin \omega t$ acts on m_2.

6 If $m_1 = 1$, $m_2 = 3$, $k_1 = 1$, $k_2 = k_3 = 3$ for the system shown in Fig. 7.18*b*, find the natural frequencies of the system. What are the relative amplitudes of the two masses when the system is vibrating at each of these frequencies?

7 Work Exercise 6 if $m_1 = m_2 = 1$, $k_1 = 1$, $k_2 = 3$, and $k_3 = 9$.

8 In Exercise 6 find the displacements of m_1 and m_2 if the system begins to move from rest with $y_1 = 1$ and $y_2 = 0$.

9 In Exercise 7 find the displacements of m_1 and m_2 if the system begins to move from its equilibrium position, m_1 having velocity 1 and m_2 having velocity -1.

10 If $m_1 = m_2$ and $k_1 = k_3$ in the system shown in Fig. 7.18*b*, find the natural frequencies of the system as functions of m_1, k_1, and k_2. What are the relative amplitudes of the two masses when the system is vibrating at each of these frequencies? How does the quantity $y_1 + y_2$ vary with time? How does the quantity $y_1 - y_2$ vary with time? Under what conditions, if any, would you expect this system to exhibit the phenomenon of beats?

***11** Find the equations of motion of the system shown in Fig. 7.18*b* if $m_1 = m_2 = \frac{1}{9}$, $k_1 = k_3 = 9$, $k_2 = \frac{20}{9}$, and if the system begins to move from rest in the position where $y_1 = 2$ and $y_2 = 0$ Analyze this motion for the existence of beats.

12 Find the natural frequencies of the network shown in Fig. 7.19*a*.

13 Find the natural frequencies of the network shown in Fig. 7.19*b*.

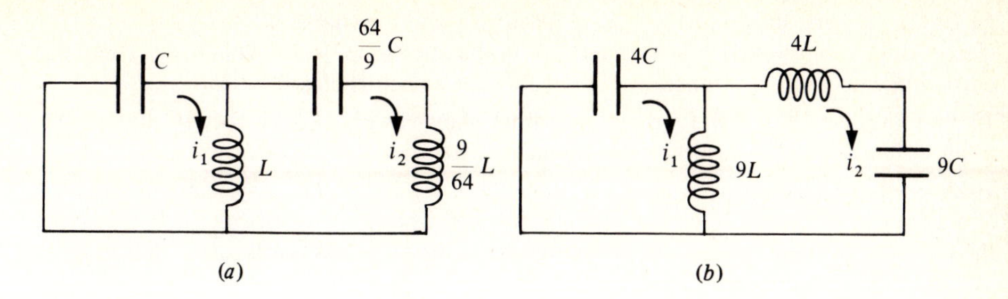

Figure 7.19

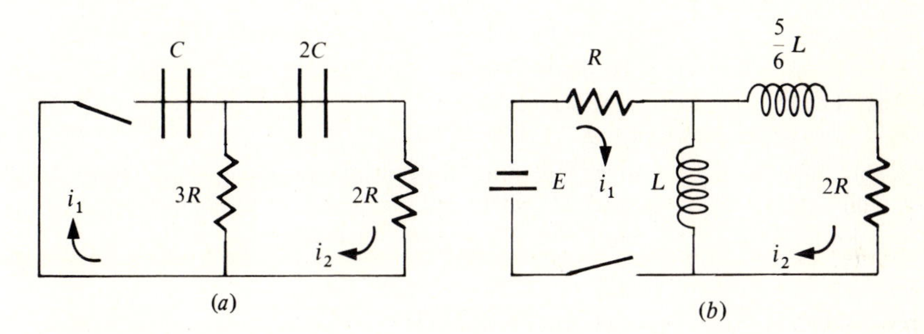

Figure 7.20

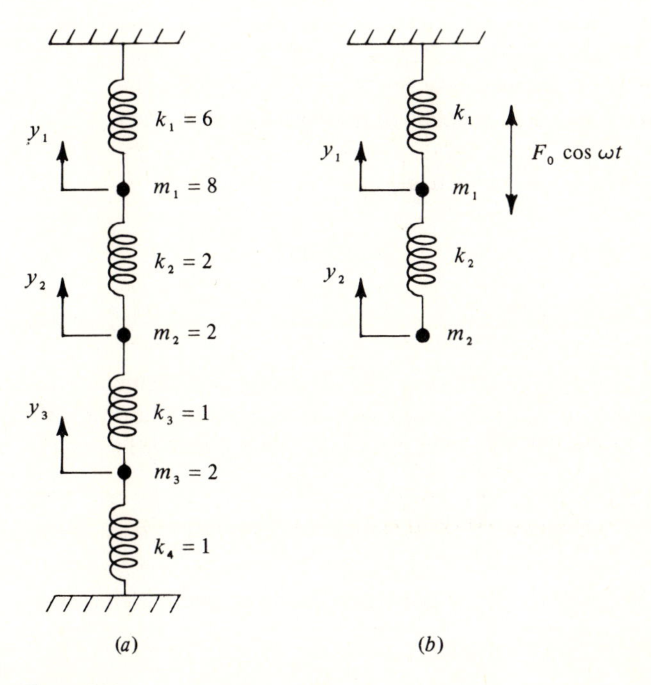

Figure 7.21

***14** In the network shown in Fig. 7.20*a*, the current and the charge on the capacitor in the closed loop are both zero, but the capacitor in the open loop bears a charge Q_0. Find the current in each loop as a function of time after the switch is closed.

***15** Find the current in each loop of the network shown in Fig. 7.20*b* if the switch is closed at an instant when all charges and currents are zero.

***16** A mass m_1 hanging from a spring of modulus k_1 constitutes a system with natural frequency $\omega = \sqrt{k_1/m_1}$. If a mass m_2 hangs from m_1 by a spring of modulus k_2, prove that the two natural frequencies of the resulting system are such that one is always less than ω and one is always greater than ω.

***17** What are the natural frequencies of the system shown in Fig. 7.21*a*? What are the relative amplitudes of the masses when the system is vibrating at each of these frequencies?

***18** In the system shown in Fig. 7.21*b*, the parameters m_1, k_1, and ω are assumed to be known. Determine k_2 and m_2 so that in the steady-state forced motion of the system the mass m_1 will remain at rest. What is the resultant amplitude of the mass m_2?

***19** Prove that for no values of the parameters m_1, m_2, k_1, k_2, and k_3 can the two natural frequencies of the system shown in Fig. 7.18*b* be equal.

***20** Set up the differential equations governing the motion of the system shown in Fig. 7.18*b*, taking into account the force of gravity and the forces due to the initial elongations of the springs. Verify that these forces cancel each other identically.

7.6 ELECTROMECHANICAL ANALOGIES

Although we have pointed out repeatedly the mathematical similarity between certain simple mechanical and electrical systems, we have not yet shown how to make the correspondence exact, so that numerical results for one system can be obtained experimentally from another. To see how this can be done, let us consider, as an illustration, the spring-suspended mass governed by the equation

$$\frac{w}{g}\frac{d^2y}{dt^2} + c\frac{dy}{dt} + ky = F_0 \cos \omega_1 t \tag{1}$$

and the series-electrical circuit governed by the equation

$$L\frac{d^2Q}{dt^2} + R\frac{dQ}{dt} + \frac{1}{C}Q = E_0 \cos \omega_2 t \tag{2}$$

At the outset, we should realize that the availability of electrical elements with specified numerical values may make it difficult to construct a quantitatively accurate electrical model of a mechanical system simply by matching the respective parameters† and putting

$$L = \frac{w}{g} = M \qquad R = c \qquad C = \frac{1}{k} \qquad E_0 = F_0 \qquad \text{and} \qquad \omega_2 = \omega_1$$

Instead, we must reduce Eqs. (1) and (2) to dimensionless form and match certain dimensionless *groups* of parameters.

† For instance, although a spring of modulus 1 lb/in is not unusual, a capacitor of capacitance 1 C/V, that is, 1 F, is highly unusual since most capacitors are in the microfarad range.

To do this for Eq. (1), let $v_1(>0)$ be an arbitrary frequency, let $s(>0)$ be an arbitrary distance, and let X and T be dimensionless variables defined by the relations

$$X = \frac{y}{s} \qquad \text{and} \qquad T = v_1 t$$

Then

$$\frac{dy}{dt} = \frac{d(sX)}{dt} = \frac{d(sX)}{dT}\frac{dT}{dt} = sv_1 \frac{dX}{dT}$$

and, similarly,

$$\frac{d^2y}{dt^2} = sv_1^2 \frac{d^2X}{dT^2}$$

Under these substitutions Eq. (1) becomes (writing M in place of w/g)

$$Msv_1^2 \frac{d^2X}{dT^2} + csv_1 \frac{dX}{dT} + ksX = F_0 \cos \frac{\omega_1}{v_1} T$$

or, dividing by Msv_1^2,

$$\frac{d^2X}{dT^2} + \frac{c}{Mv_1}\frac{dX}{dT} + \frac{k}{Mv_1^2} X = \frac{F_0}{Msv_1^2} \cos \frac{\omega_1}{v_1} T \tag{3}$$

In this form the equation is completely dimensionless, for not only are the variables X and T dimensionless, but so too are the coefficients. In fact,

F_0 = force = mass × acceleration has dimensions $M\dfrac{L}{T^2} = ML/T^2$

k = force per unit distance has dimensions $\dfrac{ML/T^2}{L} = M/T^2$

c = force per unit velocity has dimensions $\dfrac{ML/T^2}{L/T} = M/T$

s = distance has dimensions L

ω_1, v_1 = frequencies are of dimension $1/T$

and thus, checking the dimensions of the various coefficients, we find

$$\left[\frac{c}{Mv_1}\right]\dagger = \frac{M/T}{M(1/T)} = [0]$$

$$\left[\frac{k}{Mv_1^2}\right] = \frac{M/T^2}{M(1/T^2)} = [0]$$

$$\left[\frac{F_0}{Msv_1^2}\right] = \frac{ML/T^2}{ML(1/T^2)} = [0]$$

$$\omega_1/v_1 = \frac{1/T}{1/T} = [0]$$

† The dimensions of a physical quantity are frequently indicated by enclosing the quantity in square brackets. In this notation, a dimensionless quantity is indicated by the symbol [0].

To reduce Eq. (2) to dimensionless form, let $q(>0)$ be an arbitrary charge, and let $\nu_2(>0)$ be an arbitrary frequency. Then, in terms of the dimensionless variables

$$X = \frac{Q}{q} \qquad \text{and} \qquad T = \nu_2 t$$

we have

$$\frac{dQ}{dt} = q\nu_2 \frac{dX}{dT} \qquad \text{and} \qquad \frac{d^2Q}{dt^2} = q\nu_2^2 \frac{d^2X}{dT^2}$$

Making these substitutions in Eq. (2) and then dividing by $Lq\nu_2^2$ leads at once to the equation

$$\frac{d^2X}{dT^2} + \frac{R}{L\nu_2}\frac{dX}{dT} + \frac{1}{CL\nu_2^2}X = \frac{E_0}{Lq\nu_2^2}\cos\frac{\omega_2}{\nu_2}T \tag{4}$$

This, too, is a completely dimensionless equation for, from Ohm's law and the other relations defining the voltage differences across capacitors and inductors (Sec. 4.6), we have

$$[q] = \text{charge} = \text{amperes} \times T$$

$$[R] = \text{volts per ampere}$$

$$[C] = \text{charge per volt} = (\text{amperes} \times T) \text{ per volt}$$

$$[L] = \text{volts per unit rate of change of current}$$

$$= \text{volts per } (\text{amperes}/T) = (\text{volts} \times T) \text{ per ampere}$$

$$[\omega_2, \nu_2] = \text{frequencies} = 1/T$$

and from these it is easy to verify that the coefficients in Eq. (4) are dimensionless.

Now suppose that we have a mass-spring system for which we desire to make an exact electrical model with charge corresponding to displacement. This means that we are given the mechanical parameters M, c, k, F_0, and ω_1 and can choose as convenient, or necessary, the electrical parameters L, R, C, E_0, and ω_2 and the arbitrary scale factors s, ν_1 and q, ν_2.

Comparing Eqs. (3) and (4), it is clear that they will be *identical* if and only if the corresponding dimensionless coefficients are equal:

$$\begin{aligned} \frac{c}{M\nu_1} &= \frac{R}{L\nu_2} \\ \frac{k}{M\nu_1^2} &= \frac{1}{CL\nu_2^2} \\ \frac{F_0}{Ms\nu_1^2} &= \frac{E_0}{Lq\nu_2^2} \\ \frac{\omega_1}{\nu_1} &= \frac{\omega_2}{\nu_2} \end{aligned} \tag{5}$$

Of the nine quantities apparently at our disposal, namely,

$$L,\ R,\ C,\ E_0,\ \omega_2,\ s,\ v_1,\ q,\ v_2$$

only seven are essentially arbitrary, because s and q enter only as the ratio s/q, and v_1 and v_2 enter only as the ratio v_1/v_2. However, since these seven quantities (L, R, C, E_0, ω_2, and the ratios s/q and v_1/v_2) need satisfy only the four relations (5), it is clear that we have ample opportunity for suiting our convenience (e.g., availability of electrical elements in the laboratory) in constructing our model.

Now suppose that the electrical counterpart of the given mechanical system has been built and that the appropriate initial conditions have been determined from the relations

$$\frac{y_0}{s} \leftarrow X(0) \rightarrow \frac{Q_0}{q}$$

$$\frac{1}{sv_1}\frac{dy}{dt}\bigg|_{t=0} = \frac{1}{sv_1}v_0 \leftarrow \frac{dX}{dT}\bigg|_{T=0} \rightarrow \frac{1}{qv_2}\frac{dQ}{dt}\bigg|_{t=0} = \frac{1}{qv_2}i_0$$

If the charge on the capacitor in the electrical circuit is now measured and plotted dimensionlessly as

$$X = \frac{Q}{q} \qquad \text{vs.} \qquad T = v_2 t$$

the resulting graph will be identical with the dimensionless plot of

$$X = \frac{y}{s} \qquad \text{vs.} \qquad T = v_1 t$$

But from the dual interpretations

$$\frac{y}{s} \leftarrow X \rightarrow \frac{Q}{q} \qquad \text{and} \qquad v_1 t \leftarrow T \rightarrow v_2 t$$

it is clear that

$$y = \frac{s}{q}Q$$

and

$$t\,\bigg|_{\substack{\text{mechanical}\\ \text{system}}} = \frac{v_2}{v_1}\,t\,\bigg|_{\substack{\text{electrical}\\ \text{system}}}$$

Hence, having plotted Q vs. t for the electrical system, it is only necessary to regraduate the axes so that

$$1 \text{ unit on } Q \text{ scale} = \frac{s}{q} \text{ units on } y \text{ scale}$$

and

$$1 \text{ unit on electrical } t \text{ scale} = \frac{v_2}{v_1} \text{ units on mechanical } t \text{ scale}$$

in order to obtain an exact plot of the displacement y. The velocity $v = dy/dt$ can be found either as the derivative of the displacement y or directly from a plot of the current i by regraduating the axes appropriately.

A system with a single degree of freedom is so simple to analyze mathematically that there is no practical reason for constructing an electrical model of it. This is not the case, however, when there are several degrees of freedom, especially if the behavior of a system for a number of values of its parameters is to be determined. In such cases the ease with which electrical components can be connected and disconnected, and the ease with which currents and voltages can be measured and recorded automatically, often make it convenient to study a complicated mechanical system by making experimental measurements on an electrical analog.

The theory of this procedure is a direct extension of that which we have just developed. The most difficult part of the process is usually diagramming the equivalent circuit, not setting up the governing differential equations or reducing them to a dimensionless form. Lacking space for a detailed treatment of this topic, we conclude our discussion with a few general observations and an example or two.

In constructing a series model of a mass-spring system, we always have the correspondences

$$M \leftrightarrow L \qquad c \leftrightarrow R \qquad k \leftrightarrow \frac{1}{C} \qquad F_0 \leftrightarrow E_0$$

and the problem is to connect the electrical elements in such a way that a circuit mathematically equivalent to the given mechanical system will be obtained. The

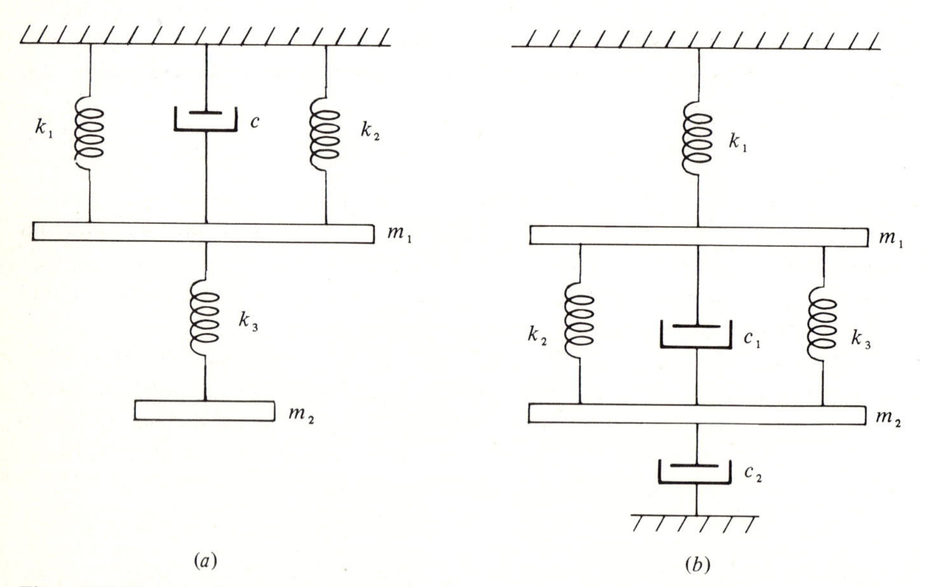

Figure 7.22 Two simple mass-spring systems.

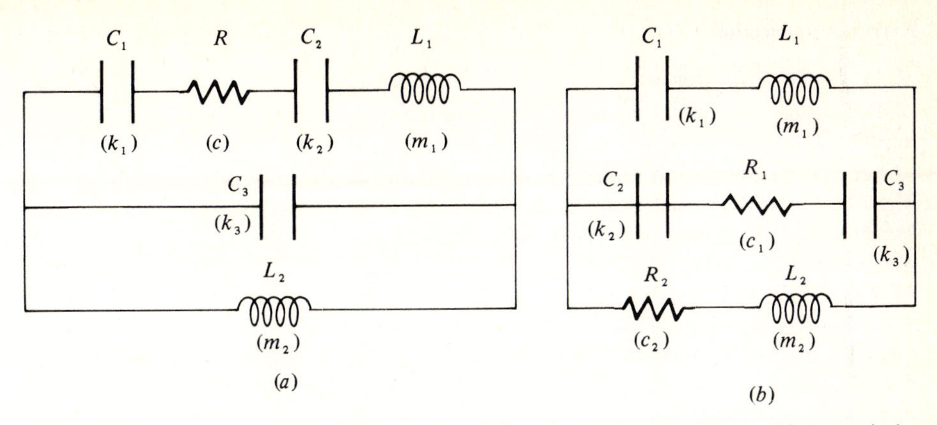

Figure 7.23 The electrical analogs of the mechanical systems shown in Figs. 7.22*a* and *b*, respectively.

fact that velocity corresponds to current indicates that components of a mechanical system between whose ends there is the same velocity difference (and hence the same displacement difference) should correspond to electrical elements between whose ends the same current flows. Now elements in apparent *mechanical parallel*, such as the springs k_1 and k_2 and the dashpot c in Fig. 7.22*a*, or the springs k_2 and k_3 and the dashpot c_1 in Fig. 7.22*b*, have the same velocity (and displacement) difference between their terminals. Hence they correspond to electrical elements through which the same current flows, i.e., to electrical elements in *series*. Moreover, elements in apparent *mechanical series* experience velocities (and displacements) totaling the velocity (or displacement) across the entire combination. Therefore the current through their analogs, as a whole, is the sum of the currents through the individual analogs. Such elements must then be in electrical *parallel*.

To diagram the series circuit equivalent to the system shown in Fig. 7.22*a*, we note that m_1, k_1, k_2 and c all experience the same velocities.† Hence their analogs must be elements in series. Also, the velocity of m_2 is the sum of the velocity of m_1 and the velocity difference across k_3. Hence the analogs of m_2, k_3, and the combination (m_1, k_1, k_2, c) must be in parallel. Figure 7.23*a* shows a circuit meeting these requirements. The elements in parentheses indicate the mechanical components to which the electrical elements correspond.

In the system shown in Fig. 7.22*b*, the dashpot c_1 experiences the velocity difference of k_2 and k_3 rather than that of k_1, and the dashpot c_2 experiences the same velocity as m_2. Hence in the overall parallel combination, the analog of c_1 is in series with the analogs of k_2 and k_3 and the analog of c_2 is in series with the analog of m_2. Figure 7.23*b* shows a circuit meeting these requirements.

† As usual, we assume that unspecified constraints prevent m_1 and m_2 from swinging or rotating, even though $k_1 \neq k_2$.

Exercises for Section 7.6

1 Verify that the coefficients in Eq. (4) are dimensionless.

2 Reduce the differential equation governing the torsional system with one degree of freedom [Eq. (2), Sec. 7.2] to dimensionless form.

3 Set up the differential equations describing the behavior of the system shown in Fig. 7.22*a* and then reduce them to dimensionless form by introducing the dimensionless variables $X_1 = y_1/s_1$ and $X_2 = y_2/s_2$, where s_1 and s_2 are arbitrary lengths, and $T = v_1 t$, where v_1 is an arbitrary frequency.

4 Set up the differential equations describing the behavior of the system shown in Fig. 7.23*a* and then reduce them to dimensionless form by introducing the dimensionless variables $X_1 = Q_1/q_1$ and $X_2 = Q_2/q_2$, where q_1 and q_2 are arbitrary charges, and $T = v_2 t$, where v_2 is an arbitrary frequency. Verify that these equations are identical with those obtained in Exercise 3.

5 Work Exercise 3 for the system shown in Fig. 7.22*b*.

6 Work Exercise 4 for the system shown in Fig. 7.23*b*.

7 Diagram an electrical network equivalent to the mechanical system shown in Fig. 7.24*a*.

8 Diagram an electrical network equivalent to the mechanical system shown in Fig. 7.24*b*.

9 Set up the differential equations for the system shown in Fig. 7.24*a* and for its electrical analog and reduce them to dimensionless form.

10 Set up the differential equations for the system shown in Fig. 7.24*b* and for its electrical analog and reduce them to dimensionless form.

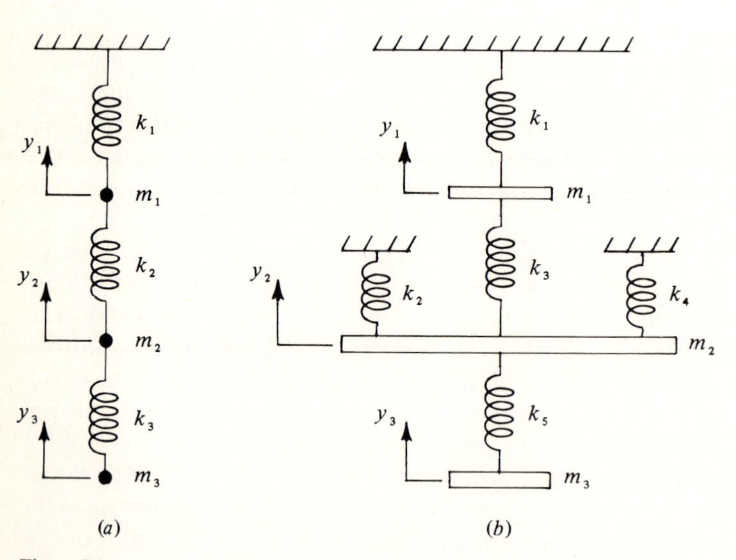

Figure 7.24

CHAPTER

EIGHT

FOURIER SERIES

8.1 INTRODUCTION

In Chap. 3 we learned that nonhomogeneous differential equations of the form

$$ay'' + by' + cy = A \cos \omega t + B \sin \omega t$$

can easily be solved for all values of the constants a, b, c, A, B, and ω. Then in Chap. 7 we discovered that such equations are fundamental in the study of physical systems subjected to certain periodic excitations. In many cases, however, the forces, torques, voltages, or currents which act on a system, although periodic, are by no means so simple that they can be represented as pure sine and cosine functions. For instance, the voltage impressed on an electrical circuit might consist of a series of pulses, as shown in Fig. 8.1*a*, or the disturbing influence acting on a mechanical system might be a force of constant magnitude whose direction is periodically and instantaneously reversed, as in Fig. 8.1*b*.

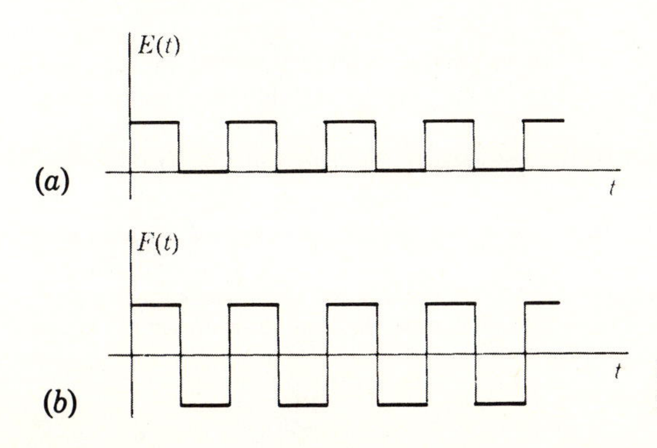

Figure 8.1 Typical periodical forcing functions.

This raises the question of whether a general periodic function† can be expressed as a series of sine and cosine terms. Specifically, since for all integral values of n

$$\cos \frac{n\pi(t+2p)}{p} = \cos\left(\frac{n\pi t}{p} + 2n\pi\right) = \cos \frac{n\pi t}{p}$$

and

$$\sin \frac{n\pi(t+2p)}{p} = \sin\left(\frac{n\pi t}{p} + 2n\pi\right) = \sin \frac{n\pi t}{p}$$

it is natural to ask whether coefficients $\{a_n\}$ and $\{b_n\}$ can be found such that a series of the form

$$\tfrac{1}{2}a_0\ddagger + a_1 \cos \frac{\pi t}{p} + a_2 \cos \frac{2\pi t}{p} + \cdots + a_n \cos \frac{n\pi t}{p} + \cdots$$
$$+ b_1 \sin \frac{\pi t}{p} + b_2 \sin \frac{2\pi t}{p} + \cdots + b_n \sin \frac{n\pi t}{p} + \cdots$$

can represent an arbitrary function $f(t)$ of the same period, $2p$. If this is the case, then by applying to the individual terms the concepts of magnification ratio and phase shift, in a mechanical problem, or complex impedance, in an electrical problem, we will be able to determine the response of a system to an arbitrary periodic disturbance. The possibility of such expansions and their determination when they exist are the subject matter of **Fourier analysis**§ to which we shall devote this chapter.

8.2 THE EULER COEFFICIENTS

To obtain formulas for the coefficients a_n and b_n in the expansion

$$f(t) = \frac{1}{2}a_0 + a_1 \cos \frac{\pi t}{p} + a_2 \cos \frac{2\pi t}{p} + \cdots + a_n \cos \frac{n\pi t}{p} + \cdots$$
$$+ b_1 \sin \frac{\pi t}{p} + b_2 \sin \frac{2\pi t}{p} + \cdots + b_n \sin \frac{n\pi t}{p} + \cdots \tag{1}$$

assuming, of course, that it exists and represents $f(t)$, we shall need the following definite integrals, which are valid for all values of d, provided m and n are integers

† A function $f(t)$ is said to be **periodic** if there exists a positive constant $2p$ with the property that $f(t+2p) = f(t)$ for all values of t. If there is a smallest number $2p$ (> 0) for which this identity holds, then $2p$ is called the **period** of the function.

‡ The introduction of the factor $\frac{1}{2}$ in this term is a conventional device to render the final formulas for the coefficients more symmetric.

§ Named for J. B. J. Fourier (1768–1830), French mathematician and confidant of Napoleon, who first undertook the systematic study of such expansions in a memorable monograph, "Theorie analytique de la chaleur," published in 1822. The use of such series in particular problems, however, dates from the time of Daniel Bernoulli (1700–1782), who used them to solve certain problems concerning vibrating strings.

satisfying the given restrictions.

$$\int_d^{d+2p} \cos\frac{n\pi t}{p}\,dt = 0 \qquad n \neq 0 \tag{2}$$

$$\int_d^{d+2p} \sin\frac{n\pi t}{p}\,dt = 0 \tag{3}$$

$$\int_d^{d+2p} \cos\frac{m\pi t}{p}\cos\frac{n\pi t}{p}\,dt = 0 \qquad m \neq n \tag{4}$$

$$\int_d^{d+2p} \cos^2\frac{n\pi t}{p}\,dt = p \qquad n \neq 0 \tag{5}$$

$$\int_d^{d+2p} \cos\frac{m\pi t}{p}\sin\frac{n\pi t}{p}\,dt = 0 \tag{6}$$

$$\int_d^{d+2p} \sin\frac{m\pi t}{p}\sin\frac{n\pi t}{p}\,dt = 0 \qquad m \neq n \tag{7}$$

$$\int_d^{d+2p} \sin^2\frac{n\pi t}{p}\,dt = p \qquad n \neq 0 \tag{8}$$

With these integrals available, the determination of a_n and b_n proceeds as follows.

To find a_0, we assume that the series (1) can legitimately be integrated term by term from $t = d$ to $t = d + 2p$.† Then

$$\int_d^{d+2p} f(t)\,dt = \frac{a_0}{2}\int_d^{d+2p} dt + a_1\int_d^{d+2p} \cos\frac{\pi t}{p}\,dt + \cdots$$
$$+ a_n\int_d^{d+2p} \cos\frac{n\pi t}{p}\,dt + \cdots + b_1\int_d^{d+2p} \sin\frac{\pi t}{p}\,dt + \cdots$$
$$+ b_n\int_d^{d+2p} \sin\frac{n\pi t}{p}\,dt + \cdots$$

The integral on the left can always be evaluated, since $f(t)$ is a known function which is assumed to be integrable. At worst, some method of approximate integration will be required. The first term on the right is simply

$$\tfrac{1}{2}a_0 t\Big|_d^{d+2p} = pa_0$$

By Eq. (2) all integrals with a cosine in the integrand vanish, and by Eq. (3) all integrals containing a sine vanish. Hence the integrated result reduces to

$$\int_d^{d+2p} f(t)\,dt = pa_0$$

or

$$a_0 = \frac{1}{p}\int_d^{d+2p} f(t)\,dt \tag{9}$$

† A sufficient condition for a series of integrable functions to be integrable term by term is that it be uniformly convergent.

To find a_n $(n = 1, 2, 3, \ldots)$, we multiply each side of (1) by $\cos(n\pi t/p)$ and then integrate from d to $d + 2p$, assuming again that term-by-term integration is justified. This gives

$$\int_d^{d+2p} f(t)\cos\frac{n\pi t}{p}\,dt = \tfrac{1}{2}a_0\int_d^{d+2p}\cos\frac{n\pi t}{p}\,dt + a_1\int_d^{d+2p}\cos\frac{\pi t}{p}\cos\frac{n\pi t}{p}\,dt + \cdots$$

$$+ a_n\int_d^{d+2p}\cos^2\frac{n\pi t}{p}\,dt + \cdots + b_1\int_d^{d+2p}\sin\frac{\pi t}{p}\cos\frac{n\pi t}{p}\,dt + \cdots$$

$$+ b_n\int_d^{d+2p}\sin\frac{n\pi t}{p}\cos\frac{n\pi t}{p}\,dt + \cdots$$

Again, the integral on the left is completely determined. By Eqs. (2) and (4), all integrals on the right containing only cosine terms vanish except the one involving $\cos^2(n\pi t/p)$, which, by Eq. (5), is equal to p. Finally, by Eq. (6), every integral which contains a sine is zero. Hence

$$\int_d^{d+2p} f(t)\cos\frac{n\pi t}{p}\,dt = pa_n$$

or

$$a_n = \frac{1}{p}\int_d^{d+2p} f(t)\cos\frac{n\pi t}{p}\,dt \tag{10}$$

To determine b_n, we continue essentially the same procedure. We multiply (1) by $\sin(n\pi t/p)$ and then integrate from d to $d + 2p$, getting

$$\int_d^{d+2p} f(t)\sin\frac{n\pi t}{p}\,dt = \frac{1}{2}a_0\int_d^{d+2p}\sin\frac{n\pi t}{p}\,dt + a_1\int_d^{d+2p}\cos\frac{\pi t}{p}\sin\frac{n\pi t}{p}\,dt + \cdots$$

$$+ a_n\int_d^{d+2p}\cos\frac{n\pi t}{p}\sin\frac{n\pi t}{p}\,dt + \cdots + b_1\int_d^{d+2p}\sin\frac{\pi t}{p}\sin\frac{n\pi t}{p}\,dt + \cdots$$

$$+ b_n\int_d^{d+2p}\sin^2\frac{n\pi t}{p}\,dt + \cdots$$

As before, every integral on the right vanishes but one, leaving

$$\int_d^{d+2p} f(t)\sin\frac{n\pi t}{p}\,dt = pb_n$$

or

$$b_n = \frac{1}{p}\int_d^{d+2p} f(t)\sin\frac{n\pi t}{p}\,dt \tag{11}$$

Formulas (9), (10), and (11) are known as the **Euler** or **Euler-Fourier formulas,** and the series (1), when its coefficients have these values, is known as the **Fourier series** of $f(t)$. In most applications, the interval over which the coefficients are computed is either $(-p, p)$ or $(0, 2p)$; so the value of d in the Euler formulas is usually either $-p$ or 0. Actually, the formula for a_0 need not be listed separately, since it can be obtained from the general expression for a_n by putting $n = 0$.† It was to achieve this that we wrote the constant term as $\frac{1}{2}a_0$ in the original expansion.

We must be careful at this stage not to delude ourselves with the belief that we have proved that every periodic function $f(t)$ has a Fourier expansion which converges to it. What our analysis has shown is merely that *if* a function $f(t)$ has an expansion of the form (1) for which term-by-term integration is valid, *then* the coefficients in that series must be given by the Euler formulas. Questions concerning the convergence of Fourier series and (if they converge) the conditions under which they will represent the functions which generated them are many and difficult. These problems are primarily of theoretical interest, however, for almost any conceivable practical application is covered by the famous **theorem of Dirichlet.**‡

Theorem 1 If $f(t)$ is a bounded periodic function which in any one period has at most a finite number of local maxima and minima and a finite number of points of discontinuity, then the Fourier series of $f(t)$ converges to $f(t)$ at all points where $f(t)$ is continuous and converges to the average of the right- and left-hand limits of $f(t)$ at each point where $f(t)$ is discontinuous.

The conditions of Theorem 1, usually referred to as the **Dirichlet conditions,** make it clear that a function need not be continuous in order to possess a valid Fourier expansion. This means that a function may have a graph consisting of a number of disjointed arcs of different curves, each defined by a different formula, and still be representable by a Fourier series. In using the Euler formulas to find the coefficients in the expansion of such a function, it will therefore be necessary to break up the range of integration $(d, d + 2p)$ to correspond to the various segments of the function. Thus in Fig. 8.2, the function $f(t)$ is defined by three

† It is not necessarily the case, however, that the value of a_0 in a particular problem can be obtained by putting $n = 0$ in the *integrated* formula for a_n. For instance, in Example 2 the integrated formula for a_n is indeterminate when $n = 0$, and evaluation of the indeterminacy yields -3 instead of the correct value 3, which is obtained by putting $n = 0$ *before* integrating.

‡ Named for the German mathematician Peter Gustave Lejeune Dirichlet (1805–1859). For a proof of this theorem see, for instance, Philip Franklin, A Simple Discussion of the Representation of Functions by Fourier Series, in "Selected Papers on Calculus," Mathematical Association of America, 1969, pp. 357–61.

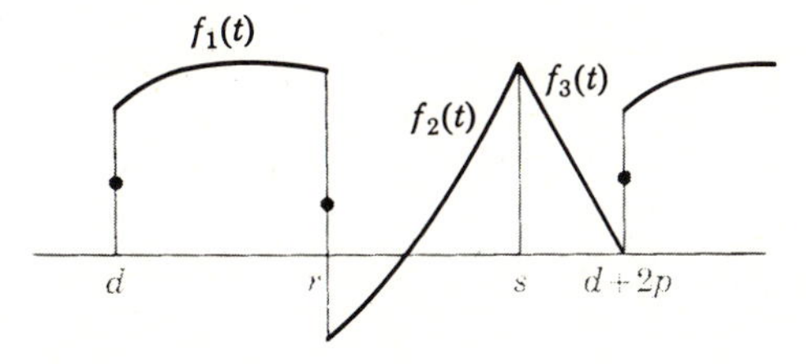

Figure 8.2 A periodic function defined by different formulas over different portions of a period.

different expressions, $f_1(t)$, $f_2(t)$, $f_3(t)$ on successive portions of the period interval $d \le t \le d + 2p$. Hence it is necessary to write the Euler formulas as

$$a_n = \frac{1}{p}\int_d^{d+2p} f(t)\cos\frac{n\pi t}{p}\,dt$$

$$= \frac{1}{p}\int_d^r f_1(t)\cos\frac{n\pi t}{p}\,dt + \frac{1}{p}\int_r^s f_2(t)\cos\frac{n\pi t}{p}\,dt + \frac{1}{p}\int_s^{d+2p} f_3(t)\cos\frac{n\pi t}{p}\,dt$$

$$b_n = \frac{1}{p}\int_d^{d+2p} f(t)\sin\frac{n\pi t}{p}\,dt$$

$$= \frac{1}{p}\int_d^r f_1(t)\sin\frac{n\pi t}{p}\,dt + \frac{1}{p}\int_r^s f_2(t)\sin\frac{n\pi t}{p}\,dt + \frac{1}{p}\int_s^{d+2p} f_3(t)\sin\frac{n\pi t}{p}\,dt$$

Incidentally, according to Theorem 1, the Fourier series of the function shown in Fig. 8.2 will converge to the average values, indicated by dots, at the discontinuities at d, r, and $d + 2p$, regardless of the definition (or lack of definition) of the function at these points.

Example 1 What is the Fourier expansion of the periodic function whose definition in one period is

$$f(t) = \begin{cases} 0 & -\pi \le t \le 0 \\ \sin t & 0 < t \le \pi \end{cases}$$

In this case the half-period of the given function is $p = \pi$. Hence, taking $d = -\pi$ in the Euler formulas, we have

$$a_n = \frac{1}{\pi}\int_{-\pi}^{\pi} f(t)\cos nt\,dt = \frac{1}{\pi}\int_{-\pi}^{0} 0\cos nt\,dt + \frac{1}{\pi}\int_0^{\pi}\sin t\cos nt\,dt$$

$$= \frac{1}{\pi}\left[-\frac{1}{2}\left\{\frac{\cos(1-n)t}{1-n} + \frac{\cos(1+n)t}{1+n}\right\}\right]_0^{\pi}$$

$$= -\frac{1}{2\pi}\left[\frac{\cos(\pi - n\pi)}{1-n} + \frac{\cos(\pi + n\pi)}{1+n} - \left(\frac{1}{1-n} + \frac{1}{1+n}\right)\right]$$

$$= -\frac{1}{2\pi}\left(\frac{-\cos n\pi}{1-n} + \frac{-\cos n\pi}{1+n} - \frac{2}{1-n^2}\right)$$

$$= \frac{1+\cos n\pi}{\pi(1-n^2)} \qquad n \ne 1$$

$$a_1 = \frac{1}{\pi}\int_0^{\pi}\sin t\cos t\,dt = \frac{\sin^2 t}{2\pi}\bigg|_0^{\pi} = 0$$

$$b_n = \frac{1}{\pi}\int_{-\pi}^{\pi} f(t)\sin nt\,dt = \frac{1}{\pi}\int_{-\pi}^{0} 0\sin nt\,dt + \frac{1}{\pi}\int_0^{\pi}\sin t\sin nt\,dt$$

$$= \frac{1}{\pi}\left[\frac{1}{2}\left\{\frac{\sin(1-n)t}{1-n} - \frac{\sin(1+n)t}{1+n}\right\}\right]_0^{\pi} = 0 \qquad n \ne 1$$

$$b_1 = \frac{1}{\pi}\int_0^{\pi}\sin^2 t\,dt = \frac{1}{\pi}\left[\frac{t}{2} - \frac{\sin 2t}{4}\right]_0^{\pi} = \frac{1}{2}$$

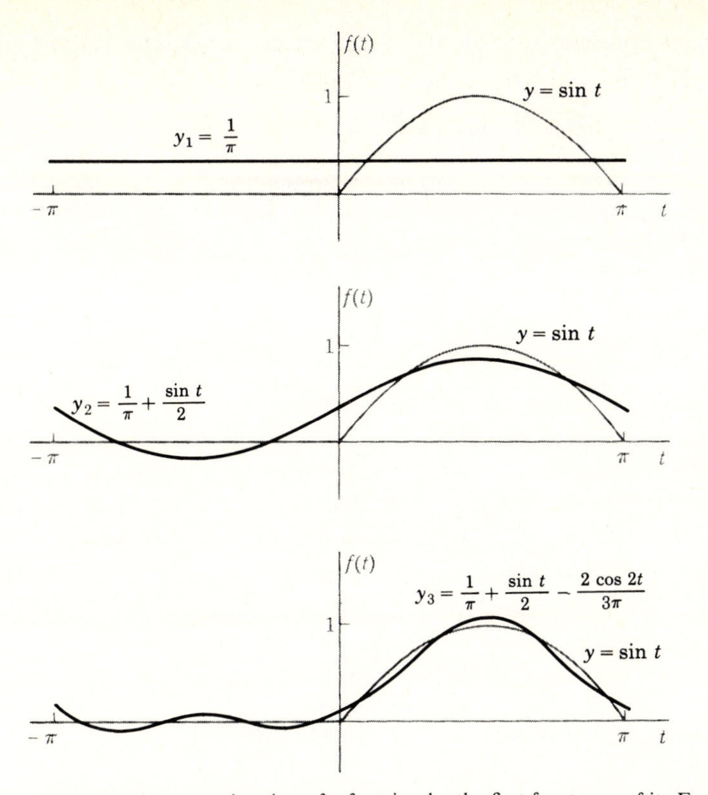

Figure 8.3 The approximation of a function by the first few terms of its Fourier expansion.

Hence, evaluating the coefficients for $n = 0, 1, 2, \ldots$, we have

$$f(t) = \frac{1}{\pi} + \frac{\sin t}{2} - \frac{2}{\pi}\left(\frac{\cos 2t}{3} + \frac{\cos 4t}{15} + \frac{\cos 6t}{35} + \frac{\cos 8t}{63} + \cdots\right)$$

Plots showing the accuracy with which the first n terms of this series represent the given function are shown in Fig. 8.3 for $n = 1, 2, 3$. For $n = 4, 5, 6, \ldots$ the graphs of the partial sums are almost indistinguishable from the graph of $f(t)$.

Interesting numerical series can often be obtained from Fourier series by evaluating them at particular points. For instance, if we set $t = \pi/2$ in the above expansion, we find

$$f\left(\frac{\pi}{2}\right) = 1 = \frac{1}{\pi} + \frac{1}{2} - \frac{2}{\pi}\left(-\frac{1}{3} + \frac{1}{15} - \frac{1}{35} + \frac{1}{63} - \cdots\right)$$

or

$$\frac{1}{1 \cdot 3} - \frac{1}{3 \cdot 5} + \frac{1}{5 \cdot 7} - \frac{1}{7 \cdot 9} + \cdots = \frac{\pi - 2}{4}$$

Example 2 Find the expansion of the periodic function whose definition in one period is

$$f(t) = \begin{cases} -t & -3 \le t \le 0 \\ t & 0 < t \le 3 \end{cases}$$

In this case the period of the function is 6. Hence $p = 3$, and, from (10) and (11), taking $d = -3$, we have

$$a_n = \frac{1}{3}\int_{-3}^{0} -t \cos \frac{n\pi t}{3}\, dt + \frac{1}{3}\int_{0}^{3} t \cos \frac{n\pi t}{3}\, dt$$

$$= -\frac{1}{3}\left[\frac{9}{n^2\pi^2} \cos \frac{n\pi t}{3} + \frac{3t}{n\pi} \sin \frac{n\pi t}{3}\right]_{-3}^{0} + \frac{1}{3}\left[\frac{9}{n^2\pi^2} \cos \frac{n\pi t}{3} + \frac{3t}{n\pi} \sin \frac{n\pi t}{3}\right]_{0}^{3}$$

$$= -\frac{3}{n^2\pi^2}(1 - \cos n\pi) + \frac{3}{n^2\pi^2}(\cos n\pi - 1)$$

$$= \frac{6}{n^2\pi^2}(\cos n\pi - 1) \qquad n \neq 0$$

$$a_0 = \frac{1}{3}\int_{-3}^{0} -t\, dt + \frac{1}{3}\int_{0}^{3} t\, dt = -\frac{t^2}{6}\bigg|_{-3}^{0} + \frac{t^2}{6}\bigg|_{0}^{3} = \frac{3}{2} + \frac{3}{2} = 3$$

$$b_n = \frac{1}{3}\int_{-3}^{0} -t \sin \frac{n\pi t}{3}\, dt + \frac{1}{3}\int_{0}^{3} t \sin \frac{n\pi t}{3}\, dt$$

$$= -\frac{1}{3}\left[\frac{9}{n^2\pi^2} \sin \frac{n\pi t}{3} - \frac{3t}{n\pi} \cos \frac{n\pi t}{3}\right]_{-3}^{0} + \frac{1}{3}\left[\frac{9}{n^2\pi^2} \sin \frac{n\pi t}{3} - \frac{3t}{n\pi} \cos \frac{n\pi t}{3}\right]_{0}^{3}$$

$$= \frac{3}{n\pi} \cos(-n\pi) - \frac{3}{n\pi} \cos n\pi = 0$$

Substituting these coefficients into the series (1), we obtain

$$f(t) = \frac{3}{2} - \frac{12}{\pi^2}\left(\frac{1}{1} \cos \frac{\pi t}{3} + \frac{1}{9} \cos \frac{3\pi t}{3} + \frac{1}{25} \cos \frac{5\pi t}{3} + \cdots\right)$$

Exercises for Section 8.2

Determine the Fourier expansions of the periodic functions whose definitions in one period are:

1 $f(t) = \begin{cases} 1 & 0 \le t \le 1 \\ 0 & 1 < t < 2\dagger \end{cases}$

2 $f(t) = \begin{cases} 1 & 0 \le t \le 1 \\ -1 & 1 < t < 2 \end{cases}$

3 $f(t) = \begin{cases} 0 & -2 \le t < -1 \\ 1 & -1 \le t < 0 \\ -1 & 0 \le t < 1 \\ 0 & 1 \le t \le 2 \end{cases}$

4 $f(t) = \begin{cases} 1 & -2 \le t < -1 \\ 0 & -1 \le t < 0 \\ -1 & 0 \le t < 1 \\ 0 & 1 \le t < 2 \end{cases}$

5 $f(t) = t \qquad 0 \le t < 1$

6 $f(t) = t \qquad -1 \le t < 1$

7 $f(t) = \begin{cases} 2 & 0 \le t < 2\pi/3 \\ 1 & 2\pi/3 \le t < 4\pi/3 \\ 0 & 4\pi/3 \le t < 2\pi \end{cases}$

8 $f(t) = \begin{cases} 0 & -2 \le t < -1 \\ 1 + t & -1 \le t < 0 \\ 1 - t & 0 \le t < 1 \\ 0 & 1 \le t \le 2 \end{cases}$

9 $f(t) = e^{-t} \qquad 0 \le t < 2$

10 $f(t) = \sin t \qquad 0 \le t \le \pi$

† The equality sign cannot be included at $t = 2$ because since the period of f is 2, it follows that $f(2) = f(0 + 2) = f(0) = 1$, and not 0.

11 $f(t) = \begin{cases} 0 & -\pi \le t < 0 \\ t & 0 \le t < \pi \end{cases}$ **12** $f(t) = \cos t \qquad -\frac{\pi}{2} \le t \le \frac{\pi}{2}$

***13** $f(t) = t - t^3 \qquad -1 \le t \le 1$ ***14** $f(t) = \begin{cases} 0 & -\pi \le t < 0 \\ t^2 & 0 \le t < \pi \end{cases}$

***15** Establish the following numerical results:

$$1 + \frac{1}{2^2} + \frac{1}{3^2} + \frac{1}{4^2} + \frac{1}{5^2} + \cdots = \frac{\pi^2}{6}$$

$$1 - \frac{1}{2^2} + \frac{1}{3^2} - \frac{1}{4^2} + \frac{1}{5^2} - \cdots = \frac{\pi^2}{12}$$

$$1 + \frac{1}{3^2} + \frac{1}{5^2} + \frac{1}{7^2} + \frac{1}{9^2} + \cdots = \frac{\pi^2}{8}$$

Hint: Use the results of Exercise 14.

16. If $f(t)$ is a continuous function of period $2p$, show that

$$a_n = -\frac{1}{n\pi}\int_{-p}^{p} f'(t)\sin\frac{n\pi t}{p}\,dt$$

and

$$b_n = \frac{1}{n\pi}\int_{-p}^{p} f'(t)\cos\frac{n\pi t}{p}\,dt$$

17. If $f(t)$ is a function of period $2p$, whose derivative is everywhere continuous, show that

$$a_n = -\frac{p}{n^2\pi^2}\int_{-p}^{p} f''(t)\cos\frac{n\pi t}{p}\,dt$$

and

$$b_n = -\frac{p}{n^2\pi^2}\int_{-p}^{p} f''(t)\sin\frac{n\pi t}{p}\,dt$$

18 If $f(t)$ is a periodic function of period $2p$ which is continuous on $(-p, p)$, show that

$$a_n = -\frac{1}{n\pi}\int_{-p}^{p} f'(t)\sin\frac{n\pi t}{p}\,dt$$

and

$$b_n = \frac{(-1)^{n+1}}{n\pi}[f(p^-) - f(-p^+)] + \frac{1}{n\pi}\int_{-p}^{p} f'(t)\cos\frac{n\pi t}{p}\,dt$$

8.3 HALF-RANGE EXPANSIONS

When $f(t)$ possesses certain symmetry properties, the coefficients in its Fourier expansion become especially simple. This was illustrated in Example 2 of the last section, where the given function was symmetric in the y axis and its expansion contained only cosine terms, i.e., only terms which themselves were symmetric in the y axis. In this section we shall investigate in detail just what effect the symmetry of $f(t)$ has on the coefficients in the Fourier series for $f(t)$.

Suppose first of all that $f(t)$ is an **even function;** i.e., suppose that $f(-t) = f(t)$ for all t or, geometrically, that the graph of $f(t)$ is symmetric in the vertical axis.

Taking $d = -p$ in the formula for a_n [Eq. (10), Sec. 8.2], we can write

$$a_n = \frac{1}{p}\int_{-p}^{p} f(t)\cos\frac{n\pi t}{p}\,dt = \frac{1}{p}\int_{-p}^{0} f(t)\cos\frac{n\pi t}{p}\,dt + \frac{1}{p}\int_{0}^{p} f(t)\cos\frac{n\pi t}{p}\,dt \quad (1)$$

Now, in the integral from $-p$ to 0 let us make the substitution

$$t = -s \qquad dt = -ds$$

Then since $t = -p$ implies $s = p$, and since $t = 0$ implies $s = 0$, this integral becomes

$$\frac{1}{p}\int_{p}^{0} f(-s)\cos\frac{-n\pi s}{p}\,(-ds) \quad (2)$$

But $f(-s) = f(s)$, from the hypothesis that $f(t)$ is an even function. Moreover, the cosine is also an even function; i.e.,

$$\cos\frac{-n\pi s}{p} = \cos\frac{n\pi s}{p}$$

Finally, the negative sign associated with ds in (2) can be eliminated by changing the limits back to the normal order, 0 to p. The integral (2) then becomes

$$\frac{1}{p}\int_{0}^{p} f(s)\cos\frac{n\pi s}{p}\,ds$$

and thus, substituting this into (1), a_n can be written

$$a_n = \frac{1}{p}\int_{0}^{p} f(s)\cos\frac{n\pi s}{p}\,ds + \frac{1}{p}\int_{0}^{p} f(t)\cos\frac{n\pi t}{p}\,dt$$

$$= \frac{2}{p}\int_{0}^{p} f(t)\cos\frac{n\pi t}{p}\,dt$$

since the two integrals are identical, except for the dummy variable of integration, which is immaterial.

Similarly, we can write

$$b_n = \frac{1}{p}\int_{-p}^{0} f(t)\sin\frac{n\pi t}{p}\,dt + \frac{1}{p}\int_{0}^{p} f(t)\sin\frac{n\pi t}{p}\,dt$$

Again, putting $t = -s$ and $dt = -ds$ in the first integral, we find

$$b_n = \frac{1}{p}\int_{p}^{0} f(-s)\sin\frac{-n\pi s}{p}\,(-ds) + \frac{1}{p}\int_{0}^{p} f(t)\sin\frac{n\pi t}{p}\,dt$$

But, by hypothesis, $f(-s) = f(s)$; and $\sin(-n\pi s/p) = -\sin(n\pi s/p)$ from the familiar properties of the sine function. Hence, reversing the limits on the first integral, as before, we have

$$b_n = -\frac{1}{p}\int_{0}^{p} f(s)\sin\frac{n\pi s}{p}\,ds + \frac{1}{p}\int_{0}^{p} f(t)\sin\frac{n\pi t}{p}\,dt = 0$$

since, except for the irrelevant dummy variable of integration, the two integrals are identical in all but sign. Thus (assuming the Dirichlet theorem) we have established the following useful result.

Theorem 1 If $f(t)$ is an even periodic function which satisfies the Dirichlet conditions, the coefficients in the Fourier series of $f(t)$ are given by the formulas

$$a_n = \frac{2}{p}\int_0^p f(t)\cos\frac{n\pi t}{p}\,dt \qquad b_n \equiv 0$$

where $2p$ is the period of $f(t)$.

If $f(t)$ is an **odd** function, i.e., if $f(t)$ has the property that $f(-t) = -f(t)$ for all values of t, then by an almost identical argument we can establish the following companion result.

Theorem 2 If $f(t)$ is an odd periodic function which satisfies the Dirichlet conditions, the coefficients in the Fourier series of $f(t)$ are given by the formulas

$$a_n \equiv 0 \qquad b_n = \frac{2}{p}\int_0^p f(t)\sin\frac{n\pi t}{p}\,dt$$

where $2p$ is the period of $f(t)$.

The observations we have just made about the Fourier coefficients of odd and even functions serve to reduce by half the labor of expanding such functions. However, their chief value is that they allow us to meet the requirements of certain problems† in which expansions containing *only* cosine terms or expansions containing *only* sine terms must be constructed.

Suppose, for instance, that the conditions of a problem require us to consider the values of a function f *only* in the interval from 0 to p. In other words, conditions of periodicity are irrelevant to the problem, and what the function may be outside the interval $[0, p]$ is immaterial. This being the case, we can extend f into a new function F whose domain is the entire t axis in the following way: On $[0, p]$ let F be equal to f, on $(-p, 0)$ let F be defined arbitrarily (subject to the Dirichlet conditions, of course), and for all other values of t let $F(t)$ be defined by the periodicity condition $F(t + 2p) = F(t)$. Thus

$$F(t) = \begin{cases} \phi(t) & -p < t < 0 \qquad \phi \text{ arbitrary} \\ f(t) & 0 \le t \le p \\ F(t + 2p) = F(t) \end{cases}$$

Between $-p$ and 0, the Fourier expansion of $F(t)$, of course, converges to $\phi(t)$; but *irrespective of the extension* ϕ, this series will converge to $f(t)$ between 0 and p, since on this interval $F(t) = f(t)$.

In particular, if we extend the function f from 0 to $-p$ by reflecting it in the vertical axis, so that $\phi(-t) = f(t)$, the original function together with its extension

† Examples of such problems will be found in Secs. 11.4 and 11.5.

is even; hence the Fourier expansion of the periodic continuation F will contain only cosine terms [including, of course, the constant term $\frac{1}{2}a_0 = \frac{1}{2}a_0 \cos(0\pi t/p)$] whose coefficients, as we showed above, are

$$a_n = \frac{2}{p}\int_0^p F(t)\cos\frac{n\pi t}{p}\,dt = \frac{2}{p}\int_0^p f(t)\cos\frac{n\pi t}{p}\,dt$$

On the other hand, if we extend the same function f from 0 to $-p$ by reflecting it in the origin, so that $\phi(-t) = -f(t)$, the extended function is odd and hence the Fourier series of its periodic continuation F will contain only sine terms, whose coefficients will be given by

$$b_n = \frac{2}{p}\int_0^p F(t)\sin\frac{n\pi t}{p}\,dt = \frac{2}{p}\int_0^p f(t)\sin\frac{n\pi t}{p}\,dt$$

Thus, simply by imagining the appropriate extension of a function f originally defined only for $0 \le t \le p$, we can obtain expansions representing f on this interval and containing only cosine terms or only sine terms, as we please. Moreover, the formulas for the coefficients in these expansions involve only the values of f on the original interval $[0, p]$ and do not depend in any way upon the extensions we used to obtain those formulas. Such series are known as **half-range expansions.**

Closely associated with half-range cosine and sine series is another special expansion, obtained in the following way: Given a function f defined originally only for $0 \le t \le p$, let us extend it from p to $2p$ by reflecting it in the line $t = p$. As Fig. 8.4 illustrates, the extended function F is defined by the rule

$$F(t) = \begin{cases} f(t) & 0 \le t \le p \\ f(2p - t) & p < t \le 2p \end{cases}$$

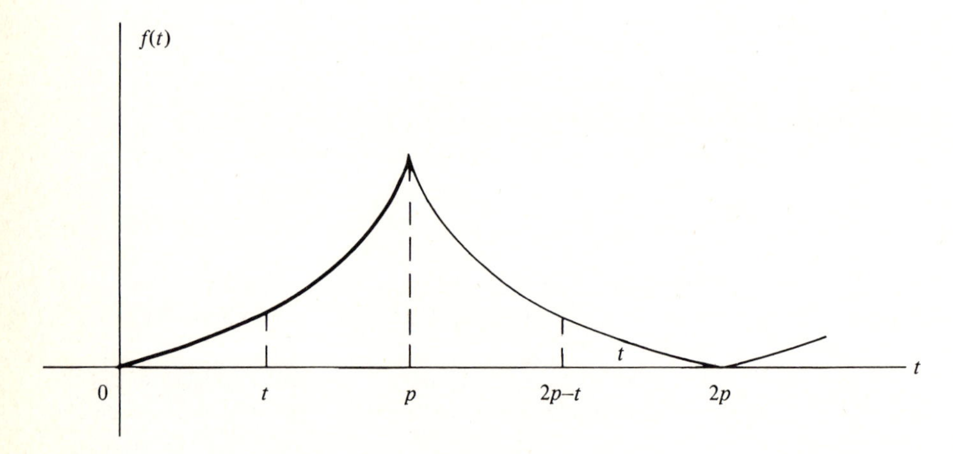

Figure 8.4 The extended function discussed in Theorem 3.

For the half-range sine expansion of the extended function F we have, by Theorem 2,

$$\begin{aligned} b_n &= \frac{2}{2p}\int_0^{2p} F(t)\sin\frac{n\pi t}{2p}\,dt \\ &= \frac{1}{p}\int_0^{p} f(t)\sin\frac{n\pi t}{2p}\,dt + \frac{1}{p}\int_p^{2p} f(2p-t)\sin\frac{n\pi t}{2p}\,dt \end{aligned} \tag{3}$$

If we make the substitutions $s = 2p - t$ and $ds = -dt$ in the second of the integrals in (3), we obtain

$$\frac{1}{p}\int_p^0 f(s)\sin\left[\frac{n\pi}{2p}(2p-s)\right](-ds) = \frac{1}{p}\int_0^p f(s)\sin\left(n\pi - \frac{n\pi s}{2p}\right)ds \tag{4}$$

Now $\quad \sin\left(n\pi - \dfrac{n\pi s}{2p}\right) = \sin n\pi \cos\dfrac{n\pi s}{2p} - \cos n\pi \sin\dfrac{n\pi s}{2p} = -(-1)^n \sin\dfrac{n\pi s}{2p}$

Hence the integral (4) becomes

$$-(-1)^n\frac{1}{p}\int_0^p f(s)\sin\frac{n\pi s}{2p}\,ds$$

Except for the dummy variable s, this is identical with the first integral in (3) if n is odd and is the negative of the first integral in (3) if n is even. Thus, combining the integrals in the two cases, we have established the following theorem.

Theorem 3 If $f(t)$, originally defined for $0 \le t \le p$, is extended over the interval $(p, 2p)$ by reflection in the line $t = p$, then for the half-range sine expansion of the extended function,

$$b_n = \begin{cases} 0 & n \text{ even} \\ \dfrac{2}{p}\displaystyle\int_0^p f(t)\sin\frac{n\pi t}{2p}\,dt & n \text{ odd} \end{cases}$$

The special series described in Theorem 3 is important because there are problems† in which a function given on an interval $(0, p)$ must be expanded in terms of sines whose angles must all be *odd* multiples of a fundamental angle of the form $\pi t/(2p)$.

Example 1 Find the half-range expansions of the function

$$f(t) = t - t^2 \qquad 0 \le t \le 1$$

The half-range cosine expansion is obtained by first extending $t - t^2$ from the given interval $[0, 1]$ to the interval $[-1, 0]$ by reflection in the y axis and then taking the function thus defined from -1 to 1 as one period of a periodic function of period

† Examples of such problems will be found in Sec. 11.5.

$2p = 2$. However, once we understand the reasoning underlying the procedure, we need give no thought to the extension but can write immediately, on the basis of Theorem 1,

$$b_n \equiv 0$$

$$a_n = \frac{2}{1}\int_0^1 (t - t^2) \cos \frac{n\pi t}{1}\, dt$$

$$= 2\left[\left(\frac{\cos n\pi t}{n^2\pi^2} + \frac{t}{n\pi}\sin n\pi t\right) - \left(\frac{2t}{n^2\pi^2}\cos n\pi t - \frac{2}{n^3\pi^3}\sin n\pi t + \frac{t^2}{n\pi}\sin n\pi t\right)\right]_0^1$$

$$= 2\left(\frac{\cos n\pi - 1}{n^2\pi^2} - \frac{2\cos n\pi}{n^2\pi^2}\right)$$

$$= -\frac{2(1 + \cos n\pi)}{n^2\pi^2} \qquad n \neq 0$$

$$a_0 = \frac{2}{1}\int_0^1 (t - t^2)\, dt = 2\left[\frac{t^2}{2} - \frac{t^3}{3}\right]_0^1 = \frac{1}{3}$$

Hence it is possible to represent $f(t) = t - t^2$ for $0 \le t \le 1$ by the series

$$f(t) = \frac{1}{6} - \frac{4}{\pi^2}\left(\frac{\cos 2\pi t}{4} + \frac{\cos 4\pi t}{16} + \frac{\cos 6\pi t}{36} + \frac{\cos 8\pi t}{64} + \cdots\right) \tag{5}$$

Similarly, the half-range sine expansion is obtained by first extending the given function, $t - t^2$, to the interval $[-1, 0]$ by reflection in the origin and then extending periodically the function thus defined over $[-1, 1]$. However, all we actually need to do to obtain the expansion is to note that, according to Theorem 2,

$$a_n \equiv 0$$

and

$$b_n = \frac{2}{1}\int_0^1 (t - t^2) \sin \frac{n\pi t}{1}\, dt$$

$$= 2\left[\left(\frac{\sin n\pi t}{n^2\pi^2} - \frac{t}{n\pi}\cos n\pi t\right) - \left(\frac{2t}{n^2\pi^2}\sin n\pi t + \frac{2}{n^3\pi^3}\cos n\pi t - \frac{t^2}{n\pi}\cos n\pi t\right)\right]_0^1$$

$$= 2\left[\left(-\frac{\cos n\pi}{n\pi}\right) - \left(\frac{2(\cos n\pi - 1)}{n^3\pi^3} - \frac{\cos n\pi}{n\pi}\right)\right]$$

$$= \frac{4(1 - \cos n\pi)}{n^3\pi^3}$$

Hence it is also possible to represent $f(t)$ for $0 < t < 1$ by the series

$$f(t) = \frac{8}{\pi^3}\left(\frac{\sin \pi t}{1} + \frac{\sin 3\pi t}{27} + \frac{\sin 5\pi t}{125} + \frac{\sin 7\pi t}{343} + \cdots\right) \tag{6}$$

Series (5) and (6) are by no means the only Fourier series that will represent $t - t^2$ on the interval $(0, 1)$. They are merely the most convenient or most useful ones. In fact, with every possible extension of $t - t^2$ from 0 to -1 there is associated a series yielding $t - t^2$ for $0 < t < 1$. For instance, a third such series might be obtained by letting the extension be simply the one defined by $t - t^2$ itself for $-1 < t < 0$. In this case

$$a_n = \frac{1}{1}\int_{-1}^{1}(t - t^2)\cos\frac{n\pi t}{1}\,dt$$

$$= \left[\left(\frac{\cos n\pi t}{n^2\pi^2} + \frac{t}{n\pi}\sin n\pi t\right) - \left(\frac{2t}{n^2\pi^2}\cos n\pi t - \frac{2}{n^3\pi^3}\sin n\pi t + \frac{t^2}{n\pi}\sin n\pi t\right)\right]_{-1}^{1}$$

$$= -\frac{4\cos n\pi}{n^2\pi^2} \qquad n \neq 0$$

$$a_0 = \frac{1}{1}\int_{-1}^{1}(t - t^2)\,dt = \left[\frac{t^2}{2} - \frac{t^3}{3}\right]_{-1}^{1} = -\frac{2}{3}$$

$$b_n = \frac{1}{1}\int_{-1}^{1}(t - t^2)\sin\frac{n\pi t}{1}\,dt$$

$$= \left[\left(\frac{\sin n\pi t}{n^2\pi^2} - \frac{t}{n\pi}\cos n\pi t\right) - \left(\frac{2t}{n^2\pi^2}\sin n\pi t + \frac{2}{n^3\pi^3}\cos n\pi t - \frac{t^2}{n\pi}\cos n\pi t\right)\right]_{-1}^{1}$$

$$= -\frac{2\cos n\pi}{n\pi}$$

Hence, for $0 \leq t < 1$ it is also possible to write

$$f(t) = -\frac{1}{3} + \frac{4}{\pi^2}\left(\frac{\cos \pi t}{1} - \frac{\cos 2\pi t}{4} + \frac{\cos 3\pi t}{9} - \frac{\cos 4\pi t}{16} + \cdots\right)$$
$$+ \frac{2}{\pi}\left(\frac{\sin \pi t}{1} - \frac{\sin 2\pi t}{2} + \frac{\sin 3\pi t}{3} - \frac{\sin 4\pi t}{4} + \cdots\right) \tag{7}$$

Figure 8.5a, b, and c shows the extended periodic functions represented, respectively, by the series (5), (6), and (7).

(a)

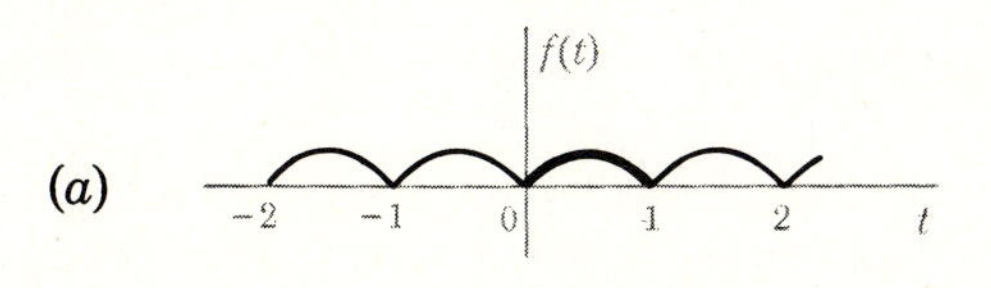

(b)

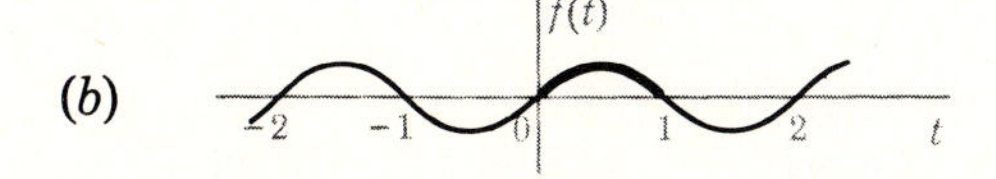

(c)

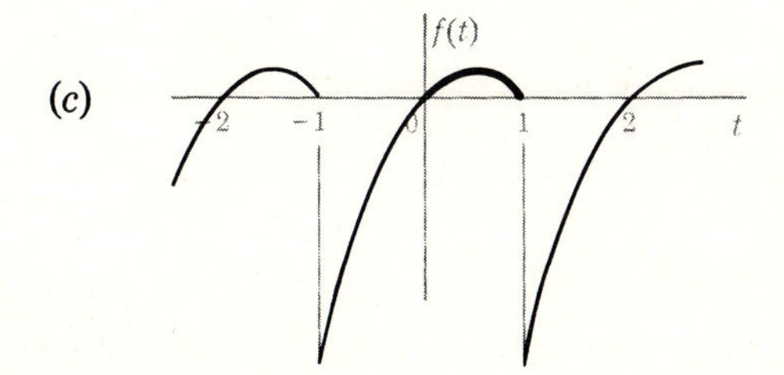

Figure 8.5 Different periodic functions coinciding over the interval (0, 1).

Figure 8.5 and the associated expansions illustrate another interesting and important fact. In Fig. 8.5*c* the graph as a whole is not continuous but has jumps at $t = \pm 1, \pm 3, \ldots$. In the corresponding series (7), the coefficients (of the sine terms) decrease only at a rate proportional to $1/n$. On the other hand, the graph in Fig. 8.5*a* is everywhere continuous but has corners, or points where the tangent changes direction discontinuously. In the corresponding series (5), the coefficients become small much more rapidly than in (7); in fact they decrease at a rate proportional to $1/n^2$. Finally, the graph in Fig. 8.5*b* not only is continuous but also has a continuously turning tangent; i.e., there are no points where the tangent changes direction abruptly. This smoother behavior of the function is reflected in the coefficients in the corresponding series (6), which in this case approach zero at a rate proportional to $1/n^3$. These observations are summed up and generalized in the following theorem, which we cite without proof.†

Theorem 4 As n becomes infinite, the coefficients a_n and b_n in the Fourier expansion of a periodic function satisfying the Dirichlet conditions always approach zero at least as rapidly as c/n,‡ where c is a constant independent of n. If the function has one or more points of discontinuity, then either a_n or b_n, and in general both, can decrease no faster than this. In general, if a function $f(t)$ and its first $k-1$ derivatives satisfy the Dirichlet conditions and are everywhere continuous, then as n becomes infinite, the coefficients a_n and b_n in the Fourier series of $f(t)$ tend to zero at least as rapidly as c/n^{k+1}. If, in addition, the kth derivative of $f(t)$ is not everywhere continuous, then either a_n or b_n, and in general both, can tend to zero no faster than c/n^{k+1}.

More concisely, though less accurately, Theorem 4 asserts that the smoother the function, the faster its Fourier expansion converges. Conversely, by observing the rate at which the terms in the Fourier series of an otherwise unknown function approach zero, we can obtain useful information about the degree of smoothness of the function.

Closely associated with the last result are the following observations, which we also state without proof.

Theorem 5§ The integral of any periodic function which satisfies the Dirichlet conditions can be found by term-by-term integration of the Fourier series of the function.

† See, for instance, H. S. Carslaw, "Fourier Series," Dover, New York, 1930, pp. 269–71.

‡ This does not mean that the coefficients are proportional to $1/n$. What it does mean is that there exists a number c, independent of n, such that both a_n and b_n are less than c/n for all values of n. For example, $2n/(n^2+9)$ approaches zero at least as rapidly as c/n because

$$\frac{2n}{n^2+9} < \frac{2n}{n^2} = \frac{2}{n} \qquad \text{for all values of } n$$

but it does not approach zero as rapidly as c/n^2 because there is no value of c such that

$$\frac{2n}{n^2+9} < \frac{c}{n^2} \qquad \text{for all values of } n$$

§ See, for instance, E. C. Titchmarsh, "Theory of Functions," Oxford, New York, 1939, pp. 419–21.

Theorem 6† If $f(t)$ is a periodic function which satisfies the Dirichlet conditions and is everywhere continuous, and if $f'(t)$ also satisfies the Dirichlet conditions, then wherever it exists, $f'(t)$ can be found by term-by-term differentiation of the Fourier series of $f(t)$.

Exercises for Section 8.3

1 Prove Theorem 2.

2 By considering the identity

$$f(t) = \frac{f(t) + f(-t)}{2} + \frac{f(t) - f(-t)}{2}$$

show that any function defined over an interval which is symmetric with respect to the origin can be written as the sum of an even function and an odd function.

Obtain the half-range sine and cosine expansions of each of the following functions.

3 $f(t) = 1 \qquad 0 \le t \le 2$ **4** $f(t) = t \qquad 0 \le t \le p$

5 $f(t) = \begin{cases} 0 & 0 \le t < 1 \\ 1 & 1 \le t \le 2 \end{cases}$ **6** $f(t) = \begin{cases} t & 0 \le t < 1 \\ 2 - t & 1 \le t \le 2 \end{cases}$

7 $f(t) = t^2 \qquad 0 \le t \le p$ **8** $f(t) = \cos t \qquad 0 \le t \le 2\pi$

9 $f(t) = \sin t \qquad 0 \le t \le 2\pi$ **10** $f(t) = e^{-at} \qquad 0 \le t \le \pi$

***11** $f(t) = \sin at \qquad 0 \le t \le \pi$, a not an integer

***12** $f(t) = \cos at \qquad 0 \le t \le \pi$, a not an integer

***13** By setting $t = \pi$ in the half-range cosine expansion of $\cos at$ obtained in Exercise 12, show that

$$\cot a\pi = \frac{1}{a\pi} - \frac{2a}{\pi} \sum_{n=1}^{\infty} \frac{1}{n^2 - a^2} = \frac{1}{\pi} \sum_{n=-\infty}^{\infty} \frac{1}{n + a}$$

***14** Obtain a series, different from the half-range sine expansion, which will represent $t - t^2$ for $0 \le t \le 1$ and which has coefficients decreasing as $1/n^3$.

***15** Is it possible to obtain a series representing $t - t^2$ for $0 \le t \le 1$ whose coefficients will decrease as $1/n^4$?

***16** Find a function whose half-range cosine series will have coefficients which decrease as $1/n^4$. Determine the expansion.

***17** Determine how $f(t)$, originally defined only for $0 \le t \le p$, must be extended from p to $2p$ if the half-range cosine expansion of the extended function is to contain no terms of the form $\cos(n\pi t/p)$, n even. Derive a formula for the nonzero coefficients.

****18** If

$$f(t) = \begin{cases} 1 & 0 \le t \le a \\ \dfrac{t - 1}{a - 1} & a < t < 1 \\ 0 & 1 \le t \le 2 \end{cases}$$

† See, for instance, E. T. Whittaker and G. N. Watson, "Modern Analysis," Macmillan, New York, 1943, pp. 168–69.

and if a is only slightly less than 1, discuss the behavior of the coefficients in the half-range cosine expansion of $f(t)$ for small and medium values of n as well as for $n \to \infty$.

***19** Without calculating the coefficients, determine how fast the coefficients in the half-range expansions of each of the following functions will decrease:

(a) $f(t) = t(1-t)^2 \qquad 0 \le t \le 1$

(b) $f(t) = \begin{cases} 10t^4 - 15t^3 + 6t & 0 \le t < 1 \\ 1 & 1 \le t \le 2 \end{cases}$

(c) $f(t) = \begin{cases} t^4 - 3t^3 + 2t^2 + t & 0 \le t < 1 \\ 2t - t^2 & 1 \le t \le 2 \end{cases}$

(d) $f(t) = 1/(2 + \cos t) \qquad 0 \le t \le 2\pi$

***20** Prove Theorem 4 for the special case $k = 1$. Under what conditions, if any, can either a_n or b_n decrease faster than c/n^2? *Hint:* Assuming that $f'(t)$ has a single point of discontinuity in each period, apply integration by parts to the formulas for a_n and b_n, taking u in the formula $\int u\,dv = uv - \int v\,du$ to be $f(t)$.

8.4 ALTERNATIVE FORMS OF FOURIER SERIES

The original form of the Fourier series of a function, as derived in Sec. 8.2, can be converted into several other trigonometric forms and into one in which imaginary exponentials appear instead of real trigonometric functions. For instance, in the series

$$f(t) = \frac{a_0}{2} + \sum_{n=1}^{\infty} \left(a_n \cos \frac{n\pi t}{p} + b_n \sin \frac{n\pi t}{p} \right)$$

we can apply to each pair of terms of the same frequency the usual procedure for reducing the sum of a sine and a cosine of the same angle to a single term:

$$f(t) = \frac{a_0}{2} + \sum_{n=1}^{\infty} \sqrt{a_n^2 + b_n^2} \left(\frac{a_n}{\sqrt{a_n^2 + b_n^2}} \cos \frac{n\pi t}{p} + \frac{b_n}{\sqrt{a_n^2 + b_n^2}} \sin \frac{n\pi t}{p} \right)$$

If we now define the angles γ_n and δ_n from the triangle shown in Fig. 8.6 and set

$$A_0 = \frac{a_0}{2} \qquad \text{and} \qquad A_n = \sqrt{a_n^2 + b_n^2}$$

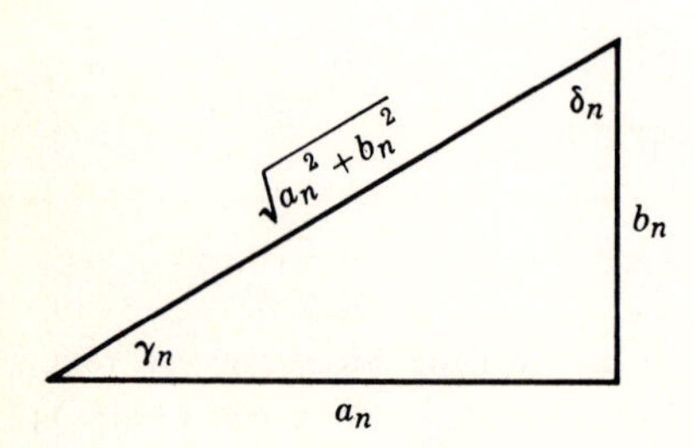

Figure 8.6 The triangle defining the phase angles γ_n and δ_n for the resultant of the terms of frequency $n\pi/p$ in a Fourier series.

the last series can be written

$$f(t) = A_0 + \sum_{n=1}^{\infty} A_n\left(\cos\frac{n\pi t}{p}\cos\gamma_n + \sin\frac{n\pi t}{p}\sin\gamma_n\right)$$

$$= A_0 + \sum_{n=1}^{\infty} A_n\cos\left(\frac{n\pi t}{p} - \gamma_n\right)$$

or, equally well,

$$f(t) = A_0 + \sum_{n=1}^{\infty} A_n\left(\cos\frac{n\pi t}{p}\sin\delta_n + \sin\frac{n\pi t}{p}\cos\delta_n\right)$$

$$= A_0 + \sum_{n=1}^{\infty} A_n\sin\left(\frac{n\pi t}{p} + \delta_n\right)$$

In either of these forms, the quantity $A_n = \sqrt{a_n^2 + b_n^2}$ is the resultant amplitude of the components of frequency $n\pi/p$, that is, the amplitude of the n**th harmonic** in the expansion. The phase angles

$$\gamma_n = \tan^{-1}\frac{b_n}{a_n} \qquad \text{and} \qquad \delta_n = \tan^{-1}\frac{a_n}{b_n} = \frac{\pi}{2} - \gamma_n$$

measure the lag or lead of the nth harmonic with reference to a standard cosine or standard sine wave of the same frequency.

The complex exponential form of a Fourier series is obtained by substituting the exponential equivalents of the cosine and sine terms into the original form of the series:

$$f(t) = \frac{a_0}{2} + \sum_{n=1}^{\infty}\left(a_n\frac{e^{ni\pi t/p} + e^{-ni\pi t/p}}{2} + b_n\frac{e^{ni\pi t/p} - e^{-ni\pi t/p}}{2i}\right)$$

Collecting terms on the various exponentials and noting that $1/i = -i$, we obtain

$$f(t) = \frac{a_0}{2} + \sum_{n=1}^{\infty}\left(\frac{a_n - ib_n}{2}e^{ni\pi t/p} + \frac{a_n + ib_n}{2}e^{-ni\pi t/p}\right)$$

If we now define

$$c_0 = \frac{a_0}{2} \qquad c_n = \frac{a_n - ib_n}{2} \qquad c_{-n} = \frac{a_n + ib_n}{2}$$

the last series can be written in the more symmetric form

$$f(t) = \sum_{n=-\infty}^{\infty} c_n e^{ni\pi t/p} \tag{1}$$

When it is used at all, this exponential form is used as a basic form in its own right; i.e., it is not obtained by transformation from the trigonometric form but is constructed directly from the given function. To do this requires that expressions be available for the direct evaluation of the coefficients c_n. These can easily be found from the definitions of c_0, c_n, and c_{-n}. For

$$c_0 = \frac{1}{2} a_0 = \frac{1}{2p} \int_d^{d+2p} f(t)\, dt$$

$$c_n = \frac{a_n - ib_n}{2} = \frac{1}{2}\left[\frac{1}{p}\int_d^{d+2p} f(t) \cos \frac{n\pi t}{p}\, dt - i\frac{1}{p}\int_d^{d+2p} f(t) \sin \frac{n\pi t}{p}\, dt\right]$$

$$= \frac{1}{2p}\int_d^{d+2p} f(t)\left(\cos \frac{n\pi t}{p} - i \sin \frac{n\pi t}{p}\right) dt$$

$$= \frac{1}{2p}\int_d^{d+2p} f(t) e^{-ni\pi t/p}\, dt$$

$$c_{-n} = \frac{a_n + ib_n}{2} = \frac{1}{2}\left[\frac{1}{p}\int_d^{d+2p} f(t) \cos \frac{n\pi t}{p}\, dt + i\frac{1}{p}\int_d^{d+2p} f(t) \sin \frac{n\pi t}{p}\, dt\right]$$

$$= \frac{1}{2p}\int_d^{d+2p} f(t)\left(\cos \frac{n\pi t}{p} + i \sin \frac{n\pi t}{p}\, dt\right)$$

$$= \frac{1}{2p}\int_d^{d+2p} f(t) e^{ni\pi t/p}\, dt$$

Clearly, whether the index n is positive, negative, or zero, c_n is correctly given by the single formula

$$c_n = \frac{1}{2p}\int_d^{d+2p} f(t) e^{-ni\pi t/p}\, dt \tag{2}$$

As usual, d will almost always be either $-p$ or 0.

Example 1 Find the complex form of the Fourier series of the periodic function whose definition in one period is $f(t) = e^{-t}$, $-1 \le t < 1$.

Since $p = 1$, we have from (2), taking $d = -1$,

$$c_n = \frac{1}{2}\int_{-1}^{1} e^{-t} e^{-ni\pi t}\, dt = \frac{1}{2}\left[\frac{e^{-(1+ni\pi)t}}{-(1+ni\pi)}\right]_{-1}^{1}$$

$$= \frac{e^{-(1+ni\pi)} - e^{(1+ni\pi)}}{-2(1+ni\pi)}$$

$$= \frac{ee^{ni\pi} - e^{-1}e^{-ni\pi}}{2(1+ni\pi)}$$

Now $e^{i\pi} = \cos \pi + i \sin \pi = -1$, and thus $e^{ni\pi} = e^{-ni\pi} = (-1)^n$. Hence

$$c_n = \frac{(-1)^n}{(1+ni\pi)} \frac{e - e^{-1}}{2} = \frac{(-1)^n(1 - ni\pi) \sinh 1}{1 + n^2\pi^2}$$

The expansion of $f(t)$ is therefore

$$f(t) = \sum_{n=-\infty}^{\infty} (-1)^n \frac{(1 - ni\pi) \sinh 1}{1 + n^2\pi^2} e^{ni\pi t}$$

This, of course, can be converted into the real trigonometric form without difficulty, for we have, by definition,

$$c_n = \frac{a_n - ib_n}{2} \qquad \text{and} \qquad c_{-n} = \frac{a_n + ib_n}{2}$$

and thus, by adding and subtracting these expressions, we obtain

$$a_n = c_n + c_{-n} \qquad \text{and} \qquad b_n = i(c_n - c_{-n})$$

Therefore in this problem

$$a_n = \frac{(-1)^n(1 - ni\pi)\sinh 1}{1 + n^2\pi^2} + \frac{(-1)^n(1 + ni\pi)\sinh 1}{1 + n^2\pi^2} = \frac{(-1)^n 2\sinh 1}{1 + n^2\pi^2}$$

$$b_n = i\left[\frac{(-1)^n(1 - ni\pi)\sinh 1}{1 + n^2\pi^2} - \frac{(-1)^n(1 + ni\pi)\sinh 1}{1 + n^2\pi^2}\right] = \frac{(-1)^n 2n\pi \sinh 1}{1 + n^2\pi^2}$$

Hence we can also write

$$f(t) = \sinh 1 - 2\sinh 1\left(\frac{\cos \pi t}{1 + \pi^2} - \frac{\cos 2\pi t}{1 + 4\pi^2} + \frac{\cos 3\pi t}{1 + 9\pi^2} - \cdots\right)$$

$$- 2\pi \sinh 1\left(\frac{\sin \pi t}{1 + \pi^2} - \frac{2\sin 2\pi t}{1 + 4\pi^2} + \frac{3\sin 3\pi t}{1 + 9\pi^2} - \cdots\right)$$

Exercises for Section 8.4

What is the amplitude of the resultant term of frequency $n\pi/p$ in the Fourier series of the periodic functions whose definitions in one period are the following?

1 $f(t) = t + t^2 \qquad -1 \le t < 1$

***2** $f(t) = \begin{cases} -t & -1 \le t < 0 \\ t^2 - t & 0 \le t < 1 \end{cases}$

3 $f(t) = e^{-t} \qquad -1 \le t < 1$

4 Show that the amplitude of the nth harmonic in the Fourier series of a function $f(t)$ is equal to $2\sqrt{c_n c_{-n}}$.

Find the complex form of the Fourier series of the periodic functions whose definitions in one period are:

5 $f(t) = \begin{cases} 1 & 0 \le t < 1 \\ 0 & 1 \le t < 2 \end{cases}$ **6** $f(t) = \begin{cases} 1 & 0 \le t < 1 \\ -1 & 1 \le t < 2 \end{cases}$

7 $f(t) = t \qquad 0 \le t < 1$ **8** $f(t) = t \qquad -1 \le t < 1$

***9** $f(t) = \cos t \qquad -\pi/2 \le t \le \pi/2$. *Hint:* Use the fact that $\cos\theta = \frac{1}{2}(e^{i\theta} + e^{-i\theta})$.

***10** $f(t) = \sin t \qquad 0 \le t \le \pi$ ***11** $f(t) = \cosh t \qquad -1 \le t < 1$

***12** $f(t) = \sinh t \qquad -1 \le t < 1$

8.5 APPLICATIONS OF FOURIER SERIES

Although we will see in Chap. 11 that Fourier series play an essential role in the application of partial differential equations to the study of such physical phenomena as the vibration of strings and the flow of heat, their most important application at the present stage of our work is in the analysis of the behavior of physical systems subjected to general periodic disturbances. We shall conclude this chapter with a discussion of two typical problems, one involving a mechanical system, the other an electrical circuit.

Example 1 Determine the steady-state forced vibrations of the system shown in Fig. 8.7*a* if the applied force $F(t)$ is the periodic force shown in Fig. 8.7*b*.

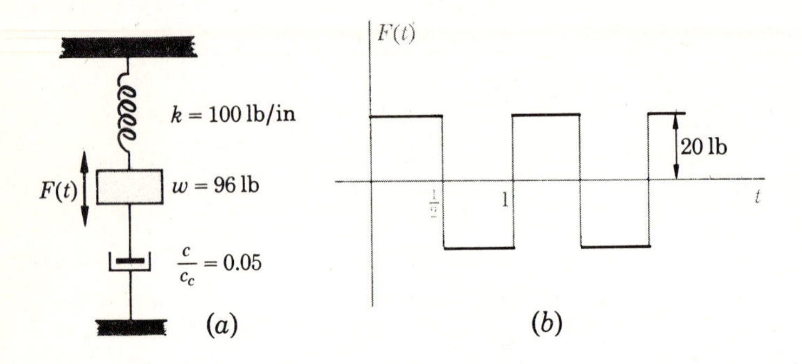

Figure 8.7 A spring-mass system acted upon by an alternating square-wave force.

Since the concepts we developed in our study of forced vibrations in Chap. 7 are applicable only to periodic functions which are simple sines and cosines, our first step must be to express the driving force $F(t)$ in terms of such functions; i.e., our first step must be to determine the Fourier expansion of $F(t)$. By extending $F(t)$ as an odd function of t, no cosine terms can be present in its expansion, and thus we need only compute b_n:

$$b_n = \frac{2}{\frac{1}{2}} \int_0^{1/2} 20 \sin \frac{n\pi t}{\frac{1}{2}}\, dt = 80 \left[-\frac{\cos 2n\pi t}{2n\pi} \right]_0^{1/2}$$

$$= 40 \frac{1 - \cos n\pi}{n\pi}$$

$$= \begin{cases} 0 & n \text{ even} \\ \dfrac{80}{n\pi} & n \text{ odd} \end{cases}$$

Hence $$F(t) = \frac{80}{\pi}\left(\sin 2\pi t + \frac{\sin 6\pi t}{3} + \frac{\sin 10\pi t}{5} + \frac{\sin 14\pi t}{7} + \cdots \right)$$

and this is the expression that would appear on the right-hand side of the differential equation describing the motion of the system if we were to set up the equation. Now we are asked only to find the *steady-state* forced motion of the system; hence we need to determine only the particular integral corresponding to $F(t)$. Moreover, since the relevant differential equation (even though we have not set it up) is obviously linear, the required particular integral can be found very simply by using the ideas of Sec. 7.3. In fact, all we need do is to apply the proper magnification and phase shift to each component of the driving force $F(t)$ and add the results. Preparatory to this, we must

Table 8.1

Term	δ_{st}	$\dfrac{\omega}{\omega_N}$	$M = \dfrac{1}{\sqrt{\left(1-\dfrac{\omega^2}{\omega_N^2}\right)^2+\left(2\dfrac{c}{c_c}\dfrac{\omega}{\omega_N}\right)^2}}$	$\alpha = \tan^{-1}\dfrac{2\dfrac{c}{c_c}\dfrac{\omega}{\omega_N}}{1-\dfrac{\omega^2}{\omega_N^2}}$	Steady-state term = $\delta_{st} M \sin(\omega t - \alpha)$
1	$\dfrac{4}{5\pi}$	$\dfrac{2\pi}{20}$	1.11	0.035 rad	$0.28 \sin(2\pi t - 0.035)$
2	$\dfrac{4}{15\pi}$	$\dfrac{6\pi}{20}$	6.83	0.701 rad	$0.58 \sin(6\pi t - 0.701)$
3	$\dfrac{4}{25\pi}$	$\dfrac{10\pi}{20}$	0.68	3.035 rad	$0.03 \sin(10\pi t - 3.035)$
4	$\dfrac{4}{35\pi}$	$\dfrac{14\pi}{20}$	0.26	3.084 rad	$0.01 \sin(14\pi t - 3.084)$
...	...	...	...	...	...

determine the static deflections that would be produced in the system by steady forces having the magnitudes of the various terms of $F(t)$. These are equal to

$$(\delta_{st})_n = \frac{80/(n\pi)\text{ lb}}{100\text{ lb/in}} = \frac{4}{5n\pi}\text{ in} \qquad n \text{ odd}$$

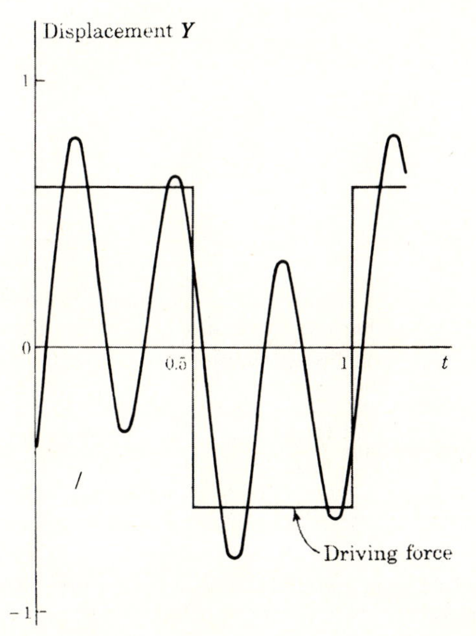

Figure 8.8 A response of apparent frequency greater than that of the excitation producing it.

Then we must calculate the undamped natural frequency of the system:

$$\omega_N{}^\dagger = \sqrt{\frac{kg}{w}} = \sqrt{\frac{100 \times 384}{96}} = 20 \text{ rad/s}$$

The rest of the work can best be presented in tabular form (see Table 8.1). Figure 8.8 shows the steady-state motion of the system plotted as a function of time.

This example illustrates an exceedingly important but sometimes misunderstood characteristic of forced vibrations. If the driving force is not a pure sine or cosine function, its Fourier expansion will contain terms whose frequencies are above the fundamental, or apparent, frequency of the excitation. If the frequency of one of these harmonics happens to be close to the natural, or resonant, frequency of the system, and if the amount of friction in the system is small, the corresponding magnification ratio will be large and its value may offset many times the smaller amplitude of that harmonic and make the resultant term the dominant part of the entire response. If and when this happens, the response will appear to be of a higher frequency than the force which produces it. Figure 8.8 shows this clearly, for, although the force alternates only once per second, the weight appears to move up and down 3 times per second.

It is interesting to note that although the driving force in this example is discontinuous, both the displacement and the velocity it produces are continuous. This is suggested by the plot of the displacement shown in Fig. 8.8 and confirmed by an application of Theorem 4, Sec. 8.3. In fact, since the frequency of the general term in the Fourier expansion of the driving force $F(t)$ is $(2n - 1)2\pi \approx 4n\pi$, it follows, by neglecting all but the highest power of n in the denominator of the magnification ratio M, that for n sufficiently large, M is arbitrarily close to

$$\frac{1}{\omega^2/\omega_N^2} \approx \frac{1}{(4n\pi)^2/(20)^2} = \frac{25}{n^2\pi^2}$$

Therefore, since the static deflection corresponding to the general term in the expansion of $F(t)$ is

$$(\delta_{st})_n = \frac{4}{5(2n-1)\pi} \approx \frac{2}{5n\pi}$$

it follows that as n becomes infinite, the coefficient of the general term in the expansion of the steady-state displacement, namely, $(\delta_{st})_n M$, tends to zero as

$$\frac{2}{5n\pi}\frac{25}{n^2\pi^2} = \frac{10}{n^3\pi^3}$$

Thus, according to Theorem 4, Sec. 8.3, the steady-state displacement $Y(t)$ and the steady-state velocity dY/dt are continuous, but the acceleration d^2Y/dt^2 is discontinuous. Of course, independent of this reasoning, the acceleration must be discontinuous since, from Newton's law,

$$\text{Acceleration} = \frac{\text{force}}{\text{mass}}$$

and in this problem the given force is discontinuous.

† We must remember that here the subscript N in ω_N stands for *natural* and is in no way connected with the parameter n which identifies the general term in the Fourier expansion of $F(t)$.

Example 2 Find the steady-state current produced in the circuit shown in Fig. 8.9*a* by the periodic voltage shown in Fig. 8.9*b*.

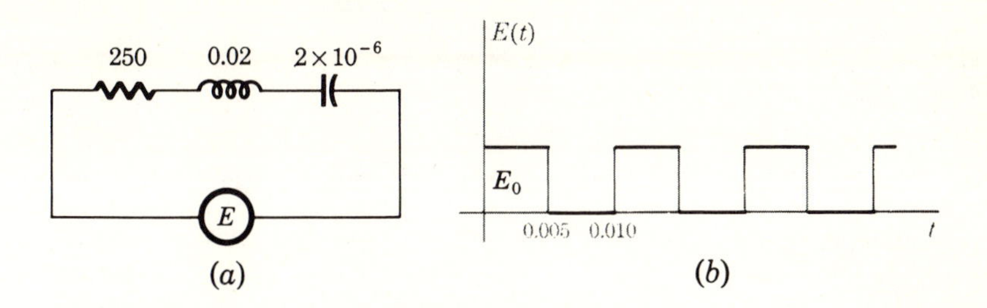

Figure 8.9 A series circuit driven by a square-wave voltage.

As in Example 1, our first step here is to determine the Fourier expansion of the impressed voltage. However, since we plan to use the complex impedance to find the steady-state current, we use the complex exponential rather than the real trigonometric form of the Fourier series. Hence we compute

$$c_n = \frac{1}{0.01}\int_0^{0.005} E_0 e^{-ni\pi t/0.005}\,dt = 100E_0 \left.\frac{e^{-ni\pi t/0.005}}{-ni\pi/0.005}\right|_0^{0.005}$$

$$= E_0 \frac{1 - e^{-ni\pi}}{2ni\pi}$$

$$= \begin{cases} 0 & n \text{ even}, n \neq 0 \\ \dfrac{E_0}{ni\pi} = -\dfrac{iE_0}{n\pi} & n \text{ odd} \end{cases}$$

$$c_0 = \frac{1}{0.01}\int_0^{0.005} E_0\,dt = \frac{E_0}{2}$$

Therefore $$E(t) = E_0\left(\cdots + \frac{ie^{-600i\pi t}}{3\pi} + \frac{ie^{-200i\pi t}}{\pi} + \frac{1}{2} - \frac{ie^{200i\pi t}}{\pi} - \frac{ie^{600i\pi t}}{3\pi} - \cdots\right)$$

In Sec. 7.4 we showed that the steady-state current produced by a voltage of the form $Ae^{i\omega t}$ can be found simply by dividing the voltage by the complex impedance

$$Z(\omega) = R + i\left(\omega L - \frac{1}{\omega C}\right)$$

Using the data of the present problem, we have

$$Z(\omega) = 250 + i\left(0.02\omega - \frac{10^6}{2\omega}\right)$$

or, since $$\omega = 200n\pi \qquad n \text{ odd}$$

we have $$Z(\omega) \equiv Z_n = 250 + i\left(4n\pi - \frac{2{,}500}{n\pi}\right) \qquad n \text{ odd}$$

Hence, dividing each term in the expansion of the voltage $E(t)$ by the value of Z for the corresponding frequency, i.e., the corresponding value of n, we find

$$I(t) = \sum_{n=-\infty}^{\infty} D_n e^{200ni\pi t} \qquad n \text{ odd}\dagger$$

where

$$D_n = \frac{c_n}{Z_n} = -\frac{iE_0}{n\pi} \frac{1}{250 + i[4n\pi - 2{,}500/(n\pi)]} = \frac{-iE_0}{250n\pi + i(4n^2\pi^2 - 2{,}500)}$$

If we want the real trigonometric form of this expansion, namely,

$$I(t) = \frac{a_0}{2} + a_1 \cos 200\pi t + a_3 \cos 600\pi t + \cdots + b_1 \sin 200\pi t + b_3 \sin 600\pi t + \cdots$$

we have at once

$$a_n = D_n + D_{-n} = -iE_0 \left[\frac{1}{250n\pi + i(4n^2\pi^2 - 2{,}500)} + \frac{1}{-250n\pi + i(4n^2\pi^2 - 2{,}500)} \right]$$

$$= -\frac{2E_0(4n^2\pi^2 - 2{,}500)}{(250n\pi)^2 + (4n^2\pi^2 - 2{,}500)^2} \qquad n \text{ odd}$$

$$b_n = i(D_n - D_{-n}) = E_0 \left[\frac{1}{250n\pi + i(4n^2\pi^2 - 2{,}500)} - \frac{1}{-250n\pi + i(4n^2\pi^2 - 2{,}500)} \right]$$

$$= \frac{500n\pi E_0}{(250n\pi)^2 + (4n^2\pi^2 - 2{,}500)^2} \qquad n \text{ odd}$$

Exercises for Section 8.5

1 In Example 2, show that the current $I(t)$ is continuous even though the voltage $E(t)$ is discontinuous. Is the charge $Q(t)$ on the capacitor continuous?

2 In Example 1, determine the first four terms in the steady-state response of the system if the amount of friction is doubled and the spring is changed to one of modulus 144 lb/in.

3 In Example 2, determine the steady-state current if

$$E(t) = \begin{cases} E_0 & 0 < t < 0.005 \\ -E_0 & 0.005 < t < 0.01 \end{cases}$$

and $E(t)$ is periodic with period $2p = 0.01$.

4 In Example 1, determine the first four terms in the steady-state response of the system if $k = 28$ g/cm, $w = 35$ g, $c/c_c = 0.1$, and the magnitude of F is 20 g.

5 Determine the steady-state current in the circuit shown in Fig. 8.10, $E(t)$ being periodic with period $2p = 0.02$.

† Because of the presence of the capacitor, the impedance Z_0 for the dc component, or component of zero frequency, is infinite. Hence the term $E_0/2$ in the expansion of $E(t)$ makes no contribution to the steady-state current.

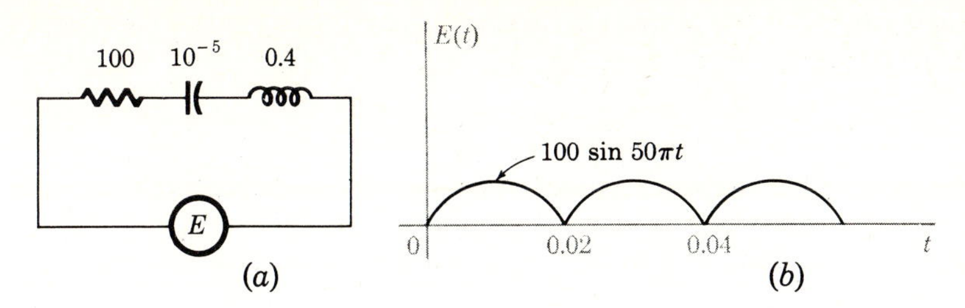

Figure 8.10

***6** Find the steady-state motion of the system shown in Fig. 8.11 if the periodic force F is applied to the upper weight. Discuss the possibility of resonance and approximate resonance in the system.

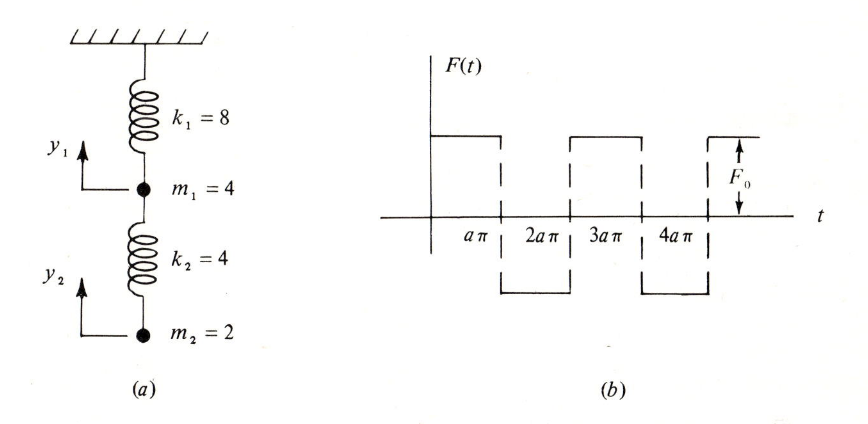

Figure 8.11

***7** Work Exercise 6 if the force F is applied to the lower weight.

CHAPTER

NINE

SERIES SOLUTIONS OF LINEAR DIFFERENTIAL EQUATIONS

9.1 INTRODUCTION

With the exception of the Euler-Cauchy equation which we studied in Sec. 4.5, linear differential equations of order two or higher with variable coefficients can rarely be solved in terms of elementary functions. For such equations, the only thing resembling a general solution procedure is the so-called **method of Frobenius,**† which bears a strong resemblance to the method of undetermined coefficients that we studied in Sec. 4.3. In this method, the product of an undetermined power of $|x - x_0|$ and a Taylor series in $x - x_0$ is assumed as a tentative solution. This series is then substituted into the given differential equation, like terms are collected, and finally the coefficient of each power is equated to zero. This gives an infinite set of conditions from which, sometimes more easily than one might expect, the successive coefficients in the assumed solution can be found. Rarely will the resultant series be the expansions of known functions. Instead, they become the definitions of new functions which, in particular cases, may be investigated in detail and eventually tabulated.

To educate ourselves in these new ideas at a very simple level, let us suppose that *in complete ignorance of the trigonometric functions and their properties* we are to solve the equation

$$y'' + y = 0 \tag{1}$$

To apply the method of Frobenius to this equation (assuming that we are seeking a solution around the origin), we begin by assuming an infinite series solution of

† Named for the German mathematician F. G. Frobenius (1849–1917).

the form

$$y = a_0 + a_1 x + a_2 x^2 + \cdots + a_n x^n + \cdots = \sum_{n=0}^{\infty} a_n x^n \tag{2}$$

Then we substitute this series and its second derivative

$$\sum_{n=2\dagger}^{\infty} n(n-1)a_n x^{n-2}$$

found by termwise differentiation, into Eq. (1), getting

$$\sum_{n=2}^{\infty} n(n-1)a_n x^{n-2} + \sum_{n=0}^{\infty} a_n x^n = 0 \tag{3}$$

Now if we change the index of summation in the first sum in (3) from n to m by the substitution $n = m + 2$, we obtain

$$\sum_{m=0}^{\infty} (m+2)(m+1)a_{m+2} x^m$$

This sum has the same limits as the second sum in (3). Moreover, if the dummy index m is replaced by n, its general term will contain the same power of x as the general term in the second sum in (3). Hence, if we make this change of notation, we can combine the two sums into a single sum, getting

$$\sum_{n=0}^{\infty} [(n+2)(n+1)a_{n+2} + a_n]x^n = 0 \tag{4}$$

Since the various powers of x in (4) are linearly independent, this series can vanish identically if and only if the coefficient of each power of x is zero. Thus we obtain the infinite set of equations

$$(n+2)(n+1)a_{n+2} + a_n = 0$$

or

$$a_{n+2} = -\frac{a_n}{(n+2)(n+1)} \qquad n = 0, 1, 2, \ldots \tag{5}$$

A relation, such as (5), which holds between two or more members of an infinite sequence whenever their indices differ by prescribed integers is called a **recurrence relation** or, more explicitly, a **linear recurrence relation.**

If n is even, say $n = 2m$, then from (5) we find, successively,

$$a_2 = -\frac{a_0}{2 \cdot 1} = -\frac{a_0}{2!}$$

$$a_4 = -\frac{a_2}{4 \cdot 3} = \frac{a_0}{4!}$$

$$a_6 = -\frac{a_4}{6 \cdot 5} = -\frac{a_0}{6!}$$

† Note that $n(n-1)a_n x^{n-2} = 0$ for $n = 0$ and $n = 1$.

and, by an obvious induction,

$$a_{2m} = (-1)^m \frac{a_0}{(2m)!} \qquad m = 0, 1, 2, \ldots$$

Similarly, if n is odd, say $n = 2m + 1$, we find from (5)

$$a_3 = -\frac{a_1}{3 \cdot 2} = -\frac{a_1}{3!}$$

$$a_5 = -\frac{a_3}{5 \cdot 4} = \frac{a_1}{5!}$$

and in general

$$a_{2m+1} = (-1)^m \frac{a_1}{(2m+1)!} \qquad m = 0, 1, 2, \ldots$$

Substituting these values of the a's into the assumed series (2) and collecting terms on a_0 and a_1, we obtain the solution

$$y = a_0 \sum_{m=0}^{\infty} (-1)^m \frac{x^{2m}}{(2m)!} + a_1 \sum_{m=0}^{\infty} (-1)^m \frac{x^{2m+1}}{(2m+1)!} \tag{6}$$

If, in truth, we knew nothing about the trigonometric functions, the two series in (6) could be used to define new functions, say

$$C(x) = 1 - \frac{x^2}{2!} + \frac{x^4}{4!} - \cdots = \sum_{m=0}^{\infty} (-1)^m \frac{x^{2m}}{(2m)!}$$

and $$S(x) = x - \frac{x^3}{3!} + \frac{x^5}{5!} - \cdots = \sum_{m=0}^{\infty} (-1)^m \frac{x^{2m+1}}{(2m+1)!}$$

which, by the ratio test, are defined for all real values of x. With $C(x)$ and $S(x)$ thus defined, we could study them by utilizing various properties of their series representations, together with the fact that each function satisfies the differential equation (1).

For instance, it is immediately clear that

$$C(0) = 1 \qquad \text{and} \qquad S(0) = 0$$

By differentiating each series term-by-term† and inspecting the results, it is also clear that

$$\frac{dC}{dx} = -S(x) \qquad \text{and} \qquad \frac{dS}{dx} = C(x)$$

Actually, all the analytic properties of the trigonometric functions can be developed from this approach, although we shall leave the derivation of further properties of $C(x)$ (actually $\cos x$) and $S(x)$ (actually $\sin x$) to the exercises.

† This is justified, since the ratio and comparison tests can be used to verify that both $C(x)$ and $S(x)$ converge absolutely and uniformly for all values of x.

In subsequent sections we shall investigate the application of the method of Frobenius to linear differential equations for which we have no alternative solution procedures. Then as an illustration of the general theory, we shall make a moderately detailed study of one of the most important linear, second-order, variable-coefficient differential equations, namely Bessel's equation,

$$x^2y'' + xy' + (x^2 - \nu^2)y = 0$$

This will lead us to a new (for us) class of functions known as Bessel functions, some of whose properties and applications we shall investigate.

Exercises for Section 9.1

1 Using the method of Frobenius, find the series solution $y = E(x)$ of the equation $y' - y = 0$ which satisfies the condition $y(0) = 1$.

2 Without solving the equation or using any of the properties of the exponential function, show that $y = E(\lambda x)$ is a solution of the equation $y' - \lambda y = 0$. *Hint:* Change the independent variable of the given equation from x to t by the substitution $t = \lambda x$.

***3** Using the result of Exercise 2, establish the identity $E(ax)E(bx) = E[(a + b)x]$. *Hint:* Combine the differential equations satisfied by $E(ax)$ and $E(bx)$.

4 Prove that $C(-x) = C(x)$ and that $S(-x) = -S(x)$.

***5** Prove that $C^2(x) + S^2(x) = 1$. *Hint:* Recall Abel's identity.

***6** Without solving the equation or using any of the properties of the trigonometric functions, show that $y = C(\lambda x)$ and $y = S(\lambda x)$ are solutions of the equation $y'' + \lambda^2 y = 0$. *Hint:* Note the suggestion given in Exercise 2.

***7** Without using any of the properties of the trigonometric functions, prove that $S(x + a) = C(a)S(x) + S(a)C(x)$. *Hint:* Note that $C(x)$, $S(x)$, and $S(x + a)$ all satisfy the equation $y'' + y = 0$. Then use Theorem 4, Sec. 3.2.

***8** Prove that $C(x + a) = C(a)C(x) - S(a)S(x)$.

***9** Prove that there is at least one positive value of x for which $C(x) = 0$.

***10** Prove that there is at least one positive value of x for which $S(x) = 0$.

****11** If $x = p$ is the smallest positive value of x for which $S(x) = 0$, show that $C(p) = -1$, $C(p/2) = 0$, $S(p/2) = 1$, $S(2p) = 0$, $C(2p) = 1$.

****12** Show that $C(x)$ and $S(x)$ are periodic functions of x of period $2p$, where p is the smallest positive zero of $S(x)$.

9.2 THEORETICAL PRELIMINARIES

The equation $y'' + y = 0$ which we used in the last section to introduce the method of Frobenius is atypically simple; in fact, since it is a linear differential equation with constant coefficients, there was no need to use series for its solution at all. Typical serious applications of the method of Frobenius are concerned with series solutions of equations of the form

$$y'' + P(x)y' + Q(x)y = 0 \tag{1}$$

in some neighborhood of a point $x = x_0$ at which the coefficient functions $P(x)$ and $Q(x)$ may not even be continuous. In general, series of the familiar Maclaurin or Taylor type, which sufficed for the solution of the equation $y'' + y = 0$ in the last section, are inadequate for the solution of an equation like (1). Instead, we must assume a series representation for y of the form

$$y = |x - x_0|^r[a_0 + a_1(x - x_0) + a_2(x - x_0)^2 + \cdots] \qquad a_0 \neq 0 \qquad (2a)$$

where $x = x_0$ is the point around which a solution is sought and the exponent r need not be a positive integer. Since r may not be an integer, the absolute value of $x - x_0$ is necessary in the factor which multiplies the power series in (2a) to ensure that y is real when $x - x_0$ is negative. For simplicity, however, we shall usually drop absolute values, assume for y a series of the form

$$y = \sum_{n=0}^{\infty} a_n(x - x_0)^{n+r} \qquad a_0 \neq 0 \qquad (2b)$$

and seek solutions only over an interval of the form $x_0 < x < R$.

The analysis involves a consideration of several cases, depending upon the behavior of the coefficient functions $P(x)$ and $Q(x)$ at the point $x = x_0$ around which we propose to expand the solution y. In most of our work, the variables x and y and the coefficient functions $P(x)$ and $Q(x)$ will all be real. This is not a necessary restriction, however, and in the basic definitions and theorems we shall introduce in this section, x, y, $P(x)$, and $Q(x)$ may take on either real or complex values.

The first concept we shall need is that of an **analytic function.**

Definition 1 A function $f(x)$ is said to be **analytic** at a point $x = x_0$ if and only if it has a Taylor series at $x = x_0$ which represents the function in some neighborhood of $x = x_0$.

In particular, polynomial functions are analytic everywhere, and rational functions are analytic at all points where they are defined, i.e., at all points except the zeros of their denominators.

Using the concept of an analytic function, we can now classify the point x_0 at which we seek a solution of (1).

Definition 2 If the coefficient functions $P(x)$ and $Q(x)$ in the equation $y'' + P(x)y' + Q(x)y = 0$ are both analytic at $x = x_0$, then x_0 is said to be an **ordinary point** of the equation.

If at least one of the functions $P(x)$ and $Q(x)$ is not analytic at $x = x_0$ but if the functions defined by the products $(x - x_0)P(x)$ and $(x - x_0)^2Q(x)$ are analytic at $x = x_0$, then x_0 is said to be a **regular singular point** of the equation.

If at least one of the products $(x - x_0)P(x)$ and $(x - x_0)^2Q(x)$ is not analytic at $x = x_0$, then x_0 is said to be an **irregular singular point** of the equation.

In our work we shall be concerned exclusively with the expansion of solutions of Eq. (1) around ordinary points and regular singular points.

Example 1 For the differential equation

$$y'' + \frac{2}{x}y' + \frac{3}{x(x-1)^3}y = 0$$

$x = 0$ and $x = 1$ are singular points, since at $x = 0$ both $P(x)$ and $Q(x)$ are undefined, while at $x = 1$, although $P(x)$ is analytic, $Q(x)$ is not. All other points are ordinary points. The point $x = 0$ is a regular singular point since each of the products

$$xP(x) = 2$$

and $$x^2Q(x) = \frac{3x}{(x-1)^3} = -3x(1-x)^{-3} = -3x(1 + 3x + 6x^2 + \cdots) \qquad |x| < 1$$

is analytic at $x = 0$, that is, can be represented around $x = 0$ by a series of positive integral powers of x. The point $x = 1$ is an irregular singular point, however, because although the product

$$(x-1)P(x) = \frac{2(x-1)}{x} = 2(x-1)[1 + (x-1)]^{-1}$$

$$= 2(x-1)[1 - (x-1) + (x-1)^2 - \cdots] \qquad |x-1| < 1$$

is analytic at $x = 1$, the product

$$(x-1)^2Q(x) = \frac{3}{x(x-1)}$$

is undefined at $x = 1$ and hence is not analytic at $x = 1$.

The importance of the classification of values of x into ordinary and singular points is apparent from the following theorems, which are proved in more advanced treatments of the theory of differential equations.†

Theorem 1 At an ordinary point $x = x_0$ of the differential equation

$$y'' + P(x)y' + Q(x)y = 0$$

every solution is analytic, i.e., can be represented by a series of the form

$$y = a_0 + a_1(x - x_0) + a_2(x - x_0)^2 + \cdots$$

Moreover, the radius of convergence of each series solution is not less than the distance from x_0 to the nearest singular point of the equation.

Theorem 2 At a regular singular point $x = x_0$ of the differential equation

$$y'' + P(x)y' + Q(x)y = 0$$

there is at least one solution which possesses an expansion of the form

$$y = |x - x_0|^r[a_0 + a_1(x - x_0) + a_2(x - x_0)^2 + \cdots]$$

† See, for instance, E. T. Whittaker and G. N. Watson, "Modern Analysis," Macmillan, New York, 1943, pp. 194–203.

and this series will converge for $0 < |x - x_0| < R$, where R is not less than the distance from x_0 to the nearest of the other singular points of the equation.

Theorem 3 At an irregular singular point $x = x_0$ of the differential equation

$$y'' + P(x)y' + Q(x)y = 0$$

there are in general no solutions with expansions consisting solely of powers of $x - x_0$.

In using Theorems 1 and 2 to infer the minimum radius of convergence of power-series solutions of Eq. (1), it must be borne in mind that the singular point nearest to, but distinct from, the point of expansion may be complex, even though the point around which we are expanding is real. For instance, for the differential equation

$$y'' + \frac{1}{1 + x^2} y' + y = 0$$

the coefficient functions

$$P(x) = \frac{1}{1 + x^2} \qquad \text{and} \qquad Q(x) = 1$$

are analytic for all real values of x. However, $P(x)$ fails to be analytic at $x = \pm i$; hence, these two points are singular points of the differential equation. Therefore, a series solution around the ordinary point $x = 2$, say, would have radius of convergence $R = \sqrt{5}$ at least, since in the complex plane the distance from the point of expansion $x = 2$ to the nearest singular point $x = i$ (or $x = -i$) is $\sqrt{5}$ (Fig. 9.1).

Before the method of Frobenius is used to obtain series solutions of Eq. (1) around an ordinary point or a regular singular point, it is convenient to translate axes, if necessary, so that the point of expansion $x = x_0$ becomes $x = 0$.

If $x = 0$ is an ordinary point of Eq. (1), then $P(x)$ and $Q(x)$ are analytic at the origin, and according to Theorem 1, every solution around the origin can be

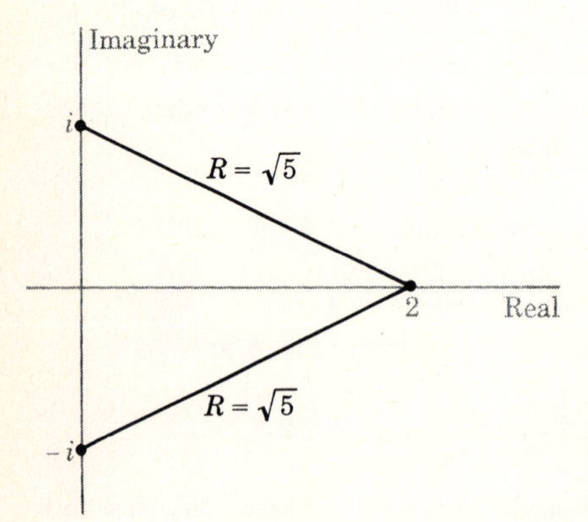

Figure 9.1 The radius of convergence as the distance to a complex singular point.

represented by a series of the Maclaurin form:

$$y = \sum_{n=0}^{\infty} a_n x^n$$

Therefore, in this case there is no need to include the factor $|x|^r$ in applying the method of Frobenius. (This is illustrated by the example of Sec. 9.1.)

If $x = 0$ is a regular singular point of Eq. (1), then both $xP(x)$ and $x^2Q(x)$ are analytic at the origin, and we can write

$$xP(x) = p_0 + p_1 x + p_2 x^2 + \cdots$$

$$x^2Q(x) = q_0 + q_1 x + q_2 x^2 + \cdots$$

Therefore, multiplying Eq. (1) by x^2 and then substituting these series for $xP(x)$ and $x^2Q(x)$, we have

$$x^2y'' + x(p_0 + p_1 x + p_2 x^2 + \cdots)y' + (q_0 + q_1 x + q_2 x^2 + \cdots)y = 0 \tag{3}$$

Now we assume a solution of the form (2), with $x_0 = 0$,

$$y = x^r(a_0 + a_1 x + a_2 x^2 + \cdots) \qquad a_0 \neq 0 \tag{4}$$

If we substitute this into Eq. (1), now written in the form (3), we have

$$\begin{aligned} x^2[a_0 r(r-1)x^{r-2} &+ a_1(r+1)rx^{r-1} + a_2(r+2)(r+1)x^r + \cdots] \\ &+ x(p_0 + p_1 x + p_2 x^2 + \cdots)[a_0 r x^{r-1} + a_1(r+1)x^r + a_2(r+2)x^{r+1} + \cdots] \\ &+ (q_0 + q_1 x + q_2 x^2 + \cdots)(a_0 x^r + a_1 x^{r+1} + a_2 x^{r+2} + \cdots) = 0 \end{aligned}$$

or, collecting terms on the various powers of x,

$$\begin{aligned} a_0[r(r-1) &+ p_0 r + q_0]x^r \\ &+ \{a_1[(r+1)r + p_0(r+1) + q_0] + a_0(p_1 r + q_1)\}x^{r+1} \\ &+ \{a_2[(r+2)(r+1) + p_0(r+2) + q_0] \\ &+ a_1[p_1(r+1) + q_1] + a_0(p_2 r + q_2)\}x^{r+2} + \cdots = 0 \end{aligned} \tag{5}$$

Equation (5) will be an identity if and only if the coefficient of each power of x is zero, and thus we obtain the set of equations

$$\begin{aligned} &a_0[r(r-1) + p_0 r + q_0] = 0 \\ &a_1[(r+1)r + p_0(r+1) + q_0] + a_0(p_1 r + q_1) = 0 \\ &a_2[(r+2)(r+1) + p_0(r+2) + q_0] \\ &\qquad + a_1[p_1(r+1) + q_1] + a_0(p_2 r + q_2) = 0 \\ &\cdots\cdots\cdots\cdots\cdots\cdots\cdots\cdots\cdots\cdots \end{aligned} \tag{6}$$

Since $a_0 \neq 0$, it follows from the first of these equations that

$$r^2 + (p_0 - 1)r + q_0 = 0 \tag{7}$$

This quadratic equation in r is known as the **indicial equation** of the differential equation relative to the point of expansion, and its roots r_1 and r_2 are known as the **exponents** of the regular singular point under consideration ($x = 0$). For each of these values there is, in general, a series solution of Eq. (1) of the form (4). Moreover, the coefficients in these expansions can be determined, one by one, from the successive equations in the set (6), which express each of the a's in terms of the a's which precede it in the series (4).

Example 2 Find series solutions around the origin for the equation

$$y'' + x^2y' + 2xy = 0$$

In this equation both $P(x) = x^2$ and $Q(x) = 2x$ are analytic everywhere, and therefore $x = 0$, like all other points, is an ordinary point of the given equation. Hence, according to Theorem 1, every solution around the origin will possess a Maclaurin expansion, and these series will converge for all values of x.

To obtain a solution, we assume for y a series of the form

$$y = \sum_{n=0}^{\infty} a_n x^n \tag{8}$$

and substitute it into the given equation, getting

$$\begin{aligned}(2 \cdot 1a_2 + 3 \cdot 2a_3x + 4 \cdot 3a_4x^2 + \cdots) + x^2(a_1 + 2a_2x + \cdots)\\ + 2x(a_0 + a_1x + a_2x^2 + \cdots) = 0\end{aligned} \tag{9}$$

The terms common to the three series in (9) are those which contain the second and higher powers of x. Hence it is convenient to write Eq. (9) in the form

$$\left[2a_2 + 6a_3x + \sum_{m=0}^{\infty}(m+4)(m+3)a_{m+4}x^{m+2}\right] + \sum_{m=0}^{\infty}(m+1)a_{m+1}x^{m+2} + \left(2a_0x + \sum_{m=0}^{\infty}2a_{m+1}x^{m+2}\right) = 0$$

and then combine the three sums into one, getting

$$2a_2 + (6a_3 + 2a_0)x + \sum_{m=0}^{\infty}[(m+4)(m+3)a_{m+4} + (m+3)a_{m+1}]x^{m+2} = 0 \tag{10}$$

Equation (10) will be an identity if and only if the coefficient of each power of x is zero. Hence we have

$$a_2 = 0 \qquad 6a_3 + 2a_0 = 0 \qquad \text{or} \qquad a_3 = -\frac{a_0}{3}$$

and, in general, the recurrence relation

$$(m+4)(m+3)a_{m+4} + (m+3)a_{m+1} = 0$$

or

$$a_{m+4} = -\frac{a_{m+1}}{m+4} \qquad m = 0, 1, 2, \ldots \tag{11}$$

From (11) we find at once that

$$a_2 = a_5 = a_8 = \cdots = a_{3k-1} = 0$$

$$a_3 = -\frac{a_0}{3}, \; a_6 = -\frac{a_3}{6} = \frac{a_0}{3^2 2!}, \; a_9 = -\frac{a_6}{9} = -\frac{a_0}{3^3 3!}, \ldots, \; a_{3k} = (-1)^k \frac{a_0}{3^k k!}$$

$$a_4 = -\frac{a_1}{1 \cdot 4}, \; a_7 = -\frac{a_4}{7} = \frac{a_1}{1 \cdot 4 \cdot 7}, \; a_{10} = -\frac{a_7}{10} = -\frac{a_1}{1 \cdot 4 \cdot 7 \cdot 10}, \ldots .$$

$$a_{3k+1} = (-1)^k \frac{a_1}{1 \cdot 4 \cdot 7 \cdot 10 \cdots (3k+1)}$$

Substituting the values of the a's into the series (8) and collecting terms on a_0 and a_1, we obtain the solution

$$y = a_0\left(1 - \frac{x^3}{3} + \frac{x^6}{3^2 2!} - \frac{x^9}{3^3 3!} + \cdots\right)$$
$$+ a_1\left(x - \frac{x^4}{1 \cdot 4} + \frac{x^7}{1 \cdot 4 \cdot 7} - \frac{x^{10}}{1 \cdot 4 \cdot 7 \cdot 10} + \cdots\right)$$

The power series

$$y_1 = \sum_{k=0}^{\infty} (-1)^k \frac{x^{3k}}{3^k k!} \qquad \text{and} \qquad y_2 = \sum_{k=0}^{\infty} (-1)^k \frac{x^{3k+1}}{1 \cdot 4 \cdot 7 \cdot 10 \cdots (3k+1)}$$

are two linearly independent particular solutions of the given equation whose complete solution is therefore

$$y = a_0 y_1 + a_1 y_2$$

The fact, guaranteed by Theorem 1, that the series defining y_1 and y_2 converge for all values of x can, of course, be confirmed by the ratio test.

Exercises for Section 9.2

1 Why is it no restriction to assume that $a_0 \neq 0$ in the series (2) and (4)?

2 If (inefficiently) a solution of the form $y = x^r \sum_{n=0}^{\infty} a_n x^n$ is assumed at an ordinary point of Eq. (1), show that the roots of the indicial equation are always $r = 0$ and $r = 1$.

3 Find the singular points of each of the following equations and determine whether they are regular or irregular.

(*a*) $y'' + xy' + y = 0$ (*b*) $e^x y'' + 2y' - xy = 0$
(*c*) $x^2 y'' - \lambda^2 y = 0$ (*d*) $x^2 y'' + y' + y = 0$
(*e*) $(1 - x^2)y'' + y' + y = 0$ (*f*) $x^2(1 - x)y'' + (1 - x)y' + y = 0$

4 Find the indicial equation relative to each of the singular points of each of the equations in Exercise 3.

Find two independent power-series solutions around the origin for each of the following equations.

5 $y'' + xy' + y = 0$ **6** $y'' + x^2 y' + 3xy = 0$
7 $y'' + x^2 y' + xy = 0$ ***8** $y'' + 2y' + y = 0$
***9** $y'' - 3y' + 2y = 0$ **10** $(1 + x^2)y'' - 2y = 0$

*11 Find the first three nonzero terms in the two particular power-series solutions of the equation $y'' + y' + [1/(1 - x)]y = 0$ around the origin.

*12 Find two linearly independent power-series solutions around $x = 1$ for the equation $y'' + xy' + y = 0$.

13 What do you think would be appropriate definitions for the terms *ordinary point, regular singular point, irregular singular point*, and *indicial equation* for the general, linear, third-order differential equation $y''' + P(x)y'' + Q(x)y' + R(x)y = 0$?

*14 What is the indicial equation relative to the origin for the third-order differential equation

$$y''' + P(x)y'' + Q(x)y' + R(x)y = 0$$

given that $xP(x)$, $x^2Q(x)$, and $x^3R(x)$ are each analytic at the origin?

*15 The point at infinity is said to be an ordinary point or a singular point of the differential equation $y'' + P(x)y' + Q(x)y = 0$ according as the equation obtained from this by the substitution $x = 1/u$ has an ordinary point or a singular point at $u = 0$. Show that under this substitution the original equation becomes

$$u^4 \frac{d^2y}{du^2} + \left[2u^3 - u^2 P\left(\frac{1}{u}\right)\right]\frac{dy}{du} + Q\left(\frac{1}{u}\right)y = 0$$

and use this result to determine the nature of the point at infinity for the equation $(x^2 + 1)y'' + y' + y = 0$.

*16 Verify that under the change of dependent variable defined by the substitution $y = z \exp\left[-\frac{1}{2}\int P(x)\,dx\right]$ the differential equation $y'' + P(x)y' + Q(x)y = 0$ becomes

$$\frac{d^2z}{dx^2} + R(x)z = 0 \qquad \text{where} \qquad R(x) = Q(x) - \frac{1}{2}\frac{dP(x)}{dx} - \frac{1}{4}P^2(x)$$

*17 Using the result of Exercise 16, determine conditions on a_1, b_1, a_2, b_2, and c_2 which will ensure that the equation

$$y'' + (a_1 + b_1 x)y' + (a_2 + b_2 x + c_2 x^2)y = 0$$

can be solved in terms of elementary functions.

*18 If the origin is an irregular singular point of the equation $y'' + P(x)y' + Q(x)y = 0$ and if in the neighborhood of the origin $xP(x)$ and $x^2Q(x)$ have expansions of the form

$$xP(x) = x^{-\alpha}(p_0 + p_1 x + \cdots)$$

and

$$x^2Q(x) = x^{-\beta}(q_0 + q_1 x + \cdots)$$

where at least one of the numbers (α, β) is positive, show that if there is an indicial equation relative to the origin, then it is of the first degree at most.

9.3 SERIES SOLUTIONS AROUND A REGULAR SINGULAR POINT

When the difference between the roots of the indicial equation relative to a regular singular point of the differential equation

$$y'' + P(x)y' + Q(x)y = 0$$

is not an integer, the equation always possesses two linearly independent series solutions of the form

$$y = |x - x_0|^r \sum_{n=0}^{\infty} a_n(x - x_0)^n \tag{1}$$

At least one, and usually both, of these will involve fractional powers of $x - x_0$ and hence will not be a Taylor series.

If the roots of the indicial equation are equal, it is obvious that two series solutions of the form (1) cannot be obtained. It is also true (though not obvious) that if the roots of the indicial equation differ by an integer, two (independent) solutions of the form (1) may not exist. In this section we shall first investigate the use of the method of Frobenius to solve Eq. (1) around a regular singular point when the roots of the indicial equation do not differ by an integer. Then we shall consider the modifications of the method that may be required when the difference between the roots of the indicial equation is an integer.

Example 1 The equation $2x^2y'' + (x^2 - x)y' + y = 0$ has a regular singular point at the origin. Hence, by Theorem 2, Sec. 9.2, it has at least one solution of the form $y = x^r \sum_{n=0}^{\infty} a_n x^n$. Find all such solutions.

Since the given equation can be written in the form

$$x^2y'' + x(-\tfrac{1}{2} + x/2)y' + \tfrac{1}{2}y = 0$$

it follows from Eq. (3), Sec. 9.2, that $p_0 = -\frac{1}{2}$ and $q_0 = \frac{1}{2}$. Hence, the indicial equation is

$$r^2 - \tfrac{3}{2}r + \tfrac{1}{2} = 0$$

and its roots are $r = 1, \frac{1}{2}$. Since the difference between these roots is not an integer, it is clear that there are two linearly independent solutions of the required form.

If $r = 1$, we have

$$y = \sum_{n=0}^{\infty} a_n x^{n+1} \qquad y' = \sum_{n=0}^{\infty} (n+1)a_n x^n \qquad y'' = \sum_{n=1}^{\infty} (n+1)na_n x^{n-1}$$

Hence, substituting these into the given equation, we obtain

$$\sum_{n=1}^{\infty} 2(n+1)na_n x^{n+1} + \left[\sum_{n=0}^{\infty} (n+1)a_n x^{n+2} - \sum_{n=0}^{\infty} (n+1)a_n x^{n+1}\right] + \sum_{n=0}^{\infty} a_n x^{n+1} = 0$$

or, combining the first, third, and fourth sums,

$$\sum_{n=1}^{\infty} (2n^2 + n)a_n x^{n+1} + \sum_{n=0}^{\infty} (n+1)a_n x^{n+2} = 0$$

These two sums can also be combined, if the index of summation in the second is changed from n to $n + 1$. Doing this, we have

$$\sum_{n=1}^{\infty} (2n^2 + n)a_n x^{n+1} + \sum_{n=1}^{\infty} na_{n-1}x^{n+1} = \sum_{n=1}^{\infty} [(2n^2 + n)a_n + na_{n-1}]x^{n+1} = 0$$

The last equation will be identically satisfied if and only if the coefficient of each power of x is zero. Hence, we obtain the recurrence relation

$$(2n^2 + n)a_n + na_{n-1} = 0 \qquad \text{or} \qquad a_n = -\frac{a_{n-1}}{2n+1} \qquad n = 1, 2, 3, \ldots \tag{2}$$

From (2) we find that

$$a_1 = -\frac{a_0}{3},\ a_2 = -\frac{a_1}{5} = \frac{a_0}{3 \cdot 5},\ a_3 = -\frac{a_2}{7} = -\frac{a_0}{3 \cdot 5 \cdot 7}, \ldots$$

Hence, taking $a_0 = 1$, we have as one particular solution

$$y_1 = \sum_{n=0}^{\infty} (-1)^n \frac{x^{n+1}}{1 \cdot 3 \cdot 5 \cdots (2n+1)}$$

Since $x = 0$ is the only singular point of the given equation, it follows that the series defining y_1 converges for all values of x.

If $r = \frac{1}{2}$, we have, similarly,

$$y = \sum_{n=0}^{\infty} a_n x^{n+1/2} \qquad y' = \sum_{n=0}^{\infty} (n + \tfrac{1}{2})a_n x^{n-1/2}$$

$$y'' = \sum_{n=0}^{\infty} (n + \tfrac{1}{2})(n - \tfrac{1}{2})a_n x^{n-3/2} \qquad x > 0$$

Hence, substituting into the given equation, we obtain

$$\sum_{n=0}^{\infty} 2(n + \tfrac{1}{2})(n - \tfrac{1}{2})a_n x^{n+1/2}$$

$$+ \left[\sum_{n=0}^{\infty} (n + \tfrac{1}{2})a_n x^{n+3/2} - \sum_{n=0}^{\infty} (n + \tfrac{1}{2})a_n x^{n+1/2}\right] + \sum_{n=0}^{\infty} a_n x^{n+1/2} = 0$$

or, combining the first, third, and fourth sums,

$$\sum_{n=1}^{\infty} (2n^2 - n)a_n x^{n+1/2} + \sum_{n=0}^{\infty} (n + \tfrac{1}{2})a_n x^{n+3/2} = 0$$

If we change the index of summation in the last sum from n to $n + 1$ and then combine the two sums, we have

$$\sum_{n=1}^{\infty} (2n^2 - n)a_n x^{n+1/2} + \sum_{n=1}^{\infty} (n - \tfrac{1}{2})a_{n-1} x^{n+1/2}$$

$$= \sum_{n=1}^{\infty} [(2n^2 - n)a_n + (n - \tfrac{1}{2})a_{n-1}]x^{n+1/2} = 0$$

The last equation will be identically satisfied if and only if the a's satisfy the recurrence relation

$$(2n^2 - n)a_n + (n - \tfrac{1}{2})a_{n-1} = 0 \qquad \text{or} \qquad a_n = -\frac{a_{n-1}}{2n} \tag{3}$$

From (3) we find

$$a_1 = -\frac{a_0}{2},\ a_2 = -\frac{a_1}{4} = \frac{a_0}{2^2 2!},\ a_3 = -\frac{a_2}{6} = -\frac{a_0}{2^3 3!}, \ldots$$

Hence, taking $a_0 = 1$, we have a second, linearly independent solution

$$y_2 = \sum_{n=0}^{\infty}(-1)^n \frac{x^{n+1/2}}{2^n n!} = \sqrt{x}\, e^{-x/2} \qquad x > 0$$

Had we begun with the assumption $y = \sqrt{-x} \sum_{n=0}^{\infty} a_n x^n$, $x < 0$, we would have obtained the same recurrence relation (3), the same values for the a's, and the solution

$$y_3 = \sqrt{-x} \sum_{n=0}^{\infty}(-1)^n \frac{x^n}{2^n n!} \qquad x < 0$$

Hence, combining the particular solutions y_2 and y_3, we have as a second solution, valid for all values of x,

$$y_4 = \sqrt{|x|} \sum_{n=0}^{\infty}(-1)^n \frac{x^n}{2^n n!} = \sqrt{|x|}\, e^{-x/2}$$

When the roots of the indicial equation differ by an integer, there is usually only one solution of the form $|x - x_0|^r \sum_{n=0}^{\infty} a_n(x - x_0)^n$ around a regular singular point, and other methods must be used to find a second, independent solution. One way to do this is to use the procedure developed in Sec. 3.3. Having one series solution y_1, we can assume $y = \phi(x)y_1$ and then attempt to determine $\phi(x)$ so that the product $\phi(x)y_1$ will satisfy the given equation. In another method, the series which becomes y_1 when r is given the proper value is also considered as a function of r, and a second solution is obtained from the partial derivative of this series with respect to r. This extension of the method of Frobenius can best be made clear through an example.

Example 2 Find two linearly independent solutions, valid near the origin, for the equation $xy'' + y' + y = 0$.

Since the given equation has $P(x) = Q(x) = 1/x$, it follows that the origin is a regular singular point. Hence, by Theorem 2, Sec. 9.2, there is at least one solution with an expansion of the form

$$y = \sum_{n=0}^{\infty} a_n x^{n+r} \qquad a_0 \neq 0 \text{ (possibly } x > 0) \tag{4}$$

Substituting this series and its first two derivatives into the given differential equation gives us

$$\sum_{n=0}^{\infty}(n + r)(n + r - 1)a_n x^{n+r-1} + \sum_{n=0}^{\infty}(n + r)a_n x^{n+r-1} + \sum_{n=0}^{\infty} a_n x^{n+r} = 0$$

or, combining the first two series,

$$\sum_{n=0}^{\infty}(n + r)^2 a_n x^{n+r-1} + \sum_{n=0}^{\infty} a_n x^{n+r} = 0 \tag{5}$$

If the term corresponding to $n = 0$ is detached from the first series in (5), and if the index of summation is changed from n to $n - 1$ in the second series, the two series can be combined, giving us

$$r^2 a_0 x^{r-1} + \sum_{n=1}^{\infty} (n+r)^2 a_n x^{n+r-1} + \sum_{n=1}^{\infty} a_{n-1} x^{n+r-1}$$

$$= r^2 a_0 x^{r-1} + \sum_{n=1}^{\infty} [(n+r)^2 a_n + a_{n-1}] x^{n+r-1} = 0 \tag{6}$$

The last equation will be identically satisfied if and only if

$$r^2 a_0 = 0 \tag{7}$$

and the a's satisfy the recurrence relation

$$(n+r)^2 a_n + a_{n-1} = 0 \qquad \text{or} \qquad a_n = -\frac{a_{n-1}}{(n+r)^2} \tag{8}$$

From (7) we infer that the indicial equation is $r^2 = 0$. However, to set the stage for the new procedure that will furnish us with a second, linearly independent solution around $x = 0$, we shall continue for a while with r as a parameter before we finally replace it with the value 0. Doing this, we find from the recurrence relation (8) that

$$a_1 = -\frac{a_0}{(r+1)^2} \qquad a_2 = -\frac{a_1}{(r+2)^2} = \frac{a_0}{(r+1)^2(r+2)^2}$$

and, by an obvious induction,

$$a_n = (-1)^n \frac{a_0}{(r+1)^2(r+2)^2 \cdots (r+n)^2} \qquad n = 1, 2, \ldots$$

With a_0 arbitrary, and all the other a's now determined in terms of a_0 and r, Eq. (4) may be written as

$$y_r = a_0 x^r \left[1 - \frac{x}{(r+1)^2} + \frac{x^2}{(r+1)^2(r+2)^2} - \frac{x^3}{(r+1)^2(r+2)^2(r+3)^2} + \cdots\right] \tag{9}$$

Since the a's have all been determined so as to satisfy the recurrence relation (8), every term in (6) except the first, namely $r^2 a_0 x^{r-1}$, is equal to zero. Hence y_r satisfies not the given homogeneous differential equation but the nonhomogeneous equation

$$xy'' + y' + y = r^2 a_0 x^{r-1} \tag{10}$$

Of course, if we put $r = 0$, Eq. (10) reduces to the given differential equation; and y_r, with $a_0 = 1$, particularizes to the corresponding solution

$$y_1 = 1 - \frac{x}{(1!)^2} + \frac{x^2}{(2!)^2} - \frac{x^3}{(3!)^2} + \cdots = \sum_{n=0}^{\infty} (-1)^n \frac{x^n}{(n!)^2}$$

Since the indicial equation, namely $r^2 = 0$, has only the one root $r = 0$, it is evident that a second solution of the form (4) cannot be found by using the roots of the indicial equation. In an attempt to find a second solution, let us substitute y_r into (10), thus obtaining the identity

$$xy_r'' + y_r' + y_r = r^2 x^{r-1}$$

in x and r (a_0 having been divided out). Now differentiate both members of this identity partially with respect to r, remembering, of course, that x is independent of r.

The result is

$$x\frac{\partial(y_r'')}{\partial r}+\frac{\partial(y_r')}{\partial r}+\frac{\partial(y_r)}{\partial r}=2rx^{r-1}+r^2x^{r-1}\ln|x| \qquad x\neq 0$$

When the order of differentiation with respect to x and r is interchanged, this becomes

$$x\left(\frac{\partial y_r}{\partial r}\right)''+\left(\frac{\partial y_r}{\partial r}\right)'+\frac{\partial y_r}{\partial r}=r(2+r\ln|x|)x^{r-1} \tag{11}$$

If we set $r=0$ in this equation, the right-hand side becomes zero, which shows that $\partial y_r/\partial r|_{r=0}$ satisfies the original differential equation and is, presumably, a second, linearly independent solution.

Carrying out the indicated differentiation of y_r, as given by (9) with $a_0=1$ for convenience, we have

$$\begin{aligned}\frac{\partial y_r}{\partial r}&=x^r\ln|x|\left[1-\frac{x}{(r+1)^2}+\frac{x^2}{(r+1)^2(r+2)^2}-\frac{x^3}{(r+1)^2(r+2)^2(r+3)^2}+\cdots\right]\\&\quad+x^r\left\{-x\frac{-2}{(r+1)^3}+x^2\left[\frac{-2}{(r+1)^3}\frac{1}{(r+2)^2}+\frac{1}{(r+1)^2}\frac{-2}{(r+2)^3}\right]\right.\\&\qquad-x^3\left[\frac{1}{(r+2)^2(r+3)^2}\frac{-2}{(r+1)^3}+\frac{1}{(r+1)^2(r+3)^2}\frac{-2}{(r+2)^3}\right.\\&\qquad\left.\left.+\frac{1}{(r+1)^2(r+2)^2}\frac{-2}{(r+3)^3}\right]+\cdots\right\}\end{aligned}$$

Finally, letting $r=0$, we have as a second solution of the original equation

$$\begin{aligned}y_2=\left.\frac{\partial y_r}{\partial r}\right|_{r=0}&=\ln|x|\left[1-\frac{x}{(1!)^2}+\frac{x^2}{(2!)^2}-\frac{x^3}{(3!)^2}+\cdots\right]\\&\quad+2\left[\frac{x}{(1!)^2}-\frac{x^2}{(2!)^2}\left(1+\frac{1}{2}\right)+\frac{x^3}{(3!)^2}\left(1+\frac{1}{2}+\frac{1}{3}\right)-\cdots\right]\\&=y_1\ln|x|-2\sum_{n=1}^{\infty}(-1)^n\frac{x^n}{(n!)^2}H_n \qquad x\neq 0\end{aligned}$$

where $H_n=\sum_{k=1}^{n}1/k$ is the nth partial sum of the harmonic series.

Since y_2 contains the term $\ln|x|$ while y_1 does not, it is clear that y_1 and y_2 cannot be linearly dependent. In other words, they are linearly independent particular solutions, and a complete solution of the original equation is $y=c_1y_1+c_2y_2$.

If we review thoughtfully the work of Example 2, it should be clear that

$$\left.\frac{\partial y_r}{\partial r}\right|_{r=0}$$

turned out to be a second solution of the given differential equation because the right-hand side of Eq. (11) vanished for $r=0$. This, in turn, was true because the right-hand side of Eq. (10) contained r^2 as a factor or, in more general terms, because the indicial equation had a double root. If an indicial equation has roots differing by a nonzero integer k, say $r=r_1$ and $r=r_1+k$, then the method of

Example 2 leads to a nonhomogeneous differential equation of the form (10) with $(r - r_1)(r - r_1 - k)x^{r-1}$ as its nonhomogeneous term. The derivative of this expression with respect to r does not vanish for either $r = r_1$ or $r = r_1 + k$. Thus the nonhomogeneous equation does not reduce to the given differential equation, and the method of Example 2 appears to fail. However, it can be shown† that there is nonetheless a second solution of the form

$$y_2 = y_1(x) \ln |x| + x^{r_2} \sum_{n=0}^{\infty} b_n x^n \tag{12}$$

as the results of Example 2 suggest. Hence, once y_1 has been found, the coefficients $\{b_n\}$ in the series in the expression (12) can be determined by substituting y_2 into the given differential equation and equating to zero the coefficient of each power of x, as usual. The details in this procedure are straightforward but very tedious, and we shall not pursue them further.

Exercises for Section 9.3

1 Show that when the tentative solution (12) is substituted into the differential equation $y'' + P(x)y' + Q(x)y = 0$, the logarithmic terms all drop out and only powers of x remain.

Find two linearly independent solutions near the origin for each of the following equations.

2 $2x^2y'' + (2x^2 + x)y' - y = 0$

3 $9x^2y'' + 9(x^2 + x)y' + (12x - 1)y = 0$

4 $2x^2y'' + (2x^2 - 3x)y' + (x + 2)y = 0$

5 $3x^2y'' + (3x^2 + 5x)y' + (6x - 1)y = 0$

***6** $4x^2y'' + (4x + 1)y = 0$ ***7** $xy'' + y' + xy = 0$

***8** $xy'' + (1 + x)y' + y = 0$ ***9** $x^2y'' + (x^2 - x)y' + y = 0$

***10** $x^2y'' - xy' + (1 + x)y = 0$ ***11** $x^2y'' + 3xy' + (1 + x)y = 0$

***12** By assuming that $y = (-x)^{1/2} \sum_{n=0}^{\infty} a_n x^n$, obtain a solution of the equation of Example 1 corresponding to the exponent $r = \frac{1}{2}$ which will be valid for negative values of x.

****13** If the roots of the indicial equation differ by 1, show that there will be two linearly independent solutions of the form $y = x^r \sum_{n=0}^{\infty} a_n x^n$ if and only if $p_1 r_1 + q_1 = 0$, where r_1 is the smaller of the two roots.

Show that each of the following equations has two linearly independent solutions of the form $y = x^r \sum_{n=0}^{\infty} a_n x^n$, and find these solutions.

***14** $x^2y'' - x(2 - x)y' + (2 - x)y = 0$

***15** $x^2y'' - x(4 - x)y' + (6 - 2x)y = 0$

***16** $x^2y'' - 2xy' + (2 - x^2)y = 0$

***17** $x^2y'' - 2xy' + (2 - x^3)y = 0$

† See, for instance, Garrett Birkhoff and Gian-Carlo Rota, "Ordinary Differential Equations," 2d ed., Blaisdell, Waltham, Mass., 1969, p. 261.

9.4 THE SERIES SOLUTION OF BESSEL'S EQUATION

One of the most important of all variable-coefficient differential equations is

$$x^2\frac{d^2y}{dx^2} + x\frac{dy}{dx} + (\lambda^2x^2 - \nu^2)y = 0 \qquad \nu \geq 0 \tag{1}$$

which is known as **Bessel's equation† of order ν with a parameter λ**. This arises in a great variety of problems, several of which we shall examine later in this chapter. We will also encounter other, more typical, applications of this equation when we study partial differential equations in Chap. 11.

As a preliminary step to make the series solution of Eq. (1) a little simpler, it is convenient to eliminate the parameter λ (temporarily) by the change of independent variable defined by the substitution

$$t = \lambda x \tag{2}$$

Since $dy/dx = \lambda\, dy/dt$ and $d^2y/dx^2 = \lambda^2 d^2y/dt^2$, this substitution changes Eq. (1) into the equation

$$t^2\frac{d^2y}{dt^2} + t\frac{dy}{dt} + (t^2 - \nu^2)y = 0 \tag{3}$$

which is known simply as **Bessel's equation of order ν**.

For Eq. (3) it is clear that in the notation of Eq. (1), Sec. 9.2,

$$P(t) = \frac{1}{t} \qquad \text{and} \qquad Q(t) = \frac{t^2 - \nu^2}{t^2}$$

Hence, the origin is a regular singular point of the equation, and all other values of t are ordinary points. At the origin, around which we propose to obtain series solutions of (3), the indicial equation [Eq. (7), Sec. 9.2] is $r^2 - \nu^2 = 0$, and therefore, by the theory of the preceding sections we are led to try a series solution of the form

$$y_\nu = \sum_{k=0}^{\infty} a_k t^{\nu+k} \tag{4}$$

corresponding to the root $r = \nu$; the root $r = -\nu$ we shall consider later.

Substituting this into Eq. (3), we obtain

$$t^2\sum_{k=0}^{\infty} a_k(\nu + k)(\nu + k - 1)t^{\nu+k-2} + t\sum_{k=0}^{\infty} a_k(\nu + k)t^{\nu+k-1}$$
$$+ (t^2 - \nu^2)\sum_{k=0}^{\infty} a_k t^{\nu+k} = 0$$

† Named for the German mathematician and astronomer Friedrich Wilhelm Bessel (1784–1846), although special cases of this equation had been studied earlier by Jakob Bernoulli (1703), Daniel Bernoulli (1732), and Leonhard Euler (1764).

or, bringing the coefficients into the respective sums and then combining the sums,

$$\sum_{k=0}^{\infty} a_k[(v+k)(v+k-1)+(v+k)-v^2]t^{v+k}+\sum_{k=0}^{\infty} a_k t^{v+k+2}=0$$

or

$$\sum_{k=1}^{\infty} a_k k(2v+k)t^{v+k}+\sum_{k=0}^{\infty} a_k t^{v+k+2}=0$$

If the term corresponding to $k = 1$ is detached from the first sum and if the index of summation in the second sum is changed from k to $k + 2$, the two series can be combined into one, as follows:

$$\left[a_1(2v+1)t^{v+1}+\sum_{k=2}^{\infty} a_k k(2v+k)t^{v+k}\right]+\sum_{k=2}^{\infty} a_{k-2}t^{v+k}$$

$$=a_1(2v+1)t^{v+1}+\sum_{k=2}^{\infty}[k(2v+k)a_k+a_{k-2}]t^{v+k}=0$$

The last equation will be satisfied identically if and only if the coefficient of each power of x is zero, that is, if and only if

$$a_1(2v+1)=0 \qquad (5)$$

and the a's satisfy the recurrence relation

$$k(2v+k)a_k+a_{k-2}=0 \qquad \text{or} \qquad a_k=-\frac{a_{k-2}}{k(2v+k)} \qquad k=2,3,\ldots \qquad (6)$$

From (5) and the restriction that $v \geq 0$, it is clear that $a_1 = 0$. From (6) it then follows that $a_3 = a_5 = \cdots = a_{2m+1} = \cdots = 0$. Likewise, from (6) it follows that

$$a_2=-\frac{a_0}{2(2v+2)}=-\frac{a_0}{2^2\cdot 1!\,(v+1)}$$

$$a_4=-\frac{a_2}{4(2v+4)}=-\frac{a_2}{2^2\cdot 2(v+2)}=\frac{a_0}{2^4\cdot 2!\,(v+2)(v+1)}$$

$$a_6=-\frac{a_4}{6(2v+6)}=-\frac{a_4}{2^2\cdot 3(v+3)}$$

$$=-\frac{a_0}{2^6\cdot 3!\,(v+3)(v+2)(v+1)}$$

and in general

$$a_{2m}=\frac{(-1)^m a_0}{2^{2m}m!\,(v+m)(v+m-1)\cdots(v+3)(v+2)(v+1)} \qquad m=1,2,3,\ldots$$

Now a_{2m} is the coefficient of t^{v+2m} in the series (4) for y_v. Hence it would probably be convenient if a_{2m} contained the factor 2^{v+2m} in its denominator instead of just 2^{2m}. To achieve this, we multiply and divide a_{2m} by 2^v, getting

$$a_{2m}=\frac{(-1)^m}{2^{v+2m}m!\,(v+m)\cdots(v+2)(v+1)}(2^v a_0)$$

Furthermore, the factors $(v + m) \cdots (v + 2)(v + 1)$ in the denominator of a_{2m} suggest a factorial. In fact, if v were a positive integer, a factorial could be created by multiplying the numerator and denominator of a_{2m} by $(v!)$. However, since v is not necessarily an integer, we must use not $(v!)$ but its generalization, $\Gamma(v + 1)$ (see Sec. 6.4), for this purpose. Doing this, we can write

$$a_{2m} = \frac{(-1)^m}{2^{v+2m} m!\,(v + m) \cdots (v + 2)(v + 1)\Gamma(v + 1)} [2^v \Gamma(v + 1) a_0]$$

Since the gamma function satisfies the recurrence relation

$$(v + j)\Gamma(v + j) = \Gamma(v + j + 1)$$

the factors $(v + 1)$, $(v + 2)$, ..., $(v + m)$ can be successively telescoped into the gamma function, and the expression for a_{2m} can be written

$$a_{2m} = \frac{(-1)^m}{2^{v+m} m!\,\Gamma(v + m + 1)} [2^v \Gamma(v + 1) a_0]$$

In this formula a_0 is still arbitrary, and since we are looking only for particular solutions, it is convenient to choose

$$a_0 = \frac{1}{2^v \Gamma(v + 1)}$$

so that, finally,

$$a_{2m} = \frac{(-1)^m}{2^{v+2m} m!\,\Gamma(v + m + 1)} \qquad m = 0, 1, 2, \ldots$$

With a_k thus determined for even values of k and $a_k = 0$ for odd values of k, substitution into (4) gives a solution y_v for each $v \geq 0$. For each v, the function y_v is called a **Bessel function of the first kind of order** v and is denoted by the symbol $J_v(t)$. Thus

$$J_v(t) = t^v \left[\frac{1}{2^v \Gamma(v + 1)} - \frac{t^2}{2^{v+2} \Gamma(v + 2)} + \frac{t^4}{2^{v+4} 2!\,\Gamma(v + 3)} - \cdots \right]$$

$$= \sum_{m=0}^{\infty} \frac{(-1)^m t^{v+2m}}{2^{v+2m} m!\,\Gamma(v + m + 1)} \tag{7}$$

Since Bessel's equation of order v has no finite singular points except the origin, it follows from Theorem 2, Sec. 9.2, that the series (7) converges for all $t \geq 0$. To ensure convergence to a real-valued function for *all* values of t, the factor t^v in (7) must, of course, be replaced by $|t|^v$. The graphs of $J_0(t)$ and $J_1(t)$ are shown in Fig. 9.2. Their resemblance to the graphs of $\cos t$ and $\sin t$ is interesting. In particular, they illustrate the important fact that for each value of v the equation $J_v(t) = 0$ has infinitely many roots.

Let us now consider the series arising from the other root of the indicial equation, namely, $r = -v$. We could, of course, begin again with a series analo-

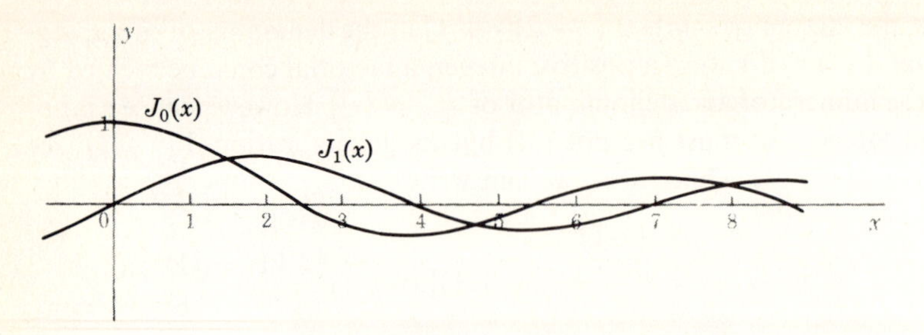

Figure 9.2 The Bessel functions of the first kind $J_0(x)$ and $J_1(x)$.

gous to (4) and determine its coefficients one by one, just as we did for $J_\nu(t)$, but there is no need to go to this trouble. In fact, since ν enters into Bessel's equation only in the form of a square, it follows that the series obtained from (7) by replacing ν by $-\nu$ will satisfy Bessel's equation, provided only that the gamma functions appearing in the denominators of the various terms are all defined. This is necessarily the case unless ν is an integer; hence when ν is not an integer, the function

$$J_{-\nu}(t) = \sum_{m=0}^{\infty} \frac{(-1)^m t^{-\nu+2m}}{2^{-\nu+2m} m!\,\Gamma(-\nu+m+1)} \tag{8}$$

is a second particular solution of Bessel's equation of order ν. Moreover, since $J_{-\nu}(t)$ contains negative powers of t while $J_\nu(t)$ does not, it is obvious that in the neighborhood of the origin $J_{-\nu}(t)$ is unbounded while $J_\nu(t)$ remains finite. Hence, when ν is not an integer, $J_\nu(t)$ and $J_{-\nu}(t)$ are two linearly independent solutions of Bessel's equation. According to Theorem 4, Sec. 3.2, a complete solution of Bessel's equation when ν is not an integer is then

$$y(t) = c_1 J_\nu(t) + c_2 J_{-\nu}(t) \tag{9}$$

From the symmetric way in which ν enters into Eq. (9), it is evident that our earlier restriction that $\nu \geq 0$ can now be removed.

For many purposes, it is convenient to take the linear combination

$$Y_\nu(t) = \frac{\cos \nu\pi J_\nu(t) - J_{-\nu}(t)}{\sin \nu\pi} \tag{10}$$

instead of $J_{-\nu}(t)$ as a second, independent solution of Bessel's equation. Using $Y_\nu(t)$, which is known as the **Bessel function of the second kind of order** ν, we can thus write a complete solution of Bessel's equation in the alternative form

$$y_\nu(t) = c_1 J_\nu(t) + c_2 Y_\nu(t) \qquad \nu \text{ not an integer} \tag{11}$$

It is interesting to note that (9) and (11) are correct expressions for the general solution of Eq. (3) even when ν is an odd multiple of $\frac{1}{2}$ and the roots of the indicial equation $r^2 - \nu^2 = 0$ differ by an integer. In the last section we pointed out that when this happens, a second, independent series solution of the form (4) will usually not exist. It *may* exist, however, and this is one of the instances when it actually does.

If ν is an integer, say $\nu = n$, the situation is somewhat different. Again the roots of the indicial equation differ by an integer, namely, $2n$, and it is to be expected that a second solution of the form (4) will not exist. In fact, when we consider $J_{-\nu}(t)$ as the limit of $J_\nu(t)$ as ν approaches $-n$ and remember that the value of the gamma function becomes infinite when its argument approaches any nonpositive integer, then it follows that as ν approaches $-n$, the first n terms in the series (7) approach zero and the series effectively begins with the term for which $m = n$:

$$J_{-n}(t) = \sum_{m=n}^{\infty} \frac{(-1)^m t^{-n+2m}}{2^{-n+2m} m!\,\Gamma(-n+m+1)}$$

In this, let the variable of summation be changed from m to j by the substitution $m = j + n$. Then

$$\begin{aligned} J_{-n}(t) &= \sum_{j=0}^{\infty} \frac{(-1)^{j+n} t^{-n+2(j+n)}}{2^{-n+2(j+n)}(j+n)!\,\Gamma[-n+(j+n)+1]} \\ &= \sum_{j=0}^{\infty} \frac{(-1)^n(-1)^j t^{n+2j}}{2^{n+2j}(j+n)!\,\Gamma(j+1)} \\ &= (-1)^n \sum_{j=0}^{\infty} \frac{(-1)^j t^{n+2j}}{2^{n+2j}\Gamma(n+j+1)j!} = (-1)^n J_n(t) \end{aligned}$$

Thus, when ν is an integer, the function $J_{-\nu}(t)$ is proportional to $J_\nu(t)$. These two solutions are therefore not independent, and the linear combination $c_1 J_\nu(t) + c_2 J_{-\nu}(t)$ is no longer a complete solution of Bessel's equation. Moreover, without additional definitions, Eq. (11) no longer provides a complete solution, since $Y_\nu(t)$, as defined by (10), assumes the indeterminate form 0/0 when ν is an integer.

A complete solution when ν is an integer can be found in any of several ways. One is to use the method developed in Sec. 3.3 for finding a second solution of a linear, second-order differential equation when one solution is known. The result, as given by Corollary 2, Theorem 1, Sec. 3.3, with $y_1(t) = J_n(t)$ and $P(t) = 1/t$, is

$$y(t) = c_1 J_n(t) + c_2 J_n(t) \int \frac{dt}{t J_n^2(t)}$$

Another is to use the procedure suggested in the last section for finding a second series solution when the roots of the indicial equation differ by a positive integer. Still another procedure is to evaluate the limit of $Y_\nu(t)$ as $\nu \to n$. This limit, which can be proved to exist and which turns out to be independent of $J_n(t)$ for all values of n, is commonly denoted by $Y_n(t)$; that is,

$$Y_n(t) = \lim_{\nu \to n} Y_\nu(t) = \lim_{\nu \to n} \frac{\cos \nu\pi J_\nu(t) - J_{-\nu}(t)}{\sin \nu\pi} \tag{12}$$

With formula (12), we can now eliminate from (11) the restriction that ν not be an integer and use it for all values of ν, integral as well as nonintegral. Plots of

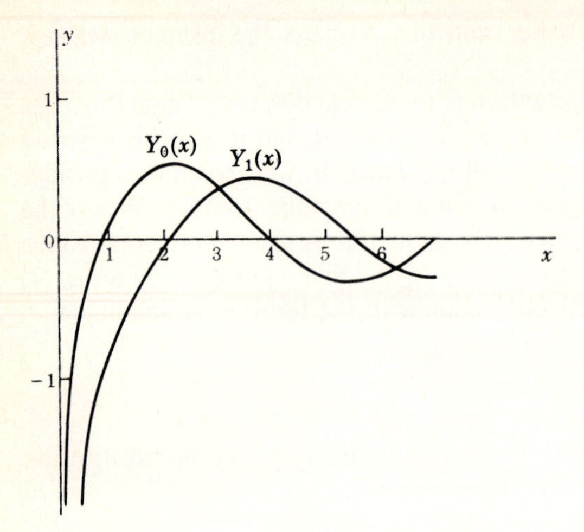

Figure 9.3 The Bessel functions of the second kind $Y_0(x)$ and $Y_1(x)$.

$Y_0(t)$ and $Y_1(t)$ are shown in Fig. 9.3. Among other things, they illustrate the important facts that for all values of v, $Y_v(t)$ is unbounded in the neighborhood of the origin and, like $J_v(t)$ and $J_{-v}(t)$, has infinitely many real zeros.

Reversing the transformation (2) which we used to eliminate the parameter λ from the general form of Bessel's equation (1), we can now summarize the results of the preceding discussion in the following theorem.

Theorem 1 For all real values of v and λ, a complete solution of Bessel's equation of order v with a parameter λ,

$$xy'' + xy' + (\lambda^2 x^2 - v^2)y = 0$$

is

$$y(x) = c_1 J_v(\lambda x) + c_2 Y_v(\lambda x)$$

If v is not an integer, a complete solution is also defined by

$$y(x) = c_1 J_v(\lambda x) + c_2 J_{-v}(\lambda x)$$

$J_v(\lambda x)$, $J_{-v}(\lambda x)$, and $Y_v(\lambda x)$ all have infinitely many real zeros. If $v \geq 0$, $J_v(\lambda x)$ is finite for all values of x but $J_{-v}(\lambda x)$ and $Y_v(\lambda x)$ are unbounded in the neighborhood of the origin.

Exercises for Section 9.4

1 If y_1 and y_2 are any two solutions of Bessel's equation of order v, show that $y_1 y_2' - y_1' y_2 = c/x$, where c is a suitable constant. *Hint:* Recall Abel's identity from Sec. 3.2.

2 If y_1 and y_2 are two independent solutions of Bessel's equation of order v, show that there is no value of x for which y_1 and y_2 are simultaneously zero. *Hint:* Use the result of Exercise 1.

***3** By determining the coefficient of $1/x$ on the left-hand side, show that if ν is not an integer,

$$J_\nu(x)J'_{-\nu}(x) - J'_\nu(x)J_{-\nu}(x) = -\frac{2}{\pi x}\sin \nu\pi$$

Is this result correct if ν is an integer? *Hint:* Use the result of Exercise 1 and the fact that $\Gamma(x)\Gamma(1-x) = \pi/(\sin \pi x)$ if x is not an integer.

***4** Show that $J_n(x)Y'_n(x) - J'_n(x)Y_n(x) = 2/(\pi x)$.

5 Show that except possibly at the origin, no nontrivial solution of Bessel's equation can have a double root.

6 If ν is not an integer, show that

$$Y = \frac{\pi}{2\sin \nu\pi}\left[J_\nu(x)\int_a^x f(s)J_{-\nu}(s)\,ds - J_{-\nu}(x)\int_a^x f(s)J_\nu(s)\,ds\right]$$

is a particular integral of the nonhomogeneous Bessel equation

$$x^2y'' + xy' + (x^2 - \nu^2)y = xf(x)$$

What is the corresponding result if ν is an integer? *Hint:* Use the method of variation of parameters and the result of Exercise 3.

***7** Show that under the change of dependent variable defined by the substitution $y = u/\sqrt{t}$, Bessel's equation of order ν becomes

$$\frac{d^2u}{dt^2} + \left(1 + \frac{1-4\nu^2}{4t^2}\right)u = 0$$

Hence show that for large values of t, solutions of Bessel's equation are described approximately by expressions of the form

$$c_1\frac{\sin t}{\sqrt{t}} + c_2\frac{\cos t}{\sqrt{t}}$$

[More precisely, it can be shown that

$$J_\nu(t) \sim \sqrt{\frac{2}{\pi t}}\cos\left(t - \frac{\pi}{4} - \frac{\nu\pi}{2}\right)$$

$$Y_\nu(t) \sim \sqrt{\frac{2}{\pi t}}\sin\left(t - \frac{\pi}{4} - \frac{\nu\pi}{2}\right)$$

where the symbol $\sim$ means that the limit of the ratio of the two quantities connected by it approaches 1 as t becomes infinite.]

***8** Show that

$$\int \frac{dx}{xJ_\nu^2(x)} = \frac{\pi}{2}\frac{Y_\nu(x)}{J_\nu(x)}$$

9 What is

$$\int \frac{dx}{xY_\nu^2(x)}$$

***10** What is

$$\frac{d}{dx}\left[\ln\frac{Y_\nu(x)}{J_\nu(x)}\right]$$

9.5 MODIFIED BESSEL FUNCTIONS

Certain equations closely resembling Bessel's equation occur so often that their solutions are also named and studied as functions in their own right. The most important of these is

$$x^2y'' + xy' - (x^2 + v^2)y = 0 \tag{1}$$

which is known as the **modified Bessel equation of order** v. Since this can be written in the form

$$x^2y'' + xy' + (i^2x^2 - v^2)y = 0$$

it is evident that this is nothing but Bessel's equation of order v with the imaginary parameter $\lambda = i$. However, in actual applications, to write the complete solution of (1) in the form

$$y = c_1 J_v(ix) + c_2 Y_v(ix)$$

and retain the imaginaries would be about as awkward as to take the solution of

$$y'' - y = 0$$

to be
$$y = c_1 \cos ix + c_2 \sin ix$$

and use this complex expression instead of resorting to real exponentials or hyperbolic functions. Accordingly, we seek modifications of $J_v(ix)$ and $Y_v(ix)$ which will be real functions of real variables.

Now,
$$J_v(ix) = \sum_{k=0}^{\infty} \frac{(-1)^k(ix)^{v+2k}}{2^{v+2k}k!\,\Gamma(v+k+1)}$$

$$= i^v \sum_{k=0}^{\infty} \frac{x^{v+2k}}{2^{v+2k}k!\,\Gamma(v+k+1)}$$

Moreover, $J_v(ix)$ multiplied by any constant will also be a solution of the equation we are considering. Hence, in particular, we can multiply it by i^{-v}, getting

$$i^{-v}J_v(ix) = \sum_{k=0}^{\infty} \frac{x^{v+2k}}{2^{v+2k}k!\,\Gamma(v+k+1)}$$

This is a completely real function, identical with $J_v(x)$ except that its terms, instead of alternating in sign, are all positive. This new function, which is related to $J_v(x)$ in the same way that cosh x and sinh x are related to cos x and sin x, is known as the **modified Bessel function of the first kind of order** v and is customarily denoted by $I_v(x)$.† If v is not an integer, the function $I_{-v}(x)$ obtained from $I_v(x)$ by replacing v by $-v$ throughout is a second, independent solution of Eq. (1), whose complete solution can therefore be written

$$y = c_1 I_v(x) + c_2 I_{-v}(x)$$

† A few authors and some handbooks continue to use the original designation $i^{-v}J_v(ix)$ to denote this function.

On the other hand, instead of using $I_{-\nu}(x)$, one may take the second solution of the modified Bessel equation to be the linear combination

$$K_\nu(x) = \frac{\pi}{2} \frac{I_{-\nu}(x) - I_\nu(x)}{\sin \nu\pi}$$

which is known as the **modified Bessel function of the second kind of order** ν. If ν is not an integer, this is a well-defined solution which is clearly independent of $I_\nu(x)$. If ν is an integer n, this assumes the indeterminate form 0/0, but a tedious evaluation by L'Hospital's rule leads to a limiting expression

$$K_n(x) = \lim_{\nu \to n} K_\nu(x) = \lim_{\nu \to n} \frac{\pi}{2} \frac{I_{-\nu}(x) - I_\nu(x)}{\sin \nu\pi}$$

which is a solution independent of $I_n(x)$. This is a useful result because, as we might expect, $I_\nu(x)$ and $I_{-\nu}(x)$ are not independent when ν is an integer. In fact, when $\nu = n$, we have the identity

$$(-1)^n J_{-n}(ix) = J_n(ix)$$

and then, by obvious steps,

$$(i^2)^n J_{-n}(ix) = J_n(ix)$$

$$i^n J_{-n}(ix) = i^{-n} J_n(ix)$$

$$I_{-n}(x) = I_n(x)$$

Plots of $I_0(x)$ and $I_1(x)$ are shown in Fig. 9.4; plots of $K_0(x)$ and $K_1(x)$ in Fig. 9.5. As these graphs illustrate, the modified Bessel functions have no real

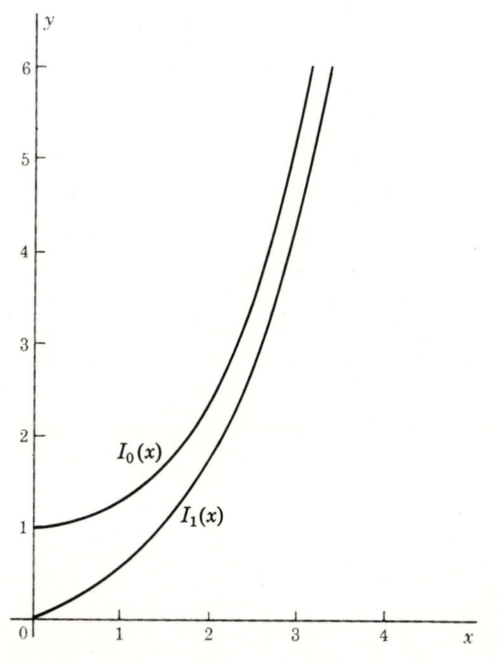

Figure 9.4 The modified Bessel function of the first kind $I_0(x)$ and $I_1(x)$.

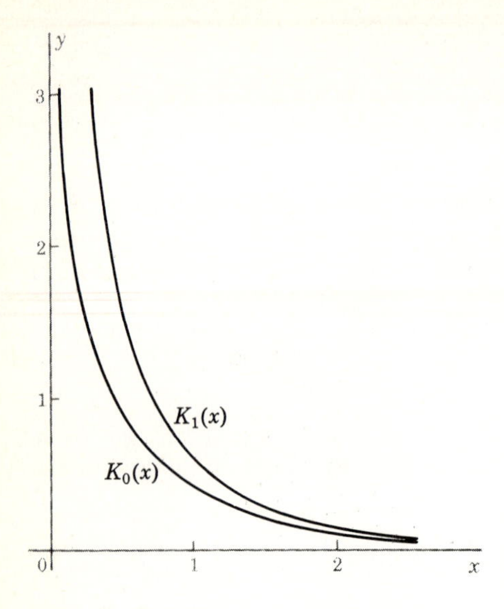

Figure 9.5 The modified Bessel functions of the second kind $K_0(x)$ and $K_1(x)$.

zeros except possibly at $x = 0$. They also illustrate that for $\nu \geq 0$, $I_\nu(x)$ is finite at the origin but $K_\nu(x)$, like $I_{-\nu}(x)$, becomes infinite as x approaches zero.

Like the ordinary Bessel equation, the modified Bessel equation frequently occurs in a form containing a parameter λ:

$$x^2y'' + xy' - (\lambda^2x^2 + \nu^2)y = 0 \tag{2}$$

A complete solution of this is, of course,

$$y = c_1 I_\nu(\lambda x) + c_2 K_\nu(\lambda x) \qquad \nu \text{ unrestricted}$$

If ν is not an integer, we have the alternative form

$$y = c_1 I_\nu(\lambda x) + c_2 I_{-\nu}(\lambda x)$$

Exercises for Section 9.5

1 If y_1 and y_2 are any two solutions of the modified Bessel equation of order ν, show that $y_1 y_2' - y_1' y_2 = c/x$, where c is a suitable constant.

2 By determining the coefficient of $1/x$ on the left-hand side, show that if ν is not an integer,

$$I_\nu(x)I'_{-\nu}(x) - I'_\nu(x)I_{-\nu}(x) = -\frac{2}{\pi x}\sin \nu\pi$$

Is this result correct if ν is an integer?

3 Show that $I_\nu(x)K'_\nu(x) - I'_\nu(x)K_\nu(x) = -1/x$.

4 Show that

$$\int \frac{dx}{xI_\nu^2(x)} = -\frac{K_\nu(x)}{I_\nu(x)}$$

5 What is

$$\int \frac{dx}{xK_\nu^2(x)}$$

6 Show that under the change of dependent variable defined by the substitution $y = u/\sqrt{x}$, the modified Bessel equation of order ν becomes

$$\frac{d^2u}{dx^2} - \left(1 + \frac{4\nu^2 - 1}{4x^2}\right)u = 0$$

Hence show that for large values of x, solutions of the modified Bessel equation are described approximately by expressions of the form

$$c_1 \frac{e^x}{\sqrt{x}} + c_2 \frac{e^{-x}}{\sqrt{x}}$$

[More precisely, it can be shown that

$$I_\nu(x) \sim \frac{e^x}{\sqrt{2\pi x}} \qquad \text{and} \qquad K_\nu(x) \sim \sqrt{\frac{\pi}{2x}}\, e^{-x}$$

as x becomes infinite.]

9.6 EQUATIONS SOLVABLE IN TERMS OF BESSEL FUNCTIONS

There are many differential equations whose solutions can be expressed in terms of Bessel functions, in particular, the large and important family described in the following theorem.

Theorem 1 If $(1 - a)^2 \geq 4c$ and if neither d nor p nor q is zero, then, except in the obvious special cases when it reduces to Euler's equation, the differential equation

$$x^2y'' + x(a + 2bx^p)y' + [c + dx^{2q} + b(a + p - 1)x^p + b^2x^{2p}]y = 0$$

has as a complete solution

$$y = x^\alpha e^{-\beta x^p}[c_1 J_\nu(\lambda x^q) + c_2 Y_\nu(\lambda x^q)]$$

where $\qquad \alpha = \dfrac{1-a}{2} \qquad \beta = \dfrac{b}{p} \qquad \lambda = \dfrac{\sqrt{|d|}}{q} \qquad \nu = \dfrac{\sqrt{(1-a)^2 - 4c}}{2q}$

If $d < 0$, J_ν and Y_ν are to be replaced by I_ν and K_ν, respectively. If ν is not an integer, Y_ν and K_ν can be replaced by $J_{-\nu}$ and $I_{-\nu}$ if desired.

The proof of this theorem, while straightforward, is lengthy and involved, and we shall not present it here. It consists in transforming the given equation by means of the substitutions

$$y = x^{(1-a)/2}e^{-(b/p)x^p}Y \qquad \text{and} \qquad x = \left(\frac{qX}{\sqrt{|d|}}\right)^{1/q}$$

and verifying that when the parameters are properly identified, the result is precisely Bessel's equation in terms of the new variables X and Y.

One special case of Theorem 1 is useful enough to be stated as a corollary.

Corollary 1 If $(1-r)^2 \geq 4b$, if $a \neq 0$, and if either $r - 2 < s$ or $b = 0$, then, except in the obvious special case when it reduces to Euler's equation,† the differential equation

$$(x^r y')' + (ax^s + bx^{r-2})y = 0$$

has as a complete solution

$$y = x^\alpha[c_1 J_\nu(\lambda x^\gamma) + c_2 Y_\nu(\lambda x^\gamma)]$$

where $$\alpha = \frac{1-r}{2} \qquad \gamma = \frac{2-r+s}{2} \qquad \lambda = \frac{2\sqrt{|a|}}{2-r+s} \qquad \nu = \frac{\sqrt{(1-r)^2 - 4b}}{2-r+s}$$

If $a < 0$, J_ν and Y_ν are to be replaced by I_ν and K_ν, respectively. If ν is not an integer, Y_ν and K_ν can be replaced by $J_{-\nu}$ and $I_{-\nu}$ if desired.

Example 1 Find a complete solution of the equation

$$x^2 y'' + x(4x^4 - 3)y' + (4x^8 - 5x^2 + 3)y = 0$$

Clearly, this is a special case of the equation of Theorem 1 with $a = -3$, $b = 2$, $p = 4$, $c = 3$, $d = -5$, and $q = 1$. Hence,

$$\alpha = 2 \qquad \beta = \tfrac{1}{2} \qquad \lambda = \sqrt{|-5|} = \sqrt{5} \qquad \text{and} \qquad \nu = 1$$

A complete solution is therefore

$$y = x^2 e^{-x^4/2}[c_1 I_1(\sqrt{5}\,x) + c_2 K_1(\sqrt{5}\,x)]$$

Example 2 What is a complete solution of the equation $y'' + y = 0$?

Obviously, one possibility is

$$y = c_1 \cos x + c_2 \sin x$$

However, $y'' + y = 0$ is also a special case of the equation of Corollary 1, with $r = 0$, $s = 0$, $a = 1$, and $b = 0$. Hence

$$\alpha = \tfrac{1}{2} \qquad \gamma = 1 \qquad \lambda = 1 \qquad \nu = \tfrac{1}{2}$$

and so we can also write

$$y = \sqrt{x}\,[d_1 J_{1/2}(x) + d_2 J_{-1/2}(x)]$$

It follows, therefore, from Theorem 4, Sec. 3.2, that for proper choice of the constants c_1 and c_2, each of the particular solutions

$$\sqrt{x}\,J_{1/2}(x) \qquad \text{and} \qquad \sqrt{x}\,J_{-1/2}(x)$$

must be expressible in the form $c_1 \cos x + c_2 \sin x$.

Now, since $\Gamma(\frac{3}{2}) = \frac{1}{2}\Gamma(\frac{1}{2}) = \frac{1}{2}\sqrt{\pi}$, the series for $J_{1/2}(x)$ begins with the term

$$\frac{x^{1/2}}{2^{1/2}\Gamma(\frac{3}{2})} = \sqrt{\frac{2x}{\pi}}$$

† Equation (1), Sec. 4.5.

Hence the series for $\sqrt{x}\,J_{1/2}(x)$ begins with the term $\sqrt{2/\pi}\,x$. Therefore, if we write

$$\sqrt{x}\,J_{1/2}(x) = \sqrt{\frac{2}{\pi}}\,x - \cdots = c_1 \cos x + c_2 \sin x$$

$$= c_1\left(1 - \frac{x^2}{2} + \cdots\right) + c_2\left(x - \frac{x^3}{6} + \cdots\right)$$

and put $x = 0$ in this identity, we find $c_1 = 0$. Subsequently, by equating the coefficients of x, we find

$$c_2 = \sqrt{\frac{2}{\pi}}$$

We have thus established the interesting and important result that

$$\sqrt{x}\,J_{1/2}(x) = \sqrt{\frac{2}{\pi}} \sin x \qquad \text{or} \qquad J_{1/2}(x) = \sqrt{\frac{2}{\pi x}} \sin x$$

In a similar fashion it can be shown that

$$J_{-1/2}(x) = \sqrt{\frac{2}{\pi x}} \cos x$$

Exercises for Section 9.6

1 Verify that Corollary 1 is a special case covered by Theorem 1.

Solve each of the following equations in terms of Bessel functions.

2 Example 2, Sec. 9.3 **3** $4x^2y'' + (4x + 1)y = 0$

4 Exercise 10, Sec. 9.3 **5** Exercise 11, Sec. 9.3

6 Can Exercises 8 and 9, Sec. 9.3, be solved in terms of Bessel functions by means of Theorem 1 and its corollary?

Find a complete solution of each of the following equations.

7 $xy'' + y' + x^2y = 0$

8 $(x^2y')' + [x^2 - n(n - 1)]y = 0$

9 $y'' + x^m y = 0$ **10** $xy'' + 2y' + 4xy = 0$

11 $xy'' - y' + 4x^5y = 0$ **12** $x^2y'' + 3xy' + (1 + x)y = 0$

13 $x^2y'' + 2x^2y' + (x^4 + x^2 - 2)y = 0$

14 $x^2y'' + (2x^2 + x)y' + (x^2 + 3x - 1)y = 0$

15 $[(a + bx)y']' - y = 0$. *Hint:* Introduce a new independent variable via the substitution $t = a + bx$.

16 $[(1 + x)^2y']' + (a + bx)y = 0 \qquad a - b < \frac{1}{4}$

***17** $y'' + ae^{mx}y = 0$. *Hint:* Introduce a new independent variable via the substitution $t = e^{mx}$.

***18** $y'' + ay' + (b + ce^{mx})y = 0$, $am > 0$ and $a^2 \geq 4b$. *Hint:* Introduce a new independent variable via the substitution $t = e^{ax}$.

19 Show that $I_{1/2}(x) = \sqrt{2/(\pi x)} \sinh x$ and $I_{-1/2}(x) = \sqrt{2/(\pi x)} \cosh x$.

****20** Show that any solution of

$$(x^{m-1}y')' + kx^{m-2}y = 0 \qquad \text{or} \qquad (x^{m-1}y')' - kx^{m-2}y = 0$$

will also satisfy the equation $(x^m y'')'' - k^2 x^{m-2} y = 0$.

21 Find a complete solution of $(x^2y'')'' - 9y = 0$.

***22** Find a complete solution of $x^2y^{\text{iv}} + 8xy''' + 12y'' - y = 0$.

9.7 IDENTITIES FOR THE BESSEL FUNCTIONS

The Bessel functions are related by an amazing array of identities. Fundamental among these are the consequences of the following pair of theorems.

Theorem 1 $$\frac{d[x^\nu J_\nu(x)]}{dx} = x^\nu J_{\nu-1}(x)$$

PROOF To prove this theorem, we first multiply the series for $J_\nu(x)$ by x^ν and then differentiate it term by term:

$$J_\nu(x) = \sum_{k=0}^{\infty} \frac{(-1)^k x^{\nu+2k}}{2^{\nu+2k}k!\,\Gamma(\nu+k+1)}$$

$$x^\nu J_\nu(x) = \sum_{k=0}^{\infty} \frac{(-1)^k x^{2\nu+2k}}{2^{\nu+2k}k!\,\Gamma(\nu+k+1)}$$

$$\begin{aligned}
\frac{d[x^\nu J_\nu(x)]}{dx} &= \sum_{k=0}^{\infty} \frac{(-1)^k 2(\nu+k)x^{2\nu+2k-1}}{2^{\nu+2k}k!\,(\nu+k)\Gamma(\nu+k)} \\
&= \sum_{k=0}^{\infty} \frac{(-1)^k x^\nu x^{\nu-1+2k}}{2^{\nu-1+2k}k!\,\Gamma(\nu+k)} \\
&= x^\nu \sum_{k=0}^{\infty} \frac{(-1)^k x^{\nu-1+2k}}{2^{\nu-1+2k}k!\,\Gamma(\nu-1+k+1)} \\
&= x^\nu J_{\nu-1}(x)
\end{aligned}$$

as asserted.

Theorem 2 $$\frac{d[x^{-\nu} J_\nu(x)]}{dx} = -x^{-\nu} J_{\nu+1}(x)$$

PROOF This theorem can be proved in essentially the same manner as Theorem 1, but it is easier and perhaps more instructive to proceed as follows. By applying Corollary 1 of Theorem 1, Sec. 9.6, to the equation

$$\frac{d(x^{-1-2\nu}y')}{dx} + x^{-1-2\nu}y = 0$$

with $r = s = -1 - 2\nu$ and $a = 1$, $b = 0$, it is clear that $x^{1+\nu}J_{1+\nu}(x)$ is a particular solution. Hence, substituting this for y in the given equation, we have

$$\frac{d\{x^{-1-2\nu}[x^{1+\nu}J_{1+\nu}(x)]'\}}{dx} + x^{-1-2\nu}[x^{1+\nu}J_{1+\nu}(x)] = 0$$

Now, using Theorem 1 to compute the derivative of the quantity $x^{1+\nu}J_{1+\nu}(x)$, we have further

$$\frac{d\{x^{-1-2\nu}[x^{1+\nu}J_\nu(x)]\}}{dx} = -x^{-\nu}J_{\nu+1}(x)$$

or finally,

$$\frac{d[x^{-\nu}J_\nu(x)]}{dx} = -x^{-\nu}J_{\nu+1}(x)$$

as asserted.

By using its definition in terms of $J_\nu(x)$ and $J_{-\nu}(x)$, one can readily show that *the Bessel function of the second kind also satisfies the identities of Theorems 1 and 2.* Furthermore, by arguments similar to those we have just used, the following theorems can be established.

Theorem 3 $\dfrac{d[x^\nu I_\nu(x)]}{dx} = x^\nu I_{\nu-1}(x)$

Theorem 4 $\dfrac{d[x^{-\nu} I_\nu(x)]}{dx} = x^{-\nu} I_{\nu+1}(x)$

Theorem 5 $\dfrac{d[x^\nu K_\nu(x)]}{dx} = -x^\nu K_{\nu-1}(x)$

Theorem 6 $\dfrac{d[x^{-\nu} K_\nu(x)]}{dx} = -x^{-\nu} K_{\nu+1}(x)$

Performing the indicated differentiations in the identities of Theorems 1 and 2, we obtain, respectively,

$$x^\nu J'_\nu(x) + \nu x^{\nu-1}J_\nu(x) = x^\nu J_{\nu-1}(x)$$
$$x^{-\nu} J'_\nu(x) - \nu x^{-\nu-1}J_\nu(x) = -x^{-\nu} J_{\nu+1}(x)$$

Dividing the first of these by x^ν and multiplying the second by x^ν and solving for $J'_\nu(x)$ in each case gives

$$J'_\nu(x) = J_{\nu-1}(x) - \frac{\nu}{x}J_\nu(x) \tag{1}$$

$$J'_\nu(x) = \frac{\nu}{x}J_\nu(x) - J_{\nu+1}(x) \tag{2}$$

Adding these and dividing by 2, we obtain a third formula for $J'_\nu(x)$:

$$J'_\nu(x) = \frac{J_{\nu-1}(x) - J_{\nu+1}(x)}{2} \tag{3}$$

Subtracting (2) from (1) gives the important recurrence formula

$$J_{\nu-1}(x) + J_{\nu+1}(x) = \frac{2\nu}{x} J_\nu(x)$$

Written as

$$J_{\nu+1}(x) = \frac{2\nu}{x} J_\nu(x) - J_{\nu-1}(x) \tag{4}$$

this formula serves to express Bessel functions of higher orders in terms of functions of lower orders, frequently a useful manipulation. Written as

$$J_{\nu-1}(x) = \frac{2\nu}{x} J_\nu(x) - J_{\nu+1}(x) \tag{5}$$

it serves similarly to express Bessel functions of large negative orders (for instance) in terms of Bessel functions whose orders are numerically smaller.

Example 1 Express $J_4(ax)$ in terms of $J_0(ax)$ and $J_1(ax)$.
Taking $\nu = 3$ in (4), we first have

$$J_4(ax) = \frac{6}{ax} J_3(ax) - J_2(ax)$$

Applying (4) again to $J_3(ax)$ and then to $J_2(ax)$, we have further

$$\begin{aligned} J_4(ax) &= \frac{6}{ax}\left[\frac{4}{ax} J_2(ax) - J_1(ax)\right] - J_2(ax) \\ &= \left(\frac{24}{a^2x^2} - 1\right) J_2(ax) - \frac{6}{ax} J_1(ax) \\ &= \left(\frac{24}{a^2x^2} - 1\right)\left[\frac{2}{ax} J_1(ax) - J_0(ax)\right] - \frac{6}{ax} J_1(ax) \\ &= \left(\frac{48}{a^3x^3} - \frac{8}{ax}\right) J_1(ax) - \left(\frac{24}{a^2x^2} - 1\right) J_0(ax) \end{aligned}$$

Example 2 Show that $d[xJ_\nu(x)J_{\nu+1}(x)]/dx = x[J_\nu^2(x) - J_{\nu+1}^2(x)]$.
Performing the indicated differentiation, we have

$$\frac{d[xJ_\nu(x)J_{\nu+1}(x)]}{dx} = J_\nu(x)J_{\nu+1}(x) + xJ'_\nu(x)J_{\nu+1}(x) + xJ_\nu(x)J'_{\nu+1}(x)$$

Then, substituting for $xJ'_\nu(x)$ from (2) and for $xJ'_{\nu+1}(x)$ from (1), we have

$$\begin{aligned} \frac{d[xJ_\nu(x)J_{\nu+1}(x)]}{dx} &= J_\nu(x)J_{\nu+1}(x) + J_{\nu+1}(x)[\nu J_\nu(x) - xJ_{\nu+1}(x)] \\ &\quad + J_\nu(x)[xJ_\nu(x) - (\nu + 1)J_{\nu+1}(x)] \\ &= x[J_\nu^2(x) - J_{\nu+1}^2(x)] \end{aligned}$$

The basic differentiation identities of Theorems 1 and 2, when written as integration formulas

$$\int x^{\nu} J_{\nu-1}(x)\,dx = x^{\nu} J_{\nu}(x) + c \tag{6}$$

$$\int x^{-\nu} J_{\nu+1}(x)\,dx = -x^{-\nu} J_{\nu}(x) + c \tag{7}$$

suffice for the integration of numerous simple expressions involving Bessel functions. For example, taking $\nu = 1$ in (6), we have

$$\int x J_0(x)\,dx = x J_1(x) + c$$

Similarly, taking $\nu = 0$ in (7), we find

$$\int J_1(x)\,dx = -J_0(x) + c$$

Usually, however, integration by parts must be used in addition to formulas (6) and (7).

Example 3 What is $\int J_3(x)\,dx$?

If we multiply and divide the integrand by x^2, we have

$$\int x^2[x^{-2} J_3(x)]\,dx$$

and so, integrating by parts with

$$u = x^2 \qquad dv = x^{-2} J_3(x)\,dx$$

$$du = 2x\,dx \qquad v = -x^{-2} J_2(x) \qquad \text{by (7), with } \nu = 2$$

we have

$$\begin{aligned} \int J_3(x)\,dx &= -J_2(x) + 2\int x^{-1} J_2(x)\,dx \\ &= -J_2(x) - 2x^{-1} J_1(x) + c \qquad \text{by (7), with } \nu = 1 \end{aligned}$$

Example 4 What is

$$\int \frac{J_2(3x)}{x^2}\,dx$$

Here it is convenient to multiply the numerator and denominator by $9x^2$, getting

$$\frac{1}{9}\int (3x)^2 J_2(3x)\frac{dx}{x^4}$$

Now, integrating by parts with

$$u = (3x)^2 J_2(3x) \qquad dv = \frac{dx}{x^4}$$

$$du = (3x)^2 J_1(3x)\,3dx \qquad v = -\frac{1}{3x^3}$$

we have

$$\int \frac{J_2(3x)}{x^2}\,dx = \frac{1}{9}\left[-\frac{3J_2(3x)}{x} + 3\int 3xJ_1(3x)\frac{dx}{x^2}\right]$$

Again using integration by parts, with

$$u = 3xJ_1(3x) \qquad dv = \frac{dx}{x^2}$$

$$du = 3xJ_0(3x)\,3dx \qquad v = -\frac{1}{x}$$

we have further

$$\int \frac{J_2(3x)}{x^2}\,dx = \frac{1}{9}\left\{-\frac{3J_2(3x)}{x} + 3\left[-3J_1(3x) + 9\int J_0(3x)\,dx\right]\right\}$$

$$= -\frac{J_2(3x)}{3x} - J_1(3x) + 3\int J_0(3x)\,dx$$

The residual integral $\int J_0(3x)\,dx$ cannot be evaluated in finite form in terms of any of the Bessel functions we have encountered.

In general, an integral of the form

$$\int x^m J_n(x)\,dx$$

where m and n are integers such that $m + n \geq 0$, can be completely integrated if $m + n$ is odd but will ultimately depend upon the residual integral $\int J_0(x)\,dx$ if $m + n$ is even. For this reason $\int_0^x J_0(x)\,dx$ has been tabulated.†

Example 5 Show that $\int_0^{\pi/2} J_0(x \cos \phi) \cos \phi \, d\phi = (\sin x)/x$.

As a first step, let us replace $J_0(x \cos \phi)$ by its equivalent infinite series and then perform the indicated integration term by term. This gives us

$$\int_0^{\pi/2} J_0(x \cos \phi) \cos \phi \, d\phi = \int_0^{\pi/2} \sum_{k=0}^{\infty} \frac{(-1)^k (x \cos \phi)^{2k}}{2^{2k}(k!)^2} \cos \phi \, d\phi$$

$$= \sum_{k=0}^{\infty} \left[\frac{(-1)^k x^{2k}}{2^{2k}(k!)^2} \int_0^{\pi/2} \cos^{2k+1} \phi \, d\phi\right]$$

The integral which occurs in the general term is given immediately by the first of **Wallis's formulas,**‡ namely,

$$\int_0^{\pi/2} \cos^{2k+1} \phi \, d\phi = \frac{2 \cdot 4 \cdot 6 \cdot \cdots \cdot (2k)}{1 \cdot 3 \cdot 5 \cdot \cdots \cdot (2k+1)} = \frac{2^{2k}(k!)^2}{(2k+1)!}$$

† "Handbook of Mathematical Functions," Superintendent of Documents, GPO, Washington, D.C., 1965.

‡ Named for the English mathematician John Wallis (1616–1703). Wallis' second formula is given in the hint to Exercise 14.

Using this, the last series becomes

$$\sum_{k=0}^{\infty}\left[\frac{(-1)^k x^{2k}}{2^{2k}(k!)^2}\cdot\frac{2^{2k}(k!)^2}{(2k+1)!}\right]=\sum_{k=0}^{\infty}\frac{(-1)^k x^{2k}}{(2k+1)!}=\frac{1}{x}\sum_{k=0}^{\infty}\frac{(-1)^k x^{2k+1}}{(2k+1)!}$$

Since the last series is just the Maclaurin expansion of $\sin x$, the given integral is equal to $(\sin x)/x$, as asserted.

Exercises for Section 9.7

1 Express $J_5(x)$ in terms of $J_0(x)$ and $J_1(x)$.

2 Express $J_{3/2}(x)$ and $J_{-3/2}(x)$ in terms of $\sin x$ and $\cos x$.

3 What is $d[x^2 J_3(2x)]/dx$? **4** What is $d[xJ_0(x^2)]/dx$?

5 What is $\int x^3 J_2(3x)\,dx$? **6** What is $\int (1/x^2)J_3(2x)\,dx$?

7 Show that
$$\frac{d[x^2 J_{\nu-1}(x)J_{\nu+1}(x)]}{dx}=2x^2 J_\nu(x)\frac{d[J_\nu(x)]}{dx}$$

8 Show that

(*a*) $4J_\nu''(x)=J_{\nu-2}(x)-2J_\nu(x)+J_{\nu+2}(x)$

(*b*) $8J_\nu'''(x)=J_{\nu-3}(x)-3J_{\nu-1}(x)+3J_{\nu+1}(x)-J_{\nu+3}(x)$

9 Show that
$$J_\nu''(x)=\left[\frac{\nu(\nu+1)}{x^2}-1\right]J_\nu(x)-\frac{1}{x}J_{\nu-1}(x)$$

10 Show that
$$\frac{d[x^\nu Y_\nu(x)]}{dx}=x^\nu Y_{\nu-1}(x)$$

Hint: Use Eq. (10), Sec. 9.4.

11 Show that
$$\frac{d[x^{-\nu}Y_\nu(x)]}{dx}=-x^{-\nu}Y_{\nu+1}(x)$$

Hint: Use Eq. (10), Sec. 9.4.

12 Prove Theorem 2 by using the series expansion for $J_\nu(x)$.

***13** Assuming Theorem 2, prove Theorem 1 by beginning with the differential equation
$$\frac{d[x^{1+2\nu}y']}{dx}+x^{1+2\nu}y=0$$

***14** By expressing the exponential as an infinite series and then integrating term by term, show that $I_0(x)=(1/\pi)\int_0^\pi \exp(x\cos\phi)\,d\phi$. *Hint:* It will be helpful to use **Wallis's second formula,**
$$\int_0^{\pi/2}\cos^{2k}\phi\,d\phi=\frac{1\cdot 3\cdot 5\cdot\cdots\cdot(2k-1)}{2\cdot 4\cdot 6\cdot\cdots\cdot 2k}\cdot\frac{\pi}{2}=\frac{(2k)!}{2^{2k}(k!)^2}\cdot\frac{\pi}{2}$$

***15** Show that $\int_0^{\pi/2} J_1(x\cos\phi)\,d\phi=(1-\cos x)/x$.

****16** What is $\int_0^{\pi/2} J_2(x\cos\phi)\cos\phi\,d\phi$?

17 Show that $\int J_0(x)\,dx=2[J_1(x)+J_3(x)+J_5(x)+\cdots]$. *Hint:* Use Formula (3).

18 Show that

$$\int J_0(x)\,dx = J_1(x) + \int \frac{J_1(x)}{x}\,dx$$

$$= J_1(x) + \frac{J_2(x)}{x} + 1 \cdot 3 \int \frac{J_2(x)}{x^2}\,dx$$

$$= J_1(x) + \frac{J_2(x)}{x} + \frac{1 \cdot 3}{x^2} J_3(x) + 1 \cdot 3 \cdot 5 \int \frac{J_3(x)}{x^3}\,dx$$

....

$$= J_1(x) + \frac{J_2(x)}{x} + \frac{1 \cdot 3}{x^2} J_3(x) + \cdots + \frac{(2n-2)!}{2^{n-1}(n-1)!\,x^{n-1}} J_n(x)$$

$$+ \frac{(2n)!}{2^n n!} \int \frac{J_n(x)}{x^n}\,dx$$

Hint: Use repeated integration by parts, each time taking $dv = x^{k+1}J_k(x)\,dx$.

19 Show that

$$J_\nu(x)J_{-(\nu+1)}(x) + J_{\nu+1}(x)J_{-\nu}(x) = -\frac{2 \sin \nu\pi}{\pi x}$$

Hint: Recall the result of Exercise 3, Sec. 9.4.

20 What is $J_\nu(x)Y_{\nu+1}(x) + J_{\nu+1}(x)Y_\nu(x)$? *Hint:* Recall the result of Exercise 4, Sec. 9.4.

21 What is $I_\nu(x)I_{-(\nu+1)}(x) - I_{\nu+1}(x)I_{-\nu}(x)$? *Hint:* Recall the result of Exercise 2, Sec. 9.5.

22 What is $I_\nu(x)K_{\nu+1}(x) + I_{\nu+1}(x)K_\nu(x)$? *Hint:* Recall the result of Exercise 3, Sec. 9.5.

***23** Verify that

$$J_2(x) = x^2\left(\frac{1}{x}\frac{d}{dx}\right)\left(\frac{1}{x}\frac{d}{dx}\right)J_0(x)$$

***24** Verify that

$$J_n(x) = (-1)^n x^n \left(\frac{1}{x}\frac{d}{dx}\right)^n J_0(x)$$

***25** Show that $\int J_0(x) \cos x\,dx = xJ_0(x) \cos x + xJ_1(x) \sin x + c$.

***26** Show that $\int J_0(x) \sin x\,dx = xJ_0(x) \sin x - xJ_1(x) \cos x + c$.

***27** What is $\int J_1(x) \cos x\,dx$? ***28** What is $\int J_1(x) \sin x\,dx$?

***29** Show that

$$\int xJ_n^2(x)\,dx = \frac{x^2}{2}[J_n^2(x) - J_{n-1}(x)J_{n+1}(x)] + c$$

Hint: After integrating by parts, the result of Exercise 7 may be helpful.

***30** Show that

$$\int [J_{\nu-1}^2(x) - J_{\nu+1}^2(x)]x\,dx = 2\nu J_\nu^2(x) + c$$

****31** What is:

(*a*) $\int xJ_0(x) \cos x\,dx$ (*b*) $\int xJ_1(x) \sin x\,dx$

****32** What is:

(*a*) $\int xJ_1(x) \cos x \, dx$ (*b*) $\int xJ_0(x) \sin x \, dx$

***33** Show that:

(*a*) $\int xJ_0(x)\,dx = xJ_1(x) + c$
(*b*) $\int x^2J_0(x)\,dx = x^2J_1(x) + xJ_0(x) - \int J_0(x)\,dx + c$
(*c*) $\int x^3J_0(x)\,dx = (x^3 - 4x)J_1(x) + 2x^2J_0(x) + c$
(*d*) $\int x^4J_0(x)\,dx = (x^4 - 9x^2)J_1(x) + (3x^3 - 9x)J_0(x) + 9\int J_0(x)\,dx + c$

***34** Show that:

(*a*) $\displaystyle\int \frac{J_1(x)}{x}\,dx = -J_1(x) + \int J_0(x)\,dx + c$

(*b*) $\int J_1(x)\,dx = -J_0(x) + c$
(*c*) $\int xJ_1(x)\,dx = -xJ_0(x) + \int J_0(x)\,dx + c$
(*d*) $\int x^2J_1(x)\,dx = 2xJ_1(x) - x^2J_0(x) + c$
(*e*) $\int x^3J_1(x)\,dx = 3x^2J_1(x) - (x^3 - 3x)J_0(x) - 3\int J_0(x)\,dx + c$
(*f*) $\int x^4J_1(x)\,dx = (4x^3 - 16x)J_1(x) - (x^4 - 8x^2)J_0(x) + c$

****35** By replacing $J_0(\lambda x)$ by its infinite series and then integrating termwise, show that $\int_0^\infty e^{-ax}J_0(\lambda x)\,dx = 1/\sqrt{\lambda^2 + a^2}$.

***36** Using the result of Exercise 35, determine the value of $\int_0^\infty e^{-ax}J_1(\lambda x)\,dx$.

****37** What is:

(*a*) $\int_0^\infty xe^{-ax}J_1(\lambda x)\,dx$ (*b*) $\int_0^\infty xe^{-ax}J_0(\lambda x)\,dx$

Hint: Use the result of Exercise 35.

***38** What is:

(*a*) $\int_0^\infty e^{-ax}I_0(\lambda x)\,dx$ (*b*) $\int_0^\infty e^{-ax}I_1(\lambda x)\,dx$

39 What is:

(*a*) $\int J_0(\sqrt{x})\,dx$ (*b*) $\int J_2(\sqrt{x})\,dx$

40 What is $\int xJ_2(1 - x)\,dx$?

41 What is $\int x \ln xJ_0(\lambda x)\,dx$?

42 What is:

(*a*) $\int xJ_0(\sqrt{x})\,dx$ (*b*) $\int xJ_2(\sqrt{x})\,dx$

***43** (*a*) Prove Theorem 3. (*b*) Prove Theorem 4.

44 Show that:

(*a*) $I'_\nu(x) = I_{\nu-1}(x) - \dfrac{\nu}{x}I_\nu(x)$ (*b*) $I'_\nu(x) = \dfrac{\nu}{x}I_\nu(x) + I_{\nu+1}(x)$

(*c*) $I'_\nu(x) = \frac{1}{2}[I_{\nu-1}(x) + I_{\nu+1}(x)]$ (*d*) $I_{\nu-1}(x) - I_{\nu+1}(x) = \dfrac{2\nu}{x}I_\nu(x)$

45 What is:

(*a*) $\int xI_0(x)\,dx$ (*b*) $\int x^2I_0(x)\,dx$
(*c*) $\int xI_1(x)\,dx$ (*d*) $\int x^2I_1(x)\,dx$

***46** Show that

$$x^2J_2(x) = \frac{1}{2}\int_0^x t(x^2 - t^2)J_0(t)\,dt$$

Hint: Observe that $x^2J_2(x) = \int_0^x t^2J_1(t)\,dt = \int_0^x t\int_0^t sJ_0(s)\,ds\,dt$ and then change the order of integration in the repeated integral.

***47** Show that

$$x^{n+1}J_{n+1}(x) = \frac{1}{2^n n!}\int_0^x t(x^2 - t^2)^n J_0(t)\,dt$$

***48** By substituting $x \sin\theta$ for t in the formula of Exercise 47, show that

$$J_{n+1}(x) = \frac{x^{n+1}}{2^n n!}\int_0^{\pi/2} \sin\theta \cos^{2n+1}\theta J_0(x\sin\theta)\,d\theta$$

****49** Verify that $y_1 = J_0(\lambda x)$, $y_2 = Y_0(\lambda x)$, $y_3 = xJ_1(\lambda x)$, and $y_4 = xY_1(\lambda x)$ are four particular solutions of the equation

$$\left(\frac{d^2}{dx^2} + \frac{1}{x}\frac{d}{dx} + \lambda^2\right)\left(\frac{d^2}{dx^2} + \frac{1}{x}\frac{d}{dx} + \lambda^2\right)y = 0$$

****50** Find a complete solution of the equation

$$\left(\frac{d^2}{dx^2} + \frac{1}{x}\frac{d}{dx} - \lambda^2\right)\left(\frac{d^2}{dx^2} + \frac{1}{x}\frac{d}{dx} - \lambda^2\right)y = 0$$

9.8 APPLICATIONS OF BESSEL FUNCTIONS

The most important applications of Bessel functions are probably those that arise in connection with partial differential equations, and we shall investigate a number of these in Chap. 11. There are other applications, however, and we shall conclude this chapter by discussing two problems, one in heat flow and one in Laplace transforms, in which Bessel functions are associated with ordinary differential equations.

Example 1 A metal fin of triangular cross section is attached to a wall to help carry off heat from the latter. Assuming dimensions and coordinates as shown in Fig. 9.6, find the steady-state temperature distribution from the base of the fin to its tip if the wall temperature is u_w and if the fin cools freely into air of constant temperature u_0.

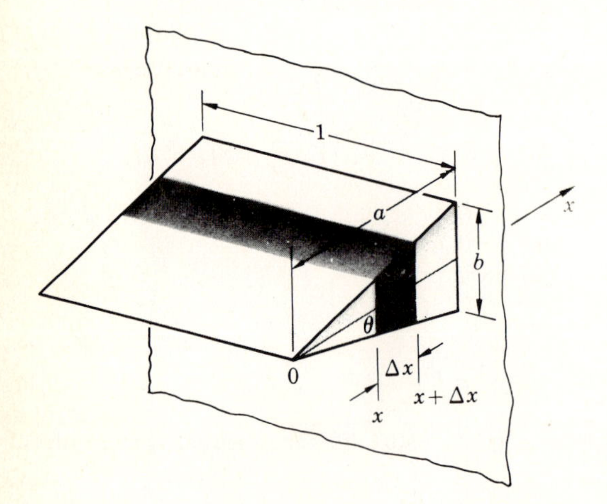

Figure 9.6 A portion of a triangular cooling fin attached to a flat wall.

We shall base our analysis on a unit length of the fin and shall assume that the fin is so thin that temperature variations parallel to the base can be neglected. In other words, we assume that the temperature is the same at all points of any particular cross section parallel to the wall. The physical laws which we will need for the formulation of the problem are **Fourier's law of heat conduction:**

The rate at which heat flows through an area is proportional to the area and to the temperature gradient in the direction perpendicular to the area.

and the following form of **Newton's law of cooling:**

The rate at which heat is lost from an area on the surface of a body is proportional to the area and to the difference between the surface temperature of the body and the temperature of the surrounding medium.

The proportionality constants in these laws are known, respectively, as the **thermal conductivity** and the **surface conductivity.**

Now consider the heat balance in the infinitesimal portion of the fin between x and $x + \Delta x$. This element gains heat by internal flow through its right face (the face nearest to the wall) and loses heat by internal flow through its left face and also by cooling through its upper and lower surfaces. Through the right face, the gain of heat per unit time is, by Fourier's law,

$$\text{Area} \times \text{thermal conductivity} \times \text{temperature gradient}$$

or

$$\left[\left(1\frac{bx}{a}\right)k\frac{du}{dx}\right]_{x+\Delta x} = \left(\frac{bkx}{a}\frac{du}{dx}\right)_{x+\Delta x}$$

Similarly, through the left face the element loses heat at the rate

$$\left(\frac{bkx}{a}\frac{du}{dx}\right)_x$$

By Newton's law, the element loses heat through its upper and lower surfaces at the rate

$$\text{Area} \times \text{surface conductivity} \times (\text{surface temperature} - \text{air temperature})$$

or

$$2\left(1\frac{\Delta x}{\cos\theta}\right)h(u - u_0) = \frac{2h(u - u_0)\,\Delta x}{\cos\theta}$$

Under steady-state conditions the rate of gain of heat must equal the rate of loss, and thus we have

$$\left(\frac{bkx}{a}\frac{du}{dx}\right)_{x+\Delta x} = \left(\frac{bkx}{a}\frac{du}{dx}\right)_x + \frac{2h(u - u_0)\,\Delta x}{\cos\theta}$$

Writing this as

$$\frac{(x\,du/dx)_{x+\Delta x} - (x\,du/dx)_x}{\Delta x} - \frac{2ah}{bk\cos\theta}(u - u_0) = 0$$

and letting $\Delta x \to 0$, we obtain the differential equation

$$\frac{d(xu')}{dx} - \frac{2ah}{bk \cos \theta}(u - u_0) = 0$$

If we set $U = u - u_0$ and $\alpha^2 = 2ah/(bk \cos \theta) = [h/(k \sin \theta)]$, this becomes

$$\frac{d(xU')}{dx} - \alpha^2 U = 0$$

This can be solved immediately by means of the corollary of Theorem 1, Sec. 9.6, and we have

$$U = u - u_0 = c_1 I_0(2\alpha\sqrt{x}) + c_2 K_0(2\alpha\sqrt{x})$$

Since $K_0(2\alpha\sqrt{x})$ is infinite when $x = 0$, c_2 must be zero, leaving

$$u - u_0 = c_1 I_0(2\alpha\sqrt{x})$$

Furthermore, $u = u_w$ when $x = a$; hence

$$u_w - u_0 = c_1 I_0(2\alpha\sqrt{a}) \qquad \text{or} \qquad c_1 = \frac{u_w - u_0}{I_0(2\alpha\sqrt{a})}$$

Therefore

$$u = u_0 + (u_w - u_0)\frac{I_0(2\alpha\sqrt{x})}{I_0(2\alpha\sqrt{a})}$$

Example 2 What is $\mathscr{L}\{t^\nu J_\nu(\lambda t)\}$ if $\nu \geq 0$?

It is possible to determine the required transform by expressing $t^\nu J_\nu(\lambda t)$ as an infinite series and then taking the transform term by term. However, it is more instructive to proceed as follows.

From Corollary 1, Theorem 1, Sec. 9.6, it is clear that $y = t^\nu J_\nu(\lambda t)$ is a solution of the differential equation

$$(t^{1-2\nu}y')' + \lambda^2 t^{1-2\nu} y = 0$$

that is,

$$ty'' + (1 - 2\nu)y' + \lambda^2 ty = 0$$

If we take the Laplace transformation of this equation, recalling Theorem 5, Sec. 6.5, we obtain

$$-\frac{d}{ds}(s^2\mathscr{L}\{y\} - sy_0 - y_0') + (1 - 2\nu)(s\mathscr{L}\{y\} - y_0) - \lambda^2 \frac{d}{ds}\mathscr{L}\{y\}$$

$$= -s^2\frac{d\mathscr{L}\{y\}}{ds} - 2s\mathscr{L}\{y\} + y_0 + (1 - 2\nu)(s\mathscr{L}\{y\} - y_0) - \lambda^2\frac{d\mathscr{L}\{y\}}{ds}$$

$$= -(s^2 + \lambda^2)\frac{d\mathscr{L}\{y\}}{ds} - (1 + 2\nu)s\mathscr{L}\{y\} + 2\nu y_0 = 0$$

Now if $\nu \geq 0$, the term $2\nu y_0$ vanishes identically, because either $\nu = 0$ or else

$$y_0 \equiv t^\nu J_\nu(\lambda t)\Big|_{t=0} = 0$$

Hence, the last equation reduces to the separable differential equation

$$\frac{d\mathscr{L}\{y\}}{\mathscr{L}\{y\}} + (1 + 2\nu)\frac{s\,ds}{s^2 + \lambda^2} = 0$$

Integrating this, we have

$$\ln |\mathscr{L}\{y\}| + \frac{1+2v}{2} \ln (s^2 + \lambda^2) = \ln |c|$$

and therefore

$$\mathscr{L}\{y\} \equiv \mathscr{L}\{t^v J_v(\lambda t)\} = \frac{c}{(s^2 + \lambda^2)^{(1+2v)/2}}$$

To determine c, we compare the leading terms on the two sides of the last equation, using the definition of $J_v(\lambda t)$ and the formula for $\mathscr{L}\{t^{2v}\}$ on the left side and the binomial expansion (Item 1, Appendix B.1) on the right side:

$$\mathscr{L}\left\{t^v \left[\frac{\lambda^v t^v}{2^v \Gamma(v+1)} - \cdots\right]\right\} = c(s^2 + \lambda^2)^{-(2v+1)/2}$$

$$\frac{\lambda^v}{2^v \Gamma(v+1)} \mathscr{L}\{t^{2v}\} - \cdots = c\left(\frac{1}{s^{2v+1}} - \cdots\right)$$

$$\frac{\lambda^v}{2^v \Gamma(v+1)} \left[\frac{\Gamma(2v+1)}{s^{2v+1}}\right] - \cdots = c\left(\frac{1}{s^{2v+1}} - \cdots\right)$$

Hence, since this must be an identity, it follows that

$$c = \frac{\lambda^v \Gamma(2v+1)}{2^v \Gamma(v+1)}$$

and so

$$\mathscr{L}\{t^v J_v(\lambda t)\} = \frac{\lambda^v \Gamma(2v+1)}{2^v \Gamma(v+1)(s^2 + \lambda^2)^{(2v+1)/2}} \qquad v \geq 0 \tag{1}$$

Numerous other transforms can be obtained by using (1). For instance, since, from (1),

$$\mathscr{L}\{J_0(\lambda t)\} = \frac{1}{\sqrt{s^2 + \lambda^2}} \qquad \text{and since} \qquad \frac{dJ_0(\lambda t)}{dt} = -\lambda J_1(\lambda t)$$

it follows that

$$\mathscr{L}\{J_1(\lambda t)\} = -\frac{1}{\lambda} \mathscr{L}\left\{\frac{dJ_0(\lambda t)}{dt}\right\} = -\frac{1}{\lambda}[s\mathscr{L}\{J_0(\lambda t)\} - J_0(0)]$$

$$= -\frac{1}{\lambda}\left(\frac{s}{\sqrt{s^2 + \lambda^2}} - 1\right) = \frac{1}{\lambda} \frac{\sqrt{s^2 + \lambda^2} - s}{\sqrt{s^2 + \lambda^2}}$$

$$= \frac{\lambda}{\sqrt{s^2 + \lambda^2}(s + \sqrt{s^2 + \lambda^2})}$$

Other results will be found among the exercises.

Exercises for Section 9.8

1 What is $\mathscr{L}\{tJ_0(\lambda t)\}$? **2** What is $\mathscr{L}\{t^2 J_0(\lambda t)\}$?

3 What is $\mathscr{L}\{J_2(\lambda t)\}$? *Hint:* Recall from Eq. (3), Sec. 9.7, that

$$J_2(\lambda t) = J_0(\lambda t) - 2\frac{dJ_1(\lambda t)}{d(\lambda t)}$$

***4** What is $\mathscr{L}\{J_n(\lambda t)\}$?

5 Show that $\int_0^\infty J_0(\lambda t)\,dt = 1/\lambda$. *Hint:* Consider the integral defining the Laplace transform of $J_0(\lambda t)$.

6 What is:

(a) $\int_0^\infty J_1(\lambda t)\,dt$ (b) $\int_0^\infty tJ_0(\lambda t)\,dt$ (c) $\int_0^\infty tJ_1(\lambda t)\,dt$

7 What is $\int_0^\infty e^{-at}J_0(\lambda t)\,dt$? **8** What is $\int_0^\infty J_n(\lambda t)\,dt$?

9 What is $\mathscr{L}^{-1}\{1/\sqrt{s^2 + 4s + 13}\}$?

10 What is $\mathscr{L}^{-1}\{1/(s^2 + 2s + 10)^{3/2}\}$?

11 What is $\mathscr{L}^{-1}\{1/[(s + a)\sqrt{s^2 + b^2}]\}$?

***12** What is $\mathscr{L}\{J_n(\lambda t)/t\}$? *Hint:* Recall Theorem 6, Sec. 6.5.

****13** Show that $\mathscr{L}\{I_0(\lambda t)\} = 1/\sqrt{s^2 - \lambda^2}$.

14 What is:

(a) $\mathscr{L}\{tI_0(\lambda t)\}$ (b) $\mathscr{L}\{tI_1(\lambda t)\}$

15 What is $\mathscr{L}\{I_1(\lambda t)\}$?

16 What is $\mathscr{L}^{-1}\{1/\sqrt{s(s - 1)}\}$?

17 Show that $\int_0^t J_0(\lambda)J_0(t - \lambda)\,d\lambda = \sin t$. *Hint:* Recall the convolution theorem.

18 What is $\mathscr{L}^{-1}\{1/\sqrt{s^4 - a^4}\}$?

19 What is $\mathscr{L}^{-1}\{1/\sqrt{s^4 + 5s^2 + 4}\}$?

***20** Find a particular integral of the equation $y'' + y = J_0(t)$.

21 Show that $\int_0^t \sin(t - \lambda)J_0(\lambda)\,d\lambda = tJ_1(t)$.

22 What is $\int_0^t \sin(t - \lambda)J_1(\lambda)\,d\lambda$?

***23** Show that

$$I_0(t) = \frac{e^{-t}}{\pi}\int_0^t \frac{e^{2\lambda}}{\sqrt{\lambda(t - \lambda)}}\,d\lambda$$

Hint: Combine Formula 4, Sec. 6.4, for the case $n = -\frac{1}{2}$ with Theorem 4, Sec. 6.5, and then apply the convolution theorem to the result of Exercise 13.

24 Show that $\int_0^t \cos(t - \lambda)J_0(\lambda)\,d\lambda = tJ_0(t)$.

25 Show that $\int_0^t J_0(t - \lambda)J_1(\lambda)\,d\lambda = J_0(t) - \cos t$.

26 What is $\int_0^t J_1(t - \lambda)J_1(\lambda)\,d\lambda$?

***27** Prove that

$$J_{m+n}(t) = n\int_0^t J_m(t - \lambda)J_n(\lambda)\frac{d\lambda}{\lambda} = m\int_0^t J_m(t - \lambda)J_n(\lambda)\frac{d\lambda}{t - \lambda}$$

***28** Prove that

$$J_{m+n}(t) = (m + k)\int_0^t J_{m+k}(\lambda)J_{n-k}(t - \lambda)\frac{d\lambda}{\lambda}$$

$$= (n - k)\int_0^t J_{m+k}(\lambda)J_{n-k}(t - \lambda)\frac{d\lambda}{t - \lambda}$$

****29** In Example 1, verify that all the heat that enters the fin is lost from its surface. What fraction of the heat entering the fin is lost from the section between $x = 0$ and $x = \lambda a$?

****30** Work Example 1 if instead of a fin, a right circular cone of height h and radius a is attached to the wall to carry off heat.

****31** The moment of inertia of the cross sections of a cantilever beam of length L is proportional to x, where x is the distance from the free end of the beam. An oblique tensile force F, whose direction makes an angle θ with the undeflected direction of the beam, acts at the free end of the beam. Find the equation of the deflection curve of the beam. *Hint:* Recall Example 9, Sec. 4.6.

****32** Work Exercise 31 if the force is an oblique compressive force.

****33** A body whose mass varies according to the law $m(t) = m_0(1 + \alpha t)$ moves along the x axis under the influence of a force of attraction which varies directly as the distance from the origin. Determine the equation of motion of the body if it starts from rest at the point $x = x_0$. *Hint:* Recall the general form of Newton's second law, $d(mv)/dt = F$.

****34** Work Exercise 33 if the mass of the body varies according to the law

$$m(t) = m_0(1 + \alpha t)^{-1}$$

****35** Work Exercise 33 if the force is directed away from the origin.

****36** Work Exercise 34 if the force is directed away from the origin.

***37** Show that the steady-state radial temperature distribution in a thin fin of rectangular cross section, thickness w, and outer radius R which completely encircles a heated cylinder of radius r satisfies the differential equation

$$\frac{d(x\,du/dx)}{dx} - \frac{2hx(u - u_0)}{kw} = 0$$

where x is measured radially outward from the axis of the cylinder and the other parameters have the same significance as in Example 1. *Hint:* Note that the fin resembles a thin washer with the cylinder running snugly through it.

***38** Work Exercise 37 if the fin is of triangular cross section, w being the thickness of the fin where it is attached to the cylinder. *Hint:* Note the suggestion made in Exercise 37.

CHAPTER

TEN

FURTHER THEORY OF LINEAR DIFFERENTIAL EQUATIONS

10.1 INTRODUCTION

In Chap. 3 we explored some of the elementary theory of linear differential equations and encountered such concepts as the wronskian of a set of solutions, Abel's identity for the wronskian, reduction of order, and variation of parameters. Subsequently, in Chap. 4 we used these ideas as a basis for solving linear differential equations with constant coefficients, both homogeneous and nonhomogeneous. Now we shall return to the theory of linear differential equations and undertake a modest extension of the work of Chap. 3. This time, however, the fruits of our investigation will not be new or more refined methods of solution but rather properties of solutions that can be inferred without finding the solutions themselves. In particular, we shall investigate such things as the adjoint equation of a linear differential equation, the existence and distribution of the zeros of solutions, the orthogonality of solutions (which makes possible a far-reaching generalization of Fourier series), and Green's functions. Although these ideas are very general, they can, of course, be applied to specific differential equations and, in particular, we shall illustrate them with occasional references to Bessel's equation and its solutions.

10.2 THE ADJOINT EQUATION

Suppose that we are given the differential equation

$$a_0(x)y' + a_1(x)y = 0 \tag{1}$$

where $a_0(x)$ and $a_1(x)$ are continuous and $a_0(x)$ possesses a continuous derivative on the interval I: $a \le x \le b$. Can the left-hand side of Eq. (1) be converted into a product of the form

$$b(x)y$$

on I by multiplying it by a suitable integrating factor $z = z(x)$? This will be possible if and only if the unknown functions z and $b(x)$ satisfy the equation

$$z[a_0(x)y' + a_1(x)y] = [b(x)y]' = b(x)y' + b'(x)y$$

which, by comparing the coefficients of y and y' on each side, implies that

$$a_0(x)z = b(x) \tag{2a}$$

$$a_1(x)z = b'(x) \tag{2b}$$

If we eliminate $b(x)$ between Eq. (2a) and Eq. (2b), we obtain

$$[a_0(x)z]' - [a_1(x)z] = 0 \tag{3a}$$

or, performing the indicated differentiation and collecting terms,

$$a_0(x)z' + [a_0'(x) - a_1(x)]z = 0 \tag{3b}$$

After z has been found from this linear (and separable), first-order equation, $b(x)$ can be found at once from Eq. (2a). Incidentally, this is essentially the process by which we found an integrating factor for the general, linear, first-order differential equation in Sec. 1.6.

Now suppose that we are given the second-order equation

$$a_0(x)y'' + a_1(x)y' + a_2(x)y = 0 \tag{4}$$

where the coefficient function $a_i(x)$ satisfies the differentiability condition

$$a_i(x) \text{ possesses a continuous } (2 - i)\text{th derivative} \qquad i = 0, 1, 2$$

Is it possible to convert the left-hand side of this equation into an exact derivative by multiplying it by a suitable factor $z = z(x)$? Since (4) contains three terms, each of which involves one and only one of the three quantities y, y', y'', it is clear that it cannot be converted into the derivative of a simple product. It may be possible, however, to convert it into the derivative of an expression of the form

$$b_0(x)y' + b_1(x)y$$

This will be the case if and only if z, $b_0(x)$, and $b_1(x)$ satisfy the relation

$$\begin{aligned} z[a_0(x)y'' + a_1(x)y' + a_2(x)y] &= [b_0(x)y' + b_1(x)y]' \\ &= b_0(x)y'' + [b_0'(x) + b_1(x)]y' + b_1'(x)y \end{aligned}$$

and this will be true if and only if

$$\begin{aligned} a_0(x)z &= b_0(x) \\ a_1(x)z &= b_0'(x) + b_1(x) \\ a_2(x)z &= b_1'(x) \end{aligned} \tag{5}$$

If $b_0(x)$ is eliminated from Eqs. (5) by differentiating the first of these equations and substituting into the second, and if $b_1(x)$ is eliminated in the same fashion between the new second equation and the third, we obtain

$$[a_0(x)z]'' - [a_1(x)z]' + a_2(x)z = 0 \tag{6a}$$

or

$$a_0(x)z'' + [2a_0'(x) - a_1(x)]z' + [a_0''(x) - a_1'(x) + a_2(x)]z = 0 \tag{6b}$$

Clearly, when z is determined from Eq. (6*b*) the unknown function $b_0(x)$ can be found at once from the first equation in (5), and $b_1(x)$ can then be found from the second equation in (5).

The similarity between Eq. (6*a*) and Eq. (3*a*) is apparent. As a matter of definition, either form of Eq. (3) is said to be the **adjoint equation** of Eq. (1), and either form of Eq. (6) is said to be the **adjoint equation** of Eq. (4). The concept of the adjoint equation can be extended to linear differential equations of all orders, but we shall leave this to the exercises.

It appears at first glance that the corresponding adjoint may be useful in solving Eq. (1) or Eq. (4). In fact, if Eq. (1) is multiplied by the function z found from Eq. (3*b*), its left member becomes the derivative of $b(x)y$ and can be integrated at once. Similarly, if Eq. (4) is multiplied by the function z found from Eq. (6*b*), its left member becomes the derivative of $b_0(x)y' + b_1(x)y$ and can be integrated at once. Then y can be found by solving the resulting first-order equation

$$b_0(x)y' + b_1(x)y = C$$

In practice, however, the adjoint equation is almost always as difficult to solve as the original equation, and as a consequence is of little help as a tool for solving linear differential equations. It is of fundamental importance, however, in the theory of linear equations, and we will see some of its uses later in this chapter. Meanwhile we shall conclude this section by establishing several interesting properties of the adjoint of the general, linear, second-order differential equation.

Theorem 1 The adjoint of the adjoint of a linear, second-order differential equation is the equation itself.

PROOF As we have seen [Eq. (6*b*)], the adjoint of the equation $a_0(x)y'' + a_1(x)y' + a_2(x)y = 0$ is

$$a_0(x)z'' + [2a_0'(x) - a_1(x)]z' + [a_0''(x) - a_1'(x) + a_2(x)]z = 0$$

Using Eq. (6*a*), the adjoint of this equation is (using y as the name of the dependent variable in the new adjoint)

$$[a_0(x)y]'' - \{[2a_0'(x) - a_1(x)]y\}' + [a_0''(x) - a_1'(x) + a_2(x)]y = 0$$

or, performing the differentiations,

$$a_0(x)y'' + 2a_0'(x)y' + a_0''(x)y - [2a_0'(x)y' + 2a_0''(x)y - a_1(x)y' - a_1'(x)y]$$
$$+ a_0''(x)y - a_1'(x)y + a_2(x)y = 0$$

When terms are collected, this reduces to

$$a_0(x)y'' + a_1(x)y' + a_2(x)y = 0$$

which is the original equation, as asserted.

Theorem 2 The equation $a_0(x)y'' + a_1(x)y' + a_2(x)y = 0$ is self-adjoint, that is, is its own adjoint, if and only if $a_0'(x) = a_1(x)$.

PROOF By Eq. (6*b*), the adjoint of the given equation is

$$a_0(x)z'' + [2a_0'(x) - a_1(x)]z' + [a_0''(x) - a_1'(x) + a_2(x)]y = 0$$

For this to have the same form as the given equation, it is necessary and sufficient that simultaneously

$$2a_0'(x) - a_1(x) = a_1(x) \qquad \text{and} \qquad a_0''(x) - a_1'(x) + a_2(x) = a_2(x)$$

The first of these conditions implies that $a_0'(x) = a_1(x)$, and since it follows from this that $a_0''(x) = a_1'(x)$, the second condition is also satisfied and the theorem is established.

Corollary 1 If the equation $a_0(x)y'' + a_1(x)y' + a_2(x)y = 0$ is self-adjoint, it can be written in the form $[a_0(x)y']' + a_2(x)y = 0$.

PROOF If the indicated differentiation is performed, we have

$$a_0(x)y'' + a_0'(x)y' + a_2(x)y = 0$$

which is equal to the given equation if the condition for self-adjointness, namely $a_0'(x) = a_1(x)$, is satisfied.

Theorem 3 Over any interval on which $a_0(x) \neq 0$, the equation $a_0(x)y'' + a_1(x)y' + a_2(x)y = 0$ can be made self-adjoint by multiplying it by the factor

$$\frac{1}{a_0(x)} \exp\left[\int \frac{a_1(x)}{a_0(x)}\,dx\right]$$

PROOF Multiplying the given equation by the indicated factor gives

$$\exp\left[\int \frac{a_1(x)}{a_0(x)}\,dx\right]y'' + \frac{a_1(x)}{a_0(x)}\exp\left[\int \frac{a_1(x)}{a_0(x)}\,dx\right]y' + \frac{a_2(x)}{a_0(x)}\exp\left[\int \frac{a_1(x)}{a_0(x)}\,dx\right]y = 0$$

Since the coefficient of y' is obviously the derivative of the coefficient of y'', the equation is self-adjoint, by Theorem 2, and can be put in the form

$$\left\{\exp\left[\int \frac{a_1(x)}{a_0(x)}\,dx\right]y'\right\}' + \frac{a_2(x)}{a_0(x)}\exp\left[\int \frac{a_1(x)}{a_0(x)}\,dx\right]y = 0$$

Theorem 4 Over any interval where $a_0(x) \neq 0$, Abel's identity for the self-adjoint equation $[a_0(x)y']' + a_2(x)y = 0$ becomes

$$a_0(x)W(y_1, y_2) = C \qquad \text{where } C \text{ is a constant}$$

PROOF In expanded form, the given equation is

$$a_0(x)y'' + a_0'(x)y' + a_2(x)y = 0 \qquad \text{or} \qquad y'' + \frac{a_0'(x)}{a_0(x)}y' + \frac{a_2(x)}{a_0(x)}y = 0$$

For this equation, Abel's identity in its original form [Lemma 1, Sec. 3.2] is

$$W(y_1, y_2) = C \exp\left[-\int \frac{a_0'(x)}{a_0(x)}\,dx\right] = C \exp\left[-\ln|a_0(x)|\right] = \frac{C}{a_0(x)}$$

Cross-multiplying now gives the assertion of the theorem.

Exercises for Section 10.2

Find the adjoint of each of the following differential equations.

1 $xy'' + xy' + 2y = 0$ **2** $(1 + x^2)y'' + 4xy' + 2y = 0$

3 $xy'' + y' + xy = 0$ **4** $(\cos x)y'' + (\sin x)y' + y = 0$

Reduce each of the following equations to self-adjoint form.

5 $y'' + 2y' + 5y = 0$ **6** $(\sin x)y'' + (\cos x)y' + y = 0$

7 $(\sin x)y'' - (\cos x)y' + y = 0$ **8** $x^2y'' + xy' + (x^2 - 1)y = 0$

Find functions $z(x)$, $b_0(x)$, and $b_1(x)$ such that

9 $z(x)(y'' + 4y) = [b_0(x)y' + b_1(x)y]'$

10 $z(x)(2x^2y'' + 7xy' + 2y) = [b_0(x)y' + b_1(x)y]'$

11 If $r(x)$ is different from zero on an interval I and if x_0 and x are points of I, show that

$$\sin\left[\int_{x_0}^{x} \frac{dt}{r(t)}\right] \qquad \text{and} \qquad \cos\left[\int_{x_0}^{x} \frac{dt}{r(t)}\right]$$

are linearly independent solutions of the self-adjoint equation

$$[r(x)y']' + \frac{1}{r(x)}y = 0$$

12 Use the result of Exercise 11 to solve the equation

$$y'' + y' + e^{-2x}y = 0$$

Hint: First put the equation in self-adjoint form.

13 Verify the identities

$$a_2 zy = a_2 zy \qquad a_1 zy' = \frac{d}{dx}[(a_1 z)y] - (a_1 z)'y$$

$$a_0 zy'' = \frac{d}{dx}[(a_0 z)y' - (a_0 z)'y] + (a_0 z)''y$$

Then, by adding these equations, show that $z(a_0 y'' + a_1 y' + a_2 y)$ is an exact derivative if and only if $(a_0 z)'' - (a_1 z)' + a_2 z = 0$.

***14** Extend the procedure of Exercise 13 and show that $z(a_0 y''' + a_1 y'' + a_2 y' + a_3 y)$ is an exact derivative if and only if z satisfies the adjoint equation

$$(a_0 z)''' - (a_1 z)'' + (a_2 z)' - a_3 z = 0$$

***15** Show that the adjoint of the adjoint of a third-order, linear differential equation is the equation itself.

***16** What conditions must be satisfied for a third-order equation to be self-adjoint?

10.3 ZEROS OF SOLUTIONS

In Exercise 14, Sec. 3.2, we showed, by a simple application of the wronskian, that *if y_1 and y_2 are two linearly independent solutions of the equation $a_0(x)y'' + a_1(x)y' + a_2(x)y = 0$, there is no value of x for which y_1 and y_2 are simultaneously zero.* In this section we shall investigate further properties of the zeros of solutions of the general, linear, second-order equation, on an interval I, either finite or infinite, where the coefficient functions are continuous and the leading coefficient is different from zero.

Our first theorem is a generalization of the familiar fact that between any two consecutive zeros of either $\sin x$ or $\cos x$ there is exactly one zero of the other function.

Theorem 1 (The Sturm Separation Theorem†) If y_1 and y_2 are linearly independent solutions of the equation $a_0(x)y'' + a_1(x)y' + a_2(x)y = 0$ over an interval I on which $a_0(x) \neq 0$, then between any two consecutive zeros of either of these solutions in I there is exactly one zero of the other.

PROOF Let y_1 and y_2 be two linearly independent solutions of the equation $a_0(x)y'' + a_1(x)y' + a_2(x)y = 0$; let x_1 and x_2 be two consecutive zeros of one of these solutions, say y_2; and suppose, contrary to the theorem, that the other solution, y_1, does not vanish for any value of x between x_1 and x_2. Now consider the function

$$f = y_2/y_1$$

From the properties of y_2 it is clear that $f(x_1) = f(x_2) = 0$. Moreover, since y_1 and y_2 are differentiable and $y_1 \neq 0$ on $[x_1, x_2]$, it follows that f satisfies all the hypotheses of Rolle's theorem. Hence f' must equal zero for at least one value of x, say $\bar{x}$, between x_1 and x_2. Now

$$f' = \frac{y_1 y_2' - y_2 y_1'}{y_1^2} = \frac{W(y_1, y_2)}{y_1^2}$$

Therefore, by Rolle's theorem, $W(y_1, y_2)|_{x=\bar{x}} = 0$. But this is impossible because, by hypothesis, y_1 and y_2 are linearly independent on the interval $[x_1, x_2]$ and thus their wronskian cannot vanish at any point of $[x_1, x_2]$. The only way to avoid this contradiction is for y_1 to vanish at least once between x_1 and x_2, so that the conditions for Rolle's theorem are not fulfilled. There cannot be more than one zero of y_1 between two consecutive zeros of y_2 because if there were, a repetition of the argument with the roles of y_1 and y_2 interchanged would show that there was a zero of y_2 between these two zeros of y_1 and hence x_1 and x_2 would not be *consecutive* zeros of y_2. Thus the theorem is established.

† Named for the Swiss mathematician J. C. F. Sturm (1803–1855).

Our next theorem extends the separation properties established in the last theorem from solutions of one equation to solutions of related equations.

Theorem 2 (The Sturm Comparison Theorem) On the interval I: $[a, b]$, either finite or infinite, let $r(x)$ be a positive-valued function with a continuous derivative and let $q_1(x)$ and $q_2(x)$ be continuous, nonidentical functions such that $q_2(x) \geq q_1(x)$ on I. If y_1 is any solution of the equation $[r(x)y']' + q_1(x)y = 0$ and if y_2 is any solution of the equation $[r(x)y']' + q_2(x)y = 0$, then between any two consecutive zeros of y_1 in $[a, b]$ there is at least one zero of y_2.

PROOF Let y_1 be any solution of the equation $[r(x)y']' + q_1(x)y = 0$; let y_2 be any solution of the equation $[r(x)y']' + q_2(x)y = 0$; let x_1 and x_2 be consecutive zeros of y_1 in $[a, b]$; and suppose, contrary to the theorem, that y_2 is not zero for any value of x between x_1 and x_2. Then since neither y_1 nor y_2 vanishes on the open interval (x_1, x_2), there is no loss of generality in supposing that y_1 and y_2 are both positive for $x_1 < x < x_2$.

Since y_1 and y_2 are solutions of the given equations, we have

$$[r(x)y_1']' + q_1(x)y_1 = 0 \tag{1}$$

$$[r(x)y_2']' + q_2(x)y_2 = 0 \tag{2}$$

for all x in $[a, b]$. If Eq. (2) is multiplied by y_1 and subtracted from y_2 times Eq. (1), we have

$$y_2[r(x)y_1']' - y_1[r(x)y_2']' = [q_2(x) - q_1(x)]y_1 y_2 \tag{3}$$

Now by performing the indicated differentiations, it is easy to verify that the left-hand side of Eq. (3) is equal to

$$\frac{d}{dx}[r(x)(y_1' y_2 - y_2' y_1)]$$

Hence, substituting this for the left-hand side of Eq. (3) and then integrating from x_1 to x_2, we obtain

$$r(x)(y_1' y_2 - y_2' y_1)\Big|_{x_1}^{x_2} = \int_{x_1}^{x_2} [q_2(x) - q_1(x)]y_1 y_2 \, dx$$

or, performing the evaluation on the left and remembering that $y_1(x_1) = y_1(x_2) = 0$,

$$r(x_2)y_1'(x_2)y_2(x_2) - r(x_1)y_1'(x_1)y_2(x_1) = \int_{x_1}^{x_2} [q_2(x) - q_1(x)]y_1 y_2 \, dx \tag{4}$$

Now y_1 is zero at x_1 and at x_2 and positive between these points. Hence the slope of y_1 must be nonnegative at x_1 and nonpositive at x_2; that is, $y_1'(x_1) \geq 0$ and $y_1'(x_2) \leq 0$. Also, since y_2 is positive on the open interval (x_1, x_2), it follows from the continuity of y_2 that $y_2(x_1) \geq 0$ and $y_2(x_2) \geq 0$. Hence, since $r(x)$ is positive at both x_1 and x_2, the left side of Eq. (4) must be equal to or less than zero. However, since y_1 and y_2 are both positive on (x_1, x_2), and since $q_1(x)$ and $q_2(x)$ are *different* continuous functions such that $q_2(x) \geq q_1(x)$ at all points of (x_1, x_2), it follows that the right-hand side of Eq. (4) is positive. Thus we have reached a contradiction which overthrows the possibility that y_2 does not vanish between x_1 and x_2, and the theorem is established.

Example 1 If $m > n \geq 0$, show that there is at least one zero of $J_n(x)$ between any two consecutive positive zeros of $J_m(x)$.

To establish this result, we first write the Bessel equations of orders m and n in the self-adjoint form to which Theorem 2 applies:

$$x^2y'' + xy' + (x^2 - m^2)y = 0 \rightarrow (xy')' + \left(x - \frac{m^2}{x}\right)y = 0$$

$$x^2y'' + xy' + (x^2 - n^2)y = 0 \rightarrow (xy')' + \left(x - \frac{n^2}{x}\right)y = 0$$

For these equations

$$r(x) = x \qquad q_1(x) = x - \frac{m^2}{x} \qquad \text{and} \qquad q_2 = x - \frac{n^2}{x}$$

Clearly, if r_1 is the smallest positive zero of $J_m(x)$, then in the interval $[r_1/2, \infty)$ these functions satisfy the conditions of Theorem 2, with $q_2(x) > q_1(x)$. Hence, by Theorem 2, there is at least one zero of $J_n(x)$ between any two consecutive positive zeros of $J_m(x)$ if $m > n \geq 0$.

Example 2 If $q(x)$ is a nonnegative, continuous function which is not identically zero, show that no (nontrivial) solution of the equation $y'' - q(x)y = 0$ can have more than one real zero.

This result follows immediately by applying Theorem 2 to the equations $y'' - q(x)y = 0$ and $y'' = 0$. In this case $q_1(x) = -q(x) \leq 0$, $q_2(x) = 0$, and therefore $q_2(x) \geq q_1(x)$. Moreover, $q_2(x) \not\equiv q_1(x)$. Hence if any solution of the equation $y'' - q(x)y = 0$ had as many as two zeros, every solution of the equation $y'' = 0$ would have to vanish at least once between those zeros. However, $y = 1$ is one solution of the equation $y'' = 0$, and it is never zero. Thus we have a contradiction if any solution of $y'' - q(x)y = 0$ has more than one zero.

If a solution of one differential equation is known to have infinitely many zeros and if, by means of the comparison theorem (Theorem 2), we are able to show that between successive zeros of that solution there is at least one zero of every solution of another equation, then, obviously, the solutions of the second equation also have infinitely many zeros. The next theorem provides us with a sufficient condition for solutions of an equation to have infinitely many zeros which does not involve comparing the given equation with a second equation. Preparatory to establishing this test, we find it convenient to prove the following lemma.

Lemma 1 If y_1 and y_2 are any two linearly independent solutions of the equation $[r(x)y']' + q(x)y = 0$ over an interval where $[r(x)y']$ is differentiable, $q(x)$ is continuous, and $r(x) \neq 0$, then functions $u(x)$ and $v(x)$ exist such that

$$y_1 = u(x) \sin v(x) \qquad y_2 = u(x) \cos v(x) \tag{5}$$

and

$$r(x)u^2(x)v'(x) = K \qquad K \text{ a nonzero constant} \tag{6}$$

PROOF First we define

$$u(x) = \sqrt{y_1^2 + y_2^2} \tag{7}$$

noting that $u(x)$ can never be zero since y_1 and y_2 cannot vanish for the same value of x. Then

$$\left(\frac{y_1}{u(x)}\right)^2 + \left(\frac{y_2}{u(x)}\right)^2 = 1$$

and hence it is possible to define a function $v(x)$ so that

$$\frac{y_1}{u(x)} = \sin v(x) \qquad \frac{y_2}{u(x)} = \cos v(x)$$

that is, so that $y_1 = u(x) \sin v(x)$ and $y_2 = u(x) \cos v(x)$. Furthermore, from Abel's identity, as given for self-adjoint equations by Theorem 4, Sec. 10.2, namely,

$$r(x)(y_1 y_2' - y_2 y_1') = C$$

we have, on substituting the values of y_1 and y_2,

$$r(x)\{u(x) \sin v(x)[u'(x) \cos v(x) - u(x)v'(x) \sin v(x)] \\ - u(x) \cos v(x)[u'(x) \sin v(x) + u(x)v'(x) \cos v(x)]\} = C$$

or, simplifying,

$$r(x)u^2(x)v'(x) = -C = K$$

Finally, $K \neq 0$ because $r(x) \neq 0$ and y_1 and y_2 are linearly independent solutions.

We can now establish the following theorem.

Theorem 3 In the equation $[r(x)y']' + q(x)y = 0$, let $r(x)$ and $q(x)$ be continuous and let $r(x)$ be positive for $0 < x < \infty$. If the two improper integrals

$$\int_1^\infty \frac{dx}{r(x)} \qquad \int_1^\infty q(x)\,dx \tag{8}$$

diverge to $+\infty$, then every solution of the given equation has infinitely many zeros in the interval $1 < x < \infty$.

PROOF We note first that if one solution has infinitely many zeros, then so does every solution. If the solutions are linearly independent, this follows from Theorem 1; if the solutions are linearly dependent, it follows because such solutions have identical zeros. Let us suppose, then, contrary to the theorem, that some solution, say y_1, vanishes only a finite number of times for $x > 1$. In other words, suppose that there is some value of x, say $x = x_0$, such that $y_1 \neq 0$ for $x \geq x_0$. Now let us introduce a new variable z defined by the substitution

$$z = -\frac{r(x)y_1'}{y_1} \qquad x \geq x_0 \tag{9}$$

Then
$$z' = -\frac{[r(x)y_1']'}{y_1} + r(x)y_1' \frac{y_1'}{y_1^2}$$

or, since $[r(x)y_1']' = -q(x)y_1$ and $r(x)(y_1')^2/y_1^2 = z^2/r(x)$,

$$z' = q(x) + \frac{1}{r(x)} z^2$$

If we now integrate this from x_0 to x, we obtain

$$\int_{x_0}^{x} z'\,dx = z(x) - z(x_0) = \int_{x_0}^{x} q(x)\,dx + \int_{x_0}^{x} \frac{z^2}{r(x)}\,dx$$

The second of the conditions (8) shows that $z \to +\infty$ as $x \to +\infty$. Therefore we conclude from (9) that y_1 and y_1' must have opposite signs for sufficiently large values of x. But if $y_1 < 0$ and $y_1' > 0$, it follows that y_1 is a monotone, increasing function which is bounded above. Similarly, if $y_1 > 0$ and $y_1' < 0$, then y_1 is a monotone, decreasing function which is bounded below. Hence, in either case, y_1 must approach a finite limit. Now if one solution, such as y_1, has only a finite number of zeros on $1 < x < \infty$, it follows from Theorem 1 that every solution has at most a finite number of zeros on $1 < x < \infty$. Therefore, by a repetition of the preceding argument, any other solution, say y_2, must also approach a finite limit as x becomes infinite. Hence, from Eq. (7), $u(x)$ must approach a finite limit as x becomes infinite. Likewise, if $y_1 > 0$ and $y_2 > 0$ for $x = x_0$ (as we may suppose without loss of generality), then, from (5), $v(x)$ must be a first-quadrant angle such that

$$v(x) = \mathrm{Tan}^{-1}\left(\frac{y_1}{y_2}\right)$$

Hence $v(x)$ also approaches a finite limit as x becomes infinite. However, from Eq. (6) we have, by solving for $v'(x)$ and integrating,

$$v(x) = K \int_{x_0}^{x} \frac{dx}{r(x)u^2(x)} + v(x_0) \qquad K \neq 0$$

Since $u(x)$ approaches a finite limit, as we have seen, it follows from the first of the conditions (8) that $v(x)$ becomes either positively or negatively infinite (depending on the sign of K) as x becomes infinite. This contradicts the fact that we showed earlier that $v(x)$ has a finite limit as x becomes infinite, and thus the theorem is established.

Example 3 Show that $J_0(x)$ has infinitely many zeros on $1 < x < \infty$.

As we know, $J_0(x)$ is a solution of the Bessel equation of order zero, that is, $x^2y'' + xy' + (x^2 - 0^2)y = 0$ or, in self-adjoint form,

$$(xy')' + xy = 0$$

For this equation, $r(x) = x$ and $q(x) = x$. Hence

$$\int_1^{\infty} \frac{dx}{r(x)} = \int_1^{\infty} \frac{dx}{x} = \ln x \Big|_1^{\infty} = +\infty$$

and

$$\int_1^{\infty} q(x)\,dx = \int_1^{\infty} x\,dx = \frac{1}{2}x^2 \Big|_1^{\infty} = +\infty$$

Therefore, by Theorem 3, $J_0(x)$ has infinitely many positive zeros.

Exercises for Section 10.3

1 In the proof of Theorem 2, why is there no loss of generality in assuming that $y_1(x)$ and $y_2(x)$ are both positive for $x_1 < x < x_2$?

2 Prove Theorem 2 under the assumption that both $y_1(x)$ and $y_2(x)$ are negative for $x_1 < x < x_2$.

3 Show in two different ways that every solution of the equation $y'' + (1 + x)y = 0$ has infinitely many positive roots.

4 If $r(x) > 0$ and $q(x) \geq 0$ for $x > 0$, show that no (nontrivial) solution of the equation $[r(x)y']' - q(x)y = 0$ has more than one positive root.

5 Prove that the modified Bessel function $I_\nu(x)$ has no positive zeros.

6 Do solutions of the equation $(1 + x^2)y'' + y = 0$ have infinitely many positive zeros?

7 Show that between any two consecutive zeros of $\sin x + 2 \cos x$ there is exactly one zero of $2 \sin x + \cos x$.

***8** Show that the existence of infinitely many zeros of solutions of an Euler equation of the second order cannot be established by a direct application of Theorem 3.

***9** Show (as in Exercise 7, Sec. 9.4) that under the substitution $y = z/\sqrt{x}$ the Bessel equation of order ν becomes the equation

$$z'' + \left(1 + \frac{1 - 4\nu^2}{4x^2}\right)z = 0$$

Hence show that if $\nu = \frac{1}{2}$, the zeros of $J_\nu(x)$ occur at intervals of exactly π; that if $0 < \nu < \frac{1}{2}$, the interval between successive zeros is less than π; and that if $\nu > \frac{1}{2}$, the interval between successive zeros is more than π.

10.4 ORTHOGONALITY PROPERTIES OF SOLUTIONS

In Chap. 8 our derivation of the Euler formulas for the coefficients in the Fourier series of a given function made use of a number of formulas [Eqs. (2) to (8), Sec. 8.2] involving integrals of products of sines and cosines. We observed that these were all formulas from elementary calculus, and had we been asked to verify them, we no doubt would have done so by using standard techniques of integration. The use of elementary integration, coupled with the familiar properties of the sine and cosine, is obviously sufficient for the derivation of the formulas we needed in Sec. 8.2, but actually it is not necessary.

To illustrate this, let us attempt, specifically, to establish the formula

$$\int_0^\pi \sin mx \sin nx \, dx = 0 \qquad m \neq n;\ m, n \text{ positive integers}$$

using only the fact that for all integral values of n, $y = \sin nx$ satisfies the differential equation $y'' + n^2y = 0$ and vanishes at $x = 0$ and at $x = \pi$. If $y_m = \sin mx$ and $y_n = \sin nx$, then

$$y_m(0) = y_m(\pi) = 0 \qquad y_n(0) = y_n(\pi) = 0 \tag{1}$$

and

$$y_m'' + m^2 y_m = 0 \tag{2a}$$

$$y_n'' + n^2 y_n = 0 \tag{2b}$$

If Eq. (2*a*) is multiplied by y_n and subtracted from y_m times Eq. (2*b*), the result is

$$y_m y_n'' - y_n y_m'' = (m^2 - n^2) y_m y_n \tag{3}$$

Now $y_m y_n'' - y_n y_m'' = d(y_m y_n' - y_n y_m')/dx$. Hence, substituting this into the left member of Eq. (3) and integrating from 0 to π, we have

$$y_m y_n' - y_n y_m' \Big|_0^\pi = (m^2 - n^2) \int_0^\pi y_m y_n \, dx \tag{4}$$

Because of the properties (1), the left side of Eq. (4) is zero, and thus, when $m \neq n$, it follows that

$$\int_0^\pi y_m y_n \, dx \equiv \int_0^\pi \sin mx \sin nx \, dx = 0 \tag{4a}$$

To obtain the value of the integral when $m = n$, it is instructive to reason as follows. Although we are interested primarily in the case in which m and n are positive integers, our derivation of the antiderivative relation asserted by Eq. (4) is correct regardless of the values of m and n. Hence, in the indefinite integral in (4), namely,

$$\int y_m y_n \, dx = \frac{y_m y_n' - y_n y_m'}{m^2 - n^2} = \frac{\sin mx(n \cos nx) - \sin nx(m \cos mx)}{m^2 - n^2}$$

we may regard m as a continuous variable and investigate the limit of the right-hand side as $m \to n$. Applying L'Hospital's rule, that is, differentiating the numerator and the denominator of the last fraction *with respect to m* and taking the limit of the resulting fraction as $m \to n$, we obtain

$$\lim_{m \to n} \frac{x \cos mx(n \cos nx) - \sin nx(\cos mx - mx \sin mx)}{2m}$$

$$= \frac{nx \cos^2 nx - \sin nx \cos nx + nx \sin^2 nx}{2n} = \frac{nx - \sin nx \cos nx}{2n}$$

Thus, evaluating the indefinite integral we have now obtained, we find

$$\int_0^\pi y_n^2 \, dx \equiv \int_0^\pi \sin^2 nx \, dx = \frac{nx - \sin nx \cos nx}{2n} \Big|_0^\pi = \frac{\pi}{2}$$

In a similar fashion, all the formulas in the list in Sec. 8.2 can be checked.

Since $y'' + \lambda^2 y = 0$ is a self-adjoint equation, the preceding discussion suggests that perhaps solutions of a general, self-adjoint, second-order differential equation which satisfy suitable boundary conditions may have integral properties analogous to those of the sine and cosine and that there may be generalizations of Fourier series involving such solutions. This is indeed the case, but it will be easier to investigate the matter if we first introduce several definitions.

Definition 1 If a sequence of real functions

$$\{\phi_n(x)\} \qquad n = 1, 2, 3, \ldots$$

which are defined over some interval (a, b), finite or infinite, has the property that

$$\int_a^b \phi_m(x)\phi_n(x)\,dx \begin{cases} = 0 & m \neq n \\ \neq 0 & m = n \end{cases}$$

then the functions are said to form an **orthogonal set** on that interval.

From the condition when $m = n$, it follows that no member of an orthogonal set of functions can be identically zero.

Definition 2 If the functions of an orthogonal set $\{\phi_n(x)\}$ have the property that

$$\int_a^b \phi_n^2(x)\,dx = 1 \qquad \text{for all values of } n$$

then the functions are said to be **orthonormal** on the interval (a, b).

Any set of orthogonal functions can easily be converted into an orthonormal set. In fact, if the functions of the set $\{\phi_n(x)\}$ are orthogonal, and if k_n is the (necessarily positive) value of $\int_a^b \phi_n^2(x)\,dx$, then the functions

$$\frac{\phi_1(x)}{\sqrt{k_1}}, \frac{\phi_2(x)}{\sqrt{k_2}}, \frac{\phi_3(x)}{\sqrt{k_3}}, \ldots$$

are clearly orthonormal. It is therefore no specialization to assume that an orthogonal set of functions is also orthonormal.

Definition 3 If a sequence of real functions $\{\phi_n(x)\}$ has the property that over some interval (a, b), finite or infinite,

$$\int_a^b p(x)\phi_m(x)\phi_n(x)\,dx \begin{cases} = 0 & m \neq n \\ \neq 0 & m = n \end{cases}$$

then the functions are said to be **orthogonal with respect to the weight function** $p(x)$ on that interval.

Any set of functions orthogonal with respect to a weight function $p(x)$ can be converted into a set of functions orthogonal in the first sense (Definition 1) simply by multiplying each member of the set by $\sqrt{p(x)}$ if, as we shall suppose, $p(x) \geq 0$ on the interval of orthogonality.

With respect to any set of functions $\{\phi_n(x)\}$ orthogonal over an interval (a, b), an arbitrary function $f(x)$ has a formal expansion analogous to a Fourier expansion, for we can write

$$f(x) = a_1\phi_1(x) + a_2\phi_2(x) + a_3\phi_3(x) + \cdots + a_n\phi_n(x) + \cdots \tag{5}$$

Then, multiplying by $\phi_n(x)$ and integrating formally between the appropriate limits, a and b, we have

$$\int_a^b f(x)\phi_n(x)\,dx = a_1 \int_a^b \phi_1(x)\phi_n(x)\,dx + a_2 \int_a^b \phi_2(x)\phi_n(x)\,dx + \cdots + a_n \int_a^b \phi_n^2(x)\,dx + \cdots$$

From the property of orthogonality, all integrals on the right are zero except the one which contains a square in its integrand. Hence we can solve at once for a_n as the quotient of two known integrals:

$$a_n = \frac{\int_a^b f(x)\phi_n(x)\,dx}{\int_a^b \phi_n^2(x)\,dx}$$

To illustrate how sets of orthogonal functions arise, it may be helpful to reconsider the equation

$$y'' + \lambda^2 y = 0 \tag{6}$$

and recall the process by which solutions of this equation satisfying the conditions

$$y(0) = 0 \qquad y(\pi) = 0 \tag{7}$$

are found. First, we know that a complete solution of Eq. (6) is

$$y = A \cos \lambda x + B \sin \lambda x$$

If the solution is to vanish when $x = 0$, we must have

$$0 = A \cos 0 + B \sin 0 \qquad \text{Hence} \qquad A = 0$$

Furthermore, if this solution is to vanish when $x = \pi$, we must have

$$0 = B \sin \lambda\pi$$

This will be true, of course, if $B = 0$; but in this case, with A already known to be zero, the solution for y is identically zero and cannot be included in an orthogonal set. The only alternative is for $\sin \lambda\pi$ to be zero, which implies that

$$\lambda = \pm 1, \pm 2, \pm 3, \ldots, \pm n, \ldots$$

Thus, nontrivial solutions of Eq. (6) satisfying the conditions (7) exist for the specific values $\lambda = \pm 1, \pm 2, \pm 3, \ldots$ and no others. Finally, because it was possible to prove [see Eq. (4a)] that the linearly independent solutions

$$\sin x, \sin 2x, \sin 3x, \ldots, \sin nx, \ldots$$

were orthogonal on the interval $[0, \pi]$, that is, on the interval determined by the points at which the boundary conditions (7) were given, an arbitrary function can be expanded, at least formally, in a series of these functions.

Generalizing the preceding discussion, we say that the problem of finding solutions of a differential equation involving a parameter, such as λ in Eq. (6), which satisfy given conditions at two different points is a **two-point characteristic value problem.** The values of the parameter (the values $\lambda = 1, 2, 3, \ldots$ in the preceding discussion) for which there exist nontrivial solutions satisfying the boundary conditions [the conditions (7) in the preceding discussion] are called the **characteristic values**† of the problem because they identify or *characterize* the only nontrivial solutions meeting the requirements of the problem. The equation from which the admissible values of the parameter are found (the equation $\sin \lambda\pi = 0$ in the preceding discussion) is called the **characteristic equation** of the problem. The solutions which correspond to the various admissible values of the parameter (the functions $\sin x$, $\sin 2x$, $\sin 3x, \ldots$ in the preceding discussion) are called the **characteristic functions** or **eigenfunctions** of the problem. The study of two-point characteristic value problems and the expansion of functions in series of the characteristic functions of such problems, after the orthogonality of the characteristic functions has been established, is of fundamental importance in large areas of both pure and applied mathematics.

We are now in a position to prove the following fundamental theorem, known as the **Sturm-Liouville orthogonality theorem,**‡ which guarantees that under suitable conditions the characteristic functions of a characteristic value problem are indeed orthogonal.

Theorem 1 Consider the differential equation

$$\frac{d[r(x)y']}{dx} + [q(x) + \lambda p(x)]y = 0$$

where $r(x)$ and $p(x)$ are continuous on the closed interval $a \leq x \leq b$ and $q(x)$ is continuous at least on the open interval $a < x < b$. Let $\lambda_1, \lambda_2, \lambda_3, \ldots$ be the values of the parameter λ for which there exist nontrivial solutions of this equation on $[a, b]$ which possess continuous first derivatives and satisfy the boundary conditions

$$a_1 y(a) - a_2 y'(a) = 0$$

$$b_1 y(b) - b_2 y'(b) = 0$$

where a_1, a_2, b_1, b_2 are any constants such that a_1 and a_2 are not both zero and b_1 and b_2 are not both zero. Finally, let $y_1, y_2, y_3, \ldots$ be the solutions corresponding to these values of λ. Then the functions $\{y_n(x)\}$ form a system orthogonal with respect to the weight function $p(x)$ over the interval (a, b).

Proof To prove this, let y_m and y_n be nontrivial solutions of the given equation associated with two distinct values of λ, say λ_m and λ_n. This means that

$$\frac{d(ry_m')}{dx} + (q + \lambda_m p)y_m = 0$$

$$\frac{d(ry_n')}{dx} + (q + \lambda_n p)y_n = 0$$

† Some writers graft the German word *eigen* meaning *own*, *peculiar*, or *proper* onto the word *values* and use the hybrid term *eigenvalues* to refer to what we have called characteristic values. Others use the full German term *eigenwerte*.

‡ Conamed for the French mathematician Joseph Liouville (1809–1882).

If y_m times the second of these equations is subtracted from y_n times the first, we obtain

$$(\lambda_m - \lambda_n)py_m y_n = y_m(ry_n')' - y_n(ry_m')' \tag{8}$$

Now, by performing the indicated differentiations, it is easy to verify that

$$y_m(ry_n')' - y_n(ry_m')' = \frac{d}{dx}[r(y_m y_n' - y_m' y_n)]$$

Hence, substituting the last expression for the right-hand side of Eq. (8) and then integrating from a to b, we have

$$(\lambda_m - \lambda_n)\int_a^b py_m y_n\,dx = [r(y_m y_n' - y_m' y_n)]_a^b$$

$$= r(b)[w(y_m, y_n)]_{x=b} - r(a)[w(y_m, y_n)]_{x=a} \tag{9}$$

Now y_m and y_n are not merely solutions of the given differential equation. For all values of the integers m and n, they also satisfy the boundary conditions

$$a_1 y(a) = a_2 y'(a) \qquad \text{and} \qquad b_1 y(b) = b_2 y'(b)$$

From the boundary condition at $x = b$ we have

$$b_1 y_m(b) - b_2 y_m'(b) = 0$$

$$b_1 y_n(b) - b_2 y_n'(b) = 0$$

in which, by hypothesis, b_1 and b_2 are not both zero. Thus these equations constitute a system of two homogeneous, linear equations with a nontrivial solution, and therefore the determinant of their coefficients must be zero. Since this determinant is precisely $w(y_m, y_n)|_{x=b}$, it follows that the antiderivative in (9) vanishes at the upper limit, $x = b$. Similarly, it follows that the antiderivative in (9) is zero at $x = a$. Moreover, if $r(a) = 0$, then the first boundary condition becomes irrelevant; i.e., the integrated terms vanish at $x = a$ without the need of any condition on the solutions y_m and y_n. Likewise, if $r(b) = 0$, the second boundary condition is irrelevant.† We have thus shown that under the conditions of the theorem

$$(\lambda_m - \lambda_n)\int_a^b py_m y_n\,dx = 0$$

Since λ_m and λ_n were any two *distinct* values of λ, the difference $\lambda_m - \lambda_n$ cannot vanish. Hence

$$\int_a^b py_m y_n\,dx = 0 \qquad m \neq n$$

and the theorem is established.

The orthogonality of the characteristic functions of the boundary-value problem covered by Theorem 1 guarantees that (at least formally) an arbitrary function admits of a generalized expansion of the form (5). However, before the

† If $r(x)$ is zero at either a or b, then $x = a$ or $x = b$ is a singular point of the given differential equation (Sec. 9.2), and it is to be expected that one of the solutions of the equation will be unbounded in the neighborhood of that point. When this is the case, the condition that $y(x)$ be bounded replaces the more explicit condition of Theorem 1.

numerical values of the coefficients in such an expansion can be found, the value of the integral of the weighted square of the general characteristic function,

$$\int_a^b p(x)y_n^2(x)\,dx \tag{10}$$

must be known. The evaluation of this integral in general is very difficult, but for one important class of equations (all are special cases of the equation covered by Corollary 1, Theorem 1, Sec. 9.6) an antiderivative can be found by an extension of the process by which we evaluated $\int \sin^2 nx\,dx$ earlier in this section. The following theorem gives the precise result.

Theorem 2 If $y_n(x) \equiv y(\lambda_n x)$ is a solution of the equation

$$(x^u y')' + (cx^{u-2} + \lambda^2 x^u)y = 0$$

when $\lambda = \lambda_n$, then

$$\int x^u y^2(\lambda_n x)\,dx = \frac{x^u}{2\lambda_n}$$

$$\times \left[\frac{x}{\lambda_n}\left(\frac{dy(\lambda_n x)}{dx}\right)^2 + \left(\frac{u-1}{\lambda_n}\right)y(\lambda_n x)\frac{dy(\lambda_n x)}{dx} + \left(\frac{c}{\lambda_n x} + \lambda_n x\right)y^2(\lambda_n x)\right]$$

PROOF At the outset, our expectation is that any solution of the given equation when $\lambda = \lambda_n$ will be of the form $y(\lambda_n, x)$ and not of the very special form $y(\lambda_n x)$. To establish the remarkable fact that λ_n occurs in the solution y_n only through the product $\lambda_n x$, let us transform the given equation by the substitution $t = \lambda_n x$. Then

$$y' \equiv \frac{dy}{dx} = \frac{dy}{dt}\frac{dt}{dx} = \lambda_n \frac{dy}{dt} \qquad \lambda_n \neq 0$$

and hence the given equation becomes

$$\lambda_n \frac{d}{dt}\left[\left(\frac{t}{\lambda_n}\right)^u \lambda_n \frac{dy}{dt}\right] + \left[c\left(\frac{t}{\lambda_n}\right)^{u-2} + \lambda_n^2\left(\frac{t}{\lambda_n}\right)^u\right]y = 0$$

Each term in this equation now contains λ^{2-u} as a factor; hence the equation is equivalent to

$$\frac{d}{dt}\left(t^u \frac{dy}{dt}\right) + (ct^{u-2} + t^u)y = 0$$

Clearly, since λ_n has been completely eliminated from the equation, all solutions are functions of t with no occurrences of λ_n except implicitly in t. Thus all solutions of the original equation are functions of the product argument $\lambda_n x$, as implied by the statement of the theorem.

Now let λ_m and λ_n be two distinct values of λ and let $y_m \equiv y(\lambda_m x)$ and $y_n \equiv y(\lambda_n x)$ be solutions corresponding to these values of λ. Then from the proof of Theorem 1, we have the formula

$$\int p(x)y_m y_n\,dx = \frac{r(x)}{\lambda_m - \lambda_n}(y_m y_n' - y_m' y_n)$$

or in the present case (noting that λ^2 now plays the role of λ),

$$\int x^u y(\lambda_m x) y(\lambda_n x)\,dx = \frac{x^u}{\lambda_m^2 - \lambda_n^2}\left[y(\lambda_m x)\frac{dy(\lambda_n x)}{dx} - y(\lambda_n x)\frac{dy(\lambda_m x)}{dx}\right]$$

$$= \frac{x^u}{\lambda_m^2 - \lambda_n^2}\left[y(\lambda_m x)\lambda_n \frac{dy(\lambda_n x)}{d(\lambda_n x)} - y(\lambda_n x)\lambda_m \frac{dy(\lambda_m x)}{d(\lambda_m x)}\right]$$

We now propose to use L'Hospital's rule to evaluate the indeterminacy which arises in the last expression when $\lambda_m \to \lambda_n$. We do this, of course, by first differentiating the numerator and the denominator independently with respect to λ_m, remembering that

$$\frac{d}{d\lambda_m} = x\frac{d}{d(\lambda_m x)}$$

The result is

$$\frac{x^u}{2\lambda_m}\left[x\frac{dy(\lambda_m x)}{d(\lambda_m x)}\lambda_n\frac{dy(\lambda_n x)}{d(\lambda_n x)} - y(\lambda_n x)\frac{dy(\lambda_m x)}{d(\lambda_m x)} - y(\lambda_n x)\lambda_m x\frac{d^2 y(\lambda_m x)}{d(\lambda_m x)^2}\right]$$

As $\lambda_m \to \lambda_n$, this becomes

$$\frac{x^u}{2\lambda_n}\left[\lambda_n x\left(\frac{dy(\lambda_n x)}{d(\lambda_n x)}\right)^2 - y(\lambda_n x)\frac{dy(\lambda_n x)}{d(\lambda_n x)} - \lambda_n x y(\lambda_n x)\frac{d^2 y(\lambda_n x)}{d(\lambda_n x)^2}\right]$$

or

$$\frac{x^u}{2\lambda_n}\left[\lambda_n x\left(\frac{dy(\lambda_n x)}{\lambda_n\,dx}\right)^2 - y(\lambda_n x)\frac{dy(\lambda_n x)}{\lambda_n\,dx} - \lambda_n x y(\lambda_n x)\frac{d^2 y(\lambda_n x)}{\lambda_n^2\,dx^2}\right] \tag{11}$$

Now from the given differential equation we have, on performing the indicated differentiation,

$$x^u\frac{d^2 y}{dx^2} + ux^{u-1}\frac{dy}{dx} + (cx^{u-2} + \lambda^2 x^u)y = 0$$

Hence, dividing by x^{u-2} and then solving for $d^2y/dx^2 \equiv d^2y(\lambda x)/dx^2$, we have

$$\frac{d^2 y(\lambda x)}{dx^2} = -\frac{1}{x^2}\left[ux\frac{dy(\lambda x)}{dx} + (c + \lambda^2 x^2)y(\lambda x)\right]$$

When this is evaluated for $\lambda = \lambda_n$ and then substituted for the second derivative in (11), we obtain

$$\int x^u y^2(\lambda_n x)\,dx = \frac{x^u}{2\lambda_n}\left\{\frac{x}{\lambda_n}\left(\frac{dy(\lambda_n x)}{dx}\right)^2 - y(\lambda_n x)\frac{1}{\lambda_n}\frac{dy(\lambda_n x)}{dx}\right.$$

$$\left. - \frac{x}{\lambda_n}y(\lambda_n x)\left[-\frac{u}{x}\frac{dy(\lambda_n x)}{dx} - \left(\frac{c}{x^2} + \lambda_n^2\right)y(\lambda_n x)\right]\right\}$$

$$= \frac{x^u}{2\lambda_n}\left[\frac{x}{\lambda_n}\left(\frac{dy(\lambda_n x)}{dx}\right)^2 + \frac{u-1}{\lambda_n}y(\lambda_n x)\frac{dy(\lambda_n x)}{dx}\right.$$

$$\left. + \left(\frac{c}{\lambda_n x} + \lambda_n x\right)y^2(\lambda_n x)\right]$$

as asserted. With this antiderivative available, the evaluation of the definite integral in (10) can be carried out once the boundary conditions are known.

For Bessel's equation of order ν, which is covered by Theorem 2 with $u = 1$ and $c = -\nu^2$, we have the antiderivative given by the following corollary.

Corollary 1 If $y_n \equiv y(\lambda_n x)$ is a solution of Bessel's equation of order ν, then

$$\int xy^2(\lambda_n x)\,dx = \frac{1}{2\lambda_n^2}\left[x^2\left(\frac{dy(\lambda_n x)}{dx}\right)^2 + (\lambda_n^2 x^2 - \nu^2)y^2(\lambda_n x)\right]$$

Example 1 Expand the function $f(x) = 4 - x^2$ on the interval $(0, 2)$ in terms of the Bessel functions of order zero which satisfy the boundary condition $y(2) = 0$ and are bounded at the origin.

The general solution of Bessel's equation of order zero with a parameter is

$$y = c_1 J_0(\lambda x) + c_2 Y_0(\lambda x)$$

Since $Y_0(\lambda x)$ is unbounded in the neighborhood of the origin, it follows that $c_2 = 0$. The boundary condition at $x = 2$ then requires that $c_1 J_0(2\lambda) = 0$. If $c_1 = 0$, the entire solution is identically zero and of no use to us. Hence the parameter λ must satisfy the equation

$$J_0(2\lambda) = 0$$

This is, then, the characteristic equation of our problem. Now the roots of the equation $J_0(z) = 0$ are†

$$z_1 \doteq 2.4048,\ z_2 \doteq 5.5201,\ z_3 \doteq 8.6537,\ \ldots$$

Hence, since $z = 2\lambda$, the characteristic values of the problem are

$$\lambda_1 \doteq 1.2024,\ \lambda_2 \doteq 2.7600,\ \lambda_3 \doteq 4.3268,\ \ldots$$

The characteristic functions are then

$$y_1 = J_0(\lambda_1 x) \doteq J_0(1.2024x),\ y_2 = J_0(\lambda_2 x) \doteq J_0(2.7600x),$$
$$y_3 = J_0(\lambda_3 x) \doteq J_0(4.3268x),\ \ldots$$

and we must now determine the values of the coefficients in the proposed expansion

$$4 - x^2 = a_1 J_0(\lambda_1 x) + a_2 J_0(\lambda_2 x) + \cdots + a_n J_0(\lambda_n x) + \cdots \tag{12}$$

Now it follows from Theorem 1 applied to Bessel's equation, written in the self-adjoint form $(xy')' + (\lambda^2 x - \nu^2/x)y = 0$, that when they satisfy suitable boundary conditions on an interval I: $a \le x \le b$, Bessel functions in general, and the functions $\{J_0(\lambda_n x)\}$ in particular, are orthogonal with respect to the weight function $p(x) = x$ on the interval I. Therefore, to find a_n in the series (12), we multiply the series by $xJ_0(\lambda_n x)$ and then integrate term by term from $x = 0$ to $x = 2$, that is, over the interval determined by the boundary conditions. From the orthogonality of the Bessel functions $\{J_0(\lambda_n x)\}$, it follows that every term on the right except the nth is equal to zero, and we have

$$\int_0^2 (4 - x^2)xJ_0(\lambda_n x)\,dx = \int_0^2 (4x - x^3)J_0(\lambda_n x)\,dx = a_n \int_0^2 xJ_0^2(\lambda_n x)\,dx \tag{13}$$

† See, for instance, Eugene Jahnke, Fritz Emde, and Friedrich Losch, "Tables of Higher Functions," 6th ed., McGraw-Hill, New York, 1960, p. 192; or "Handbook of Mathematical Functions," GPO, Washington, D.C., 1965, p. 409.

Breaking up the integral on the left and using formula (6), Sec. 9.7, together with the fact that $J_1(0) = 0$, we have for the first term

$$\int_0^2 4xJ_0(\lambda_n x)\,dx = \frac{4}{\lambda_n^2}\int_0^2 (\lambda_n x)J_0(\lambda_n x)\,d(\lambda_n x) = \frac{4}{\lambda_n^2}[(\lambda_n x)J_1(\lambda_n x)]_0^2$$

$$= \frac{8}{\lambda_n}J_1(2\lambda_n)$$

For the second term on the left in (13), we have, using Exercise 33*c*, Sec. 9.7,

$$-\int_0^2 x^3 J_0(\lambda_n x)\,dx = -\frac{1}{\lambda_n^4}\int_0^2 (\lambda_n x)^3 J_0(\lambda_n x)\,d(\lambda_n x)$$

$$= -\frac{1}{\lambda_n^4}[\{(\lambda_n x)^3 - 4(\lambda_n x)\}J_1(\lambda_n x) + 2(\lambda_n x)^2 J_0(\lambda_n x)]_0^2$$

$$= -\left(\frac{8}{\lambda_n} - \frac{8}{\lambda_n^3}\right)J_1(2\lambda_n)$$

since, from the characteristic equation, $J_0(2\lambda_n) = 0$ for all n.

To determine the integral on the right in (13), we must evaluate the antiderivative given by Corollary 1:

$$\int_0^2 xJ_0^2(\lambda_n x)\,dx = \frac{1}{2\lambda_n^2}\left[x^2\left(\frac{dJ_0(\lambda_n x)}{dx}\right)^2 + (\lambda_n^2 x^2 - 0^2)J_0^2(\lambda_n x)\right]_0^2$$

$$= \frac{1}{2\lambda_n^2}\left\{4\left[\left.\frac{dJ_0(\lambda_n x)}{dx}\right|_{x=2}\right]^2\right\} \qquad \text{since } J_0(2\lambda_n) = 0$$

Furthermore, from Theorem 2, Sec. 9.7, we have

$$\frac{dJ_0(\lambda_n x)}{dx} = -\lambda_n J_1(\lambda_n x)$$

Hence the value of the last integral is

$$\frac{2}{\lambda_n^2}[-\lambda_n J_1(2\lambda_n)]^2 = 2J_1^2(2\lambda_n)$$

Finally, substituting our various partial results into (13), we find

$$\frac{8}{\lambda_n}J_1(2\lambda_n) - \left(\frac{8}{\lambda_n} - \frac{8}{\lambda_n^3}\right)J_1(2\lambda_n) = a_n 2J_1^2(2\lambda_n)$$

Therefore $\quad a_n = \dfrac{4}{\lambda_n^3 J_1(2\lambda_n)} \quad$ and $\quad 4 - x^2 = \displaystyle\sum_{n=1}^{\infty} \frac{4}{\lambda_n^3 J_1(2\lambda_n)} J_0(\lambda_n x)$

The degree to which just the first term of this series approximates $y = 4 - x^2$ is shown in Fig. 10.1. With the scales of Fig. 10.1, a plot of the first two terms is almost indistinguishable from the plot of $y = 4 - x^2$.

Exercises for Section 10.4

1 Work Example 1 if the right-hand boundary condition is $y(3) = 0$.

2 Work Example 1 if the right-hand boundary condition is $y'(2) = 0$.

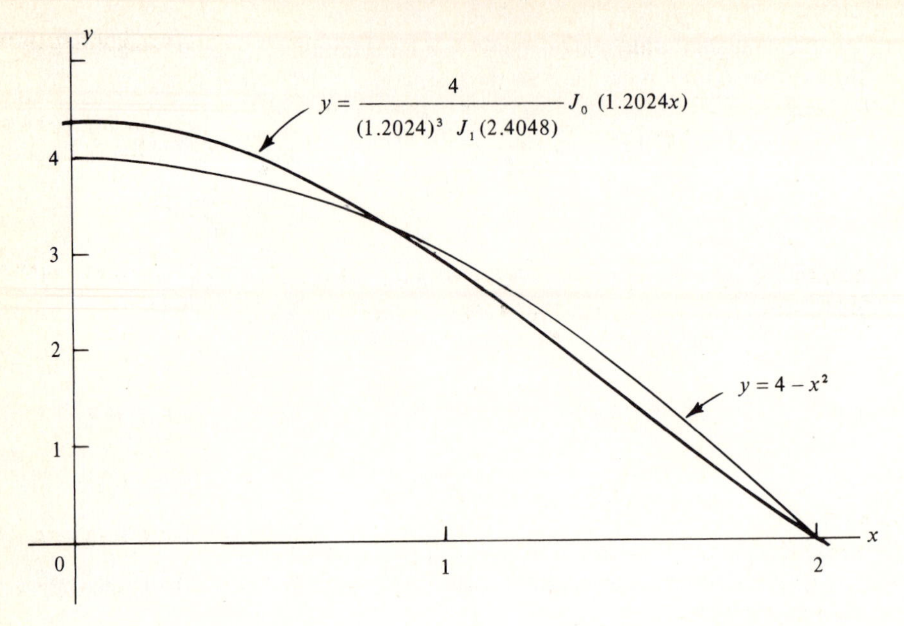

Figure 10.1 The approximation of a function by the first term of a Bessel function expansion.

3 Work Example 1 if the right-hand boundary condition is $y(2) = y'(2)$.

4 Expand $f(x) = x$ over the interval $(0, 2)$ in terms of the Bessel functions of order 1 which satisfy the boundary condition $y(2) = 0$.

5 Expand $f(x) = 1$ over the interval $(0, 3)$ in terms of the Bessel functions of order 2 which satisfy the boundary condition $y'(2) = 0$.

6 Is it possible to expand a function in terms of modified Bessel functions? Explain.

What is the characteristic equation of each of the following boundary-value problems.

7 $y'' + \lambda^2 y = 0 \qquad y(0) = 0 \qquad y'(L) = 0$

8 $y'' + \lambda^2 y = 0 \qquad y(0) = 0 \qquad y(\pi) = y'(\pi)$

9 $(xy')' + \dfrac{\lambda^2}{x} y = 0 \qquad y(1) = 0 \qquad y(2) = 0$

10 Expand $f(x) = x$ in terms of the characteristic functions of the boundary-value problem

$$y'' + \lambda^2 y = 0 \qquad y(0) = y(2\pi) \qquad y'(0) = y'(2\pi)$$

11 In Exercise 8, calculate the first two characteristic values correct to two decimal places.

10.5 GREEN'S FUNCTIONS

Not only are the mathematical objects known as *Green's functions*† of fundamental importance in the theory of boundary-value problems, but they are also mathematical characterizations of important physical concepts. They can be

† Named for the English mathematical physicist George Green (1793–1841).

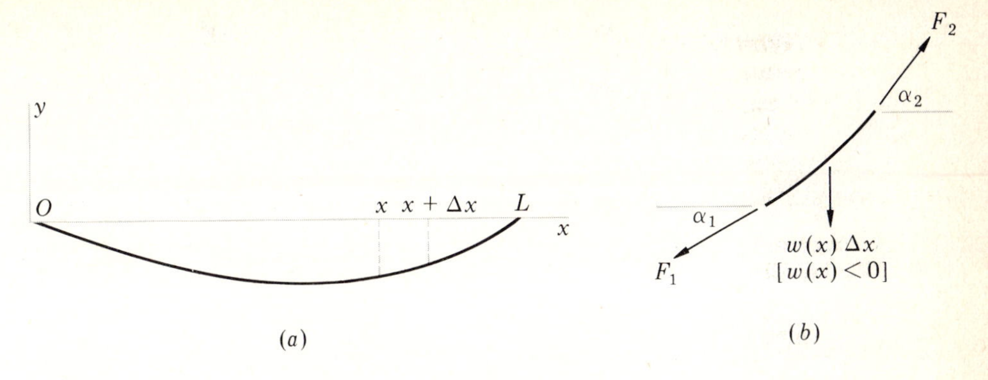

Figure 10.2 A stretched string deflected by a distributed load.

introduced from either point of view, of course, but the motivation for their study is more striking, perhaps, when they are developed via their physical counterparts. Accordingly, we shall begin this section by investigating a simple problem in mechanics whose solution will lead us quickly to what in all but name is a Green's function.

Consider a perfectly flexible elastic string stretched to a length l under tension T. We assume, as a first possibility, that the string bears a distributed load per unit length $w(x)$ which includes the weight of the string itself. Furthermore, we shall suppose that the static deflections produced by this load are all perpendicular to the original, undeflected position of the string and are all in the same plane (Fig. 10.2*a*). Hence, given any two values of x in $[0, l]$, the load acting on the portion of the string between these points is the same before and after the string deflects.

On an arbitrary element of the string we then have the forces shown in Fig. 10.2*b*. Since the deflected string is in equilibrium, the net horizontal force and the net vertical force on the element must both be zero. Hence

$$F_1 \cos \alpha_1 = F_2 \cos \alpha_2 \tag{1}$$

$$F_2 \sin \alpha_2 = F_1 \sin \alpha_1 - w(x)\, \Delta x \tag{2}$$

The first of these equations tells us that the horizontal component of the force in the string is a constant, and we shall further assume that the deflections are so small that this constant horizontal component does not differ appreciably from the tension T in the string before it is loaded. Then, dividing the respective terms in Eq. (2) by the "equal" quantities $F_2 \cos \alpha_2$, $F_1 \cos \alpha_1$, and T, we have

$$\tan \alpha_2 = \tan \alpha_1 - \frac{w(x)\, \Delta x}{T} \tag{3}$$

Now $\tan \alpha_2$ is the slope of the deflection curve at the point $x + \Delta x$, and $\tan \alpha_1$ is

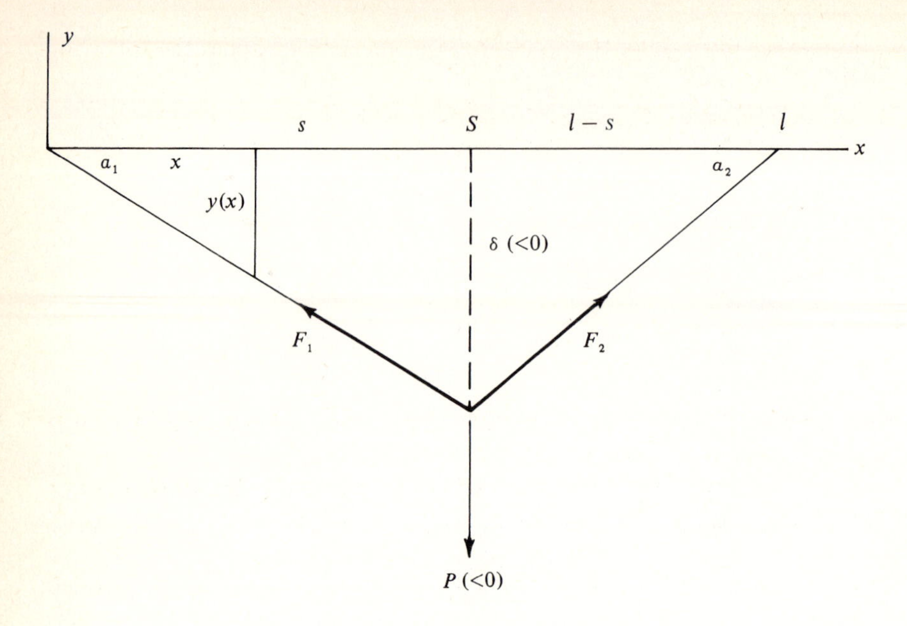

Figure 10.3 A stretched string deflected by a concentrated load.

the slope of the deflection curve at the point x. Hence Eq. (3) can be rewritten in the form

$$\frac{y'(x + \Delta x) - y'(x)}{\Delta x} = -\frac{w(x)}{T}$$

In the limit as $\Delta x \to 0$, we thus obtain

$$Ty'' = -w(x) \tag{4}$$

as the differential equation satisfied by the deflection curve of the string.

Let us now determine the deflection of the string under the influence of a concentrated rather than a distributed load. A concentrated load is, of course, a mathematical fiction which cannot be realized physically, since any nonzero load concentrated at a single point implies an infinite pressure which would immediately cut through the string. Nonetheless, the use of concentrated loads in analyzing physical systems, such as beams and strings, is both common and fruitful.

At the outset, we note from Eq. (4) that y'' is zero at all points of the string where there is no distributed load.† Hence, since $y'' = 0$ implies that y is a linear function, it follows that the deflection curve of the string under the influence of a single concentrated load P consists of two linear sections, as shown in Fig. 10.3.

† Of course, the concept of a weightless string is also a mathematical fiction.

As in our earlier discussion, the equilibrium of the string implies the following conditions

$$F_1 \cos \alpha_1 = F_2 \cos \alpha_2 = T$$

$$F_1 \sin \alpha_1 + F_2 \sin \alpha_2 = -P$$

From these we obtain, as above, the equation

$$\tan \alpha_1 + \tan \alpha_2 = \frac{-P}{T}$$

or $$\frac{-\delta}{s} + \frac{-\delta}{l-s} = \frac{-P}{T} \qquad \text{and} \qquad \delta = \frac{P(l-s)s}{Tl}$$

where δ denotes the transverse deflection of the point of the string that was initially a distance s from the origin.

With the deflection δ known, it is a simple matter to use similar triangles to find the deflection of the string at any point x. The results are

$$y(x, s) = \begin{cases} \dfrac{P(l-s)x}{Tl} & 0 \le x \le s \\[2ex] \dfrac{P(l-x)s}{Tl} & s \le x \le l \end{cases} \tag{5}$$

The notation $y(x, s)$, rather than just $y(x)$, is used, of course, to indicate that the deflection y depends on both the point s where the load is applied and the point x where the deflection is observed. These formulas give the deflection which a load P applied to the string at a point s produces at a point x. Conversely, if the roles of x and s are interchanged, these formulas give the deflection which a load P applied at the point x produces at the point s. For instance, if $x < s$, the deflection at x due to a load P at s is given by the first of the formulas (5) and is simply

$$\frac{P(l-s)x}{Tl}$$

If the roles of x and s are interchanged, the load is now at x and the deflection is measured at s ($> x$). Its value is therefore given by the second formula in (5) with x and s interchanged, of course, and hence is

$$\frac{P(l-s)x}{Tl}$$

as before. Thus it is clear that *the deflection of a string at a point x due to a concentrated load P applied at a point s is the same as the deflection produced at the point s by an equal load applied at the point x.* When P is a unit load, it is customary to use the notation $G(x, s)$ as a new name for the corresponding function $y(x, s)$ defined by (5). The preceding observation then asserts that $G(x, s)$ is symmetric in the two variables x and s; that is,

$$G(x, s) = G(s, x)$$

$G(x, s)$ is often referred to as an **influence function,** since it describes the *influence* which a unit load concentrated at the point s has at any point x of the string.

It is interesting and important to note that by means of the influence function $G(x, s)$ an expression for the deflection of a string under an arbitrary distributed load can be found without solving Eq. (4). To do this, we reason as follows. Let the interval $[0, l]$ be subdivided into n subintervals by the points $s_0 = 0, s_1, s_2, \ldots, s_n = l$; let $\Delta s_i = s_i - s_{i-1}$; and let ξ_i be an arbitrary point in Δs_i. Further, let the portion of the distributed load acting on the subinterval Δs_i, namely, $w(\xi_i)\, \Delta s_i$, be regarded as concentrated at the point $s = \xi_i$. The deflection produced at the point x by this load is the magnitude of the load multiplied by the deflection produced at x by a *unit* load at the point $s = \xi_i$, namely,

$$[w(\xi_i)\, \Delta s_i]G(x, \xi_i)$$

If we add up all the deflections produced at the point x by the various small concentrated forces which together approximate the actual distributed load, we obtain the sum

$$\sum_{i=1}^{n} w(\xi_i)G(x, \xi_i)\, \Delta s_i$$

In the limit, as each $\Delta s_i \to 0$, this sum becomes an integral, and for the deflection at an arbitrary point x we have the formula

$$y(x) = \int_0^l w(s)G(x, s)\, ds \tag{6}$$

Thus, once the function $G(x, s)$ is known, the deflection of the string under *any* piecewise continuous distributed load is given immediately by the integral (6).

We have already observed that the influence function

$$G(x, s) = \begin{cases} \dfrac{(l-s)x}{Tl} & 0 \le x \le s \\[2ex] \dfrac{(l-x)s}{Tl} & s \le x \le l \end{cases} \tag{7}$$

associated with the differential equation $Ty'' = -w(x)$ is a symmetric function of the two arguments x and s. Other properties of $G(x, s)$ are also worth noting. In the first place, it is obvious that $G(x, s)$ satisfies the boundary conditions of the problem; i.e., just as the string satisfies the conditions that its deflection is zero when $x = 0$ and when $x = l$, so it is also true that $G(0, s) = G(l, s) = 0$ for all values of s such that $0 \le s \le l$. It is also easy to verify that $G(x, s)$ is a continuous function of x on the interval $[0, l]$. This is obvious, except possibly at the point $x = s$, where we have for the left- and right-hand limits of $G(x, s)$

$$\lim_{x \to s^-} G(x, s) = \lim_{x \to s^-} \frac{(l-s)x}{Tl} = \frac{(l-s)s}{Tl}$$

$$\lim_{x \to s^+} G(x, s) = \lim_{x \to s^+} \frac{(l-x)s}{Tl} = \frac{(l-s)s}{Tl}$$

and these are clearly equal and equal to $G(s, s)$. On the other hand, the derivative of $G(x, s)$ with respect to x is discontinuous at the point $x = s$, where, in fact, it has a (downward) jump of $-1/T$. To verify this, we note first that $G(x, s)$ is obviously differentiable at all points of the interval $[0, l]$ except possibly at $x = s$. There we observe that

$$\lim_{x \to s^-} G_x(x, s) = \lim_{x \to s^-} \frac{l - s}{Tl} = \frac{l - s}{Tl}$$

$$\lim_{x \to s^+} G_x(x, s) = \lim_{x \to s^+} \left(-\frac{s}{Tl}\right) = -\frac{s}{Tl}$$

These limiting values are not equal, and their difference is

$$-\frac{s}{Tl} - \frac{l - s}{Tl} = -\frac{1}{T}$$

as asserted. Finally, we note that since $G(x, s)$ consists of two linear expressions, it satisfies the homogeneous differential equation $Ty'' = 0$ at all points of the interval $[0, l]$ except at $x = s$. In fact, at $x = s$ the second derivative $G_{xx}(x, s)$ does not exist since, as we have just observed, $G_x(x, s)$ is discontinuous at that point.

The properties which we have just noted are not accidental characteristics of the influence function of one particular problem. Instead, as the following definition makes clear, they identify an important class of functions associated with linear differential equations with variable as well as constant coefficients.

Definition 1 Consider the differential equation

$$a_0(x)y'' + a_1(x)y' + a_2(x)y = 0$$

and the homogeneous boundary conditions $\alpha_1 y(a) = \alpha_2 y'(a)$, $\beta_1 y(b) = \beta_2 y'(b)$, where α_1 and α_2 are not both zero, and β_1 and β_2 are not both zero. A function $G(x, s)$ with the property that

1. $G(x, s)$ satisfies the differential equation for $a \leq x < s$ and for $s < x \leq b$.
2. $\alpha_1 G(a, s) = \alpha_2 G_x(a, s)$, $\beta_1 G(b, s) = \beta_2 G_x(b, s)$ for $a \leq s \leq b$.
3. $G(x, s)$ is a continuous function of x for $a \leq x \leq b$.
4. $G_x(s, x)$ is continuous for $a \leq x < s$ and for $s < x \leq b$ but has a step discontinuity of magnitude $-1/a_0(s)$ at $x = s$.

is called the **Green's function** of the problem defined by the given differential equation and its boundary conditions.†

Example 1 Using Definition 1, construct the Green's function for the differential equation $y'' + k^2y = 0$ with the boundary conditions $y(0) = y(b) = 0$.

† The property of symmetry, i.e., the property that $G(x, s) = G(s, x)$, is not a part of the definition of a Green's function but is a consequence of the other properties when the differential system to which it corresponds is self-adjoint. (See Exercise 11.)

Since any solution of the equation $y'' + k^2y = 0$ is of the form $y = A \cos kx + B \sin kx$, it follows from property 1 that the required function $G(x, s)$ must be defined by expressions of the form

$$G(x, s) = \begin{cases} A_1 \cos kx + B_1 \sin kx & 0 \le x \le s \\ A_2 \cos kx + B_2 \sin kx & s \le x \le b \end{cases}$$

In order for the left-hand boundary condition to be met, as required by property 2, it is necessary that $A_1 = 0$. Similarly, in order for the boundary condition at $x = b$ to be met, it is necessary that $A_2 \cos kb + B_2 \sin kb = 0$, which will be the case if we take $A_2 = C \sin kb$ and $B_2 = -C \cos kb$, where C is arbitrary. Thus $G(x, s)$ is restricted to the form

$$G(x, s) = \begin{cases} B_1 \sin kx & 0 \le x \le s \\ C(\sin kb \cos kx - \cos kb \sin kx) = C \sin k(b - x) & s \le x \le b \end{cases}$$

Further, in order for $G(x, s)$ to be continuous at $x = s$, as required by property 3, it is necessary that $B_1 \sin ks = C \sin k(b - s)$, from which it follows that $B_1 = E \sin k(b - s)$ and $C = E \sin ks$, where E is arbitrary. Thus $G(x, s)$ is further reduced to the form

$$G(x, s) = \begin{cases} E \sin k(b - s) \sin kx & 0 \le x \le s \\ E \sin ks \sin k(b - x) & s \le x \le b \end{cases}$$

Finally, to satisfy property 4, we must have

$$\lim_{x \to s^+} G_x(x, s) - \lim_{x \to s^-} G_x(x, s) = -1 \qquad [\text{since } a_0(x) = 1]$$

or

$$\lim_{x \to s^+} [-kE \sin ks \cos k(b - x)] - \lim_{x \to s^-} [kE \sin k(b - s) \cos kx] = -1$$

$$-kE[\sin ks \cos k(b - s) + \sin k(b - s) \cos ks] = -1$$

$$-kE \sin kb = -1$$

$$E = \frac{1}{k \sin kb}$$

With E known, the Green's function $G(x, s)$ is completely determined, and we have

$$G(x, s) = \begin{cases} \dfrac{\sin kx \sin k(b - s)}{k \sin kb} & 0 \le x \le s \\ \dfrac{\sin ks \sin k(b - x)}{k \sin kb} & s \le x \le b \end{cases}$$

provided, of course, that $kb \neq n\pi$.† It is interesting to note that in this example $G(x, s) = G(s, x)$ even though we did not impose this condition in the course of our derivation.

† It is worth noting that if $kb = n\pi$, there is a nontrivial function meeting the boundary conditions $y(0) = y(b) = 0$ and satisfying the equation $y'' + k^2y = 0$ at *all* points of the interval $[0, b]$. In fact, beginning with the general solution

$$y = A \cos kx + B \sin kx$$

it is clear that the conditions $y(0) = y(b) = 0$ will be met if and only if $A = 0$ and $B \sin kb = 0$. Since $kb = n\pi$, the second condition is satisfied for all values of B; that is, B need not be zero. This illustrates the important general result that *for equations containing a parameter* (such as k in the equation $y'' + k^2y = 0$) *Green's function fails to exist for any value of the parameter for which there is a nontrivial solution of the differential equation which satisfies the boundary conditions of the problem.*

Green's functions not only are closely related to the influence functions which arise in many practical problems but also have much in common with the results of the method of variation of parameters, discussed in Sec. 3.4. To explore this matter, let us consider again the differential equation

$$a_0(x)y'' + a_1(x)y' + a_2(x)y = f(x)$$

or, equivalently,

$$y'' + \frac{a_1(x)}{a_0(x)}y' + \frac{a_2(x)}{a_0(x)}y = \frac{f(x)}{a_0(x)} \qquad a_0(x) \neq 0 \text{ for } a \leq x \leq b$$

If $y_1(x)$ and $y_2(x)$ are two linearly independent solutions of the related homogeneous equation, then, according to the method of variation of parameters [Eqs. (8), Sec. 3.4], $y = u_1 y_1 + u_2 y_2$ will be a solution of the given nonhomogeneous equation provided

$$u_1' = -\frac{y_2}{y_1 y_2' - y_1' y_2}\frac{f}{a_0} \qquad u_2' = \frac{y_1}{y_1 y_2' - y_1' y_2}\frac{f}{a_0}$$

Let us now solve for u_1 and u_2 by integrating their derivatives between x and a and between x and b, respectively. Then, recalling that $y_1 y_2' - y_1' y_2$ is the wronskian W of the two solutions y_1 and y_2, we have, using s as a dummy variable of integration,

$$u_1 = \int_x^a \frac{-y_2(s)}{W(s)}\frac{f(s)}{a_0(s)}\,ds = \int_a^x \frac{y_2(s)}{W(s)}\frac{f(s)}{a_0(s)}\,ds$$

$$u_2 = \int_x^b \frac{y_1(s)}{W(s)}\frac{f(s)}{a_0(s)}\,ds$$

and $$y = u_1 y_1 + u_2 y_2 = y_1(x)\int_a^x \frac{y_2(s)}{W(s)}\frac{f(s)}{a_0(s)}\,ds + y_2(x)\int_x^b \frac{y_1(s)}{W(s)}\frac{f(s)}{a_0(s)}\,ds$$

We now ask under what conditions, if any, the solution defined by the last equation can satisfy the boundary conditions $y(a) = y(b) = 0$. Putting $x = a$ makes the first integral vanish; hence $y(a)$ will be zero for arbitrary f if and only if the particular solution $y_2(x)$ is zero when $x = a$. At $x = b$, the second integral is zero, and y will be zero for arbitrary f if and only if $y_1(x) = 0$ when $x = b$. Assuming that y_1 and y_2 are chosen to satisfy these conditions, and moving $y_1(x)$ and $y_2(x)$ into the respective integrals, we have

$$y = \int_a^x \frac{y_1(x)y_2(s)}{W(s)a_0(s)}f(s)\,ds + \int_x^b \frac{y_1(s)y_2(x)}{W(s)a_0(s)}f(s)\,ds$$

which is of the form

$$y = \int_a^b G(x, s)f(s)\,ds \tag{8}$$

where
$$G(x, s) = \begin{cases} \dfrac{y_1(s)y_2(x)}{W(s)a_0(s)} & x \le s \le b, \text{ that is, } a \le x \le s \\ \dfrac{y_1(x)y_2(s)}{W(s)a_0(s)} & a \le s \le x, \text{ that is, } s \le x \le b \end{cases}$$

From the way in which $y_1(x)$ and $y_2(x)$ were selected, it is clear that $G(x, s)$ satisfies the boundary conditions of the problem. It is also evident that $G(x, s)$ is a continuous function of x for each value of s, since $y_1(x)$ and $y_2(x)$ are continuous functions of x and at $x = s$ the right- and left-hand limits $G(s^+, s)$ and $G(s^-, s)$ are equal. Furthermore, except at $x = s$, $G(x, s)$ satisfies the homogeneous form of the given differential equation since $y_1(x)$ and $y_2(x)$ are solutions of this equation. Finally, for $G_x(x, s)$ we have

$$G_x(x, s) = \begin{cases} \dfrac{y_1(s)y_2'(x)}{W(s)a_0(s)} & a \le x < s \\ \dfrac{y_1'(x)y_2(s)}{W(s)a_0(s)} & s < x \le b \end{cases}$$

Hence
$$G_x(s^+, s) - G_x(s^-, s) = \frac{y_1'(s)y_2(s)}{W(s)a_0(s)} - \frac{y_1(s)y_2'(s)}{W(s)a_0(s)}$$
$$= -\frac{y_1(s)y_2'(s) - y_1'(s)y_2(s)}{W(s)a_0(s)} = -\frac{1}{a_0(s)}$$

which shows that $G_x(x, s)$ has a jump of the amount prescribed by condition 4 of Definition 1 at $x = s$. Thus $G(x, s)$ has the four properties necessary to make it the Green's function for the given problem. Moreover, from Eq. (8) it is clear that the solution of the general, nonhomogeneous boundary-value problem is given by a formula just like the one we derived for the particular case of the loaded string, Eq. (6). In the present case, however, there is no guarantee that $G(x, s) = G(s, x)$.

As a final example of a more sophisticated application of a Green's function, let us consider the following problem. A flexible elastic string of weight per unit length $\rho(x)$ is stretched under tension T between two points a distance l apart. Determine the frequencies at which the string can perform free vibrations and the shape of the curve of maximum deflection corresponding to each natural frequency.

This problem is just like the one we discussed earlier except that now the load per unit length, instead of being a known static load, is the (unknown) dynamic load arising from the inertia forces of the moving string itself. To obtain an expression for this load, consider an element of the string of length Δx. The mass of this element is $[\rho(x)\,\Delta x]/g$, and when the string is vibrating, the acceleration of this element is $\ddot{y}$. Hence, the inertia force due to the motion of this element is

$$-\frac{\rho(x)\,\Delta x}{g}\ddot{y}\dagger$$

† The minus sign here indicates that when the acceleration is negative, the string is deflected as though it carried a positive load, and vice versa.

and the corresponding load *per unit length* is

$$-\frac{\rho(x)}{g}\ddot{y}$$

Now if the string is vibrating at a single natural frequency in the absence of any damping forces, all points of the string must move harmonically with that frequency and with phase differences which are either 0 or 180°. In other words, during the vibration, the curve of the string is defined by an equation of the form

$$y(x, t) = \phi(x) \sin \omega t$$

where ω is the (as yet unknown) frequency of the vibrations and $\phi(x)$ is the (as yet unknown) curve of maximum displacements. Thus the load per unit length of the string is of the form

$$w(x, t) = -\frac{\rho(x)}{g}[-\omega^2\phi(x) \sin \omega t]$$

and Eq. (6) gives us

$$y(x, t) \equiv \phi(x) \sin \omega t = \int_0^l G(x, s)\omega^2 \frac{\rho(s)}{g} \phi(s) \sin \omega t \, ds$$

The factor $\sin \omega t$ is independent of the variable of integration s and hence can be removed from the integral and canceled from the equation. Thus the amplitude function $\phi(x)$ satisfies the *integral equation*†

$$\phi(x) = \frac{\omega^2}{g} \int_0^l G(x, s)\rho(s)\phi(s) \, ds$$

In a typical problem, this equation would probably have to be solved approximately in the following way. Make some reasonable guess as to the nature of the function $\phi(x)$, say $\phi_1(x)$; substitute this into the integrand; and compute the integral as a function of x. If the result is proportional to the "input" $\phi_1(x)$, we have a solution and the unknown frequency ω can be inferred from the proportionality constant. Of course, it is highly unlikely that we would ever hit upon the solution by guessing, and so the integrated result, or "output," say $\phi_2(x)$, will not be proportional to $\phi_1(x)$ and we shall not have found a solution. However, the process can be repeated, $\phi_2(x)$ can be substituted into the integral to give a new output $\phi_3(x)$, and so on. It can be shown that in a large class of cases the sequence of functions $\{\phi_n(x)\}$ determined by this procedure will converge to a function $\phi(x)$ which is a solution, and the ratio $\phi_{n-1}(x)/\phi_n(x)$ will approach a value which is independent of x. Thus by repeating the iteration a sufficient number of times, $\phi(x)$ and ω can be approximated with satisfactory accuracy.

† An equation involving the integral of an unknown function is called an **integral equation.** As this problem illustrates, Green's functions play an important role in connecting the theory of differential equations with the theory of integral equations.

In particular, if $\rho(x)$ is a constant, say ρ, then for a uniform string of length l we have the integral equation

$$\phi(x) = \frac{\omega^2 \rho}{g} \int_0^l G(x, s)\phi(s)\, ds$$

or, substituting the Green's function for a string from Eq. (7),

$$\begin{aligned}\phi(x) &= \frac{\omega^2\rho}{g} \int_0^x \frac{(l-x)s}{Tl} \phi(s)\, ds + \frac{\omega^2\rho}{g} \int_x^l \frac{(l-s)x}{Tl} \phi(s)\, ds \\ &= \frac{\omega^2\rho(l-x)}{gTl} \int_0^x s\phi(s)\, ds + \frac{\omega^2\rho x}{gTl} \int_x^l (l-s)\phi(s)\, ds\end{aligned}$$

A reasonable guess for $\phi(x)$ might be $\phi_1(x) = A \sin(n\pi x/l)$, since this is a simple function with the property that for each value of n it is zero for $x = 0$ and for $x = l$, as the deflection of a string with fixed ends should be. Using this, we have for the integrals on the right

$$\begin{aligned}\frac{A\omega^2\rho(l-x)}{gTl} \int_0^x s \sin\frac{n\pi s}{l}\, ds &+ \frac{A\omega^2\rho x}{gTl} \int_x^l (l-s) \sin\frac{n\pi s}{l}\, ds \\ &= \frac{A\omega^2\rho(l-x)}{gTl} \left[\frac{l^2}{n^2\pi^2} \sin\frac{n\pi s}{l} - \frac{ls}{n\pi} \cos\frac{n\pi s}{l}\right]_0^x \\ &\quad + \frac{A\omega^2\rho x}{gTl} \left[-\frac{(l-s)l}{n\pi} \cos\frac{n\pi s}{l} - \frac{l^2}{n^2\pi^2} \sin\frac{n\pi s}{l}\right]_x^l \\ &= \frac{A\omega^2\rho}{gT} \frac{l^2}{n^2\pi^2} \sin\frac{n\pi x}{l}\end{aligned}$$

This will be equal to the input, $A \sin(n\pi x/l)$, if and only if

$$\frac{\omega^2\rho}{gT} \frac{l^2}{n^2\pi^2} = 1$$

Thus there are infinitely many natural frequencies, given by the formula

$$\omega_n = \frac{n\pi}{l} \sqrt{\frac{gT}{\rho}}$$

with corresponding deflection curves

$$y_n(x, t) = A_n \sin\frac{n\pi x}{l} \sin\omega_n t$$

It is interesting to note that we shall obtain these results by an entirely different method in Sec. 11.4.

Exercises for Section 10.5

1 Construct the Green's function for a string of length l using the method of variation of parameters.

2 Determine the deflection curve of a string of length l bearing equal concentrated loads P at $x = l/4$ and at $x = l/2$.

3 Determine the deflection curve of a string of length l bearing concentrated loads P at $x = l/4$, $-2P$ at $x = l/2$, and $3P$ at $x = 3l/4$.

4 Find the deflection curve of a string of length l bearing a load per unit length $w(x) = -x$, first by solving the differential equation (4) and then by using the Green's function for the string.

5 Work Exercise 4 if the load per unit length is

$$w(x) = \begin{cases} 0 & 0 \le x < l/2 \\ -1 & l/2 < x \le l \end{cases}$$

6 Find the deflection curve of a string of length l bearing a load per unit length $w(x) = -x$ and a concentrated load P at $x = l/4$.

7 Construct the Green's function for the equation $y'' + 2y' + 2y = 0$ with the boundary conditions $y(0) = 0$, $y(\pi/2) = 0$. Is this Green's function symmetric? What is the Green's function for the equation $e^{2x}y'' + 2e^{2x}y' + 2e^{2x}y = 0$? Is this Green's function symmetric?

8 Find the Green's function for the equation $y'' + k^2y = 0$ if the boundary conditions are

(*a*) $y(0) = y'(b) = 0$ (*b*) $y'(0) = y(b) = 0$
(*c*) $y'(a) = y'(b) = 0$ (*d*) $y(a) = y'(a)$, $y(b) = 0$

9 Work Exercise 8 for the equation $y'' - k^2y = 0$.

10 Find the Green's function for the equation $(y'/x^2)' + 2y/x^4 = 0$ with the boundary conditions $y(0) = y'(1) = 0$. *Hint:* The given differential equation is an Euler equation, after the indicated differentiation is carried out and the resulting equation cleared of fractions.

11 Show that the Green's function for a differential system of the form $[r(x)y']' + p(x)y = 0$ with boundary conditions of the type prescribed by Definition 1 is symmetric. *Hint:* Recall Abel's identity for the wronskian of a second-order differential equation.

12 The angle of twist θ produced in a uniform shaft of length l by a torque T is given by the formula $\theta = Tl/E_sJ$, where E_s is the modulus of elasticity in shear of the material of the shaft and J is the polar moment of inertia of the cross-sectional area of the shaft about the center of gravity of the cross section. Using this formula, find the influence function which gives the angle of twist at a point x due to a unit torque applied at a point s of a shaft rigidly clamped at the left end and free at the right end.

***13** Find the influence function which gives the deflection of a uniform cantilever beam at a point x due to a unit load applied at the point s. Verify that this influence function $G(x, s)$ has the following properties, the primes denoting differentiation with respect to x:

(*a*) $G(x, s)$ satisfies the differential equation $EIy^{iv} = 0$ for $0 \le x < s$ and $s < x \le l$.

(*b*) $G(0, s) = G'(0, s) = G''(l, s) = G'''(l, s) = 0$; that is, $G(x, s)$ satisfies the end conditions for a cantilever beam.

(*c*) $G(x, s)$, $G'(x, s)$, and $G''(x, s)$ are continuous for $0 \le x \le l$.

(*d*) $G'''(x, s)$ has a jump of $-1/EI$ at $x = s$.

Is $G(x, s)$ a symmetric function of x and s?

14 Using the influence function obtained in Exercise 13, find the deflection curve of a uniform cantilever beam bearing concentrated loads P at $x = l/2$ and at $x = l$.

***15** Find the deflection curve of a uniform cantilever beam of length l bearing a load per unit length $w(x) = x$, first by solving the differential equation $EIy^{iv} = -w(x)$ and then by using the influence function obtained in Exercise 13.

CHAPTER

ELEVEN

PARTIAL DIFFERENTIAL EQUATIONS

11.1 INTRODUCTION

At various points in our previous work, most notably in Chap. 7, we have seen how the analysis of mechanical and electrical systems containing lumped parameters often leads to ordinary differential equations in which the time t is the (only) independent variable. However, the assumption that all masses exist as conceptualized mass points; that all springs are weightless; or that elements of an electrical circuit are concentrated in ideal resistors, capacitors, and inductors, rather than continuously distributed, is frequently not sufficiently accurate. In such cases, a more realistic approach must take into account the fact that the dependent variables depend not only on t, but also on one or more space variables. Because there is more than one independent variable, the formulation of such problems leads to partial, rather than ordinary, differential equations. In this chapter we shall discuss such equations as they commonly arise in applied mathematics. We shall begin by examining in some detail the derivation from physical principles of several important partial differential equations. Then, knowing the forms of most common occurrence, we shall investigate methods of solution and their application to specific problems.

11.2 THE DERIVATION OF EQUATIONS

One of the first physical problems to be attacked through the use of partial differential equations was that of the vibration of a stretched, flexible string. Today, after nearly 250 years, it is still an excellent initial example to use as an introduction to the study of partial differential equations.†

† We have already considered one aspect of this problem in Sec. 10.5. There, through the use of a Green's function, we were able to determine the natural frequencies and the corresponding deflection curves of a uniform string vibrating transversely. Now, through the use of partial differential equations, we are able to make an even more detailed study of vibrating strings.

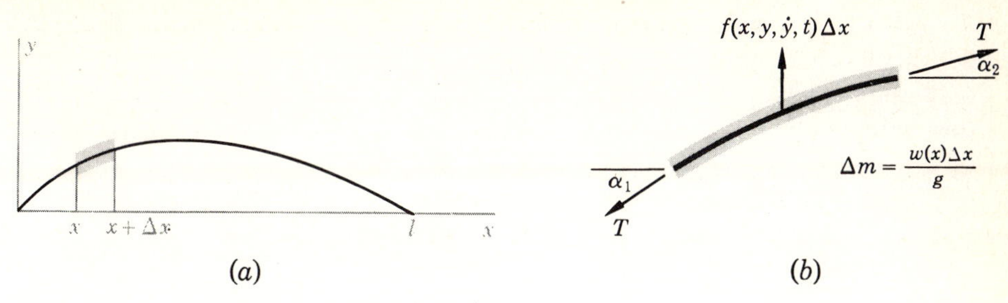

Figure 11.1 A typical element of a vibrating string.

Let us consider, then, an elastic string, stretched under a tension T between two points on the x axis (Fig. 11.1a). The weight of the string per unit length after it is stretched we suppose to be a known function $w(x)$. Besides the elastic and inertia forces inherent in the system, the string may also be acted upon by a distributed load whose magnitude per unit length we assume to be a known function of x, y, t, and the transverse velocity $\dot{y}$, say $f(x, y, \dot{y}, t)$. In formulating the problem, we assume that

1. The motion takes place entirely in one plane, and in this plane each particle moves at right angles to the equilibrium position of the string.
2. The deflection of the string during the motion is so small that the resulting change in length of the string has no effect on the tension T.
3. The string is perfectly flexible, i.e., can transmit force only in the direction of its length.
4. The slope of the deflection curve of the string is at all points and at all times so small that with satisfactory accuracy $\sin \alpha$ can be replaced by $\tan \alpha$, where α is the inclination angle of the tangent to the deflection curve.

Gravitational and frictional forces, if any, we suppose to be taken into account in the expression for the load per unit length $f(x, y, \dot{y}, t)$.

With these assumptions in mind, let us consider a general infinitesimal segment of the string as a free body (Fig. 11.1b). By assumption 1, the mass of such an element is $\Delta m = w(x)\, \Delta x/g$. By assumption 2, the forces which act at the ends of the element are the same, namely, T. By assumption 3, these forces are directed along the respective tangents to the deflection curve; and, by assumption 4, their transverse components are

$$T \sin \alpha_2 = T \sin \alpha \Big|_{x+\Delta x} \doteq T \tan \alpha \Big|_{x+\Delta x}$$

and

$$T \sin \alpha_1 = T \sin \alpha \Big|_{x} \doteq T \tan \alpha \Big|_{x}$$

The acceleration produced in Δm by these forces and by the portion of the distributed load $f(x, y, \dot{y}, t)\, \Delta x$ which acts over the interval Δx is approximately

$\partial^2 y/\partial t^2$, where y is the ordinate of an arbitrary point of the element. The time derivative is here written as a partial derivative because of course y depends not only upon t but upon x as well. Applying Newton's law to the element, we can thus write

$$\frac{w(x)\,\Delta x}{g}\frac{\partial^2 y}{\partial t^2} = T \tan \alpha\Big|_{x+\Delta x} - T \tan \alpha\Big|_x + f(x, y, \dot{y}, t)\,\Delta x \tag{1}$$

or, dividing by Δx,

$$\frac{w(x)}{g}\frac{\partial^2 y}{\partial t^2} = T\left(\frac{\tan \alpha|_{x+\Delta x} - \tan \alpha|_x}{\Delta x}\right) + f(x, y, \dot{y}, t)$$

The fraction on the right consists of the difference between the values of $\tan \alpha$ at $x + \Delta x$ and at x, divided by the difference Δx. In other words, it is precisely the difference quotient for the function $\tan \alpha$. Hence its limit as $\Delta x \to 0$ is the derivative of $\tan \alpha$ with respect to x, that is, $(\partial \tan \alpha)/\partial x$. But since $\tan \alpha = \partial y/\partial x$, this can be written simply as $\partial^2 y/\partial x^2$. Our final result, then, is that the deflection $y(x, t)$ of a stretched string satisfies the partial differential equation†

$$\frac{\partial^2 y}{\partial t^2} = \frac{Tg}{w(x)}\frac{\partial^2 y}{\partial x^2} + \frac{g}{w(x)} f(x, y, \dot{y}, t) \tag{2}$$

In most applications the weight of the string per unit length $w(x)$ is a constant, and external forces are negligible; i.e., with satisfactory accuracy $f(x, y, \dot{y}, t)$ can be assumed to be identically zero. When this is the case, Eq. (2) reduces to the **one-dimensional wave equation**

$$\frac{\partial^2 y}{\partial t^2} = a^2 \frac{\partial^2 y}{\partial x^2} \qquad a^2 = \frac{Tg}{w} \tag{3}$$

The dimensions of a^2 are

$$\frac{\text{Force} \times \text{acceleration}}{\text{Weight/unit length}} = \frac{(ML/T^2)(L/T^2)}{(ML/T^2)(1/L)} = \frac{L^2}{T^2}$$

that is, a has the dimensions of velocity. The significance of this will become apparent in Sec. 11.3 when we discuss the d'Alembert solution of the wave equation.

† The question of what constitutes a satisfactory derivation of the partial differential equation describing a given physical system is not a simple one. To attempt to give a careful limiting argument is, in effect, "to strain at a gnat and swallow a camel," since, being ultimately atomic, no physical system is continuous. Perhaps our purported derivations should be regarded merely as plausibility arguments suggesting that certain partial differential equations be accepted as the axioms of a theoretical or "rational" study of applied mathematics, whose practical importance, in contrast to its purely mathematical interest, is to be judged by how well its conclusions describe past observations and predict new ones.

The one-dimensional wave equation is a very important equation because it describes the behavior of numerous physical systems besides the vibrating string. For instance, when a uniform shaft vibrates torsionally, the angle of twist θ of a typical cross section at a distance x from one end of the shaft satisfies the equation

$$\frac{\partial^2 \theta}{\partial t^2} = a^2 \frac{\partial^2 \theta}{\partial x^2} \qquad a^2 = \frac{E_s g}{\rho}$$

where E_s is the shear modulus of the material of the shaft, ρ is its weight per unit volume, and g is the acceleration of gravity. Similarly, when a bar of uniform cross section vibrates longitudinally, the displacement u of a typical cross section at a distance x from one end of the bar satisfies the equation

$$\frac{\partial^2 u}{\partial t^2} = a^2 \frac{\partial^2 u}{\partial x^2} \qquad a^2 = \frac{Eg}{\rho}$$

where E is Young's modulus for the material of the bar and ρ is its weight per unit volume. Likewise, when the air in an organ pipe of constant cross section vibrates, the condensation factor

$$\sigma = \frac{\rho - \rho_0}{\rho_0}$$

where ρ_0 is the initial and ρ the instantaneous density of the air, satisfies the equation

$$\frac{\partial^2 \sigma}{\partial t^2} = a^2 \frac{\partial^2 \sigma}{\partial x^2}$$

where a depends in a complicated way on the initial pressure and density of the air and the ratio of the specific heats of air at constant pressure and constant volume. In each of these equations, a has the dimensions of velocity.

Closely related to the elastic string is the elastic membrane, stretched across some simple closed curve in the xy plane. In deriving the equation governing the vibration of a membrane, we assume that:

1. The motion is such that each particle of the membrane moves at right angles to the xy plane.
2. The deflection of the membrane is so small that it has no effect on the tension per unit length T acting across an arbitrary curve in the membrane.
3. The membrane is perfectly flexible, i.e., can transmit force only in directions which are tangent to its instantaneous position.
4. The slope of any tangent to the deflected membrane is so small that with satisfactory accuracy $\sin \theta$ can be replaced by $\tan \theta$, where θ is the acute angle which the tangent makes with the xy plane.
5. Gravitational and frictional forces, if not negligible, are included in a distributed force whose magnitude per unit area is $f(x, y, z, \dot{z}, t)$.

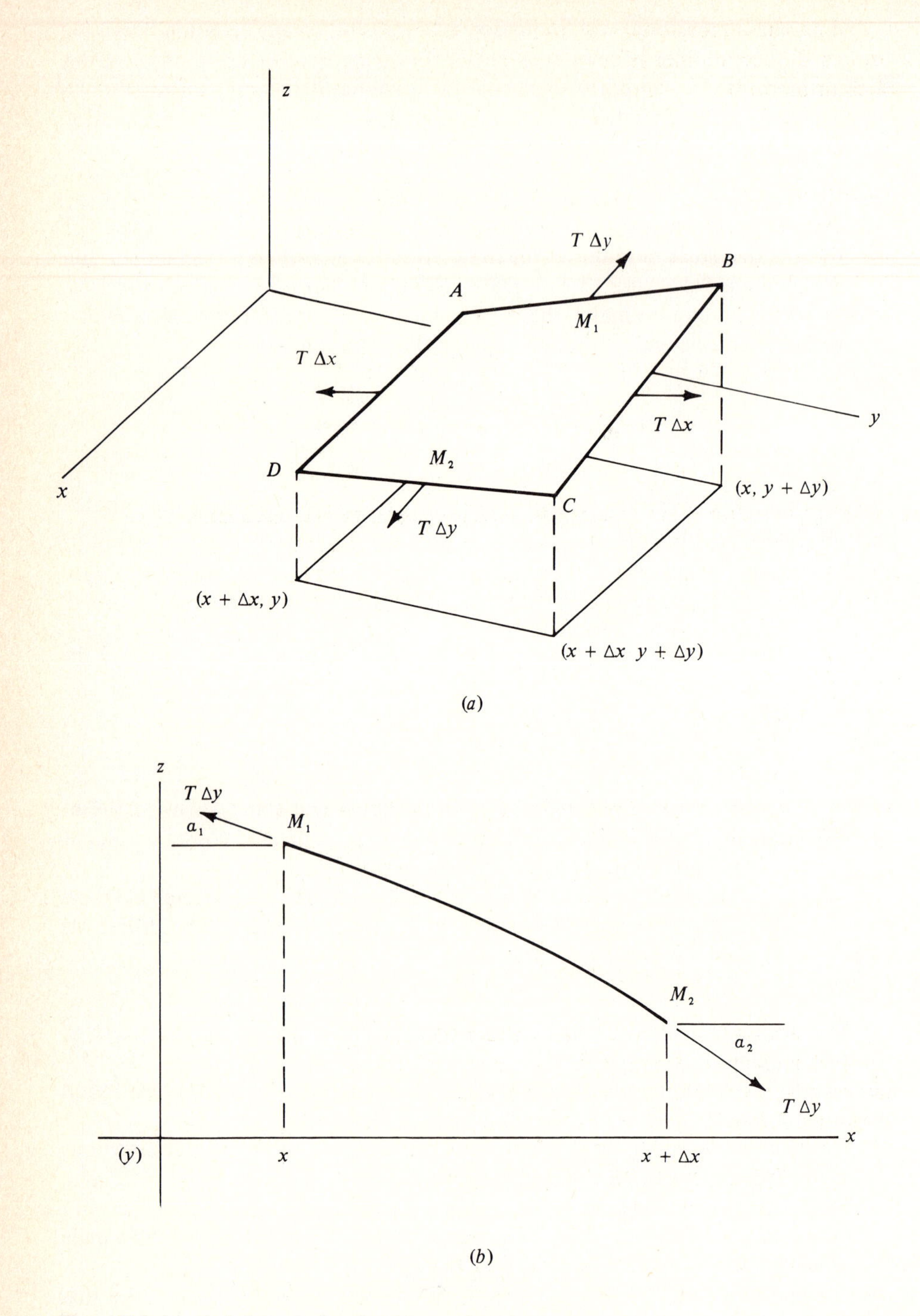

Figure 11.2 A typical element of a vibrating membrane.

With these assumptions in mind, let us consider a general infinitesimal portion of the membrane, as suggested by Fig. 11.2*a*. By assumption 1, the mass of such an element is

$$\Delta m = \frac{w(x, y)\,\Delta x\,\Delta y}{g}$$

where $w(x, y)$ is the weight per unit area of the membrane in its equilibrium position. By assumption 2, the forces applied to the element across the edges AB and DC are each equal to $T\,\Delta y$. These forces, being perpendicular to the edges, are (approximately) parallel to the xz plane, and we shall consider them concentrated at the midpoints

$$M_1\colon \left[x,\ y + \frac{\Delta y}{2},\ z\left(x,\ y + \frac{\Delta y}{2}\right)\right]$$

and

$$M_2\colon \left[x + \Delta x,\ y + \frac{\Delta y}{2},\ z\left(x + \Delta x,\ y + \frac{\Delta y}{2}\right)\right]$$

of the respective edges. By assumption 3, these forces act in the directions of the tangents to the deflection surface at M_1 and M_2 which are parallel to the xz plane. If α_1 and α_2 are, respectively, the inclination angles of these tangents, then, by 4, the transverse components of the forces are (Fig. 11.2*b*)

$$(T\,\Delta y)\sin\alpha_2 = (T\,\Delta y)\sin\alpha\Big|_{\substack{x+\Delta x\\ y+\Delta y/2}} \doteq (T\,\Delta y)\tan\alpha\Big|_{\substack{x+\Delta x\\ y+\Delta y/2}}$$

and

$$(T\,\Delta y)\sin\alpha_1 = (T\,\Delta y)\sin\alpha\Big|_{\substack{x\\ y+\Delta y/2}} \doteq (T\,\Delta y)\tan\alpha\Big|_{\substack{x\\ y+\Delta y/2}}$$

Similarly, the transverse components of the forces which act on the element across the edges BC and AD are

$$(T\,\Delta x)\sin\beta_2 = (T\,\Delta x)\sin\beta\Big|_{\substack{x+\Delta x/2\\ y+\Delta y}} = (T\,\Delta x)\tan\beta\Big|_{\substack{x+\Delta x/2\\ y+\Delta y}}$$

and

$$(T\,\Delta x)\sin\beta_1 = (T\,\Delta x)\sin\beta\Big|_{\substack{x+\Delta x/2\\ y}} = (T\,\Delta x)\tan\beta\Big|_{\substack{x+\Delta x/2\\ y}}$$

The acceleration produced by these forces and by the portion of the distributed load $f(x, y, z, \dot{z}, t)\,\Delta x\,\Delta y$ which acts on Δm is approximately $\partial^2 z/\partial t^2$, where z is the deflection at an arbitrary point in the area $\Delta x\,\Delta y$. Applying Newton's law to Δm, we then have, approximately,

$$\begin{aligned}\frac{w(x, y)\,\Delta x\,\Delta y}{g}\frac{\partial^2 z}{\partial t^2} &= (T\,\Delta y)\left(\tan\alpha\Big|_{\substack{x+\Delta x\\ y+\Delta y/2}} - \tan\alpha\Big|_{\substack{x\\ y+\Delta y/2}}\right)\\ &\quad + (T\,\Delta x)\left(\tan\beta\Big|_{\substack{x+\Delta x/2\\ y+\Delta y}} - \tan\beta\Big|_{\substack{x+\Delta x/2\\ y}}\right)\\ &\quad + f(x, y, z, \dot{z}, t)\,\Delta x\,\Delta y\end{aligned}$$

From this, by dividing by $\Delta x\ \Delta y$, noting the two difference quotients that arise on the right-hand side, and then taking limits as Δx and Δy approach zero, we obtain

$$\frac{w(x, y)}{g}\frac{\partial^2 z}{\partial t^2} = T\left(\frac{\partial \tan \alpha}{\partial x} + \frac{\partial \tan \beta}{\partial y}\right) + f(x, y, z, \dot{z}, t) \tag{4}$$

Finally, we recognize that $\tan \alpha$ is the slope of a curve cut from the deflection surface by a plane parallel to the xz plane, and therefore

$$\tan \alpha = \frac{\partial z}{\partial x}$$

Similarly, we recognize that

$$\tan \beta = \frac{\partial z}{\partial y}$$

Hence, making these substitutions in Eq. (4), we find that the deflection of the membrane, z, satisfies the equation

$$\frac{\partial^2 z}{\partial t^2} = \frac{Tg}{w(x, y)}\left(\frac{\partial^2 z}{\partial x^2} + \frac{\partial^2 z}{\partial y^2}\right) + \frac{g}{w(x, y)} f(x, y, z, \dot{z}, t) \tag{5}$$

If the weight per unit area of the membrane is the same at all points, and if external forces are negligible, i.e., if $w(x, y)$ is a constant and if $f(x, y, z, \dot{z}, t)$ is effectively zero, then Eq. (5) reduces to the **two-dimensional wave equation**

$$\frac{\partial^2 z}{\partial t^2} = a^2\left(\frac{\partial^2 z}{\partial x^2} + \frac{\partial^2 z}{\partial y^2}\right) \qquad a^2 = \frac{Tg}{w} \tag{6}$$

Here, as in the case of the vibrating string, the parameter a has the dimensions of velocity.

An entirely different class of problems leading to partial differential equations is encountered in the study of the flow of heat in thermally conducting regions. To obtain the equation governing this phenomenon, we must make use of the following experimental facts:

1. Heat flows in the direction of decreasing temperature.
2. The rate at which heat flows through an area is proportional to the area and to the temperature gradient (see item 9, Appendix B.3), in degrees per unit distance, in the direction perpendicular to the area.
3. The quantity of heat gained or lost by a body when its temperature changes is proportional to the mass of the body and to the temperature change.

The proportionality constant in 2 is called the **thermal conductivity** of the material k. The proportionality constant in 3 is called the **specific heat** of the material c.

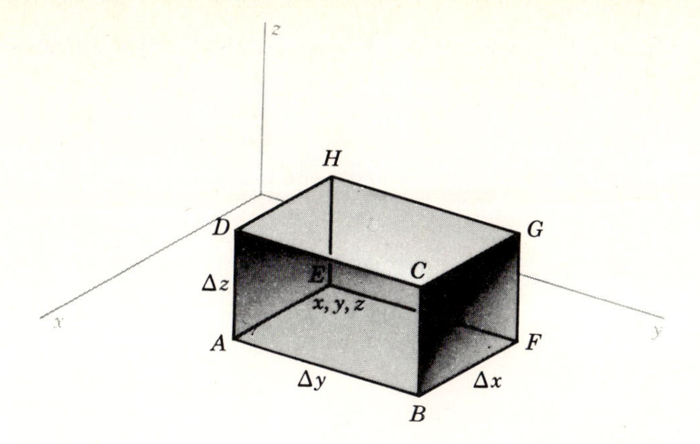

Figure 11.3 A typical volume element in a region of three-dimensional heat flow.

Let us now consider the thermal conditions in a small element $\Delta x\ \Delta y\ \Delta z$ of a conducting solid (Fig. 11.3). If the weight of the conducting material per unit volume is ρ, the mass of such an element is

$$\Delta m = \frac{\rho\ \Delta x\ \Delta y\ \Delta z}{g}$$

Furthermore, if u is the temperature at any point at any time, and if Δu is the temperature change which occurs in the element in the time interval Δt, then the quantity of heat stored in the element in this time is, by 3,

$$\Delta H = c\ \Delta m\ \Delta u = \frac{c\rho\ \Delta x\ \Delta y\ \Delta z\ \Delta u}{g}$$

and the rate at which heat is being stored is approximately

$$\frac{\Delta H}{\Delta t} = \frac{c\rho}{g}\ \Delta x\ \Delta y\ \Delta z\ \frac{\Delta u}{\Delta t} \tag{7}$$

The heat which produces the temperature change Δu comes from two sources. First, heat may be generated throughout the body, by electrical or chemical means for instance, at a known rate per unit volume, say $f(x, y, z, t)$. The rate at which heat is being received by the element from this source is then

$$f(x, y, z, t)\ \Delta x\ \Delta y\ \Delta z \tag{8}$$

Second, the element may also gain heat by virtue of heat transfer through its various faces.

In particular, the rate at which heat flows into the element through the rear face $EFGH$ is, by 2, approximately

$$-k\ \Delta y\ \Delta z\ \left.\frac{\partial u}{\partial x}\right|_{\substack{x \\ y+\Delta y/2 \\ z+\Delta z/2}}$$

where, as an average figure, we have used the temperature gradient $\partial u/\partial x$ at the center $(x, y + \frac{1}{2}\Delta y, z + \frac{1}{2}\Delta z)$ of the face $EFGH$. The minus sign is necessary because the element *gains* heat through the rear face if the normal temperature gradient, i.e., the rate of change of temperature in the x direction, is *negative.* Similarly, the element gains heat through the front face $ABCD$ at the approximate rate

$$k\,\Delta y\,\Delta z\,\frac{\partial u}{\partial x}\bigg|_{\substack{x+\Delta x\\ y+\Delta y/2\\ z+\Delta z/2}}$$

The sum of these two expressions is the net rate at which the element gains heat because of heat flow in the x direction.

In the same way, we find that the rates at which the element gains heat because of flow in the y and z directions are, respectively,

$$-k\,\Delta x\,\Delta z\,\frac{\partial u}{\partial y}\bigg|_{\substack{x+\Delta x/2\\ y\\ z+\Delta z/2}} \qquad +k\,\Delta x\,\Delta z\,\frac{\partial u}{\partial y}\bigg|_{\substack{x+\Delta x/2\\ y+\Delta y\\ z+\Delta z/2}}$$

and

$$-k\,\Delta x\,\Delta y\,\frac{\partial u}{\partial z}\bigg|_{\substack{x+\Delta x/2\\ y+\Delta y/2\\ z}} \qquad +k\,\Delta x\,\Delta y\,\frac{\partial u}{\partial z}\bigg|_{\substack{x+\Delta x/2\\ y+\Delta y/2\\ z+\Delta z}}$$

Now the rate at which heat is being stored in the element, Eq. (7), must equal the rate at which heat is being produced in the element as given by (8), plus the rate at which heat is flowing into the element from the rest of the region. Hence we have the approximate relation

$$\begin{aligned}\frac{c\rho}{g}\,\Delta x\,\Delta y\,\Delta z\,\frac{\Delta u}{\Delta t} = f(x, y, z, t)\,\Delta x\,\Delta y\,\Delta z + k\,\Delta y\,\Delta z\left(\frac{\partial u}{\partial x}\bigg|_{\substack{x+\Delta x\\ y+\Delta y/2\\ z+\Delta z/2}} - \frac{\partial u}{\partial x}\bigg|_{\substack{x\\ y+\Delta y/2\\ z+\Delta z/2}}\right)\\ + k\,\Delta x\,\Delta z\left(\frac{\partial u}{\partial y}\bigg|_{\substack{x+\Delta x/2\\ y+\Delta y\\ z+\Delta z/2}} - \frac{\partial u}{\partial y}\bigg|_{\substack{x+\Delta x/2\\ y\\ z+\Delta z/2}}\right)\\ + k\,\Delta x\,\Delta y\left(\frac{\partial u}{\partial z}\bigg|_{\substack{x+\Delta x/2\\ y+\Delta y/2\\ z+\Delta z}} - \frac{\partial u}{\partial z}\bigg|_{\substack{x+\Delta x/2\\ y+\Delta y/2\\ z}}\right)\end{aligned}$$

Finally, dividing by $k\,\Delta x\,\Delta y\,\Delta z$ and letting Δx, Δy, Δz, and Δt approach zero, we obtain the equation of heat conduction

$$a^2\frac{\partial u}{\partial t} = \frac{\partial^2 u}{\partial x^2} + \frac{\partial^2 u}{\partial y^2} + \frac{\partial^2 u}{\partial z^2} + \frac{1}{k}f(x, y, z, t) \qquad a^2 = \frac{c\rho}{kg} \tag{9}$$

The parameter a in this equation does not have the dimensions of velocity.

In many important cases, heat is neither generated nor lost in the body, and we are interested only in the limiting steady-state temperature distribution which exists when all change of temperature with time has ceased. Under these conditions both $f(x, y, z, t)$ and $\partial u/\partial t$ are identically zero, and Eq. (9) becomes simply

$$\frac{\partial^2 u}{\partial x^2} + \frac{\partial^2 u}{\partial y^2} + \frac{\partial^2 u}{\partial z^2} = 0 \qquad (10)$$

This exceedingly important equation, which arises in many applications besides steady-state heat flow, is known as **Laplace's equation,** and is often written in the abbreviated form

$$\nabla^2 u = 0 \qquad (11)$$

It is important to note that nowhere in the derivation of any of the preceding equations was any use made of boundary conditions. For instance, the same partial differential equation is satisfied by the deflections of a membrane whether the membrane is round or square. Likewise, the flow of heat in a body is described by the same equation whether the surface is maintained at a constant temperature, insulated against heat loss, or allowed to cool freely by radiation to the surrounding medium. In general, as we shall soon see, the role of boundary conditions, e.g., permanent conditions of constraint or of temperature, is to determine the *form* of those solutions which are relevant to a particular problem. Subsequent to this, initial conditions of displacement, velocity, or temperature, say, determine specific values for the arbitrary constants appearing in these solutions.

There are many important partial differential equations besides the ones we have derived, and we shall refer to several of these in the exercises. However, the wave equation, the heat equation, and Laplace's equation provide ample material to illustrate solution techniques and typical applications, and we shall concentrate our attention on these equations.

From our work in ordinary differential equations we know that in general the solutions of such equations contain arbitrary constants whose values are subsequently determined to fit given conditions on the dependent variable and its derivatives. In contrast to this, the solutions of partial differential equations involve arbitrary *functions* whose forms are subsequently determined to fit whatever conditions the dependent variable and its derivatives are required to satisfy. We shall explore these matters in some detail in the following sections, but to set the stage for this work, we conclude this section with several introductory examples.

Example 1 If f is an arbitrary twice-differentiable function of a single variable, show that $z = f(\alpha x + \beta y + at)$ is a solution of the two-dimensional wave equation for all values of α and β which satisfy the relation $\alpha^2 + \beta^2 = 1$.

Observing that $z = f(\alpha x + \beta y + at)$ is equivalent to the chain of relations $z = f(w)$ and $w = \alpha x + \beta y + at$, we have

$$\frac{\partial z}{\partial x} = \frac{df}{dw}\frac{\partial w}{\partial x} = \alpha \frac{df}{dw} \qquad \text{and} \qquad \frac{\partial^2 z}{\partial x^2} = \alpha \frac{d^2 f}{dw^2}\frac{\partial w}{\partial x} = \alpha^2 \frac{d^2 f}{dw^2}$$

Similarly, $$\frac{\partial^2 z}{\partial y^2} = \beta^2 \frac{d^2 f}{dw^2} \qquad \text{and} \qquad \frac{\partial^2 z}{\partial t^2} = a^2 \frac{d^2 f}{dw^2}$$

Hence, substituting into the wave equation (6), we have

$$a^2 \frac{d^2 f}{dw^2} = a^2 \left(\alpha^2 \frac{d^2 f}{dw^2} + \beta^2 \frac{d^2 f}{dw^2} \right) = a^2 \frac{d^2 f}{dw^2} (\alpha^2 + \beta^2)$$

and this will be an identity if and only if $\alpha^2 + \beta^2 = 1$, as asserted.

Example 2 Determine for what values of α, if any, the arbitrary twice-differentiable function $u = f(\alpha x + y)$ is a solution of the equation

$$\frac{\partial^2 u}{\partial x^2} + \frac{\partial^2 u}{\partial x \, \partial y} - 6 \frac{\partial^2 u}{\partial y^2} = 0$$

Using the chain rule, as in Example 1, we have at once

$$\frac{\partial^2 u}{\partial x^2} = \alpha^2 f'' \qquad \frac{\partial^2 u}{\partial x \, \partial y} = \alpha f'' \qquad \frac{\partial^2 u}{\partial y^2} = f''$$

where f'' denotes the second derivative of f with respect to its entire argument $\alpha x + y$. Hence, substituting into the given equation, we must have

$$\alpha^2 f'' + \alpha f'' - 6f'' = 0$$

This will be true if and only if

$$\alpha^2 + \alpha - 6 = (\alpha + 3)(\alpha - 2) = 0$$

Hence $\alpha = -3, 2$, and thus if f is any twice-differentiable function, $u = f(-3x + y)$ and $u = f(2x + y)$ are solutions of the given equation.

Exercises for Section 11.2

1 What is the form of the heat equation if the thermal conductivity k and the specific heat c vary from point to point in the body?

***2** Derive the partial differential equation satisfied by the concentration u of a liquid diffusing through a porous solid. *Hint:* The rate at which liquid diffuses through an area is proportional to the area and to the concentration gradient (see item 5, Appendix B.3) in the direction perpendicular to the area.

***3** Consider a region of space filled with a moving fluid. Let the density of the fluid at the point (x, y, z) at time t be $\rho(x, y, z, t)$, and let the particle instantaneously at the point (x, y, z) have velocity components v_x, v_y, and v_z, respectively, in the directions of the coordinate axes. By considering the flow through the boundaries of an infinitesimal region of dimensions Δx, Δy, Δz, show that the velocity components satisfy the so-called **equation of continuity**

$$\frac{\partial(\rho v_x)}{\partial x} + \frac{\partial(\rho v_y)}{\partial y} + \frac{\partial(\rho v_z)}{\partial z} + \frac{\partial \rho}{\partial t} = 0$$

*4 The flow of electricity through a cable is governed by the so-called **telephone equations**

$$\frac{\partial^2 e}{\partial x^2} = LC\frac{\partial^2 e}{\partial t^2} + (RC + GL)\frac{\partial e}{\partial t} + RGe$$

and

$$\frac{\partial^2 i}{\partial x^2} = LC\frac{\partial^2 i}{\partial t^2} + (RC + GL)\frac{\partial i}{\partial t} + RGi$$

where $e(x, t)$ and $i(x, t)$ are, respectively, the instantaneous voltage and current at a distance x from the sending end of the cable, and R, C, L, G describe electrical properties of the cable which need not be specified here. In one important case $RC = GL$. Assuming this condition, and putting $a^2 = RG$ and $v^2 = 1/(LC)$, show that if $e(x, t)$ or, equally well, $i(x, t)$ is written in the form $e(x, t) = \varepsilon^{-avt}y(x, t)$, then the function y satisfies the wave equation

$$v^2\frac{\partial^2 y}{\partial x^2} = \frac{\partial^2 y}{\partial t^2}$$

Note: To avoid confusion with the voltage, ε is used here in place of e to denote the base of natural logarithms.

5 Verify that each of the following equations has the indicated solution.

(*a*) $\dfrac{\partial^2 u}{\partial x\,\partial y} = 0 \qquad u = f(x) + g(y)$ (*b*) $u\dfrac{\partial^2 u}{\partial x\,\partial y} = \dfrac{\partial u}{\partial x}\dfrac{\partial u}{\partial y} \qquad u = f(x)g(y)$

(*c*) $a\dfrac{\partial u}{\partial x} + b\dfrac{\partial u}{\partial y} = 0 \qquad u = f(ay - bx)$

(*d*) $\dfrac{\partial^2 y}{\partial x^2} = \dfrac{\partial^2 y}{\partial t^2} \qquad y = f(x - t) + g(x + t)$

(*e*) $\dfrac{\partial^3 u}{\partial x^3} - 6\dfrac{\partial^3 u}{\partial y\,\partial x^2} + 11\dfrac{\partial^3 u}{\partial y^2\,\partial x} - 6\dfrac{\partial^3 u}{\partial y^3} = 0$

$$u = f(x + y) + g(2x + y) + h(3x + y)$$

6 Explain how the method of undetermined coefficients can be used to obtain a particular solution of the equation

$$a\frac{\partial^2 u}{\partial x^2} + b\frac{\partial^2 u}{\partial x\,\partial y} + c\frac{\partial^2 u}{\partial y^2} = \phi(x, y)$$

if

(*a*) $\phi(x, y) = e^{mx+ny}$ (*b*) $\phi(x, y) = \sin(mx + ny)$
(*c*) $\phi(x, y) = \cos(mx + ny)$ (*d*) $\phi(x, y) = px^2 + qxy + ry^2$
(*e*) $\phi(x, y)$ is a homogeneous polynomial of degree k in x and y.

7 Find a particular solution of each of the following equations.

(*a*) $\dfrac{\partial^2 u}{\partial x^2} + \dfrac{\partial^2 u}{\partial y^2} = \cos(x + 2y)$ (*b*) $\dfrac{\partial^2 u}{\partial x^2} - \dfrac{\partial u}{\partial y} = 2e^{2x+3y}$

(*c*) $\dfrac{\partial^2 u}{\partial x^2} + 3\dfrac{\partial^2 u}{\partial x\,\partial y} + 2\dfrac{\partial^2 u}{\partial y^2} = 2x - y$

8 (*a*) Show that Laplace's equation in three dimensions

$$\frac{\partial^2 u}{\partial x^2} + \frac{\partial^2 u}{\partial y^2} + \frac{\partial^2 u}{\partial z^2} = 0$$

is satisfied by the function

$$u = \frac{1}{\sqrt{(x-a)^2 + (y-b)^2 + (z-c)^2}}$$

for all values of the constants a, b, c.

(*b*) Determine whether or not Laplace's equation in two dimensions

$$\frac{\partial^2 u}{\partial x^2} + \frac{\partial^2 u}{\partial y^2} = 0$$

is satisfied by the function

$$u = \frac{1}{\sqrt{(x-a)^2 + (y-b)^2}}$$

9 Show that Laplace's equation in two dimensions is satisfied by the function $u = \ln [(x-a)^2 + (y-b)^2]$ for all values of the constants a and b.

10 Show that under the substitution $u(x, y) = w(x, y)e^{-(bx+ay)}$ the equation

$$\frac{\partial^2 u}{\partial x\, \partial y} + a\frac{\partial u}{\partial x} + b\frac{\partial u}{\partial y} + cu = 0$$

becomes

$$\frac{\partial^2 w}{\partial x\, \partial y} + (c - ab)w = 0$$

11 By assuming $u(x, t) = \phi(x) \sin t$, find a solution of the equation

$$\frac{\partial^2 u}{\partial x^2} - \frac{\partial^2 u}{\partial t^2} = x \sin t$$

satisfying the conditions $u(0, t) = 0$ and $u(l, t) = 0$, $l \neq n\pi$.

12 Show that if $u_1(x, y)$ and $u_2(x, y)$ are solutions of the equation

$$p_1(x, y)\frac{\partial^2 u}{\partial x^2} + p_2(x, y)\frac{\partial^2 u}{\partial x\, \partial y} + p_3(x, y)\frac{\partial^2 u}{\partial y^2} + q_1(x, y)\frac{\partial u}{\partial x} + q_2(x, y)\frac{\partial u}{\partial y} + r_1(x, y)u = 0$$

then for all values of the constants c_1 and c_2, the expression $c_1 u_1(x, y) + c_2 u_2(x, y)$ is also a solution.

11.3 THE D'ALEMBERT SOLUTION OF THE WAVE EQUATION

Each of the equations we derived in the last section can be solved by a method of considerable generality known as *separation of variables*. For the one-dimensional wave equation, however, there is also an elegant, special method known as

d'Alembert's solution† which, because of the importance of this equation, we shall examine in some detail before developing more general techniques.

The whole matter is very simple. In fact, if f is a function possessing a second derivative, then, by the chain rule,

$$\frac{\partial f(x-at)}{\partial t} = -af'(x-at) \qquad \frac{\partial f(x-at)}{\partial x} = f'(x-at)$$

$$\frac{\partial^2 f(x-at)}{\partial t^2} = a^2 f''(x-at) \qquad \frac{\partial^2 f(x-at)}{\partial x^2} = f''(x-at)$$

and from these results it is evident that $y = f(x-at)$ satisfies the equation

$$\frac{\partial^2 y}{\partial t^2} = a^2 \frac{\partial^2 y}{\partial x^2} \tag{1}$$

It is an equally simple matter to prove that if g is an arbitrary, twice-differentiable function, then $g(x + at)$ is likewise a solution of (1). Hence, since (1) is a linear equation, it follows (see Exercise 12, Sec. 11.2) that the sum

$$y = f(x-at) + g(x+at) \tag{2}$$

is also a solution. In fact, it can be shown (see Exercise 9) that if f and g are arbitrary, twice-differentiable functions, then (2) is a *complete* solution of (1); that is, *any* solution of (1) can be expressed in the form (2).

This form of the solution of the wave equation is especially useful for revealing the significance of the parameter a and its dimensions of velocity. Suppose, specifically, that we consider the vibrations of a uniform string‡ stretching from $-\infty$ to ∞. If its transverse displacement is given by (2), we have in fact two waves traveling in opposite directions along the string, each with velocity a. Consider the function $f(x-at)$. At $t = 0$, it defines the curve $y = f(x)$, and at any later time $t = t_1$, it defines the curve $y = f(x - at_1)$. But these curves are identical except that the latter is translated to the right a distance equal to at_1. Thus the entire configuration moves along the string without distortion a distance of at_1 in t_1 units of time. The velocity with which the wave is propagated is therefore

$$v = \frac{at_1}{t_1} = a$$

Similarly, the function $g(x + at)$ defines a configuration which moves to the left

† Named for the French mathematician Jean Le Rond d'Alembert (1717–1783). The d'Alembert solution is actually not a special method but a special application of a general procedure known as the **method of characteristics.** Unfortunately, this cannot be applied with comparable simplicity to problems involving the heat equation and Laplace's equation, and so, despite its theoretical interest, we shall not discuss it here. An introduction to the theory can be found in Arnold Sommerfeld, "Partial Differential Equations in Physics," Academic, New York, 1949, pp. 36–43.

‡ The use of the string as an illustration is purely a matter of convenience, and *any* quantity satisfying the wave equation possesses the mathematical properties developed for the string.

along the string with constant velocity a. The total displacement of the string is, of course, the algebraic sum of these two traveling waves.

To carry the solution through in detail, let us suppose that the initial displacement of the string at any point x is given by $\phi(x)$ and that the initial velocity of the string at any point is $\theta(x)$. Then, as conditions to determine the form of f and g, we have, from (2) and its first derivative with respect to t,

$$y(x, 0) = \phi(x) = \Big[f(x - at) + g(x + at)\Big]_{t=0} = f(x) + g(x) \tag{3}$$

$$\left.\frac{\partial y}{\partial t}\right|_{x,0} = \theta(x) = \Big[-af'(x - at) + ag'(x + at)\Big]_{t=0} = -af'(x) + ag'(x) \tag{4}$$

Dividing Eq. (4) by a and then integrating with respect to x, we find

$$-f(x) + g(x) = \frac{1}{a}\int_{x_0}^{x} \theta(x)\, dx$$

Combining this with Eq. (3) and introducing the dummy variable s in the integrals, we obtain

$$f(x) = \frac{1}{2}\left[\phi(x) - \frac{1}{a}\int_{x_0}^{x} \theta(s)\, ds\right] \qquad g(x) = \frac{1}{2}\left[\phi(x) + \frac{1}{a}\int_{x_0}^{x} \theta(s)\, ds\right]$$

With the forms of f and g known, we can now write

$$\begin{aligned} y &= f(x - at) + g(x + at) \\ &= \left[\frac{\phi(x - at)}{2} - \frac{1}{2a}\int_{x_0}^{x-at} \theta(s)\, ds\right] + \left[\frac{\phi(x + at)}{2} + \frac{1}{2a}\int_{x_0}^{x+at} \theta(s)\, ds\right] \end{aligned}$$

or, combining the integrals,

$$y(x, t) = \frac{\phi(x - at) + \phi(x + at)}{2} + \frac{1}{2a}\int_{x-at}^{x+at} \theta(s)\, ds \tag{5}$$

Example 1 A string stretching to infinity in both directions is given the initial displacement

$$\phi(x) = \frac{1}{1 + 8x^2}\,\dagger$$

and released from rest. Determine its subsequent motion.

† The initial deflection curve $y = \phi(x)$ clearly violates assumption 4, Sec. 11.2, since at $x = -\frac{1}{4}$ (for instance), $\phi'(x) \equiv \tan \alpha = \frac{16}{9} = 1.78$ while $\sin \alpha = 0.87$. This difficulty can easily be overcome, however, by assuming instead of $\phi(x)$ a new deflection curve

$$\phi^*(x) = \frac{\phi(x)}{k}$$

where k is a sufficiently large constant, say $k = 10^{10}$. Using $\phi(x)$ instead of $\phi^*(x)$ in this and similar problems is just a convenient way of eliminating the constant factor $1/k$ at each step of our work.

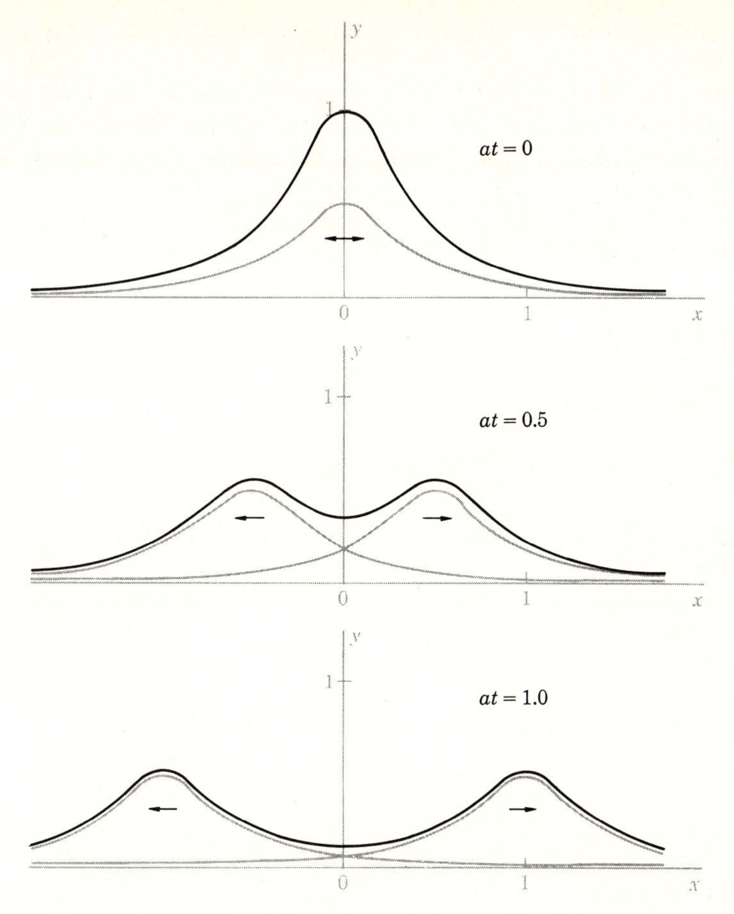

Figure 11.4 The propagation of a disturbance along a two-way infinite string.

Since $\theta(x) \equiv 0$, we have from (5) simply

$$y(x, t) = \frac{\phi(x - at) + \phi(x + at)}{2} = \frac{1}{2}\left[\frac{1}{1 + 8(x - at)^2} + \frac{1}{1 + 8(x + at)^2}\right]$$

The deflection of the string when $at = 0.0, 0.5$, and 1.0 is shown in Fig. 11.4. In this case the string does not vibrate but merely collapses back to its equilibrium position as the two traveling waves move off to infinity in opposite directions.

The motion of a semi-infinite string whose finite end is fixed can be regarded as the motion of one half of a two-way infinite string having a fixed point, or **node,** located at some finite point, say the origin. To capitalize on this fact, we need only imagine the actual string, stretching from 0 to ∞, to be extended in the opposite direction to $-\infty$. The initial conditions of displacement and velocity for the new portion of the string we define to be equal in magnitude but opposite in sign to those given for the actual string.† The solution for the resulting two-way infinite string can be written down at once, using Eq. (5). In the nature of the extended initial conditions, the displacement at the origin due to the wave traveling to the right from the left half of the string will always be equal but opposite in sign to the displacement at the origin due to the wave traveling to the left from the right half of the string. Hence the string will always remain at rest at the origin, and the solution for the right half of the extended string will be precisely the solution of the original problem.

Example 2 A semi-infinite string is given the displacement shown in Fig. 11.5*a* and released from rest. Determine its subsequent motion.

We first imagine the string extended to $-\infty$ and released from rest in the extended initial configuration shown in Fig. 11.5*b*. Since $\theta(x) \equiv 0$, we have from (5)

$$y(x, t) = \frac{\phi(x - at) + \phi(x + at)}{2}$$

where $\phi(x)$ is the displacement function shown in Fig. 11.5*b*.‡ We thus have two displacement waves, each of shape defined by $\frac{1}{2}\phi(x)$, one traveling to the right and one traveling to the left along the string. Plots of these waves are shown in Fig. 11.6. An inspection of these configurations reveals the important fact that a displacement wave is reflected from a fixed end without distortion but with reversal of sign.

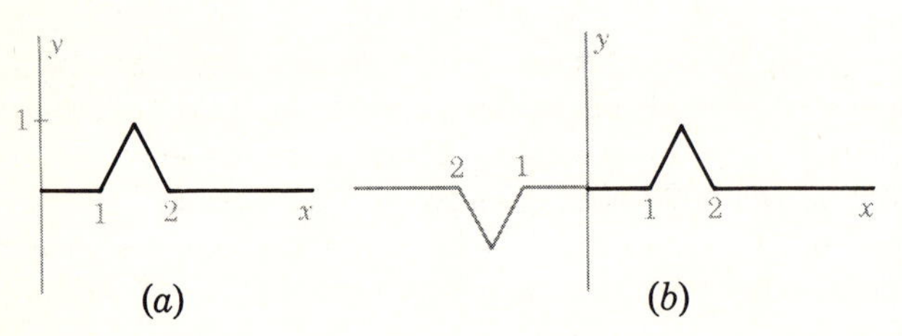

Figure 11.5 A semi-infinite string and its conceptual extension.

† This method of extending the initial conditions is sufficient but not necessary (see Exercise 5).

‡ If, as suggested by Fig. 11.5*a*, the graph of $\phi(x)$ has one or more corner points, then, strictly speaking, $\phi(x)$ does not describe an admissible initial displacement function. In fact, in the derivation of Eq. (5) both $f(x)$ and $g(x)$ were assumed to be twice differentiable, and therefore $\phi(x)$ must also be twice differentiable, which is not the case if there are points where the derivative of $\phi(x)$ is undefined. The apparent solutions obtained from Eq. (5) by overlooking this fact are therefore at best only formal solutions and are to be viewed with suspicion unless and until it is verified directly that they satisfy the given partial differential equation and its accompanying boundary and initial conditions. Questions concerning the existence and uniqueness of solutions of partial differential equations are quite difficult, and in our work we shall be concerned mainly with techniques for obtaining formal solutions. For an extended discussion of the problem of establishing the validity of solutions derived by purely formal means see, for instance, R. V. Churchill, "Fourier Series and Boundary Value Problems," 2d ed., McGraw-Hill, New York, 1963, pp. 126–63.

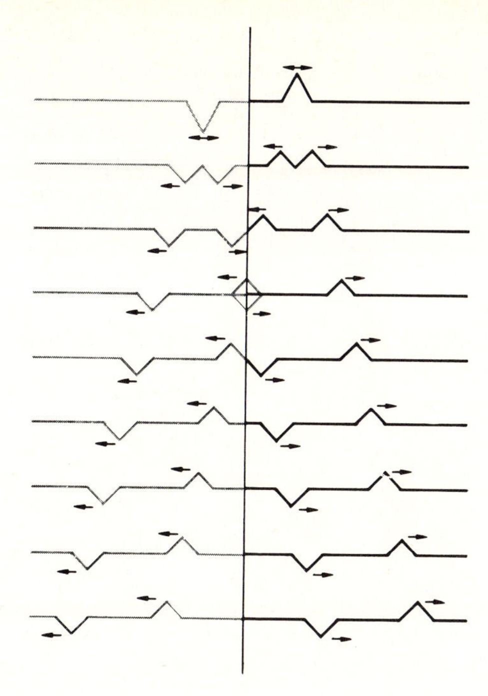

Figure 11.6 The propagation of a disturbance along a semi-infinite string.

The motion of a finite string can be obtained as the motion of a segment of an infinite string with suitably defined displacement and velocity. If the string is given between 0 and l, say, we first imagine that it is extended from 0 to $-l$ with initial conditions which are equal but opposite in sign to those for the actual string. Then we extend the string to infinity in each direction subject to initial conditions which duplicate with period $2l$ the initial configuration between $-l$ and l.† Finally, after we obtain the solution for the infinite string we have thus created, its behavior for $0 \leq x \leq l$ will be a complete description of the motion of the actual finite string in which we are really interested.

Example 3 A string of length l is given the initial displacement shown in Fig. 11.7 and released from rest. Determine its subsequent motion.

A suitable extension of the string and one-half cycle of its motion are shown in Fig. 11.8. An inspection of Fig. 11.8 shows that the period of the motion, i.e., the least time for its return to its initial state, is just the time for either of the traveling waves to traverse a distance $2l$. In other words, since the velocity of the waves is a, the period is

† This, of course, is essentially the procedure we used in Sec. 8.3 to obtain the half-range sine expansion of a function originally defined only over a finite portion of the real axis. The relation of Fourier series to the problem of the vibrating string, and to the solution of the wave equation in general, will become clear in the next section when we develop the method of separation of variables.

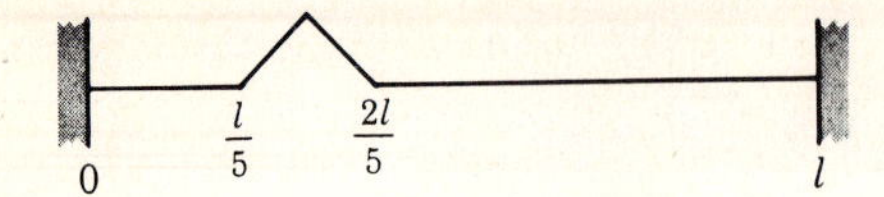

Figure 11.7 A finite string with initial displacement.

$2l/a$. The frequency of the vibrations is therefore $a/(2l)$. We shall encounter this formula again when we solve the wave equation by the method of separation of variables in the next section.

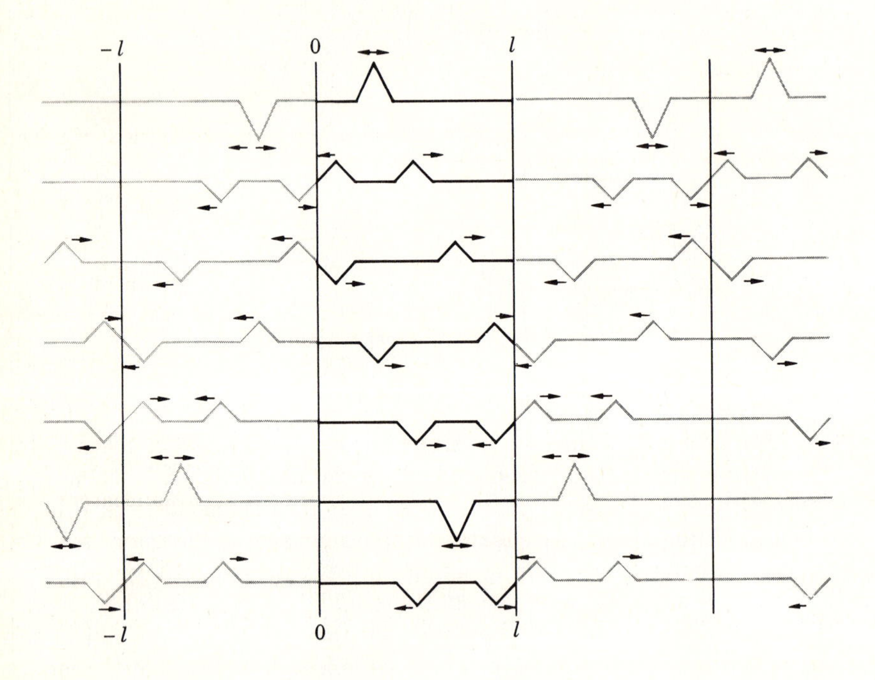

Figure 11.8 One-half cycle of the motion of a finite string.

Example 4 The initial displacement of a two-way infinite string is

$$y(x, 0) = \frac{1}{1 + x^2} \qquad x^2 < \infty$$

Show that if the string is given the proper initial velocity, its subsequent motion will consist solely of a wave traveling to the right.

By Eq. (5), the motion of the string is described by the equation

$$y(x, t) = \frac{\phi(x - at) + \phi(x + at)}{2} + \frac{1}{2a}[\Theta(x + at) - \Theta(x - at)]$$

where $\Theta(s)$ is any antiderivative of $\theta(s)$. In order that there should be no wave traveling to the left, it is necessary that the terms which describe such waves, that is, those terms which are functions of $x + at$, be missing from $y(x, t)$. Thus we must have

$$\frac{\phi(x + at)}{2} + \frac{1}{2a}\Theta(x + at) = 0$$

or, specifically,

$$\Theta(x + at) = -\frac{a}{1 + (x + at)^2}$$

Differentiating this with respect to $x + at$ gives us

$$\Theta'(x + at) = \theta(x + at) = \frac{2a(x + at)}{[1 + (x + at)^2]^2}$$

Hence if $\theta(x) = \dot{y}(x, 0) = 2ax/[1 + x^2]^2$, the motion of the string will consist of a single wave traveling to the right, as required.

Exercises for Section 11.3

1 A uniform string stretching from $-\infty$ to ∞ is given the initial displacement

$$y(x, 0) = \begin{cases} 1 - |x| & x^2 < 1 \\ 0 & x^2 \geq 1 \end{cases}$$

and released from rest. Find the displacement of the string as a function of x and t. What is the transverse velocity of the string at $x = 0$ as a function of t? *Hint:* It will be convenient to express the initial-displacement condition in terms of unit step functions (see Fig. 6.3).

2 A uniform string stretching from $-\infty$ to ∞ is given the initial displacement

$$y(x, 0) = \begin{cases} \cos x & x^2 < \pi^2/4 \\ 0 & x^2 \geq \pi^2/4 \end{cases}$$

and released from rest. Find its displacement as a function of x and t. What is its velocity at $x = 0$?

3 A uniform string stretching from $-\infty$ to ∞ while at rest in its equilibrium position is struck in such a way that the portion of the string between $x = -1$ and $x = 1$ is given a velocity of 1. Find its displacement as a function of x and t. What is its velocity at $x = 0$?

4 A uniform string stretching from 0 to ∞ is initially displaced to coincide with the curve $y = xe^{-x}$ and released from rest in that position. Find its displacement as a function of x and t. What is its velocity at $x = 1$?

5 A uniform string stretching from 0 to ∞ begins its motion with initial displacement $\phi(x)$ and initial velocity $\theta(x)$. Show that its motion can be found as the motion of the right half of a two-way infinite string provided merely that the initial displacement $\phi(-x)$ and the initial velocity $\theta(-x)$ for the negative extension of the string satisfy the condition

$$\phi(x) + \phi(-x) = -\frac{1}{a}\int_{-x}^{x} \theta(s)\, ds$$

6 If a semi-infinite string begins its motion with initial displacement $\phi(x) = (\sin x)/a$ and initial velocity $\theta(x) = 1$, and if the negative extension of the string is imagined to have the initial displacement $\phi(-x) = 0$, find the necessary initial velocity for the extended portion of the string.

7 The initial displacement of a two-way infinite string is $y(x, 0) = e^{-x^2}$. With what velocity must the string start to move if its subsequent motion is to consist solely of a wave traveling to the right? of a wave traveling to the left?

8 The initial velocity of a two-way infinite string is

$$y(x, 0) = \begin{cases} \sin x & x^2 < \pi^2 \\ 0 & x^2 \geq \pi^2 \end{cases}$$

From what initial displacement must the string start to move if its subsequent motion is to consist solely of a wave traveling to the right?

9 Show that under the substitutions $u = x - at$ and $v = x + at$, the equation

$$\frac{\partial^2 y}{\partial t^2} = a^2 \frac{\partial^2 y}{\partial x^2}$$

becomes $\partial^2 y/(\partial u\, \partial v) = 0$. Hence show that $y = f(x - at) + g(x + at)$ is the most general solution of the one-dimensional wave equation.

10 Discuss the possibility of finding solutions of the form $u = f(x + \lambda y)$ for the equation

$$A\frac{\partial^2 u}{\partial x^2} + 2B\frac{\partial^2 u}{\partial x\, \partial y} + C\frac{\partial^2 u}{\partial y^2} = 0 \qquad A, B, C \text{ constants}$$

and show that according as $B^2 - AC$ is greater than, equal to, or less than zero, there will be two, one, or no (real) values of λ for which such solutions exist. (The given equation is said to be **hyperbolic, parabolic,** or **elliptic** in the respective cases, and the nature of its solutions and their properties is significantly different in each case.)

11.4 SEPARATION OF VARIABLES

We are now ready to consider the solution of partial differential equations by the method of separation of variables. Although this method is not universally applicable, it suffices for most of the partial differential equations encountered in elementary applications and leads directly to the heart of the branch of mathematics which deals with *boundary-value problems.*

The idea behind the method is the familiar mathematical strategem of reducing a new problem to dependence upon an old one. In this case we attempt to convert the given partial differential equation into several ordinary differential equations, hoping that what we know about the latter will prove adequate for a successful continuation of the search for solutions.

To illustrate the details of the procedure, let us again consider the finite string, whose transverse displacements $y(x, t)$ we know satisfy the one-dimensional wave equation

$$\frac{\partial^2 y}{\partial t^2} = a^2 \frac{\partial^2 y}{\partial x^2}$$

As a working hypothesis, we assume that solutions for y exist as products of a function of x alone and a function of t alone:

$$y(x, t) = X(x)T(t)$$

If this is the case, partial differentiation of y with respect to either x or t amounts to total differentiation of one or the other of the factors of y, and we have

$$\frac{\partial^2 y}{\partial x^2} = X''T \qquad \text{and} \qquad \frac{\partial^2 y}{\partial t^2} = XT''$$

Substituting these into the wave equation, we obtain

$$XT'' = a^2 X''T$$

Formal division by the product XT (assuming that neither X nor T is identically zero) then gives

$$\frac{T''}{T} = a^2 \frac{X''}{X} \tag{1}$$

as a necessary condition that $y(x, t) = X(x)T(t)$ should be a solution.

Now the left member of (1) is clearly independent of x. Hence (in spite of its appearance) the right-hand side of (1) must also be independent of x, since it is identically equal to the expression on the left. Similarly, each member of (1) must be independent of t. Therefore, being independent of both x and t, each side of (1) must be a constant, say μ, and we can write

$$\frac{T''}{T} = a^2 \frac{X''}{X} = \mu$$

Thus the determination of solutions of the original partial differential equation has been reduced to the determination of solutions of the two ordinary differential equations

$$T'' = \mu T \qquad \text{and} \qquad X'' = \frac{\mu}{a^2} X$$

Assuming that we need consider only real values of μ, there are three cases to investigate:

$$\mu > 0 \qquad \mu = 0 \qquad \mu < 0$$

If $\mu > 0$, we can write $\mu = \lambda^2$ $(\lambda > 0)$. In this case the two differential equations and their solutions are

$$T'' = \lambda^2 T \qquad\qquad X'' = \frac{\lambda^2}{a^2} X$$

$$T = Ae^{\lambda t} + Be^{-\lambda t} \qquad\qquad X = Ce^{\lambda x/a} + De^{-\lambda x/a}$$

But a solution of the form

$$y(x, t) = X(x)T(t) = (Ce^{\lambda x/a} + De^{-\lambda x/a})(Ae^{\lambda t} + Be^{-\lambda t})$$

cannot describe the undamped vibrations of a system because it is not perodic, i.e., does not repeat itself periodically as time increases. Hence, although product solutions of the differential equation exist for $\mu > 0$, they have no significance in relation to the problem we are considering.

If $\mu = 0$, the equations and their solutions are

$$T'' = 0 \qquad X'' = 0$$

$$T = At + B \qquad X = Cx + D$$

But, again, a solution of the form

$$y(x, t) = X(x)T(t) = (Cx + D)(At + B)$$

cannot describe a periodic motion. Hence the alternative $\mu = 0$ must also be rejected.

Finally, if $\mu < 0$, we can write $\mu = -\lambda^2$ $(\lambda > 0)$. Then the component differential equations and their solutions are

$$T'' = -\lambda^2 T \qquad X'' = -\frac{\lambda^2}{a^2} X$$

$$T = A \cos \lambda t + B \sin \lambda t \qquad X = C \cos \frac{\lambda}{a} x + D \sin \frac{\lambda}{a} x$$

In this case the solution

$$y(x, t) = X(x)T(t) = \left(C \cos \frac{\lambda}{a} x + D \sin \frac{\lambda}{a} x\right)(A \cos \lambda t + B \sin \lambda t) \qquad (2)$$

is clearly periodic, repeating itself every time t increases by $2\pi/\lambda$. In other words, $y(x, t)$ represents a vibratory motion with period $2\pi/\lambda$ or frequency $\lambda/(2\pi)$.

It remains now to determine the value or values of λ and the constants A, B, C, and D by imposing suitable auxiliary conditions. Let us assume, specifically, that the ends of the string are fixed, so that their displacements are always zero. In other words, for all values of t, we have

$$y(0, t) = 0 \qquad \text{and} \qquad y(l, t) = 0$$

where l is the length of the string.

If the first of these conditions is imposed on the general solution (2), we have

$$0 = C(A \cos \lambda t + B \sin \lambda t)$$

This will be true, of course, if $A = B = 0$; but in this case the function T is identically zero and so, too, is the product $y(x, t) = X(x)T(t)$. This corresponds to the trivial solution in which the string remains permanently at rest. Therefore, since we are interested in solutions which describe vibratory motion, we must have $C = 0$.

When we impose the right-hand end condition, namely $y(l, t) = 0$ for all values of t (remembering that C is now zero), we obtain

$$0 = D \sin \frac{\lambda l}{a} (A \cos \lambda t + B \sin \lambda t)$$

Again, and for the same reason, we must reject the possibility that $A = B = 0$. Furthermore, we note that if $D = 0$, the solution is also trivial since, with C and D both zero, the function X and hence the product $y = X(x)T(t)$ are identically zero. The only possibility that can lead to a nontrivial solution capable of describing periodic motion is then

$$\sin \frac{\lambda l}{a} = 0$$

This will be true, and D will not have to be zero, if and only if

$$\frac{\lambda l}{a} = n\pi$$

From the continuous infinity of values of the parameter λ for which periodic product solutions of the wave equation exist, we have thus been forced to reject all but the values

$$\lambda_n = \frac{n\pi a}{l} \qquad n = 1, 2, 3, \ldots \tag{3}$$

These and only these values of λ (still infinite in number, however) yield solutions which, in addition to being periodic, also satisfy the end, or boundary, conditions of the problem at hand. In the language of Sec. 10.4, these are the *characteristic values* of the problem, and the equation from which we obtained them, namely, $\sin (\lambda l/a) = 0$, is the *characteristic equation.* With the product solutions which correspond to these admissible values of λ, we must now attempt to construct a solution which will satisfy the remaining conditions of the problem, namely, the initial conditions which assert that the string starts its motion at $t = 0$ with a known displacement $y(x, 0) = \phi(x)$ and a known velocity $\partial y/\partial t\,|_{x,0} = \theta(x)$ at every point x.

Now the wave equation is linear, and thus if we have several solutions, their sum is also a solution (see Exercise 12, Sec. 11.2). Hence, writing the solution associated with the nth value of λ in the form

$$\begin{aligned} y_n(x, t) &= \sin \frac{\lambda_n}{a} x \, (A_n \cos \lambda_n t + B_n \sin \lambda_n t) \\ &= \sin \frac{n\pi x}{l} \left(A_n \cos \frac{n\pi a t}{l} + B_n \sin \frac{n\pi a t}{l} \right)\dagger \end{aligned}$$

† The constants A and B now bear subscripts to indicate that they are not necessarily the same in the solutions associated with the different values of λ. The constant D which appears in the solution for $X(x)$ can, of course, be absorbed into the constants A and B and need not be explicitly included.

it is natural enough (though perhaps optimistic, in view of the questions of convergence that are raised) to ask if an *infinite* series of *all* the y_n's, say

$$y(x, t) = \sum_{n=1}^{\infty} y_n(x, t) = \sum_{n=1}^{\infty} \sin \frac{n\pi x}{l} \left(A_n \cos \frac{n\pi at}{l} + B_n \sin \frac{n\pi at}{l} \right) \tag{4}$$

can be made to yield a solution fitting the initial conditions of transverse displacement and velocity.

This can be done, and in fact in this case the determination of the coefficients A_n and B_n requires nothing more than a simple application of Fourier series, as developed in Chap. 8. For if we set $t = 0$ in $y(x, t)$, we obtain from Eq. (4) and the given initial-displacement condition the requirement that

$$y(x, 0) \equiv \phi(x) = \sum_{n=1}^{\infty} A_n \sin \frac{n\pi x}{l}$$

The problem of determining the A_n's so that this will be true is nothing but the problem of expanding a given function $\phi(x)$ in a half-range sine series over the interval $(0, l)$. Using Theorem 2, Sec. 8.3, we have explicitly

$$A_n = \frac{2}{l} \int_0^l \phi(x) \sin \frac{n\pi x}{l} \, dx$$

To determine the B_n's, we note, further, that

$$\frac{\partial y}{\partial t} = \sum_{n=1}^{\infty} \sin \frac{n\pi x}{l} \left(-A_n \sin \frac{n\pi at}{l} + B_n \cos \frac{n\pi at}{l} \right) \frac{n\pi a}{l} \tag{5}$$

Hence, putting $t = 0$, we have, from the initial velocity condition,

$$\left. \frac{\partial y}{\partial t} \right|_{x,0} \equiv \theta(x) = \sum_{n=1}^{\infty} \left(\frac{n\pi a}{l} B_n \right) \sin \frac{n\pi x}{l}$$

This, again, merely requires that the B_n's be determined so that the quantities

$$\frac{n\pi a}{l} B_n$$

will be the coefficients in the half-range sine expansion of the known function $\theta(x)$. Thus

$$\frac{n\pi a}{l} B_n = \frac{2}{l} \int_0^l \theta(x) \sin \frac{n\pi x}{l} \, dx \qquad \text{or} \qquad B_n = \frac{2}{n\pi a} \int_0^l \theta(x) \sin \frac{n\pi x}{l} \, dx$$

Aside from questions of convergence, our problem is now solved. We know that a uniform string of weight per unit length w, stretched under tension T to a length l, can vibrate at any of an infinite number of natural frequencies,

$$\frac{\lambda_n}{2\pi} = \frac{na}{2l} \text{ cycles/unit time}\dagger \qquad n = 1, 2, 3, \ldots; \ a^2 = \frac{Tg}{w}$$

† These, of course, are the frequencies we obtained in Sec. 10.5 through the use of the Green's function for a uniform string.

If and when the string vibrates at a single one of these frequencies, we know that its transverse displacement varies periodically between extreme values proportional to

$$\sin \frac{n\pi x}{l} \qquad 0 \le x \le l$$

Finally, assuming any initial conditions of displacement and velocity which satisfy the Dirichlet conditions (Sec. 8.2), we know how to determine, at least formally,† the actual displacement curve of the string as the superposition of an infinite series of the displacement curves associated with the respective natural frequencies $\lambda_n/2\pi$.

The functions $\sin(n\pi x/l)$, whose orthogonality properties make possible the series representation of the given initial conditions $y(x, 0) = \phi(x)$ and $\dot{y}(x, 0) = \theta(x)$, are the *characteristic functions* of the problem. The frequency $\lambda_1/(2\pi) = a/(2l)$ is the lowest natural frequency, i.e., the *fundamental tone* of the string. The other frequencies

$$\frac{\lambda_n}{2\pi} = \frac{na}{2l} \qquad n = 2, 3, 4, \ldots$$

are the *harmonics*, whose combined presence gives the string its total musical quality.

Example 1 A uniform string stretched between the points (0, 0) and (*l*, 0) is given the initial displacement

$$y(x, 0) = \begin{cases} x & 0 \le x \le l/2 \\ l - x & l/2 \le x \le l \end{cases}$$

and then released from rest. Find its subsequent displacement as a function of x and t.

From the discussion in the text we know that the displacement and velocity of the string are given by the series

$$y(x, t) = \sum_{n=1}^{\infty} \sin \frac{n\pi x}{l} \left(A_n \cos \frac{n\pi at}{l} + B_n \sin \frac{n\pi at}{l} \right) \tag{4}$$

$$\dot{y}(x, t) = \sum_{n=1}^{\infty} \sin \frac{n\pi x}{l} \left(-A_n \sin \frac{n\pi at}{l} + B_n \cos \frac{n\pi at}{l} \right) \frac{n\pi a}{l} \tag{5}$$

From the fact that the initial velocity is identically zero we have, by setting $t = 0$ in the series (5),

$$\dot{y}(x, 0) = \theta(x) \equiv 0 = \sum_{n=1}^{\infty} \left(\frac{n\pi a}{l} B_n \right) \sin \frac{n\pi x}{l}$$

† See the footnote to Example 2, Sec. 11.3.

where, from the properties of half-range sine expansions,

$$\frac{n\pi a}{l} B_n = \frac{2}{l}\int_0^l \theta(x) \sin \frac{n\pi x}{l}\, dx = \frac{2}{l}\int_0^l 0 \sin \frac{n\pi x}{l}\, dx = 0$$

Thus $B_n = 0$ for all values of n.

From the given initial displacement we have, similarly,

$$y(x, 0) = \phi(x) = \sum_{n=1}^{\infty} A_n \sin \frac{n\pi x}{l}$$

where
$$A_n = \frac{2}{l}\int_0^l \phi(x) \sin \frac{n\pi x}{l}\, dx$$

$$= \frac{2}{l}\int_0^{l/2} x \sin \frac{n\pi x}{l}\, dx + \frac{2}{l}\int_{l/2}^{l} (l - x) \sin \frac{n\pi x}{l}\, dx$$

$$= \frac{2}{l}\left[-\frac{xl}{n\pi} \cos \frac{n\pi x}{l} + \frac{l^2}{n^2\pi^2} \sin \frac{n\pi x}{l}\right]_0^{l/2}$$

$$+ \frac{2}{l}\left[-\frac{l^2}{n\pi} \cos \frac{n\pi x}{l}\right]_{l/2}^{l}$$

$$- \frac{2}{l}\left[-\frac{xl}{n\pi} \cos \frac{n\pi x}{l} + \frac{l^2}{n^2\pi^2} \sin \frac{n\pi x}{l}\right]_{l/2}^{l}$$

A straightforward evaluation and simplification of the last expression give

$$A_n = \frac{4l}{n^2\pi^2} \sin \frac{n\pi}{2} \qquad \text{or} \qquad A_n = \begin{cases} \dfrac{4l}{n^2\pi^2} & n = 1, 5, 9, \ldots \\ 0 & n = 2, 4, 6, \ldots \\ -\dfrac{4l}{n^2\pi^2} & n = 3, 7, 11, \ldots \end{cases}$$

The fact that A_n is zero for all even values of n tells us that when the string is set in motion in this particular way, all even-ordered harmonics are missing from its composite tone.

Exercises for Section 11.4

1 Verify that the solutions of the wave equation obtained in this section can all be written in the form

$$y(x, t) = f(x - at) + g(x + at)$$

as required by the d'Alembert theory.

2 Which of the following equations can be solved by the method of separation of variables?

(a) $a\dfrac{\partial^2 u}{\partial x\, \partial y} + bu = 0$ (b) $x^2\dfrac{\partial^2 u}{\partial x^2} + y\dfrac{\partial^2 u}{\partial y^2} = 0$

(c) $a\dfrac{\partial^2 u}{\partial x^2} + b\dfrac{\partial^2 u}{\partial x\, \partial y} + c\dfrac{\partial u}{\partial y} = 0$ (d) $a\dfrac{\partial^2 u}{\partial x^2} + b\dfrac{\partial^2 u}{\partial y^2} + c\dfrac{\partial^2 u}{\partial z^2} = 0$

(e) $a\dfrac{\partial^2 u}{\partial x^2} + b\dfrac{\partial^2 u}{\partial x\, \partial y} + c\dfrac{\partial^2 u}{\partial y^2} = 0$ (f) $a\dfrac{\partial^2 u}{\partial x^2} + b\dfrac{\partial^2 u}{\partial y^2} + c\dfrac{\partial u}{\partial x} + d\dfrac{\partial u}{\partial y} = 0$

*3 In Example 1, show that the displacement of the midpoint of the string varies between its successive maximum and minimum values as a linear function of at. *Hint:* After evaluating the displacement for $l/2$, recall the result of Example 2, Sec. 8.2.

4 While in its equilibrium position, a uniform string stretched between the points $(0, 0)$ and $(l, 0)$ is given the initial velocity

$$\dot{y}(x, 0) = \theta(x) = \begin{cases} ax/l & 0 \leqslant x \leqslant l/2 \\ a(l - x)/l & l/2 \leqslant x \leqslant l \end{cases}$$

Find its subsequent displacement as a function of x and t.

**5 While in its equilibrium position, a uniform string stretched between the points $(0, 0)$ and $(l, 0)$ is given the initial velocity

$$\dot{y}(x, 0) = \theta(x) = \begin{cases} 0 & 0 \leqslant x < (l - k)/2 \\ a/k & (l - k)/2 < x < (l + k)/2 \\ 0 & (l + k)/2 < x \leqslant l \end{cases}$$

Find its subsequent displacement as a function of x and t. Does your answer appear to have a meaningful limit as $k \to 0$? If so, to what problem do you think it is the answer?

6 A uniform string stretched between the points $(0, 0)$ and $(l, 0)$ is given the following initial displacement and initial velocity:

$$y(x, 0) = \phi(x) = \sin \pi x/l \qquad 0 \leqslant x \leqslant l$$

$$\dot{y}(x, 0) = \theta(x) = \begin{cases} 0 & 0 \leqslant x < l/4 \\ a & l/4 < x < 3l/4 \\ 0 & 3l/4 < x \leqslant l \end{cases}$$

Find its subsequent displacement as a function of x and t.

11.5 ONE-DIMENSIONAL HEAT FLOW

The analysis of the vibrating string which we undertook in the last section illustrates adequately the process of assuming a product solution and subsequently separating a partial differential equation into several ordinary differential equations. However, it does not illustrate the variety of boundary conditions that occur in other phenomena, nor does it suggest that other expansions besides half-range sine series may be required in order to satisfy given initial conditions. Accordingly, both for its own intrinsic interest and as an illustration of features not found in the vibrating string, we shall devote this section to a discussion of the one-dimensional flow of heat. Among other things, this will lead us to half-range cosine series, Fourier series which are neither half-range sine series nor half-range cosine series, and expansions which, strictly speaking, are not Fourier series of any kind.

The problem we propose to study is the flow of heat in a long rod of constant cross section whose curved surface is perfectly insulated and whose diameter is so small that with satisfactory accuracy temperature variations over any particular cross section may be neglected. If coordinates are chosen so that the rod extends along the x axis, say from $x = 0$ to $x = l$, it follows from these assumptions that the temperature varies only with x and t. Hence, in the general heat equation [Eq. (9), Sec. 11.2] the derivatives

$$\frac{\partial^2 u}{\partial y^2} \quad \text{and} \quad \frac{\partial^2 u}{\partial z^2}$$

are both zero, and therefore, under the assumption that there are no heat sources, the equation we have to solve is simply

$$\frac{\partial^2 u}{\partial x^2} = a^2 \frac{\partial u}{\partial t} \tag{1}$$

For heat flow in a thin rod there are three important kinds of end, or boundary, conditions:

1. *Constant temperature*, that is, the temperature at an end is maintained at a constant value at all times; or, analytically, $u = c$ at the end in question.
2. *Perfect insulation*, that is, heat flow through the end is zero at all times. Since heat will always flow if the temperature gradient is different from zero, it follows that $\partial u/\partial x = 0$ at a perfectly insulated end.
3. *Free escape*, that is, the rod loses heat by radiation into the surrounding air. It can be shown† that the analytic formulation of this condition is that at all times $\partial u/\partial x = \pm h(u - u_a)$ where h (> 0) is a thermal constant, u_a is the (constant) temperature of the air, and the minus sign applies at the right end, the plus sign at the left.

Since any one of these conditions may apply at either end of the rod, there are $3 \times 3 = 9$ possible combinations of these end conditions that might occur. We shall not investigate all these, but we shall consider examples involving end conditions of each type.

Regardless of the end conditions or the initial temperature distribution in the rod, to solve a problem in one-dimensional heat flow by the method of separation of variables, we must first assume a product solution

$$u(x, t) = X(x)T(t)$$

compute its partial derivatives

$$\frac{\partial^2 u}{\partial x^2} = X''T \quad \text{and} \quad \frac{\partial u}{\partial t} = XT'$$

substitute them into the heat equation (1), getting

$$X''T = a^2 XT'$$

† See the author's "Advanced Engineering Mathematics," 4th ed., McGraw-Hill, New York, 1975, pp. 352–53.

and then separate variables to obtain

$$\frac{X''}{X} = a^2 \frac{T'}{T}$$

Reasoning exactly as we did for the vibrating string in the last section, we observe next that the two fractions in the last equation must have a common constant value, say μ. This, in turn, means that we have the two ordinary differential equations

$$X'' = \mu X \qquad \text{and} \qquad T' = \frac{\mu}{a^2} T$$

Since the form of the solution of the first of these equations varies according as μ is greater than, equal to, or less than zero, we must now consider three cases:

1. $\mu > 0$, say $\mu = \lambda^2$ $(\lambda > 0)$
2. $\mu = 0$
3. $\mu < 0$, say $\mu = -\lambda^2$ $(\lambda > 0)$

In case 1 we have

$$X = Ae^{\lambda x} + Be^{-\lambda x} \qquad T = Ce^{\lambda^2 t/a^2}$$

This possibility must be rejected, however, because the form of T implies that as t becomes infinite, the factor T, and hence the product XT, increases exponentially for each x, which is clearly impossible in a physical system with no heat generation.

In case 2 we have

$$X = A + Bx \qquad T = C$$

and therefore, absorbing the constant C in the arbitrary constants A and B and then renaming them to avoid confusion with the solution in case 3,

$$u = XT = \frac{A_0}{2} + B_0 x \tag{2}$$

At the present stage there is no physical (or mathematical) reason for rejecting this possibility; in fact, with certain boundary conditions it becomes an essential part of the final solution.

In case 3 we have

$$X = A \cos \lambda x + B \sin \lambda x \qquad T = Ce^{-\lambda^2 t/a^2}$$

and (again absorbing C in A and B)

$$u = XT = (A \cos \lambda x + B \sin \lambda x)e^{-\lambda^2 t/a^2} \tag{3}$$

Here, also, there is no a priori reason to question the relevance of a solution of this type.

We have now proceeded as far as we can without information about the end conditions. In continuing we shall consider, in turn, four typical problems:

1. Each end of the rod is perfectly insulated, and the initial temperature distribution $u(x, 0) = f(x)$ is known.
2. The left end of the rod is maintained at the temperature $u(0, t) = 0$, the right end is perfectly insulated, and the initial temperature distribution $u(x, 0) = f(x)$ is known.
3. The left end of the rod is maintained at the temperature $u(0, t) = 50$, the right end is maintained at the temperature $u(l, t) = 100$, and the initial temperature distribution $u(x, 0) = f(x)$ is known.
4. The left end of the rod is maintained at the temperature $u(0, t) = 0$, the right end radiates freely into air of constant temperature zero, and the initial temperature distribution $u(x, 0) = f(x)$ is known.

Example 1 In the first problem, the insulated end conditions (see condition 2, above) imply that for the temperature function u, we have

$$\left.\frac{\partial u}{\partial x}\right|_{0,t} = 0 \qquad \text{and} \qquad \left.\frac{\partial u}{\partial x}\right|_{l,t} = 0$$

From (2) we obtain

$$\frac{\partial u}{\partial x} = B_0$$

and this will be zero at $x = 0$ and at $x = l$ if and only if $B_0 = 0$, leaving

$$u = u_0 = \frac{A_0}{2}$$

Similarly, from (3),

$$\frac{\partial u}{\partial x} = \lambda(-A \sin \lambda x + B \cos \lambda x)e^{-\lambda 2t/a2} \tag{4}$$

At $x = 0$ this becomes

$$\lambda B e^{-\lambda 2t/a2}$$

and for this to be zero, it is necessary that $B = 0$. To satisfy the right-hand end condition, we must have

$$-\lambda A(\sin \lambda l)e^{-\lambda 2t/a^2} = 0$$

If $A = 0$, the entire solution is trivial since B is already zero. Hence we must have

$$\sin \lambda l = 0 \qquad \lambda l = n\pi \qquad \text{and} \qquad \lambda_n = \frac{n\pi}{l} \qquad n = 1, 2, 3, \ldots$$

Therefore, returning to Eq. (3), we have the infinite family of product solutions

$$u_n(x, t) = A_n(\cos \lambda_n x)e^{-\lambda_n^2 t/a^2} = A_n\left(\cos \frac{n\pi x}{l}\right)e^{-n2\pi 2t/(a2l2)} \qquad n = 1, 2, 3, \ldots$$

In general, none of these solutions, by itself, will reduce to the given initial condition $u(x, 0) = f(x)$, $0 \le x \le l$ when $t = 0$. Hence, as we did in the string problem, we form an infinite series involving all of them (including $u_0 = A_0/2$, of course) in the hope that we may be able to make this series satisfy the initial condition. This gives us

$$u(x, t) = \frac{A_0}{2} + \sum_{n=1}^{\infty} u_n(x, t) = \frac{A_0}{2} + \sum_{n=1}^{\infty} A_n\left(\cos\frac{n\pi x}{l}\right)e^{-n^2\pi^2 t/(a^2 l^2)} \tag{5}$$

To determine A_n, we first set $t = 0$ in Eq. (5), getting

$$u(x, 0) = f(x) = \frac{A_0}{2} + \sum_{n=1}^{\infty} A_n \cos\frac{n\pi x}{l}$$

This we recognize as just the half-range cosine expansion of the given function $f(x)$. Hence, by Theorem 1, Sec. 8.3, we have

$$A_n = \frac{2}{l}\int_0^l f(x)\cos\frac{n\pi x}{l}\,dx \qquad n = 0, 1, 2, 3, \ldots$$

With the values of the A's determined, the formal solution of our problem is complete for any initial temperature distribution $u(x, 0) = f(x)$ which satisfies the Dirichlet conditions on $[0, l]$.

It is clear from Eq. (5) that as $t \to \infty$, every term in the series approaches zero because of the negative exponential factor. Thus at every point of the rod, the temperature $u(x, t)$ approaches the constant value $A_0/2$ as time goes on, which certainly accords with our expectation for a body which can neither gain nor lose heat.

Example 2 In the second problem the end conditions are

$$u(0, t) = 0 \qquad \text{and} \qquad \left.\frac{\partial u}{\partial x}\right|_{l,t} = 0$$

Imposing the first of these on the solution (2), we find $A_0 = 0$. Imposing the second on (2), we find $B_0 = 0$. Hence this particular product solution contributes nothing to the final result.

Imposing the first end condition on the solution (3), we obtain

$$Ae^{-\lambda^2 t/a^2} = 0$$

and therefore $A = 0$. Imposing the second end condition on the derivative (4), we find that

$$-\lambda B(\cos \lambda l)e^{-\lambda^2 t/a^2} = 0$$

If $B = 0$, the entire solution is trivial, since A is already zero; hence we must have

$$\cos \lambda l = 0 \qquad \lambda l = \frac{\pi}{2} + n\pi \qquad \text{and} \qquad \lambda_n = \frac{(2n+1)\pi}{2l} \qquad n = 0, 1, 2, \ldots$$

Corresponding to these characteristic values we therefore have from (3) the infinite family of product solutions

$$\begin{aligned} u_n(x, t) &= B_n \exp\left[-\frac{\lambda_n^2 t}{a^2}\right]\sin \lambda_n x \\ &= B_n \exp\left[-\frac{(2n+1)^2\pi^2 t}{4a^2 l^2}\right]\sin\frac{(2n+1)\pi x}{2l} \qquad n = 0, 1, 2, \ldots \end{aligned}$$

In general, none of these product solutions will reduce to the given initial condition $u(x, 0) = f(x)$ when $t = 0$. Hence, as before, we first combine them into an infinite series and then attempt to make the series itself satisfy the initial condition. This gives us

$$u(x, t) = \sum_{n=0}^{\infty} u_n(x, t) = \sum_{n=0}^{\infty} B_n \exp\left[-\frac{(2n+1)^2\pi^2 t}{4a^2 l^2}\right] \sin\frac{(2n+1)\pi x}{2l} \tag{6}$$

Setting $t = 0$ in this expression, we obtain

$$u(x, 0) = f(x) = \sum_{n=0}^{\infty} B_n \sin\frac{(2n+1)\pi x}{2l}$$

Obviously, this is not a half-range cosine expansion. On the other hand, because of the form of the angles in the various terms, it is not the usual half-range sine expansion. It is, however, the special sine expansion covered by Theorem 3, Sec. 8.3. Hence

$$B_n = \frac{2}{l}\int_0^l f(x) \sin\frac{(2n+1)\pi x}{2l}\,dx \qquad n = 0, 1, 2, \ldots$$

With the B's determined, the formal solution of this problem is now complete for any initial temperature distribution $u(x, 0) = f(x)$ which satisfies the Dirichlet conditions on $[0, l]$.

Since every term in the series (6) contains a negative exponential factor, it follows that the temperature throughout the rod approaches zero as t becomes infinite, regardless of the initial temperature distribution. This, of course, is what we expect as the heat "drains" out of the rod through the end which is kept at the temperature zero.

Example 3 In the third problem, the end conditions are

$$u(0, t) = 50 \qquad \text{and} \qquad u(l, t) = 100 \tag{7}$$

Imposing the first of these conditions on the solution (2), we find that $A_0/2 = 50$. Imposing the second condition on (2), we obtain

$$50 + B_0 l = 100 \qquad \text{or} \qquad B_0 = \frac{50}{l}$$

Hence we have

$$u_0(x, t) = 50 + \frac{50}{l}x$$

as one possible product solution satisfying the boundary conditions.

If, thoughtlessly, we also attempt to impose the end conditions (7) on the general product solution (3), it appears that we must have

$$50 = A \exp\left[-\frac{\lambda^2 t}{a^2}\right] \qquad \text{and} \qquad 100 = (A\cos\lambda l + B\sin\lambda l)\exp\left[-\frac{\lambda^2 t}{a^2}\right]$$

and, clearly, there are no values of the constants A and B for which these can hold for all values of t. A moment's reflection, however, should convince us that we actually want the product solutions given by (3) to be zero at $x = 0$ and at $x = l$. In fact, since we already have one solution which takes the value 50 at $x = 0$ and the value 100 at $x = l$, and since we must eventually form a series of this and all the other product solutions, it is clear that if all the latter are zero at each end of the rod, then at $x = 0$ and at $x = l$ the entire series will reduce to

$$u(0, t) = 50 + 0 + 0 + 0 + \cdots$$

$$u(l, t) = 100 + 0 + 0 + 0 + \cdots$$

as required.

Imposing the new conditions, namely $u(0, t) = 0$ and $u(l, t) = 0$, which we have found that the product solutions (3) must satisfy, we obtain

$$A \exp\left[-\frac{\lambda^2 t}{a^2}\right] = 0 \qquad \text{and} \qquad (A \cos \lambda l + B \sin \lambda l) \exp\left[-\frac{\lambda^2 t}{a^2}\right] = 0$$

The first of these implies that $A = 0$. The second implies that if we are to avoid a trivial solution, we must take

$$\sin \lambda l = 0 \qquad \text{and} \qquad \lambda_n = \frac{n\pi}{l} \qquad n = 1, 2, 3, \ldots$$

Using these characteristic values in (3), we obtain the family of product solutions

$$u_n(x, t) = B_n \exp\left[-\frac{n^2\pi^2 t}{a^2 l^2}\right] \sin \frac{n\pi x}{l} \qquad n = 1, 2, 3, \ldots$$

Hence, forming a series of all our product solutions, preparatory to imposing the initial temperature condition, we have

$$u(x, t) = \sum_{n=0}^{\infty} u_n(x, t) = 50 + \frac{50}{l}x + \sum_{n=1}^{\infty} B_n \exp\left[-\frac{n^2\pi^2 t}{a^2 l^2}\right] \sin \frac{n\pi x}{l} \tag{8}$$

Finally, putting $t = 0$ in this expression, we obtain

$$u(x, 0) = f(x) = 50 + \frac{50}{l}x + \sum_{n=1}^{\infty} B_n \sin \frac{n\pi x}{l}$$

or

$$f(x) - \left(50 + \frac{50}{l}x\right) = \sum_{n=1}^{\infty} B_n \sin \frac{n\pi x}{l}$$

The series we are dealing with here is just the half-range sine expansion not of $f(x)$ but of the difference $f(x) - [50 + (50/l)x]$. Hence, by Theorem 2, Sec. 8.3,

$$B_n = \frac{2}{l}\int_0^l \left[f(x) - \left(50 + \frac{50}{l}x\right)\right] \sin \frac{n\pi x}{l}\, dx$$

With the B's determined by this formula, Eq. (8) becomes the formal solution to our problem for any initial temperature distribution $u(x, 0) = f(x)$ which satisfies the Dirichlet conditions on $[0, l]$.

An inspection of Eq. (8) shows that as $t \to \infty$, the temperature in the rod approaches the linear function

$$u(x, t) = 50 + \frac{50}{l}x$$

as a limiting, steady-state condition.

Example 4 In our fourth and final problem in one-dimensional heat flow, the end conditions are (since $u_a = 0$)

$$u(0, t) = 0 \qquad \text{and} \qquad \left.\frac{\partial u}{\partial x}\right|_{l,t} = -hu$$

Imposing the first of these on the solution (2) gives $A_0/2 = 0$. Imposing the second gives

$$B = -h(Bl) \qquad \text{or} \qquad (1 + hl)B = 0$$

Since h and l are positive constants, $1 + hl$ can never be zero, and therefore $B = 0$. Thus no (nontrivial) solution of the form (2) can satisfy the end conditions of this problem.

When the left end condition is imposed on the product solution (3), we find, exactly as in Examples 2 and 3, that $A = 0$. To satisfy the right end condition, we must have

$$\lambda B \exp\left[-\frac{\lambda^2 t}{a^2}\right] \cos \lambda l = -hB \exp\left[-\frac{\lambda^2 t}{a^2}\right] \sin \lambda l$$

or, dividing out the nonzero exponential factor and collecting terms,

$$B(h \sin \lambda l + \lambda \cos \lambda l) = 0$$

If $B = 0$, the solution (3) becomes trivial, since A is already known to be zero. Hence, to obtain solutions of any significance in the problem, we must have

$$h \sin \lambda l + \lambda \cos \lambda l = 0 \qquad \text{or} \qquad \tan \lambda l = -\frac{\lambda}{h} = -\frac{\lambda l}{hl}$$

or finally

$$\tan z = -\alpha z$$

where

$$z = \lambda l \qquad \text{and} \qquad \alpha = \frac{1}{hl}$$

The equation $\tan z = -\alpha z$ is not like the simple equations

$$\sin \lambda l = 0 \qquad \text{and} \qquad \cos \lambda l = 0$$

which determined the admissible values of λ in Examples 1, 2, and 3, and its roots cannot be found by inspection. To determine them, it is convenient to consider the graphs of the two functions

$$y_1 = \tan z \qquad \text{and} \qquad y_2 = -\alpha z$$

The abscissas of the points of intersection of these curves (Fig. 11.9), being values of z for which $y_1 = y_2$, are then roots of the equation $\tan z = -\alpha z$. Clearly, there are infinitely many roots z_n. However, unlike the roots of $\sin \lambda l = 0$ and $\cos \lambda l = 0$, they are not evenly spaced, although as Fig. 11.9 indicates, the length of the interval between successive values of z_n *approaches* π as n becomes infinite.

From each root z_n, we obtain at once the corresponding value of λ,

$$\lambda_n = \frac{z_n}{l}$$

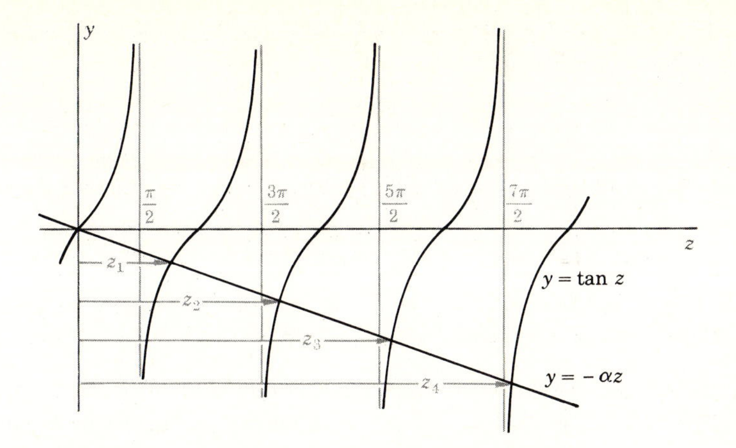

Figure 11.9 A graphical solution of the equation $\tan z = -\alpha z$.

and the corresponding product solution

$$u_n(x, t) = B_n \exp\left[-\frac{\lambda_n^2 t}{a^2}\right] \sin \lambda_n x$$

As in each of the preceding examples, we must now form an infinite series of these product solutions

$$u(x, t) = \sum_{n=1}^{\infty} u_n(x, t) = \sum_{n=1}^{\infty} B_n \exp\left[-\frac{\lambda_n^2 t}{a^2}\right] \sin \lambda_n x \tag{9}$$

and attempt to determine the constants B_n so that the function defined by this series will satisfy the initial temperature condition $u(x, 0) = f(x)$. Finally, putting $t = 0$ in (9), we find that this requires that

$$u(x, 0) = f(x) = \sum_{n=1}^{\infty} B_n \sin \lambda_n x \tag{10}$$

Because of the incommensurate angles of the various terms, the series (10) is not a Fourier series, and therefore the B_n's cannot be determined by the theory of Chap. 8. However, the functions $X_n(x) = \sin \lambda_n x$ are solutions of a characteristic value problem of the Sturm-Liouville type, and using this fact, we can determine the required expansion (10). Specifically, for the product solution $u(x, t) = X(x)T(t)$, the condition that $u(0, t) = 0$ for all values of t is equivalent to $X(0)T(t) = 0$, or simply

$$X(0) = 0 \tag{11}$$

Likewise, the other end condition, namely

$$\left.\frac{\partial u}{\partial x}\right|_{l,t} = -hu(l, t)$$

is equivalent to $X'(l)T(t) = -hX(l)T(t)$, or simply

$$hX(l) + X'(l) = 0 \tag{12}$$

Moreover, the values $\{\lambda_n\}$ which we determined above are just the values of the parameter λ for which the differential equation

$$X'' + \lambda^2 X = 0 \tag{13}$$

has nontrivial solutions X satisfying (11) and (12). Thus Eq. (13) and the boundary conditions (11) and (12) constitute a characteristic value problem covered by the Sturm-Liouville theorem (Theorem 1, Sec. 10.4) with

$$r(x) = 1 \qquad q(x) = 0 \qquad p(x) = 1$$

$$a = 0 \qquad b = l$$

$$a_1 = 1 \qquad a_2 = 0 \qquad b_1 = h \qquad b_2 = -1$$

and λ^2 and X written in place of λ and y. Hence the characteristic functions $X_n(x) = \sin \lambda_n x$ are orthogonal with respect to the weight function $p(x) = 1$ on the interval $(0, l)$.

Now that the orthogonality of the functions $\sin \lambda_n x$ has been established, the coefficients B_n can be determined by the same process that we used to derive the Euler formulas in Sec. 8.2 and used in Example 1, Sec. 10.4 to expand the function $4 - x^2$ in a series of Bessel functions. Thus we multiply both members of the series (10) by $\sin \lambda_n x$ and integrate from $x = 0$ to $x = l$. Then, noting that because of the orthogonality of the characteristic functions, every integral but one on the right-hand side is zero, we have

$$\int_0^l f(x) \sin \lambda_n x \, dx = B_n \int_0^l \sin^2 \lambda_n x \, dx \tag{14}$$

The integral on the right can be evaluated by Theorem 2, Sec. 10.4, but it is easier to evaluate it directly, as follows:

$$\int_0^l \sin^2 \lambda_n x \, dx = \int_0^l \frac{1 - \cos 2\lambda_n x}{2} \, dx = \frac{x}{2} - \frac{\sin 2\lambda_n x}{4\lambda_n}\Bigg|_0^l = \frac{l}{2} - \frac{\sin 2\lambda_n l}{4\lambda_n}$$

Now the values of $\lambda_n l \equiv z_n$ satisfy the characteristic equation $\tan z = -\alpha z$ or $\sin z = -\alpha z \cos z$. Hence, using the double-angle sine formula and then substituting for $\sin z_n$, we find for the value of the integral

$$\frac{l}{2} - \frac{\sin \lambda_n l \cos \lambda_n l}{2\lambda_n} = \frac{l}{2}\left(1 - \frac{\sin z_n \cos z_n}{z_n}\right) = \frac{l}{2}(1 + \alpha \cos^2 z_n)$$

Therefore, from (14)

$$B_n = \frac{2}{l(1 + \alpha \cos^2 z_n)} \int_0^l f(x) \sin \lambda_n x \, dx$$

With the coefficients in the series (9) determined so that the expansion (10) is correct, the last condition of the problem is met, and our solution is complete.

Exercises for Section 11.5

1 Evaluate the integral on the right in (14) by means of Theorem 2, Sec. 10.4.

2 In Example 4, verify by direct integration that the functions $\sin \lambda_n x$ are orthogonal on $(0, l)$.

***3** In Example 4, compute the values of z_1 and z_2 correct to two decimal places, assuming that $\alpha = 1$. Using these values, compute B_1 and B_2 if $f(x) = x$.

4 Complete Example 1 by finding the value of A_n for the initial temperature distribution $u(x, 0) = 100$.

5 Complete Example 2 by finding the value of B_n if $u(x, 0) = x$.

6 Work Example 3 if the left end is maintained at the constant temperature $u(0, t) = 50$, the right end is maintained at the constant temperature $u(l, t) = 0$, and $u(x, 0) = 25$.

7 Work Example 3 if both ends of the rod are maintained at the constant temperature 50 and $u(x, 0) = 0$.

****8** If both ends of a thin rod of length l radiate freely into air of constant temperature $u_a = 0$, find the characteristic equation of the problem and show that it has infinitely many roots. What are the characteristic functions of the problem? Are they orthogonal?

9 A thin rod of length l is initially at the temperature $u(x, 0) = 100$ throughout. At $t = 0$ the temperature at each end is suddenly reduced to zero and maintained at that value thereafter. Find the temperature distribution in the rod at any subsequent time.

****10** In Exercise 9, find the temperature at the midpoint of the rod and show that for arbitrarily small positive values of t it is different from 100. What does this appear to say about the velocity with which thermal disturbances are propagated? Do you agree with this conclusion? *Hint:* Consider $u(l/2, t)$ as a power series in the quantity $w = \exp[-\pi^2 t/(a^2 l^2)]$ and recall that a power series cannot converge to a constant over any interval unless it converges to that same constant at *every* point of its interval of convergence.

11.6 FURTHER APPLICATIONS

Boundary-value problems leading to the expansion of a given function in terms of a set of orthogonal functions arise in a great variety of physical problems. We shall conclude this chapter by examining several of these, two dealing with two-dimensional heat flow and two involving vibrating systems.

Example 1 A thin sheet of metal coincides with the square in the xy plane whose vertices are (0, 0), (1, 0), (1, 1), and (0, 1). The two faces of the sheet are perfectly insulated, so that heat flow in the sheet is purely two-dimensional. The edges parallel to the y axis are maintained at the constant temperature zero, the lower edge is perfectly insulated, and along the upper edge the temperature distribution $u(x, 1) = f(x)$ is maintained. Find the steady-state temperature distribution $u(x, y)$ in the sheet.

Since we are asked to find the *steady-state* temperature in the sheet, it follows that $\partial u/\partial t \equiv 0$, and our problem is to solve Laplace's equation

$$\frac{\partial^2 u}{\partial x^2} + \frac{\partial^2 u}{\partial y^2} = 0 \qquad (1)$$

subject to the boundary conditions

$$u(0, y) = u(1, y) = 0 \qquad \left.\frac{\partial u}{\partial y}\right|_{x,0} = 0 \qquad u(x, 1) = f(x)$$

Assuming a product solution $u(x, y) = X(x)Y(y)$, substituting into Eq. (1), and separating variables, we obtain

$$\frac{X''}{X} = -\frac{Y''}{Y} \tag{2}$$

As before, we reason that these two fractions must have a common constant value, say μ, which, as far as we know at the outset, may be positive, zero, or negative.

If $\mu > 0$, say $\mu = \lambda^2$ $(\lambda > 0)$, we have from (2)

$$X'' = \lambda^2 X \qquad Y'' = -\lambda^2 Y$$

$$X = A \cosh \lambda x + B \sinh \lambda x \qquad Y = C \cos \lambda y + D \sin \lambda y$$

Considering first the conditions along the vertical edges, because they appear to be simpler and more explicit, we have along the left edge $u(0, y) = X(0)Y(y) = AY(y) = 0$. Unless Y is identically zero, this will be true only if $A = 0$. From the condition along the right edge, we have $u(1, y) = X(1)Y(y) = B \sinh \lambda \; Y(y) = 0$ which implies that $B = 0$. Hence no (nontrivial) solution arising from the possibility that $\mu > 0$ can satisfy the left- and right-hand edge conditions.

If $\mu = 0$, we have from (2)

$$X'' = 0 \qquad Y'' = 0$$

$$X = A + Bx \qquad Y = C + Dy$$

Again imposing the conditions which must hold along the vertical edges, we have $u(0, y) = X(0)Y(y) = AY(y) = 0$, which implies that $A = 0$, and $u(1, y) = X(1)Y(y) = BY(y) = 0$, which implies that $B = 0$. Thus in this case also, we have only a trivial solution.

If $\mu < 0$, say $\mu = -\lambda^2$ $(\lambda > 0)$, we have from (2)

$$X'' = -\lambda^2 X \qquad Y'' = \lambda^2 Y$$

$$X = A \cos \lambda x + B \sin \lambda x \qquad Y = C \cosh \lambda y + D \sinh \lambda y$$

The left edge condition requires that $u(0, y) = X(0)Y(y) = AY(y) = 0$, and therefore $A = 0$. The right edge condition requires that $u(1, y) = X(1)Y(y) = B \sin \; \lambda \; Y(y) = 0$. If $B = 0$, the entire solution is trivial, since A is already zero. Hence we must have

$$X(1) = \sin \lambda = 0$$

which yields the characteristic values

$$\lambda_n = n\pi \qquad n = 1, 2, 3, \ldots$$

Thus, at this stage of the problem, the relevant product solutions have been restricted to the form

$$u_n(x, y) = \sin n\pi x(C_n \cosh n\pi y + D_n \sinh n\pi y)$$

where the constant B has been absorbed in the arbitrary constants C_n and D_n.

To satisfy the condition of perfect insulation along the lower edge, we must have

$$\left.\frac{\partial u}{\partial y}\right|_{x,0} = X(x)Y'(0) = \sin n\pi x(n\pi D_n) = 0$$

which implies that $D_n = 0$. Thus when $D_n = 0$, the product solutions

$$u_n(x, y) = C_n \sin n\pi x \cosh n\pi y \qquad n = 1, 2, 3, \ldots \tag{3}$$

satisfy the given boundary conditions along the vertical edges and the lower edge. In general, however, they will not satisfy the nonconstant temperature condition along the upper edge.

Guided by our past experience, we now form a series of the product solutions (3), getting

$$u(x, y) = \sum_{n=1}^{\infty} u_n(x, y) = \sum_{n=1}^{\infty} C_n \sin n\pi x \cosh n\pi y \tag{4}$$

and attempt to determine the C_n's so that the function defined by this series will reduce to the given temperature distribution along the upper edge. This requires that

$$u(x, 1) = f(x) = \sum_{n=1}^{\infty} (C_n \cosh n\pi) \sin n\pi x$$

which we recognize as just the half-range sine expansion of the given function $f(x)$ over the interval (0, 1). Therefore, by Theorem 2, Sec. 8.3,

$$C_n \cosh n\pi = \int_0^1 f(x) \sin n\pi x \, dx \qquad \text{and} \qquad C_n = \frac{1}{\cosh n\pi} \int_0^1 f(x) \sin n\pi x \, dx$$

With the C_n's determined, the series (4) is the (formal) solution to our problem.

Example 2 A circular sheet of metal of radius b has its two faces perfectly insulated, so that heat flow in the sheet is purely two-dimensional. At $t = 0$, with the temperature throughout the sheet equal to 100, the temperature around the edge of the sheet is suddenly reduced to zero and maintained thereafter at that value. Find the temperature distribution in the sheet at any subsequent time.

From the statement of the problem, it is clear that the temperature in the sheet will vary with time, as heat is lost from the hot interior through the cold boundary. Thus it appears that we must solve the two-dimensional (non-steady-state) heat equation

$$\frac{\partial^2 u}{\partial x^2} + \frac{\partial^2 u}{\partial y^2} = a^2 \frac{\partial u}{\partial t} \tag{5}$$

However, because of the curved boundary and its relatively complicated cartesian equation $x^2 + y^2 = b^2$, this problem is essentially impossible to solve in terms of the cartesian coordinates which appear in Eq. (5). Instead, we shall work in polar coordinates, where the equation of the boundary is simply $r = b$. Naturally, this will involve solving the partial differential equation into which Eq. (5) is transformed by the familiar substitutions $x = r \cos \theta$ and $y = r \sin \theta$. By a straightforward but moderately complicated application of standard techniques of partial differentiation, it can be shown that in polar coordinates Eq. (5) becomes

$$\frac{\partial^2 u}{\partial r^2} + \frac{1}{r}\frac{\partial u}{\partial r} + \frac{1}{r^2}\frac{\partial^2 u}{\partial \theta^2} = a^2 \frac{\partial u}{\partial t} \tag{6}$$

Hence, our problem is to solve this equation, subject to the boundary condition

$u(b, \theta, t) = 0$, the initial condition $u(r, \theta, 0) = 100$, and, of course, the condition that u remain finite at all points of the sheet.

Our first observation is that this problem possesses circular symmetry, for both the boundary condition and the initial condition are independent of θ, and hence the temperature distribution throughout the sheet will depend only on r. Thus

$$\frac{\partial^2 u}{\partial \theta^2} \equiv 0$$

and the equation we must solve is simply

$$\frac{\partial^2 u}{\partial r^2} + \frac{1}{r}\frac{\partial u}{\partial r} = a^2 \frac{\partial u}{\partial t} \tag{7}$$

subject to the boundary condition $u(b, t) = 0$ and the initial condition $u(r, 0) = 100$.

As usual, we begin by assuming a product solution

$$u(r, t) = R(r)T(t)$$

substituting it into Eq. (7), and separating variables:

$$R''T + \frac{1}{r}R'T = a^2 RT'$$

$$\frac{R''}{R} + \frac{1}{r}\frac{R'}{R} = a^2 \frac{T'}{T} \tag{8}$$

If the common value of the two members of Eq. (8) is a positive constant, say λ^2, the differential equation for the factor T is

$$T' = \frac{\lambda^2}{a^2} T$$

which predicts an exponential increase in temperature as t increases. Since this is impossible, this case must be rejected.

If the common value of the expressions in Eq. (8) is zero, we have the equations

$$\frac{R''}{R} + \frac{1}{r}\frac{R'}{R} = 0 \qquad \text{or} \qquad R'' + \frac{1}{r}R' = 0 \qquad \text{and} \qquad T' = 0$$

The equation for R is a separable, ordinary differential equation† which can be solved at once:

$$\frac{dR'}{R'} + \frac{dr}{r} = 0 \qquad \ln R' + \ln r = \ln A \qquad R'r = A$$

$$dR = \frac{A}{r}\,dr \qquad R = A \ln r + B$$

Since the temperature must be bounded at the center of the sheet, where $r = 0$, it follows that $A = 0$. Then, to satisfy the condition that the temperature on the boundary is zero, it is necessary that $B = 0$. Hence, only a trivial solution results.

† It can also be converted into an Euler-Cauchy equation (see Sec. 4.5) by multiplying through by r^2.

If the separation constant in Eq. (8) is a negative constant, say $-\lambda^2$ ($\lambda > 0$), we have the two ordinary differential equations

$$\frac{R''}{R} + \frac{1}{r}\frac{R'}{R} = -\lambda^2 \qquad \text{or} \qquad r^2R'' + rR' + \lambda^2 r^2 R = 0 \qquad \text{and} \qquad T' = -\frac{\lambda^2}{a^2}T$$

The solution for T is, of course,

$$T = Ce^{-\lambda^2 t/a^2} \tag{9}$$

The equation for R is just Bessel's equation of order zero, with parameter λ, and therefore the solution for R is

$$R = AJ_0(\lambda r) + BY_0(\lambda r) \tag{10}$$

Since $Y_0(\lambda r)$ is unbounded in the neighborhood of $r = 0$, which of course is part of the metal sheet, it is necessary that $B = 0$. To meet the boundary condition that $u(b, t) = R(b)T(t) = AJ_0(\lambda b)T(t) = 0$, it is necessary that

$$J_0(\lambda b) = 0$$

Thus the characteristic values of the problem, $\lambda_1, \lambda_2, \lambda_3, \ldots$, are determined by the positive roots of the equation†

$$J_0(z) = 0$$

via the relation

$$\lambda_n = \frac{z_n}{b}$$

and from these values we obtain from (9) and (10) the infinite family of product solutions

$$u_n(r, t) = R_n(r)T_n(t) = A_n J_0(\lambda_n r) \exp(-\lambda_n^2 t/a^2) \qquad n = 1, 2, 3, \ldots \tag{11}$$

Our last major step is to form a series of the product solutions (11),

$$u(r, t) = \sum_{n=1}^{\infty} u_n(r, t) = \sum_{n=1}^{\infty} A_n J_0(\lambda_n r) \exp(-\lambda_n^2 t/a^2) \tag{12}$$

and determine the A_n's so that this series will reduce to the given initial temperature distribution $u(r, 0) = 100$ when $t = 0$. Setting $t = 0$ in Eq. (12), we see that this requires that

$$100 = \sum_{n=1}^{\infty} A_n J_0(\lambda_n r) \tag{13}$$

To determine the A_n's, we first recall (Theorem 1, Sec. 10.4) that solutions of Bessel's equation are orthogonal with respect to r over the interval $(0, b)$ provided they are bounded at $r = 0$ and satisfy a condition of the form

$$b_1 R(b) - b_2 R'(b) = 0$$

† See, for instance, Eugene Jahnke, Fritz Emde, and Friedrich Losch, "Tables of Higher Functions," 6th ed., McGraw-Hill, New York, 1960, p. 192; or "Handbook of Mathematical Functions," GPO, Washington, D.C., 1965, p. 409.

at $r = b$. Since these conditions are met in our problem, we next multiply the series (13) by $rJ_0(\lambda_n r)$ and integrate from $r = 0$ to $r = b$. Because of the property of orthogonality, all but one of the integrals on the right are zero, and we have

$$\int_0^b 100rJ_0(\lambda_n r)\,dr = A_n \int_0^b rJ_0^2(\lambda_n r)\,dr \tag{14}$$

For the integral on the left in (14), we have

$$\begin{aligned}\int_0^b 100rJ_0(\lambda_n r)\,dr &= \frac{100}{\lambda_n^2}\int_0^b (\lambda_n r)J_0(\lambda_n r)\,d(\lambda_n r)\\ &= \frac{100}{\lambda_n^2}[(\lambda_n r)J_1(\lambda_n r)]_0^b \qquad \text{[by Eq. (6), Sec. 9.7]}\\ &= \frac{100b}{\lambda_n}J_1(\lambda_n b)\end{aligned}$$

For the integral on the right in (14), we have, by Corollary 1, Theorem 2, Sec. 10.4, and the identity $dJ_0(\lambda_n x)/dx = -\lambda_n J_1(\lambda_n x)$,

$$\frac{b^2}{2}J_1^2(\lambda_n b)$$

Hence, from (14),

$$A_n = \frac{200}{b\lambda_n J_1(\lambda_n b)} \qquad n = 1, 2, 3, \ldots$$

With the values of A_n determined, Eq. (12) provides the solution to our problem.

Example 3 What are the natural frequencies of a uniform circular membrane of radius b, fixed along its circumference, and vibrating in such a way that its deflection surface possesses circular symmetry?

Since we are concerned with the free vibration of a uniform membrane, our first thought is that we must solve the two-dimensional wave equation, Eq. (6), Sec. 11.2,

$$a^2\left(\frac{\partial^2 z}{\partial x^2} + \frac{\partial^2 z}{\partial y^2}\right) = \frac{\partial^2 z}{\partial t^2}$$

However, for the same reasons that led us to use polar coordinates in the last example, we must also use them here. Since the derivatives involving the space variables x and y are the same in the wave equation and Laplace's equation, it follows from the result cited in Example 2 that in polar coordinates the two-dimensional wave equation is

$$a^2\left(\frac{\partial^2 z}{\partial r^2} + \frac{1}{r}\frac{\partial z}{\partial r} + \frac{1}{r^2}\frac{\partial^2 z}{\partial \theta^2}\right) = \frac{\partial^2 z}{\partial t^2}$$

However, in the present problem we are told that the deflection function z possesses circular symmetry, so that z is independent of θ, just as it was in Example 2. Therefore

$$\frac{\partial^2 z}{\partial \theta^2} \equiv 0$$

and we need be concerned only with the simpler equation

$$a^2\left(\frac{\partial^2 z}{\partial r^2}+\frac{1}{r}\frac{\partial z}{\partial r}\right)=\frac{\partial^2 z}{\partial t^2} \tag{15}$$

Our first step, of course, is to assume a product solution $z(r, t) = R(r)T(t)$, substitute it into Eq. (15), and separate variables:

$$a^2\left(R''T+\frac{1}{r}R'T\right)=RT''$$

$$\frac{R''}{R}+\frac{1}{r}\frac{R'}{R}=\frac{1}{a^2}\frac{T''}{T}$$

If the common value of the two members of the last equation is positive or zero, the solutions of the resulting equation for the factor T are nonoscillatory and therefore incapable of describing vibratory motion. Hence we must suppose that the separation constant is negative, say $-\lambda^2$ $(\lambda > 0)$. Thus we have

$$\frac{R''}{R}+\frac{1}{r}\frac{R'}{R}=-\lambda^2 \qquad T''=-a^2\lambda^2 T$$

$$r^2R''+rR'+\lambda^2r^2R=0$$

$$R=AJ_0(\lambda r)+BY_0(\lambda r) \qquad T=C\cos a\lambda t+D\sin a\lambda t$$

and $$z(r,t)=R(r)T(t)=[AJ_0(\lambda r)+BY_0(\lambda r)][C\cos a\lambda t+D\sin a\lambda t]$$

Since $Y_0(\lambda r)$ is unbounded in the neighborhood of $r = 0$, whereas the deflection z is necessarily finite at all points of the membrane, it follows that $B = 0$. Furthermore, since the membrane is obviously held in place around its edge, where $r = b$, it follows that

$$z(b,t)=AJ_0(\lambda b)(C\cos a\lambda t+D\sin a\lambda t)=0$$

If $A = 0$, the entire solution is trivial and all we have is a description of the membrane remaining permanently at rest. Hence we must have

$$J_0(\lambda b)=0$$

Corresponding to the roots $z_1 \doteq 2.4048$, $z_2 \doteq 5.5201$, $z_3 \doteq 8.6537, \ldots$ of the equation $J_0(z) = 0$, we thus have the values

$$\lambda_1=\frac{z_1}{b}, \qquad \lambda_2=\frac{z_2}{b}, \qquad \lambda_3=\frac{z_3}{b}, \ldots$$

For each of these, *and for no others*, the wave equation has periodic solutions which are independent of θ, bounded in the neighborhood of $r = 0$, and zero on the circle $r = b$. From the time factor

$$T=C\cos a\lambda t+D\sin a\lambda t$$

it is clear, finally, that the natural frequencies at which vibrations can occur are

$$\omega_n=\frac{a\lambda_n}{2\pi}=\frac{az_n}{2\pi b} \qquad \text{cycles per unit time}$$

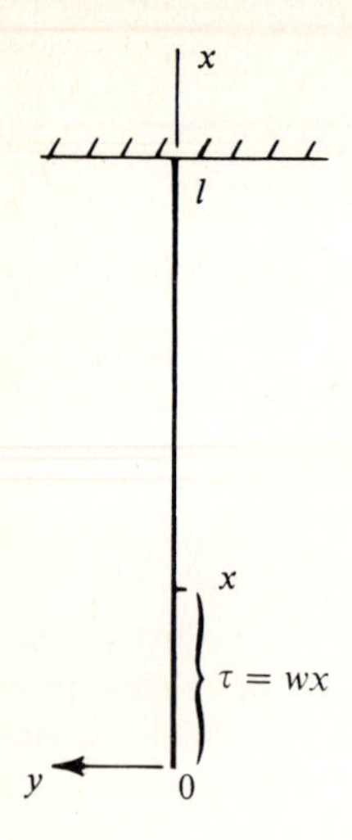

Figure 11.10 A hanging cable.

Example 4 A uniform, perfectly flexible cable of weight per unit length w is suspended from a frictionless support, as shown in Fig. 11.10. What are the frequencies at which it can vibrate if it is displaced slightly from its equilibrium position?

This is essentially the problem of the vibrating string which we discussed in Sec. 11.2 and again in Secs. 11.3 and 11.4, with two important differences. Here, instead of being constant, the tension τ at a general point of the cable is equal to the weight wx of the cable below that point, and now one end of the cable is free, whereas before both ends were fixed. Hence, because $\tau = wx$ is variable, Eq. (1), Sec. 11.2, becomes in the limit

$$\frac{w}{g}\frac{\partial^2 y}{\partial t^2} = \frac{\partial[wx(\partial y/\partial x)]}{\partial x} \tag{16}$$

As usual, we assume a product solution $y(x, t) = X(x)T(t)$ and attempt to separate variables. Thus, substituting, we have

$$\frac{w}{g}T''X = w\frac{\partial(xX'T)}{\partial x} = wT(xX')' \qquad \text{or} \qquad \frac{(xX')'}{X} = \frac{T''}{gT}$$

The common value of these two fractions must be a negative constant, say $-\lambda^2$ $(\lambda > 0)$, for otherwise T will not be a periodic function of the time t, as we know it must. Hence for the factor T we have

$$T'' = -\lambda^2 gT \qquad \text{and} \qquad T = C\cos\lambda\sqrt{g}\,t + D\sin\lambda\sqrt{g}\,t \tag{17}$$

For the factor X we have the differential equation

$$(xX')' + \lambda^2 X = 0 \tag{18}$$

From (18), using Corollary 1, Theorem 1, Sec. 9.6, with

$$r = 1 \qquad a = \lambda^2 \qquad b = 0 \qquad s = 0 \qquad \alpha = 0 \qquad \gamma = \tfrac{1}{2}$$

$$\lambda(\text{of the corollary}) = 2\lambda[\text{of Eq. (18)}] \qquad \nu = 0$$

we obtain the solution

$$X = AJ_0(2\lambda\sqrt{x}) + BY_0(2\lambda\sqrt{x})$$

Therefore

$$y(x, t) = X(x)T(t) = [AJ_0(2\lambda\sqrt{x}) + BY_0(2\lambda\sqrt{x})][C \cos \lambda\sqrt{g}t + D \sin \lambda\sqrt{g}t]$$

Since the deflection $y(0, t) = X(0)T(t)$ at the free end of the cable remains finite, whereas $Y_0(2\lambda\sqrt{x})$ is unbounded in the neighborhood of $x = 0$, it follows that $B = 0$. Also, since the cable is fixed at $x = l$, we must have

$$y(l, t) = X(l)T(t) = AJ_0(2\lambda\sqrt{l})T(t) = 0$$

If $A = 0$, we have only the trivial solution corresponding to the cable hanging permanently motionless. Therefore it follows that

$$J_0(2\lambda\sqrt{l}) = 0$$

since $T(t)$ is not to vanish identically.

If $z_1, z_2, z_3, \ldots$ are the roots of the equation $J_0(z) = 0$, then

$$\lambda_1 = \frac{z_1}{2\sqrt{l}}, \qquad \lambda_2 = \frac{z_2}{2\sqrt{l}}, \qquad \lambda_3 = \frac{z_3}{2\sqrt{l}}, \ldots$$

For these values, *and for these values only*, Eq. (16) has periodic solutions which are bounded in the neighborhood of $x = 0$ and zero at $x = l$. Finally, from the time factor T given by Eq. (17), it is clear that the natural frequencies of the cable are given by the formula

$$\omega_n = \frac{\lambda_n\sqrt{g}}{2\pi} = \frac{z_n}{4\pi}\sqrt{\frac{g}{l}} \qquad \text{cycles per unit time}$$

It is interesting to note that if the flexible cable were replaced by a rigid bar of the same length and weight per unit length, the latter would swing about one end with frequency

$$\omega = \frac{1}{2\pi}\sqrt{\frac{3g}{2l}}$$

The ratio of the lowest frequency of the cable to the single frequency of an equivalent bar is thus (since $z_1 = 2.4048$)

$$(z_1/2)\sqrt{g/l}\Bigg/\sqrt{\frac{3}{2}g/l} \doteq \frac{1.2024}{1.2247} \doteq 0.9818$$

Exercises for Section 11.6

1 In Example 1, find $u(x, y)$ if $f(x) = x - x^2$.

2 Work Example 1 if the boundary conditions are

$$u(0, y) = u(x, 0) = u(1, y) = 0 \qquad u(x, 1) = f(x)$$

3 Work Example 1 if the boundary conditions are

$$u(0, y) = u(x, 0) = 0 \qquad \left.\frac{\partial u}{\partial x}\right|_{1, y} = 0 \qquad u(x, 1) = f(x)$$

4 Work Example 1 if the boundary conditions are

$$u(0, y) = u(x, 0) = 0 \qquad u(x, 1) = 100 \qquad u(1, y) = 100y$$

5 Work Example 1 if the boundary conditions are

$$u(0, y) = 0 \qquad \left.\frac{\partial u}{\partial y}\right|_{x,0} = \left.\frac{\partial u}{\partial y}\right|_{x,1} = 0 \qquad u(1, y) = f(y)$$

***6** Extend Example 3 by finding the deflection of the membrane as a function of r and t if the initial deflection is $z(r, t) = b^2 - r^2$.

***7** Work Example 4 if the weight per unit length of the cable is $(k + 1)x^k$, $k > 0$.

***8** A thin, circular sheet of radius b has its two faces perfectly insulated, so that heat flow in the sheet is purely two-dimensional. The upper half of the boundary of the sheet is maintained at the temperature $u(b, \theta) = 100$ $(0 < \theta < \pi)$. The lower half of the boundary is maintained at the temperature $u(b, \theta) = 0$ $(\pi < \theta < 2\pi)$. Find the steady-state temperature distribution in the sheet.

***9** Work Exercise 8 if the sheet is semicircular and the temperature along the curved portion of the boundary is 100 and the temperature along the diametral edge is 0.

***10** Work Example 2 if the entire boundary of the sheet is perfectly insulated and the initial temperature distribution is $u(r, 0) = f(r)$.

***11** A right circular cylinder of radius b and height h has its upper and lower bases maintained at the constant temperature zero. The curved surface is maintained at the temperature distribution $u(b, z) = f(z)$. Determine the steady-state temperature distribution throughout the cylinder.

***12** Work Exercise 11 if the lower base and curved surface of the cylinder are maintained at the temperature zero and the upper base is maintained at the temperature 100.

***13** A thin sheet of metal bounded by the x axis and the lines $x = 0$ and $x = 1$ and stretching to infinity in the y direction has its two faces perfectly insulated, so that heat flow in the sheet is purely two-dimensional. The vertical edges of the sheet are maintained at the temperature zero, and the base is maintained at the temperature 100. Find the steady-state temperature distribution in the sheet.

***14** Work Exercise 13 if the boundary conditions are

(*a*) $\left.\dfrac{\partial u}{\partial x}\right|_{0,y} = \left.\dfrac{\partial u}{\partial x}\right|_{1,y} = 0,\ u(x, 0) = 100$

(*b*) $u(0, y) = 0,\ u(1, y) = 100,\ u(x, 0) = 100x$

(*c*) $u(0, y) = 0,\ \left.\dfrac{\partial u}{\partial x}\right|_{1,y} = 0,\ u(x, 0) = 100$

(*d*) $u(0, y) = 0,\ u(x, 0) = 0,\ u(1, y) = 100$

****15** It can be shown that the free transverse vibrations of a uniform beam are governed by the equation

$$a^2 \frac{\partial^4 y}{\partial x^4} = -\frac{\partial^2 y}{\partial t^2}$$

where a is a combination of physical constants of the beam. The three common end conditions of a beam, namely *built-in*, *simply supported* or *hinged*, and *free*, are characterized by the following conditions on the deflection function $y(x, t)$, primes denoting differentiation with respect to x:

Built-in	$y = y' = 0$
Simply supported	$y = y'' = 0$
Free	$y'' = y''' = 0$

Find the equation whose roots determine the natural frequencies of each of the following beams:

(*a*) Simply supported at $x = 0$ and $x = l$
(*b*) Built-in at $x = 0$ and $x = l$
(*c*) Free at $x = 0$ and $x = l$
(*d*) Simply supported at $x = 0$, built-in at $x = l$
(*e*) Simply supported at $x = 0$, free at $x = l$
(*f*) Built-in at $x = 0$, free at $x = l$

CHAPTER

TWELVE

THE NUMERICAL SOLUTION OF DIFFERENTIAL EQUATIONS

12.1 INTRODUCTION

Although we have spent several hundred pages developing methods for the exact solution of differential equations, all we are really able to do is solve a relatively few important, but essentially routine problems. The equations that arise at the research frontier, whether in pure mathematics or in advanced applications, are almost always beyond our power to solve exactly, and computer-assisted numerical methods must be used to approximate their solutions. The study of such methods, and the errors they involve, is one of the major topics in the field of *numerical analysis*. We cannot go deeply into these matters, but we shall present two methods of great utility for approximating the solutions of ordinary differential equations. One of these is the *Runge-Kutta method*; the other is *Milne's method*. Questions concerning the theoretical errors in such methods, as well as the errors due to round-off in the computations themselves, we shall leave to courses in numerical analysis.

12.2 THE RUNGE-KUTTA METHOD

The fundamental problem in the numerical solution of ordinary differential equations is the solution of the first-order equation

$$\frac{dy}{dx} = f(x, y) \tag{1}$$

subject to the initial condition that $y = y_0$ when $x = x_0$. We do not, of course, expect to find y as an explicit function of x. Instead, our object is to obtain satisfactory approximations to the values of the solution $y(x)$ on a specified set of x values, $x_1, x_2, x_3, \ldots$. In our discussion of methods for doing this, we shall denote approximations to the exact values $y(x_1), y(x_2), y(x_3), \ldots$ by $y_1, y_2, y_3, \ldots$, respectively.

The first step in the pointwise solution of the initial-value problem $dy/dx = f(x, y)$, $y(x_0) = y_0$, is to approximate the value of y at $x_1 = x_0 + \Delta x$, that is, $y(x_1) = y_0 + \Delta y$. The simplest way to do this is to approximate Δy by the usual differential estimate for the true increment Δy, namely,

$$\Delta y \doteq dy = y'(x_0)\,\Delta x \tag{2}$$

The differential equation itself gives us the value of the derivative at the point (x_0, y_0) on the required solution curve, for we have from (1)

$$y'(x_0) = f(x_0, y_0)$$

Hence, from (2),

$$\Delta y \doteq f(x_0, y_0)\,\Delta x$$

and therefore

$$y(x_1) = y_0 + \Delta y \doteq y_0 + f(x_0, y_0)\,\Delta x = y_1 \tag{3}$$

Once y_1 is determined as an approximation to $y(x_1)$, the same procedure can be repeated at (x_1, y_1), and so on as far as required. The geometric interpretation of this process, which is known as **Euler's method**, is shown in Fig. 12.1.

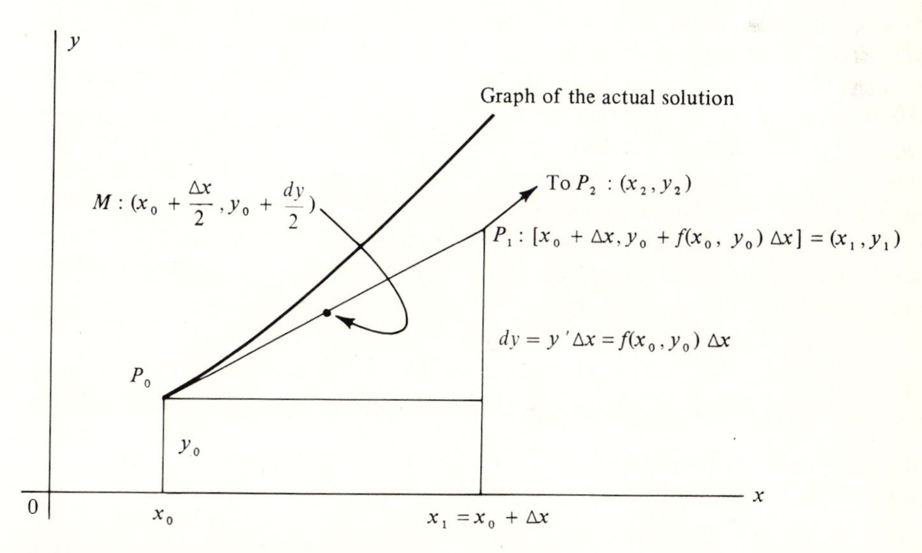

Figure 12.1 The Euler method of solving $dy/dx = f(x, y)$.

On the other hand, having obtained y_1 as a first approximation to $y(x_1)$ by Euler's method, one can use the differential equation (1) to compute y' at the new point (x_1, y_1) and then use the average of the derivatives at (x_0, y_0) and (x_1, y_1) to obtain a (presumably) more accurate estimate of Δy and hence of $y(x_1)$ before attempting to approximate $y(x_2)$. This method yields the value

$$\Delta y \doteq \tfrac{1}{2}[y'(x_0) + y'(x_1)]\,\Delta x$$

and, from this, using (1) to obtain the necessary values of the derivatives, we find

$$y_0 + \Delta y \doteq y_0 + \tfrac{1}{2}[f(x_0, y_0) + f(x_1, y_1)]\,\Delta x = (y_1)_2 \tag{4}$$

as a second approximation to $y(x_1)$. This process is known as the **modified Euler method.**

Still another possibility, after determining y_1 as a first estimate of $y(x_1)$ by Euler's method, is to reapproximate Δy and $y(x_1)$ using not the average of the derivatives at P_0: (x_0, y_0) and P_1: (x_1, y_1), but the derivative at the midpoint of the segment P_0P_1 (Fig. 12.1), namely,

$$M\colon \left(\frac{x_0 + x_1}{2}, \frac{y_0 + y_1}{2}\right)$$

This gives the (presumably) improved estimate

$$\Delta y \doteq f\left(\frac{x_0 + x_1}{2}, \frac{y_0 + y_1}{2}\right)\Delta x = f\left[x_0 + \frac{\Delta x}{2}, y_0 + \frac{1}{2}f(x_0, y_0)\,\Delta x\right]\Delta x$$

and, as a third approximation to $y(x_1)$,

$$(y_1)_3 = y_0 + f\left[x_0 + \frac{\Delta x}{2}, y_0 + \frac{1}{2}f(x_0, y_0)\,\Delta x\right]\Delta x \tag{5}$$

The solution process based on formula (5) is known as **Runge's method.**

Neither the Euler method, based on Eq. (3), nor the modified Euler method, based on Eq. (4), nor Runge's method, based on Eq. (5), is an effective way of solving the differential equation (1) because unless Δx is very small, the errors implicit in the various approximations for Δy build up significantly as successive values of y are calculated. The Runge-Kutta method is essentially a generalization of these three simple procedures in which at each stage three or more preliminary estimates of Δy are calculated. The value which is then used for Δy to compute the next value of y is a linear combination of these estimates in which the constants of combination are chosen to make the error as small as possible.

Specifically, in Kutta's third-order method, the three estimates of Δy are

$$(\Delta y)_1 = f(x_0, y_0)\,\Delta x \tag{6}$$

which is just the estimate used in Euler's method,

$$(\Delta y)_2 = f[x_0 + p\,\Delta x, y_0 + p(\Delta y)_1]\,\Delta x \qquad 0 < p < 1 \tag{7}$$

which is just like the estimate used in Runge's method except that instead of being evaluated at the midpoint of the segment P_0P_1 (i.e., with $p = \frac{1}{2}$), the derivative is evaluated at a point $[x_0 + p\,\Delta x,\ y_0 + p(\Delta y)_1]$ yet to be determined; and

$$(\Delta y)_3 = f[x_0 + q\,\Delta x,\ y_0 + r(\Delta y)_2 + \overline{q - r}(\Delta y)_1]\,\Delta x \qquad 0 < q, r < 1 \tag{8}$$

where q and r are yet to be determined. Finally the value which is actually used for Δy in the calculation of y_1 is taken to be

$$(\Delta y)_4 = a(\Delta y)_1 + b(\Delta y)_2 + c(\Delta y)_3 \tag{9}$$

where a, b, c are parameters which, like the parameters p, q, r, are to be chosen to ensure the highest possible accuracy in estimating Δy.

The determination of a, b, c, p, q, r is accomplished in the following way. First, Δy is expanded in a power series in Δx:

$$\Delta y = y(x_1) - y_0 = y'(x_0)\,\Delta x + \frac{1}{2!}y''(x_0)(\Delta x)^2 + \frac{1}{3!}y'''(x_0)(\Delta x)^3 + \cdots \tag{10}$$

using implicit differentiation of the given equation (1) to compute the derivatives of y which must be evaluated to obtain the coefficients in the series. Then, in a similar fashion, the approximations $(\Delta y)_1$, $(\Delta y)_2$, and $(\Delta y)_3$, which are functions of Δx defined by Eqs. (6), (7), and (8), respectively, are expressed as power series in Δx. These three expansions are then substituted into (9), so that $(\Delta y)_4$, the final estimate of Δy, is also expressed as a power series in Δx. Finally, a, b, c, p, q, r are chosen to make the series for Δy and the series for $(\Delta y)_4$ identical as far as the terms involving $(\Delta x)^3$. When this is accomplished, the difference between the true value of Δy, as given by the series (10), and the final estimate $(\Delta y)_4$, as obtained from (9), is of the order of $(\Delta x)^4$. In other words, the error inherent in the method is proportional to the fourth power of the tabular interval Δx.

When the coefficients of Δx and $(\Delta x)^2$ in the two series are equated, we obtain the conditions

$$a + b + c = 1 \qquad pb + qc = \tfrac{1}{2} \tag{11}$$

Two more conditions,

$$p^2b + q^2c = \tfrac{1}{3} \qquad prc = \tfrac{1}{6} \tag{12}$$

arise when the coefficients of $(\Delta x)^3$ are equated. Equations (11) and the first of Eqs. (12) can easily be solved for a, b, and c in terms of p and q. The second equation in (12) can then be used to express r in terms of p and q also. The results are

$$a = \frac{6pq - 3(p+q) + 2}{6pq} \qquad b = \frac{2 - 3q}{6p(p-q)}$$
$$c = \frac{2 - 3p}{6q(q-p)} \qquad r = \frac{q(q-p)}{p(2-3p)} \tag{13}$$

Since p and q are arbitrary, we thus have a two-parameter family of formulas which can be used for the step-by-step solution of the equation $y' = f(x, y)$ with an error which is of the order of $(\Delta x)^4$.

Two special cases of Kutta's third-order method are worthy of note. In listing them, we shall for convenience introduce the following conventional notation:

$$\Delta x = h \qquad (\Delta y)_1 = k_1 \qquad (\Delta y)_2 = k_2 \qquad (\Delta y)_3 = k_3$$

Case 1: $a = \frac{1}{4} \qquad b = 0 \qquad c = \frac{3}{4} \qquad p = \frac{1}{3} \qquad q = r = \frac{2}{3}$

$$\Delta y \doteq (\Delta y)_4 = \tfrac{1}{4}(k_1 + 3k_3)$$

where

$$k_1 = f(x_0, y_0)h$$
$$k_2 = f(x_0 + \tfrac{1}{3}h, y_0 + \tfrac{1}{3}k_1)h$$
$$k_3 = f(x_0 + \tfrac{2}{3}h, y_0 + \tfrac{2}{3}k_2)h$$

Case 2: $a = \frac{1}{4} \qquad b = c = \frac{3}{8} \qquad p = q = r = \frac{2}{3}$

$$\Delta y \doteq (\Delta y)_4 = \tfrac{1}{8}(2k_1 + 3k_2 + 3k_3)$$

where

$$k_1 = f(x_0, y_0)h$$
$$k_2 = f(x_0 + \tfrac{2}{3}h, y_0 + \tfrac{2}{3}k_1)h$$
$$k_3 = f(x_0 + \tfrac{2}{3}h, y_0 + \tfrac{2}{3}k_2)h$$

The values of the parameters in case 2 cannot be obtained from Eqs. (13), since $p = q$, but can be checked directly in Eqs. (11) and (12).

The preceding discussion can be extended without difficulty (except in detail!) to yield step-by-step solution procedures in which the error is of the order of $h^5 = (\Delta x)^5$. In particular, the following two sets of formulas are quite useful.

Case 3: $\Delta y \doteq (\Delta y)_5 = \frac{1}{6}(k_1 + 2k_2 + 2k_3 + k_4)$ where

$$k_1 = f(x_0, y_0)h$$
$$k_2 = f(x_0 + \tfrac{1}{2}h, y_0 + \tfrac{1}{2}k_1)h$$
$$k_3 = f(x_0 + \tfrac{1}{2}h, y_0 + \tfrac{1}{2}k_2)h$$
$$k_4 = f(x_0 + h, y_0 + k_3)h$$

Case 4: $\Delta y \doteq (\Delta y)_5 = \frac{1}{8}(k_1 + 3k_2 + 3k_3 + k_4)$ where

$$k_1 = f(x_0, y_0)h$$
$$k_2 = f(x_0 + \tfrac{1}{3}h, y_0 + \tfrac{1}{3}k_1)h$$
$$k_3 = f(x_0 + \tfrac{2}{3}h, y_0 - \tfrac{1}{3}k_1 + k_2)h$$
$$k_4 = f(x_0 + h, y_0 + k_1 - k_2 + k_3)h$$

The solution process based on Case 3 is usually referred to specifically as the **Runge-Kutta method.**

Example 1 Tabulate the solution of $y' = x^2 + y$ at $x = 0.1, 0.2, 0.3, 0.5$, and 1.0, given that $y = -1$ when $x = 0$.

Using the Runge-Kutta formulas of Case 3 for the first increment, we have

$$k_1 = f(x_0, y_0)h = [(0)^2 - 1](0.1) = -0.10000$$

$$k_2 = f(x_0 + \tfrac{1}{2}h, y_0 + \tfrac{1}{2}k_1)h = [(0.05000)^2 - 1.05000](0.1) = -0.10475$$

$$k_3 = f(x_0 + \tfrac{1}{2}h, y_0 + \tfrac{1}{2}k_2)h = [(0.05000)^2 - 1.05238](0.1) = -0.10500$$

$$k_4 = f(x_0 + h, y_0 + k_3)h = [(0.10000)^2 - 1.10500](0.1) = -0.10950$$

and
$$\begin{aligned} \Delta y &\doteq (\Delta y)_5 \\ &= \tfrac{1}{6}(k_1 + 2k_2 + 2k_3 + k_4) \\ &= \tfrac{1}{6}(-0.10000 - 0.20950 - 0.21000 - 0.10950) \\ &= -0.10483 \end{aligned}$$

Hence
$$y_1 = y_0 + (\Delta y)_5 = -1.10483$$

For the second increment, we have similarly

$$k_1 = -0.10948 \qquad k_2 = -0.11371 \qquad k_3 = -0.11392 \qquad k_4 = -0.11788$$

and
$$(\Delta y)_5 = -0.11377$$

Hence
$$y_2 = y_1 + (\Delta y)_5 = -1.21860$$

For the third increment, we have

$$k_1 = -0.11786 \qquad k_2 = -0.12150 \qquad k_3 = -0.12169 \qquad k_4 = -0.12503$$

and
$$(\Delta y)_5 = -0.12154$$

Hence
$$y_3 = y_2 + (\Delta y)_5 = -1.34014$$

For the fourth increment (noting that this time $h = \Delta x = 0.2$), we have

$$k_1 = -0.25003 \qquad k_2 = -0.26103 \qquad k_3 = -0.26213 \qquad k_4 = -0.27045$$

and
$$(\Delta y)_5 = -0.26113$$

Hence
$$y_4 = y_3 + (\Delta y)_5 = -1.60127$$

For the fifth increment (noting that now $h = \Delta x = 0.5$), we have

$$k_1 = -0.67564 \qquad k_2 = -0.68830 \qquad k_3 = -0.69146 \qquad k_4 = -0.64636$$

and
$$(\Delta y)_5 = -0.68025$$

Hence
$$y_5 = y_4 + (\Delta y)_5 = -2.28152$$

The differential equation $y' = x^2 + y$ is so simple that it can be solved exactly without recourse to numerical methods, and by the methods of Chap. 1 or Chap. 3 we find at once that the required solution is

$$y = e^x - x^2 - 2x - 2$$

For $x = 0.1, 0.2, 0.3, 0.5$, and 1.0, this gives us the correct values, correct to five decimal places,

$$y(x_1) = y(0.1) \doteq -1.10483 \qquad y(x_2) = y(0.2) \doteq -1.21860$$

$$y(x_3) = y(0.3) \doteq -1.34014 \qquad y(x_4) = y(0.5) \doteq -1.60128$$

$$y(x_5) = y(1.0) \doteq -2.28172$$

The values we computed for y_1, y_2, and y_3 agree with these to five decimal places; the value we computed for y_4 differs from the correct value by 1 in the fifth place; and our value for y_5 differs from the correct value by only 20 in the fifth place, i.e., by 2 in the fourth place.

Among other things, this example illustrates the important fact that the value of $h \equiv \Delta x$ need not be constant throughout the process but may be changed, as time and circumstances may dictate.

Any of the Runge-Kutta formulas can be used to solve simultaneous differential equations. For instance, using Case 3, we can tabulate the solution of the initial-value problem

$$\frac{dy}{dx} = f(x, y, z) \qquad \frac{dz}{dx} = g(x, y, z) \qquad y = y_0,\ z = z_0 \text{ when } x = x_0$$

at intervals of $\Delta x \equiv h$ (either constant or variable) by computing

$$k_1 \equiv (\Delta y)_1 = f(x_0, y_0, z_0)h$$

$$l_1 \equiv (\Delta z)_1 = g(x_0, y_0, z_0)h$$

$$k_2 \equiv (\Delta y)_2 = f(x_0 + \tfrac{1}{2}h, y_0 + \tfrac{1}{2}k_1, z_0 + \tfrac{1}{2}l_1)h$$

$$l_2 \equiv (\Delta z)_2 = g(x_0 + \tfrac{1}{2}h, y_0 + \tfrac{1}{2}k_1, z_0 + \tfrac{1}{2}l_1)h$$

$$k_3 \equiv (\Delta y)_3 = f(x_0 + \tfrac{1}{2}h, y_0 + \tfrac{1}{2}k_2, z_0 + \tfrac{1}{2}l_2)h$$

$$l_3 \equiv (\Delta z)_3 = g(x_0 + \tfrac{1}{2}h, y_0 + \tfrac{1}{2}k_2, z_0 + \tfrac{1}{2}l_2)h$$

$$k_4 \equiv (\Delta y)_4 = f(x_0 + h, y_0 + k_3, z_0 + l_3)h$$

$$l_4 \equiv (\Delta z)_4 = g(x_0 + h, y_0 + k_3, z_0 + l_3)h$$

and then, using the formulas,

$$\Delta y \doteq (\Delta y)_5 = \tfrac{1}{6}(k_1 + 2k_2 + 2k_3 + k_4)$$

$$\Delta z \doteq (\Delta z)_5 = \tfrac{1}{6}(l_1 + 2l_2 + 2l_3 + l_4)$$

Since any differential equation of the form

$$y^{(n)} = f(x, y, y', \ldots, y^{(n-1)}) \qquad n \geq 2$$

and any equation of the form $g(x, y, y', \ldots, y^{(n-1)}, y^{(n)}) = 0$ which can be solved for $y^{(n)}$, can be written as a system of simultaneous, first-order differential equations, Runge-Kutta methods also suffice for the solution of such equations. For example, under the substitution $dy/dx \equiv y' = z$, the initial-value problem

$$\frac{d^2y}{dx^2} = g(x, y, y') \qquad y = y_0, y' = y'_0 \qquad \text{when } x = x_0$$

becomes $$\frac{dy}{dx} = z \qquad \frac{dz}{dx} = g(x, y, z) \qquad y = y_0, z = y'_0 \qquad \text{when } x = x_0$$

which is just like the system we discussed above with $f(x, y, z) \equiv z$.

Exercises for Section 12.2

1 In Example 1, calculate $y(0.4)$ and $y(0.6)$.

2 For the problem of Example 1, find $y(0.1)$, $y(0.2)$, and $y(0.3)$ using Euler's method.

3 Work Example 1 using the modified Euler method.

4 Work Example 1 using Runge's method.

5 Using Kutta's third-order approximation 1, compute $y(1.1)$, $y(1.2)$, and $y(1.3)$ if $y' = x - y$ and $y = 1$ when $x = 1$. Carry four decimal places in your work, and compare your answers with the exact solution.

6 Using Kutta's third-order approximation 2, compute $y(0.1)$, $y(0.2)$, and $y(0.3)$ if $y' = x + y$ and $y = 1$ when $x = 0$. Carry four decimal places in your work, and compare your answers with the exact solution.

7 Using the Runge-Kutta method 3, tabulate the function $y = e^{-x^2}$ for $x = 0.0$, 0.1, 0.2, 0.3, 0.4, and 0.5. How do your answers compare with the exact values of this function? *Hint:* Find a differential equation satisfied by y.

8 Using the Runge-Kutta method 3, evaluate $\int_0^x e^{-t^2}\,dt$ for $x = 0.0$, 0.2, 0.4, 0.6, 0.8, and 1.0.

9 Using the Runge-Kutta method 3, tabulate the solution of the system

$$\frac{dy}{dx} = x + z \qquad \frac{dz}{dx} = x - y \qquad y = 0, z = 1 \qquad \text{when } x = 0$$

at intervals of $h = 0.1$ from $x = 0.0$ to $x = 0.5$.

10 Set up the Kutta third-order approximation corresponding to the values $p = \frac{1}{2}$, $q = 1$ and show that it reduces to Simpson's rule when $f(x, y)$ is independent of y.

12.3 FINITE DIFFERENCE FORMULAS

The Runge-Kutta method, which we discussed in the last section, is what is known as a **one-step method** for the numerical solution of an ordinary differential equation. By this we mean that given a differential equation $y' = f(x, y)$ and a single pair of values (x_0, y_0), the method enables us to approximate the value of y at $x_0 + h$ without additional information. In contrast to this are **multistep methods** which require a knowledge of several pairs of x, y values before they can be used to extend the solution. One important multistep method is *Milne's method*, which we shall discuss in the next section. Preparatory to this, we must introduce a few simple ideas from the *calculus of finite differences.* Among these are the finite difference operators Δ and E, the Gregory-Newton interpolation formula, and several formulas for numerical differentiation. Some of these will also be utilized in the work of the next chapter, where we shall study *difference equations.*

Suppose that we have a function $y = f(x)$ given in tabular form at a sequence of equally spaced values of x, say x_0, $x_1 = x_0 + h$, $x_2 = x_0 + 2h$, $x_3 = x_0 + 3h$, and so on.

x	$f(x)$
x_0	$f(x_0)$
x_1	$f(x_1)$
x_2	$f(x_2)$
x_3	$f(x_3)$
$\vdots$	$\vdots$

If x_k and x_{k+1} are consecutive values of x in such a tabulation, the **first forward differences** of $f(x)$ are defined by the formula

$$\Delta f(x_k) = f(x_{k+1}) - f(x_k) \tag{1a}$$

Still more simply, if f_k is used as an abbreviation of $f(x_k)$, the first forward differences of $f(x)$ are given by the formula

$$\Delta f_k = f_{k+1} - f_k \tag{1b}$$

Similarly, if Δf_k and Δf_{k+1} are consecutive first differences of $f(x)$, the second forward differences of $f(x)$ are defined by the formula

$$\Delta^2 f_k = \Delta(\Delta f_k) = \Delta f_{k+1} - \Delta f_k \tag{2}$$

In the same way, we have for the third differences of $f(x)$

$$\Delta^3 f_k = \Delta(\Delta^2 f_k) = \Delta^2 f_{k+1} - \Delta^2 f_k \tag{3}$$

Higher forward differences of $f(x)$ are defined in a similar fashion as the difference between consecutive differences of one lower order, and we have in general

$$\Delta^n f_k = \Delta(\Delta^{n-1} f_k) = \Delta^{n-1} f_{k+1} - \Delta^{n-1} f_k \tag{4}$$

In many applications it is convenient to have the differences of a function prominently displayed. This is usually done by constructing a **difference table** in which each difference is entered in the appropriate column midway between the elements in the preceding column from which it is constructed:†

† For instance, for convenience in performing proportional-part interpolation, many tables of the elementary functions contain a column of the differences between consecutive entries in the table.

x	$f(x)$	Δ	Δ^2	Δ^3	$\cdot$
x_0	f_0				
		Δf_0			
x_1	f_1		$\Delta^2 f_0$		
		Δf_1		$\Delta^3 f_0$	
x_2	f_2		$\Delta^2 f_1$		$\cdot$
		Δf_2		$\Delta^3 f_1$	
x_3	f_3		$\Delta^2 f_2$	$\cdot$	
		Δf_3			
x_4	f_4		$\cdot$		
		$\cdot$			
$\cdot$	$\cdot$				

or in a specific numerical example

x	$f(x) = x^3$	Δ	Δ^2	Δ^3	Δ^4
0	0				
		1			
1	1		6		
		7		6	
2	8		12		0
		19		6	
3	27		18		0
		37		6	
4	64		24		0
		61		6	
5	125		30		$\cdot$
		91		$\cdot$	
6	216		$\cdot$		
		$\cdot$			
$\cdot$	$\cdot$				

From the definition of the differences of a function it follows that Δ is a linear operator (see item 12, Appendix B.2); that is, the differences of a function have the properties described by the following two theorems, whose proofs we leave as exercises.

Theorem 1 $\Delta(cf) = c\,\Delta f$ c a constant

Theorem 2 $\Delta(f \pm g) = \Delta f \pm \Delta g$

The operator Δ also has the simple and familiar exponential property described by Theorem 3.

Theorem 3 $\Delta^m(\Delta^n f) = \Delta^{m+n} f \qquad m, n$ nonnegative integers

The fact that the expression defining the first forward differences of a function is just the numerator of the difference quotient whose limit defines the derivative of the function suggests that in some respects the properties of the differences of a function and the properties of the derivatives of a function may be analogous. For instance, the properties of the difference operator Δ asserted by Theorems 1, 2, and 3 are also properties of the derivative operator D. Still another common property, suggested by the difference table for x^3 which we presented above, is contained in the following theorem.

Theorem 4 The nth differences of a polynomial of degree n are constant.

Proof Because of Theorems 1 and 2, it is clearly sufficient to prove the assertion of the theorem for the special polynomial x^n. To do this, we observe first that $x_{k+1} = x_k + h$. Hence

$$\Delta x_k^n = x_{k+1}^n - x_k^n = (x_k + h)^n - x_k^n$$

Therefore, applying the binomial expansion to $(x_k + h)^n$, we have

$$\begin{aligned}\Delta x_k^n &= \left[x_k^n + n x_k^{n-1} h + \frac{n(n-1)}{2} x_k^{n-2} h^2 + \cdots + n x_k h^{n-1} + h^n\right] - x_k^n \\ &= nh x_k^{n-1} + \frac{n(n-1)}{2} h^2 x_k^{n-2} + \cdots + n h^{n-1} x_k + h^n\end{aligned}$$

This shows that the first difference of any positive integral power of x is a polynomial in x_k of degree $n - 1$. The second difference of x^n is, of course, the first difference of this polynomial; and by Theorems 1 and 2 and what we have just proved, this difference is a polynomial of degree $n - 2$. Continuing in this way, we see that at each stage the operation of differencing leads to a polynomial of one lower degree. Clearly, after n steps, the degree of the resultant polynomial will thus be reduced to zero. In other words, the nth difference of x^n will be a constant, and the theorem is proved.

Closely associated with Δ is the shift operator E, which is defined to be the operator which increases the argument of a function by one tabular interval. Thus

$$Ef(x_k) = Ef_k = f(x_k + h) = f(x_{k+1}) = f_{k+1}$$

Applying E a second time again increases the argument of $f(x)$ by h; that is,

$$E^2 f(x_k) = E[Ef(x_k)] = Ef(x_k + h) = f(x_k + 2h) = f_{k+2}$$

Similarly, applying E a third time, we have

$$E^3 f(x_k) = E[E^2 f(x_k)] = Ef_{k+2} = f_{k+3}$$

Clearly, it is meaningful to speak of the operation of changing the argument of a function by *any* multiple of the tabular interval. Hence, in general, we define

$$E^r f(x_k) = f(x_k + rh) = f_{k+r} \tag{5}$$

for *any* real number r, positive, negative, or zero. From its definition, it is easy to verify that E obeys the laws

$$E(cf_k) = cEf_k \qquad c \text{ a constant}$$

$$E(f_k + g_k) = Ef_k + Eg_k$$

$$E^m(E^n f_k) = E^{m+n} f_k \qquad m, n \text{ any real numbers}$$

Two operators with the property that when they are applied to the same function, they yield the same result are said to be **operationally equivalent.** Now from the definition of Δf_k, we have

$$\Delta f_k = f_{k+1} - f_k = Ef_k - f_k$$

or, symbolically,

$$\Delta f_k = (E - 1) f_k$$

Hence we have the operational equivalences

$$\Delta = E - 1 \tag{6}$$

$$E = 1 + \Delta \tag{7}$$

By means of Eq. (6), we can express the various forward differences of a function in terms of successive entries in the table of the function. Using (6), we can write

$$\Delta^n f_k = (E - 1)^n f_k$$

Then, using the binomial expansion, we have

$$\begin{aligned}
\Delta^n f_k &= \left[E^n - nE^{n-1} + \frac{n(n-1)}{2} E^{n-2} - \cdots + (-1)^{n-1} nE + (-1)^n\right] f_k \\
&= E^n f_k - nE^{n-1} f_k + \frac{n(n-1)}{2} E^{n-2} f_k - \cdots + (-1)^{n-1} nEf_k + (-1)^n f_k \\
&= f_{k+n} - nf_{k+n-1} + \frac{n(n-1)}{2} f_{k+n-2} - \cdots \\
&\qquad + (-1)^{n-1} nf_{k+1} + (-1)^n f_k
\end{aligned}$$

Specifically, taking $k = 0$ and $n = 1, 2, 3, 4$, we have

$$\begin{aligned}
\Delta f_0 &= f_1 - f_0 \\
\Delta^2 f_0 &= f_2 - 2f_1 + f_0 \\
\Delta^3 f_0 &= f_3 - 3f_2 + 3f_1 - f_0 \\
\Delta^4 f_0 &= f_4 - 4f_3 + 6f_2 - 4f_1 + f_0
\end{aligned} \tag{8}$$

Using Eq. (7), we can now derive the interpolation formula on which Milne's method for the numerical solution of differential equations is based. The fundamental problem of interpolation, of course, is to obtain the value of a tabulated function at a point which is not one of the tabular points. Specifically, we desire a formula for $f(x)$ where $x = x_0 + rh$ and r is not an integer. Now, by Eq. (5),

$$f(x) = f(x_0 + rh) = E^r f_0$$

and by Eq. (7) this is the same as

$$f(x) = (1 + \Delta)^r f_0$$

Hence, using the binomial expansion, we have

$$f(x) = f(x_0 + rh) = \left[1 + r\Delta + \frac{r(r-1)}{2!}\Delta^2 + \frac{r(r-1)(r-2)}{3!}\Delta^3 + \frac{r(r-1)(r-2)(r-3)}{4!}\Delta^4 + \cdots\right] f_0$$

$$= f_0 + r\,\Delta f_0 + \frac{r(r-1)}{2!}\Delta^2 f_0 + \frac{r(r-1)(r-2)}{3!}\Delta^3 f_0 + \frac{r(r-1)(r-2)(r-3)}{4!}\Delta^4 f_0 + \cdots \qquad (9)$$

Equation (9) is known as the **Gregory-Newton interpolation formula.**†

If $f(x)$ is a polynomial of degree n, we know from Theorem 4 that its differences beyond the nth are all zero. Hence the series in (9) is finite, and the value of $f(x)$ is given exactly by this formula which, in fact, is an identity. If $f(x)$ is not a polynomial, its differences never become zero. In this case, Eq. (9) is a finite sum and represents $f(x)$ exactly only if r is a positive integer. For all other values of r, the series in (9) is an infinite series, and an error term estimating the remainder after n terms should be obtained. This is done in courses in numerical analysis, but in our work we shall neglect this theoretical refinement and simply assume that we have extended the interpolation series to the point where the error term is either zero or at least negligibly small.

Example 1 Compute $\sqrt{50.2}$ from the following data:

x	$\sqrt{x}$	Δ	Δ^2	Δ^3
50	7.07107			
		0.07036		
51	7.14143		−0.00069	
		0.06967		0.00003
52	7.21110		−0.00066	
		0.06901		0.00001
53	7.28011		−0.00065	
		0.06836		0.00002
54	7.34847		−0.00063	
		0.06773		
55	7.41620			

† All operational derivations are suggestive rather than rigorous, and ours is no exception. For a more satisfactory development of the Gregory-Newton interpolation formula see, for instance, the author's "Advanced Engineering Mathematics," 4th ed., McGraw-Hill, New York, 1975, pp. 117–20.

The calculation of the various differences of $\sqrt{x}$ presents no difficulty, and we obtain easily the values listed to the right of the original data. Since $h = 1$ and since $50.2 = 50 + 0.2h$, it follows that $r = 0.2$. Thus the Gregory-Newton formula gives

$$\sqrt{50.2} \doteq 7.07107 + (0.2)(0.07036) + \frac{(0.2)(-0.8)}{2}(-0.00069)$$

$$+ \frac{(0.2)(-0.8)(-1.8)}{6}(0.00003) + \cdots$$

$$= 7.07107 + 0.01407 + 0.00006 + 0.00000 = 7.08520$$

The exact value of $\sqrt{50.2}$, correct to five decimal places, is also 7.08520.

Our immediate interest in the Gregory-Newton interpolation formula is that by differentiating it with respect to r, we can obtain formulas for the various derivatives of a tabulated function. Specifically, if we perform the indicated multiplications in the coefficients in Eq. (9), we have

$$f(x_0 + rh) = f_0 + r\,\Delta f_0 + \frac{r^2 - r}{2}\Delta^2 f_0 + \frac{r^3 - 3r^2 + 2r}{6}\Delta^3 f_0$$

$$+ \frac{r^4 - 6r^3 + 11r^2 - 6r}{24}\Delta^4 f_0 + \cdots$$

Then if we differentiate this with respect to r, we obtain

$$hf'(x_0 + rh) = \Delta f_0 + \frac{2r - 1}{2}\Delta^2 f_0 + \frac{3r^2 - 6r + 2}{6}\Delta^3 f_0$$

$$+ \frac{2r^3 - 9r^2 + 11r - 3}{12}\Delta^4 f_0 + \cdots \tag{10}$$

$$h^2 f''(x_0 + rh) = \Delta^2 f_0 + (r - 1)\,\Delta^3 f_0 + \frac{6r^2 - 18r + 11}{12}\Delta^4 f_0 + \cdots \tag{11}$$

. . . .

Usually we are concerned with the values of the derivatives of $f(x)$ at the tabular points themselves. These are obtained from Eqs. (10), (11), ... simply by putting $r = 0$ and dividing by $h, h^2, \ldots$:

$$f'(x_0) = \frac{1}{h}\left(\Delta f_0 - \frac{1}{2}\Delta^2 f_0 + \frac{1}{3}\Delta^3 f_0 - \frac{1}{4}\Delta^4 f_0 + \cdots\right) \tag{12}$$

$$f''(x_0) = \frac{1}{h^2}\left(\Delta^2 f_0 - \Delta^3 f_0 + \frac{11}{12}\Delta^4 f_0 - \cdots\right) \tag{13}$$

. . . .

Example 2 From the data in the following table, compute the first two derivatives of $\ln x$ at $x = 200$.

x	$\ln x$	Δ	Δ^2	Δ^3
200	5.29831737			
		0.00498754		
201	5.30330491		−0.00002475	
		0.00496279		0.00000024
202	5.30826770		−0.00002451	
		0.00493828		0.00000024
203	5.31320598		−0.00002427	
		0.00491401		0.00000025
204	5.31811999		−0.00002402	
		0.00488999		
205	5.32300998			

The calculation of the various differences is an easy matter; then, recognizing that $h = 1$ and $x_0 = 200$, we have from Eqs. (12) and (13)

$$f'(200) = \tfrac{1}{1}[0.00498754 - \tfrac{1}{2}(-0.00002475) + \tfrac{1}{3}(0.00000024) - \cdots]$$

$$= 0.00500000$$

$$f''(200) = \frac{1}{1^2}[-0.00002475 - 0.00000024 + \cdots]$$

$$= -0.00002499$$

The value we have obtained for $f'(200)$ is correct to eight places. The value of $f''(200)$ differs from the correct value by 1 in the eighth place.

Exercises for Section 12.3

1 Prove Theorem 1. **2** Prove Theorem 2.

3 Construct the difference table for $f(x) = x^3$ for values of x from 0 to 4 with $h = 0.5$.

4 (*a*) Apply Eq. (9) to the difference table for x^3 given in the text, taking $x_0 = 0$, and verify that the result is an identity.

(*b*) Work part (*a*) with $x_0 = 1$.

5 Fit a polynomial of minimum degree to the following data:

x	0	1	2	3	4	5
$f(x)$	17	15	13	17	33	67

Hint: Construct a difference table for the function and then set up the Gregory-Newton formula, letting $r = x$.

6 (*a*) Compute $\sqrt{50.2}$ from the data of Example 1, taking $x_0 = 51$.
(*b*) Compute $\sqrt{50.5}$ from the data of Example 1.

7 Find the first and second derivatives of $\sqrt{x}$ at $x = 2.5$ from the following data:

x	2.50	2.55	2.60	2.65	2.70
$\sqrt{x}$	1.58114	1.59687	1.61245	1.62788	1.64317

8 (*a*) Show that the approximation provided by the first two terms of Eq. (9) is equivalent to proportional-part interpolation, that is, is equivalent to reading the value of $f(x)$ at $x = x_0 + rh$ from the chord joining the points (x_0, f_0) and $(x_0 + h, f_1)$ rather than from the actual arc of $y = f(x)$.

*(*b*) Find the equation of the parabola of the family $y = a + bx + cx^2$ which passes through the three points $(0, f_0)$, (h, f_1), and $(2h, f_2)$. Then find the ordinate of this parabola at $x = rh$, and show that this value is the approximation to $f(x)$ provided by the first three terms of Eq. (9).

9 The function $(x)^{(n)} = x(x-1)(x-2)\cdots(x-n+1)$ (n a positive integer) is called the **positive factorial function.** Show that $\Delta(x)^{(n)} = n(x)^{(n-1)}$.

10 The function $(x)^{-(n)} = 1/[(x+1)(x+2)\cdots(x+n)]$ (n a positive integer) is called the **negative factorial function.** Show that $\Delta(x)^{-(n)} = -n(x)^{-(n+1)}$.

11 By substituting from Eqs. (8) into Eqs. (12) and (13), express $f'(x_0)$ and $f''(x_0)$ in terms of successive values of $f(x)$.

12 Introduce the operators E and D in Maclaurin's expansion

$$f(x+h) = f(x) + hf'(x) + \frac{h^2}{2!}f''(x) + \frac{h^3}{3!}f'''(x) + \cdots$$

and thus establish the operational equivalences

$$E = e^{hD} \qquad \Delta = e^{hD} - 1 \qquad D = \frac{1}{h}\ln(1+\Delta)$$

Use the last of these equivalences to give another operational derivation of Eq. (12).

12.4 MILNE'S METHOD

To develop Milne's method for the numerical solution of the differential equation

$$y' = f(x, y) \tag{1}$$

we begin with Eq. (10), Sec. 12.3, written in terms of y rather than f, and evaluate it for $r = 1, 2, 3$, and 4, getting

$$y'_1 = \frac{1}{h}\left(\Delta y_0 + \frac{1}{2}\Delta^2 y_0 - \frac{1}{6}\Delta^3 y_0 + \frac{1}{12}\Delta^4 y_0 + \cdots\right)$$

$$y'_2 = \frac{1}{h}\left(\Delta y_0 + \frac{3}{2}\Delta^2 y_0 + \frac{1}{3}\Delta^3 y_0 - \frac{1}{12}\Delta^4 y_0 + \cdots\right)$$

$$y'_3 = \frac{1}{h}\left(\Delta y_0 + \frac{5}{2}\Delta^2 y_0 + \frac{11}{6}\Delta^3 y_0 + \frac{1}{4}\Delta^4 y_0 + \cdots\right)$$

$$y'_4 = \frac{1}{h}\left(\Delta y_0 + \frac{7}{2}\Delta^2 y_0 + \frac{13}{3}\Delta^3 y_0 + \frac{25}{12}\Delta^4 y_0 + \cdots\right)$$

Next, we neglect all differences beyond the fourth, and those we retain we replace with their equivalents in terms of the successive values of y according to Eqs. (8), Sec. 12.3, with f again replaced by y. For y'_1, this gives

$$y'_1 \doteq \frac{1}{h}\left[(y_1 - y_0) + \frac{1}{2}(y_2 - 2y_1 + y_0) - \frac{1}{6}(y_3 - 3y_2 + 3y_1 - y_0) + \frac{1}{12}(y_4 - 4y_3 + 6y_2 - 4y_1 + y_0)\right]$$

$$= \frac{1}{12h}(-3y_0 - 10y_1 + 18y_2 - 6y_3 + y_4) \qquad (2)$$

and, similarly,

$$y'_2 \doteq \frac{1}{12h}(y_0 - 8y_1 + 8y_3 - y_4) \qquad (3)$$

$$y'_3 \doteq \frac{1}{12h}(-y_0 + 6y_1 - 18y_2 + 10y_3 + 3y_4) \qquad (4)$$

$$y'_4 \doteq \frac{1}{12h}(3y_0 - 16y_1 + 36y_2 - 48y_3 + 25y_4) \qquad (5)$$

Now if we subtract Eq. (3) from twice the sum of Eq. (2) and Eq. (4) and solve the resulting equation for y_4, we obtain

$$y_4 = y_0 + \frac{4h}{3}(2y'_1 - y'_2 + 2y'_3) \qquad (6)$$

Since any value of x can be interpreted as x_0, Eq. (6) really expresses a relation between the value of y at an arbitrary point, the value of y at the fourth preceding tabular point, and the values of y' at the three intervening points. Hence Eq. (6) can be written in more general terms as

$$y_{n+1} = y_{n-3} + \frac{4h}{3}(2y'_{n-2} - y'_{n-1} + 2y'_n) \qquad (7)$$

If we know the values of y and y' at x_{n-3}, x_{n-2}, x_{n-1}, and x_n, Eq. (7) enables us to "reach out" one step further and compute y_{n+1}. With y_{n+1} known, we can return to the given differential equation (1) and compute y'_{n+1}. Then using Eq. (7) again, with n increased by 1 throughout, we can find y_{n+2}, and so on, step by step, until the solution has been tabulated over the desired set of x values. All that remains is to determine enough values of y and y' to get the process started.

One way to do this is to begin with the given initial values (x_0, y_0) and use the Runge-Kutta method to compute y_1, y_2, y_3 and the corresponding derivatives y'_1, y'_2, y'_3. With this information, Milne's method can now be applied and $y_4, y_5, \ldots$ can be computed, using Eqs. (6) and (7).

Another possibility is to begin the tabulation of y by expanding it in a Taylor series about the point $x = x_0$, assuming that such an expansion exists:

$$y = y_0 + y'_0(x - x_0) + \frac{y''_0}{2!}(x - x_0)^2 + \frac{y'''_0}{3!}(x - x_0)^3 + \cdots \tag{8}$$

The value of y_0 is, of course, given. The value of y'_0 can be found at once by substituting x_0 and y_0 into the given differential equation (1). To find the second derivative, we need only differentiate the given equation, getting

$$y'' = \frac{\partial f}{\partial x} + \frac{\partial f}{\partial y} y' \tag{9}$$

Since $f(x, y)$ is a given function, its partial derivatives are known and become definite numbers when x_0 and y_0 are substituted into them. Moreover, the value of y' at (x_0, y_0) has already been found, and thus (9) furnishes the value of y''_0. Similarly, differentiating (9) and evaluating the result at (x_0, y_0) will give y'''_0, and so on, assuming that the appropriate derivatives of $f(x, y)$ exist at (x_0, y_0). In this way the first few terms in the expansion of y around $x = x_0$ can be constructed. In especially favorable cases, an explicit formula for the general term of the series (8) can be found and the interval of convergence established. When this happens, (8) is the required solution of Eq. (1), and we need look no further. In general, however, successive differentiation of $f(x, y)$ becomes too complicated to continue, or the resulting series converges too slowly to be of practical value in computing y at tabular points relatively far from x_0.

On the other hand, for tabular values of x relatively close to x_0 and with (8) available as a representation of y in some neighborhood of $x = x_0$, we can set $x = x_0 + h \equiv x_1$ and calculate y_1. Similarly, setting $x = x_0 + 2h$ and $x_0 + 3h$, we can find y_2 and y_3. Then substituting (x_1, y_1), (x_2, y_2), and (x_3, y_3) into the given differential equation, we can compute y'_1, y'_2, and y'_3 without difficulty. With these values we are then in a position to begin the step-by-step solution of the differential equation by means of Eq. (7).

From the preceding discussion it would seem that Eq. (7) is, in general, adequate for the step-by-step solution of $y' = f(x, y)$. However, as a precaution against errors of various kinds, it is desirable to have a second, independent formula which can be used to check each computed value of y_{n+1}. To obtain such an expression, we return to Eqs. (2) through (5), add 4 times Eq. (4) to the sum of Eqs. (3) and (5), and solve the resulting equation for y_4, getting

$$y_4 = y_2 + \frac{h}{3}(y'_2 + 4y'_3 + y'_4)$$

or, in more general terms,

$$y_{n+1} = y_{n-1} + \frac{h}{3}(y'_{n-1} + 4y'_n + y'_{n+1}) \tag{10}$$

This formula cannot be used as a formula of extrapolation, since it involves y'_{n+1}, which cannot be found unless y_{n+1} is already known. However, after y_{n+1} has been calculated by means of (7), y'_{n+1} can be calculated, using Eq. (1), and enough information is then available to permit the use of (10) as another means of computing y_{n+1}. If the value of y_{n+1}, as given by (10), agrees with the value found from (7), we are ready to move on to the calculation of y_{n+2}. On the other hand, if the two values of y_{n+1} do not agree, we then use the second value of y_{n+1} to compute a new value of y'_{n+1}, which we substitute into (10) to find still another value for y_{n+1}. This process is continued until two successive values of y_{n+1} are in agreement. Once this happens, we are ready to continue the tabulation of y by returning to Eq. (7) and determining an initial estimate of y_{n+2}.

Formulas like (7), which express a new value exclusively in terms of quantities already found, are known as **open formulas** or **predictor formulas.** Those, like (10), which express a new value in terms of one or more additional new quantities and which, therefore, can be used only for purposes of checking and refining are known as **closed formulas** or **corrector formulas.**

The method of Milne is readily extended to the solution of simultaneous and higher-order equations. For instance, if we have the equations

$$y' = f(x, y, z) \qquad \text{and} \qquad z' = g(x, y, z) \tag{11}$$

with the initial conditions $y = y_0$, $z = z_0$ when $x = x_0$, and if by independent means we have calculated (y_1, y_2, y_3), (z_1, z_2, z_3), and the related quantities (y'_1, y'_2, y'_3) and (z'_1, z'_2, z'_3), then, using (7) and an identical version of it with z replacing y, we can compute y_4 and z_4. After that, we can compute y'_4 and z'_4 from the given differential equations and again use (7) to obtain y_5 and z_5, and so on, as far as desired. Of course, the closed formula (10) can be used to check and correct both y_{n+1} and z_{n+1} if and when this is deemed necessary.

The application of Milne's method to equations of higher order is immediate, since (as we have several times observed) such an equation can always be replaced by a system of simultaneous, first-order equations. For instance, $y'' = g(x, y, y')$ is equivalent to the system

$$y' = z \qquad z' = g(x, y, z)$$

which is just a special case, with $f(x, y, z) \equiv z$, of the general problem of two simultaneous, first-order equations.

Example 1 Tabulate the solution of $y'' + y^2 = x$ at intervals of $h = 0.1$ from $x = 0$ to $x = 0.6$, if $y = 0$ and $y' = 1$ when $x = 0$.

If we are to use either the Runge-Kutta method or Milne's method, it is first necessary to convert the given second-order equation into a pair of first-order equations by putting $z = y'$. This gives us the equivalent system

$$y' = z$$
$$z' = x - y^2 \qquad y = 0, z = 1 \quad \text{when } x = 0 \tag{12}$$

Beginning with this information, the Runge-Kutta method can be used to find $y_1, y_2, y_3, z_1, z_2, z_3$, and the corresponding derivatives which we must know before Milne's method can be applied. However, to illustrate the use of Taylor series, we shall obtain the necessary starting values from the expansions of y and z in terms of powers of x.

With a general pair of simultaneous, first-order differential equations such as (11), it is necessary to construct the series expansion of y and, independently, the series expansion of z. However, when a second-order equation of the form $y'' = f(x, y, y')$ is reduced to a pair of first-order equations, one of the equations is always $y' = z$. Hence in such cases the series for z can be found by simply differentiating the series for y after the latter has been constructed.

To determine the successive terms in the power series expansion of y in a neighborhood of $x = 0$, we need the values of the first few derivatives of y at $x = 0$. The values $y_0 = 0$ and $y'_0 = 1$ are given. That $y''_0 = 0$ follows immediately from the given differential equation. The values of $y'''_0, y^{iv}_0, y^{v}_0, y^{vi}_0, \ldots$ can be obtained by repeated differentiation of the given equation and evaluation of the results:

$$y''' = 1 - 2yy'$$
$$y^{iv} = -2yy'' - 2(y')^2$$
$$y^{v} = -2yy''' - 2y'y'' - 4y'y'' = -2yy''' - 6y'y''$$
$$y^{vi} = -2yy^{iv} - 2y'y''' - 6y'y''' - 6(y'')^2$$
$$= -2yy^{iv} - 8y'y''' - 6(y'')^2$$

....

Evaluating these derivatives at $x = x_0 = 0$, remembering that $y_0 = 0$, $y'_0 = 1$, and $y''_0 = 0$, we obtain

$$y'''_0 = 1, \; y^{iv}_0 = -2, \; y^{v}_0 = 0, \; y^{vi}_0 = -8, \ldots$$

Hence, substituting these values into Eq. (8), we find

$$y = x + \frac{1}{3!}x^3 - \frac{2}{4!}x^4 - \frac{8}{6!}x^6 + \cdots$$
$$= x + \tfrac{1}{6}x^3 - \tfrac{1}{12}x^4 - \tfrac{1}{90}x^6 + \cdots \tag{13}$$

and, by differentiating Eq. (13),

$$z = y' = 1 + \tfrac{1}{2}x^2 - \tfrac{1}{3}x^3 - \tfrac{1}{15}x^5 + \cdots \tag{14}$$

When the series (13) and (14) are evaluated for $x_1 = 0.1$, $x_2 = 0.2$, $x_3 = 0.3$, and the corresponding derivatives of y and z are computed from (12), we obtain the following starting values:

$$\begin{array}{llll} y_0 = 0.0000 & y'_0 = 1.0000 & z_0 = 1.0000 & z'_0 = 0.0000 \\ y_1 = 0.1002 & y'_1 = 1.0047 & z_1 = 1.0047 & z'_1 = 0.0900 \\ y_2 = 0.2012 & y'_2 = 1.0173 & z_2 = 1.0173 & z'_2 = 0.1595 \\ y_3 = 0.3038 & y'_3 = 1.0358 & z_3 = 1.0358 & z'_3 = 0.2077 \end{array}$$

From these, using Eq. (7) (and its counterpart with y replaced by z), we find

$$y_4 = 0.4085 \qquad \text{and} \qquad z_4 = 1.0581$$

Corresponding to these values, we find from (12) that

$$y'_4 = 1.0581 \qquad \text{and} \qquad z'_4 = 0.2331$$

When we recompute y_4 and y_5 using the closed formula (10), we obtain the same values for y_4 and z_4, so we accept them as correct.

Continuing the process, we return to Eq. (7) and compute y_5 and z_5, getting

$$y_5 = 0.5155 \qquad \text{and} \qquad z_5 = 1.0817$$

From these we find that

$$y'_5 = 1.0817 \qquad \text{and} \qquad z'_5 = 0.2343$$

Checking the values of y_5 and z_5 by recalculating them from Eq. (10), we obtain

$$y_5 = 0.5155 \qquad \text{and} \qquad z_5 = 1.0816$$

A second application of Eq. (10) repeats these values, which we therefore accept as correct.

Another application of Eq. (7) leads to the values

$$y_6 = 0.6248 \qquad \text{and} \qquad z_6 = 1.1041$$

which are again confirmed by Eq. (10). Continuing in this fashion, the solution of the given initial-value problem can be tabulated as far as desired.

Exercises for Section 12.4

1 Continue Example 1 by calculating y_7 and y_8.

2 Using the values of y_0, y_1, y_2, and y_3 found in Example 1, Sec. 12.2, calculate y_4 and y_5 by Milne's method.

3 Given the equation $y' = x - y$ and the starting values

x	y
0.0	1.0000
0.1	0.9097
0.2	0.8375
0.3	0.7816

find $y(0.4)$, $y(0.5)$, and $y(0.6)$ (*a*) using only Eq. (7); (*b*) using Eq. (10) to correct the values found from Eq. (7). Compare your answers in each case with those given by the exact solution.

4 Work Exercise 3, given the equation $y' = x^2 - y$ and the starting values

x	y
0.0	0.0000
0.1	0.0003
0.2	0.0025
0.3	0.0084

5 Using Milne's method, and the Maclaurin series for y to obtain the necessary starting values, tabulate the solution of the equation $y' = x + y^2$ at intervals of $h = 0.1$ from $x = 0$ to $x = 0.5$ if $y_0 = 0$.

6 Show that the use of Maclaurin's expansion alone is sufficient to solve the initial-value problem $y'' = xy$, $y_0 = 1$, $y'_0 = 0$.

7 Work Exercise 3 using the predictor formula

$$y_{n+1} = y_n + \frac{h}{12}(23y'_n - 16y'_{n-1} + 5y'_{n-2})$$

and the corrector formula

$$y_{n+1} = y_n + \frac{h}{12}(5y'_{n+1} + 8y'_n - y'_{n-1})$$

(These equations constitute the so-called **Adams-Moulton method** for the numerical solution of differential equations.)

8 Assuming the values of y_0, y_1, and y_2 of Example 1, continue the solution using the predictor formula

$$y_{n+1} = 2y_n - y_{n-1} + h^2(y''_n + \tfrac{1}{12}\Delta^2 y''_{n-2})$$

and the corrector formula

$$y_{n+1} = 2y_n - y_{n-1} + h^2(y''_n + \tfrac{1}{12}\Delta^2 y''_{n-1})$$

***9** Show that the error in Milne's predictor formula is $\frac{14}{45}h^5 y_0^{\text{v}}$. In other words, show that the dominant term in the difference between y_4, as given by its Maclaurin expansion, and y_4, as determined by combining the Maclaurin series for y'_1, y'_2, and y'_3 according to Milne's predictor formula, is $\frac{14}{45}h^5 y_0^{\text{v}}$.

***10** What is the error in Milne's corrector formula?

CHAPTER

THIRTEEN

LINEAR DIFFERENCE EQUATIONS

13.1 INTRODUCTION

The similarities between the difference operator Δ, defined in Sec. 12.3, and the derivative operator D,† and the fact that derivatives can be approximated by expressions involving differences, suggest that there may be a theory of *difference equations* roughly paralleling the theory of differential equations; and this is indeed the case. However, in the study of difference equations, we do not ordinarily consider equations of the form

$$f(\Delta)y = \phi(x)$$

as might be expected by analogy with the differential equation

$$f(D)y = \phi(x)$$

but rather equations of the form

$$F(E)y = \phi(x) \tag{1}$$

where E is the shift operator, also defined in Sec. 12.3. This, of course, is simply a matter of notational convenience, since, by using the operational equivalence $\Delta = E - 1$ [Eq. (6), Sec. 12.3], any function of Δ can be transformed at once into a function of E, and vice versa. In this chapter we shall investigate techniques for the solution of Eq. (1) when $F(E)$ is a polynomial in E with constant coefficients. Then, having learned how to solve such equations, we shall examine some of their simpler applications, which, interestingly enough, are often closely related to differential equations and systems of differential equations.

† The properties described in Theorem 4 and in Exercises 9 and 10, Sec. 12.3, are particularly suggestive.

Exercises for Section 13.1

Express each of the following equations as an equation of the form $F(E)y = \phi(x)$.

1 $\Delta^2 y - 3\,\Delta y + 2y = 0$ **2** $\Delta^3 y - 6\,\Delta^2 y + 11\,\Delta y - 6y = x^2$

3 $(\Delta^2 + 2\,\Delta + 2)y = 3^x$ **4** $y_{n+2} = y_{n+1} + y_n$

5 If $F(E)$ is the expression obtained from $f(\Delta)$ by the substitution $\Delta = E - 1$, what is the relation between the roots of the equations $f(\Delta) = 0$ and $F(E) = 0$?

13.2 HOMOGENEOUS, LINEAR DIFFERENCE EQUATIONS

By a **linear, constant-coefficient difference equation** we mean an equation of the form

$$(a_0 E^r + a_1 E^{r-1} + \cdots + a_{r-1} E + a_r)y = \phi(x) \qquad a_0, a_1, \ldots, a_r \text{ constants} \quad (1)$$

Since the substitution $t = hx$ will transform a function of t tabulated at intervals of h into a function of x tabulated at unit intervals, it is clearly no restriction to assume $h = 1$, so that invariably $Ef(x) = f(x + 1)$, and we shall do this throughout this chapter.

In Eq. (1), if both a_0 and a_r are different from zero, as we shall henceforth suppose, the positive integer r is called the **order** of the equation. If $\phi(x)$ is identically zero, Eq. (1) is said to be **homogeneous.**† If $\phi(x)$ is not identically zero, Eq. (1) is said to be **nonhomogeneous.** By a **solution** of Eq. (1) we mean a function of x with the property that when substituted into (1), it reduces the equation to an identity for $x = x_0 \pm n, n = \ldots, -2, -1, 0, 1, 2, \ldots, x_0$ arbitrary. In this section we shall be concerned with developing methods for the solution of homogeneous, linear difference equations with constant coefficients. The solution of nonhomogeneous equations we will take up in the next section.

Example 1 Verify that for all values of the constants c_1 and c_2, the expression $y = c_1 + c_2 2^x$ is a solution of the equation

$$(E^2 - 3E + 2)y = 0$$

Since $(E^2 - 3E + 2)y$ means $E^2 y - 3Ey + 2y$, substitution of $y = c_1 + c_2 2^x$ yields

$$(c_1 + c_2 2^{x+2}) - 3(c_1 + c_2 2^{x+1}) + 2(c_1 + c_2 2^x)$$

$$= c_1(1 - 3 + 2) + c_2 2^x(2^2 - 3 \cdot 2 + 2) = c_1 \cdot 0 + c_2 \cdot 0 = 0$$

as asserted.

It is interesting to note that the function $y = c_1 + c_2 2^x$ is by no means the only solution of the equation of Example 1. In fact, since $Ef(x) = f(x + 1) = f(x)$ for

† A homogeneous, linear difference equation with constant coefficients is sometimes referred to as a **linear recurrence formula,** since it is simply a repeating, or *recurring*, relation between the values of y at *any* $r + 1$ consecutive values of x.

any function $f(x)$ of period 1, it follows that if $f_1(x)$ and $f_2(x)$ are arbitrary functions of period 1, then both $c_1 f_1(x)$ and $c_2 f_2(x)$ behave as constants as far as the operator E is concerned. Therefore

$$y = c_1 f_1(x) + c_2 f_2(x)2^x$$

is also a solution of the equation $(E^2 - 3E + 2)y = 0$. In practical problems, however, one is seldom interested in x and y as continuous variables connected by Eq. (1), and in our work we shall attempt no more than the determination of y as a function of x on the domain set $x = \ldots, -2, -1, 0, 1, 2, \ldots$. Hence, we shall consistently ignore the possibility of including functions of period 1 in the coefficients of solutions such as $y = c_1 + c_2 2^x$. Moreover, we shall henceforth emphasize the discrete character of the domain set $\ldots, -2, -1, 0, 1, 2, \ldots$ by using n rather than x as our independent variable.

It is important to note that when suitable starting values of y are given, the exact solution of any homogeneous, linear difference equation can be tabulated in a step-by-step fashion over any desired range of x values. In fact, given the equation

$$(a_0 E^r + a_1 E^{r-1} + \cdots + a_{r-1} E + a_r)y = 0 \tag{2}$$

we can always divide by a_0 and rename the coefficients, getting

$$(E^r - p_1 E^{r-1} - \cdots - p_{r-1} E - p_r)y = 0$$

Then, setting $E^k y = y_k$, we may solve for $E^r y \equiv y_r$, getting

$$y_r = p_1 y_{r-1} + p_2 y_{r-2} + \cdots + p_{r-1} y_1 + p_r y_0 \tag{3}$$

Thus if we are given the values of $y(n)$ for $n = 0, 1, 2, \ldots, r - 1$,† say $y_0, y_1, y_2, \ldots, y_{r-1}$, we can compute $y(r) = y_r$ immediately. Then, noting that Eq. (3) is simply a relation connecting the values of y at *any* set of $r + 1$ consecutive, equally spaced values of x, we can advance each subscript by 1 and thus obtain y_{r+1}:

$$y_{r+1} = p_1 y_r + p_2 y_{r-1} + \cdots + p_{r-1} y_2 + p_r y_1$$

Since y_r has just been found, this equation gives us the value of y_{r+1}. Continuing in this fashion, the solution for y can be tabulated as far as required.‡

Example 2 If $(E^2 - E - 6)y = 0$ and $y_0 = 0$, $y_1 = 1$, determine $y_2, y_3, \ldots, y_8$.

Clearly, the given equation is equivalent to the recurrence relation $y_{n+2} = y_{n+1} + 6y_n$, $n = 0, 1, 2, 3, \ldots$. Hence, beginning with $n = 0$ and substituting the given starting values, we have

$$y_2 = y_1 + 6y_0 = 1 + 6 \cdot 0 = 1$$

† This, of course, is analogous to being given the values of $y_0, y_0', \ldots, y_0^{(r-1)}$ in an initial-value problem involving a linear differential equation of order r.

‡ The tabulation process we have just described can be applied equally well to Eq. (3) in the case when the coefficients are suitably defined functions of n, provided that the values of $y(n)$ for $n = 0, 1, 2, \ldots, r - 1$ are given.

Then, taking $n = 1$, we have

$$y_3 = y_2 + 6y_1 = 1 + 6 \cdot 1 = 7$$

and, continuing in this fashion,

$$y_4 = 13 \qquad y_5 = 55 \qquad y_6 = 133 \qquad y_7 = 463 \qquad y_8 = 1{,}261$$

From the preceding discussion and the accompanying example, it is evident that the *tabulation* of the solution of a linear difference equation is a simple matter. What we really want, however, is a closed *formula* for a complete solution containing arbitrary constants which can be determined to fit prescribed starting values. For linear difference equations with constant coefficients, the determination of a complete solution is not difficult and, in fact, is similar to the methods we have learned for the solution of linear differential equations with constant coefficients. For second-order equations, the process is based on the following theorems, which are analogous, respectively, to Theorem 1, Sec. 1.2, and Theorem 4, Sec. 3.2.

Theorem 1 If $y_1(n)$ and $y_2(n)$ are any two solutions of the homogeneous equation $(a_0 E^2 + a_1 E + a_2)y = 0$, then $c_1 y_1(n) + c_2 y_2(n)$, where c_1 and c_2 are arbitrary constants, is also a solution.

Theorem 2 If $y_1(n)$ and $y_2(n)$ are two solutions of the homogeneous equation $(a_0 E^2 + a_1 E + a_2)y = 0$ for which

$$C[y_1(n), y_2(n)]\dagger = \begin{vmatrix} y_1(n) & y_2(n) \\ Ey_1(n) & Ey_2(n) \end{vmatrix} \neq 0$$

then any solution $y_3(n)$ of the homogeneous equation can be written in the form $y_3(n) = c_1 y_1(n) + c_2 y_2(n)$ where c_1 and c_2 are suitable constants.

The proof of Theorem 1 and the proof of Theorem 2 when the coefficients are constants are simple, and we shall leave them as exercises. The proof of Theorem 2 when the coefficients are nonconstant functions of n we shall leave to more complete texts on difference equations.‡

To find particular nontrivial solutions of the homogeneous equation

$$(a_0 E^2 + a_1 E + a_2)y = 0 \tag{4}$$

where the coefficients a_0, a_1, a_2 are constants, we might try, as with the analogous differential equation, the substitution

$$y = e^{mn} \tag{5}$$

† The function $C[y_1(n), y_2(n)]$ is customarily referred to as **Casorati's determinant,** after the Italian mathematician Felice Casorati (1835–1890). Its resemblance to the wronskian $W[y_1(x), y_2(x)]$ (see Sec. 3.2) is apparent.

‡ See, for instance, L. M. Milne-Thompson, "The Calculus of Finite Differences," Macmillan and Co., London, 1933, pp. 354–55.

However, it is more convenient to assume

$$y = M^n \qquad M \neq 0 \tag{6}$$

which is clearly equivalent to (5) with $M = e^m$. Substituting this into Eq. (4) and recalling our agreement that $Ef(n) = f(n+1)$, we obtain

$$a_0 M^{n+2} + a_1 M^{n+1} + a_2 M^n = 0$$

or, dividing out M^n, which is not zero,

$$a_0 M^2 + a_1 M + a_2 = 0 \tag{7}$$

Naturally enough, Eq. (7) is called the **characteristic equation** of the difference equation (4).

If the roots M_1 and M_2 of (7) are distinct, then

$$C(M_1^n, M_2^n) = M_1^n M_2^{n+1} - M_2^n M_1^{n+1} = M_1^n M_2^n (M_2 - M_1) \neq 0\dagger$$

and hence, by Theorem 2, a complete solution of Eq. (4) is

$$y = c_1 M_1^n + c_2 M_2^n \tag{8}$$

If M_1 and M_2 are real, this is a completely acceptable form of the solution. However, if M_1 and M_2 are complex, then (8) is inconvenient for most purposes, and it is desirable that we reduce it to a more useful form. To do this, let the roots of the characteristic equation be

$$M_1, M_2 = p \pm iq = re^{\pm i\theta}$$

where $\qquad r = \sqrt{p^2 + q^2} \qquad$ and $\qquad \tan \theta = \dfrac{q}{p}$ ‡

Then we can write

$$\begin{aligned} y &= c_1(re^{i\theta})^n + c_2(re^{-i\theta})^n \\ &= r^n(c_1 e^{i\theta n} + c_2 e^{-i\theta n}) \\ &= r^n[c_1(\cos \theta n + i \sin \theta n) + c_2(\cos \theta n - i \sin \theta n)] \\ &= r^n[(c_1 + c_2) \cos \theta n + i(c_1 - c_2) \sin \theta n)] \end{aligned}$$

or, renaming the constants,

$$y = r^n(A \cos \theta n + B \sin \theta n) \tag{9}$$

If $M_1 = M_2$, clearly $C(M_1^n, M_2^n) = 0$, and we must find a second, independent solution before we can construct a complete solution of Eq. (4). Again, by analogy with differential equations, we are led to try

$$y = nM_1^n$$

† Since $a_2 \neq 0$ [or else the difference equation (4) would be of order less than 2, contrary to hypothesis], it follows from (7) that neither M_1 nor M_2 can be zero.

‡ For a discussion of the exponential form of a complex number, see item 2, Appendix B.1.

Table 13.1

Difference equation: $(a_0 E^2 + a_1 E + a_2)y = 0 \qquad a_0, a_2 \neq 0$
Characteristic equation: $a_0 M^2 + a_1 M + a_2 = 0$

Nature of the roots of the characteristic equation	Condition on the coefficients of the characteristic equation	Complete solution of the difference equation
Real and unequal $M_1 \neq M_2$	$a_1^2 - 4a_0 a_2 > 0$	$y = c_1 M_1^n + c_2 M_2^n$
Real and equal $M_1 = M_2$	$a_1^2 - 4a_0 a_2 = 0$	$y = c_1 M_1^n + c_2 n M_1^n$
Conjugate complex $M_1 = p + iq$, $M_2 = p - iq$	$a_1^2 - 4a_0 a_2 < 0$	$y = r^n(A \cos \theta n + B \sin \theta n)$, $r = \sqrt{p^2 + q^2}$, $\tan \theta = q/p$

And we find by direct substitution that this is indeed a solution when the characteristic equation (7) has equal roots, for we have, on substituting,

$$\begin{aligned} &a_0(n + 2)M_1^{n+2} + a_1(n + 1)M_1^{n+1} + a_2 n M_1^n \\ &= nM_1^n(a_0 M_1^2 + a_1 M_1 + a_2) + M_1^{n+1}(2a_0 M_1 + a_1) \end{aligned}$$

and this is identically zero since the coefficient of nM_1^n vanishes because, in any case, M_1 satisfies the characteristic equation (7); and the coefficient of M_1^{n+1} vanishes because when the characteristic equation has equal roots, their common value is $M_1 = -a_1/(2a_0)$. Moreover, for the solutions $y_1 = M_1^n$ and $y_2 = nM_1^n$, we have

$$C(M_1^n, nM_1^n) = M_1^n(n + 1)M_1^{n+1} - nM_1^n M_1^{n+1} = M_1^{2n+1} \neq 0$$

Hence, according to Theorem 2, a complete solution of Eq. (4) when the characteristic equation has equal roots is

$$y = c_1 M_1^n + c_2 n M_1^n \tag{10}$$

The results of the preceding discussion are summarized in Table 13.1.

Example 3 Find a complete solution of the difference equation

$$(E^2 + 2E + 4)y = 0$$

The characteristic equation in this case is $M^2 + 2M + 4 = 0$, and its roots are $M_1, M_2 = -1 \pm i\sqrt{3}$. Since

$$r = \sqrt{(-1)^2 + (\sqrt{3})^2} = 2 \qquad \text{and} \qquad \theta = \tan^{-1} \frac{\sqrt{3}}{-1} = \frac{2\pi}{3}$$

we have as a complete solution

$$y = 2^n\left(A \cos \frac{2\pi n}{3} + B \sin \frac{2\pi n}{3}\right)$$

Example 4 Find a closed formula for the particular solution of the equation $(E^2 - E - 6)y = 0$ that was tabulated in Example 2.

The characteristic equation of the difference equation $(E^2 - E - 6)y = 0$ is

$$M^2 - M - 6 = 0$$

and its roots are $M_1 = -2$ and $M_2 = 3$. Hence a complete solution of the equation is

$$y = c_1(-2)^n + c_2 3^n$$

Since we are given the starting values $y_0 = 0$ and $y_1 = 1$, the coefficients c_1 and c_2 must satisfy the conditions

$$c_1 + c_2 = 0$$

$$-2c_1 + 3c_2 = 1$$

Hence $c_1 = -\frac{1}{5}$, $c_2 = \frac{1}{5}$, and the required solution is

$$y = -\tfrac{1}{5}(-2)^n + \tfrac{1}{5}3^n$$

Example 5 Show that the equation $(E^2 - 2\lambda E + 1)y = 0$ has the indicated solution in each of the following cases:

a $\lambda < -1$ $\quad y = (-1)^n(A \cosh \mu n + B \sinh \mu n) \quad \cosh \mu = -\lambda$
b $\lambda = -1$ $\quad y = (-1)^n(A + Bn)$
c $-1 < \lambda < 1$ $\quad y = A \cos \mu n + B \sin \mu n \quad \cos \mu = \lambda$
d $\lambda = 1$ $\quad y = A + Bn$
e $\lambda > 1$ $\quad y = A \cosh \mu n + B \sinh \mu n \quad \cosh \mu = \lambda$

Following our general procedure, we first set up the characteristic equation

$$M^2 - 2\lambda M + 1 = 0 \tag{11}$$

and determine its roots

$$M_1, M_2 = \lambda \pm \sqrt{\lambda^2 - 1} \tag{12}$$

The structure of this formula for the roots makes it clear why we have to consider five separate cases. The individual cases are handled as follows.

a Since $\lambda < -1$ implies $-\lambda > 1$, it follows that it is possible to define a quantity μ by the equation

$$\cosh \mu = -\lambda$$

Under this substitution the roots (12) become

$$\begin{aligned} M_1, M_2 &= -\cosh \mu \pm \sqrt{\cosh^2 \mu - 1} = -\cosh \mu \pm \sinh \mu \\ &= -(\cosh \mu + \sinh \mu), -(\cosh \mu - \sinh \mu) \\ &= -e^{\mu}, -e^{-\mu} \end{aligned}$$

Therefore $y = c_1(-e^{\mu})^n + c_2(-e^{-\mu})^n = (-1)^n(c_1 e^{\mu n} + c_2 e^{-\mu n})$

$$\begin{aligned} &= (-1)^n[c_1(\cosh \mu n + \sinh \mu n) + c_2(\cosh \mu n - \sinh \mu n)] \\ &= (-1)^n[(c_1 + c_2)\cosh \mu n + (c_1 - c_2)\sinh \mu n] \\ &= (-1)^n(A \cosh \mu n + B \sinh \mu n) \end{aligned}$$

where $A = c_1 + c_2$ and $B = c_1 - c_2$.

b In this case, $\lambda = -1$ so that $M_1 = M_2 = -1$ and therefore

$$y = A(-1)^n + Bn(-1)^n = (-1)^n(A + Bn)$$

c Since $-1 < \lambda < 1$, it is possible to define a quantity μ by the equation

$$\cos \mu = \lambda \qquad 0 < \mu < \pi$$

Under this substitution the roots (12) become

$$\begin{aligned} M_1, M_2 &= \cos \mu \pm \sqrt{\cos^2 \mu - 1} = \cos \mu \pm \sqrt{-\sin^2 \mu} \\ &= \cos \mu + i \sin \mu, \cos \mu - i \sin \mu \\ &= e^{i\mu}, e^{-i\mu} \end{aligned}$$

Therefore
$$\begin{aligned} y &= c_1(e^{i\mu})^n + c_2(e^{-i\mu})^n \\ &= c_1(\cos \mu n + i \sin \mu n) + c_2(\cos \mu n - i \sin \mu n) \\ &= (c_1 + c_2) \cos \mu n + i(c_1 - c_2) \sin \mu n \\ &= A \cos \mu n + B \sin \mu n \end{aligned}$$

where $A = c_1 + c_2$ and $B = i(c_1 - c_2)$.

d In this case, $\lambda = 1$ so that $M_1 = M_2 = 1$, and therefore

$$y = A + Bn$$

e Since $\lambda > 1$, it is possible to define a quantity μ by the equation

$$\cosh \mu = \lambda$$

Under this substitution the roots (12) become

$$M_1, M_2 = \cosh \mu \pm \sqrt{\cosh^2 \mu - 1} = \cosh \mu \pm \sinh \mu = e^{\mu}, e^{-\mu}$$

Therefore
$$\begin{aligned} y &= c_1(e^{\mu})^n + c_2(e^{-\mu})^n \\ &= c_1(\cosh \mu n + \sinh \mu n) + c_2(\cosh \mu n - \sinh \mu n) \\ &= (c_1 + c_2) \cosh \mu n + (c_1 - c_2) \sinh \mu n \\ &= A \cosh \mu n + B \sinh \mu n \end{aligned}$$

where, as in **a**, $A = c_1 + c_2$ and $B = c_1 - c_2$.

Exercises for Section 13.2

Verify that each of the following equations has the indicated solution.

1 $(E^2 + 7E + 12)y = 0$ $\qquad y = c_1(-3)^n + c_2(-4)^n$

2 $(E^3 + 2E^2 - 5E - 6)y = 0$ $\qquad y = c_1(-1)^n + c_2 2^n + c_3(-3)^n$

3 $(E^2 + E + 1)y = 0$ $\qquad y = c_1 \cos (2\pi n/3) + c_2 \sin (2\pi n/3)$

Tabulate from $n = 0$ to $n = 10$ the solution of each of the following equations determined by the given starting values.

4 $(E^2 - 2E + 2)y = 0$ $\qquad y_0 = 1, y_1 = -1$

5 $(E^3 - E^2 + E - 1)y = 0$ $\qquad y_0 = 1, y_1 = 0, y_2 = 2$

6 $(E^2 - E + 2)y = 0$ $\qquad y_0 = 1, \Delta y_0 = 3$

7 Find a complete solution of each of the following equations.

(*a*) $(E^2 - 7E + 10)y = 0$ (*b*) $(E^2 + 6E + 9)y = 0$
(*c*) $(E^2 + 2E + 2)y = 0$ (*d*) $(\Delta^2 - 3\,\Delta + 2)y = 0$
(*e*) $(E^2 + 9)y = 0$ (*f*) $(E^3 - 9E^2 + 26E - 24)y = 0$

8 Find the solution of each of the following equations which satisfies the given conditions.

(*a*) $(E^2 - 2E - 3)y = 0$ $y_0 = 0,\ y_1 = -4$
(*b*) $(E^2 + 4E + 4)y = 0$ $y_0 = 2,\ y_1 = -2$
(*c*) $(E^2 + 9)y = 0$ $y_0 = 1,\ y_1 = -3$
(*d*) $(E^3 - 6E^2 + 11E - 6)y = 0$ $y_0 = 4,\ \Delta y_0 = 1,\ \Delta^2 y_0 = -1$

9 Prove Theorem 1.

10 Discuss the solution of each of the following equations:

(*a*) $Ey = 0$ (*b*) $(a_0 E^2 + a_1 E)y = 0$ $a_0,\ a_1$ constants

***11** If $y_1(n)$ and $y_2(n)$ are any two solutions of the general, linear, second-order difference equation

$$[a_0(n)E^2 + a_1(n)E + a_2(n)]y = 0$$

show that Casorati's determinant $C(y_1, y_2)$ satisfies the relation

$$[a_0(n)E - a_2(n)]C = 0$$

Hint: Write down the conditions that both $y_1(n)$ and $y_2(n)$ satisfy the given equation; then eliminate the terms $Ey_1(n)$ and $Ey_2(n)$.

***12** Prove Theorem 2 in the special case where the coefficients are constants. *Hint:* Recall the proof of Theorem 4, Sec. 3.2, and then use the result of Exercise 11.

13.3 NONHOMOGENEOUS, LINEAR DIFFERENCE EQUATIONS

The solution of nonhomogeneous, linear difference equations is based on the following theorem, whose resemblance to Theorem 1, Sec. 3.4, is unmistakable.

Theorem 1 If $Y(n)$ is any solution of the nonhomogeneous equation

$$(a_0 E^r + a_1 E^{r-1} + \cdots + a_{r-1} E + a_r)y = \phi(n)$$

and if $c_1 y_1(n) + c_2 y_2(n) + \cdots + c_r y_r(n)$ is a complete solution of the homogeneous equation obtained from this by deleting the term $\phi(n)$, then any solution $y(n)$ of the nonhomogeneous equation can be written in the form

$$y(n) = c_1 y_1(n) + c_2 y_2(n) + \cdots + c_r y_r(n) + Y(n)$$

where $c_1, c_2, \ldots, c_r$ are suitable constants.

As in the case of differential equations, the homogeneous equation which results when $\phi(n)$ is deleted from the nonhomogeneous equation is called the **related homogeneous equation,** and the complete solution of the related homogeneous equation is called the **complementary function** of the nonhomogeneous equation. By analogy, the particular solution $Y(n)$ is often referred to as a **particular**

Table 13.2

Difference equation: $(a_0 E^2 + a_1 E + a_2)y = \phi(n)$

$\phi(n)$†	Necessary choice for particular solution Y‡
1. α (constant)	A
2. αn^k (k a positive integer)	$A_0 n^k + A_1 n^{k-1} + \cdots + A_{k-1} n + A_k$
3. αk^n	Ak^n
4. $\alpha \cos kn$	$A \cos kn + B \sin kn$
5. $\alpha \sin kn$	
6. $\alpha n^k l^n \cos mn$	$(A_0 n^k + \cdots + A_{k-1} n + A_k) l^n \cos mn$
7. $\alpha n^k l^n \sin mn$	$+ (B_0 n^k + \cdots + B_{k-1} n + B_k) l^n \sin mn$

† When $\phi(n)$ consists of a sum of several terms, the appropriate choice for Y is the sum of the Y expressions corresponding to these terms individually.

‡ Whenever a term in any of the Y's listed in this column duplicates a term in the complementary function, all terms in that Y expression must be multiplied by the lowest positive integral power of n sufficient to eliminate all such duplications.

integral of the nonhomogeneous equation, even though no integration is involved in its determination. In our work we will almost always be concerned with nonhomogeneous equations of the second order.

To solve the nonhomogeneous equation

$$(a_0 E^2 + a_1 E + a_2)y = \phi(n) \qquad a_0, a_1, a_2 \text{ constants} \tag{1}$$

by means of Theorem 1, we must add a particular solution of (1) to a complete solution of the related homogeneous equation. To find the necessary particular solution Y, we use the method of undetermined coefficients very much as we did in solving nonhomogeneous differential equations in Sec. 4.3. And here, as in Sec. 4.3, this method is able to handle only equations in which $\phi(n)$ is a linear combination of terms or products of terms of the form†

$$k^n \qquad \cos kn \qquad \sin kn \qquad\qquad k \text{ a constant}$$

and

$$n^k \qquad\qquad k \text{ a nonnegative integer}$$

We begin by assuming for Y an arbitrary linear combination of all the terms which arise from $\phi(n)$ by repeatedly applying the operator E. As in the case of differential equations, if any term in the initial choice for Y duplicates a term in the complementary function, it and all terms associated with it in Y (see Table 13.2) must be multiplied by the lowest positive integral power of n which will eliminate all duplications. When the form of Y is thus determined, Y is substituted into the difference equation and the arbitrary constants in Y are chosen to make the resulting equation an identity. The procedure is summarized in Table 13.2.

Example 1 Find a complete solution of the equation $(E^2 + 2E - 8)y = 5n + 3^n$.

The characteristic equation in this case is $M^2 + 2M - 8 = 0$, and from its roots $M_1 = 2$ and $M_2 = -4$ we can immediately construct the complementary function

$$c_1 2^n + c_2(-4)^n$$

According to lines 2 and 3 in Table 13.2, the particular integral we would normally try is

$$Y = An + B + C3^n$$

and since none of the terms in this expression duplicates, i.e., is linearly dependent upon, a term in the complementary function, we proceed with this trial solution without modification. Substituting Y into the given equation yields

$$[A(n+2) + B + C3^{n+2}] + 2[A(n+1) + B + C3^{n+1}] - 8[An + B + C3^n] = 5n + 3^n$$

or, collecting like terms,

$$-5An + (4A - 5B) + 7C3^n = 5n + 3^n$$

† In Sec. 4.3 we identified these functions as those which possess only a finite number of linearly independent derivatives. Here they appear as the functions which yield only a finite number of linearly independent terms under repeated applications of the operator E.

This will be an identity if and only if

$$-5A = 5 \qquad 4A - 5B = 0 \qquad 7C = 1$$

Hence

$$A = -1 \qquad B = -\tfrac{4}{5} \qquad C = \tfrac{1}{7}$$

Therefore $Y = -n - \frac{4}{5} + \frac{1}{7}3^n$, and the required solution is

$$y = c_1 2^n + c_2(-4)^n - n - \tfrac{4}{5} + \tfrac{1}{7}3^n$$

Example 2 Find a complete solution of the equation $(E^2 - 5E + 6)y = 2^n$.

In this case the characteristic equation is $M^2 - 5M + 6 = 0$, and from its roots $M_1 = 2$ and $M_2 = 3$ we obtain at once the complementary function

$$c_1 2^n + c_2 3^n$$

According to line 3 in Table 13.2, we would normally try $Y = A2^n$ as a particular integral. However, since $A2^n$ and the term $c_1 2^n$ in the complementary function are linearly dependent, we must modify Y, according to the second footnote in the table, and continue instead with

$$Y = An2^n$$

Substituting this into the given equation, we have

$$A(n+2)2^{n+2} - 5A(n+1)2^{n+1} + 6An2^n = 2^n$$

or, dividing by 2^n and collecting terms,

$$-2A2^n = 2^n$$

Hence $A = -\frac{1}{2}$, $Y = -\frac{1}{2}n2^n$, and the required solution is

$$y = c_1 2^n + c_2 3^n - \tfrac{1}{2}n2^n$$

Example 3 Find a complete solution of the equation $(E^2 - 4E + 4)y = 5 \sin (\pi n/2)$.

In this case the characteristic equation is $M^2 - 4M + 4 = 0$, and its roots are $M_1 = M_2 = 2$. Hence the complementary function is

$$c_1 2^n + c_2 n 2^n$$

As a particular integral we would normally try

$$Y = A \cos \frac{\pi n}{2} + B \sin \frac{\pi n}{2}$$

and since neither of these terms is linearly dependent upon anything in the complementary function, we continue with it without modification. Substituting Y into the given equation, we obtain

$$\left[A \cos \frac{\pi(n+2)}{2} + B \sin \frac{\pi(n+2)}{2}\right] - 4\left[A \cos \frac{\pi(n+1)}{2} + B \sin \frac{\pi(n+1)}{2}\right]$$
$$+ 4\left[A \cos \frac{\pi n}{2} + B \sin \frac{\pi n}{2}\right] = 5 \sin \frac{\pi n}{2}$$

Hence, simplifying the trigonometric functions and collecting terms, we have

$$\left[-A\cos\frac{\pi n}{2} - B\sin\frac{\pi n}{2}\right] - 4\left[-A\sin\frac{\pi n}{2} + B\cos\frac{\pi n}{2}\right] + 4\left[A\cos\frac{\pi n}{2} + B\sin\frac{\pi n}{2}\right] = 5\sin\frac{\pi n}{2}$$

and $$(3A - 4B)\cos\frac{\pi n}{2} + (4A + 3B)\sin\frac{\pi n}{2} = 5\sin\frac{\pi n}{2}$$

The last equation will be an identity if and only if

$$3A - 4B = 0 \qquad \text{and} \qquad 4A + 3B = 5$$

Hence, solving these simultaneously,

$$A = \frac{4}{5} \qquad B = \frac{3}{5} \qquad \text{and} \qquad Y = \frac{4}{5}\cos\frac{\pi n}{2} + \frac{3}{5}\sin\frac{\pi n}{2}$$

The required solution is therefore

$$y = c_1 2^n + c_2 n 2^n + \frac{4}{5}\cos\frac{\pi n}{2} + \frac{3}{5}\sin\frac{\pi n}{2}$$

Exercises for Section 13.3

Find a complete solution of each of the following equations.

1 $(E^2 - 3E - 10)y = 3(2^n)$

2 $(E^2 - 6E + 9)y = 6$

3 $(\Delta^2 - 7\Delta + 12)y = 3^n$

4 $(E^3 + 2E^2 - 4E - 8)y = 3^n$

5 $(E^2 - E - 2)y = 2^{2n}$

6 $(E^2 + 2E - 3)y = 2$

7 $(E^2 - E - 6)y = 6n^2$

8 $(E^3 - 4E^2 + 6E - 4)y = n + 1$

9 $(E^2 + 2E + 1)y = 1$

10 $(E^2 - 3E + 2)y = 2^n + 2^{-n}$

***11** $(E^2 + 4)y = \cos n$

12 $(E^2 - 4E + 4)y = 2^n$

13 $(E^4 + 8E^2 - 9)y = 20$

14 $(E^4 + 10E^2 + 9)y = \cos\frac{\pi n}{4}$

15 Prove Theorem 1.

13.4 APPLICATIONS OF LINEAR DIFFERENCE EQUATIONS

There are numerous problems in which the values of a function defined on a set of integers can be connected by a linear recurrence relation; and when this is the case, a formula for the function can be found by solving the corresponding difference equation. Our first two examples involve problems of this sort.

Example 1 If D_n is the nth-order determinant

$$\begin{vmatrix} 1+a^2 & a & 0 & \cdots & 0 & 0 \\ a & 1+a^2 & a & \cdots & 0 & 0 \\ 0 & a & 1+a^2 & \cdots & 0 & 0 \\ \cdots & \cdots & \cdots & \cdots & \cdots & \cdots \\ 0 & 0 & 0 & \cdots & 1+a^2 & a \\ 0 & 0 & 0 & \cdots & a & 1+a^2 \end{vmatrix}$$

show that $D_n = (1 + a^2)D_{n-1} - a^2 D_{n-2}$ and determine D_n as a function of n.

The form of the given recurrence relation suggests that we should begin by expanding D_n in terms of the elements in its first row (or column). Doing this, we obtain

$$D_n = (1 + a^2)\begin{vmatrix} 1+a^2 & a & \cdots & 0 & 0 \\ a & 1+a^2 & \cdots & 0 & 0 \\ \cdots & \cdots & \cdots & \cdots & \cdots \\ 0 & 0 & \cdots & 1+a^2 & a \\ 0 & 0 & \cdots & a & 1+a^2 \end{vmatrix}$$

$$-a\begin{vmatrix} a & a & \cdots & 0 & 0 \\ 0 & 1+a^2 & \cdots & 0 & 0 \\ \cdots & \cdots & \cdots & \cdots & \cdots \\ 0 & 0 & \cdots & 1+a^2 & a \\ 0 & 0 & \cdots & a & 1+a^2 \end{vmatrix}$$

The first of the two determinants on the right is identical in structure to D_n except that it contains only $n - 1$ rows and $n - 1$ columns; in other words, it is just D_{n-1}. The second determinant on the right does not have the form of D_n, but if we expand it in terms of the elements in its first column, we obtain a times a determinant of order $n - 2$ which is of the same structure as D_n, that is, is just D_{n-2}. Thus we have verified the asserted recurrence relation

$$D_n = (1 + a^2)D_{n-1} - a^2 D_{n-2}$$

or, equivalently,

$$[E^2 - (1 + a^2)E + a^2]D_n = 0 \qquad n = 1, 2, 3, \ldots$$

The characteristic equation of this difference equation is $M^2 - (1 + a^2)M + a^2 = 0$, and from its roots $M_1 = 1$ and $M_2 = a^2$ we can immediately construct the complete solution

$$D = c_1 + c_2(a^2)^n \qquad a \neq \pm 1 \tag{1}$$

To determine c_1 and c_2, we return to the definition of D_n and note that

$$D_1 = 1 + a^2 \qquad \text{and} \qquad D_2 = \begin{vmatrix} 1+a^2 & a \\ a & 1+a^2 \end{vmatrix} = 1 + a^2 + a^4$$

Hence, substituting these into the complete solution (1), we obtain the conditions

$$1 + a^2 = c_1 + c_2 a^2 \qquad \text{and} \qquad 1 + a^2 + a^4 = c_1 + c_2 a^4$$

Solving these simultaneously, we find that

$$c_1 = \frac{1}{1 - a^2} \qquad \text{and} \qquad c_2 = -\frac{a^2}{1 - a^2} \qquad a \neq \pm 1$$

Therefore $\quad D_n = \dfrac{1}{1-a^2} - \dfrac{a^2}{1-a^2}(a^2)^n = \dfrac{1-a^{2n+2}}{1-a^2} = 1 + a^2 + a^4 + \cdots + a^{2n}$

If $a = \pm 1$, the recurrence relation becomes

$$(E^2 - 2E + 1)D_n = 0 \qquad n = 1, 2, 3, \ldots$$

and its characteristic equation has the repeated root $M = 1$. Hence, in this case,

$$D_n = c_3 + c_4 n$$

When $a = \pm 1$, the values of D_1 and D_2 are 2 and 3, respectively. Therefore c_3 and c_4 must satisfy the equations

$$c_3 + c_4 = 2$$

$$c_3 + 2c_4 = 3$$

and we find that $c_3 = c_4 = 1$. The solution when $a = \pm 1$ is thus

$$D_n = 1 + n$$

which is just what the answer in general becomes when $a = \pm 1$.

Example 2 Find a formula for the sum of the series

$$s_n = \sum_{k=1}^{k=n} kr^k \qquad r \neq 1$$

Clearly, the difference between s_{n+1} and s_n is just the last term in s_{n+1}, namely, $(n+1)r^{n+1}$. Hence s_n satisfies the nonhomogeneous difference equation

$$s_{n+1} - s_n = (n+1)r^{n+1} \qquad \text{or} \qquad (E-1)s_n = (n+1)r^{n+1}$$

Moreover, although Tables 13.1 and 13.2 appear to be concerned only with second-order equations, the procedures they outline are correct for difference equations of order 1 as well as for equations of higher order. Hence we set up the characteristic equation $M - 1 = 0$, and from its root we construct the complementary function

$$s_n = c_1(1)^n = c_1$$

To find a particular integral, we assume

$$S_n = (An + B)r^{n+1}$$

Then, substituting, we must have

$$[A(n+1) + B]r^{n+2} - (An + B)r^{n+1} = (n+1)r^{n+1}$$

or, dividing out r^{n+1} and collecting terms,

$$n(Ar - A) + (Ar + Br - B) = n + 1$$

This will be an identity in the variable n if and only if

$$A(r-1) = 1 \qquad \text{and} \qquad Ar + Br - B = 1$$

or, since $r \neq 1$,

$$A = \frac{1}{r-1} \qquad \text{and} \qquad B = -\frac{1}{(r-1)^2}$$

Hence
$$S_n = \left[\frac{n}{r-1} - \frac{1}{(r-1)^2}\right] r^{n+1}$$

and a complete solution is

$$s_n = c_1 + S_n = c_1 + \frac{n(r-1)-1}{(r-1)^2} r^{n+1}$$

To determine c_1, we use the obvious fact that $s_n = r$ when $n = 1$. Thus we must have

$$r = c_1 + \frac{r-2}{(r-1)^2} r^2 \qquad \text{or} \qquad c_1 = \frac{r}{(r-1)^2}$$

Hence, finally,

$$s_n = \frac{r + [n(r-1)-1]r^{n+1}}{(r-1)^2}$$

Linear difference equations also occur in many important physical problems. In particular, they should be anticipated whenever one is investigating a configuration, either mechanical or electrical, which consists of a series of identical elements connected to one another in the same way. Our next two examples are of this sort. Example 3 is an electrical problem in which difference equations simplify the solution of a system of linear algebraic equations arising from the analysis of a network of identical loops. Example 4 involves a mechanical system consisting of a series of identical spring-connected masses. In it, both differential equations and difference equations are involved.

Example 3 In the system shown in Fig. 13.1, the point P_0 is kept at the constant potential $V = V_0$ with respect to the ground, which is assumed to be at the potential

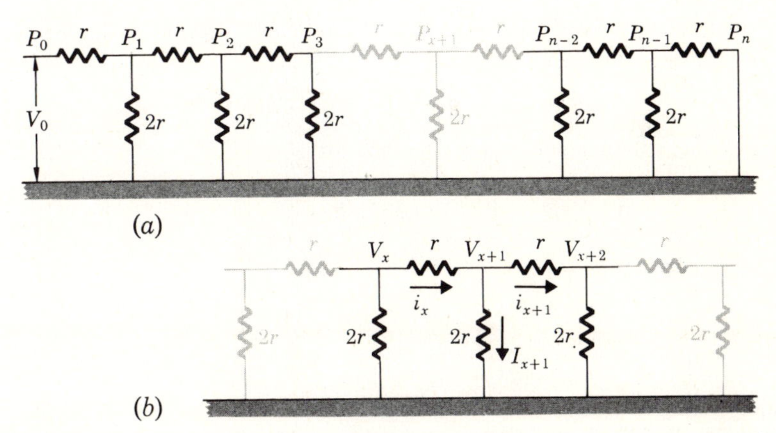

Figure 13.1 A ladder-type network with identical loops. (Although the network shown in Fig. 13.1*a* appears to contain exactly seven loops, the number of loops is actually indefinite. This is implied by the fact that the central portion of the figure is drawn with lighter lines.)

zero. The point P_N is connected to the ground by a wire of zero resistance, so that it remains at the potential $V_N = 0$. What is the potential $V_1, V_2, \ldots, V_{N-1}$ at the respective points $P_1, P_2, \ldots, P_{N-1}$?

According to Kirchhoff's first law, the sum of the currents flowing toward any junction in a network must equal the sum of the currents flowing away from that junction. Hence at a general point P_{n+1} (Fig. 13.1*b*) we have

$$i_n = i_{n+1} + I_{n+1}$$

If we now replace each current by its equivalent according to Ohm's law, $I = V/R$, where V is the potential difference between the ends of the wire through which the current I is flowing, we obtain

$$\frac{V_n - V_{n+1}}{r} = \frac{V_{n+1} - V_{n+2}}{r} + \frac{V_{n+1} - 0}{2r}$$

or, simplifying and collecting terms,

$$V_{n+2} - \tfrac{5}{2}V_{n+1} + V_n = 0 \qquad n = 0, 1, 2, \ldots, N-2 \tag{2}$$

Equation (2) constitutes a system of $n - 1$ linear algebraic equations from which the unknown potentials $V_1, V_2, \ldots, V_{N-1}$ can be found by completely elementary, though very tedious, steps for any particular value of n. However, it is much simpler and much more elegant to regard Eq. (2) as a second-order difference equation

$$(E^2 - \tfrac{5}{2}E + 1)V_n = 0 \tag{3}$$

subject to the end conditions $V_0 = V$ and $V_N = 0$, which will serve to determine the values of the arbitrary constants appearing in any complete solution of (3).

Taking this point of view, we first set up the characteristic equation of Eq. (3), namely,

$$M^2 - \tfrac{5}{2}M + 1 = 0$$

From its roots, $M_1 = \frac{1}{2}$ and $M_2 = 2$, we then construct a complete solution of Eq. (3):

$$V_n = A(\tfrac{1}{2})^n + B(2^n) \tag{4}$$

When the boundary conditions at $n = 0$ and $n = N$ are imposed on the complete solution (4), we obtain the equations

$$A + B = V_0 \qquad \text{and} \qquad A(\tfrac{1}{2})^N + B(2^N) = 0$$

When these are solved simultaneously, we find that

$$A = \frac{2^{2N}}{2^{2N} - 1} V_0 \qquad \text{and} \qquad B = -\frac{1}{2^{2N} - 1} V_0$$

The final solution is therefore

$$V_n = \left(\frac{2^{2N}}{2^n} - 2^n\right)\frac{V_0}{2^{2N} - 1} \qquad n = 0, 1, 2, \ldots, N$$

Example 4 A system consists of N equal masses connected by identical springs as shown in Fig. 13.2. Neglecting friction, find the natural frequencies at which the system can vibrate. Determine also, for each natural frequency, a set of numbers proportional to the amplitudes through which the masses vibrate at that frequency.

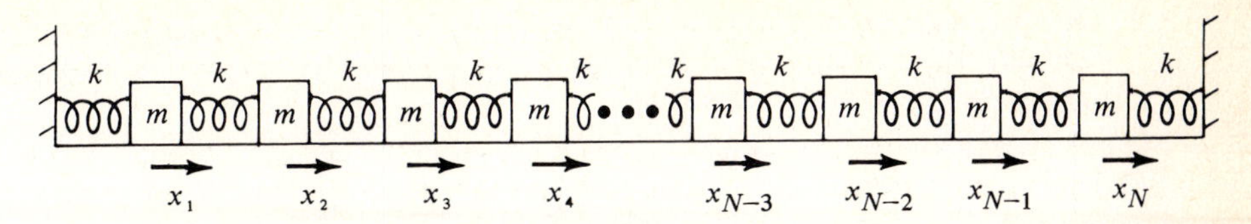

Figure 13.2 A system of N spring-connected masses.

Let x_n be the instantaneous displacement of the nth mass from its equilibrium position. Then the change in length of the nth spring is

$$x_1 - 0 = x_1 \qquad n = 1$$

$$x_n - x_{n-1} \qquad n = 2, 3, \ldots, N$$

$$0 - x_N = -x_N \qquad n = N + 1$$

and the force exerted by each spring is equal to its change in length multiplied by the spring modulus k. Applying Newton's law to each mass, exactly as we did in Example 1, Sec. 7.5, we obtain the following system of homogeneous, linear, constant-coefficient, second-order differential equations

$$\begin{aligned}
m\ddot{x}_1 &= -k(x_1 - 0) + k(x_2 - x_1) \\
m\ddot{x}_2 &= -k(x_2 - x_1) + k(x_3 - x_2) \\
m\ddot{x}_3 &= -k(x_3 - x_2) + k(x_4 - x_3) \\
&\cdots\cdots\cdots\cdots\cdots\cdots\cdots\cdots \\
m\ddot{x}_{N-2} &= -k(x_{N-2} - x_{N-3}) + k(x_{N-1} - x_{N-2}) \\
m\ddot{x}_{N-1} &= -k(x_{N-1} - x_{N-2}) + k(x_N - x_{N-1}) \\
m\ddot{x}_N &= -k(x_N - x_{N-1}) + k(0 - x_N)
\end{aligned}$$

or, letting $D^2 = d^2/dt^2$ and then collecting terms,

$$\begin{aligned}
(mD^2 + 2k)x_1 \qquad - kx_2 \qquad\qquad &= 0 \\
-kx_1 + (mD^2 + 2k)x_2 - kx_3 \qquad &= 0 \\
-kx_2 + (mD^2 + 2k)x_3 - kx_4 \qquad &= 0 \\
\cdots\cdots\cdots\cdots\cdots\cdots\cdots\cdots\cdots \qquad & \\
-kx_{N-3} + (mD^2 + 2k)x_{N-2} - kx_{N-1} &= 0 \\
-kx_{N-2} + (mD^2 + 2k)x_{N-1} - kx_N \qquad &= 0 \\
-kx_{N-1} + (mD^2 + 2k)x_N &= 0
\end{aligned} \qquad (5)$$

Since only derivatives of even order occur in these equations (because friction in the system was assumed to be negligible), we assume solutions of the form

$$x_n = A_n \cos \omega t\dagger \qquad n = 1, 2, \ldots, N$$

† The assumption that $x_n = A_n \sin \omega t$ would work equally well, of course.

where ω (> 0) is the unknown frequency of the response and the A_n's are arbitrary constants. Substituting these expressions for the x_n's into Eqs. (5), dividing each equation by $-k \cos \omega t$, and then setting

$$\frac{m\omega^2}{k} = \alpha^2 \qquad \alpha > 0$$

we obtain the algebraic equations

$$\begin{aligned}
-(2 - \alpha^2)A_1 + A_2 &= 0 \\
A_1 - (2 - \alpha^2)A_2 + A_3 &= 0 \\
A_2 - (2 - \alpha^2)A_3 + A_4 &= 0 \\
\cdots\cdots\cdots\cdots\cdots\cdots\cdots\cdots\cdots & \\
A_{N-3} - (2 - \alpha^2)A_{N-2} + A_{N-1} &= 0 \\
A_{N-2} - (2 - \alpha^2)A_{N-1} + A_N &= 0 \\
A_{N-1} - (2 - \alpha^2)A_N &= 0
\end{aligned} \tag{6}$$

In order for this system of N homogeneous linear equations to have a nontrivial solution for the N unknown amplitudes $A_1, A_2, \ldots, A_N$, that is, in order for Eqs. (6) to describe a response of the system which is not identically zero, it is necessary that the determinant of the coefficients in these equations be zero. However, in this problem the determinant of the coefficients is of the Nth order, and to expand it and then solve the resulting Nth-degree equation in the unknown frequency parameter $\alpha^2 = m\omega^2/k$ for even a moderately large value of N would be prohibitively time-consuming. Hence it is much better to proceed in the following way: With the exception of the first and last equations, every equation in the set (6) is of the form

$$A_n - (2 - \alpha^2)A_{n+1} + A_{n+2} = 0$$

In other words, for $n = 1, 2, \ldots, N - 2$, the A's satisfy the second-order difference equation

$$[E^2 - (2 - \alpha^2)E + 1]A_n = 0 \tag{7}$$

The first and last equations in the set (6), which clearly do not fit the pattern of Eq. (7), are, of course, the two conditions necessary for the determination of the two arbitrary constants which will appear in the complete solution of the difference equation (7).

If we write Eq. (7) in the form

$$\left[E^2 - 2\left(1 - \frac{\alpha^2}{2}\right)E + 1\right]A_n = 0 \tag{8}$$

it is clear that with $1 - \alpha^2/2 = \lambda$, it is precisely the equation which we considered at length in Example 5, Sec. 13.2. Hence, since ω, and therefore α, is unknown, it appears that there are five cases to be considered, namely,

$$\lambda = 1 - \frac{\alpha^2}{2} < -1 \qquad\qquad \lambda = 1 - \frac{\alpha^2}{2} = -1$$

$$-1 < \lambda = 1 - \frac{\alpha^2}{2} < 1$$

$$\lambda = 1 - \frac{\alpha^2}{2} = 1 \qquad\qquad \lambda = 1 - \frac{\alpha^2}{2} > 1$$

The last two possibilities can be ruled out immediately, since they imply that

$$\alpha^2 = \frac{M\omega^2}{k} \leq 0$$

whereas M, ω, and k are all intrinsically positive quantities. The first two possibilities can also be ruled out, though with somewhat more difficulty, and we shall leave them as exercises.

Continuing, then, with the third possibility, we put

$$\lambda = 1 - \frac{\alpha^2}{2} = \cos\mu$$

which implies that

$$2 - \alpha^2 = 2\cos\mu \qquad \alpha = \sqrt{2(1-\cos\mu)} = 2\sin\frac{\mu}{2} \tag{9}$$

From the results of Example 5, Sec. 13.2, we know that a complete solution of Eq. (8) in this case is

$$A_n = c_1 \cos n\mu + c_2 \sin n\mu \qquad n = 1, 2, \ldots, N \tag{10}$$

To determine the constants c_1 and c_2, we first evaluate Eq. (10) for $n = 1$ and $n = 2$ and substitute the results into the first of the equations in (6). This gives us the condition

$$-2\cos\mu(c_1\cos\mu + c_2\sin\mu) + (c_1\cos 2\mu + c_2\sin 2\mu) = 0$$

or $$c_1(-2\cos^2\mu + \cos 2\mu) + c_2(-2\cos\mu\sin\mu + \sin 2\mu) = 0$$

Finally, since $\cos 2\mu = 2\cos^2\mu - 1$ and $\sin 2\mu = 2\sin\mu\cos\mu$, this reduces to

$$-c_1 = 0 \tag{11}$$

Similarly, to impose the second boundary condition, we evaluate the complete solution (10) for $n = N - 1$ and $n = N$ and substitute the resulting expressions into the last of the equations in (6), remembering, of course, that we now know that $c_1 = 0$. This gives us

$$c_2 \sin(N-1)\mu - 2\cos\mu(c_2\sin N\mu) = 0$$

or, expanding $\sin(N-1)\mu$ and collecting terms,

$$c_2[(\sin N\mu\cos\mu - \cos N\mu\sin\mu) - 2\cos\mu\sin N\mu]$$
$$= -c_2(\sin N\mu\cos\mu + \cos N\mu\sin\mu) = -c_2\sin(N+1)\mu = 0$$

Since c_1 is already known to be zero, the entire solution becomes trivial if c_2 is also zero, and in this uninteresting case Eqs. (10) merely describe the system remaining permanently at rest. To obtain a solution describing possible motions of the system, we must therefore have

$$\sin(N+1)\mu = 0 \tag{12}$$

that is, the parameter μ must have one of the values

$$\mu_j = \frac{j\pi}{N+1} \qquad j = 1, 2, \ldots, N \tag{13}$$

These values of μ define the N natural frequencies of the system since, from (9),

$$\alpha = \sqrt{\frac{m\omega^2}{k}} = 2 \sin \frac{\mu}{2}$$

Hence the required frequencies are given by the formula

$$\omega_j = 2\sqrt{\frac{k}{m}} \sin \frac{j\pi}{2(N+1)} \qquad j = 1, 2, \ldots, N \tag{14}$$

Finally, it follows from Eqs. (10) and (13) that when the system is vibrating at one of the natural frequencies $\omega = \omega_j$ given by Eq. (14), the amplitudes with which the successive masses vibrate are proportional to the numbers

$$\sin \frac{nj\pi}{N+1} \qquad n = 1, 2, \ldots, N$$

Exercises for Section 13.4

1 Verify that if j takes on integral values greater than N in (14), no new values of the frequency ω are obtained.

2 Show that when the system of Example 4 is vibrating at the natural frequency ω_j corresponding to the value μ_j, the amplitudes of the masses are related by the formula

$$A_n = (-1)^{j+1} A_{N-n+1}$$

3 Taking $c_2 = 1$ in Eq. (10), plot the successive amplitudes for each of the first four natural frequencies by drawing vertical lines of the appropriate height at $n = 1, 2, \ldots, N = 12$.

4 Verify the assertion of Example 4 that the case $1 - \alpha^2/2 = -1$ can be ruled out by showing that it leads only to $c_1 = c_2 = 0$.

****5** Verify the assertion of Example 4 that the case $1 - \alpha^2/2 < -1$ can be ruled out by showing that it leads only to $c_1 = c_2 = 0$. *Hint:* The work is very much like the work of Example 4 except that it involves hyperbolic functions and their identities rather than trigonometric functions.

****6** Work Example 4 if the spring connecting the Nth mass to the wall is removed.

****7** Work Example 4 if neither the first mass nor the last mass is connected to the adjacent wall.

***8** Work Example 3 if there is a resistor of resistance $2r$ in the wire connecting P_N to the ground.

***9** Work Example 3 if both P_0 and P_N are maintained at the constant potential V_0.

***10** Work Example 3 if the common value of the resistance in the vertical branches is kr, $k > 0$.

****11** Show that the integral

$$I_n(\lambda) = \int_0^\pi \frac{\cos nt - \cos n\lambda}{\cos t - \cos \lambda}\, dt \qquad n = 2, 3, 4, \ldots$$

satisfies the equation $[E^2 - (2\cos\lambda)E + 1]I_n = 0$. Solve this equation and find an explicit expression for I_n, given $I_0 = 0$, $I_1 = \pi$.

12 Express each of the following nth-order determinants as an explicit function of n.

(a) $\begin{vmatrix} 5 & 2 & 0 & \cdots & 0 & 0 \\ 3 & 5 & 2 & \cdots & 0 & 0 \\ 0 & 3 & 5 & \cdots & 0 & 0 \\ \cdots & \cdots & \cdots & \cdots & \cdots & \cdots \\ 0 & 0 & 0 & \cdots & 5 & 2 \\ 0 & 0 & 0 & \cdots & 3 & 5 \end{vmatrix}$ (b) $\begin{vmatrix} 5 & 1 & 0 & \cdots & 0 & 0 \\ 6 & 5 & 1 & \cdots & 0 & 0 \\ 0 & 6 & 5 & \cdots & 0 & 0 \\ \cdots & \cdots & \cdots & \cdots & \cdots & \cdots \\ 0 & 0 & 0 & \cdots & 5 & 1 \\ 0 & 0 & 0 & \cdots & 6 & 5 \end{vmatrix}$

(c) $\begin{vmatrix} 4 & 4 & 0 & \cdots & 0 & 0 \\ 1 & 4 & 4 & \cdots & 0 & 0 \\ 0 & 1 & 4 & \cdots & 0 & 0 \\ \cdots & \cdots & \cdots & \cdots & \cdots & \cdots \\ 0 & 0 & 0 & \cdots & 4 & 4 \\ 0 & 0 & 0 & \cdots & 1 & 4 \end{vmatrix}$ (d) $\begin{vmatrix} 2 & 2 & 0 & \cdots & 0 & 0 \\ 1 & 2 & 2 & \cdots & 0 & 0 \\ 0 & 1 & 2 & \cdots & 0 & 0 \\ \cdots & \cdots & \cdots & \cdots & \cdots & \cdots \\ 0 & 0 & 0 & \cdots & 2 & 2 \\ 0 & 0 & 0 & \cdots & 1 & 2 \end{vmatrix}$

***13** If $a > 2$, show that the value of the nth-order determinant

$$\begin{vmatrix} a & 1 & 0 & \cdots & 0 & 0 \\ 1 & a & 1 & \cdots & 0 & 0 \\ 0 & 1 & a & \cdots & 0 & 0 \\ \cdots & \cdots & \cdots & \cdots & \cdots & \cdots \\ 0 & 0 & 0 & \cdots & a & 1 \\ 0 & 0 & 0 & \cdots & 1 & a \end{vmatrix}$$

is $[\sinh (n + 1)\mu]/\sinh \mu$ where $\cosh \mu = a/2$. What is the value of the determinant if $a = 2$? $-2 < a < 2$? $a = -2$? $a < -2$?

14 Find a formula for the sum of each of the following series.

(a) $\sum_{i=1}^{n} i$ (b) $\sum_{i=1}^{n} i^2$ (c) $\sum_{i=1}^{n} (2i - 1)^2$

***15** Find a formula for the sum of each of the following series.

(a) $\sum_{i=1}^{n} i^2 r^i$ (b) $\sum_{i=1}^{n} \sin ir$ (c) $\sum_{i=1}^{n} \cos ir$

13.5 DIFFERENCE EQUATIONS AND THE NUMERICAL SOLUTION OF DIFFERENTIAL EQUATIONS

One of the many uses of difference equations is in analyzing the accuracy of methods for the numerical solution of differential equations. Although a detailed discussion of this topic must be left to more advanced texts on numerical analysis, we can make an introductory investigation of this problem by considering the solution of the simple equation

$$y' = Ay \qquad A \text{ a constant} \tag{1}$$

by the predictor-corrector formulas used in Milne's method.

Milne's method, as we know, makes use of the predictor formula [Eq. (7), Sec. 12.4]

$$y_{n+1} = y_{n-3} + \frac{4h}{3}(2y'_{n-2} - y'_{n-1} + 2y'_n)$$

to obtain a first estimate of y_{n+1} and then iterates the corrector formula [Eq. (10), Sec. 12.3]

$$y_{n+1} = y_{n-1} + \frac{h}{3}(y'_{n-1} + 4y'_n + y'_{n+1})$$

to determine the value of y_{n+1} which is finally accepted. Applied to the particular differential equation (1), Milne's corrector-formula becomes

$$y_{n+1} = y_{n-1} + \frac{h}{3}(Ay_{n-1} + 4Ay_n + Ay_{n+1})$$

which is simply a linear recurrence relation connecting the values of y at three arbitrary consecutive values of x. Hence, *the numerical solution given by Milne's method is just a particular solution of the second-order difference equation*

$$\left(1 - \frac{Ah}{3}\right)y_{n+1} - \frac{4Ah}{3}y_n - \left(1 + \frac{Ah}{3}\right)y_{n-1} = 0 \qquad 1 - \frac{Ah}{3} \neq 0 \tag{2}$$

The restriction $1 - Ah/3 \neq 0$ is not serious, since our concern is with the properties of Milne's method when h is sufficiently small, say $h < 3/|A|$. If, for convenience, we put $Ah/3 = \lambda$, the characteristic equation of Eq. (2) becomes

$$(1 - \lambda)M^2 - 4\lambda M - (1 + \lambda) = 0$$

and its roots are

$$M_1, M_2 = \frac{2\lambda \pm \sqrt{1 + 3\lambda^2}}{1 - \lambda} \qquad \lambda \neq 1$$

Hence, the tabular function y_n, which is our approximation to the solution of (1), is contained in the complete solution

$$y_n = c_1 M_1^n + c_2 M_2^n$$

and as $h \to 0$, this particular solution should, presumably, approach the exact solution of (1), namely,

$$y = y_0 e^{A(x - x_0)} \tag{3}$$

To determine under what conditions, if any, this will be the case, it is convenient to expand M_1 and M_2 in terms of powers of λ, using the binomial theorem, and then retain only the lowest power of λ in the expansion of each root. Doing this, noting that each of the expansions we need is convergent if $3\lambda^2 < 1$, or $0 < h < \sqrt{3}/|A|$, we have

$$\begin{aligned}
M_1, M_2 &= [2\lambda \pm (1 + 3\lambda^2)^{1/2}](1 - \lambda)^{-1} \\
&= [2\lambda \pm (1 + \tfrac{3}{2}\lambda^2 + \cdots)](1 + \lambda + \lambda^2 + \cdots) \\
&= (2\lambda \pm 1 + \cdots)(1 + \lambda + \cdots)
\end{aligned}$$

and $$M_1 = 1 + 3\lambda + \cdots \qquad M_2 = -1 + \lambda + \cdots$$

Hence $$y_n \doteq c_1(1 + 3\lambda)^n + c_2(-1 + \lambda)^n$$

or, finally, since $\lambda = Ah/3$ and $n = (x_n - x_0)/h$,

$$y_n \doteq c_1(1 + Ah)^{(x_n - x_0)/h} + c_2(-1)^n\left(1 - \frac{Ah}{3}\right)^{(x_n - x_0)/h} \tag{4}$$

As $h \to 0$, each term in (4) contains an indeterminate of the form 1^∞. Evaluating these by L'Hospital's rule, or recalling from calculus that

$$\lim_{z\to 0} (1 + z)^{b/z} = e^b$$

we see that for small values of h we have the approximation

$$y_n = c_1 e^{A(x_n - x_0)} + c_2(-1)^n e^{-A(x_n - x_0)/3} \tag{5}$$

If our initial data were perfectly accurate, and if there were no round-off errors in our calculations, the correct values of c_1 and c_2 would be y_0 and 0, respectively, and (5) would reduce to the exact solution (3) of the differential equation $y' = Ay$ with the initial condition $y(x_0) = y_0$. However, because of inevitable errors of various kinds, c_2 will in general not be zero. Thus the numerical solution we obtain is actually the sum of an approximation to the exact solution plus a so-called **parasitic solution** which for small values of h is approximately $c_2(-1)^n e^{-A(x_n - x_0)/3}$.

If $A > 0$, the parasitic solution is approximately a decaying exponential function which soon becomes vanishingly small. In this case, the procedure is said to be **numerically stable.** On the other hand, if $A < 0$, the numerical value of the parasitic solution increases exponentially (while continually alternating in sign), and sooner or later, depending on the size of c_2 and A, it will become the principal part of the numerical solution we obtain. In this case, the procedure is said to be **numerically unstable.**

The analysis, leading to essentially the same conclusions, which can be made for the nonhomogeneous equation $y' = Ay + f(x)$ is similar except that the difference equation arising from Milne's corrector formula is nonhomogeneous rather than homogeneous.

When A is a constant, the equation $y' = Ay$ is so simple that numerical methods for solving it are completely unnecessary and therefore a discussion of their stability is irrelevant. On the other hand, for the general, first-order differential equation

$$y' = f(x, y) \tag{6}$$

numerical methods of solution are often required, and the question of their stability is important. Unfortunately, for Eq. (6) Milne's corrector formula does not, in general, become a linear difference equation, and in such cases the preceding analysis cannot be immediately applied. However, if $f(x, y)$ is replaced by its Taylor expansion around the point (x_0, y_0), namely,

$$y' = f(x, y) = f(x_0, y_0) + \left[\frac{\partial f}{\partial x}\bigg|_{x_0, y_0}(x - x_0) + \frac{\partial f}{\partial y}\bigg|_{x_0, y_0}(y - y_0)\right]$$

$$+ \frac{1}{2!}\left[\frac{\partial^2 f}{\partial x^2}\bigg|_{x_0, y_0}(x - x_0)^2 + 2\frac{\partial^2 f}{\partial x\,\partial y}\bigg|_{x_0, y_0}(x - x_0)(y - y_0)\right.$$

$$\left. + \frac{\partial^2 f}{\partial y^2}\bigg|_{x_0, y_0}(y - y_0)^2\right] + \cdots$$

and if powers of $x - x_0$ and $y - y_0$ higher than the first are neglected, then in a suitable neighborhood of (x_0, y_0), the differential equation $y' = f(x, y)$ becomes, approximately,

$$y' \equiv (y - y_0)' = \frac{\partial f}{\partial y}\bigg|_{x_0, y_0}(y - y_0) + \left[f(x_0, y_0) + \frac{\partial f}{\partial x}\bigg|_{x_0, y_0}(x - x_0)\right]$$

which is a nonhomogeneous differential equation of the form we have just considered. By the preceding discussion, the solution given by Milne's corrector formula will be stable if the coefficient of $y - y_0$, namely,

$$\frac{\partial f}{\partial y}\bigg|_{x_0, y_0}$$

is positive and unstable if it is negative.

If any of the other predictor-corrector methods we described in Sec. 12.4 (see Exercises 7 and 8, Sec. 12.4) is applied to Eq. (1), the corresponding corrector formula becomes a homogeneous, constant-coefficient linear difference equation and the preceding analysis can be repeated. The chief difference between the various difference equations thus obtained is that some may be of order greater than 2. This means that a characteristic equation may be encountered which has more than two roots, so that the related complementary function has the form

$$y_n = c_1 M_1^n + c_2 M_2^n + \cdots + c_k M_k^n$$

As with Milne's method, one of these terms will approach the exact solution of Eq. (1) as $h \to 0$; the others will be parasitic solutions. If the absolute values of the extraneous roots from which the parasitic solutions arise are all less than 1, then as n, or x, increases, each will decay exponentially and the process will yield a meaningful approximation to the exact solution. On the other hand, if the absolute value of even one of the extraneous roots is greater than 1, then the corresponding parasitic solution will increase exponentially and the process will be numerically unstable.

Since numerical instability is commonly observed in predictor-corrector methods, most workers prefer the Runge-Kutta method, which involves no numerical instability, at least for sufficiently small values of the tabular interval h.

Exercises for Section 13.5

1 (*a*) Show that Euler's method for the numerical solution of differential equations [Eq. (3), Sec. 12.2] is numerically stable for the differential equation $y' = Ay$ for all values of A.

(*b*) Verify that as $h \to 0$, the general solution of the difference equation involved in part (*a*) converges to the general solution of $y' = Ay$ for all values of A.

2 Work Exercise 1 for the modified Euler method [Eq. (4), Sec. 12.2].

3 (*a*) Discuss the stability of the closed formula given in Exercise 7, Sec. 12.4, in relation to the equation $y' = Ay$.

(*b*) Verify that as $h \to 0$, one of the solutions of the difference equation involved in part (*a*) converges to the general solution of $y' = Ay$.

***4** Show that the corrector formula

$$y_{n+1} = y_n + \frac{h}{24}(9y'_{n+1} + 19y'_n - y'_{n-1} + y'_{n-2})$$

is stable in relation to the equation $y' = Ay$ if $\gamma \equiv Ah/24$ is a sufficiently small negative number. *Hint:* Note that from the characteristic equation γ can easily be plotted as a function of M.

***5** Discuss the stability of Milne's predictor formula [Eq. (7), Sec. 12.4] in relation to the equation $y' = Ay$. *Hint:* Note the hint given in Exercise 4.

***6** (*a*) Show that the closed formula given in Exercise 8, Sec. 12.4, is numerically stable for the equation $y'' = Ay$ for all values of A.

(*b*) Verify that as $h \to 0$, the general solution of the difference equation involved in part (*a*) converges to the general solution of $y'' = Ay$.

CHAPTER

FOURTEEN

THE DESCRIPTIVE THEORY OF NONLINEAR DIFFERENTIAL EQUATIONS

14.1 INTRODUCTION

Up to this point, most of our work has been concerned with devising methods for solving differential equations. For several types of first-order equations and for the very important class of linear, constant-coefficient equations of all orders, we are now able to obtain solutions in finite form. For many linear differential equations with variable coefficients, we have learned how to obtain solutions in the form of infinite series. And for equations whose complexity makes exact solution difficult or impossible, we have developed methods for obtaining highly accurate numerical approximations to their solutions. Given sufficient time and computer resources, it is probably correct to say that all solutions of any solvable differential equation, or system of differential equations, can be found with acceptable accuracy. However, many questions phrased in terms of differential equations can be adequately answered without actually solving the equations. Frequently, only some descriptive property, such as the stability or periodicity of solutions, is to be investigated. In this chapter we shall undertake an introductory discussion of these matters for systems which can be described by two first-order differential equations.

14.2 THE PHASE PLANE AND CRITICAL POINTS

As a starting point, let us consider the initial-value problem

$$\frac{d^2x}{dt^2} = f\left(x, \frac{dx}{dt}\right) \qquad x = x_0, \frac{dx}{dt} = x_0' \quad \text{when } t = 0 \tag{1}$$

where f has continuous first partial derivatives with respect to x and x'. As we observed several times in our previous work, if we put $dx/dt = y$, this equation, with its accompanying conditions, can be replaced by the system

$$\frac{dx}{dt} = y \qquad \frac{dy}{dt} = f(x, y) \qquad x = x_0,\ y = y_0\ (= x_0') \quad \text{when } t = 0 \tag{2}$$

By a solution of this system we mean, as usual, a pair of differentiable functions $\{x = x(t),\ y = y(t)\}$ which, on some interval containing $t = 0$, reduce Eqs. (2) to identities and are such that $x(0) = x_0$ and $y(0) = y_0$.

If we choose, we can think of the functions $x = x(t)$ and $y = y(t)$ as the parametric equations of an arc in the xy plane which passes through the point (x_0, y_0). From this point of view, the xy plane is called the **phase plane** of the original equation (1) or of the system (2), and the arc defined parametrically by the equations $x = x(t)$ and $y = y(t)$ is called a **path** or **orbit** or **trajectory** of either (1) or (2). Assuming that the points of a trajectory are in one-to-one correspondence with the values of the parameter t, the direction in which t increases is said to be the **positive direction** on the trajectory (see Exercise 16).

If the values $x = x_0$ and $y = y_0$ $(= x_0')$ are assigned when $t = t_0$, rather than when $t = 0$, we obtain a different solution but the same trajectory, since the equations

$$x = x(t) \qquad y = y(t) \qquad \alpha < t < \beta$$

define the same arc as the equations

$$x = x(t - t_0) \qquad y = y(t - t_0) \qquad \alpha + t_0 < t < \beta + t_0$$

Thus *solution* and *trajectory* are not synonomous terms.

If the parameter t is eliminated between the equations $x = x(t)$ and $y = y(t)$, we obtain the xy equation of a curve, which may or may not be the corresponding trajectory. For example, if we eliminate the parameter t between the equations

$$x = e^t \qquad y = e^{2t} \qquad -\infty < t < \infty \tag{3}$$

we obtain $y = x^2$. However, the parabola represented by this equation is not the trajectory defined by (3) because these equations define only the portion of the parabola to the right of the y axis; that is, the trajectory is only the (open) right half of the parabola $y = x^2$.

The xy equation of a curve containing the trajectory through (x_0, y_0) can also be found by first eliminating t (in the form of the differential dt) from the system (2) by dividing the second equation by the first, getting

$$\frac{dy}{dx} = \frac{f(x, y)}{y} \qquad (y \neq 0) \qquad y = y_0 \text{ when } x = x_0$$

and then solving this equation in the variables x and y.

More generally, the preceding observations can all be applied to systems of the form

$$\begin{aligned} \frac{dx}{dt} &= g(x, y) \\ \frac{dy}{dt} &= f(x, y) \end{aligned} \qquad x = x_0,\ y = y_0 \text{ when } t = t_0 \tag{4}$$

where both f and g are functions possessing continuous first partial derivatives. Here, also, the xy plane is called the **phase plane,** and the arc defined parametrically by any solution

$$x = x(t) \qquad y = y(t)$$

is called a **path** or **orbit** or **trajectory** of the system. Systems such as (4) in which the independent variable does not appear explicitly in either f or g are said to be **autonomous.** In this chapter we shall be concerned exclusively with autonomous systems.

Example 1 Find the trajectories of the system

$$\begin{aligned} \frac{dx}{dt} &= 3x + y \\ \frac{dy}{dt} &= x + 3y \end{aligned} \tag{5}$$

Using the methods of Chap. 5, we seek solutions of the form

$$\begin{bmatrix} x \\ y \end{bmatrix} = \begin{bmatrix} A \\ B \end{bmatrix} e^{mt}$$

Substituting into (5) and dividing by e^{mt} yields the equations

$$\begin{aligned} (m - 3)A - B &= 0 \\ -A + (m - 3)B &= 0 \end{aligned} \tag{6}$$

This system of algebraic equations will have a nontrivial solution if and only if

$$\begin{vmatrix} m - 3 & -1 \\ -1 & m - 3 \end{vmatrix} = m^2 - 6m + 8 = 0$$

that is, if and only if $m = 2$ or $m = 4$. If $m = 2$, then, from (6), $-A - B = 0$, and we have the particular solution

$$\begin{bmatrix} x \\ y \end{bmatrix} = \begin{bmatrix} 1 \\ -1 \end{bmatrix} e^{2t}$$

If $m = 4$, then, from (6), $A - B = 0$, and we have the particular solution

$$\begin{bmatrix} x \\ y \end{bmatrix} = \begin{bmatrix} 1 \\ 1 \end{bmatrix} e^{4t}$$

These two solutions are linearly independent, and hence a complete solution is

$$\begin{bmatrix} x \\ y \end{bmatrix} = c_1 \begin{bmatrix} 1 \\ -1 \end{bmatrix} e^{2t} + c_2 \begin{bmatrix} 1 \\ 1 \end{bmatrix} e^{4t}$$

The trajectories of the system are thus defined parametrically by the equations

$$\begin{aligned} x &= \quad c_1 e^{2t} + c_2 e^{4t} \\ y &= -c_1 e^{2t} + c_2 e^{4t} \qquad -\infty < t < \infty \end{aligned} \tag{7}$$

To eliminate the parameter t between the equations in (7), we first add these equations and then subtract them, getting

$$\begin{aligned} x + y &= 2c_2 e^{4t} \\ x - y &= 2c_1 e^{2t} \end{aligned}$$

Finally, eliminating the exponentials between these equations, we have

$$\left(\frac{x - y}{2c_1}\right)^2 = e^{4t} = \frac{x + y}{2c_2}$$

or

$$(x - y)^2 = k(x + y) \qquad k = 2c_1^2/c_2 \tag{8}$$

If $k = 0$, this yields the line $y = x$, corresponding to $c_1 = 0$. The line $y = -x$, corresponding to $c_2 = 0$, is not included in (8) unless k is allowed to become infinite.

Alternatively, if we divide the second of the equations (5) by the first, we obtain

$$\frac{dy}{dx} = \frac{x + 3y}{3x + y}$$

This is a homogeneous first-order equation which can easily be solved by means of the substitution $y = ux$, as we saw in Sec. 1.4. The result, of course, is Eq. (8).

Equation (8) defines a family of parabolas (Fig. 14.1) which, with one exception, are all tangent at the origin to the line $x + y = 0$ and have the line $x - y = 0$ as axis. The one exception is the degenerate parabola consisting of the repeated line $x - y = 0$, corresponding to the value $k = 0$. These curves are *not* the trajectories of the given system. For instance, if $x = -1$ and $y = 2$ when $t = 0$, then, from (7), c_1 and c_2 must satisfy the conditions

$$\begin{aligned} c_1 + c_2 &= -1 \\ -c_1 + c_2 &= \quad 2 \end{aligned}$$

Hence $c_1 = -\frac{3}{2}$, $c_2 = \frac{1}{2}$, and

$$\begin{aligned} x &= -\tfrac{3}{2}e^{2t} + \tfrac{1}{2}e^{4t} \\ y &= \quad \tfrac{3}{2}e^{2t} + \tfrac{1}{2}e^{4t} \qquad -\infty < t < \infty \end{aligned}$$

The arc, or trajectory, corresponding to these equations is only the (open) half of the parabola $(x - y)^2 = 9(x + y)$ extending from the origin (which is excluded) through the point $(-1, 2)$. Since c_1 appears as a square in the parameter k in Eq. (8), it follows that the other (open) half of the parabola $(x - y)^2 = 9(x + y)$ is the trajectory

$$\begin{aligned} x &= \quad \tfrac{3}{2}e^{2t} + \tfrac{1}{2}e^{4t} \\ y &= -\tfrac{3}{2}e^{2t} + \tfrac{1}{2}e^{4t} \end{aligned}$$

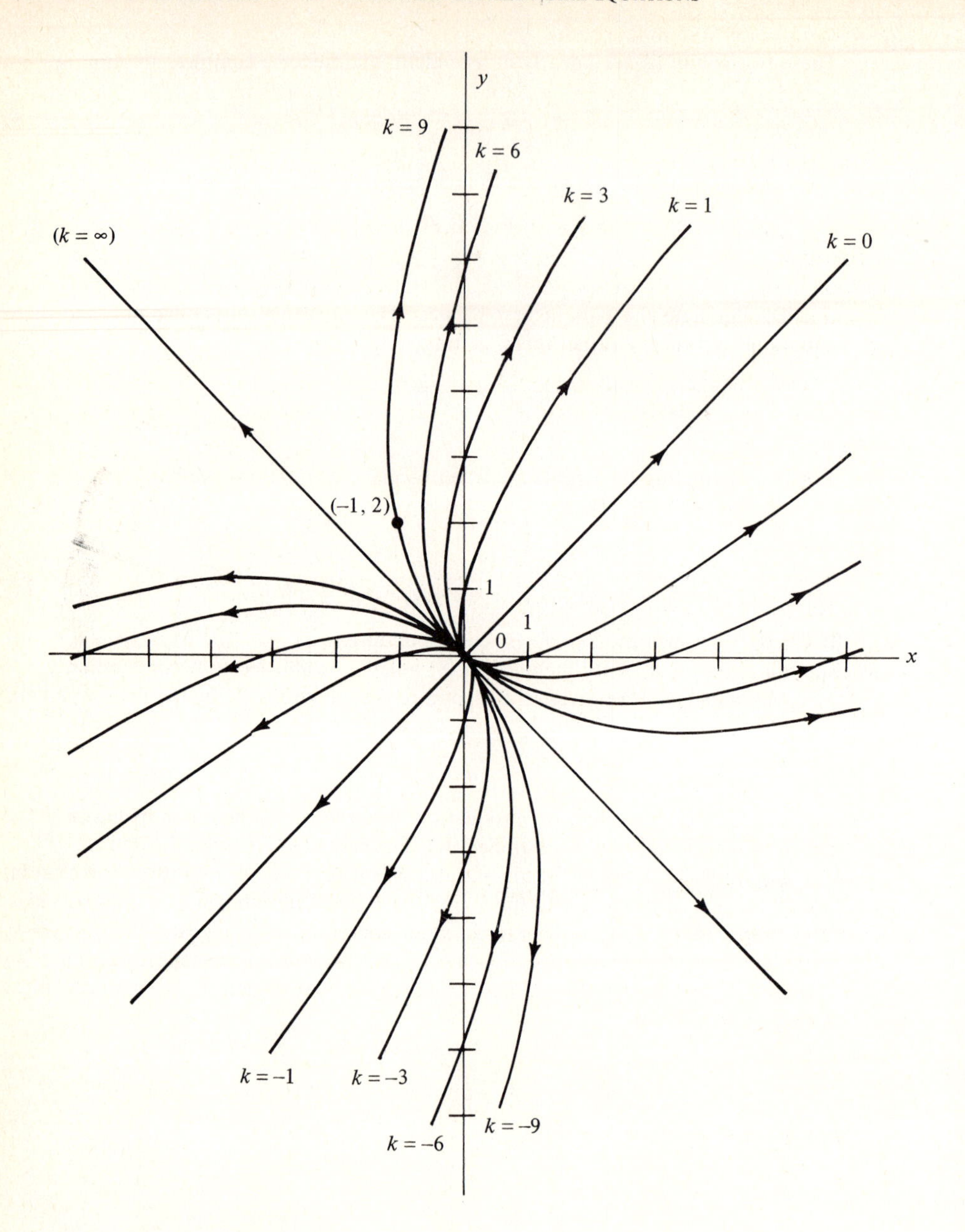

Figure 14.1 Typical trajectories of the system $dx/dt = 3x + y$, $dy/dt = x + 3y$.

More generally, every parabola of the family (8) (with the origin excluded) is the union of two trajectories. Similarly, each of the lines $y = \pm x$ (with the origin excluded) is the union of two trajectories.

In general, the equation

$$\frac{dy}{dx} = \frac{f(x, y)}{g(x, y)}$$

obtained from (4) by dividing the second equation by the first, gives the slope at the point (x, y) of the trajectory of (4) which passes through that point. However, if (x_0, y_0) is a point where $f(x, y)$ and $g(x, y)$ are simultaneously zero, then the slope at that point is an indeterminate of the form 0/0. Such a point is called a **critical point** or **equilibrium point** of the system (4). If a critical point has the property that there exists a circle which contains it but no other critical point, then that critical point is said to be **isolated.** Throughout this chapter we shall be concerned only with isolated critical points.

In the next section we shall investigate the nature of the trajectories of the system

$$\frac{dx}{dt} = ax + by$$

$$\frac{dy}{dt} = cx + ey \qquad a, b, c, e \text{ constants} \tag{9}$$

in the neighborhood of the obvious isolated critical point (0, 0). Since the equations in (9) are linear, the system can always be solved explicitly. Moreover, since the related first-order equation

$$\frac{dy}{dx} = \frac{cx + ey}{ax + by}$$

is homogeneous, it, too, can be solved in every case. However, a knowledge of the possible configurations of the trajectories of (9) in the neighborhood of (0, 0) is fundamental for the descriptive study of the system (4) in the general case when at least one of the functions f, g is nonlinear.

Exercises for Section 14.2

Determine the critical points of each of the following systems.

1 $\dfrac{dy}{dt} = x + 2y - 3$

$\dfrac{dx}{dt} = 3x + y + 1$

2 $\dfrac{dy}{dt} = xy + x - 2y + 4$

$\dfrac{dx}{dt} = 3x - y + 2$

3 $\dfrac{dy}{dt} = 9x^2 + 16y^2 - 25$

$\dfrac{dx}{dt} = 16x^2 + 9y^2 - 25$

4 $\dfrac{dy}{dt} = x - y$

$\dfrac{dx}{dt} = 3x - y - 3xy + x^3$

Find parametric equations for the trajectories of each of the following equations. In each case, find the xy equation of the family of curves containing the trajectories.

5 $x'' - 3x' + 2x = 0$ **6** $x'' + 3x' + 2x = 0$

7 $x'' + x = 0$ **8** $x'' + 4x = 0$

Find parametric equations for the trajectories of each of the following systems. In each case, find the xy equation of the family of curves containing the trajectories.

9 $$\frac{dy}{dt} = x \qquad \frac{dx}{dt} = y$$

10 $$\frac{dy}{dt} = x \qquad \frac{dx}{dt} = -y$$

11 $$\frac{dy}{dt} = y \qquad \frac{dx}{dt} = -x + 2y$$

12 $$\frac{dy}{dt} = 3x + y \qquad \frac{dx}{dt} = -x + y$$

13 Find the xy equation of the family of curves containing the trajectories of the equation $x'' + x + 2x^3 = 0$. Can parametric equations be found for the trajectories themselves? What are the trajectories?

14 For an autonomous system, how many trajectories pass through a particular point which is not a critical point? Why?

15 Find the solutions of the nonautonomous system

$$\frac{dx}{dt} = y \qquad \frac{dy}{dt} = 6t$$

which satisfy the conditions (a) $x = 1$, $y = 1$ when $t = 1$; (b) $x = 1$, $y = 1$ when $t = 2$. How many different trajectories pass through the point (1, 1)? How does this behavior compare with that of an autonomous system?

16 Determine A and B so that $x = At^2$, $y = Bt^2$ will be a solution of the system

$$\frac{dx}{dt} = 8\sqrt{y} \qquad \frac{dy}{dt} = \sqrt{x}$$

Describe the trajectory defined by this solution and show that the parameter t increases in each direction on this path.

14.3 CRITICAL POINTS AND THE TRAJECTORIES OF LINEAR SYSTEMS

In this section we shall examine the possible configurations of the trajectories of the system

$$\begin{aligned} \frac{dx}{dt} &= ax + by \\ \frac{dy}{dt} &= cx + ey \end{aligned} \qquad a, b, c, e \text{ real constants; } ae - bc \neq 0 \tag{1}$$

or of the corresponding single equation

$$\frac{dy}{dx} = \frac{cx + ey}{ax + by} \tag{1a}$$

in the neighborhood of the critical point (0, 0). The restriction $ae - bc \neq 0$ is necessary, for otherwise every point of the line $ax + by = 0$ is a critical point of (1) and (0, 0) is not an isolated critical point, as we have supposed. To see this, assume (for definiteness) that $a \neq 0$ and consider the equations

$$\begin{aligned} ax + by &= 0 \\ cx + ey &= 0 \end{aligned} \tag{2}$$

If c times the first equation is subtracted from a times the second, we obtain $(ae - bc)y = 0$. Hence (2) is equivalent to the system

$$\begin{aligned} ax + by &= 0 \\ (ae - bc)y &= 0 \end{aligned} \tag{3}$$

If $ae - bc \neq 0$, the only solution of (3), and hence of (2), is $x = y = 0$, and (0, 0) is the only critical point of the system (1). On the other hand, if $ae - bc = 0$, then the system (2) and the single equation $ax + by = 0$ have identical solutions; that is, every point of the line $ax + by = 0$ is a critical point of the system (1) and the equation (1a).

As usual, to solve the system (1), we assume $x = Ae^{mt}$, $y = Be^{mt}$, substitute, and obtain at once the characteristic equation

$$\begin{vmatrix} m - a & -b \\ -c & m - e \end{vmatrix} = m^2 - (a + e)m + ae - bc = 0 \tag{4}$$

As we learned in Chap. 5, the roots of this equation determine the form of the solutions for x and y, and these, in turn, determine the nature of the trajectories of (1) and (1a). There are three distinct cases to consider, according as the discriminant

$$\Delta = (a + e)^2 - 4(ae - bc)$$

of (4) is greater than, equal to, or less than zero. However, there are several subcases which must be distinguished. For instance, when the roots of (4) are real and distinct, the trajectories are significantly different when the roots are of like sign and when they are of unlike sign. Furthermore, when the roots are complex, the configuration of the trajectories is significantly different when the roots are pure imaginaries and when they are not.

The various possibilities are described in Table 14.1. Each of these cases can be investigated in complete generality by solving either Eqs. (1) or Eq. (1a) under the appropriate restrictions on the coefficients. However, for simplicity, we shall explain the general nature of the trajectories in each case by considering only a suitable prototype.

Table 14.1

Nature of the characteristic roots	Conditions on the coefficients in the characteristic equation $[\Delta \equiv (a+e)^2 - 4(ae - bc)]$	
1. Real, unequal, and of like sign	$\Delta > 0$, $ae - bc > 0$†	
2. Real, unequal, and of unlike sign	$\Delta > 0$, $ae - bc < 0$	
3a. Real and equal	$\Delta = 0$; b, c not both zero	
3b. Real and equal	$\Delta = 0$, $b = c = 0$	$(a = e)$
4. Pure imaginary	$\Delta < 0$, $a + e = 0$	$(ae - bc > 0)$
5. Complex but not pure imaginary	$\Delta < 0$, $a + e \neq 0$	

† From the elementary theory of quadratic equations, the constant term $ae - bc$ in the characteristic equation (4) is equal to the product of the roots. Hence the roots are of like sign when $ae - bc > 0$ and of unlike sign when $ae - bc < 0$. Since $ae - bc \neq 0$, the characteristic equation can never have zero as a root.

In discussing the various cases, it will be convenient to have the following definitions.

Definition 1 A trajectory T defined by the equations $x = x(t)$ and $y = y(t)$ is said to **approach** the critical point (0, 0) as $t \to +\infty$ if and only if

$$\lim_{t \to +\infty} x(t) = 0 \qquad \text{and} \qquad \lim_{t \to +\infty} y(t) = 0$$

T is said to **approach** the critical point (0, 0) as $t \to -\infty$ if and only if

$$\lim_{t \to -\infty} x(t) = 0 \qquad \text{and} \qquad \lim_{t \to -\infty} y(t) = 0$$

Definition 2 A trajectory T defined by the equations $x = x(t)$ and $y = y(t)$ is said to **enter** the critical point (0, 0) as $t \to +\infty$ if and only if T approaches (0, 0) as $t \to +\infty$ and

$$\lim_{t \to +\infty} \frac{y(t)}{x(t)}$$

either exists or is $\pm\infty$. T is said to **enter** the critical point (0, 0) as $t \to -\infty$ if and only if T approaches (0, 0) as $t \to -\infty$ and

$$\lim_{t \to -\infty} \frac{y(t)}{x(t)}$$

either exists or is $\pm\infty$.

Definition 3 A trajectory T defined by the equations $x = x(t)$ and $y = y(t)$ is said to recede indefinitely from the critical point (0, 0) as $t \to +\infty$ (or $t \to -\infty$) if and only if at least one of the functions $x(t)$ and $y(t)$ becomes infinite as $t \to +\infty$ (or $t \to -\infty$).

Case 1: For the system

$$\frac{dx}{dt} = \lambda x$$

$$\frac{dy}{dt} = 2\lambda y \qquad \lambda \neq 0 \tag{5}$$

we have $a = \lambda$, $b = c = 0$, $e = 2\lambda$. Hence $\Delta = \lambda^2$, $ae - bc = 2\lambda^2$, and thus the characteristic roots λ, 2λ are real, unequal, and of like sign. In this case the related equation ($1a$) is

$$\frac{dy}{dx} = \frac{2y}{x}$$

and its solution is

$$y = kx^2 \tag{6}$$

This is the equation of a family of parabolas, each tangent to the line $y = 0$ at the critical point (0, 0) and each having $x = 0$ as axis.

If the parametric equations of the trajectories are found by solving for x and y from Eqs. (5), we find

$$x = Ae^{\lambda t} \qquad y = Be^{2\lambda t} \tag{7}$$

Eliminating t between these equations leads, of course, to the family of curves (6), although in every case the trajectories are only portions of the corresponding curves. From Eqs. (7) it is clear that as $t \to +\infty$, every trajectory, including the open halves of the line $x = 0$ corresponding to $A = 0$, recedes indefinitely from the critical point if $\lambda > 0$ and approaches the critical point in a well-defined direction if $\lambda < 0$. Figure 14.2 shows the configuration of trajectories for $\lambda > 0$.

A critical point around which the configuration of trajectories resembles that shown in Fig. 14.2 is called a **node.** A node is characterized by the existence of a neighborhood of the critical point such that all trajectories in that neighborhood

1. Approach and enter the critical point as $t \to +\infty$, or
2. Approach and enter the critical point as $t \to -\infty$.

Case 2: For the system

$$\frac{dx}{dt} = \lambda x$$

$$\frac{dy}{dt} = -\lambda y \qquad \lambda \neq 0 \tag{8}$$

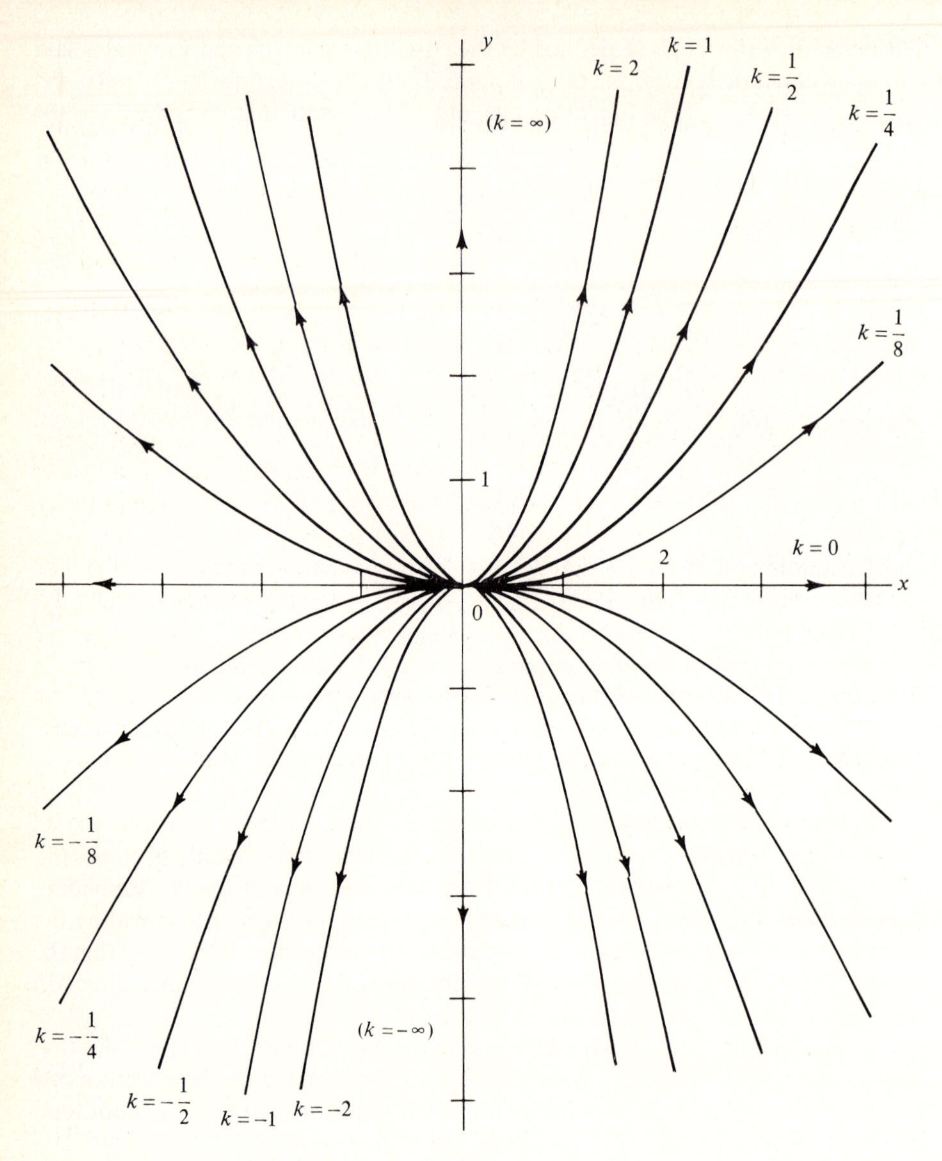

Figure 14.2 Typical trajectories around the node of the system $dx/dt = \lambda x$, $dy/dt = 2\lambda y$ ($\lambda > 0$).

we have $a = \lambda$, $b = c = 0$, $e = -\lambda$. Hence $\Delta = 4\lambda^2$, $ae - bc = -\lambda^2$, and the characteristic roots λ, $-\lambda$ are real, unequal, and of opposite sign. In this case the related equation (1a) is

$$\frac{dy}{dx} = -\frac{y}{x}$$

and its solution is

$$xy = k \tag{9}$$

This is the equation of a family of curves which, for $k \neq 0$, are hyperbolas having $x = 0$ and $y = 0$ as asymptotes. For $k = 0$ the corresponding curve consists of the asymptotes themselves. From Eqs. (8) we obtain at once the parametric equations of the trajectories

$$x = Ae^{\lambda t} \qquad y = Be^{-\lambda t} \tag{10}$$

Elimination of t between these equations leads immediately to Eq. (9). From Eqs. (10) it is clear that whether λ is positive or negative, each hyperbolic trajectory recedes indefinitely from the critical point both as $t \to +\infty$ and as $t \to -\infty$. As $t \to +\infty$, the trajectories on $x = 0$ approach and enter the critical point if $\lambda > 0$ and recede indefinitely from the critical point if $\lambda < 0$. As $t \to +\infty$, the trajectories on $y = 0$ recede indefinitely from the critical point if $\lambda > 0$ and approach and enter the critical point if $\lambda < 0$. Figure 14.3 shows the configuration of the trajectories for $\lambda > 0$.

A critical point around which the configuration of trajectories resembles that shown in Fig. 14.3 is called a **saddle point.** A saddle point is characterized by the existence of a neighborhood of the critical point in which:

1. There are (at least) two trajectories which approach and enter the critical point from opposite directions as $t \to +\infty$, and there are (at least) two other trajectories which approach and enter the critical point from opposite directions as $t \to -\infty$.
2. All other trajectories recede indefinitely from the critical point both as $t \to +\infty$ and as $t \to -\infty$.

Case 3a: For the system

$$\frac{dx}{dt} = \lambda x$$

$$\frac{dy}{dt} = \lambda x + \lambda y \qquad \lambda \neq 0 \tag{11}$$

we have $a = \lambda$, $b = 0$, $c = e = \lambda$. Hence $\Delta = 0$, $ae - bc = \lambda^2$, and the characteristic roots λ, λ are real and equal. In this case the related equation ($1a$) is

$$\frac{dy}{dx} = \frac{x + y}{x}$$

and its solution (obtained by solving it either as a homogeneous or a linear equation) is

$$y = x \ln |cx| \qquad c \neq 0 \tag{12}$$

Since $$\lim_{x \to 0} (x \ln |cx|) = 0$$

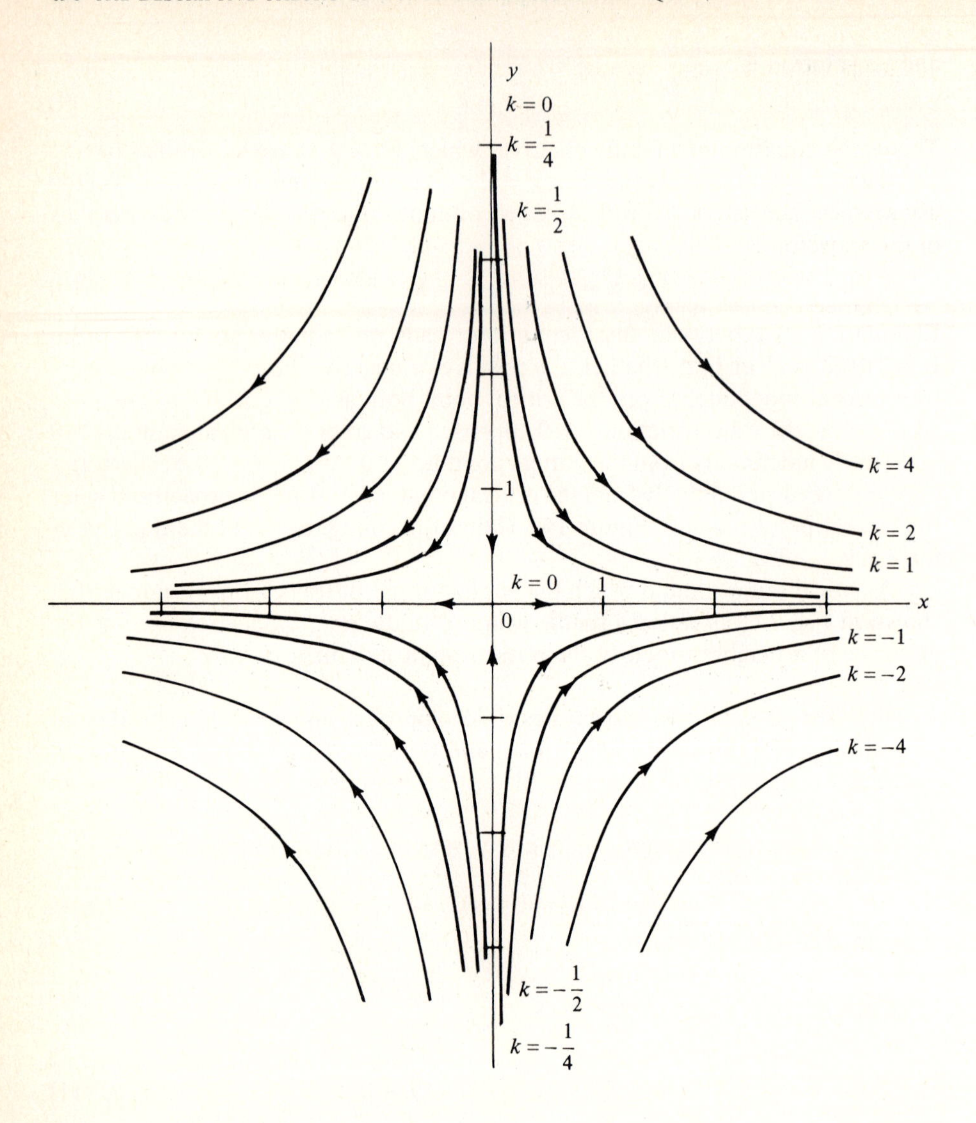

Figure 14.3 Typical trajectories around the saddle point of the system $dx/dt = \lambda x$, $dy/dt = -\lambda y$ $(\lambda > 0)$.

and since
$$\frac{d}{dx}(x \ln |cx|) = 1 + \ln |cx|$$

becomes infinite as $x \to 0$, it follows that (12) is the equation of a family of transcendental curves each of which approaches the critical point (0, 0) with slope approaching ∞ as x approaches zero. From Eqs. (11) we find that the parametric equations of the trajectories are

$$x = Ae^{\lambda t}$$

$$y = Be^{\lambda t} + A\lambda te^{\lambda t}$$

Thus if $\lambda > 0$, each trajectory recedes indefinitely from the critical point as $t \to +\infty$, while if $\lambda < 0$, each trajectory approaches and enters the critical point as $t \to +\infty$. Therefore the critical point in this case is also a node. Figure 14.4 shows the configuration of the trajectories for $\lambda > 0$.

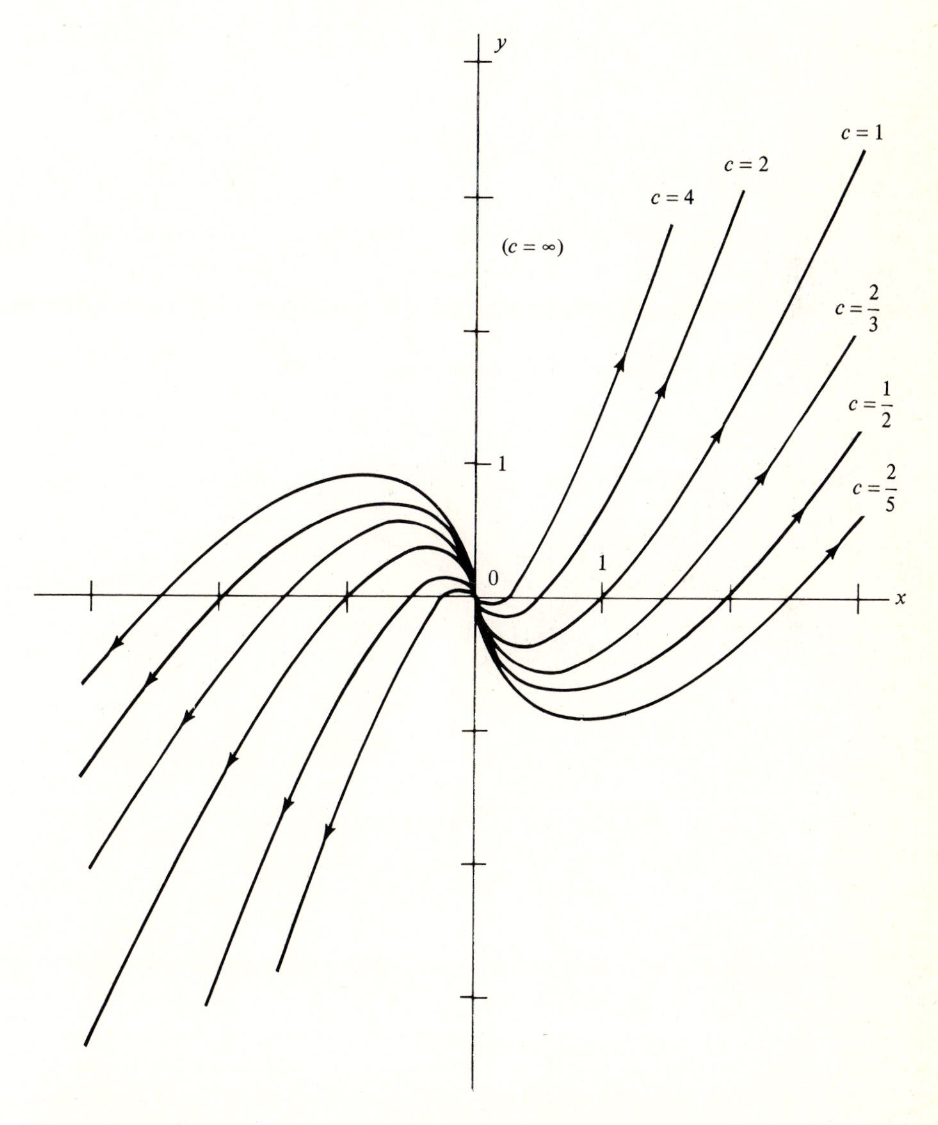

Figure 14.4 Typical trajectories around the node of the system $dx/dt = \lambda x$, $dy/dt = \lambda x + \lambda y$ ($\lambda > 0$).

Case 3b: For the system

$$\frac{dx}{dt} = \lambda x$$

$$\frac{dy}{dt} = \lambda y$$

we have $a = e = \lambda$, $b = c = 0$, $\Delta = 0$ and the equal characteristic roots λ, λ. The parametric equations of the trajectories in this case are

$$x = Ae^{\lambda t} \qquad y = Be^{\lambda t}$$

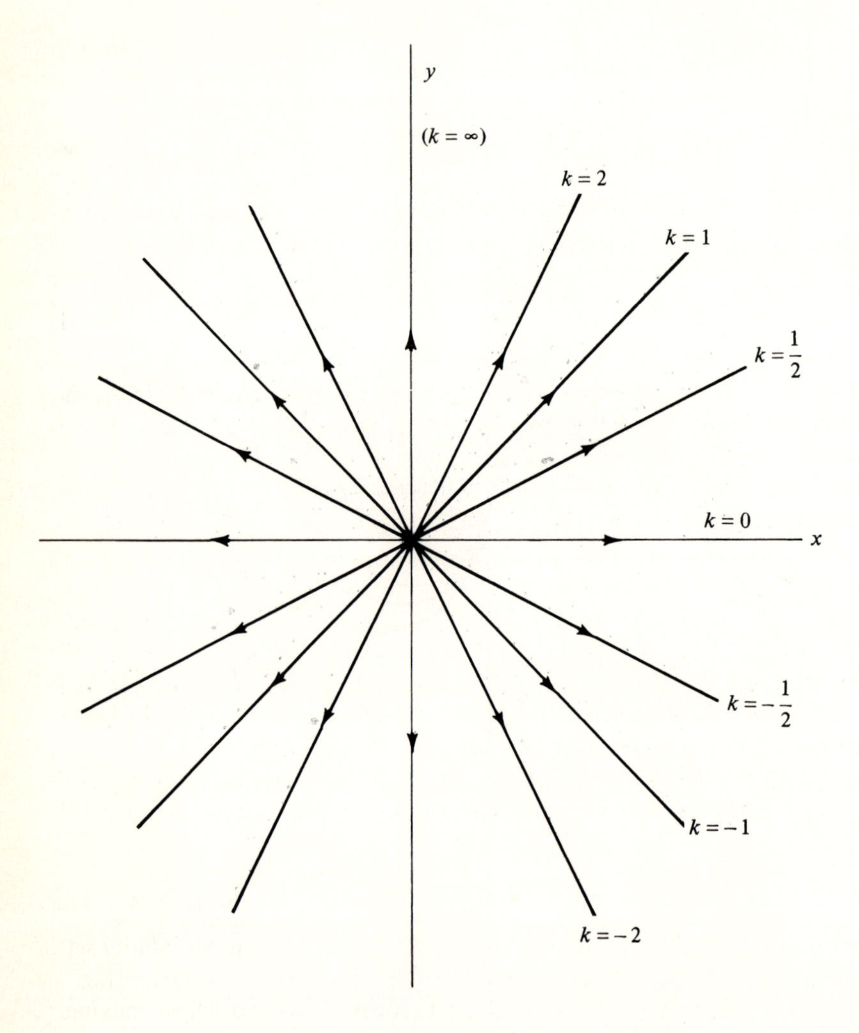

Figure 14.5 Typical trajectories around the node of the system $dx/dt = \lambda x$, $dy/dt = \lambda y$ $(\lambda > 0)$.

and the associated family of curves is the family of lines

$$y = kx$$

If $\lambda > 0$, every trajectory recedes indefinitely from the critical point as $t \to +\infty$; and if $\lambda < 0$, every trajectory approaches and enters the critical point as $t \to +\infty$. Figure 14.5 shows the configuration of trajectories for $\lambda > 0$. The critical point in this example is also a node, although it differs from the patterns of cases 1 and 3*a* of Table 14.1 in that the curves do not have a common tangent at the critical point. A node of this type is sometimes called a **proper node,** while nodes of the type occurring in cases 1 and 3*a* are called **improper nodes.**
Case 4: For the system

$$\frac{dx}{dt} = \lambda y$$

$$\frac{dy}{dt} = -\lambda x \qquad \lambda \neq 0 \tag{13}$$

we have $a = e = 0, b = \lambda, c = -\lambda$. Hence $\Delta = -4\lambda^2, a + e = 0$, and the characteristic roots $\pm i\lambda$ are pure imaginaries. The related equation (1*a*) is

$$\frac{dy}{dx} = -\frac{x}{y} \tag{14}$$

and its solution is $x^2 + y^2 = r^2$. This is the equation of a family of circles each having the critical point (0, 0) as center. Solving Eqs. (13), we find that the parametric equations of the trajectories are

$$x = A \cos \lambda t + B \sin \lambda t$$

$$y = -A \sin \lambda t + B \cos \lambda t \tag{15}$$

and these we recognize as parametric equations of the circles of the family $x^2 + y^2 = r^2$ which we obtained from (14). From Eqs. (15) it follows (see Exercise 13) that as $t \to +\infty$, all trajectories are traversed repeatedly in the clockwise direction if $\lambda > 0$ and in the counterclockwise direction if $\lambda < 0$. Figure 14.6 shows the configuration of the trajectories for $\lambda > 0$.

A critical point around which the configuration of trajectories resembles the configuration shown in Fig. 14.6 is called a **center.** Its significant characteristics are:

1. There exists a neighborhood of the critical point containing an infinite set of closed trajectories each of which contains the critical point in its interior.
2. For every $\varepsilon > 0$, there are trajectories in this neighborhood whose maximum chord is less than ε in length.

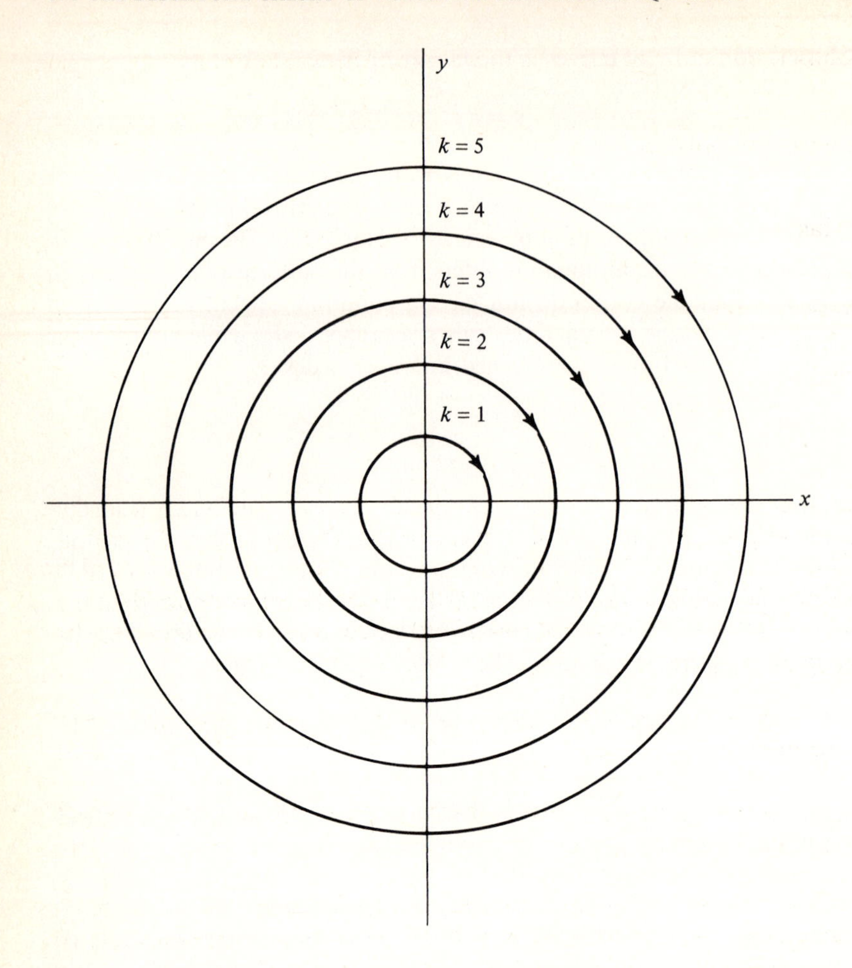

Figure 14.6 Typical trajectories around the center of the system $dx/dt = \lambda y$, $dy/dt = -\lambda x$ ($\lambda > 0$).

Case 5: For the system

$$\frac{dx}{dt} = \lambda x - y$$

$$\frac{dy}{dt} = x + \lambda y \qquad \lambda \neq 0 \tag{16}$$

we have $a = e = \lambda$, $b = -1$, $c = 1$. Hence $\Delta = -4$, $a + e = 2\lambda$, and (since $\lambda \neq 0$) the characteristic roots $\lambda \pm i$ are complex numbers which are not pure imaginaries. The related equation (1a) is

$$\frac{dy}{dx} = \frac{x + \lambda y}{\lambda x - y}$$

and its solution (found by solving it as a homogeneous equation) is

$$x^2 + y^2 = K^2 e^{2\lambda \tan^{-1}(y/x)} \qquad K \neq 0$$

or, transforming to polar coordinates,

$$r^2 = K^2 e^{2\lambda\theta}$$

or finally, taking square roots and letting k be either positive or negative,

$$r = ke^{\lambda\theta}$$

This is the equation of a family of spirals, each of which winds around the critical point (0, 0) as a limit. From Eqs. (16) we find that the parametric equations of the trajectories are

$$x = e^{\lambda t}(A \cos t + B \sin t)$$

$$y = e^{\lambda t}(A \sin t - B \cos t)$$

From these it is evident that if $\lambda > 0$, every trajectory recedes indefinitely from the critical point as $t \to +\infty$, and if $\lambda < 0$, every trajectory approaches the critical point as $t \to +\infty$. Figure 14.7 shows the configuration of the trajectories for $\lambda > 0$.

A critical point around which the configuration of trajectories resembles the configuration shown in Fig. 14.7 is called a **spiral point** or **focal point.** Its significant characteristics are:

1. There exists a neighborhood of the critical point such that every trajectory in this neighborhood approaches the critical point either as $t \to +\infty$ or as $t \to -\infty$.
2. As each trajectory approaches the critical point, it winds around the critical point an infinite number of times.

An examination of the families of trajectories appearing in Figs. 14.2 through 14.7 indicates, first of all, that only around a center are there periodic solutions of the given system, for only in this case does a trajectory containing a point (x_0, y_0) ever return to that point and hence (by the fundamental existence and uniqueness theorem) repeat the behavior it began previously at that point. The question of stability, which we shall soon consider, cannot be answered solely by an inspection of the family of trajectories because (except around a saddle point) trajectories may approach, or recede from, the critical point as t becomes infinite, depending on the directions established on the trajectories by the roots of the characteristic equation.

To aid us in discussing stability, we need both the concept of a *stable critical point* and the concept of an *asymptotically stable critical point*. To formulate these definitions, let C: (x_0, y_0) be an isolated critical point of the system

$$\frac{dx}{dt} = g(x, y)$$

$$\frac{dy}{dt} = f(x, y)$$

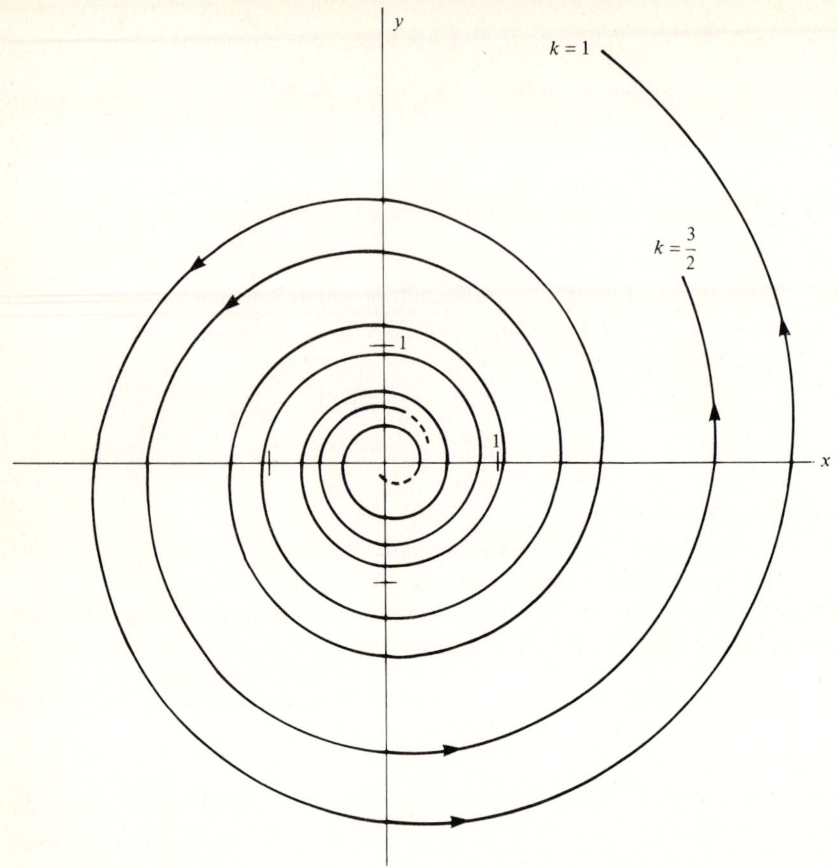

Figure 14.7 Typical trajectories around the spiral point of the system $dx/dt = \lambda x - y$, $dy/dt = x + \lambda y$ $(\lambda > 0)$.

let Γ: $\{x = x(t), y = y(t)\}$ be an arbitrary trajectory of the system, and let

$$D(t) = \sqrt{[x(t) - x_0]^2 + [y(t) - y_0]^2}$$

be the distance of an arbitrary point on Γ from the critical point.

Definition 1 The isolated critical point C is said to be **stable** if and only if for every $\varepsilon > 0$ there exists a $\delta > 0$ such that on any trajectory which contains a point $[x(t^*), y(t^*)]$ for which $D(t^*) < \delta$ the distance $D(t)$ exists and is less than ε for all $t \geq t^*$.

Stated less formally, Definition 1 says that a critical point is stable if and only if for every distance $\varepsilon > 0$ there is a distance δ (necessarily equal to or less than ε) such that any trajectory that once comes within δ of the critical point remains within ε of the critical point for all subsequent values of t. Figure 14.8a illustrates this behavior.

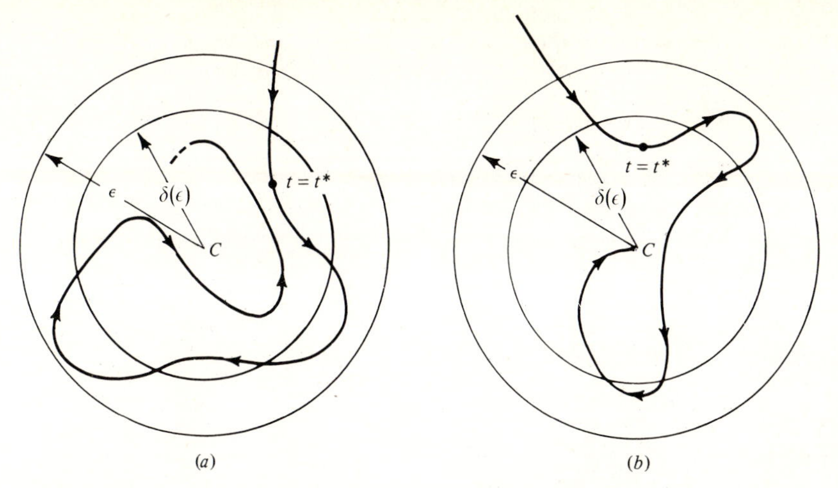

Figure 14.8 (*a*) A stable trajectory; (*b*) an asymptotically stable trajectory.

Definition 2 The isolated critical point C is said to be **asymptotically stable** if and only if, first, it is stable and, second, there exists a $\delta^* > 0$ such that if $D(t^*) < \delta^*$, then

$$\lim_{t \to +\infty} x(t) = x_0 \qquad \text{and} \qquad \lim_{t \to +\infty} y(t) = y_0$$

Less formally, Definition 2 says that a critical point C is asymptotically stable if and only if it is stable and every trajectory that comes sufficiently close to C actually approaches C. Figure 14.8*b* illustrates this behavior.

Naturally, to complete our vocabulary, we also have the following definition.

Definition 3 A critical point which is not stable is said to be **unstable.**

Table 14.2 summarizes the stability properties of the various critical points we have discussed.

The preceding results are stated still more concisely in the following theorem.

Theorem 1 Given the linear system $dx/dt = ax + by$, $dy/dt = cx + ey$ with $ae - bc \neq 0$, so that $(0, 0)$ is the only critical point of the system. If the roots of the characteristic equation $m^2 - (a + e)m + (ae - bc) = 0$ are real and negative or complex with negative real part, then the critical point is asymptotically stable. If the roots of the characteristic equation are real and positive or complex with positive real part, then the critical point is unstable. If the roots of the characteristic equation are pure imaginary, then the critical point is stable but not asymptotically stable. Only when the characteristic roots are pure imaginaries does the given system have periodic solutions.

Table 14.2 Stability properties of the critical point (0, 0) of the linear system $dx/dt = ax + by$, $dy/dt = cx + ey$, whose characteristic equation is

$$m^2 - (a + e)m + (ae - bc) = 0$$

Nature of the characteristic roots	Nature of the critical point	Stability of the critical point
1. Real, unequal, and of like sign	Node (improper)	Asymptotically stable if roots are negative; unstable if roots are positive
2. Real, unequal, and of unlike sign	Saddle point	Unstable
3. Real and equal	Node (proper or improper)	Asymptotically stable if roots are negative; unstable if roots are positive
4. Pure imaginary	Center	Stable but not asymptotically stable
5. Complex but not pure imaginary	Spiral point	Asymptotically stable if real part of roots is negative; unstable if real part of roots is positive

Exercises for Section 14.3

Determine the nature of the critical point (0, 0) for each of the following systems and tell whether it is stable, asymptotically stable, or unstable.

1 $\dfrac{dx}{dt} = 2x + 5y$

$\dfrac{dy}{dt} = x - 2y$

2 $\dfrac{dx}{dt} = 2x + 5y$

$\dfrac{dy}{dt} = -x + 5y$

3 $\dfrac{dx}{dt} = -4x + 3y$

$\dfrac{dy}{dt} = -2x + y$

4 $\dfrac{dx}{dt} = 2x + y$

$\dfrac{dy}{dt} = -x + 2y$

5 $\dfrac{dx}{dt} = x - 4y$

$\dfrac{dy}{dt} = x + 5y$

6 $\dfrac{dx}{dt} = -3x + y$

$\dfrac{dy}{dt} = -x - 3y$

7 Is the following an acceptable definition of asymptotic stability? A critical point C is asymptotically stable if every trajectory that comes sufficiently close to C actually approaches C. Justify your answer.

8 The motion of a mass-spring system is governed by the equation $m(d^2x/dt^2) + c(dx/dt) + kx = 0$. Convert this to a system of two first-order equations, show that this system has a single critical point, identify its type, and determine its stability properties.

9 Solve the system (11), and derive Eq. (12) by eliminating the parameter t between the functions thus obtained.

10 Work Exercise 9 for the system used to illustrate case 4.

11 Work Exercise 9 for the system used to illustrate case 5.

12 Show that if a trajectory of the system $dx/dt = g(x, y)$, $dy/dt = f(x, y)$ starts at a point which is not a critical point, it cannot reach a critical point in a finite length of time. *Hint:* Let $x = x(t)$, $y = y(t)$ be equations of the given trajectory, let (x_0, y_0) be the critical point, and suppose that there is a time $t = t_0$ for which $x(t_0) = x_0$ and $y(t_0) = y_0$. Then note that $x = x_0$, $y = y_0$ is a solution of the initial-value problem for which $x = x_0$ and $y = y_0$ when $t = t_0$.

13 Either by using the polar coordinate relation $\theta = \tan^{-1}(y/x)$ or by expressing x and y as single trigonometric functions, verify that as $t \to +\infty$, the trajectories described by (15) are traversed in the clockwise direction if $\lambda > 0$ and in the counterclockwise direction if $\lambda < 0$.

14.4 CRITICAL POINTS OF SYSTEMS WHICH ARE APPROXIMATELY LINEAR

Having completed our preparatory investigation of the critical points and paths of linear systems, we now turn our attention to the general, nonlinear autonomous system

$$\frac{dx}{dt} = F(x, y)$$

$$\frac{dy}{dt} = G(x, y) \tag{1}$$

As a starting point, we assume that this system has an isolated critical point which, without loss of generality, we suppose to be the origin. In this section we shall suppose further that in some neighborhood of the origin, F and G can be written in the approximately linear form

$$F(x, y) = ax + by + f(x, y)$$

$$G(x, y) = cx + ey + g(x, y) \tag{2}$$

where at least one of the functions f, g is nonlinear and f and g are small in comparison with $r = \sqrt{x^2 + y^2}$. More explicitly, the last requirement means that

$$\lim_{(x, y)\to(0, 0)} \frac{f(x, y)}{\sqrt{x^2 + y^2}} = 0 \qquad \text{and} \qquad \lim_{(x, y)\to(0, 0)} \frac{g(x, y)}{\sqrt{x^2 + y^2}} = 0$$

In particular, this will be the case if F and G possess Taylor expansions (item 17,

Appendix B.2) in which linear terms are actually present because then, remembering that by hypothesis $F(0, 0) = G(0, 0) = 0$, we have

$$F(x, y) = \left[\frac{\partial F}{\partial x}\bigg|_{0,0} x + \frac{\partial F}{\partial y}\bigg|_{0,0} y\right] + \frac{1}{2}\left[\frac{\partial^2 F}{\partial x^2}\bigg|_{0,0} x^2 + 2\frac{\partial^2 F}{\partial x\,\partial y}\bigg|_{0,0} xy + \frac{\partial^2 F}{\partial y^2}\bigg|_{0,0} y^2\right] + \cdots$$

and
$$G(x, y) = \left[\frac{\partial G}{\partial x}\bigg|_{0,0} x + \frac{\partial G}{\partial y}\bigg|_{0,0} y\right] + \frac{1}{2}\left[\frac{\partial^2 G}{\partial x^2}\bigg|_{0,0} x^2 + 2\frac{\partial^2 G}{\partial x\,\partial y}\bigg|_{0,0} xy + \frac{\partial^2 G}{\partial y^2}\bigg|_{0,0} y^2\right] + \cdots$$

With $f(x, y)$ and $g(x, y)$ interpreted to be the portions of the respective series consisting of the terms in x and y of the second degree and higher, these expressions have the form of the right-hand members of Eqs. (2). In studying these equations, we shall also suppose that

$$ae - bc = \left(\frac{\partial F}{\partial x}\bigg|_{0,0}\right)\left(\frac{\partial G}{\partial y}\bigg|_{0,0}\right) - \left(\frac{\partial F}{\partial y}\bigg|_{0,0}\right)\left(\frac{\partial G}{\partial x}\bigg|_{0,0}\right) \neq 0$$

Under the assumption that $F(x, y)$ and $G(x, y)$ possess Taylor expansions which contain linear terms, and with the accompanying definitions of $f(x, y)$ and $g(x, y)$, it follows that $f(x, y)$ and $g(x, y)$ are negligibly small in comparison with x and y in sufficiently small neighborhoods of the origin, and for this reason such systems are often said to be **almost linear.** This suggests that near the origin the system defined by (1) and (2) behaves essentially like the linear system we studied in the last section. This is true in some cases but false in others, as the following theorem, which we cite without proof, makes clear.

Theorem 1 Consider the nonlinear autonomous system

$$\frac{dx}{dt} = ax + by + f(x, y)$$

$$\frac{dy}{dt} = cx + ey + g(x, y)$$

where $ae - bc \neq 0$ and

$$\lim_{(x,y)\to(0,0)} \frac{f(x, y)}{\sqrt{x^2 + y^2}} = 0 \qquad \lim_{(x,y)\to(0,0)} \frac{g(x, y)}{\sqrt{x^2 + y^2}} = 0$$

and the corresponding linear system

$$\frac{dx}{dt} = ax + by$$

$$\frac{dy}{dt} = cx + ey$$

Let $(0, 0)$ be an isolated critical point of each system, and let m_1 and m_2 be the roots of the characteristic equation

$$m^2 - (a + e)m + (ae - bc) = 0$$

of the linear system. Then

1. If m_1 and m_2 are real, unequal, and of like sign, then the critical point $(0, 0)$ is a node of both the linear system and the nonlinear system.
2. If m_1 and m_2 are real, unequal, and of unlike sign, then the critical point $(0, 0)$ is a saddle point of both the linear system and the nonlinear system, although the trajectories which are the asymptotes may be curves rather than straight lines.
3. If m_1 and m_2 are conjugate complex numbers which are not pure imaginaries, then the critical point $(0, 0)$ is a spiral point of both the linear system and the nonlinear system.
4. If m_1 and m_2 are real and equal, then the critical point $(0, 0)$ is a node of both the linear system and the nonlinear system except when simultaneously $a = e \neq 0$ and $b = c = 0$.
5. If m_1 and m_2 are real and equal and if simultaneously $a = e \neq 0$ and $b = c = 0$, then although $(0, 0)$ is a node of the linear system, it may be either a node or a spiral point of the nonlinear system.
6. If m_1 and m_2 are pure imaginaries, then although $(0, 0)$ is a center of the linear system, it may be either a center or a spiral point of the nonlinear system.

Example 1 Determine the nature of the critical point $(0, 0)$ for the system

$$\frac{dx}{dt} = x + 2y + x \cos y$$

$$\frac{dy}{dt} = -y - \sin y$$

Our first step is to determine the related linear system. This requires that the linear terms in $x \cos y$ and $-\sin y$ be found and combined with the linear terms originally present. Hence we replace $x \cos y$ and $-\sin y$ by their Maclaurin expansions and then collect terms, getting

$$\frac{dx}{dt} = x + 2y + x\left(1 - \frac{y^2}{2!} + \frac{y^4}{4!} - \cdots\right) = 2x + 2y + \left(-\frac{y^2}{2} + \frac{y^4}{24} - \cdots\right)$$

$$\frac{dy}{dt} = -y - \left(y - \frac{y^3}{3!} + \frac{y^5}{5!} - \cdots\right) = -2y - \left(-\frac{y^3}{6} + \frac{y^5}{120} - \cdots\right)$$

The linear system we must investigate in order to apply Theorem 1 is therefore

$$\frac{dx}{dt} = 2x + 2y$$

$$\frac{dy}{dt} = -2y$$

The characteristic equation of this system is $m^2 - 4 = 0$. Since its roots, $m = \pm 2$, are real, unequal, and of unlike sign, the critical point $(0, 0)$ is a saddle point of the linear

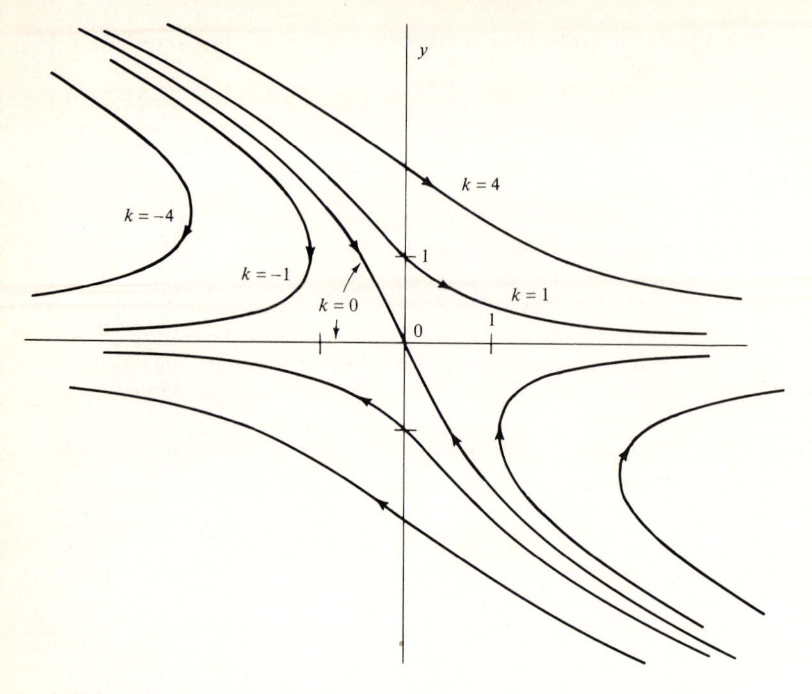

Figure 14.9*a* Typical trajectories around the critical point of the system

$$dx/dt = x + 2y + x \cos y, \; dy/dt = -y - \sin y.$$

system. Therefore, by observation 2 of Theorem 1, it is also a saddle point of the given nonlinear system.

The equation obtained by dividing the two given equations is

$$\frac{dy}{dx} = \frac{-y - \sin y}{x + 2y + x \cos y}$$

or

$$(y + \sin y)\, dx + (x + 2y + x \cos y)\, dy = 0$$

This equation happens to be exact, and hence (atypically) it can easily be solved. The solution, obtained by inspection, is

$$xy + y^2 + x \sin y = k$$

Figure 14.9*a* shows the configuration of the trajectories for the given nonlinear system, and Fig. 14.9*b* shows the trajectories of the related linear system. In this case one of the asymptotes of the trajectories of the nonlinear system is a curve and not a straight line.

Example 2 Determine the nature of the critical point (0, 0) for the system

$$\frac{dx}{dt} = y$$

$$\frac{dy}{dt} = -x - y^2$$

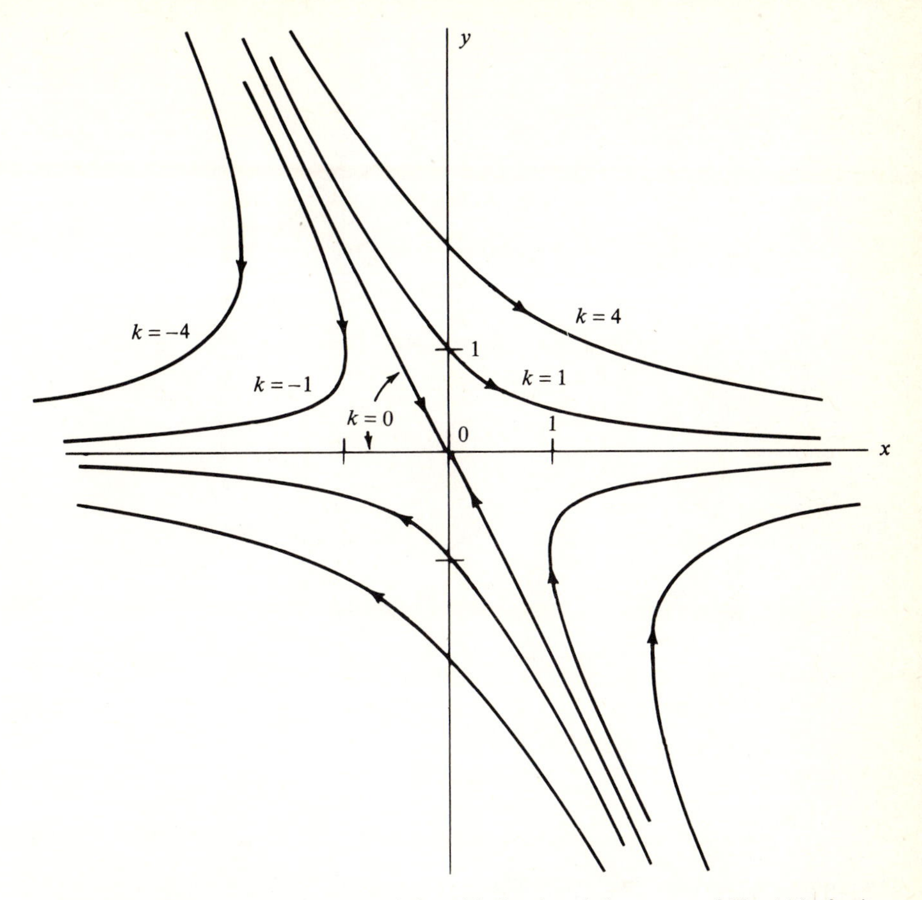

Figure 14.9*b* Typical trajectories around the critical point of the system of Fig. 14.9*a* for its related linear system $dx/dt = 2x + 2y$, $dy/dt = -2y$.

Since $\lim_{(x,y)\to(0,0)} y^2/\sqrt{x^2+y^2} = 0$, the related linear system is

$$\frac{dx}{dt} = y$$

$$\frac{dy}{dt} = -x$$

and its characteristic equation is $m^2 + 1 = 0$. Since the characteristic roots are pure imaginaries, the critical point $(0, 0)$ is a center for the linear system. Therefore, according to observation 6 of Theorem 1, $(0, 0)$ is either a center or a spiral point of the given nonlinear system, but the theorem does not tell us which.

In this case, however, the nature of the critical point can be determined, since the trajectories can be found explicitly. In fact, from the original pair of equations we find

$$\frac{dy}{dx} = -\frac{x}{y} - y \qquad \text{or} \qquad \frac{dy}{dx} + y = -\frac{x}{y}$$

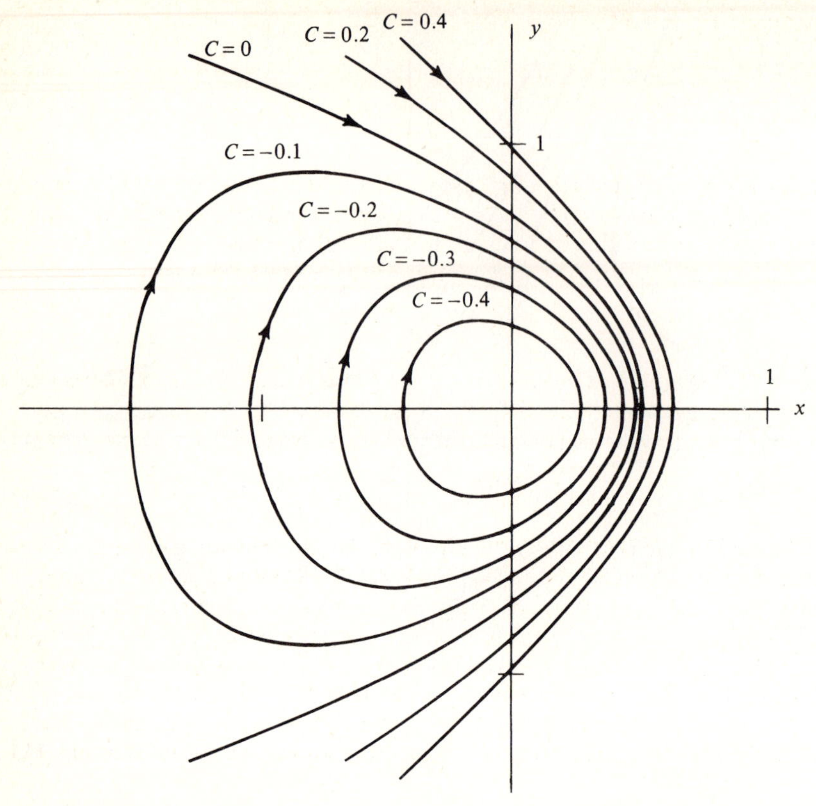

Figure 14.10 Typical trajectories around the center of the system $dx/dt = y$, $dy/dt = -x - y^2$.

which is a Bernoulli equation (Sec. 1.7) with $n = -1$. We therefore multiply the last equation by $2y$ and then put $z = y^2$, getting

$$\frac{dz}{dx} + 2z = -2x$$

Solving this as a linear first-order equation (Sec. 1.6), we find

$$z = y^2 = -x + \tfrac{1}{2} + Ce^{-2x}$$

If $C = 0$, this is the equation of a parabola. For $C > 0$, the trajectories are open curves somewhat resembling parabolas. For $-\frac{1}{2} < C < 0$, the trajectories are simple closed curves containing the critical point in their interiors. For $C = -\frac{1}{2}$, the path reduces to the single point (0, 0), and for $C < -\frac{1}{2}$, no path curves exist. From the plot of the paths shown in Fig. 14.10, it is clear that (0, 0) is a center of the given system.

Example 3 Determine the nature of the critical point (0, 0) for the system

$$\frac{dx}{dt} = y$$

$$\frac{dy}{dt} = -x - y^3$$

By inspection, the related linear system is

$$\frac{dx}{dt} = y$$

$$\frac{dy}{dt} = -x$$

which is the same approximating system we found in Example 2. Therefore, again, (0, 0) is a center of the related linear system and either a center or a spiral point of the given nonlinear system, but Theorem 1 cannot tell us which.

In this case, neither the given system nor the related equation

$$\frac{dy}{dx} = -\frac{x + y^3}{y} \tag{3}$$

can be solved exactly by any method with which we are familiar. However, from a purely descriptive consideration of the last equation, it is possible to argue convincingly that (0, 0) is a spiral point. To do this, we observe first that the cubic curves

$$m = -\frac{x + y^3}{y} \qquad \text{or} \qquad x = -my - y^3$$

are the loci of points at which solutions of Eq. (3), i.e., trajectories of the given system, have slope m. In particular, $x = -y^3$ is the locus of points at which the trajectories have slope zero; $y = 0$ is the locus of points at which their slope is infinite; $x = -y - y^3$ is the locus of points at which their slope is 1; and $x = y - y^3$ is the locus of points at which their slope is -1. For $0 < m < \infty$, the corresponding curves are monotonically decreasing cubics lying between the x axis and the curve $x = -y^3$. For $-\infty < m < 0$, the curves are nonmonotone cubics lying in the other two "sectors" determined by $y = 0$ and $x = -y^3$. These observations are illustrated in Fig. 14.11*a*.

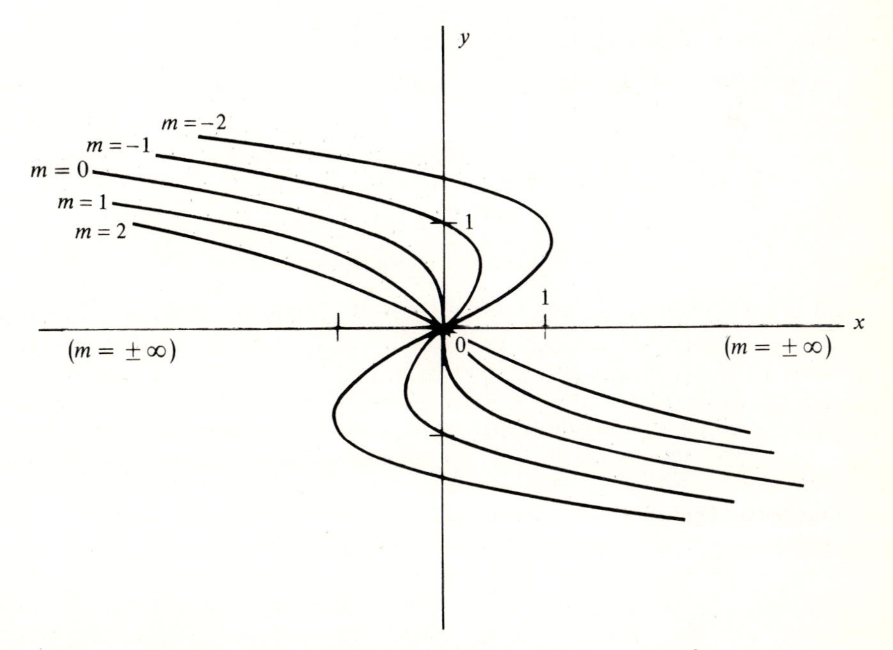

Figure 14.11*a* Typical isoclines for the system $dx/dt = y$, $dy/dt = -x - y^3$.

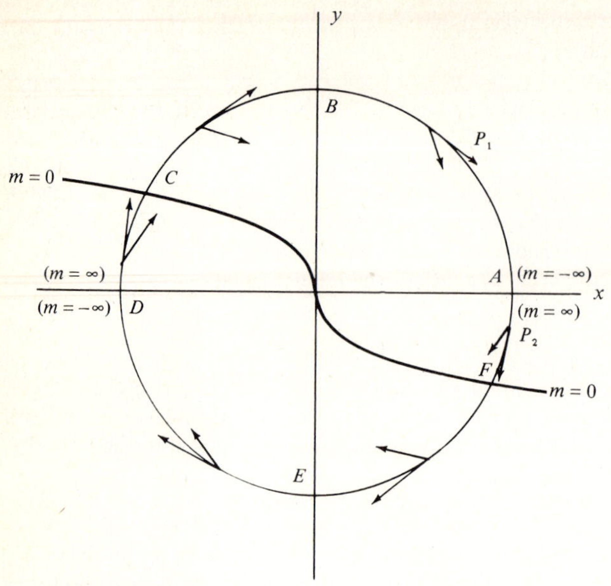

Figure 14.11*b*

Now consider an arbitrary circle $x^2 + y^2 = r^2$ (Fig. 14.11*b*). The location of the curves on which the slope assigned by Eq. (3) is $m = 1$ and $m = -1$ indicates that:

1. Along the arc AC the slope increases monotonically from $-\infty$ to 0.
2. Along the arc CD the slope increases monotonically from 0 to ∞.
3. Along the arc DF the slope increases monotonically from $-\infty$ to 0.
4. Along the arc FA the slope increases monotonically from 0 to ∞.

Now let P: $(r \cos \theta, r \sin \theta)$ be an arbitrary point on the circle distinct from A and D. At P the slope of the circle is

$$m_c = -\frac{\cos \theta}{\sin \theta}$$

and the slope assigned by Eq. (3) to the trajectory passing through P is

$$m_t = -\frac{\cos \theta + r^2 \sin^3 \theta}{\sin \theta} = -\frac{\cos \theta}{\sin \theta} - r^2 \sin^2 \theta$$

Since P is neither A nor D, the term $r^2 \sin^2 \theta$ is different from zero, and hence from the last equation it follows that $m_t < m_c$. The relative directions of the tangent to the circle at P and the tangent to the trajectory through P are therefore as shown in Fig. 14.11*b*.

Thus at all points on *any* circle $x^2 + y^2 = r^2$ (except at the ends of the horizontal diameter, where the tangent to the trajectory is also tangent to the circle) the tangent to the trajectory through that point is directed into (out of) the interior of the circle. This strongly suggests that the trajectories must be spirals approaching (receding from) the critical point as a point traverses them in the clockwise (counterclockwise) direction. Moreover, from the equation $dx/dt = y$, it follows that the trajectories are traversed in the clockwise direction as $t \to +\infty$.

When the Taylor expansions of $F(x, y)$ and $G(x, y)$ each contain linear terms, Theorem 1 gives us information about the nature of the critical points of the system

$$\frac{dx}{dt} = F(x, y)$$

$$\frac{dy}{dt} = G(x, y)$$

but it does not give us any information about the stability of the critical point. The next theorem, which we also cite without proof, provides us with useful information of this sort.

Theorem 2 Consider the nonlinear autonomous system

$$\frac{dx}{dt} = ax + by + f(x, y)$$

$$\frac{dy}{dt} = cx + ey + g(x, y)$$

where $ae - bc \neq 0$ and

$$\lim_{(x,y)\to(0,0)} \frac{f(x, y)}{\sqrt{x^2 + y^2}} = 0 \qquad \lim_{(x,y)\to(0,0)} \frac{g(x, y)}{\sqrt{x^2 + y^2}} = 0$$

and the corresponding linear system

$$\frac{dx}{dt} = ax + by$$

$$\frac{dy}{dt} = cx + ey$$

Let $(0, 0)$ be an isolated critical point of each system, and let m_1 and m_2 be the roots of the characteristic equation

$$m^2 - (a + e)m + (ae - bc) = 0$$

of the linear system. Then:

1. If m_1 and m_2 are real and negative or conjugate complex with negative real part, then $(0, 0)$ is an asymptotically stable critical point of both the linear system and the given nonlinear system.
2. If m_1 and m_2 are pure imaginary, then although $(0, 0)$ is a stable critical point of the linear system, it is not necessarily a stable critical point of the nonlinear system. In fact, the critical point $(0, 0)$ of the given nonlinear system may be asymptotically stable, stable but not asymptotically stable, or unstable.
3. If either m_1 or m_2 is real and positive, or if m_1 and m_2 are conjugate complex with positive real parts, then $(0, 0)$ is an unstable critical point for both the linear system and the given nonlinear system.

Reconsidering Example 1 in the light of Theorem 2, we note that since the characteristic roots of the related linear system are ± 2, therefore, by observation 3 of Theorem 2, the critical point (0, 0) is unstable for both the linear system and the nonlinear system. We confirmed this, of course, by determining and plotting the path curves (Fig. 14.9). In both Example 2 and Example 3 the characteristic roots of the related linear system were the conjugate imaginaries $\pm i$. Hence, according to observation 2 of Theorem 2, the critical point (0, 0) while stable for the linear system may be asymptotically stable, stable but not asymptotically stable, or unstable for the given nonlinear system. From our analysis of the path curves, we found the critical point (0, 0) to be stable but not asymptotically stable in Example 2 and asymptotically stable in Example 3.

Exercises for Section 14.4

Determine the nature and stability of the critical point (0, 0) of each of the following systems.

1 $\dfrac{dx}{dt} = 3x + 4y + x^2$, $\dfrac{dy}{dt} = 4x - 3y - 2xy$

2 $\dfrac{dx}{dt} = 6x + 10y - x^2$, $\dfrac{dy}{dt} = -4x - 6y + 2xy$

3 $\dfrac{dx}{dt} = -x - x \cos y$, $\dfrac{dy}{dt} = y + \sin y$

4 $\dfrac{dx}{dt} = x + 2y + 2 \sin y$, $\dfrac{dy}{dt} = -3y - xe^x$

5 $\dfrac{dx}{dt} = 1 + y - e^{-x}$, $\dfrac{dy}{dt} = y - \sin x$

6 $\dfrac{dx}{dt} = e^{-x+y} - \cos x$, $\dfrac{dy}{dt} = \sin (x - 3y)$

7 $\dfrac{dx}{dt} = x + x^2 - 3xy$, $\dfrac{dy}{dt} = -2x + y + 3y^2$

8 $\dfrac{dx}{dt} = -\sin (x - y)$, $\dfrac{dy}{dt} = 1 - 5y - e^x$

9 Find the equation of the family of trajectories in Exercise 1.

10 Find the equation of the family of trajectories in Exercise 3 and sketch several members of the family.

Determine the critical points of each of the following systems and discuss their nature and stability:

11 $\dfrac{dx}{dt} = 2y + x^2$, $\dfrac{dy}{dt} = -2x - 4y$

Hint: Translate axes by the substitutions $X = x + h$, $Y = y + k$, where h and k are chosen so that the critical point becomes (0, 0).

12 $\dfrac{dx}{dt} = -3y + 3xy$ $\quad\dfrac{dy}{dt} = 2x - y - 3$

13 $\dfrac{dx}{dt} = x + y^2$ $\quad\dfrac{dy}{dt} = x + y$

14 $\dfrac{dx}{dt} = 1 - xy$ $\quad\dfrac{dy}{dt} = x - y^3$

15 $\dfrac{dx}{dt} = 1 - y$ $\quad\dfrac{dy}{dt} = x^2 - y^2$

***16** If the equation for dy/dx obtained from the system

$$\frac{dx}{dt} = ax + by + f(x, y) \qquad \frac{dy}{dt} = cx + ey + g(x, y)$$

is exact (Sec. 1.5), show that the critical point (0, 0) is never asymptotically stable.

***17** Where does the geometric argument used in Example 3 to prove that the path curves are spirals break down when applied to the system of Example 2?

18 Justify the directions indicated by the arrows on the trajectories (*a*) in Fig. 14.9, (*b*) in Fig. 14.10.

14.5 SYSTEMS WHICH ARE NOT APPROXIMATELY LINEAR

Up to this point we have been concerned exclusively with autonomous systems

$$\frac{dx}{dt} = F(x, y) \qquad \frac{dy}{dt} = G(x, y)$$

in which the Taylor expansions of both $F(x, y)$ and $G(x, y)$ around a critical point contained linear terms; and we have seen that from the system resulting when only the linear terms were retained, certain properties of the actual, nonlinear system could be inferred. If the expansions of $F(x, y)$ and $G(x, y)$ around a critical point contain no linear terms, Theorems 1 and 2 of the last section are clearly inapplicable, and an investigation of the trajectories and their stability is too difficult to attempt in an elementary treatment such as ours. Hence we shall content ourselves with an example or two and then cite, without proof, several fundamental theorems that pertain to these more general autonomous systems.

As an illustration of the complicated configurations of path curves which may occur, let us consider first the system

$$\frac{dx}{dt} = y^2 - x^2 \qquad \frac{dy}{dt} = 2xy \qquad \text{or} \qquad \frac{dy}{dx} = \frac{2xy}{y^2 - x^2}$$

Clearly, (0, 0) is an isolated critical point of this system, but since neither $y^2 - x^2$

nor $2xy$ contains any linear terms, the theorems of the last section cannot be applied to help us determine its nature or stability. In this case, however, the equation for dy/dx is easy to solve, for it can be written in the form

$$3y^2\,dy = 3x^2\,dy + 6xy\,dx$$

and then integrated by inspection. This gives us

$$y^3 = 3x^2y + k$$

as the equation of the family of trajectories. Several of these curves are shown in Fig. 14.12. Evidently, the critical point (0, 0) is a kind of "super" saddle point, with three asymptotes each consisting of two trajectories.

As a second example, let us consider the system

$$\frac{dx}{dt} = x^2 \qquad \frac{dy}{dt} = 2y^2 - xy$$

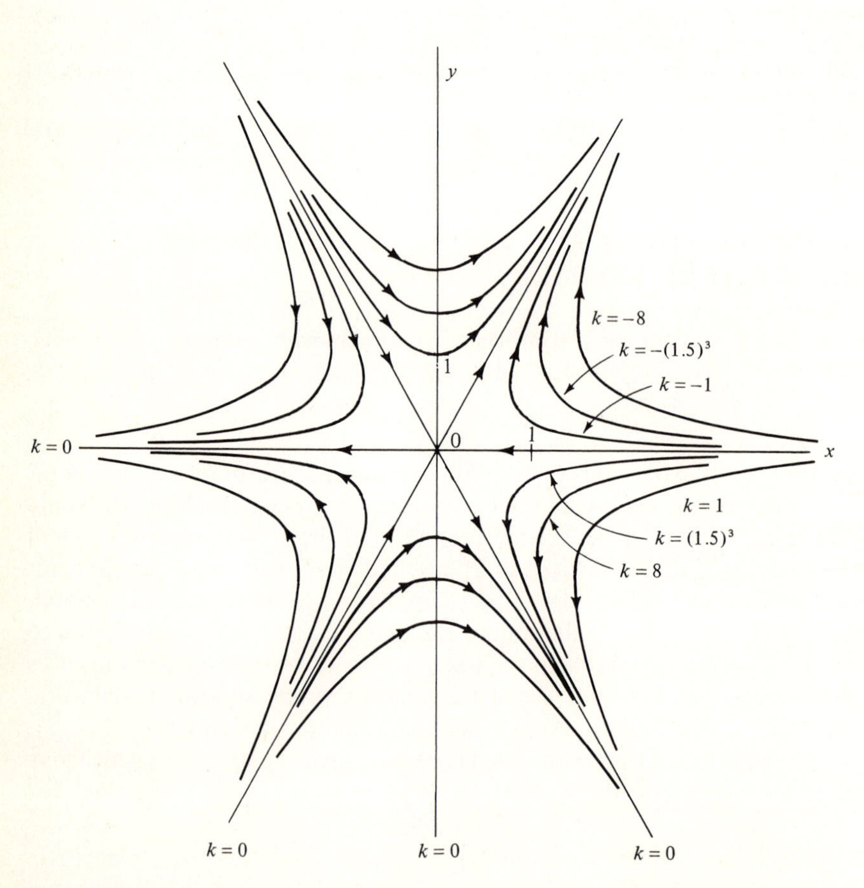

Figure 14.12 Typical trajectories around the critical point (0, 0) of the system $dx/dt = y^2 - x^2$, $dy/dt = 2xy$.

Since the associated equation

$$\frac{dy}{dx} = \frac{2y^2 - xy}{x^2}$$

is homogeneous, it is possible in this case, also, to solve for the equation of the family of trajectories. Thus, making the substitution $y = ux$ (Sec. 1.4), we obtain the separable equation

$$x\,du = 2(u^2 - u)\,dx$$

Separating variables, integrating, and then replacing u by y/x, we obtain finally

$$y = \frac{x}{1 - kx^2}$$

Several curves of this family are shown in Fig. 14.13. In this case the critical point (0, 0) appears to be a combination of a node and a saddle point.

Since there are infinitely many possibilities for the configuration of the trajectories near a critical point of a general nonlinear system, no attempt is made to describe and classify the critical points of such systems. Instead, studies of these

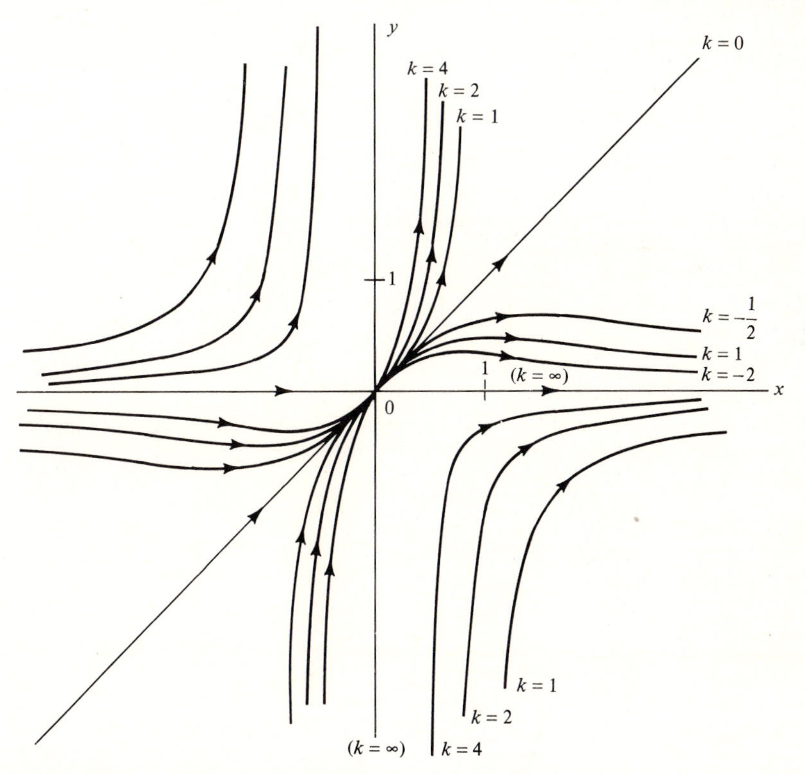

Figure 14.13 Typical trajectories around the critical point (0, 0) of the system $dx/dt = x^2$, $dy/dt = 2y^2 - xy$.

systems are concerned primarily with questions of stability and periodicity. We shall conclude this chapter by first introducing some necessary definitions and then stating without proof several of the basic theorems dealing with these matters.

Our first definition extends to functions of continuous variables the concept of *definiteness*, which we have already encountered in studying matrices and quadratic forms in algebra.

Definition 1 Let D be a domain containing the point (x_0, y_0) in its interior and let V be a function defined throughout D.

1. If $V(x_0, y_0) = 0$ and if $V(x, y) > 0$ at all other points of D, then V is said to be **positive definite** throughout D.
2. If $V(x_0, y_0) = 0$ and if $V(x, y) \geq 0$ at all other points of D, with the equality holding for at least one point other than (x_0, y_0), then V is said to be **positive semidefinite** throughout D.
3. If $V(x_0, y_0) = 0$ and if $V(x, y) < 0$ at all other points of D, then V is said to be **negative definite** throughout D.
4. If $V(x_0, y_0) = 0$ and if $V(x, y) \leq 0$ at all other points of D, with the equality holding for at least one point other than (x_0, y_0), then V is said to be **negative semidefinite** throughout D.

Example 1 If (x_0, y_0) is the point $(0, 0)$ and if D is the interior of the circle $x^2 + y^2 = 1$, then

1. The function $V = (x + y)^2 + y^4$ is positive definite in D since it is nonnegative throughout D and zero only at $(0, 0)$.
2. The function $V = y^2(x^2 + y^2)$ is positive semidefinite in D since it is nonnegative throughout D but vanishes in D not only at $(0, 0)$ but at all points of the horizontal diameter of D.
3. The function $V = (x^2 + y^2)/(x^2 + y^2 - 2)$ is negative definite in D since it is nonpositive throughout D and is zero only at $(0, 0)$.
4. The function $V = x^2/(x^2 + y^2 - 2)$ is negative semidefinite in D since it is nonpositive throughout D but vanishes in D not only at $(0, 0)$ but at all points of the vertical diameter of D.

Suppose, now, that we have a trajectory

$$\Gamma\colon x = x(t),\ y = y(t)$$

of the system

$$\frac{dx}{dt} = F(x, y)$$

$$\frac{dy}{dt} = G(x, y) \tag{1}$$

and a function V defined and differentiable at all points of Γ. This means that on the curve Γ, V is a function of t,

$$V = V[x(t), y(t)]$$

and can be differentiated with respect to t. Thus, from the familiar chain rule, we have

$$\frac{dV}{dt} \equiv \dot{V} = \frac{\partial V}{\partial x}\frac{dx}{dt} + \frac{\partial V}{\partial y}\frac{dy}{dt}$$

or, using the values of dx/dt and dy/dt provided by (1),

$$\frac{dV}{dt} \equiv \dot{V} = \frac{\partial V}{\partial x}F(x, y) + \frac{\partial V}{\partial y}G(x, y) \tag{2}$$

The function $\dot{V}$ defined by (2) is called the **derivative of V with respect to the system (1).**

We are now in a position to state two theorems, due to the Russian mathematician A. M. Liapounov (1857–1918), which deal, respectively, with the stability and the instability of the general system (1).

Theorem 1 If the autonomous system

$$\frac{dx}{dt} = F(x, y)$$

$$\frac{dy}{dt} = G(x, y)$$

has an isolated critical point at the origin, if F and G have continuous first partial derivatives in some domain D containing the origin, and if there exists a function $V = V(x, y)$ with continuous first partial derivatives which is positive definite in D, then

1. If the derivative of V with respect to the given system is negative definite in D, the origin is an asymptotically stable critical point.
2. If the derivative of V with respect to the given system is negative semidefinite in D, the origin is at least a stable critical point.

Theorem 2 Let the origin be an isolated critical point of the autonomous system

$$\frac{dx}{dt} = F(x, y)$$

$$\frac{dy}{dt} = G(x, y)$$

Let F and G have continuous first partial derivatives in some domain D containing the origin, and let V be a function which has continuous first partial derivatives in D and takes on the value 0 at the critical point (0, 0). Then the origin is an unstable critical point if either of the following conditions is satisfied:

1. In every neighborhood of the origin there is at least one point where V is positive and the derivative of V with respect to the system is positive definite in D.
2. In every neighborhood of the origin there is at least one point where V is negative and the derivative of V with respect to the system is negative definite in D.

A function $V = V(x, y)$ having either the properties described in Theorem 1 or the properties described in Theorem 2 is said to be a **Liapounov function** for the corresponding autonomous system. Unfortunately, Theorems 1 and 2 contain no clues to help us determine an appropriate function V, and it usually takes considerable ingenuity to construct one, if indeed it exists. However, the following theorem from elementary algebra is useful in identifying a class of functions from which a suitable Liapounov function $V(x, y)$ can sometimes be selected.

Theorem 3 The function $V(x, y) = Ax^2 + Bxy + Cy^2$

1. Is positive definite if and only if $A > 0$ and $B^2 - 4AC < 0$;
2. Is negative definite if and only if $A < 0$ and $B^2 - 4AC < 0$;
3. Is positive semidefinite if and only if $A > 0$ and $B^2 - 4AC = 0$;
4. Is negative semidefinite if and only if $A < 0$ and $B^2 - 4AC = 0$.

Example 2 Show that the origin is an asymptotically stable critical point of the system

$$\frac{dx}{dt} = y - 2x^3$$

$$\frac{dy}{dt} = -2x - 3y^5$$

It is not difficult to verify that the origin is the only critical point of the given system. To determine its nature, let us try, tentatively, to find a Liapounov function of the form

$$V(x, y) = Ax^2 + Cy^2$$

Using this, we see at once that

$$\begin{aligned} \dot{V} &= 2Ax(y - 2x^3) + 2Cy(-2x - 3y^5) \\ &= -4Ax^4 + (2A - 4C)xy - 6Cy^6 \end{aligned}$$

If we take $A = 2$ and $C = 1$, this becomes

$$\dot{V} = -8x^4 - 6y^6$$

which is clearly a negative definite function. Moreover, with $A = 2$ and $C = 1$, the function V becomes

$$V = 2x^2 + y^2$$

which is obviously positive definite. Hence condition 1 of Theorem 1 is satisfied, and therefore the origin is an asymptotically stable critical point, as asserted.

Example 3 Show that the origin is an asymptotically stable critical point of the system

$$\frac{dx}{dt} = -x^5 - y^3$$

$$\frac{dy}{dt} = 3x^3 - y^3$$

Clearly, the origin is the only critical point of the system. Again, to investigate it, let us begin with the tentative Liapounov function

$$V = Ax^2 + Cy^2$$

From this we find at once that

$$\begin{aligned} \dot{V} &= 2Ax(-x^5 - y^3) + 2Cy(3x^3 - y^3) \\ &= -2Ax^6 + (-2Axy^3 + 6Cx^3y) - 2Cy^4 \end{aligned}$$

As in Example 2, the first and last terms in $\dot{V}$ form a negative definite function if A and C are positive, but in this case the components of the middle term do not cancel and it is thus not clear whether $\dot{V}$ is negative definite or not.

Reconsidering the middle term in $\dot{V}$, we note that it could be eliminated by proper choice of A and C if the factor x in the first term were x^3 and the factor y in the second term were y^3. Moreover, these replacements can be accomplished by beginning with the tentative Liapounov function

$$V = Ax^4 + Cy^4$$

In fact, for this function we have

$$\begin{aligned} \dot{V} &= 4Ax^3(-x^5 - y^3) + 4Cy^3(3x^3 - y^3) \\ &= -4Ax^8 + (-4A + 12C)x^3y^3 - 4Cy^6 \end{aligned}$$

From this, by choosing $A = 3$ and $C = 1$, we obtain

$$\dot{V} = -12x^8 - 4y^6$$

and

$$V = 3x^4 + y^4$$

Thus our second choice for V is positive definite, and the related derivative $\dot{V}$ is negative definite. Hence, by Theorem 1, the origin is an asymptotically stable critical point of the given system, as asserted.

Example 4 Determine the nature of the critical point (0, 0) for the system

$$\frac{dx}{dt} = 2x^3 - 2xy^2$$

$$\frac{dy}{dt} = -5x^2y - 4y^3$$

Again trying

$$V = Ax^2 + Cy^2$$

we find

$$\begin{aligned}\dot{V} &= 2Ax(2x^3 - 2xy^2) + 2Cy(-5x^2y - 4y^3)\\ &= 4Ax^4 + (-4A - 10C)x^2y^2 - 8Cy^4\end{aligned}$$

Here, if we eliminate the middle term by taking $A = 5$ and $C = -2$, we are left with

$$\dot{V} = 20x^4 + 16y^4$$

and

$$V = 5x^2 - 2y^2$$

Obviously $\dot{V}$ is positive definite; hence Theorem 1 cannot be applied. However, it is clear that $V(0, 0) = 0$ and that $V(x, y) > 0$ at points arbitrarily close to $(0, 0)$ since in fact V is positive at all points of the x axis except the origin. Hence condition 1 of Theorem 2 is satisfied, and therefore $(0, 0)$ is an unstable critical point of the system.

Exercises for Section 14.5

Using Liapounov functions of the form $Ax^2 + By^2$, verify that for each of the following systems the origin is a critical point of the indicated type.

1 $\dfrac{dx}{dt} = -x^3 + xy^2$ $\qquad \dfrac{dy}{dt} = -4x^2y - y^3$ $\qquad$ asymptotically stable

2 $\dfrac{dx}{dt} = -x^3 + 3xy^2$ $\qquad \dfrac{dy}{dt} = x^2y - 4y^3$ $\qquad$ at least stable

3 $\dfrac{dx}{dt} = 2xy + x^3$ $\qquad \dfrac{dy}{dt} = 2x^2 - y^3$ $\qquad$ unstable

4 $\dfrac{dx}{dt} = -x - x^3 + 4xy^2$ $\qquad \dfrac{dy}{dt} = -2x - y - 4y^3$ $\qquad$ asymptotically stable

Using Liapounov functions of the form $V = Ax^2 + By^2$, determine the nature of the critical point $(0, 0)$ for each of the following systems.

5 $\dfrac{dx}{dt} = -x^3 + 3y^2$ $\qquad \dfrac{dy}{dt} = -y^3 - 2xy$

6 $\dfrac{dx}{dt} = -x^3 - 8xy^2$ $\qquad \dfrac{dy}{dt} = -2x^2y + 9y^3$

7 $\dfrac{dx}{dt} = x - 4y + x^3$ $\qquad \dfrac{dy}{dt} = -4y - y^3 + 2x^2y$

8 $\dfrac{dx}{dt} = -x + 2y - x^3$ $\qquad \dfrac{dy}{dt} = -y + 2x^2y - y^3$

9 Using the Liapounov function $V = xy$, determine the nature of the critical point $(0, 0)$ for the system

$$\frac{dx}{dt} = -2xy - y^3 \qquad \frac{dy}{dt} = -x + 2y^2$$

10 Using a Liapounov function of the form $V = Ax^2 + By^4$, determine the nature of the critical point $(0, 0)$ for the system

$$\frac{dx}{dt} = -x^3 - 3xy^4 \qquad \frac{dy}{dt} = x^2y - 2y^3 - y^5$$

11 Show that the qualification "at least" in conclusion 2 of Theorem 1 is necessary by exhibiting a positive definite Liapounov function whose derivative with respect to the system

$$\frac{dx}{dt} = -x^3 + xy^2 \qquad \frac{dy}{dt} = -3x^2y - y^3$$

is negative semidefinite and another positive definite Liapounov function whose derivative with respect to this system is negative definite.

12 Using a Liapounov function of the form $V = Ax^2 + By^2$, show that the critical point (0, 0) is asymptotically stable for the system

$$\frac{dx}{dt} = a^2y - xf(x, y) \qquad \frac{dy}{dt} = -b^2x - yf(x, y)$$

if $f(x, y)$ is positive throughout some neighborhood of the origin and unstable if $f(x, y)$ is negative throughout some neighborhood of the origin.

13 Show that (0, 0) is an asymptotically stable critical point of the system

$$\frac{dx}{dt} = y^3 - x^3f(x, y) \qquad \frac{dy}{dt} = -x^3 - y^3f(x, y)$$

if $f(x, y)$ is positive throughout some neighborhood of the origin and is an unstable critical point if $f(x, y)$ is negative throughout some neighborhood of the origin.

14 For each of the following functions, determine whether in some neighborhood of the origin it is positive definite, positive semidefinite, negative definite, negative semidefinite, or none of these.

(*a*) $x^2 - 4xy + 5y^2$ (*b*) $x^2 + x^4 + 4xy + 4y^2$ (*c*) $1 - e^{-xy}$
(*d*) $x^4 - 2x^2y^4 + y^8$ (*e*) $x^4 - 3x^2y^4 + y^8$ (*f*) $x^4 - x^2y^2 + y^8$
(*g*) $2x^4 - 4x^3y + 4x^2y^2 - 4xy^3 + 2y^4$ (*h*) $\sin(xy)$

14.6 PERIODIC SOLUTIONS AND LIMIT CYCLES

In Sec. 14.3 we observed that if the roots of the characteristic equation

$$m^2 - (a + e)m + ae - bc = 0$$

of the linear system

$$\frac{dx}{dt} = ax + by$$

$$\frac{dy}{dt} = cx + ey \tag{1}$$

were pure imaginaries, then the critical point (0, 0) was a center. This meant that the family of path curves was a continuous infinity of nested, simple closed curves each containing the critical point (0, 0) in its interior (Fig. 14.6). In this case the system had infinitely many periodic solutions, but in no other case did it have any periodic solutions at all.

In Sec. 14.4 we saw, further, that if the roots of the characteristic equation

$$m^2 - (a + e)m + ae - bc = 0$$

of the linear system associated with the approximately linear system

$$\frac{dx}{dt} = ax + by + f(x, y)$$

$$\frac{dy}{dt} = cx + ey + g(x, y) \tag{2}$$

were pure imaginaries, then the singular point (0, 0) might be a center or it might be a spiral point. This seems to suggest that for certain nonlinear systems it is also true that there are either infinitely many periodic solutions or no periodic solutions. This is not the case, however, and it is not difficult to imagine why. For all we know, the trajectories winding outward from a spiral point, instead of spiraling to infinity, may wind ever and ever nearer to some simple closed curve C enclosing the critical point and forming a barrier, so to speak, to the outward progress of the spirals. At the same time, paths outside C may spiral in from infinity and wind ever more closely around C on the outside. If this is possible, then there might be one, or perhaps several, periodic solutions forming the boundaries of annular regions around the critical point in which the members of the infinite family of spiral paths would lie.

The simplest example of a system exhibiting this behavior is probably

$$\frac{dx}{dt} = y + x - x(x^2 + y^2)$$

$$\frac{dy}{dt} = -x + y - y(x^2 + y^2) \tag{3}$$

and we shall soon see that it does indeed have a single periodic solution toward which the other solutions all spiral. As we prepare to investigate this system, it should be clear that since we are concerned with spiral behavior, it will probably be convenient to find dr/dt and $d\theta/dt$ and work in polar coordinates rather than work with Eqs. (3). In particular, from the sign of dr/dt we should be able to tell whether r increases indefinitely, decreases to zero, or approaches a limiting value $r = r_0$ as t becomes infinite. Accordingly, as a preliminary step, we begin with the familiar polar coordinate relations and compute the derivatives of r and θ with respect to t. From the equation $r^2 = x^2 + y^2 (r > 0)$ we find at once

$$r\frac{dr}{dt} = x\frac{dx}{dt} + y\frac{dy}{dt} \tag{4}$$

Then from the relation $\theta = \tan^{-1}(y/x)$ we find

$$\frac{d\theta}{dt} = \frac{1}{1 + (y/x)^2} \cdot \frac{x(dy/dt) - y(dx/dt)}{x^2} = -\frac{1}{x^2 + y^2}\left(y\frac{dx}{dt} - x\frac{dy}{dt}\right)$$

or

$$-r^2\frac{d\theta}{dt} = y\frac{dx}{dt} - x\frac{dy}{dt} \tag{5}$$

Returning to Eqs. (3), if we multiply the first equation by x and the second by y and add, we obtain

$$x\frac{dx}{dt} + y\frac{dy}{dt} = x^2 + y^2 - (x^2 + y^2)^2$$

or, using (4),

$$r\frac{dr}{dt} = r^2 - r^4$$

or, finally,

$$\frac{dr}{dt} = r(1 - r^2) \tag{6}$$

Similarly, if we multiply the first of Eqs. (3) by y and the second by x and subtract, we get

$$y\frac{dx}{dt} - x\frac{dy}{dt} = x^2 + y^2$$

or, using (5),

$$-r^2\frac{d\theta}{dt} = r^2$$

or, finally,

$$\frac{d\theta}{dt} = -1 \tag{7}$$

From (6) it is clear that $dr/dt > 0$ if $0 < r < 1$; that is, as t increases, r increases by (6), and the path curves spiral outward as long as they lie within the circle $r = 1$. It also follows from (6) that $dr/dt < 0$ if $r > 1$; that is, r decreases, and again by (6) the path curves spiral inward as long as they lie outside the circle $r = 1$. Moreover, the circle $r = 1$ is itself a trajectory, since Eqs. (6) and (7) are satisfied by

$$r = 1 \qquad \theta = -t + t_0 \qquad t_0 \text{ arbitrary}$$

More explicitly, if we separate variables in Eq. (6) and then use partial-fraction techniques, we have

$$\frac{dr}{r(1 - r^2)} = dt$$

$$\frac{1}{2}\left(\frac{2}{r} + \frac{1}{1 - r} - \frac{1}{1 + r}\right) dr = dt$$

$$\ln\frac{r^2}{|1 - r^2|} = 2t - \ln|k| \qquad k \neq 0$$

$$r^2 = \frac{e^{2t}}{k + e^{2t}} = \frac{1}{1 + ke^{-2t}}$$

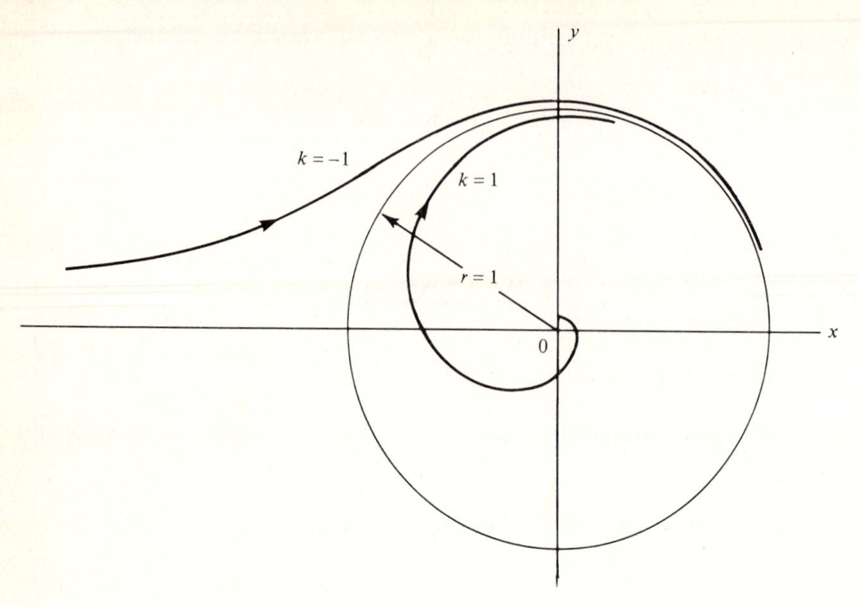

Figure 14.14 The limit cycle of the system $dx/dt = x + y - x(x^2 + y^2)$, $dy/dt = -x + y - y(x^2 + y^2)$.

or
$$r = \frac{1}{\sqrt{1 + ke^{-2t}}} \tag{8}$$

and, from (7), $\theta = -t + t_0$.

The value $k = 0$ was excluded in the derivation of (8). However, if it is now used in (8), it yields the circle $r = 1$ which, as we noted above, is one of the trajectories. If $k > 0$, the corresponding trajectories spiral toward this circle from its interior as $t \to +\infty$. If $k < 0$, the corresponding curves spiral toward the circle from the outside as $t \to +\infty$. Figure 14.14 shows the circle $r = 1$ and several of these spirals.

A simple closed curve in the phase plane which has nonclosed paths spiraling toward it, either as $t \to \infty$ or as $t \to -\infty$, is called a **limit cycle.** If all trajectories which start sufficiently close to a limit cycle, both inside and outside, spiral toward it as $t \to +\infty$, the limit cycle is said to be **stable.** If trajectories on one side of the limit cycle spiral toward it while those on the other side spiral away from it as $t \to +\infty$, the limit cycle is said to be **semistable.** If trajectories on both sides spiral away from it as $t \to +\infty$, the limit cycle is said to be **unstable.**

In the example we have just discussed, the existence of a limit cycle was established by actually solving the given system of equations. Usually, however, the equations are too difficult to solve, and we must fall back on other methods to establish the existence or nonexistence of a limit cycle, that is, a periodic solution. The fundamental results are contained in the following pair of theorems, one giving conditions sufficient for the existence of a limit cycle, the other giving conditions sufficient for the nonexistence of a limit cycle.

Theorem 1† Consider the autonomous system

$$\frac{dx}{dt} = F(x, y)$$

$$\frac{dy}{dt} = G(x, y)$$

where F and G have continuous first partial derivatives in some domain D. Let D_1 be a bounded subdomain in D, let R be the region consisting of D_1 and its boundary, and suppose that R contains no critical points of the given system. If there exists a solution $x = x(t)$, $y = y(t)$ of the system and a value t_0 such that the trajectory defined by $x = x(t)$, $y = y(t)$ remains in R for all $t \geq t_0$, then either

1. $x = x(t)$, $y = y(t)$ is a closed trajectory, i.e., a periodic solution; or
2. $x = x(t)$, $y = y(t)$ spirals toward a closed trajectory in R as $t \to \infty$.

In either case, the given system has a periodic solution.

Theorem 2‡ Let the functions F and G in the autonomous system

$$\frac{dx}{dt} = F(x, y)$$

$$\frac{dy}{dt} = G(x, y)$$

have continuous first partial derivatives in a simply connected domain D.§ If $\partial F/\partial x + \partial G/\partial y$ has the same sign throughout D, then there is no periodic solution of the given system which lies entirely in D.

If we reconsider the system (3) in the light of Theorem 1, we must first find a closed region R containing no critical points of the system, and then we must show that there is a trajectory which remains in R for all sufficiently large values of t. Since R must contain no critical points, it cannot be the interior of a circle with center at the origin. However, since $(0, 0)$ is the only critical point, R can be the closed annular region defined by two concentric circles having $(0, 0)$ as center, say $r = \frac{1}{3}$ and $r = 3$. Using the solution (8) which we found for r, it is clear that there is a solution, say the one for which $k = 3$, which remains in R for all $t \geq 0$. Had we not obtained the explicit solution (8), we could have inferred the existence of the required trajectory in R by noting from (6) that for any solution starting in R, the radius r is a decreasing function of t if $1 < r < 3$ and an increasing function of t if $\frac{1}{3} < r < 1$. Hence, any solution which starts in R must remain in R. Thus the

† This is customarily referred to as the **Poincaré-Bendixon theorem,** after the French mathematician Henri Poincaré (1854–1912) and the Swedish mathematician Ivar Bendixon (1861–1935).

‡ This theorem is usually known as **Bendixon's theorem.**

§ See item 16, Appendix B.2.

hypotheses of Theorem 1 are satisfied, and the system (3) has a periodic solution in R, namely,

$$r = 1$$

$$\theta = -t + t_0$$

as we observed earlier.

To illustrate Theorem 2, consider the system

$$\frac{dx}{dt} = x + 2xy + x^3$$

$$\frac{dy}{dt} = -y^2 + x^2 y$$

Clearly, the functions $F = x + 2xy + x^3$ and $G = -y^2 + x^2 y$ have continuous first partial derivatives at all points of the xy plane. Furthermore

$$\frac{\partial F}{\partial x} + \frac{\partial G}{\partial y} = (1 + 2y + 3x^2) + (-2y + x^2) = 1 + 4x^2$$

Since $1 + 4x^2$ is positive for all values of x (and y), Theorem 2 guarantees that the given system has no periodic solutions anywhere in the phase plane.

One of the most important applications of the ideas we have been discussing is to nonlinear, second-order differential equations, in particular to the equation

$$\frac{d^2x}{dt^2} + p(x)\frac{dx}{dt} + q(x) = 0 \tag{9}$$

which, as we have seen, is equivalent to the system

$$\frac{dx}{dt} = y$$

$$\frac{dy}{dt} = -q(x) - p(x)y$$

For particular instances of Eq. (9) that arise in important technical problems, Theorems 1 and 2 are often unable to establish either the existence or the nonexistence of periodic solutions. In such cases, the following theorem, due to the American mathematician Norman Levinson (1912–1975), is frequently applicable.

Theorem 3 Consider the second-order differential equation

$$\frac{d^2x}{dt^2} + p(x)\frac{dx}{dt} + q(x) = 0$$

and let

$$P(x) = \int_0^x p(s)\,ds \qquad \text{and} \qquad Q(x) = \int_0^x q(s)\,ds$$

If:

1. $p(x)$ is an even, continuous function for all values of x;
2. $q(x)$ is an odd function which is positive for $x > 0$ and has a continuous derivative for all values of x;
3. there exists a number x_0 such that $P(x) < 0$ for $0 < x < x_0$ and $P(x) > 0$ for $x > x_0$;
4. $P(x)$ increases monotonically for $x > x_0$ and becomes infinite as $x \to +\infty$;
5. $Q(x)$ becomes infinite as $x \to +\infty$;

then the given equation has a unique closed trajectory toward which all other trajectories (except the trivial solution $x = y \equiv dx/dt = 0$) spiral as $t \to +\infty$.

Example 1 Investigate the existence of periodic solutions of the so-called **van der Pol equation,**

$$\frac{d^2x}{dt^2} + \lambda(x^2 - 1)\frac{dx}{dt} + x = 0 \qquad \lambda > 0$$

As a first step, let us convert this equation into the equivalent system

$$\frac{dx}{dt} = y$$

$$\frac{dy}{dt} = -x - \lambda(x^2 - 1)y \tag{10}$$

and attempt to apply Theorems 1 and 2. Considering Theorem 2 first (since it is a little simpler), we note that

$$F = y \qquad G = -x - \lambda(x^2 - 1)y$$

and therefore

$$\frac{\partial F}{\partial x} + \frac{\partial G}{\partial y} = -\lambda(x^2 - 1)$$

This is positive at all points in the simply connected infinite vertical strip between the lines $x = -1$ and $x = 1$. Hence, by Theorem 2, we conclude that the given equation has no closed trajectory which lies entirely in this portion of the xy plane. Moreover, $-\lambda(x^2 - 1)$ is of constant negative sign throughout the half-plane to the left of $x = -1$ and throughout the half-plane to the right of $x = 1$. Hence, there are no closed trajectories in these regions of the phase plane. However, for all we know from Theorem 2, there may be solutions of the system (10) which define closed trajectories lying partly in one of these regions and partly in another.

In an attempt to apply Theorem 1, we next compute

$$x\frac{dx}{dt} + y\frac{dy}{dt} = r\frac{dr}{dt} = -\lambda(x^2 - 1)y^2$$

from which we conclude that

$$\frac{dr}{dt} = -\lambda(r^2 \cos^2 \theta - 1)r \sin^2 \theta$$

This shows that if $0 < r < 1$, the radius r is an increasing function of t and the path curves spiral outward. If we could show, similarly, that r is a decreasing function of t on paths that start sufficiently far from the origin [as we did in our discussion of the system (3)], we could conclude that there was an annular region in which a trajectory would have to remain, and we could apply Theorem 1. However, for $r > 1$, dr/dt is a decreasing function of t where $\cos^2 \theta > 1/r^2$ but an increasing function of t where $\cos^2 \theta < 1/r^2$, and so our attempt to invoke Theorem 1 fails.

As a final effort, let us try to apply Theorem 3. Clearly, $p(x) = \lambda(x^2 - 1)$ is an even, continuous function, and $q(x) = x$ is an odd function with a continuous derivative, as required by the theorem. Furthermore,

$$Q(x) = \int_0^x s\,ds = \frac{x^2}{2}$$

becomes infinite as $x \to \infty$. Also

$$P(x) = \int_0^x \lambda(s^2 - 1)\,ds = \lambda\left(\frac{x^3}{3} - x\right)$$

and if we take $x_0 = \sqrt{3}$, we have

$$P(x) < 0 \text{ for } 0 < x < x_0 = \sqrt{3} \qquad \text{and} \qquad P(x) > 0 \text{ for } x > x_0 = \sqrt{3}$$

Finally, since $P'(x) = \lambda(x^2 - 1) > 0$ for $x > 1$, it follows that P is a monotonically increasing function of x for $x > x_0 = \sqrt{3}$ and P becomes infinite as $x \to +\infty$. Thus, all the hypotheses of Theorem 3 are satisfied, and we can conclude that van der Pol's equation has a unique closed trajectory toward which all other trajectories spiral.

Exercises for Section 14.6

For each of the following autonomous systems, find all limit cycles and determine their stability.

1 $\dfrac{dr}{dt} = r(4 - r^2) \qquad \dfrac{d\theta}{dt} = 1$

2 $\dfrac{dr}{dt} = r(1 - r)^2 \qquad \dfrac{d\theta}{dt} = -1$

3 $\dfrac{dr}{dt} = r(r - 1)(r - 2) \qquad \dfrac{d\theta}{dt} = 1$

4 $\dfrac{dr}{dt} = r(r^2 + r - 6) \qquad \dfrac{d\theta}{dt} = 1$

5 $\dfrac{dr}{dt} = r(r - 1)(r - 2)^2(r - 3) \qquad \dfrac{d\theta}{dt} = -1$

6 $\dfrac{dr}{dt} = \sin \pi r \qquad \dfrac{d\theta}{dt} = -1$

7 Find the equation of the family of trajectories in (*a*) Exercise 1, (*b*) Exercise 2 (*c*) Exercise 3.

Show that none of the following systems has a limit cycle.

8 $\dfrac{dx}{dt} = x + 2y + x^3$

$\dfrac{dy}{dt} = -x^2 y + 2y^3$

9 $\dfrac{dx}{dt} = -2x - x \sin y$

$\dfrac{dy}{dt} = -x^2 y^3$

10 $\dfrac{dx}{dt} = x + x^2 + 2y \cos x$ **11** $\dfrac{dx}{dt} = y \sinh xy$

$\dfrac{dy}{dt} = -2xy + y^2 \sin x + y^3$ $\dfrac{dy}{dt} = y - x^2 + y \cosh x$

12 Show that the system

$$\frac{dx}{dt} = -y + \frac{x}{r} f(r) \qquad \frac{dy}{dt} = x + \frac{y}{r} f(r)$$

has limit cycles corresponding to the zeros of $f(r)$. What is the direction of motion on these curves?

Show that each of the following systems has a limit cycle.

13 $\dfrac{dx}{dt} = -2y + 4x - x(x^2 + 4y^2)$ $\dfrac{dy}{dt} = 2x + 4y - y(x^2 + 4y^2)$

14 $\dfrac{dx}{dt} = x + 3y - x(x^2 + 9y^2)$ $\dfrac{dy}{dt} = -x + 2y - y(x^2 + 9y^2)$

Show that each of the following equations has a periodic solution.

15 $\dfrac{d^2x}{dt^2} + (x^2 - 1)\dfrac{dx}{dt} + x + \sin x = 0$

16 $\dfrac{d^2x}{dt^2} + (x^4 - 1)\dfrac{dx}{dt} + x^3 = 0$

17 $\dfrac{d^2x}{dt^2} + (x^4 - 4x^2)\dfrac{dx}{dt} + x + x^3 = 0$

18 $\dfrac{d^2x}{dt^2} + (4x^2 - 1)\dfrac{dx}{dt} + \sinh x = 0$

***19** $\dfrac{d^2x}{dt^2} - (4 + 9x^2 - 5x^4)\dfrac{dx}{dt} - x + x^3 = 0$

***20** For what values of λ, if any, does the equation

$$\frac{d^2x}{dt^2} + (\cosh x - \lambda)\frac{dx}{dt} + \sinh x = 0$$

have a periodic solution?

21 What is the difference, if any, between a closed trajectory and a periodic solution?

APPENDIX

A

THE FUNDAMENTAL EXISTENCE THEOREM

A.1 THE METHOD OF SUCCESSIVE APPROXIMATIONS

In most elementary applications of differential equations, the question of the existence of a solution meeting specified initial conditions is convincingly answered by actually finding a solution and verifying that it has the required properties. The uniqueness of such a solution is usually taken for granted, on the basis of a deterministic view of the natural world and the processes of cause and effect. But what should a person do when confronted with an equation he or she is unable to solve? To continue the search for a solution when none exists is obviously futile. But to abandon a problem when it has a solution is equally unfortunate. Clearly, the existence and uniqueness of solutions of differential equations are (or ought to be) matters of concern for the applied scientist as well as the pure mathematician.

In this section and the one which follows, we shall establish a fundamental existence and uniqueness theorem for first-order differential equations. Similar theorems for systems of first-order equations and for equations of higher order can be proved by reasonably obvious modifications and extensions of the proof for first-order equations, but we shall not undertake this generalization.†

† See, for instance, Walter Leighton, "Ordinary Differential Equations," Wadsworth, Belmont, Calif., 1963, pp. 212–19.

In outline, the method of proof is the following. Given the initial-value problem

$$y' = f(x, y) \qquad y = y_0 \quad \text{when } x = x_0$$

and assuming $y = y_0$ as a first approximation to the required solution, we shall develop a substitution process by which each approximate solution generates another and (presumably) better approximation. Then we shall show that the sequence of these approximations converges to a function which is the required solution.

The general method of successive approximations is most easily understood by considering its application to the solution of an algebraic equation. Suppose we are to solve an equation that has (or can be put in) the form

$$x = g(x) \tag{1}$$

If x_1 is a root of this equation, then when we substitute it into the right-hand side of Eq. (1), we of course obtain the value x_1. On the other hand, if x_1 is not a root, then the substitution of x_1 yields a value, say x_2, which is different from x_1. Whether or not x_1 is a root, this substitution process can, in general, be continued, giving a sequence of numbers

$$\begin{aligned} x_2 &= g(x_1) \\ x_3 &= g(x_2) \\ x_4 &= g(x_3) \\ &\cdots\cdots \\ x_n &= g(x_{n-1}) \end{aligned}$$

which, under certain conditions,† will converge to the required root.

For example, to solve the equation

$$x = \frac{x^3 + x^2 + 1}{3(x^2 + 1)} \tag{2}$$

by the method of successive approximations, we might, quite arbitrarily, begin with the number $x_1 = 3$. Substituting this into the right-hand side of (2) yields the second approximation

$$x_2 = 1.2333$$

Substituting this into the fraction in (2) gives

$$x_3 = 0.5814$$

† Sufficient conditions that the sequence $x_1, x_2, x_3, \ldots$ converge to a real root r of $x = g(x)$ are that (1) x_1 lies in some interval $[a, b]$ in which r is the only root of $x = g(x)$, and (2) $0 < g'(x) < 1$ for all values of x in $[a, b]$. See, for instance, F. B. Hildebrand, "Introduction to Numerical Analysis," 2d ed., McGraw-Hill, New York, 1974, pp. 567–70.

and continuing in this fashion we obtain in succession

$$x_4 = 0.3823$$

$$x_5 = 0.3496$$

$$x_6 = 0.3460$$

$$x_7 = 0.3457$$

$$x_8 = 0.3456$$

$$x_9 = 0.3456$$

The fact that

$$x_9 = f(x_8) = f(0.3456) = 0.3456$$

shows that, to four decimal places, a root of Eq. (2) is $x = 0.3456$. Moreover, an inspection of our calculations shows that five substitutions, or iterations, gave the value of the root correct to three decimal places, while only four iterations were required to yield the root correct to two decimal places.

The process we have just described has an interesting geometric interpretation. To discover it, we note first that solving the equation $x = g(x)$ is equivalent to finding the intersection(s) of the two curves

$$y = x \qquad \text{and} \qquad y = g(x)$$

The substitution of x_1 into $g(x)$ at the first stage of the iterative procedure amounts to determining the ordinate of the curve $y = g(x)$ at the value $x = x_1$. The substitution of x_2 into $g(x)$ amounts to finding the ordinate of $y = g(x)$ at the point where the ordinate of $y = x$ is $y_1 = g(x_1) = x_2$. Similarly, the substitution of x_3 into $g(x)$ amounts to finding the ordinate of $y = g(x)$ at the point where the ordinate of $y = x$ is $y_2 = g(x_2) = x_3$, and so on. Figure A.1 shows the graphical significance of these steps for the curves involved in the example we have just discussed.

The method of successive approximations is applied to a first-order differential equation

$$y' = f(x, y) \tag{3}$$

with given initial condition

$$y(x_0) = y_0$$

in the following way. We first integrate Eq. (3) from x_0 to x, getting

$$\int_{x_0}^{x} y'(t)\, dt = \int_{x_0}^{x} f[t, y(t)]\, dt$$

or

$$y(x) = y_0 + \int_{x_0}^{x} f[t, y(t)]\, dt \tag{4}$$

A relation of this sort is called an **integral equation,** since now the unknown function y appears inside an integral. What we have done thus far has been to

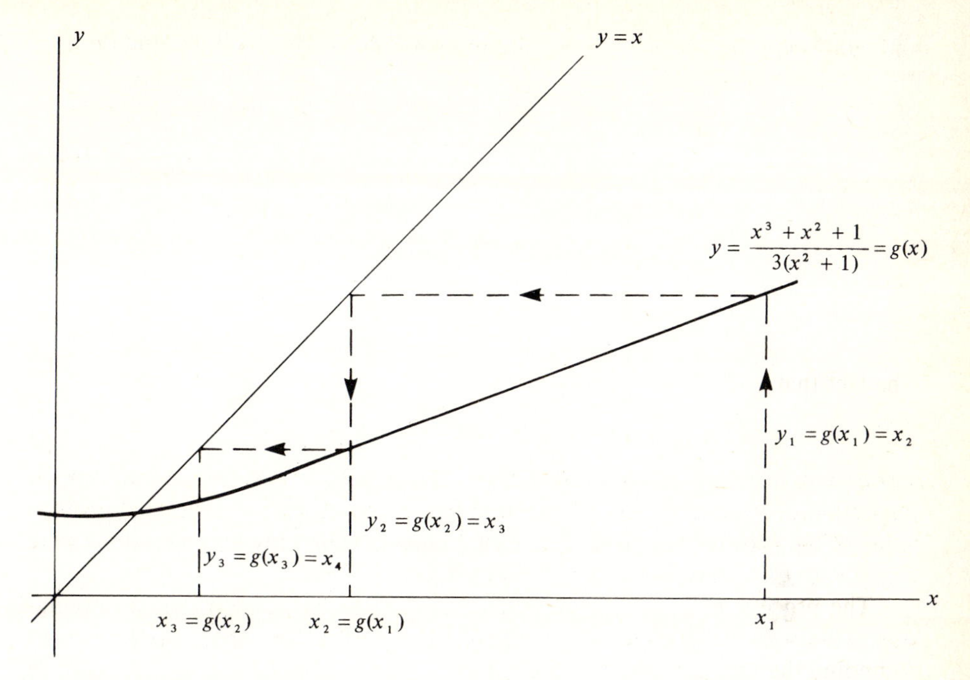

Figure A.1 The graphical significance of the iterative solution of the equation $x = g(x)$.

transform the given problem consisting of a differential equation and an initial condition into an integral equation in which the initial condition is automatically incorporated. To begin the iterative solution of the integral equation (4), we first assume an initial approximation, say y_1, to the actual solution. A simple choice for y_1 is, of course, the given constant y_0. The function y_1 is then substituted into the integrand in Eq. (4). Under this substitution, the integrand becomes a completely known function of t which can, at least in theory, be integrated. The result is the function

$$y_0 + \int_{x_0}^{x} f[t, y_1(t)]\, dt$$

which would be equal to y_1 if y_1 were indeed the required solution but which in general is a new function, say y_2. Continuing this process of substitution, we generate the sequence of functions

$$y_2 = y_0 + \int_{x_0}^{x} f[t, y_1(t)]\, dt$$

$$y_3 = y_0 + \int_{x_0}^{x} f[t, y_2(t)]\, dt$$

$$y_4 = y_0 + \int_{x_0}^{x} f[t, y_3(t)]\, dt$$

$$\cdots\cdots\cdots\cdots\cdots\cdots$$

Under appropriate conditions, this sequence will converge to a limit function $y(x)$ which is the required solution.

To illustrate this process, let us use it to solve the initial-value problem

$$y' = 2xy^2 \qquad y(0) = y_0 = -1 \tag{3.a}$$

Beginning with the assumption $y_1 = y_0 = -1$, we have

$$y_2 = -1 + \int_0^x 2t(-1)^2\,dt = -1 + x^2$$

$$y_3 = -1 + \int_0^x 2t(-1+t^2)^2\,dt = -1 + \int_0^x 2t(1 - 2t^2 + t^4)\,dt$$

$$= -1 + x^2 - x^4 + \tfrac{1}{3}x^6$$

$$y_4 = -1 + \int_0^x 2t(-1 + t^2 - t^4 + \tfrac{1}{3}t^6)^2\,dt$$

$$= -1 + \int_0^x 2t(1 - 2t^2 + 3t^4 - \tfrac{8}{3}t^6 + \tfrac{5}{3}t^8 - \tfrac{2}{3}t^{10} + \tfrac{1}{9}t^{12})\,dt$$

$$= -1 + x^2 - x^4 + x^6 - \tfrac{2}{3}x^8 + \tfrac{1}{3}x^{10} - \tfrac{1}{9}x^{12} + \tfrac{1}{63}x^{14}$$

Clearly, this iterative process can be continued indefinitely. Whether the sequence of y's will converge and, if so, what the limit will be are not entirely clear, however, although there are significant clues. Reconsidering our results, a pattern of sorts seems to be emerging, for we have

$$y_1 = -1$$

$$y_2 = -1 + x^2$$

$$y_3 = -1 + x^2 - x^4 + (\tfrac{1}{6}x^6)$$

$$y_4 = -1 + x^2 - x^4 + x^6 + (-\tfrac{2}{3}x^8 + \tfrac{1}{3}x^{10} - \tfrac{1}{9}x^{12} + \tfrac{1}{63}x^{14})$$

and it appears that although more and more terms with (apparently) unpredictable coefficients are appearing, they involve only higher and higher powers of x, while at the same time an increasing number of terms at the beginning of the series exhibit the very regular behavior

$$-1 + x^2 - x^4 + x^6 - \cdots \tag{5}$$

Since the infinite series (5) is the Maclaurin expansion of the function

$$y = -\frac{1}{1+x^2}$$

it is natural to conjecture that this is the limit of the sequence $\{y_n\}$ and, since its value is -1 when $x = 0$, that it is the required solution of the given initial-value problem. This can easily be checked by substituting into the equation (3.a) or by solving this equation by the method of separation of variables.

The example we have just worked illustrates the fact that even for very simple differential equations the method of successive approximations is usually very tedious and is therefore not of much practical importance. On the other hand, it is of great theoretical importance, for, as we shall see in the next section, it is the cornerstone of the proof that under suitable conditions the initial-value problem

$$y' = f(x, y) \qquad y(x_0) = y_0$$

has one and only one solution.

Preparatory to proving this theorem in the next section, it will be helpful to note several questions which must be answered in the course of the proof:

1. Does each of the successive approximations exist, or is it possible that at some point y_n is undefined?
2. If all members of the sequence $\{y_n\}$ exist, does the sequence converge?
3. If the sequence $\{y_n\}$ converges, does the limit function satisfy the differential equation and the given initial condition?
4. If the sequence $\{y_n\}$ converges to a solution of the problem, is it the only solution or may there be others?

To illustrate the relevance of the first of these questions, consider the iterative solution of the algebraic equation $x = g(x)$ when the graphs of $y = x$ and $y = g(x)$ are those shown in Fig. A.2*a*. It is entirely possible, as suggested by the figure, that one of the iterates is not in the domain of the function g. If this is the case, then all

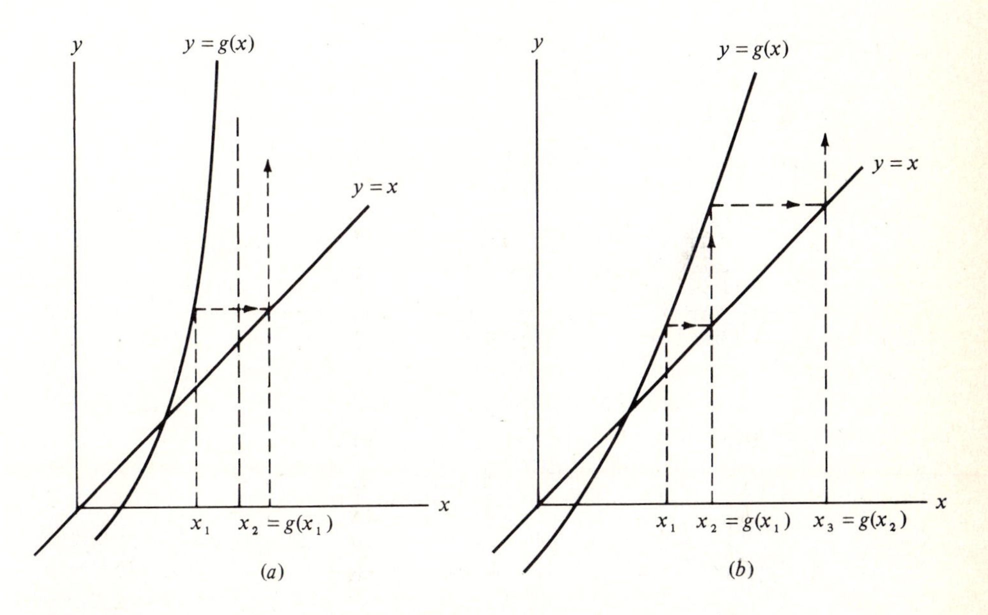

Figure A.2 Sources of trouble in the iterative solution of the equation $x = g(x)$.

subsequent iterates are undefined and the process breaks down. Similarly, the iterative solution of the differential equation $y' = f(x, y)$ may break down because one or more of the values of y_{n-1} may not be in the domain of $f(x, y)$.

The significance of question 2 is illustrated for the iterative solution of $x = g(x)$ by Fig. A.2*b*. Here, although each iterate is well defined [assuming that the domain of $g(x)$ includes all positive values of x], the process clearly diverges. And for all we know, and until we prove the contrary, the same thing may happen in the iterative solution of $y' = f(x, y)$, $y(x_0) = y_0$.

The answering of questions 3 and 4 is, of course, the central purpose of our investigation: Under what conditions does the initial-value problem $y' = f(x, y)$, $y(x_0) = y_0$ have a unique solution?

Exercises for Section A.1

Solve each of the following equations by the method of successive approximations.

1 $x = 2 - e^{-x}$ **2** $x = \dfrac{x^3 + 5}{10}$

3 $x = \ln(x + 2)$ **4** $x = \dfrac{1}{x^2 - 4x + 5}$

5 $x = \cos x$ **6** $x^3 + x - 1 = 0$

Beginning with the initial approximation $y_1 = y_0$, find y_2, y_3, and y_4 for each of the following initial-value problems. In each case compare your approximations with the exact solution.

7 $y' = y \qquad y = y_0 = 1,\ x = 0$ **8** $y' = 2xy \qquad y = y_0 = 1,\ x = 0$

9 $y' = y^2 \qquad y = y_0,\ x = 0$ **10** $y' = y + e^x \qquad y = y_0,\ x = 0$

A.2 THE FUNDAMENTAL EXISTENCE THEOREM

Our purpose in this section is to prove that under suitable conditions the general, first-order differential equation

$$y' = f(x, y) \tag{1}$$

possesses a unique solution $y(x)$ which takes on the value y_0 when x takes on the value x_0. The precise theorem, which is due to the French mathematician Émile Picard (1856–1941), is somewhat complicated to state and moderately difficult to prove. Before we undertake to establish it, it will be helpful to make several preliminary observations.

In the first place, it is natural to expect that the theorem will require certain conditions on $f(x, y)$ and perhaps on one or more of the partial derivatives of $f(x, y)$. This is indeed the case, and we shall assume that $f(x, y)$ is continuous throughout some closed rectangular region R: $|x - x_0| \le a$, $|y - y_0| \le b$ containing the point (x_0, y_0) in its interior. A second condition, sufficient for a proof of the theorem, is that f_y should be continuous throughout R. This is unnecessarily strong, however, and all that need be assumed is that the difference quotient for f_y

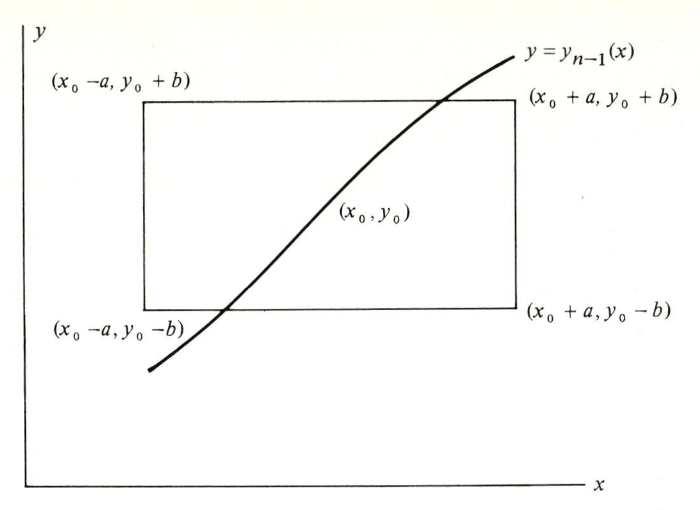

Figure A.3 An iterate which lies outside the rectangle R over a portion of the interval $[x_0 - a, x_0 + a]$.

should be bounded in R. This means that there exists a number A such that

$$\left| \frac{f(x, y_2) - f(x, y_1)}{y_2 - y_1} \right| \leq A$$

for all pairs of points (x, y_1) and (x, y_2) in R. Written in the form

$$|f(x, y_2) - f(x, y_1)| \leq A|y_2 - y_1| \dagger$$

this condition is called a **Lipschitz condition,** after the German mathematician Rudolf Lipschitz (1831–1904). The preceding Lipschitz condition does not imply the existence, much less the continuity, of f_y, but the continuity of f_y does imply the Lipschitz condition. In fact, if f_y is continuous over R, then by the law of the mean, there exists a value of y, say $y = \eta$, between y_1 and y_2 such that

$$f(x, y_2) - f(x, y_1) = f_y(x, \eta)(y_2 - y_1)$$

Furthermore, the continuity of f_y on the *closed* region R implies that f_y is bounded over R. Hence, taking absolute values and replacing $|f_y(x, \eta)|$ by the bound, say A, of $|f_y|$, we have precisely the Lipschitz condition.

The proof of the theorem of Picard is based primarily on the method of successive approximations, which we discussed and illustrated in the last section. Specifically, assuming the simple initial approximation $y_0(x) = y_0$, it involves the sequence of iterates

$$y_n = y_0 + \int_{x_0}^{x} f[t, y_{n-1}(t)]\, dt \qquad n = 1, 2, 3, \ldots \tag{2}$$

However, as we observed in the last section, it may be that this iterative process will break down at some stage. This is surely a possibility if for some values of x between $x_0 - a$ and $x_0 + a$, inclusive, the graph of one of the iterates, say y_{n-1}, lies partly outside of R (Fig. A.3). Indeed, if such is the case, then in the integral

† This condition admits of the possibility $y_1 = y_2$ which the difference quotient for f_y does not.

defining y_n it may be impossible to evaluate the integrand $f[t, y_{n-1}(t)]$ because at some points in the range of integration it does not exist.

The following lemma settles the question of the existence of the iterates by showing that there is always an interval containing x_0 over which the graph of every member of the sequence (2) lies in R, the domain of $f(x, y)$.

Lemma 1 Let $f(x, y)$ be continuous on the closed rectangle R: $|x - x_0| \le a$, $|y - y_0| \le b$; let $|f(x, y)| \le M$ in R; and let h be the smaller of the two (positive) numbers $(a, b/M)$. Then for every nonnegative integral value of n the iterates

$$y_0(x) = y_0, \; y_1(x) = y_0 + \int_{x_0}^{x} f(t, y_0)\, dt, \ldots, y_n(x) = y_0 + \int_{x_0}^{x} f[t, y_{n-1}(t)]\, dt, \ldots$$

exist, have continuous first derivatives, and satisfy the inequality

$$|y_n(x) - y_0| \le b \qquad \text{for} \quad |x - x_0| \le h$$

PROOF We note first of all that a bound M surely exists since $f(x, y)$ is assumed to be continuous on the closed region R.

We now proceed inductively. For $n = 0$ the first iterate, namely, $y_0(x) = y_0$, exists, has the continuous first derivative $y_0'(x) = 0$, and satisfies the inequality $|y_0(x) - y_0| \equiv 0 \le b$ for $|x - x_0| \le h$. Let us now prove that if $y_{n-1}(x)$ has these properties, then the same is true for y_n. Since $h \le a$, the assumption that $|y_{n-1}(x) - y_0| \le b$ for $|x - x_0| \le h$ means that $[x, y_{n-1}(x)]$ lies in R. Therefore, from the hypotheses of the lemma, $f[x, y_{n-1}(x)]$ exists and $|f[x, y_{n-1}(x)]| \le M$. Now since

$$y_n(x) = y_0 + \int_{x_0}^{x} f[t, y_{n-1}(t)]\, dt \tag{3}$$

it follows, from familiar properties of the definite integral, that for $|x - x_0| \le h$, $y_n(x)$ also exists, is continuous, and has as its derivative the continuous function $f[x, y_{n-1}(x)]$. Furthermore, from (3)

$$\begin{aligned} |y_n(x) - y_0| &= \left| \int_{x_0}^{x} f[t, y_{n-1}(t)]\, dt \right| \\ &\le \left| \int_{x_0}^{x} |f[t, y_{n-1}(t)]|\, dt \right| \dagger \\ &\le \left| \int_{x_0}^{x} M\, dt \right| = M|x - x_0| \le Mh \le b \qquad |x - x_0| \le h \end{aligned}$$

This completes the second part of the induction, and the proof is complete.

Lemma 1 shows that for each value of n, $y_n(x)$ lies within $\pm b$ of y_0 for $|x - x_0| \le h$; that is, for these values of x the graph of $y_n(x)$ lies in R. Actually, more than this can be asserted. From the relation $y_n'(x) = f[x, y_{n-1}(x)]$ it follows that $|y_n'(x)| = |f[x, y_{n-1}(x)]| \le M$. In other words, for $|x - x_0| \le h$ the slope of the graph of $y_n(x)$ is between $-M$ and M. Therefore the graph of $y_n(x)$, which of course passes through the point (x_0, y_0), must lie between the lines $y - y_0 =$

† The outer absolute values are necessary because of the possibility that $x < x_0$.

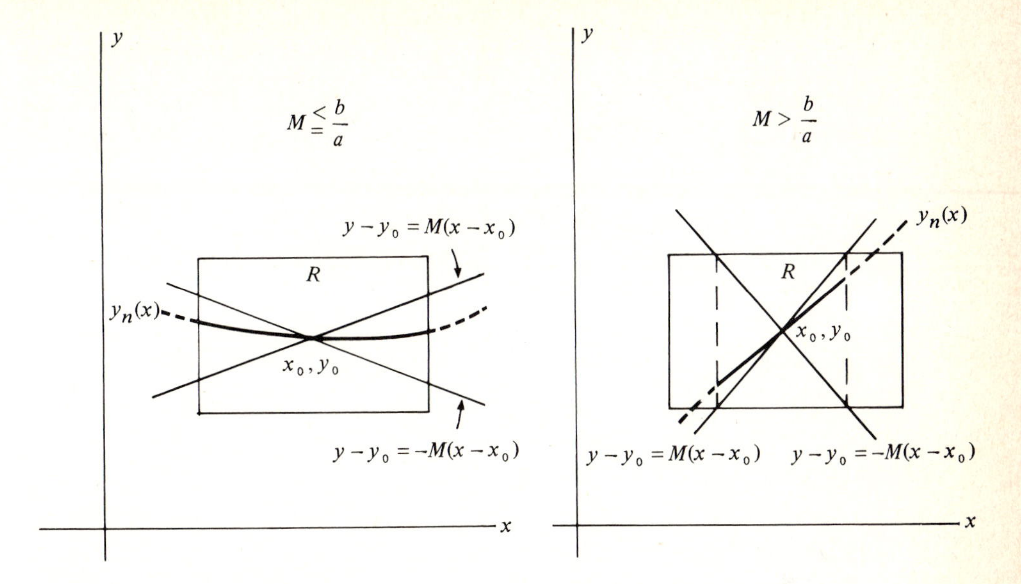

Figure A.4 The sectors in which the iterates must lie.

$\pm M(x - x_0)$. If $M \leq b/a$, these lines intersect the vertical boundaries of R, and we have the case $h = a$. If $M > b/a$, these lines intersect the horizontal boundaries of R in points whose abscissas are $x_0 \pm b/M$, and we have the case $h = b/M < a$. Figure A.4 illustrates these observations.

It is not enough to know that $y_n(x)$ exists on $[x_0 - h, x_0 + h]$ for all values of n, because the sequence $\{y_n\}$ must be convergent before there is any hope that it will lead to a solution of the given differential equation. To show that the sequence $\{y_n\}$ converges, we shall make use of the following lemma.

Lemma 2 Let $f(x, y)$ be continuous in the closed rectangle R^* defined by the inequalities $|x - x_0| \leq h$, $|y - y_0| \leq b$, let $|f(x, y)| \leq M$ in R, and let $f(x, y)$ satisfy the Lipschitz condition

$$|f(x, y_n) - f(x, y_{n-1})| \leq A|y_n - y_{n-1}|$$

in R^*. Then any two successive iterates in the sequence

$$\left\{y_n(x) = y_0 + \int_{x_0}^{x} f[t, y_{n-1}(t)]\, dt\right\}$$

satisfy the inequality

$$|y_{n+1}(x) - y_n(x)| \leq \frac{MA^n|x - x_0|^{n+1}}{(n+1)!} \leq \frac{MA^n h^{n+1}}{(n+1)!}$$

for $|x - x_0| \leq h$.

PROOF We note first that since $f(x, y)$ is continuous in the closed region R^*, it is bounded in R^*; that is, there exists a number M^* such that $|f(x, y)| \le M^*$ for all (x, y) in R^*. Moreover, since R^* is contained in R, it follows that $M^* \le M$, where M is a bound of $|f(x, y)|$ in R. Thus it is correct to write $|f(x, y)| \le M$ for all (x, y) in R^*.

Now, recalling that $y_0(x) = y_0$, we have from (2), for $n = 1$

$$
\begin{aligned}
|y_1(x) - y_0| &= \left| \int_{x_0}^{x} f(t, y_0)\,dt \right| \le \left| \int_{x_0}^{x} |f(t, y_0)|\,dt \right| \\
&\le \left| \int_{x_0}^{x} M\,dt \right| = M|x - x_0| \le Mh
\end{aligned}
$$

Likewise, when $n = 2$, we have

$$
\begin{aligned}
y_2(x) - y_1(x) &= \int_{x_0}^{x} f[t, y_1(t)]\,dt - \int_{x_0}^{x} f[t, y_0(t)]\,dt \\
&= \int_{x_0}^{x} \{f[t, y_1(t)] - f[t, y_0(t)]\}\,dt
\end{aligned}
$$

Hence, taking absolute values and using both the Lipschitz condition on $f(x, y)$ and the first of the bounds on $|y_1(x) - y_0|$ that we have just obtained, we have

$$
\begin{aligned}
|y_2(x) - y_1(x)| &\le \left| \int_{x_0}^{x} |f[t, y_1(t)] - f[t, y_0(t)]|\,dt \right| \\
&\le \left| \int_{x_0}^{x} A|y_1 - y_0|\,dt \right| \\
&\le A\left| \int_{x_0}^{x} M|t - x_0|\,dt \right| = MA\frac{|x - x_0|^2}{2} \le \frac{MAh^2}{2!}
\end{aligned}
$$

Having verified the assertion of the lemma for $n = 1$ and $n = 2$, let us now proceed inductively. To do this, we assume that $|y_n(x) - y_{n-1}(x)| \le MA^{n-1}(|x - x_0|^n/n!)$ and attempt to prove that

$$
|y_{n+1}(x) - y_n(x)| \le MA^n \frac{|x - x_0|^{n+1}}{(n+1)!} \le \frac{MA^n h^{n+1}}{(n+1)!}
$$

We have, of course,

$$
\begin{aligned}
|y_{n+1}(x) - y_n(x)| &= \left| \int_{x_0}^{x} \{f[t, y_n(t)] - f[t, y_{n-1}(t)]\}\,dt \right| \\
&\le \left| \int_{x_0}^{x} |f[t, y_n(t)] - f[t, y_{n-1}(t)]|\,dt \right| \\
&\le \left| \int_{x_0}^{x} A|y_n - y_{n-1}|\,dt \right| \\
&\le \left| \int_{x_0}^{x} A\left\{MA^{n-1}\frac{|t - x_0|^n}{n!}\right\}dt \right| \\
&= MA^n \frac{|x - x_0|^{n+1}}{(n+1)!} \le \frac{MA^n h^{n+1}}{(n+1)!}
\end{aligned}
$$

and the induction is complete.

We are now in a position to state and prove the fundamental existence and uniqueness theorem of Picard.

Theorem 1 Consider the differential equation $y' = f(x, y)$ and the initial condition $y(x_0) = y_0$ where $f(x, y)$ is continuous in the closed rectangle R: $|x - x_0| \leq a$, $|y - y_0| \leq b$ and satisfies the Lipschitz condition $|f(x, y_i) - f(x, y_j)| \leq A|y_i - y_j|$ in R. If $|f(x, y)| \leq M$ in R and if h is the smaller of the two numbers $(a, b/M)$, then there exists a unique solution of the initial-value problem $y' = f(x, y)$, $y(x_0) = y_0$ on the interval $|x - x_0| \leq h$.

PROOF By hypothesis, there exist numbers M and A such that $|f(x, y)| \leq M$ and $|f(x, y_n) - f(x, y_{n-1})| \leq A|y_n - y_{n-1}|$ in R: $|x - x_0| \leq a$, $|y - y_0| \leq b$. Hence if h is the smaller of the two numbers $(a, b/M)$, it follows from Lemmas 1 and 2 that at least over the rectangle R^*: $|x - x_0| \leq h$, $|y - y_0| \leq b$ all members of the sequence

$$y_0(x) = y_0$$

$$y_1(x) = y_0 + \int_{x_0}^{x} f[t, y_0(t)]\, dt$$

$$y_2(x) = y_0 + \int_{x_0}^{x} f[t, y_1(t)]\, dt$$

$$\cdots\cdots\cdots\cdots\cdots\cdots\cdots\cdots$$

$$y_n(x) = y_0 + \int_{x_0}^{x} f[t, y_{n-1}(t)]\, dt$$

$$\cdots\cdots\cdots\cdots\cdots\cdots\cdots\cdots$$

exist and satisfy the inequality

$$|y_{n+1}(x) - y_n(x)| \leq \frac{MA^n h^{n+1}}{(n+1)!}$$

We must now prove that this sequence converges for $|x - x_0| \leq h$; that is, we must show that there is a function $y(x)$ such that

$$\lim_{n \to \infty} y_n(x) = y(x) \qquad |x - x_0| \leq h$$

We shall do this by the rather interesting device of constructing an infinite series of functions whose $(n + 1)$st partial sum is $y_n(x)$ and then proving that this series converges. We therefore consider the infinite series

$$y_0(x) + [y_1(x) - y_0(x)] + [y_2(x) - y_1(x)] + \cdots + [y_n(x) - y_{n-1}(x)] + \cdots \tag{4}$$

Because of the obvious cancellations, the sum of the first $n + 1$ terms of this series is just $y_n(x)$. Moreover, the series of the absolute values of these terms is dominated by the series of positive constants

$$|y_0(x)| + Mh + \frac{MAh^2}{2!} + \cdots + \frac{MA^{n-1}h^n}{n!} + \cdots$$

$$= |y_0(x)| + \frac{M}{A}\left\{\left[1 + Ah + \frac{(Ah)^2}{2!} + \cdots + \frac{(Ah)^n}{n!} + \cdots\right] - 1\right\} \tag{5}$$

Since this series of positive constants has the obvious sum

$$|y_0(x)| + \frac{M}{A}(e^{Ah} - 1)$$

it follows by the comparison test that on $[x_0 - h, x_0 + h]$ the series (4) converges absolutely to some function $y(x)$. Moreover, by the Weierstrass M test (item 20, Appendix B.2), this series also converges uniformly and hence its sum $y(x)$ is continuous (item 19, Appendix B.2) since each of its terms is continuous.

In what follows we will also need to know that $[x, y(x)]$ is in R for $|x - x_0| \leq h$. To verify this, it is sufficient to note that since $|y_n(x) - y_0| \leq b$ for $|x - x_0| \leq h$ and $n = 0, 1, 2, \ldots$ (by Lemma 1) and since $y(x) = \lim_{n\to\infty} y_n(x)$, therefore $|y(x) - y_0| \leq b$ for $|x - x_0| \leq h$.

To show that $y(x)$ is a solution of the given initial-value problem, it is sufficient to show that $y(x)$ satisfies the relation

$$y(x) = y_0 + \int_{x_0}^{x} f[t, y(t)]\, dt \tag{6}$$

because this relation shows immediately that $y(x_0) = y_0$ and because by differentiation [which is justified since $f\{x, y(x)\}$ is continuous in R] we obtain $y'(x) = f[x, y(x)]$.

To verify (6), we shall show that the difference between its two members is equal to zero. Now

$$\begin{aligned}
\left| y(x) - y_0 - \int_{x_0}^{x} f[t, y(t)]\, dt \right| \\
&= \left| y(x) - y_0 - \int_{x_0}^{x} f[t, y_{n-1}(t)]\, dt \right. \\
&\qquad \left. + \int_{x_0}^{x} \{f[t, y_{n-1}(t)] - f[t, y(t)]\}\, dt \right| \\
&= \left| y(x) - y_n(x) + \int_{x_0}^{x} \{f[t, y_{n-1}(t)] - f[t, y(t)]\}\, dt \right| \\
&\leq |y(x) - y_n(x)| + \left| \int_{x_0}^{x} |f[t, y_{n-1}(t)] - f[t, y(t)]|\, dt \right| \\
&\leq |y(x) - y_n(x)| + \left| \int_{x_0}^{x} A|y_{n-1}(t) - y(t)|\, dt \right|
\end{aligned}$$

Now because of the uniform convergence of the series (4), it follows that for any $\varepsilon > 0$ there exists an integer N such that $|y(x) - y_n(x)| < \varepsilon$ for all $n > N$ and all x such that $|x - x_0| \leq h$. Hence for sufficiently large values of n we have from the last inequality

$$\begin{aligned}
\left| y(x) - y_0 - \int_{x_0}^{x} f[t, y(t)]\, dt \right| &\leq \varepsilon + \left| \int_{x_0}^{x} A\varepsilon\, dt \right| \\
&= \varepsilon + A\varepsilon|x - x_0| \leq \varepsilon(1 + Ah)
\end{aligned}$$

If the left member of this inequality was some number which was not zero, say K, we could obtain a contradiction by choosing $\varepsilon < K/(1 + Ah)$. Hence we conclude that the left member must be zero, and we have proved that $y(x) = \lim_{n \to \infty} y_n(x)$ is a solution of the given initial-value problem.

To complete the proof of the theorem of Picard, we must show, finally, that $y(x)$ is the *only* solution of the given initial-value problem on the interval $[x_0 - h, x_0 + h]$. To do this, let $\phi(x)$ be *any* solution of the equation $y' = f(x, y)$ on the interval $[x_0 - h, x_0 + h]$ such that $\phi(x_0) = y_0$. Then from the continuity of $\phi(x)$ [which follows from the fact that $\phi(x)$ is differentiable] there must be some interval around x_0, say $|x - x_0| < \delta$, in which $|\phi(x) - y_0| < b$. If the difference between $\phi(x)$ and y_0 is ever equal to b, let x_1 be the value of x closest to x_0 for which this is true. Then $|\phi(x_1) - y_0| = b$ and

$$|\phi(x) - y_0| < b \qquad \text{for} \quad |x - x_0| < |x_1 - x_0|$$

Then
$$\left|\frac{\phi(x_1) - y_0}{x_1 - x_0}\right| = \frac{b}{|x_1 - x_0|} = \frac{b}{h}\frac{h}{|x_1 - x_0|} \geq M\frac{h}{|x_1 - x_0|} \tag{7}$$

However, by the mean value theorem, there is a value $x = \xi$ between x_0 and x_1 such that

$$\left|\frac{\phi(x_1) - y_0}{x_1 - x_0}\right| = \left|\frac{\phi(x_1) - \phi(x_0)}{x_1 - x_0}\right| = |\phi'(\xi)| = |f[\xi, \phi(\xi)]| \leq M$$

Hence we have a contradiction of (7) unless $|x_1 - x_0| \geq h$. In other words, for any solution $\phi(x)$ of the given initial-value problem, the inequality $|\phi(x) - y_0| < b$ holds for all values of x between $x_0 - h$ and $x_0 + h$, inclusive.

Now from the two relations

$$y(x) = y_0 + \int_{x_0}^{x} f[t, y(t)]\, dt$$

and
$$\phi(x) = y_0 + \int_{x_0}^{x} f[t, \phi(t)]\, dt$$

which hold for any two solutions of the given initial-value problem, we have, on subtracting and taking absolute values,

$$|y(x) - \phi(x)| \leq \left|\int_{x_0}^{x} |f[t, y(t)] - f[t, \phi(t)]|\, dt\right|$$

Moreover, since we have just seen that $\phi(x)$ lies within $\pm b$ of y_0 for values of x in the interval $|x - x_0| \leq h$, it follows that both $f[t, y(t)]$ and $f[t, \phi(t)]$ are evaluated at points in R^*. Hence the Lipschitz condition is fulfilled, and from the last inequality we have

$$|y(x) - \phi(x)| \leq \left|\int_{x_0}^{x} A|y(t) - \phi(t)|\, dt\right| \tag{8}$$

Furthermore, since the maximum difference between $y(t)$ and $\phi(t)$ is $2b$, we have, from Eq. (8),

$$|y(x) - \phi(x)| \leq \left|\int_{x_0}^{x} A2b\, dt\right| = 2bA|x - x_0|$$

If we now return to (8) and use this estimate of $|y(x) - \phi(x)|$ in the integrand, we obtain the improved estimate

$$|y(x) - \phi(x)| \leq \left| \int_{x_0}^{x} A(2bA|t - x_0|)\, dt \right| = 2bA^2 \frac{|x - x_0|^2}{2}$$

If we return to (8) with this new estimate, we obtain further

$$|y(x) - \phi(x)| \leq \left| \int_{x_0}^{x} A\left(2bA^2 \frac{|t - x_0|^2}{2}\right) dt \right| = 2bA^3 \frac{|x - x_0|^3}{3!}$$

Continuing this process of refining the estimate for $|y(x) - \phi(x)|$, we obtain after n steps

$$|y(x) - \phi(x)| \leq 2bA^n \frac{|x - x_0|^n}{n!} \leq \frac{2bA^n h^n}{n!}$$

Now the last member of this continued inequality is the $(n + 1)$st term of the convergent power series expansion of $2be^{Ah}$, and hence must approach zero as n becomes infinite. Thus the difference $|y(x) - \phi(x)|$ can be made arbitrarily small by taking n sufficiently large, and therefore it must be zero. It follows that on $[x_0 - h, x_0 + h]$ $y(x) = \phi(x)$, and our proof is complete.

Exercises for Section A.2

1 Under what conditions, if any, will the inequality sign hold in Eq. (7)?

2 Prove the assertion made in the proof of Theorem 1 that if $|y_n(x) - y_0| \leq b$ and if $y(x) = \lim_{n \to \infty} y_n(x)$, then $|y(x) - y_0| \leq b$.

APPENDIX

B

REFERENCE MATERIAL FROM ALGEBRA, ANALYSIS, AND APPLIED MATHEMATICS

B.1 SELECTED TOPICS FROM ALGEBRA

(1) The Binomial Expansion

$$(a + b)^n = a^n + na^{n-1}b + \frac{n(n-1)}{2!}a^{n-2}b^2 + \cdots + \frac{n(n-1)\cdots(n-k+1)}{k!}a^{n-k}b^k + \cdots$$

If n is a nonnegative integer, the expression on the right is a finite series which is equal to $(a + b)^n$ for all values of a and b. If n is a negative number or a positive number which is not an integer, the expression on the right is an infinite series which converges to $(a + b)^n$ if $|a| > |b|$ and diverges if $|a| < |b|$.

(2) Complex Numbers

A **complex number** is a number of the form $z = x + iy$ where x and y are real numbers and $i = \sqrt{-1}$. The real number x is called the **real part** of z, and the real number y (not iy) is called the **imaginary part** of z. Associated with any complex number is its **length** or **magnitude** or **absolute value**

$$r = \sqrt{x^2 + y^2}$$

and its **angle** or **amplitude** or **argument**

$$\theta = \tan^{-1}\frac{y}{x}$$

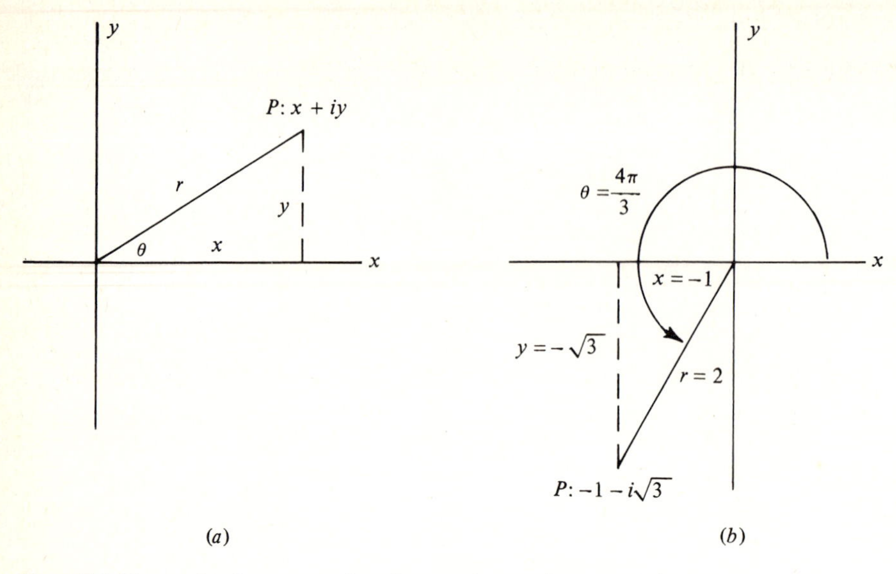

Figure B.1 The graphical representation of a complex number $z = x + iy$.

From the graphical representation of complex numbers illustrated in Fig. B.1*a*, it follows that

$$x + iy = r \cos \theta + ir \sin \theta = r(\cos \theta + i \sin \theta)$$

The last expression is called the **polar** or **trigonometric form** of the complex number. A complex number can also be written in the **exponential form**

$$x + iy = re^{i\theta}$$

By comparing the trigonometric and exponential forms of a complex number, it is evident that

$$e^{i\theta} = \cos \theta + i \sin \theta$$

This is usually referred to as **Euler's formula.**

Example 1 The length of the complex number $-1 - i\sqrt{3}$ shown in Fig. B.1*b* is

$$r = \sqrt{(-1)^2 + (-\sqrt{3})^2} = 2$$

and the angle is

$$\theta = \tan^{-1} \frac{-\sqrt{3}}{-1} = \frac{4\pi}{3}$$

The trigonometric form of the number is therefore

$$2\left(\cos \frac{4\pi}{3} + i \sin \frac{4\pi}{3}\right)$$

and the exponential form of the number is

$$2e^{4i\pi/3}$$

(3) Cramer's Rule

Given a system of n nonhomogeneous linear equations in n unknowns,

$$\begin{aligned} a_{11}x_1 + a_{12}x_2 + \cdots + a_{1n}x_n &= b_1 \\ a_{21}x_1 + a_{22}x_2 + \cdots + a_{2n}x_n &= b_2 \\ \cdots\cdots\cdots\cdots\cdots\cdots\cdots\cdots \\ a_{n1}x_1 + a_{n2}x_2 + \cdots + a_{nn}x_n &= b_n \end{aligned}$$

If the determinant of the coefficients

$$|A| = \begin{vmatrix} a_{11} & a_{12} & \cdots & a_{1n} \\ a_{21} & a_{22} & \cdots & a_{2n} \\ \cdots & \cdots & \cdots & \cdots \\ a_{n1} & a_{n2} & \cdots & a_{nn} \end{vmatrix}$$

is different from zero, then

$$x_i = \frac{|D_i|}{|A|} \qquad i = 1, 2, \ldots, n$$

where $|D_i|$ is the determinant obtained from $|A|$ by replacing the ith column of $|A|$ by the column of constants $b_1, b_2, \ldots, b_n$.

Example 2
If

$$\begin{aligned} 2x_1 + 3x_2 - x_3 &= 1. \\ x_1 \qquad\quad + 4x_3 &= -2 \\ 3x_1 - 2x_2 + 2x_3 &= 0 \end{aligned}$$

then, for example,

$$x_2 = \frac{\begin{vmatrix} 2 & 1 & -1 \\ 1 & -2 & 4 \\ 3 & 0 & 2 \end{vmatrix}}{\begin{vmatrix} 2 & 3 & -1 \\ 1 & 0 & 4 \\ 3 & -2 & 2 \end{vmatrix}} = \frac{-4}{48} = -\frac{1}{12}$$

(4) Homogeneous Linear Equations

A system of n homogeneous linear equations in n unknowns

$$\begin{aligned} a_{11}x_1 + a_{12}x_2 + \cdots + a_{1n}x_n &= 0 \\ a_{21}x_1 + a_{22}x_2 + \cdots + a_{2n}x_n &= 0 \\ \cdots\cdots\cdots\cdots\cdots\cdots\cdots\cdots \\ a_{n1}x_1 + a_{n2}x_2 + \cdots + a_{nn}x_n &= 0 \end{aligned}$$

will have a nontrivial solution, that is, a solution other than the obvious one in which the value of each x is zero, if and only if the determinant of the coefficients

$$|A| = |a_{ij}| = \begin{vmatrix} a_{11} & a_{12} & \cdots & a_{1n} \\ a_{21} & a_{22} & \cdots & a_{2n} \\ \cdots & \cdots & \cdots & \cdots \\ a_{n1} & a_{n2} & \cdots & a_{nn} \end{vmatrix}$$

is equal to zero.

If $|a_{ij}| = 0$ but if there is one row, say the kth, in $|a_{ij}|$ whose elements have cofactors which are not all zero, then every solution of the system is of the form

$$x_j = \lambda A_{kj} \qquad j = 1, 2, \ldots, n$$

where A_{kj} is the cofactor of the jth element in the kth row of $|a_{ij}|$ and λ is a constant.

Example 3 It is easy to verify that the determinant of the coefficients in the system

$$\begin{aligned} 2x_1 + 3x_2 + 4x_3 &= 0 \\ 3x_1 + 4x_2 + 5x_3 &= 0 \\ 10x_1 + 9x_2 + 8x_3 &= 0 \end{aligned} \qquad \text{namely} \qquad \begin{vmatrix} 2 & 3 & 4 \\ 3 & 4 & 5 \\ 10 & 9 & 8 \end{vmatrix}$$

is equal to zero but that the cofactor of the first element in the third row is different from zero. Hence the values of x_1, x_2, and x_3 in any solution of this system are proportional to the cofactors

$$A_{31} = \begin{vmatrix} 3 & 4 \\ 4 & 5 \end{vmatrix} = -1 \qquad A_{32} = -\begin{vmatrix} 2 & 4 \\ 3 & 5 \end{vmatrix} = 2 \qquad A_{33} = \begin{vmatrix} 2 & 3 \\ 3 & 4 \end{vmatrix} = -1$$

Thus, taking $\lambda = -1$, we have the particular solution

$$x_1 = 1 \qquad x_2 = -2 \qquad x_3 = 1$$

and all other solutions are constant multiples of this one.

(5) Partial Fractions

If $p(x)$ and $q(x)$ are polynomials and if the degree of $q(x)$ is greater than the degree of $p(x)$, then constants $A, B_1, \ldots, B_k, C, E, G_1, H_1, \ldots, G_k, H_k$ can be found such that the fraction $p(x)/q(x)$ can be expressed as a sum of partial fractions according to the following rules:

1. For each unrepeated real linear factor $x - r$ of $q(x)$, there will be a single fraction of the form

$$\frac{A}{x - r}$$

2. For each k-fold, real linear factor $(x - r)^k$ of $q(x)$, there will be a set of fractions of the form

$$\frac{B_1}{x - r} + \frac{B_2}{(x - r)^2} + \cdots + \frac{B_k}{(x - r)^k}$$

3. For each unrepeated, irreducible quadratic factor $x^2 + ax + b$ of $q(x)$, there will be a single fraction of the form

$$\frac{Cx + E}{x^2 + ax + b}$$

4. For each k-fold, irreducible quadratic factor $(x^2 + ax + b)^k$ of $q(x)$, there will be a set of fractions of the form

$$\frac{G_1 x + H_1}{x^2 + ax + b} + \frac{G_2 x + H_2}{(x^2 + ax + b)^2} + \cdots + \frac{G_k x + H_k}{(x^2 + ax + b)^k}$$

The constants can be found by putting the total set of partial fractions over a common denominator and making the numerator of the resulting fraction identical with $p(x)$.

(6) The (Determinant) Rank of a Matrix

The determinant whose elements are the elements common to any r rows and any r columns of a matrix A is called a **determinant of order** r **of the matrix** A. The matrix A is said to be of **rank** r if and only if every determinant of order $r + 1$ in A is equal to zero, but there is at least one determinant of order r in A which is different from zero.

(7) Simultaneous Linear Equations

A system of m linear equations in n unknowns

$$\begin{aligned} a_{11}x_1 + a_{12}x_2 + \cdots + a_{1n}x_n &= b_1 \\ a_{21}x_1 + a_{22}x_2 + \cdots + a_{2n}x_n &= b_2 \\ \cdots\cdots\cdots\cdots\cdots\cdots\cdots\cdots & \cdots\cdots \\ a_{m1}x_1 + a_{m2}x_2 + \cdots + a_{mn}x_n &= b_m \end{aligned}$$

will have a solution if and only if the coefficient matrix

$$\begin{bmatrix} a_{11} & a_{12} & \cdots & a_{1n} \\ a_{21} & a_{22} & \cdots & a_{2n} \\ \cdots & \cdots & \cdots & \cdots \\ a_{m1} & a_{m2} & \cdots & a_{mn} \end{bmatrix}$$

and the augmented matrix

$$\begin{bmatrix} a_{11} & a_{12} & \cdots & a_{1n} & b_1 \\ a_{21} & a_{22} & \cdots & a_{2n} & b_2 \\ \cdots & \cdots & \cdots & \cdots & \cdots \\ a_{m1} & a_{m2} & \cdots & a_{mn} & b_m \end{bmatrix}$$

have the same rank. If the common value of the rank of these two matrices is r, then the system has a family of solutions containing $n - r$ independent arbitrary constants.

In particular, even though it cannot be found by Cramer's rule [see (3) above], a nonhomogeneous system of n equations in n unknowns for which the determinant of the coefficients is equal to zero will have a solution, provided that the coefficient matrix and the augmented matrix have the same rank.

Example 4 The determinant of the coefficients of the system

$$\begin{aligned} x_1 + 2x_2 + 2x_3 &= 1 \\ 2x_1 + 3x_2 + x_3 &= 4 \\ x_1 - x_2 - 7x_3 &= 7 \end{aligned}$$

is equal to zero. Hence the solution of the system (if it exists) cannot be found by Cramer's rule. However, both the coefficient matrix

$$\begin{bmatrix} 1 & 2 & 2 \\ 2 & 3 & 1 \\ 1 & -1 & -7 \end{bmatrix}$$

and the augmented matrix

$$\begin{bmatrix} 1 & 2 & 2 & 1 \\ 2 & 3 & 1 & 4 \\ 1 & -1 & -7 & 7 \end{bmatrix}$$

are of rank $r = 2$. Therefore there is a family of solutions containing $n - r = 3 - 2 = 1$ arbitrary constant. One way to find these solutions is to reduce the given system to triangular form: Subtracting twice the first equation from the second and subtracting the first equation from the third give the equivalent system

$$\begin{aligned} x_1 + 2x_2 + 2x_3 &= 1 \\ -x_2 - 3x_3 &= 2 \\ -3x_2 - 9x_3 &= 6 \end{aligned}$$

Then subtracting three times the new second equation from the new third equation gives

$$\begin{aligned} x_1 + 2x_2 + 2x_3 &= 1 \\ -x_2 - 3x_3 &= 2 \\ 0 &= 0 \end{aligned}$$

If x_3 is given the arbitrary value λ, x_1 and x_2 can be found in terms of λ from the first two equations. Thus

$$\begin{aligned} x_1 &= 5 + 4\lambda \\ x_2 &= -2 - 3\lambda \\ x_3 &= \lambda \end{aligned} \quad \text{or} \quad \begin{bmatrix} x_1 \\ x_2 \\ x_3 \end{bmatrix} = \begin{bmatrix} 5 \\ -2 \\ 0 \end{bmatrix} + \lambda \begin{bmatrix} 4 \\ -3 \\ 1 \end{bmatrix}$$

B.2 SELECTED TOPICS FROM ANALYSIS

(1) Absolute Convergence of Improper Integrals

An improper integral of the form $\int_a^\infty f(x)\,dx$ is said to be **absolutely convergent** if the integral of the absolute value of $f(x)$, namely, $\int_a^\infty |f(x)|\,dx$, exists. Absolute convergence of an improper integral implies ordinary convergence, but not conversely.

(2) The Chain Rule

If $y = f(u)$ and $u = g(x)$, then

$$\frac{dy}{dx} = \frac{df}{du}\frac{dg}{dx}$$

If $z = f(u)$ and $u = g(x, y)$, then

$$\frac{\partial z}{\partial x} = \frac{df}{du}\frac{\partial g}{\partial x} \quad \text{and} \quad \frac{\partial z}{\partial y} = \frac{df}{du}\frac{\partial g}{\partial y}$$

(3) Continuity of Functions Defined by Improper Integrals

Let $G(s) = \int_a^\infty f(t)g(s, t)\,dt$. If $g(s, t)$ is a continuous function of s and t for $\alpha \le s \le \beta$ and $t \ge a$, if $f(t)$ is at least piecewise continuous for $t \ge a$, and if the integral defining $G(s)$ converges uniformly over the interval $\alpha \le s \le \beta$, then $G(s)$ is a continuous function of s for $\alpha \le s \le \beta$. Since the definitive property of a continuous function is that

$$\lim_{s \to s_0} G(s) = G(s_0)$$

this result asserts, in effect, that under the appropriate conditions the limit of $G(s)$ can be found by taking the limit inside the t-integral sign or, equivalently, that the order of integrating with respect to t and taking the limit with respect to s can be interchanged.

(4) The Derivative of a Determinant

If each element of an nth-order determinant $|A|$ is a differentiable function of x, the **derivative** of $|A|$ with respect to x is equal to the sum of n determinants, the ith one of which is identical with $|A|$ except for the ith row (column), which consists of the derivatives of the elements of the ith row (column) of $|A|$.

(5) The Derivative of an Improper Integral with Respect to a Parameter

Let $G(s) = \int_a^\infty f(t)g(s, t)\,dt$. If $g(s, t)$ and $g_s(s, t) \equiv \partial g(s, t)/\partial s$ are continuous functions of s and t for $\alpha \leq s \leq \beta$ and $t \geq a$, if $f(t)$ is at least piecewise continuous for $t \geq a$, if the integral defining $G(s)$ converges, and if $\int_a^\infty f(t)g_s(s, t)\,dt$ converges uniformly over the interval $\alpha \leq s \leq \beta$, then for $\alpha \leq s \leq \beta$ the derivative of $G(s)$ is given by the formula

$$G'(s) \equiv \frac{d}{ds}\int_a^\infty f(t)g(s, t)\,dt = \int_a^\infty f(t)\frac{\partial g(s, t)}{\partial s}\,dt \equiv \int_a^\infty f(t)g_s(s, t)\,dt$$

In words, this formula asserts that under the appropriate conditions, the derivative of $G(s)$ can be found by differentiating inside the t-integral sign or, equivalently, that the order of integrating with respect to t and differentiating with respect to s can be interchanged.

(6) The Greatest Lower Bound of a Set of Numbers

If S is a set of numbers and if b is a number such that $b \leq s$ for each number s in S, then b is called a **lower bound** of S. The number b is the **greatest lower bound** of S if

1. b is a lower bound of S, and
2. If $c > b$, then c is not a lower bound of S.

(7) Hyperbolic Functions

The hyperbolic sine, cosine, and tangent are defined in terms of exponential functions as follows:

$$\sinh x = \frac{e^x - e^{-x}}{2} \qquad \cosh x = \frac{e^x + e^{-x}}{2} \qquad \tanh x = \frac{\sinh x}{\cosh x} = \frac{e^x - e^{-x}}{e^x + e^{-x}}$$

Among the important properties of the hyperbolic functions are the following:

$$\sinh(-x) = -\sinh x \qquad \cosh(-x) = \cosh x \qquad \tanh(-x) = -\tanh x$$

$$\sinh 0 = 0 \qquad \cosh 0 = 1 \qquad \tanh 0 = 0$$

Sinh x and tanh x are zero if and only if $x = 0$, cosh x is never zero.

$$\lim_{x\to\infty} \sinh x = \infty \qquad \lim_{x\to\infty} \cosh x = \infty \qquad \lim_{x\to\infty} \tanh x = 1$$

The hyperbolic functions are not periodic.

$$\cosh^2 x - \sinh^2 x = 1$$

$$\sinh(x \pm y) = \sinh x \cosh y \pm \cosh x \sinh y$$

$$\cosh(x \pm y) = \cosh x \cosh y \pm \sinh x \sinh y$$

$$\sinh 2x = 2 \sinh x \cosh x$$

$$\cosh 2x = \cosh^2 x + \sinh^2 x = 2\cosh^2 x - 1 = 1 + 2\sinh^2 x$$

$$\frac{d \sinh x}{dx} = \cosh x \qquad \frac{d \cosh x}{dx} = \sinh x \qquad \frac{d \tanh x}{dx} = \frac{1}{\cosh^2 x}$$

$$\int \sinh x \, dx = \cosh x \qquad \int \cosh x \, dx = \sinh x \qquad \int \tanh x \, dx = \ln(\cosh x)$$

(8) Improper Integrals

A definite integral is said to be **improper**

1. If the integrand is unbounded in the neighborhood of one or more points in the range of integration (including the endpoints), or
2. If at least one of the limits is infinite.

If x_0 is a point in the range of integration at which $f(x)$ is infinite, then

$$\int_a^b f(x)\,dx = \lim_{\delta_1 \to 0} \int_a^{x_0 - \delta_1} f(x)\,dx + \lim_{\delta_2 \to 0} \int_{x_0 + \delta_2}^b f(x)\,dx$$

provided these limits exist. Similarly,

$$\int_a^\infty f(x)\,dx = \lim_{b \to \infty} \int_a^b f(x)\,dx$$

provided this limit exists.

(9) The Integral of an Improper Integral with Respect to a Parameter

Let $G(s) = \int_a^\infty f(t)g(s, t)\,dt$. If $g(s, t)$ is a continuous function of s and t for $\alpha \le s \le \beta$ and $t \ge a$, if $f(t)$ is at least piecewise continuous for $t \ge a$, and if the integral defining $G(s)$ converges uniformly over the interval $\alpha \le s \le \beta$, then

$$\int_\alpha^\beta G(s)\,ds \equiv \int_\alpha^\beta \left[\int_a^\infty f(t)g(s, t)\,dt \right] ds = \int_a^\infty \left[\int_\alpha^\beta f(t)g(s, t)\,ds \right] dt$$

In words, this formula asserts that under the appropriate conditions the integral of $G(s)$ can be found by integrating inside the t-integral sign or, equivalently, that the order of integrating with respect to t and with respect to s can be interchanged.

(10) Leibnitz' Rule

If

$$F(x) = \int_{a(x)}^{b(x)} f(x, t)\, dt$$

and if $\partial f(x, t)/\partial x$ is a continuous function of x and t, then

$$\frac{dF}{dx} = \int_{a(x)}^{b(x)} \frac{\partial f(x, t)}{\partial x}\, dt + f[x, b(x)]\frac{db(x)}{dx} - f[x, a(x)]\frac{da(x)}{dx}$$

(11) L'Hospital's Rule

If for $x = a$ the fraction $f(x)/g(x)$ assumes the indeterminate form 0/0 or ∞/∞, then

$$\lim_{x \to a} \frac{f(x)}{g(x)} = \lim_{x \to a} \frac{f'(x)}{g'(x)}$$

provided the limit on the right exists. Other indeterminate expressions can often be evaluated by transforming them into indeterminate forms of the type covered by L'Hospital's rule.

Example 1 What is $\lim_{t \to 0} (1 + at)^{b/t}$?

At $t = 0$, this expression is an indeterminate of the form 1^∞ and cannot be evaluated by a direct application of L'Hospital's rule. However, setting it equal to y and taking logarithms give

$$\ln y = b \frac{\ln (1 + at)}{t}$$

which is an indeterminate of the form 0/0 as $t \to 0$. Therefore, by L'Hospital's rule,

$$\lim_{t \to 0} \ln y = b \lim_{t \to 0} \frac{\ln (1 + at)}{t} = b \lim_{t \to 0} \frac{a}{1 + at} = ba$$

Hence, from the continuity of the logarithmic and exponential functions,

$$\lim_{t \to 0} (1 + at)^{b/t} = \lim_{t \to 0} y = \lim_{t \to 0} e^{\ln y}$$

$$= \exp \left[\lim_{t \to 0} \ln y\right] = e^{ab}$$

(12) Linear Operator

An operator, say L, with the property that if it can be meaningfully applied to $f_1(x)$, $f_2(x)$, and $c_1 f_1(x) + c_2 f_2(x)$, where c_1 and c_2 are arbitrary constants, then

$$L[c_1 f_1(x) + c_2 f_2(x)] = c_1 L[f_1(x)] + c_2 L[f_2(x)]$$

is called a **linear operator**.

(13) The Mean Value Theorem

If $f(x)$ is continuous over the closed interval $a \leq x \leq b$ and if $f'(x)$ exists at least over the open interval $a < x < b$, then there is at least one value of x, say $x = x_1$, between $x = a$ and $x = b$ such that

$$\frac{f(b) - f(a)}{b - a} = f'(x_1)$$

(14) Parametric Differentiation

If $x = f(t)$ and $y = g(t)$, then

$$\frac{dy}{dx} = \frac{dg}{dt} \bigg/ \frac{df}{dt} \qquad \frac{df}{dt} \neq 0$$

(15) Rolle's Theorem

If $f(x)$ is continuous over the closed interval $a \leq x \leq b$, if $f(a) = 0$ and $f(b) = 0$, and if $f'(x)$ exists at least over the open interval $a < x < b$, then there is at least one value of x between $x = a$ and $x = b$, say $x = x_1$, such that $f'(x_1) = 0$.

(16) Simply Connected Region

A region R is said to be **simply connected** if it has the property that whenever a simple closed curve C lies in R, then all points inside C are also in R. Figure B.2*a* shows a simply connected region, and Fig. B.2*b* shows a multiply connected region.

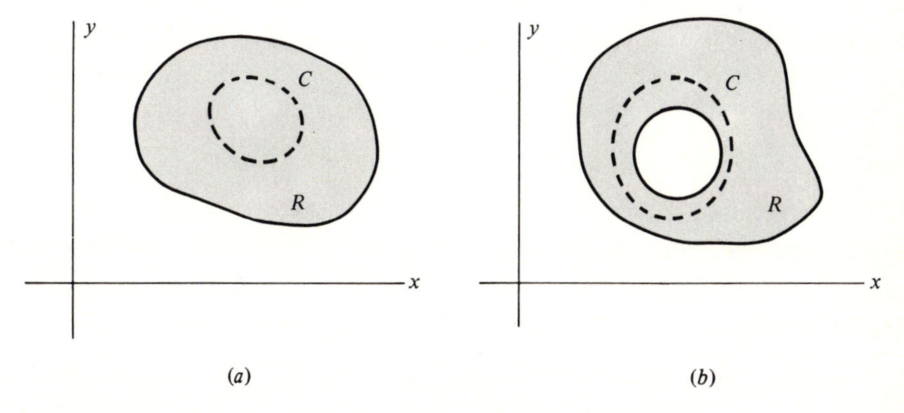

Figure B.2 (*a*) A simply connected region; (*b*) a multiply connected region.

(17) Taylor's Series for a Function of Two Variables

If $f(x, y)$ possesses partial derivatives of all orders at the point (x_0, y_0), then

$$f(x_0, y_0) + \left[\frac{\partial f}{\partial x}\bigg|_{x_0, y_0}(x - x_0) + \frac{\partial f}{\partial y}\bigg|_{x_0, y_0}(y - y_0)\right]$$

$$+ \frac{1}{2!}\left[\frac{\partial^2 f}{\partial x^2}\bigg|_{x_0, y_0}(x - x_0)^2 + 2\frac{\partial^2 f}{\partial x\,\partial y}\bigg|_{x_0, y_0}(x - x_0)(y - y_0) + \frac{\partial^2 f}{\partial y^2}\bigg|_{x_0, y_0}(y - y_0)^2\right]$$

$$+ \cdots$$

is called the **Taylor's series** of $f(x, y)$ around the point (x_0, y_0).

(18) Uniform Convergence of an Improper Integral

The improper integral $\int_a^\infty f(s, t)\,dt$ is said to **converge uniformly** for $\alpha \leq s \leq \beta$ if given any $\varepsilon > 0$, there exists a number N, depending on ε but not on s, such that

$$\left|\int_b^\infty f(s, t)\,dt\right| < \varepsilon$$

for all values of $b > N$ and for all values of s in the interval $\alpha \leq s \leq \beta$.

(19) Uniform Convergence of an Infinite Series

The infinite series

$$S(x) = u_1(x) + u_2(x) + \cdots + u_n(x) + \cdots$$

whose nth partial sum is

$$S_n(x) = u_1(x) + u_2(x) + \cdots + u_n(x)$$

is said to **converge uniformly** for $\alpha \leq x \leq \beta$ if given any $\varepsilon > 0$, there exists a number N, depending on ε but not on x, such that

$$|S(x) - S_n(x)| < \varepsilon$$

for all values of $n > N$ and all values of x in the interval $\alpha \leq x \leq \beta$. If the terms of a uniformly convergent series are continuous, then the sum to which the series converges is also continuous. This is not necessarily true if the convergence is not uniform.

(20) The Weierstrass M Test

If a sequence of positive constants $\{M_n\}$ exists such that $|f_n(x)| \leq M$ for all positive integers n and for all values of x in a region D, and if the series $M_1 + M_2 + M_3 + \cdots + M_n + \cdots$ converges, then the series

$$f_1(x) + f_2(x) + f_3(x) + \cdots + f_n(x) + \cdots$$

converges uniformly in D.

B.3 SELECTED TOPICS FROM APPLIED MATHEMATICS

(1) Abbreviations of Units

amperes, A
centimeters, cm
cubic centimeters, cm^3
degrees Celsius, °C
farads, F
gallons, gal
henries, H
kilograms, kg
ohms, Ω
seconds, s
candles (unit of light intensity), Cd
coulombs, C
cycles per second (or Hertz), Hz
degrees Fahrenheit, °F
feet, ft
grams, g
inches, in
milligrams, mg
pounds, lb
volts, V

(2) Archimedes' Principle

A body submerged, wholly or partially, in a liquid experiences an upward force equal to the weight of the liquid which it displaces.

(3) Average Heat Capacity

The amount of heat required to raise the temperature of a system by one degree is the **average heat capacity.**

(4) Btu (British Thermal Unit)

The amount of heat required to raise the temperature of one pound of water one degree Fahrenheit is one Btu.

(5) Concentration Gradient

Let $C(x, y, z)$ be the concentration of a liquid, in grams per cubic centimeter, say, at an arbitrary point P: (x, y, z) in a porous solid. Let Q be a second point in the solid and let P_1: $(x + \Delta x, y + \Delta y, z + \Delta z)$ be an arbitrary point, distinct from P, on the line PQ. Then

$$\lim_{P_1 \to P} \frac{C(x + \Delta x, y + \Delta y, z + \Delta z) - C(x, y, z)}{|PP_1|} = \lim_{\substack{\Delta x, \Delta y, \\ \Delta z \to 0}} \frac{C(x + \Delta x, y + \Delta y, z + \Delta z) - C(x, y, z)}{\sqrt{(\Delta x)^2 + (\Delta y)^2 + (\Delta z)^2}}$$

is said to be the **concentration gradient** at P in the direction PQ.

(6) Moment of Inertia

If M is an arbitrary mass and if r is the perpendicular distance from an arbitrary point of M to a given line l, then

$$\int_M r^2 \, dM$$

is said to be the **moment of inertia** of the mass M about the line, or axis, l. The moment of inertia of a circular disk of radius a and mass M about a line perpendicular to the disk at its center (usually called the *polar moment of inertia* of the disk) is $\frac{1}{2}Ma^2$. The moment of inertia of a thin uniform bar of length l and mass M about an axis perpendicular to the bar at one end is $\frac{1}{3}Ml^2$. The moment of inertia of a thin uniform bar of length l and mass M about an axis perpendicular to the bar at its midpoint is $\frac{1}{12}Ml^2$.

(7) Specific Heat

The amount of heat required to raise the temperature of a unit amount of a substance one degree is the **specific heat**. When the unit amount of the substance is one gram and the temperature increment is one degree Celsius, the specific heat is expressed in calories. When the unit amount of the substance is one pound and the temperature increment is one degree Fahrenheit, the specific heat is expressed in Btu (see Item 4).

(8) Steady State

A condition or property of a system, such as temperature, displacement, or current, which either does not vary with time or varies periodically with time.

(9) Temperature Gradient

Let $u(x, y, z)$ be the temperature at an arbitrary point P: (x, y, z) of a solid. Let Q be a second point in the solid, and let P_1: $(x + \Delta x, y + \Delta y, z + \Delta z)$ be an arbitrary point, distinct from P, on the line PQ. Then

$$\lim_{P_1 \to P} \frac{u(x + \Delta x, y + \Delta y, z + \Delta z) - u(x, y, z)}{|PP_1|}$$
$$= \lim_{\substack{\Delta x, \Delta y, \\ \Delta z \to 0}} \frac{u(x + \Delta x, y + \Delta y, z + \Delta z) - u(x, y, z)}{\sqrt{(\Delta x)^2 + (\Delta y)^2 + (\Delta z)^2}}$$

is said to be the **temperature gradient** at P in the direction PQ.

(10) Tensile Force

A force which acts to pull or stretch something, as opposed to a compressive force which acts to push or compress something, is a **tensile force**.

ANSWERS TO ODD-NUMBERED EXERCISES

Chapter 1, Sec. 1.2

1. Second order, ordinary, linear
3. Third order, ordinary, linear
5. Second order, ordinary, linear
7. Second order, partial, linear
9. Second order, partial, nonlinear
11. Yes. If c_1 and c_2 are functions of x, their derivatives will not be zero and will occur in the course of the proof with coefficients involving y_1 and y_2 which will not be identically zero.
13. $c + \ln |x|$, where $c = a + \ln |b|$
15. $\dfrac{x + B}{Cx + D}$, where $B = b/a$, $C = c/a$, $D = d/a$, and $a \neq 0$
17. $B \cos x$, where $B = 2A \cos a$
19. $A \sin 3x + B \sin x$, where $A = a - c/4$ and $B = b + 3c/4$
21. $A(x - 6y - 7) + B(3x + 4y + 5)$, where $A = a + c/2$ and $B = b + 3c/2$
31. $y'' + y' - 2y = 0$ **33.** $y'' - y = 0$
35. $y'' - 4y = 0$ **37.** $2yy'' - (y')^2 = 0$
39. The first equation is a homogeneous linear equation; the second equation is nonlinear.
41. The first equation is a homogeneous linear equation; the second equation is nonlinear.
43. Neither $f(x, y) = 4xy/(x^2 - 1)$ nor $f_y(x, y)$ is defined for $x = \pm 1$.

Chapter 1, Sec. 1.3

1. $y = ce^{-x^2}$ **3.** $y = \tan (x^3 + c)$
5. $y^2 e^{2y} = cx$ **7.** $y^2 = 1 + c(1 + x^2)$
9. $y = \ln |y + 1| + x^2 + c$ **11.** $(y - 1)e^y = c - e^{-x}$
13. $y = \dfrac{x - c(x + 1)}{1 + c(x + 1)}$ **15.** $y = \ln |\cos (c - x)| + k$

17. $y = ke^{cx}$ **19.** $y = 100e^{-2x}$

21. $y = 2(1 + x^2)$

23. $y = \begin{cases} (1 - x^2)^2 & x \leq -1 \\ 2(1 - x^2)^2 & -1 < x \leq 1 \\ 0 & 1 < x \end{cases}$

25. From calculus we know that $\int \left[\phi(u) \frac{du}{dx}\right] dx = \int \phi(u)\, du$

Hence, writing the given equation in the form $f(x) = g(y)\, dy/dx$ and integrating both sides *with respect to x*, we have

$$\int f(x)\, dx = \int \left[g(y) \frac{dy}{dx}\right] dx = \int g(y)\, dy$$

as required.

29. $y = 1 - 2x - \ln |c - x|$

31. $y = x + 1 - \dfrac{1 + 2ce^{-3x}}{1 - ce^{-3x}}$

Chapter 1, Sec. 1.4

1. Homogeneous of degree -1

3. The expression as a whole is not homogeneous, although the first major term is homogeneous of degree 1 and the second is homogeneous of degree 2.

11. $(x + y)^2 = cx$ **13.** $y = x/\ln |cx|$

15. $x^2 - 2xy - 3y^2 = c$; all the curves of this family are hyperbolas.

17. $2 \tan^{-1} (y/x) = \ln c(x^2 + y^2)$; all the curves of this family are complicated transcendental curves.

19. $2y + 2\sqrt{x^2 + y^2} = x^2$ **21.** $y^2 = \dfrac{x^2 \ln |x|}{1 + \ln |x|}$

23. $y^3 = x^3 - x^{3/2}$

25. $(y - 3)^2 + 2(y - 3)(x + 2) - (x + 2)^2 = c$

27. If $aB = bA$, then $a = \lambda A$, $b = \lambda B$, and the equation of Exercise 24 can be written in the form

$$\frac{dy}{dx} = \frac{\lambda Ax + \lambda By + c}{Ax + By + C} = \frac{\lambda(Ax + By + C) + (c - \lambda C)}{Ax + By + C} = f(Ax + By + C)$$

and the method of Exercise 27, Sec. 1.3, can be applied.

29. The equation $dy/dx = 2x/(3x - y)$ is a counterexample, since its solution $(y - x)^2 = c(y - 2x)$ defines a family of conics although $b + c = 3 \neq 0$.

31. $\theta = \ln cr$ or $\tan^{-1} (y/x) = \ln c\sqrt{x^2 + y^2}$. (The solution process here should be compared with the analysis of the same problem in Exercise 17.)

33. $\sqrt{x^2 + y^2}$

Chapter 1, Sec. 1.5

1. $xy^2 - x + \cos y = c$ **3.** $x^3 - 3x^2y - y^2 = c$

5. $x^2y^4 + x \sin y = c$ **7.** $x^3y^4 + x^3 = c$ (multiply by x^2)

9. $y = 2 \tan^{-1} (x/y) + c$ (divide by y^2)

11. $2x/y + 3 \ln |y| = c$ (homogeneous, or integrating factor $1/y^2$)
13. $x^2y^2 = x^2 - y^2 + c$ (separable, or integrating factor xy)
15. $cx^2 = y + \sqrt{x^2 + y^2}$ (homogeneous, or integrating factor $1/x^2$)
25. For the equation $xy^2\,dx + (x^2y + y)\,dy = 0$, Corollary 1 gives the solution $x^2y^2 + y^2 = k$, and this is meaningful if and only if $k \geq 0$.

Chapter 1, Sec. 1.6

1. $y = x^2 \ln |x| + cx^2$

3. $y = \sin x + c \cos x$

5. $y = \frac{1}{2} - \frac{1}{x} + \frac{c}{x^2}$

7. $y = \frac{x^2(x-1)}{2} + c(1-x)$

9. $y = e^{-x} + c\frac{e^{-x}}{x}$

11. $y = \frac{e^x}{2} + \frac{3e^{-x}}{2}$

13. $y = \frac{x^2+1}{4} + \frac{1}{1+x^2}$

15. $y = x - \sqrt{\frac{1+x^2}{2}}$

17. $x = \frac{4y^2}{5} + \frac{c}{y^3}$

Chapter 1, Sec. 1.7

1. $y = \frac{ce^{2x}}{4} - \frac{x}{2} + k$

3. $y = \frac{cx^2}{2} + k$

5. $y = \frac{x^3}{3} - \frac{x^2}{4} + c \ln |x| + k$

7. $y = -x + c\left(x + \frac{x^3}{3}\right) + k$

9. $y = \frac{x}{x+c}$

11. $\frac{1}{y} = -\frac{1}{3} + c \exp\left(-\frac{3}{2}x^2\right)$

13. $y^2 = \frac{2x-1}{2} + ce^{-2x}$

15. General solution $y = mx - 4m^3$; singular solution $y^2 = x^3/27$
17. General solution $y = mx - e^m$; singular solution $y = x \ln |x| - x$

19. $y = xy' \pm \frac{2\sqrt{3}}{9}(y')^{3/2}$

21. $y = xy' + 2\left(\frac{1-y'}{3}\right)^{3/2}$

23. $y = 2 \tan^{-1} (2x + \mathrm{Tan}^{-1} \tfrac{1}{2})$

25. $y = \frac{1}{2} - \frac{(x+2)^2}{4}$

27. $y = \begin{cases} -2x - 1 & x < -1 \\ x^2 & -1 \leq x \leq 2 \\ 4x - 4 & 2 < x \end{cases}$

29. $y = 1 + \frac{1}{1 - x + ce^{-x}}$

31. $y = 1 - x + \frac{1}{c - x}$

33. $y = x + \frac{5x}{c - x^5}$

35. $y' = y^2 + \frac{y}{x} - x^2$

Chapter 2, Sec. 2.2

1. $R = R_0 e^{-0.000436t}$, 660 years, 4.3 percent of R_0, 2.9 percent of R_0

3. 2.4 revolutions **5.** $\omega = \omega_0 e^{-kt/I}$

7. $p = \dfrac{e^{kwy} - 1}{k}$, 1.02

9. $y^{5/2} = h^{5/2} - \dfrac{5r^2h^2}{2R^2}\sqrt{2g}\,t$; $y = 0$ when $t = \dfrac{R^2}{5r^2}\sqrt{\dfrac{2h}{g}}$

11. $(2R - y)^{3/2} = R^{3/2} + \dfrac{3\pi r^2\sqrt{2g}}{4l}t$;

$y = 0$ when $t = \dfrac{4l}{3\pi r^2\sqrt{2g}}[(2R)^{3/2} - R^{3/2}]$

13. $y = kx^2$

15. $\alpha \ln \dfrac{\alpha - \sqrt{h}}{\alpha - \sqrt{y}} + \sqrt{h} - \sqrt{y} = \dfrac{a}{A}\sqrt{\dfrac{g}{2}}\,t$ where $\alpha = \dfrac{Q}{a\sqrt{2g}}$

The limiting depth is $y = \alpha^2$.

17. 120.8°

19. Loss $= \dfrac{(T_1 - T_0)2\pi k}{\ln r_1 - \ln r_0}$; $T = T_0 + (T_1 - T_0)\dfrac{\ln r - \ln r_0}{\ln r_1 - \ln r_0}$

21. $s = -16t^2 + 16t$; $s_{max} = 4$ ft (= 1764 ft above the earth); the stone strikes the ground when $t = 11$ s; its velocity of impact is -336 ft/s.

23. $y = 256t - 2{,}560(1 - e^{-t/8})$; $y_{max} = 55.0$ ft; the object strikes the ground with velocity 54.8 ft/s when $t = 3.71$ s.

25. $x = x_0 \cos\sqrt{\dfrac{k}{m}}\,t$; $v = -x_0\sqrt{\dfrac{k}{m}}\sin\sqrt{\dfrac{k}{m}}\,t$

27. $t = \sqrt{\dfrac{x_0 m}{2k}}\left(\sqrt{x_0 x - x^2} + \dfrac{x_0}{2}\cos^{-1}\dfrac{2x - x_0}{x_0}\right)$

29. $v = -\sqrt{\dfrac{2gr^2}{y_0}}\sqrt{\dfrac{y_0 - y}{y}}$;

$t = \sqrt{\dfrac{y_0}{2gr^2}}\left(\sqrt{y_0 y - y^2} + \dfrac{y_0}{2}\cos^{-1}\dfrac{2y - y_0}{y_0}\right)$

31. $s = \dfrac{g \sin\alpha}{3}t^2$ **33.** $i = \dfrac{E}{R}(1 - e^{-Rt/L})$, $t_{1/2} \doteq 0.693\dfrac{L}{R}$

35. $Q = Q_0 e^{-t/(RC)}$ **37.** $y = \dfrac{H}{w}\cosh\dfrac{wx}{H}$

39. $y^2 = 2cx + c^2$

Chapter 2, Sec. 2.3

1. $Q = 30 - \dfrac{1{,}500}{t + 50}$, $t = 50$ min

3. $Q = 90\dfrac{1 - e^{-0.00645t}}{3 - 2e^{-0.00645t}}$, $t = 44.6$ min

5. $B = \dfrac{A_0 B_0}{B_0 + (A_0 - B_0)e^{-A_0 kt}}$

7. $Q = 120\dfrac{1 - e^{-0.00314t}}{4 - 3e^{-0.00314t}}$

9. $\dfrac{4Q}{40 - Q} + \ln\left(1 - \dfrac{Q}{40}\right) = 0.0294t$

11. $Q = 200(1 - e^{-t/20})$, $t = 27.73$ min

13. $Q = 2(100 - t) - 150[(100 - t)/100]^3$; 42.3 min; $Q_{max} = 89$ lb when $t = 33\frac{1}{3}$ min

15. $y = h - (k_1/k_2\rho)(1 - e^{-k_2 t})$; $h \leq k_1/k_2\rho$, where k_1 and k_2 are the proportionality constants in the rates of evaporation and condensation.

Chapter 2, Sec. 2.4

1. Lake Erie: $t_{1/2} \doteq 1.82$ years, $t_{1/10} \doteq 6.05$ years
   Lake Ontario: $t_{1/2} \doteq 5.31$ years, $t_{1/10} \doteq 17.63$ years

3. 13,200 B.C. 5. $N = N_0 \exp(k_b - k_d)t$

7. If a is the proportionality constant in the birthrate and $bN + c$ is the proportionality factor in the death rate, then

   (a) $N = \dfrac{c}{a - b}$; $a > b$, $N_0 = \dfrac{c}{a - b}$

   (b) $N = \dfrac{c}{a - b}\dfrac{N_0}{N_0 - \left(N_0 - \dfrac{c}{a - b}\right)e^{ct}}$; $a > b$, $N_0 > \dfrac{c}{a - b}$; $t < \dfrac{1}{c}\ln\left(\dfrac{N_0}{N_0 - \dfrac{c}{a - b}}\right)$

   (c) $N = \dfrac{c}{a - b}\dfrac{N_0}{N_0 + \left(\dfrac{c}{a - b} - N_0\right)e^{ct}}$; $a > b$, $N_0 < \dfrac{c}{a - b}$

   (d) $N = N_0 e^{-ct}$, $a = b$

   (e) $N = \dfrac{c}{b - a}\dfrac{N_0}{\left(N_0 + \dfrac{c}{b - a}\right)e^{ct} - N_0}$, $a < b$

9. $N = N_0 \exp[(b_1 - d_1)t - (b_2 - d_2)t^2/2]$. The population has an extremum when $t = (b_1 - d_1)/(b_2 - d_2)$. It is a maximum if $b_2 - d_2 > 0$ and a minimum if $b_2 - d_2 < 0$. If $(b_1 - d_1)(b_2 - d_2) \geq 0$, the extremum occurs in "real time," i.e., for $t \geq 0$. As $t \to \infty$,

$$
\begin{array}{ll}
N \to 0 & \text{if } b_2 - d_2 > 0 \\
N \to \infty & \text{if } b_2 - d_2 < 0 \\
N \to 0 & \text{if } b_2 - d_2 = 0 \text{ and } b_1 - d_1 < 0 \\
N \to \infty & \text{if } b_2 - d_2 = 0 \text{ and } b_1 - d_1 > 0 \\
N = N_0 & \text{if } b_1 - d_1 = b_2 - d_2 = 0
\end{array}
$$

11. $\ln \dfrac{N}{N_0} = \dfrac{b_1}{2b_2} \ln \left| \dfrac{b_2 + t}{b_2 - t} \right| - \dfrac{d_1}{2d_2} \ln \left| \dfrac{d_2 + t}{d_2 - t} \right|$

As $t \to \infty$,

$$\ln \frac{N}{N_0} \to 0$$

which is a meaningful finite limit. This is not relevant, however, since the terms on the right in the expression for $\ln (N/N_0)$ are undefined, and the process comes to an end with N being 0, N_0, or ∞, at whichever of the times $t = |b_2|$ or $t = |d_2|$ is the smaller.

13. $Q = A - (A - Q_0)e^{-t/V}$ **15.** 24.1 min; 48.2 min

17. 66.2 min

19. $x = \dfrac{(3x_0 + y_0) + (x_0 - y_0)e^{-4kAt/(3V)}}{4}$

Chapter 2, Sec. 2.5

1. $xy = k$ **3.** $16y^3 = 9(k - x)^2$

5. $3x^2y + y^3 = k$ **7.** $y = ce^{x/y}$

9. $x^2 - y^2 = c$, $x^2 + y^2 = c^2$ **11.** $x^2 + y^2 = cx$

Chapter 2, Sec. 2.6

1. $s = \dfrac{aN(0) - [aN(0) - k]e^{-kt/N(0)}}{k}$ **3.** $x = \dfrac{N}{2}$; $x = \dfrac{kN - c}{2k}$

5. $r = r_0 \exp [\rho \pi r_0^2 x/(2W)]$ where x is the distance from the upper base to the general cross section

7. $x = -\sqrt{l^2 - y^2} + l \ln \dfrac{l + \sqrt{l^2 - y^2}}{y}$

9. 11:23 A.M.

11. $\dfrac{ds}{dt} + \dfrac{r_0 - aN(0)}{N(0)} s = -a$

Chapter 3, Sec. 3.2

3. $-3(x + 1) + (x + 2) + (2x + 1) \equiv 0$ **5.** $-6e^{2x}$

9. The functions are linearly independent.

11. The functions are linearly dependent and satisfy the equation

$$6 \ln (x - 1) + 3[2 \ln (x + 1)] - 2[3 \ln (x^2 - 1)] \equiv 0$$

13. The fractions are linearly independent.

Chapter 3, Sec. 3.3

1. $y = c_1 \sin x + c_2 \cos x$ **3.** $y = c_1 x + c_2 x^{-4}$
5. $y = c_1 e^x + c_2 xe^{-x}$ **7.** $y = c_1 e^x + c_2 xe^x + c_3 x^2 e^x$
11. $y_2 = -xe^x$ **13.** Yes
15. $y = c_1 e^x + c_2 e^{2x}$, $y = k_1(e^x + e^{2x}) + k_2(e^x - e^{2x})$

17. $y = c_1 x + c_2 \dfrac{1}{x}$, $y = k_1 x + k_2 \dfrac{x^2 + 1}{x}$

Chapter 3, Sec. 3.4

1. $y = c_1 e^x + c_2 e^{3x} + e^{-x}/8$

3. $y = c_1 e^{-x} + c_2 e^{2x} - \dfrac{e^{-x}}{9} - \dfrac{xe^{-x}}{3} = k_1 e^{-x} + c_2 e^{2x} - \dfrac{xe^{-x}}{3}$

5. $y = c_1 \sin x + c_2 \cos x - \dfrac{x \cos x}{2} + \dfrac{\sin x}{2}$

$= k_1 \sin x + c_2 \cos x - \dfrac{x \cos x}{2}$

7. $y = c_1 \sin x + c_2 \cos x - \cos x(\ln |\sec x + \tan x|)$

9. $y = c_1 x + c_2 \dfrac{1}{x} + \dfrac{x \ln |x|}{2} - \dfrac{x}{4} = k_1 x + c_2 \dfrac{1}{x} + \dfrac{x \ln |x|}{2}$

11. $y = c_1 x + c_2 x^2 + xe^x$
15. Assume $Y = u_1 y_1 + u_2 y_2 + u_3 y_3$ where u_1, u_2, u_3 are functions of x to be determined so that Y will satisfy the given differential equation. This leads to the three equations $u_1' y_1 + u_2' y_2 + u_3' y_3 = 0$, $u_1' y_1' + u_2' y_2' + u_3' y_3' = 0$, and $u_1' y_1'' + u_2' y_2'' + u_3' y_3'' = R(x)$ from which u_1', u_2', u_3' and then u_1, u_2, and u_3 can always be found provided y_1, y_2, y_3 are linearly independent solutions of the given equation.

Chapter 4, Sec. 4.1

1. Dy means the derivative of y with respect to the relevant independent variable. yD is an operator which indicates y times the derivative of a function as yet unspecified.
3. The common value of the expressions is $-21 \cos 3x - 7 \sin 3x$.
5. If and only if $r_2(x) = r_1(x) + c$

7. $r(x) = -1$ **9.** $y = c_1 x + c_2 x \displaystyle\int \frac{e^x}{x}\,dx$

Chapter 4, Sec. 4.2

1. $y = c_1 + c_2 x$ **3.** $y = c_1 e^x + c_2 xe^x$
5. $y = c_1 e^{-x} + c_2 e^{-4x}$ **7.** $y = c_1 + c_2 e^{-5x}$
9. $y = c_1 e^{2x/3} + c_2 xe^{2x/3}$ **11.** $y = e^{-5x}(A \cos x + B \sin x)$
13. $y = 2 \cos 2x + 3 \sin 2x$ **15.** $y = 0$

17. There is no solution meeting the given conditions.

19. If $x_1 = x_0 + n\pi$, the conditions can be met if and only if $y_1 = (-1)^n y_0$. In this case, the equations arising from the end conditions are dependent, and there is an infinite family of solutions meeting the given conditions.

21. $\lambda = n,\ y = A \cos nx$

25. There are no nontrivial solutions satisfying the given conditions.

31. As $m_2 \to m_1$, y becomes an indeterminate of the form 0/0. This evaluates by L'Hospital's rule to $xe^{m_1 x}$, which is a second, independent solution of the limiting equation.

37. $y = -\dfrac{ae^x + 2be^{2x}}{x^2(ae^x + be^{2x})}$

Chapter 4, Sec. 4.3

1. $y = e^{-2x}(c_1 \cos x + c_2 \sin x) + e^x/5$

3. $y = c_1 + c_2 e^{-x} + x^2/2 + x$

5. $y = c_1 \cos x + c_2 \sin x + (x \sin x)/2 - \sin 2x$

7. $y = c_1 + c_2 e^{-3x} + (\sin x - \cos x)/2$

9. $y = c_1 e^{-2x} + c_2 xe^{-2x} + xe^{-x} - 2e^{-x}$

11. $y = c_1 e^{-x} + c_2 xe^{-x} - \dfrac{3 \cos 2x}{50} + \dfrac{2 \sin 2x}{25} + \dfrac{1}{2}$

13. $y = c_1 e^{2x} + c_2 e^{3x} + \dfrac{7 \cosh x + 5 \sinh x}{24}$

15. (*b*) $A = 1$

17. $y = \frac{13}{2}e^{-2x} + 11xe^{-2x} + 2x - \frac{9}{2}$

19. $y = e^{-x}(3 \cos 2x + 4 \sin 2x) + 2 \cos x + \sin x$

21. $Y = \dfrac{(e^{\lambda t} - e^{at}) - (\lambda - a)te^{at}}{(\lambda - a)^2}$

25. $Y = x^{1/2} + \dfrac{1}{4}x^{-3/2} - \dfrac{3 \cdot 5}{4^2}x^{-7/2} + \dfrac{3 \cdot 5 \cdot 7 \cdot 9}{4^3}x^{-11/2} - \cdots$

This series diverges for all values of x.

27. (*a*) $Y = -xe^x$ (*b*) $Y = -xe^x - e^x$

29. (*a*) $Y = \dfrac{x \sin x}{2}$ (*b*) $Y = \dfrac{x \sin x}{2}$

Chapter 4, Sec. 4.4

1. $y = c_1 e^{-x} + c_2 e^{-2x} + c_3 e^{-3x} + x - 3$

3. $y = c_1 e^{-2x} + e^{2x}(c_2 \cos x + c_3 \sin x) + 3 \cos x - \sin x$

5. $y = c_1 e^x + c_2 e^{-x} + c_3 \cos 3x + c_4 \sin 3x - x^2 - \dfrac{16}{9} - \dfrac{\sin 2x}{5}$

7. $y = c_1 e^x + c_2 xe^x + c_3 e^{-2x} + c_4 xe^{-2x} + \frac{1}{18}x^2 e^x$

9. $y = e^x(c_1 \cos x + c_2 \sin x) + e^{-x}(c_3 \cos x + c_4 \sin x) + \dfrac{\cos x}{5} + \dfrac{\sin 2x}{20}$

11. $y = -2 \sin x + \cos x$ **13.** $y = \sin x - \sin 2x$

15. $Y = \frac{1}{2} \int_{x_0}^{x} f(s)(e^{x-s} - 2e^{2(x-s)} + e^{3(x-s)})\, ds$

17. $Y = \frac{1}{2} \int_{x_0}^{x} f(s)[(x-s)e^{x-s} - \sinh (x-s)]\, ds$

19. Only for values of λ which satisfy the equation $\tanh \lambda = \tan \lambda$. $y_n = A_n(\sin \lambda_n \sinh \lambda_n x - \sinh \lambda_n \sin \lambda_n x)$, where λ_n is the nth one of the roots of the equation $\tanh \lambda = \tan \lambda$.

Chapter 4, Sec. 4.5

1. $y = c_1 \dfrac{1}{x} + c_2 x$ **3.** $y = c_1 x + c_2 x \ln |x| + \dfrac{x^5}{16}$

5. $y = \dfrac{c_1}{\sqrt{x}} + \dfrac{c_2}{x} + 2 + \dfrac{x}{2}$

7. $y = c_1 x + c_2 \dfrac{1}{x} + c_3 \sin (3 \ln |x|) + c_4 \cos (3 \ln |x|) + \dfrac{1}{39x^2} - \dfrac{\ln |x|}{3}$

9. Yes

Chapter 4, Sec. 4.6

1. $y = 2 \cos 4t - 2 \sin 4t$ **3.** $y = \frac{9}{4}e^{-t} - \frac{1}{4}e^{-9t}$

5. $y = 2e^{-6t} \sin 8t$ **7.** $Y = \dfrac{-10 \cos t + 8 \sin t}{41}$

11. $Q = 10^{-3} \cos 1{,}000t$, $i = -\sin 1{,}000t$

13. $Q = -\dfrac{1}{3{,}000} \cos 1{,}000t + \dfrac{1}{750} \cos 500t$

$i = \frac{1}{3} \sin 1{,}000t - \frac{2}{3} \sin 500t$

15. $2\pi\sqrt{hw/(\rho g)}$ where ρ is the density of water

17. $r = a \cosh \omega t$

19. $y = \tan \theta \left[x - \sqrt{\dfrac{EI}{F \cos \theta}} \dfrac{\sinh \sqrt{F \cos \theta/(EI)}x}{\cosh \sqrt{F \cos \theta/(EI)}L} \right]$

where the origin is taken at the free end of the beam

21. $\omega_n = (z_n^2/L^2)\sqrt{EIg/(A\rho)}$, where z_n is the nth one of the roots of the equation $\cos z \cosh z = 1$. The deflection curve corresponding to the critical speed ω_n is

$$y_n = (\cos z_n - \cosh z_n)\left(\cos z_n \frac{x}{L} - \cosh z_n \frac{x}{L}\right) + (\sin z_n + \sinh z_n)\left(\sin z_n \frac{x}{L} - \sinh z_n \frac{x}{L}\right)$$

23. $\omega = \dfrac{1}{2\pi} \sqrt{\dfrac{3kL^2 g}{wL^2 + 3Wl^2}}$ **25.** $\omega = \dfrac{1}{2\pi} \sqrt{\dfrac{3g[(2W + w)L + 4kl^2]}{2L^2(3W + w)}}$

27. $\omega = [1/(2\pi)]\sqrt{g/l}$

29. True period $\doteq 1.18 \times$ approximate period

Chapter 5, Sec. 5.2

1. $x = c_1 e^t + c_2 e^{-t}$, $y = c_1 e^t - c_2 e^{-t}$

3. $x = c_1 e^t + 9$, $y = -\frac{3}{2}c_1 e^t + 3e^{-t} - 15$

5. $x = c_1 \cos t + c_2 \sin t - e^t$, $y = -2c_2 \cos t + 2c_1 \sin t + e^t$

7. $x = e^{-t}(c_1 \cos t + c_2 \sin t)$, $y = e^{-t}(c_2 \cos t - c_1 \sin t)$

9. $x = (-\frac{3}{2}c_1 - \frac{1}{8})e^{-t} - \frac{1}{2}c_2 e^{3t} - \frac{3}{4}te^{-t}$
$y = c_1 e^{-t} + c_2 e^{3t} + \frac{1}{2}te^{-t}$

11. $x = c_1 e^{-t} + c_2 e^t + c_3 e^{2t}$, $y = c_1 e^{-t} - c_2 e^t + 2c_3 e^{2t}$

13.
$$
\begin{aligned}
x &= c_1 e^t + c_2 e^{-t} + c_3 e^{3t} - 2e^{2t} \\
y &= -2c_2 e^{-t} + 2c_3 e^{3t} - 2e^{2t} \\
z &= -2c_1 e^t + 2c_2 e^{-t} + 2c_3 e^{3t} + 2e^{2t}
\end{aligned}
$$

15. $x = y = z = e^t$

17. $x = c_1 e^{-t} + c_2 e^{-2t} + c_3 \cos 2t + c_4 \sin 2t - 3$
$y = -c_1 e^{-t} - \frac{8}{3}c_2 e^{-2t} - 2(c_3 + c_4) \cos 2t + 2(c_3 - c_4) \sin 2t + 2$

19. (*a*) $x = 4 \sin 2t - 2$, $y = -\cos 2t - 3 \sin 2t + 1$
(*b*) $x = 3e^t + 2 \cos 2t + 2 \sin 2t - 2$
$y = -2e^t - 2 \cos 2t - \sin 2t + 1$

21. $(5D - 4)x - Dy = 0$, $3Dx + (D - 4)y = 0$

25. $Q_1 = V[(s_1 - s)e^{-at/V} + s]$

$$Q_2 = V\left[(s_2 - s)e^{-at/V} + \frac{a}{V}(s_1 - s)te^{-at/V} + s\right]$$

Q_2 will reach an extreme value when $t = (s_1 - s_2)V/[(s_1 - s)a]$ which is physically meaningful if and only if $t \geq 0$, that is, if and only if $s_1 > s_2, s$ or $s_1 < s_2, s$.

27.
$$Q_1 = 100 + \frac{200}{\sqrt{3}}e^{-3t/40} \sin\left(\frac{\sqrt{3}}{40}t - \frac{\pi}{3}\right)$$

$$Q_2 = 100 - \frac{200}{\sqrt{3}}e^{-3t/40} \sin \frac{\sqrt{3}}{40}t$$

$$Q_3 = 100 + \frac{200}{\sqrt{3}}e^{-3t/40} \sin\left(\frac{\sqrt{3}}{40}t + \frac{\pi}{3}\right)$$

29. $x = c_1 + c_2 e^t$, $y = -c_1 + c_3 e^t$

Chapter 5, Sec. 5.3

1. $x = c_1 e^{-3t} - \frac{1}{2}$, $y = c_1 e^{-3t} + \frac{1}{2}$

3. $x = -c_1 e^t - 2c_2 e^{2t} + t$, $y = c_1 e^t + c_2 e^{2t} + 1$

5.
$$
\begin{aligned}
x &= -2c_1 + c_2 e^{-5t} + 5e^t + e^{-t} \\
y &= c_1 - 3c_2 e^{-5t} - 3e^t + e^{-t}
\end{aligned}
$$

7. $x = c_1 e^t + 4c_2 e^{2t} + 5c_3 e^{3t} - 5e^{-t} + 2$
$y = -c_1 e^t - 5c_2 e^{2t} - 7c_3 e^{3t} + 3e^{-t} - 1$

9. $x = 2c_1 \cos t + 2c_2 \sin t + c_3 \cos 2t + c_4 \sin 2t + 2 \sin 3t$
$y = -3c_1 \cos t - 3c_2 \sin t - 3c_3 \cos 2t - 3c_4 \sin 2t - 3 \sin 3t$

11. $x = y = z = e^t$

17. $x_1 = \cos 2t + \sin 3t,\ x_2 = \cos 2t - \frac{2}{3}\sin 3t$

19.
$$\ddot{y}_1 = -\lambda^2(2y_1 - y_2)$$
$$\ddot{y}_2 = -\lambda^2(-y_1 + 2y_2 - y_3) \qquad \lambda^2 = \frac{4Tg}{wl}$$
$$\ddot{y}_3 = -\lambda^2(-y_2 + 2y_3)$$
$$\omega_1 = \sqrt{2 - \sqrt{2}}\lambda,\ 1 : \sqrt{2} : 1$$
$$\omega_2 = \sqrt{2}\lambda,\ 1 : 0 : -1$$
$$\omega_3 = \sqrt{2 + \sqrt{2}}\lambda,\ 1 : -\sqrt{2} : 1$$

Chapter 5, Sec. 5.4

1. $x = 2e^t(c_1 \cos 2t + c_2 \sin 2t),\ y = e^t(c_2 \cos 2t - c_1 \sin 2t)$

3. $x = c_1 e^t + 3c_2 e^{9t} - \frac{7}{20}e^{-t},\ y = -c_1 e^t + 5c_2 e^{9t} + \frac{1}{4}e^{-t}$

5. $$\begin{bmatrix} x \\ y \end{bmatrix} = c_1 \begin{bmatrix} 2 \\ -1 \end{bmatrix} e^t + c_2 \begin{bmatrix} 1 \\ 1 \end{bmatrix} e^{-2t} + \frac{1}{2}\begin{bmatrix} 2 \\ 1 \end{bmatrix} e^{-t}$$

7. $$\begin{bmatrix} x \\ y \end{bmatrix} = c_1 \begin{bmatrix} 4 \\ -3 \end{bmatrix} e^t + c_2 \begin{bmatrix} 5 \\ -3 \end{bmatrix} e^{-2t} + \begin{bmatrix} 1 \\ 0 \end{bmatrix} e^t$$

9. $$\begin{bmatrix} x \\ y \end{bmatrix} = c_1 e^{-2t}\left(\begin{bmatrix} 0 \\ 1 \end{bmatrix} \cos t - \begin{bmatrix} 1 \\ -1 \end{bmatrix} \sin t\right)$$
$$+ c_2 e^{-2t}\left(\begin{bmatrix} 1 \\ -1 \end{bmatrix} \cos t + \begin{bmatrix} 0 \\ 1 \end{bmatrix} \sin t\right) + \begin{bmatrix} 1 \\ -1 \end{bmatrix} e^t$$

11. $$\begin{bmatrix} x \\ y \end{bmatrix} = c_1 e^{-t}\left(\begin{bmatrix} 1 \\ 1 \end{bmatrix} \cos 2t - \begin{bmatrix} 2 \\ -4 \end{bmatrix} \sin 2t\right)$$
$$+ c_2 e^{-t}\left(\begin{bmatrix} 2 \\ -4 \end{bmatrix} \cos 2t + \begin{bmatrix} 1 \\ 1 \end{bmatrix} \sin 2t\right) + \frac{1}{6}\begin{bmatrix} -1 \\ 5 \end{bmatrix} e^{-t}$$

13. $$\begin{bmatrix} x \\ y \\ z \end{bmatrix} = c_1 \begin{bmatrix} 0 \\ 1 \\ -1 \end{bmatrix} e^t + c_2 \begin{bmatrix} 6 \\ 2 \\ -7 \end{bmatrix} e^{-t} + c_3 \begin{bmatrix} 3 \\ 1 \\ -1 \end{bmatrix} e^{4t}$$

15. $$\begin{bmatrix} x \\ y \\ z \end{bmatrix} = c_1 \begin{bmatrix} 1 \\ 0 \\ 0 \end{bmatrix} e^t + c_2 \begin{bmatrix} 0 \\ 1 \\ 0 \end{bmatrix} e^t + c_3 \left(\begin{bmatrix} 0 \\ 0 \\ 1 \end{bmatrix} e^t + \begin{bmatrix} 1 \\ 2 \\ 0 \end{bmatrix} te^t\right)$$

19.
$$D^r(t^3 e^{mt}) = m^r t^3 e^{mt} + 3rm^{r-1}t^2 e^{mt} + 3r(r-1)m^{r-2}te^{mt} + r(r-1)(r-2)m^{r-3}e^{mt}$$
$$p(D)(t^3 e^{mt}) = p(m)t^3 e^{mt} + 3p'(m)t^2 e^{mt} + 3p''(m)te^{mt} + p'''(m)e^{mt}$$
$$P(D)(At^3 e^{mt}) = P(m)At^3 e^{mt} + 3P'(m)At^2 e^{mt} + 3P''(m)Ate^{mt} + P'''(m)Ae^{mt}$$

23.
$$Dx_1 = x_2$$
$$Dx_2 = 14x_1 + 3x_2 + 24x_3 \qquad \text{where } x_1 = x,\ x_2 = Dx,\ x_3 = y$$
$$Dx_3 = -x_1 - x_2 - 2x_3$$

Chapter 6, Sec. 6.1

1. $y = c_1 e^{3t} + c_2 e^{2t} + \frac{1}{2}e^t$
3. $y = c_1 e^{-t} + c_2 e^{-3t} + t/3 - \frac{4}{9}$
5. $y = c_1 e^t + c_2 te^t + e^{2t}$
7. $y = c_1 e^t + c_2 e^{2t} + c_3 e^{3t} - 1$

Chapter 6, Sec. 6.2

1. (*a*) Yes, (*b*) No, (*c*) No, (*d*) Yes, (*e*) Yes, (*f*) Yes
3. (*a*) 0, (*b*) 0, (*c*) 0, (*d*) k, (*e*) k, (*f*) 0

Chapter 6, Sec. 6.3

1. $\mathscr{L}\{f^{(n)}\} = s^n \mathscr{L}\{f\} - \sum_{j=0}^{n-1} s^{n-1-j} f^{(j)}(0^+)$

7. (*a*) $\mathscr{L}\{\cos bt\} = \dfrac{s}{s^2 + b^2}$ (*b*) $\mathscr{L}\{\sin bt\} = \dfrac{b}{s^2 + b^2}$

9. $\mathscr{L}\{y\} = \dfrac{s\mathscr{L}\{f(t)\} + a_0 y_0 s}{a_0 s^2 + a_1 s + a_2}$

11. If $T(f')$ and $T(f'')$ are not to involve the evaluation of f or any of its derivatives, it is necessary that $K(s, a) = K(s, b) = 0$ and that

$$\left.\frac{\partial K(s, t)}{\partial t}\right|_{t=a} = \left.\frac{\partial K(s, t)}{\partial t}\right|_{t=b} = 0$$

If $\phi(s, t)$ is an arbitrary differentiable function which is bounded at $t = a$ and at $t = b$, these conditions are met by any kernel of the form $K(s, t) = (t - a)^2(t - b)^2\phi(s, t)$

Chapter 6, Sec. 6.4

3. $\dfrac{s}{s^2 - k^2}$ **5.** $\dfrac{1}{2}\left[\dfrac{1}{s} + \dfrac{s}{s^2 + 4b^2}\right]$

7. (*a*) e^{-3t} (*b*) $\dfrac{1}{3!}t^3$ (*c*) $\frac{1}{3}\sin 3t$

(*d*) $2\cos 3t + \sin 3t$ (*e*) $-\frac{1}{2}e^{-t} + \frac{3}{2}e^{3t}$

9. $z = 4e^{-4t} + 2e^t$
11. $y = -\frac{2}{3}\cos t - \sin t + \frac{8}{3}\cos 2t + 2\sin 2t$
$z = \frac{1}{3}\cos t + \frac{1}{3}\sin t - \frac{4}{3}\cos 2t - \frac{2}{3}\sin 2t$

13. (*a*) $\dfrac{1}{(\ln c)^{c+1}}\Gamma(c + 1)$ (*b*) $\Gamma(\frac{1}{2}) = \sqrt{\pi}$

(*c*) $\dfrac{1}{(m + 1)^{n+1}}\Gamma(n + 1)$

15. $a\sqrt{m\pi/k}$

Chapter 6, Sec. 6.5

1. $\dfrac{1}{(s-2)^2}$

3. $\dfrac{2s(s^2-27)}{(s^2+9)^3}$

5. $\dfrac{2(3s^2+12s+13)}{(s^3+6s^2+13s)^2}$

7. $\dfrac{4(s+3)}{s(s^2+6s+13)^2}$

9. $\ln \dfrac{\sqrt{s^2+9}}{s}$

11. $\dfrac{1}{3!}t^3e^{-2t}$

13. $e^{-2t} - te^{-2t}$

15. $\dfrac{1}{4} - \dfrac{1}{4}e^{-2t} - \dfrac{t}{2}e^{-2t}$

17. $\frac{1}{4}e^{-t}(1 - \cos 2t)$

19. $\dfrac{2(1-\cosh t)}{t}$

21. $-\dfrac{t}{4}\cos 2t + \dfrac{1}{8}\sin 2t$

23. $y = e^{-2t}\cos t$

25. $y = te^{-t} + \frac{1}{6}t^3e^{-t}$
27. $y = \frac{5}{8}e^t - \frac{5}{8}e^{-t} - \frac{5}{4}te^{-t} - (t^2/4)e^{-t}$
29. $x = e^{-t}(\cos t + 2\sin t)$, $y = e^{-t}(2\cos t - \sin t)$
31. (*a*) 1 (*b*) 0 (*c*) 1 (*d*) 0
33. $\ln 2$ **35.** $\frac{1}{2}[\ln(p^2+q^2) - \ln(a^2+b^2)]$
37. $y = y_0 e^{-2t} + ct^3e^{-2t}$
39. There is no solution meeting the given conditions. The solutions corresponding to $y_0 = 0$ are all of the form

$$y = c(-\tfrac{1}{2} + t + \tfrac{1}{2}e^{-4t} + te^{-4t})$$

and for each of these $y_0' = 0$.

Chapter 6, Sec. 6.6

1. $(1/s)e^{-as}$

3. $1/s$

5. $\left(\dfrac{4}{s} + \dfrac{4}{s^2} + \dfrac{2}{s^3}\right)e^{-2s}$

7. $\left(\dfrac{s\cos 1}{s^2+1} - \dfrac{\sin 1}{s^2+1}\right)e^{-s}$

9. $\dfrac{1+e^{-\pi s}}{s^2+1}$

11. $\dfrac{b}{s}(e^{-as} - 2e^{-2as} + e^{-3as})$

13. $y = \dfrac{t^2e^{-t}}{2} - \dfrac{(t-1)^2e^{-(t-1)}u(t-1)}{2}$

15. $y = e^{-t}\cos t - e^{-t}\sin t + e^{-(t-2)}\sin(t-2)u(t-2)$

Chapter 6, Sec. 6.7

1. $\frac{5}{2}e^{-t} - 9e^{-2t} + \frac{15}{2}e^{-3t}$
3. $\frac{1}{25}(3\cos t + 4\sin t - 3e^{-2t} - 10te^{-2t})$

5. $(t/4)(e^t - e^{-t})$

7. $\frac{3}{20}e^t - \frac{1}{4}e^{-t} + \frac{1}{10}(\cos t - 2 \sin t)e^{-2t}$

9. $\frac{1}{5^4}\left[\left(\frac{4}{25} + \frac{2t}{5} + \frac{2t^2}{5} + \frac{t^3}{6}\right)e^{-2t} + \left(-\frac{4}{25} + \frac{2t}{5} - \frac{2t^2}{5} + \frac{t^3}{6}\right)e^{3t}\right]$

11. $y = \frac{1}{3}e^{2t} - \frac{1}{2}e^t + \frac{1}{6}e^{-t} + \frac{1}{2}u(t-2) + \frac{1}{6}e^{2(t-2)}u(t-2) \cdot$

$\quad - \frac{1}{2}e^{t-2}u(t-2) - \frac{1}{6}e^{-(t-2)}u(t-2)$

13. $y = \left(\frac{1}{4} + \frac{t}{2} + \frac{t^2}{4}\right)e^{-t} - \frac{\cos t + \sin t}{4}$

15. $x = \frac{1}{6}(-28 + 20t - 3t^2) + \frac{1}{3}(14 + 4t)e^{-t}$

$y = \frac{1}{6}(-6 + 8t - 3t^2) + \frac{2}{3}(3 + t)e^{-t}$

19. $(t/4) \sin 2t$; yes

Chapter 6, Sec. 6.8

1. $\frac{s}{(s^2+1)^2}$ **3.** $\frac{1}{s(s^2+1)}$

5. $\frac{1}{(s+2)(s^2+1)}$ **7.** $\frac{\sin 2t}{16} - \frac{t \cos 2t}{8}$

9. $\delta(t) - 2e^{-2t}$ **11.** $\frac{1}{6}(3t \cos 3t + \sin 3t)e^{-2t}$

13. $Y = \int_0^t (t - \lambda)e^{-a(t-\lambda)}f(\lambda)\, d\lambda = \int_0^t \lambda e^{-a\lambda}f(t-\lambda)\, d\lambda$

19. se^{-st_0}

21. $k = \frac{3}{4a^3}, \frac{3}{2(as)^3}[(as - 1) + (as + 1)e^{-2as}], 1$

29. (a) $\begin{cases} 0 & t < a \\ f(t-a) & 0 \le a \le t \end{cases}$ (b) $\begin{cases} 0 & t < a \\ \int_a^t f(t-\lambda)\, d\lambda & 0 \le a \le t \end{cases}$

(c) $\frac{m!\,n!}{(m+n+1)!}t^{m+n+1}$

31. It depends on what kinds of functions are acceptable as solutions. Since $f * x = g \to \mathscr{L}\{x\} = \mathscr{L}\{g\}/\mathscr{L}\{f\}$, it is possible that $\mathscr{L}\{x\}$ may not be the transform of a "respectable" function. This will be the case, for example, if $f = g$.

35. $x = -\frac{2}{3} \sin t + \frac{4}{3} \sin 2t$

37. $x = \frac{t}{2} + \frac{\sin \sqrt{2}\,t}{2\sqrt{2}}$

39. $A(t) = \frac{1}{2} - e^{-t} + \frac{1}{2}e^{-2t}$, $h(t) = e^{-t} - e^{-2t}$

Chapter 7, Sec. 7.2

1. $\sqrt{k/I}$

3. $1/\sqrt{LC}$

Chapter 7, Sec. 7.3

5. The first integer equal to or greater than $\dfrac{\sqrt{1-(c/c_c)^2}\ \ln 2}{2\pi(c/c_c)}$
7. $Y_{ss} \doteq 0.91 \sin (15t - 0.85)$
9. $y = e^{-6t}(-2 \cos 8t - \frac{3}{2} \sin 8t) + 2$
11. $y = -e^{-6t}(\cos 8t + \frac{1}{2} \sin 8t)$
13. $y = -e^{-10t}(1 + 8t)$

23. $$x = \left(x_0 - \frac{\mu w}{k}\right) \cos \omega_n t + \frac{\mu w}{k} \qquad 0 \le t \le \frac{\pi}{\omega_n}$$
$$= -\left(x_0 - \frac{3\mu w}{k}\right) \cos \omega_n\left(t - \frac{\pi}{\omega_n}\right) - \frac{\mu w}{k} \qquad \frac{\pi}{\omega_n} \le t \le \frac{2\pi}{\omega_n}$$
$$= \left(x_0 - \frac{5\mu w}{k}\right) \cos \omega_n\left(t - \frac{2\pi}{\omega_n}\right) + \frac{\mu w}{k} \qquad \frac{2\pi}{\omega_n} \le t \le \frac{3\pi}{\omega_n}$$
....

During each cycle the amplitude decreases by the same amount, $4\mu w/k$. The particle will come to rest when the maximum amplitude is such that the spring force in that position is, for the first time, equal to or less than the static frictional force μw. Incidentally, the period of the motion with Coulomb friction is the same as the period without any damping.

25. The amplitude of the steady-state response varies between the values
$$\frac{F_1}{\omega^2 - \omega_1^2} - \frac{F_2}{\omega^2 - \omega_2^2} \quad \text{and} \quad \frac{F_1}{\omega^2 - \omega_1^2} + \frac{F_2}{\omega^2 - \omega_2^2}$$

Chapter 7, Sec. 7.4

1. $i = \frac{17}{100}e^{-500t} \sin 300t$
3. $i = \frac{1}{15}e^{-20{,}000t} - \frac{1}{15}e^{-5{,}000t}$
5. $i = -\frac{1}{8}e^{-6{,}000t} \sin 8{,}000t$
7. $i = \frac{625}{4}te^{-2{,}500t}$

Chapter 7, Sec. 7.5

1. $\omega_1 = 1,\ (a_1 : a_2 = 1 : 2);\ \omega_2 = 2,\ (a_1 : a_2 = 1 : -1)$
   $y_1 = \frac{2}{3} \sin t - \frac{1}{3} \sin 2t,\ y_2 = \frac{4}{3} \sin t + \frac{1}{3} \sin 2t$
3. $y_1 = \frac{5}{2} \sin t - 2 \sin 2t + \frac{1}{2} \sin 3t$
   $y_2 = 5 \sin t + 2 \sin 2t - 3 \sin 3t$

5. $$Y_1 = \frac{F_0}{2(\omega^2 - 1)(\omega^2 - 4)} \sin \omega t$$
$$Y_2 = \frac{(3 - \omega^2)F_0}{2(\omega^2 - 1)(\omega^2 - 4)} \sin \omega t$$

7. $\omega_1 = \sqrt{3}$, $(a_1 : a_2 = 3 : 1)$; $\omega_2 = \sqrt{13}$, $(a_1 : a_2 = 1 : -3)$

9. $y_1 = \dfrac{\sqrt{3}}{5} \sin \sqrt{3}\,t + \dfrac{2\sqrt{13}}{65} \sin \sqrt{13}\,t$

$y_2 = \dfrac{\sqrt{3}}{15} \sin \sqrt{3}\,t - \dfrac{6\sqrt{13}}{65} \sin \sqrt{13}\,t$

11. $y_1 = \cos 9t + \cos 11t$, $y_2 = \cos 9t - \cos 11t$. Since these can be rewritten in the form $y_1 = 2 \cos t \cos 10t$ and $y_2 = 2 \sin t \sin 10t$, it is clear that both y_1 and y_2 appear to vary with frequency $10[= (9 + 11)/2]$ with the slowly varying amplitudes $2 \cos t$ and $2 \sin t$, respectively. Thus the system exhibits the phenomenon of beats.

13. $\omega_1 = 1/(3\sqrt{LC},)$ $\omega_2 = 1/(12\sqrt{LC})$

15. $i_1 = \dfrac{E}{R}\left(-\dfrac{9}{17}e^{-3Rt/(5L)} - \dfrac{8}{17}e^{-4Rt/L} + 1\right)$

$i_2 = \dfrac{E}{R}\left(\dfrac{6}{17}e^{-3Rt/(5L)} - \dfrac{6}{17}e^{-4Rt/L}\right)$

17. $\omega_1 = 1/\sqrt{2}$, $a_1 : a_2 : a_3 = 1 : 2 : 2$; $\omega_2 = 1$, $a_1 : a_2 : a_3 = 1 : 0 : -2$; $\omega_3 = \sqrt{2}$, $a_1 : a_2 : a_3 = 1 : -4 : 2$

Chapter 7, Sec. 7.6

3. $$\frac{d^2X_1}{dT^2} + \frac{c}{m_1 v_1}\frac{dX_1}{dT} + \frac{k_1 + k_2 + k_3}{m_1 v_1^2}X_1 - \frac{k_3 s_2}{m_1 s_1 v_1^2}X_2 = 0$$

$$-\frac{k_3 s_1}{m_2 s_2 v_1^2}X_1 + \frac{d^2X_2}{dT^2} + \frac{k_3}{m_2 v_1^2}X_2 = 0$$

5. $$\frac{d^2X_1}{dT^2} + \frac{c_1}{m_1 v_1}\frac{dX_1}{dT} + \frac{k_1 + k_2 + k_3}{m_1 v_1^2}X_1 - \frac{c_1 s_2}{m_1 s_1 v_1}\frac{dX_2}{dT} - \frac{(k_2 + k_3)s_2}{m_1 s_1 v_1^2}X_2 = 0$$

$$-\frac{c_1 s_1}{m_2 s_2 v_1}\frac{dX_1}{dT} - \frac{(k_2 + k_3)s_1}{m_2 s_2 v_1^2}X_1 + \frac{d^2X_2}{dT^2} + \frac{(c_1 + c_2)}{m_2 v_1}\frac{dX_2}{dT} + \frac{k_2 + k_3}{m_2 v_1^2}X_2 = 0$$

7.

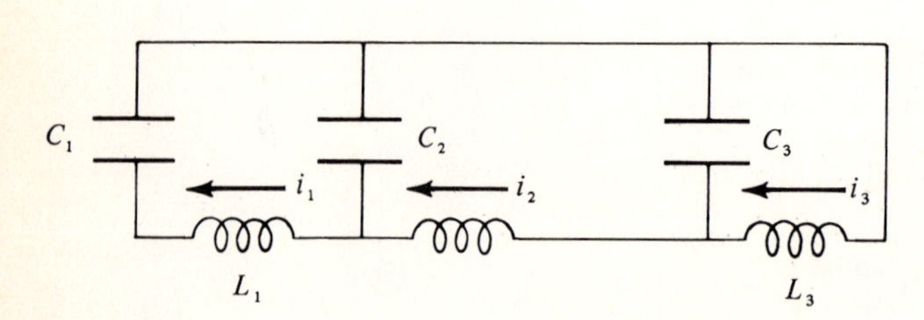

Problem 7.6-7

9. $$\frac{d^2X_1}{dT^2} + \frac{k_1 + k_2}{m_1 v_1^2} X_1 - \frac{k_2 s_2}{m_1 s_1 v_1^2} X_2 = 0$$

$$-\frac{k_2 s_1}{m_2 s_2 v_1^2} X_1 + \frac{d^2X_2}{dT^2} + \frac{k_2 + k_3}{m_2 v_1^2} X_2 - \frac{k_3 s_3}{m_2 s_2 v_1^2} X_3 = 0$$

$$-\frac{k_3 s_2}{m_3 s_3 v_1^2} X_2 + \frac{d^2X_3}{dT^2} + \frac{k_3}{m_3 v_1^2} X_3 = 0$$

$$\frac{d^2X_1}{dT^2} + \left(\frac{1}{C_1} + \frac{1}{C_2}\right)\frac{1}{L_1 v_2^2} X_1 - \frac{q_2}{C_2 q_1 L_1 v_2^2} X_2 = 0$$

$$-\frac{q_1}{C_2 q_2 L_2 v_2^2} X_1 + \frac{d^2X_2}{dT^2} + \left(\frac{1}{C_2} + \frac{1}{C_3}\right)\frac{1}{L_2 v_2^2} X_2 - \frac{q_3}{C_3 q_2 L_2 v_2^2} X_3 = 0$$

$$-\frac{q_2}{C_3 q_3 L_3 v_2^2} X_2 + \frac{d^2X_3}{dT^2} + \frac{1}{C_3 L_3 v_2^2} X_3 = 0$$

Chapter 8, Sec. 8.2

1. $f(t) = \frac{1}{2} + \frac{2}{\pi}\left(\frac{\sin \pi t}{1} + \frac{\sin 3\pi t}{3} + \frac{\sin 5\pi t}{5} + \cdots\right)$

3. $a_n \equiv 0,\ b_n = \begin{cases} -\dfrac{2}{n\pi} & n = 1, 3, 5, 7, \ldots \\ -\dfrac{4}{n\pi} & n = 2, 6, 10, 14, \ldots \\ 0 & n = 4, 8, 12, 16, \ldots \end{cases}$

5. $f(t) = \frac{1}{2} - \frac{1}{\pi}\left(\frac{\sin 2\pi t}{1} + \frac{\sin 4\pi t}{2} + \frac{\sin 6\pi t}{3} + \cdots\right)$

7. $a_0 = 2;\ a_n \equiv 0,\ n \neq 0$

$b_n = \begin{cases} \dfrac{3}{n\pi} & n = 1, 2, 4, 5, 7, 8, \ldots \\ 0 & n = 3, 6, 9, 12, \ldots \end{cases}$

9. $a_n = \dfrac{1 - e^{-2}}{1 + n^2\pi^2},\ b_n = \dfrac{n\pi(1 - e^{-2})}{1 + n^2\pi^2}$

11. $a_0 = \dfrac{\pi}{2},\ a_n = \begin{cases} -\dfrac{2}{n^2\pi^2} & n = 1, 3, 5, \ldots; \\ 0 & n = 2, 4, 6, \ldots \end{cases} \qquad b_n = \dfrac{(-1)^{n+1}}{n}$

13. $a_n \equiv 0,\ b_n = (-1)^{n+1}\dfrac{12}{n^3\pi^3}$

Chapter 8, Sec. 8.3

3. Cosine expansion: $f(t) = 1$

Sine expansion: $f(t) = \dfrac{4}{\pi}\left(\sin\dfrac{\pi t}{2} + \dfrac{1}{3}\sin\dfrac{3\pi t}{2} + \dfrac{1}{5}\sin\dfrac{5\pi t}{2} + \cdots\right)$

5. For the cosine expansion, $b_n \equiv 0$, $a_0 = 2$ and

$$a_n = \begin{cases} -\dfrac{2}{n\pi} & n = 1, 5, 9, \ldots \\ 0 & n = 2, 4, 6, \ldots \\ \dfrac{2}{n\pi} & n = 3, 7, 11, \ldots \end{cases}$$

For the sine expansion, $b_n = \begin{cases} \dfrac{2}{n\pi} & n = 1, 3, 5, \ldots \\ -\dfrac{4}{n\pi} & n = 2, 6, 10, \ldots \\ 0 & n = 4, 8, 12, \ldots \end{cases}$

7. For the cosine expansion, $b_n \equiv 0$, $a_0 = \frac{2}{3}p^2$, and

$$a_n = (-1)^n \frac{4p^2}{n^2\pi^2} \qquad n \neq 0$$

For the sine expansion, $a_n \equiv 0$ and

$$b_n = \begin{cases} \dfrac{2p^2}{n\pi}\left(1 - \dfrac{4}{n^2\pi^2}\right) & n \text{ odd} \\ -\dfrac{2p^2}{n\pi} & n \text{ even} \end{cases}$$

9. For the cosine expansion, $b_n \equiv 0$ and

$$a_n = \begin{cases} 0 & n \text{ even} \\ \dfrac{8}{\pi(4 - n^2)} & n \text{ odd} \end{cases}$$

Clearly, $f(t)$ is its own sine expansion; that is, $a_n \equiv 0$, $b_n \equiv 0$, $n \neq 2$, and $b_2 = 1$.

11. For the cosine expansion, $b_n \equiv 0$ and

$$a_n = \frac{1 - (-1)^n \cos a\pi}{\pi} \frac{2a}{a^2 - n^2}$$

For the sine expansion, $a_n \equiv 0$ and

$$b_n = \frac{(-1)^{n+1} 2n \sin a\pi}{\pi(n^2 - a^2)}$$

15. Yes. In particular, the Fourier expansion of

$$f(t) = \begin{cases} t - t^2 & 0 \le t \le 1 \\ t - t^2 - 4t^3 - 2t^4 + t^3(t + 1)^3 g(t) & -1 \le t \le 0 \end{cases}$$

where $g(t)$ is any function possessing a continuous second derivative, will converge to $t - t^2$ for $0 \le t \le 1$ and will have coefficients decreasing as $1/n^4$.

17. Extend $f(t)$ from p to $2p$ so that $f(2p - t) = -f(t)$.

$$a_n = \begin{cases} 0 & n \text{ even} \\ \dfrac{2}{p}\displaystyle\int_0^p f(t)\cos\frac{n\pi t}{2p}\,dt & n \text{ odd} \end{cases}$$

19. (*a*) Cosine expansion: $a_n \sim 1/n^2$; sine expansion: $b_n \sim 1/n^3$
(*b*) Cosine expansion: $a_n \sim 1/n^2$; sine expansion: $b_n \sim 1/n$
(*c*) Cosine expansion: $a_n \sim 1/n^2$; sine expansion: $b_n \sim 1/n^3$
(*d*) Cosine expansion: a_n decreases faster than any fixed negative power of n since the derivatives of $f(t)$ of all orders are continuous everywhere; sine expansion: $b_n \sim 1/n$

Chapter 8, Sec. 8.4

1. $A_0 = \frac{1}{3}$, $A_n = [2/(n^2\pi^2)]\sqrt{n^2\pi^2 + 4}$

3. $A_n = \dfrac{2 \sinh 1}{\sqrt{1 + n^2\pi^2}}$

5. $c_0 = \dfrac{1}{2}$; $c_n = \dfrac{1 - e^{-ni\pi}}{2ni\pi} = \dfrac{1 - (-1)^n}{2ni\pi}$

7. $c_0 = \dfrac{1}{2}$; $c_n = \dfrac{(1 + 2ni\pi)e^{-2ni\pi} - 1}{4n^2\pi^2} = \dfrac{i}{2n\pi}$

9. $c_n = \dfrac{2(-1)^n}{\pi(1 - 4n^2)}$ **11.** $c_n = \dfrac{(-1)^n \sinh 1}{1 + n^2\pi^2}$

Chapter 8, Sec. 8.5

1. Yes

3. $i_{ss} = -\displaystyle\sum_{n=-\infty}^{\infty} \frac{2iE_0 e^{200ni\pi t}}{250n\pi + i(4n^2\pi^2 - 2{,}500)}$ n odd

5. $i_{ss} = \displaystyle\sum_{n=-\infty}^{\infty} \frac{200ne^{100ni\pi t}}{(1 - 4n^2)[100n\pi + i(40n^2\pi^2 - 1{,}000)]}$

7. $Y_1 = \dfrac{2F_0 a^4}{\pi} \displaystyle\sum_{\substack{n=1 \\ n \text{ odd}}}^{\infty} \frac{1}{n(n^2 - a^2)(n^2 - 4a^2)} \sin\frac{nt}{a}$

$Y_2 = \dfrac{2F_0 a^2}{\pi} \displaystyle\sum_{\substack{n=1 \\ n \text{ odd}}}^{\infty} \frac{3a^2 - n^2}{n(n^2 - a^2)(n^2 - 4a^2)} \sin\frac{nt}{a}$

If $a \doteq n/2$ or $a \doteq n$, where n is an odd positive integer, the system will be at approximate resonance and vibrations of large amplitude will occur.

Chapter 9, Sec. 9.1

1. $y = 1 + x + \dfrac{x^2}{2!} + \dfrac{x^3}{3!} + \cdots = e^x$

Chapter 9, Sec. 9.2

1. Because if $a_0 = 0$, we can factor out the highest power of x common to all terms in the series, leaving a series whose first term is a nonzero constant.

3. (*a*) There are no singular points.
(*b*) There are no singular points.
(*c*) $x = 0$, regular (*d*) $x = 0$, irregular
(*e*) $x = 1$, regular; $x = -1$, regular
(*f*) $x = 0$, irregular; $x = 1$, regular

5. $y_1 = 1 - \dfrac{x^2}{2} + \dfrac{x^4}{2 \cdot 4} - \dfrac{x^6}{2 \cdot 4 \cdot 6} + \cdots$

$y_2 = x - \dfrac{x^3}{3} + \dfrac{x^5}{3 \cdot 5} - \dfrac{x^7}{3 \cdot 5 \cdot 7} + \cdots$

7. $y_1 = 1 - \dfrac{1^2}{3!}x^3 + \dfrac{1^2 \cdot 4^2}{6!}x^6 - \dfrac{1^2 \cdot 4^2 \cdot 7^2}{9!}x^9 + \cdots$

$y_2 = x - \dfrac{2^2}{4!}x^4 + \dfrac{2^2 \cdot 5^2}{7!}x^7 - \dfrac{2^2 \cdot 5^2 \cdot 8^2}{10!}x^{10} + \cdots$

9. $y_1 = 1 - x^2 - x^3 - \frac{7}{12}x^4 - \frac{1}{4}x^5 - \cdots = 2e^x - e^{2x}$
$y_2 = x + \frac{3}{2}x^2 + \frac{7}{6}x^3 + \frac{5}{8}x^4 + \frac{31}{120}x^5 + \cdots = -e^x + e^{2x}$

11. $y_1 = 1 - \frac{1}{2}x^2 - \frac{1}{24}x^4 - \cdots$; $y_2 = x - \frac{1}{2}x^2 - \frac{1}{24}x^4 - \cdots$

13. $x = a$ is an ordinary point if $P(x)$, $Q(x)$, and $R(x)$ are analytic at $x = a$. $x = a$ is a regular singular point if at least one of the functions $P(x)$, $Q(x)$, $R(x)$ is not analytic at $x = a$ but each of the functions $(x - a)P(x)$, $(x - a)^2Q(x)$, $(x - a)^3R(x)$ is analytic at $x = a$. $x = a$ is an irregular singular point if at least one of the functions $(x - a)P(x)$, $(x - a)^2Q(x)$, $(x - a)^3R(x)$ is not analytic at $x = a$.

15. The point at infinity is a regular singular point.

17. The conditions $b_2 = a_1 b_1/2$ and $c_2 = b_1^2/4$ are sufficient to ensure that the given equation can be solved in terms of elementary functions.

Chapter 9, Sec. 9.3

3. $y_1 = x^{1/3} \sum_{n=0}^{\infty} (-1)^n \dfrac{x^n}{n!} = x^{1/3}e^{-x}$

$y_2 = x^{-1/3}\left[1 - (3x) + \dfrac{(3x)^2}{4} - \dfrac{(3x)^3}{4 \cdot 7} + \dfrac{(3x)^4}{4 \cdot 7 \cdot 10} - \cdots\right]$

5. $y_1 = x^{1/3} \sum_{n=0}^{\infty} (-1)^n \dfrac{x^n}{n!} = x^{1/3}e^{-x}$

$y_2 = \dfrac{1}{x}\left[1 + (3x) - \dfrac{(3x)^2}{2} + \dfrac{(3x)^3}{2 \cdot 5} - \dfrac{(3x)^4}{2 \cdot 5 \cdot 8} + \cdots\right]$

7. $y_1 = \sum_{n=0}^{\infty} (-1)^n \frac{1}{(n!)^2} \left(\frac{x}{2}\right)^{2n}$

$$y_2 = y_1 \ln |x| + \left[\left(\frac{x}{2}\right)^2 - \frac{1}{(2!)^2}\left(1 + \frac{1}{2}\right)\left(\frac{x}{2}\right)^4 + \frac{1}{(3!)^2}\left(1 + \frac{1}{2} + \frac{1}{3}\right)\left(\frac{x}{2}\right)^6 - \cdots\right]$$

9. $y_1 = x \sum_{n=0}^{\infty} (-1)^n \frac{x^n}{n!} = xe^{-x}$

$$y_2 = y_1 \ln |x| + x^2\left[1 - \frac{1}{2!}\left(1 + \frac{1}{2}\right)x + \frac{1}{3!}\left(1 + \frac{1}{2} + \frac{1}{3}\right)x^2 - \cdots\right]$$

11. $y_1 = \frac{1}{x} \sum_{n=0}^{\infty} (-1)^n \frac{x^n}{(n!)^2}$

$$y_2 = y_1 \ln |x| + 2\left[1 - \frac{1}{(2!)^2}\left(1 + \frac{1}{2}\right)x + \frac{1}{(3!)^2}\left(1 + \frac{1}{2} + \frac{1}{3}\right)x^2 - \cdots\right]$$

15. $y_1 = x^2$; $y_2 = x^2 \sum_{n=1}^{\infty} (-1)^n \frac{x^n}{n!} = x^2(e^{-x} - 1)$

17. $y_1 = x + \frac{x^4}{2 \cdot 3} + \frac{x^7}{2 \cdot 3 \cdot 5 \cdot 6} + \frac{x^{10}}{2 \cdot 3 \cdot 5 \cdot 6 \cdot 8 \cdot 9} + \cdots$

$$y_2 = x^2 + \frac{x^5}{3 \cdot 4} + \frac{x^8}{3 \cdot 4 \cdot 6 \cdot 7} + \frac{x^{11}}{3 \cdot 4 \cdot 6 \cdot 7 \cdot 9 \cdot 10} + \cdots$$

Chapter 9, Sec. 9.4

3. Yes, since J_n and J_{-n} are dependent and therefore have a vanishing wronskian.

9. $\dfrac{2}{\pi x Y_\nu(x) J_\nu(x)}$

Chapter 9, Sec. 9.5

5. $I_\nu(x)/K_\nu(x)$

Chapter 9, Sec. 9.6

3. $y = \sqrt{x}[c_1 J_0(2\sqrt{x}) + c_2 Y_0(2\sqrt{x})]$
5. $y = (1/x)[c_1 J_0(2\sqrt{x}) + c_2 Y_0(2\sqrt{x})]$
7. $y = c_1 J_0(\frac{2}{3}x^{3/2}) + c_2 Y_0(\frac{2}{3}x^{3/2})$

9. $y = \sqrt{x}\left[c_1 J_{1/(2+m)}\left(\frac{2}{2+m} x^{(2+m)/2}\right) + c_2 Y_{1/(2+m)}\left(\frac{2}{2+m} x^{(2+m)/2}\right)\right]$

11. $y = x[c_1 J_{1/3}(\frac{2}{3}x^3) + c_2 J_{-1/3}(\frac{2}{3}x^3)]$
13. $y = \sqrt{x}\, e^{-x}[c_1 J_{3/4}(\frac{1}{2}x^2) + c_2 J_{-3/4}(\frac{1}{2}x^2)]$

15. $y = c_1 I_0\left(2\frac{\sqrt{a + bx}}{|b|}\right) + c_2 K_0\left(2\frac{\sqrt{a + bx}}{|b|}\right)$

17. $y = c_1 J_0(2\sqrt{a/m^2}\, e^{mx/2}) + c_2 Y_0(2\sqrt{a/m^2}\, e^{mx/2})$. If $a < 0$, J_0 and Y_0 are to be replaced by I_0 and K_0, respectively.
21. $y = c_1 J_0(2\sqrt{3x}) + c_2 Y_0(2\sqrt{3x}) + c_3 I_0(2\sqrt{3x}) + c_4 K_0(2\sqrt{3x})$

Chapter 9, Sec. 9.7

1. $J_5(x) = \left(\frac{384}{x^4} - \frac{72}{x^2} + 1\right)J_1(x) - \left(\frac{192}{x^3} - \frac{12}{x}\right)J_0(x)$

3. $-xJ_3(2x) + 2x^2J_2(2x)$
5. $\frac{1}{3}x^3J_3(3x) + c$ **21.** $-[2/(\pi x)] \sin \nu\pi$
27. $xJ_1(x) \cos x - J_0(x)(x \sin x + \cos x) + c$
31. (a) $\frac{1}{3}\{x^2[J_0(x) \cos x + J_1(x) \sin x] + xJ_1(x) \cos x\}$
(b) $\frac{1}{3}\{x^2[J_0(x) \cos x + J_1(x) \sin x] - 2xJ_1(x) \cos x\}$
37. (a) $\lambda/(a^2 + \lambda^2)^{3/2}$ (b) $a/(a^2 + \lambda^2)^{3/2}$
39. (a) $2\sqrt{x}\, J_1(\sqrt{x}) + c$ (b) $-4J_0(\sqrt{x}) - 2\sqrt{x}\, J_1(\sqrt{x}) + c$

41. $\frac{1}{\lambda} x \ln xJ_1(\lambda x) + \frac{1}{\lambda^2} J_0(\lambda x) + c$
45. (a) $xI_1(x) + c$ (b) $x^2I_1(x) - xI_0(x) + \int I_0(x)\, dx + c$
(c) $xI_0(x) + \int I_0(x)\, dx + c$ (d) $x^2I_2(x) + c$

Chapter 9, Sec. 9.8

1. $\frac{s}{(s^2 + \lambda^2)^{3/2}}$ **3.** $\frac{\lambda^2}{\sqrt{s^2 + \lambda^2}(s + \sqrt{s^2 + \lambda^2})^2}$

7. $1/\sqrt{a^2 + \lambda^2}$ **9.** $e^{-2t}J_0(3t)$

11. $e^{-at} \int_0^x e^{ax}J_0(bx)\, dx$ **15.** $\frac{\lambda}{\sqrt{s^2 - \lambda^2}(s + \sqrt{s^2 - \lambda^2})}$

19. $\int_0^t J_0(2\lambda)J_0(t - \lambda)\, d\lambda$ **29.** $\frac{\sqrt{\lambda}\, I_1(2\alpha\sqrt{\lambda a})}{I_1(2\alpha\sqrt{a})}$

31. $y(x) = \tan\theta \left[\frac{\sqrt{x}\, I_1(2a\sqrt{x})}{aI_0(2a\sqrt{\ell})} - x\right]$ where $a^2 = \frac{F \cos\theta}{Ek}$

and k is the proportionality constant in the expression for the cross-sectional moment of inertia.

33. $x(t) = \frac{x_0\{Y_1(\lambda)J_0[\lambda(1 + \alpha t)^{1/2}] - J_1(\lambda)Y_0[\lambda(1 + \alpha t)^{1/2}]\}}{J_0(\lambda)Y_1(\lambda) - J_1(\lambda)Y_0(\lambda)}$

where $\lambda = 2k/(\alpha\sqrt{m_0})$ and k^2 is the proportionality constant in the force law.

35. $x(t) = \frac{x_0\{K_1(\lambda)I_0[\lambda(1 + \alpha t)^{1/2}] + I_1(\lambda)K_0[\lambda(1 + \alpha t)^{1/2}]\}}{I_0(\lambda)K_1(\lambda) + I_1(\lambda)K_0(\lambda)}$ where $\lambda = 2k/(\alpha\sqrt{m_0})$

Chapter 10, Sec. 10.2

1. $xz'' + (2 - x)z' + z = 0$
3. $xz'' + z' + xz = 0$; i.e., the given equation is self-adjoint.

5. $(e^{2x}y')' + 5e^{2x}y = 0$ **7.** $\left(\dfrac{y'}{\sin x}\right)' + \dfrac{y}{\sin^2 x} = 0$

9. $z = b_0 = A \cos 2x + B \sin 2x$
$b_1 = 2A \sin 2x - 2B \cos 2x$

Chapter 10, Sec. 10.3

1. By hypothesis, x_1 and x_2 are consecutive zeros of y_1. Hence y_1 cannot change sign in (x_1, x_2). Also, by the counterhypothesis, y_2 is different from zero for all x in (x_1, x_2). Hence y_2 is also of constant sign in (x_1, x_2). Finally, by multiplying y_1 or y_2, or both, by -1, if necessary, each can be made positive on (x_1, x_2).

3. Consider the equations $y'' + (1 + x)y = 0$ and $y'' + y = 0$, or consider the integrals

$$\int_1^\infty \frac{dx}{r(x)} = \int_1^\infty dx \quad \text{and} \quad \int_1^\infty q(x)\,dx = \int_1^\infty (1 + x)\,dx$$

5. Consider the equations $(xy')' - (x^2 + \nu^2)y = 0$ and $y'' = 0$.

7. The functions $y_1 = \sin x + 2 \cos x$ and $y_2 = 2 \sin x + \cos x$ are linearly independent solutions of the equation $y'' + y = 0$.

Chapter 10, Sec. 10.4

1. $$4 - x^2 = \sum_{n=1}^\infty \frac{2}{3}\left(-\frac{5}{\lambda_n} + \frac{4}{\lambda_n^3}\right)\frac{J_0(\lambda_n x)}{J_1(3\lambda_n)}$$

where λ_n is the nth one of the roots of the equation $J_0(3\lambda) = 0$.

3. $$4 - x^2 = -\sum_{n=1}^\infty \frac{4}{\lambda_n^2 J_0(2\lambda_n)} J_0(\lambda_n x)$$

where λ_n is the nth one of the roots of the equation $J_0(2\lambda) + \lambda J_1(2\lambda) = 0$.

5. $$1 = -\sum_{n=1}^\infty \frac{16}{9\lambda_n^2(9\lambda_n^2 - 4)J_2(3\lambda_n)} J_2(\lambda_n x)$$

where λ_n is the nth one of the roots of the equation $3\lambda J_1(3\lambda) - 2J_2(3\lambda) = 0$.

7. $\cos \lambda L = 0$ **9.** $\sin (\lambda \ln 2) = 0$

11. $\lambda_1 \doteq 1.29$, $\lambda_2 \doteq 2.37$

Chapter 10, Sec. 10.5

3. $$y = \begin{cases} \dfrac{Px}{2T} & 0 \le x \le \dfrac{L}{4} \\[2ex] \dfrac{P(L - 2x)}{4T} & \dfrac{L}{4} \le x \le \dfrac{L}{2} \\[2ex] -\dfrac{3P(L - 2x)}{4T} & \dfrac{L}{2} \le x \le \dfrac{3L}{4} \\[2ex] \dfrac{3P(L - x)}{2T} & \dfrac{3L}{4} \le x \le L \end{cases}$$

5. $y = \begin{cases} -\dfrac{Lx}{8T} & 0 \le x \le \dfrac{L}{2} \\ \dfrac{4x^2 - 5Lx + L^2}{8T} & \dfrac{L}{2} \le x \le L \end{cases}$

7. $G(x, s) = \begin{cases} \dfrac{(e^{-s}\cos s)(e^{-x}\sin x)}{e^{-2s}} & 0 \le x \le s \\ \dfrac{(e^{-x}\cos x)(e^{-s}\sin s)}{e^{-2s}} & s \le x \le \dfrac{\pi}{2} \end{cases}$

In this case, $G(x, s)$ is not symmetric. For the equivalent equation $e^{2x}y'' + 2e^{2x}y' + 2e^{2x}y = 0$, $G(x, s)$ is the symmetric function

$$\begin{cases} (e^{-s}\cos s)(e^{-x}\sin x) & 0 \le x \le s \\ (e^{-x}\cos x)(e^{-s}\sin s) & s \le x \le \pi/2 \end{cases}$$

9. (*a*) $G(x, s) = \begin{cases} \dfrac{\cosh k(b-s)\sinh kx}{k\cosh kb} & 0 \le x \le s \\ \dfrac{\cosh k(b-x)\sinh ks}{k\cosh kb} & s \le x \le b \end{cases}$

(*b*) $G(x, s) = \begin{cases} \dfrac{\sinh k(b-s)\cosh kx}{k\cosh kb} & 0 \le x \le s \\ \dfrac{\sinh k(b-x)\cosh ks}{k\cosh kb} & s \le x \le b \end{cases}$

(*c*) $G(x, s) = \begin{cases} \dfrac{\cosh k(b-s)\cosh k(x-a)}{k\sinh k(b-a)} & a \le x \le s \\ \dfrac{\cosh k(b-x)\cosh k(s-a)}{k\sinh k(b-a)} & s \le x \le b \end{cases}$

(*d*) $G(x, s) = \begin{cases} \dfrac{\sinh k(b-s)[\sinh k(x-a) + k\cosh k(x-a)]}{k\sinh k(b-a) + k^2\cosh k(b-a)} & a \le x \le s \\ \dfrac{\sinh k(b-x)[\sinh k(s-a) + k\cosh k(s-a)]}{k\sinh k(b-a) + k^2\cosh k(b-a)} & s \le x \le b \end{cases}$

13. $G(x, s) = \begin{cases} \dfrac{x^2(x-3s)}{6EI} & 0 \le x \le s \\ \dfrac{s^2(s-3x)}{6EI} & s \le x \le L \end{cases}$

15. $y = -\dfrac{x^5}{120EI} + \dfrac{L^2x^3}{12EI} - \dfrac{L^3x^2}{6EI}$

Chapter 11, Sec. 11.2

1. $\dfrac{c\rho}{g}\dfrac{\partial u}{\partial t} = \dfrac{\partial}{\partial x}\left(k\dfrac{\partial u}{\partial x}\right) + \dfrac{\partial}{\partial y}\left(k\dfrac{\partial u}{\partial y}\right) + \dfrac{\partial}{\partial z}\left(k\dfrac{\partial u}{\partial z}\right)$

7. (*a*) $u = -\frac{1}{5}\cos(x + 2y)$ (*b*) $u = 2e^{2x}e^{3y}$
(*c*) $u = \frac{5}{6}x^3 - \frac{1}{2}x^2y$

11. $u(x, t) = \dfrac{x \sin l - l \sin x}{\sin l} \sin t \quad l \neq n\pi$

Chapter 11, Sec. 11.3

1. $y(x, t) = \frac{1}{2}(1 - |x - at|)[u(x - at + 1) - u(x - at - 1)]$

$$+ \frac{1}{2}(1 - |x + at|)[u(x + at + 1) - u(x + at - 1)]$$

$$v(0, t) = \begin{cases} -a & 0 \le at < 1 \\ 0 & 1 < at \end{cases}$$

3. $y(x, t) = \dfrac{1}{2a}[(x + at + 1)u(x + at + 1) - (x - at + 1)u(x - at + 1)$

$$- (x + at - 1)u(x + at - 1) + (x - at - 1)u(x - at - 1)]$$

$$v(0, t) = \begin{cases} 1 & 0 \le at < 1 \\ 0 & 1 < at \end{cases}$$

7. $\theta(x) = 2axe^{-x^2}$; $\theta(x) = -2axe^{-x^2}$

Chapter 11, Sec. 11.4

5. $y(x, t) = \sum_{n=1}^{\infty} B_n \sin \dfrac{n\pi x}{l} \sin \dfrac{n\pi at}{l}$

where $$B_n = \frac{2}{n\pi a} \sin \frac{n\pi}{2} \frac{\sin [n\pi k/(2l)]}{n\pi k/(2l)}$$

As $k \to 0$, $B_n \to [2/(n\pi a)] \sin (n\pi/2)$, which we infer to be the coefficient formula for the case in which the string is set in motion by a unit impulse applied at the midpoint of the string.

Chapter 11, Sec. 11.5

3. $z_1 \doteq 2.03$, $z_2 \doteq 4.91$; $B_1 \doteq 0.73l$, $B_2 \doteq -0.15l$

5. $B_n = (-1)^n 8l/[(2n + 1)^2\pi^2]$

7. $u(x, t) = 50 - \dfrac{200}{\pi} \sum_{\substack{n=1 \\ n \text{ odd}}}^{\infty} \dfrac{1}{n} \sin \dfrac{n\pi x}{l} e^{-n^2\pi^2 t/(a^2 l^2)}$

9. $u(x, t) = \dfrac{400}{\pi} \sum_{\substack{n=1 \\ n \text{ odd}}}^{\infty} \dfrac{1}{n} \sin \dfrac{n\pi x}{l} e^{-n^2\pi^2 t/(a^2 l^2)}$

Chapter 11, Sec. 11.6

1. $u(x, y) = \dfrac{8}{\pi^3} \sum_{\substack{n=1 \\ n \text{ odd}}}^{\infty} \dfrac{1}{n^3} \dfrac{\sin n\pi x \cosh n\pi y}{\cosh n\pi}$

3. $u(x, y) = \sum_{n=1}^{\infty} B_n \sin \dfrac{2n-1}{2} \pi x \sinh \dfrac{2n-1}{2} \pi y$

where $$B_n = \frac{2}{\sinh [(2n-1)\pi/2]} \int_0^1 f(x) \sin \frac{2n-1}{2} \pi x \, dx$$

5. $u(x, y) = \frac{1}{2} A_0 x + \sum_{n=1}^{\infty} A_n \sinh n\pi x \cos n\pi y$

where $A_0 = 2 \int_0^1 f(y) \, dy$ and

$$A_n = \frac{2}{\sinh n\pi} \int_0^1 f(y) \cos n\pi y \, dy \qquad n \neq 0$$

7. $J_k[2\lambda\sqrt{(k+1)l/g}] = 0$

9. $u(r, \theta) = \dfrac{400}{\pi} \sum_{\substack{n=1 \\ n \text{ odd}}}^{\infty} \dfrac{1}{n} \left(\dfrac{r}{b}\right)^n \sin n\theta$

11. $u(r, z) = \sum_{n=1}^{\infty} B_n I_0 \left(\dfrac{n\pi r}{h}\right) \sin \dfrac{n\pi z}{h}$

where $$B_n = \frac{2}{h I_0(n\pi b/h)} \int_0^h f(z) \sin \frac{n\pi z}{h} \, dz$$

13. $u(x, y) = \dfrac{400}{\pi} \sum_{n=1}^{\infty} \dfrac{1}{2n-1} \sin (2n-1)\pi x \, e^{-(2n-1)\pi y}$

15. (*a*) $\sin \sqrt{\dfrac{\lambda}{a}} l = 0$ (*b*) $\cos \sqrt{\dfrac{\lambda}{a}} l \cosh \sqrt{\dfrac{\lambda}{a}} l = 1$

(*c*) $\cos \sqrt{\dfrac{\lambda}{a}} l \cosh \sqrt{\dfrac{\lambda}{a}} l = 1$ (*d*) $\tan \sqrt{\dfrac{\lambda}{a}} l = \tanh \sqrt{\dfrac{\lambda}{a}} l$

(*e*) $\tan \sqrt{\dfrac{\lambda}{a}} l = \tanh \sqrt{\dfrac{\lambda}{a}} l$ (*f*) $\cos \sqrt{\dfrac{\lambda}{a}} l \cosh \sqrt{\dfrac{\lambda}{a}} l = -1$

Chapter 12, Sec. 12.2

1. $y(0.4) \doteq -1.4682$ $y(0.6) \doteq -1.7379$
3. $y(0.1) \doteq -1.1045$ $y(0.2) \doteq -1.2179$ $y(0.3) \doteq -1.3391$
5. $y(1.1) \doteq 1.0048$ $y(1.2) \doteq 1.0187$ $y(1.3) \doteq 1.0408$
These values are correct to four decimal places.

7. Solving the initial-value problem $y' = -2xy$, $y = 1$ when $x = 0$, yields the values

x	e^{-x^2}
0.0	1.00000
0.1	0.99005
0.2	0.96079
0.3	0.91393
0.4	0.85214
0.5	0.77880

These values are correct to five decimal places.

9.

x	y	z
0.0	0.00000	1.00000
0.1	0.10500	0.99983
0.2	0.21994	0.99866
0.3	0.34467	0.99550

Chapter 12, Sec. 12.3

3.

x	x^3	Δ	Δ^2	Δ^3
0.0	0.000			
		0.125		
0.5	0.125		0.750	
		0.875		0.750
1.0	1.000		1.500	
		2.375		0.750
1.5	3.375		2.250	
		4.625		0.750
2.0	8.000		3.000	
		7.625		0.750
2.5	15.625		3.750	
		11.375		
3.0	27.000			

5. $x^3 - 3x^2 + 17$

7. $f'(2.5) \doteq 0.3160, f''(2.5) \doteq -0.0632$

11. $f'(x_0) \doteq \frac{1}{12h}(-25f_0 + 48f_1 - 36f_2 + 16f_3 - 3f_4)$

$$f''(x_0) \doteq \frac{1}{12h^2}(35f_0 - 104f_1 + 114f_2 - 56f_3 + 11f_4)$$

Chapter 12, Sec. 12.4

1. $y_7 \doteq 0.7362$, $z_7 \doteq 1.1226$
3. $y(0.4) \doteq 0.7407$ $y(0.5) \doteq 0.7131$ $y(0.6) \doteq 0.6977$
These values are correct to four decimal places.
5.

x	y
0.0	0.0000
0.1	0.0050
0.2	0.0200
0.3	0.0451
0.4	0.0805
0.5	0.1256

Chapter 13, Sec. 13.1

1. $(E^2 - 5E + 6)y = 0$ **3.** $(E^2 + 1)y = 3^x$
5. Each of the roots of $F(E) = 0$ is one more than the corresponding root of $f(\Delta) = 0$.

Chapter 13, Sec. 13.2

5.
$y_0 = 1$ $y_6 = 2$
$y_1 = 0$ $y_7 = 3$
$y_2 = 2$ $y_8 = 1$
$y_3 = 3$ $y_9 = 0$
$y_4 = 1$ $y_{10} = 2$
$y_5 = 0$

7. (*a*) $y = c_1 2^n + c_2 5^n$ (*b*) $y = c_1(-3)^n + c_2 n(-3)^n$

(*c*) $y = 2^{n/2}\left(A \cos \frac{3\pi n}{4} + B \sin \frac{3\pi n}{4}\right)$

(*d*) $y = c_1 2^n + c_2 3^n$ (*e*) $y = 3^n\left(A \cos \frac{n\pi}{2} + B \sin \frac{n\pi}{2}\right)$

(*f*) $y = c_1 2^n + c_2 3^n + c_3 4^n$

Chapter 13, Sec. 13.3

1. $y = c_1 5^n + c_2(-2)^n - 2^{n-2}$
3. $y = c_1 4^n + c_2 5^n + \frac{1}{2} \cdot 3^n$
5. $y = c_1 2^n + c_2(-1)^n + \frac{1}{10} 4^n$
7. $y = c_1 3^n + c_2(-2)^n - n^2 - 3n - 1$
9. $y = c_1(-1)^n + c_2 n(-1)^n + \frac{1}{4}$

11. $y = 2^n\left(c_1 \cos \frac{\pi n}{2} + c_2 \sin \frac{\pi n}{2}\right) + \frac{\cos(n-2) + 4\cos n}{17 + 8\cos 2}$

13. $y = c_1 + c_2(-1)^n + 3^n\left(c_3 \cos \dfrac{\pi n}{2} + c_4 \sin \dfrac{\pi n}{2}\right) + n$

Chapter 13, Sec. 13.4

3.

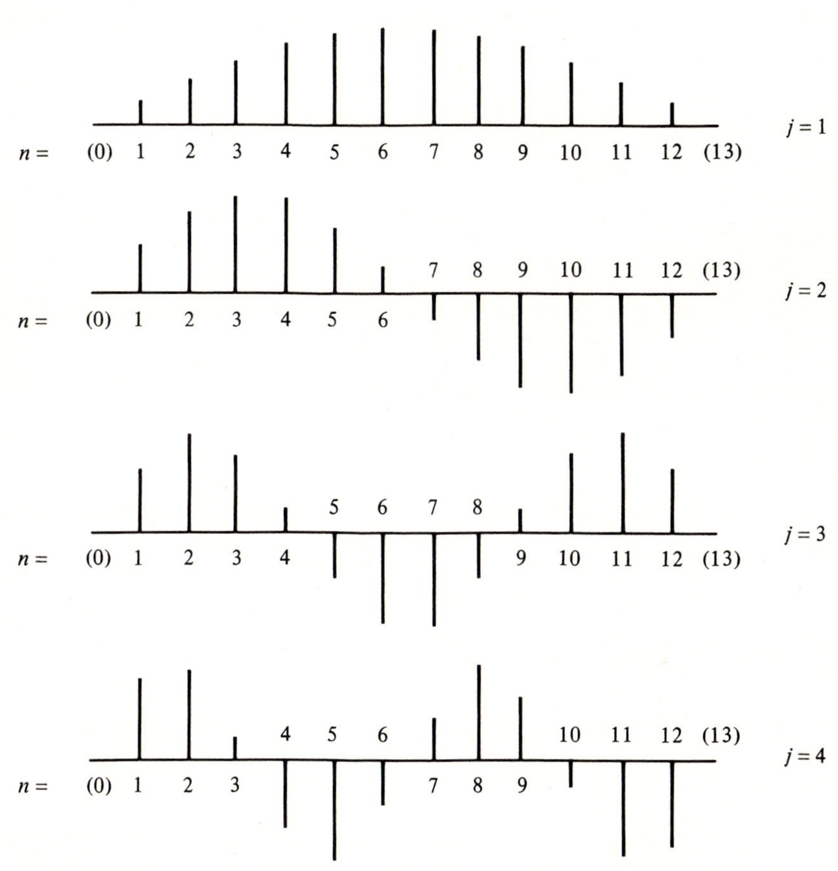

Problem 13.4-3

7. $\omega_j = 2\sqrt{\dfrac{k}{m}} \sin \dfrac{j\pi}{2N} \qquad j = 1, 2, \ldots, N$

At the natural frequency ω_j, the amplitudes with which the successive masses vibrate are proportional to $\cos\,[j\pi(2n-1)/(2N)]$, $n = 1, 2, \ldots, N$.

9. $V_n = \dfrac{V_0}{2^N + 1}(2^{N-n} + 2^n)$

11. $I_n = \dfrac{\pi \sin n\lambda}{\sin \lambda}$

13.

$$D_n = \begin{cases} \dfrac{\sinh (n+1)\mu}{\sinh \mu}; \cosh \mu = \dfrac{a}{2} & a > 2 \\ 1 + n & a = 2 \\ \dfrac{\sin (n+1)\mu}{\sin \mu}; \cos \mu = \dfrac{a}{2} & -2 < a < 2 \\ (-1)^n(1+n) & a = -2 \\ (-1)^n \dfrac{\sinh (n+1)\mu}{\sinh \mu}; \cosh \mu = -\dfrac{a}{2} & a < -2 \end{cases}$$

15. (*a*) $s_n = -\dfrac{r(r+1)}{(r-1)^3} + \left[\dfrac{n^2}{r-1} - \dfrac{2n}{(r-1)^2} + \dfrac{r+1}{(r-1)^3}\right] r^{n+1}$

(*b*) $s_n = \dfrac{\sin r + \sin rn - \sin r(n+1)}{2(1 - \cos r)}, \ r \neq 2j\pi$

(*c*) $s_n = \dfrac{-1 + \cos r + \cos rn - \cos r(n+1)}{2(1 - \cos r)}, \ r \neq 2j\pi$

Chapter 13, Sec. 13.5

1. The solution obtained by Euler's method is $y = c(1 + Ah)^{(x_n - x_0)/h}$ which approaches the true solution $y = ce^{Ax}$ for all values of A as $h \to 0$.
3. The formula is numerically stable for the equation $y' = Ay$ for all values of A.
5. The formula is numerically stable for the equation $y' = Ay$ only if $A > 0$.

Chapter 14, Sec. 14.2

1. $(-1, 2)$ **3.** $(1, 1), (1, -1), (-1, 1), (-1, -1)$
5. $x = c_1 e^t + c_2 e^{2t}, y = c_1 e^t + 2c_2 e^{2t}; x - y = k(2x - y)^2$
7. $x = c_1 \cos t + c_2 \sin t, y = -c_1 \sin t + c_2 \cos t; x^2 + y^2 = k^2$
9. $x = c_1 e^t + c_2 e^{-t}, y = c_1 e^t - c_2 e^{-t}; x^2 - y^2 = k$
11. $x = c_1 e^{-t} + c_2 e^t, y = c_2 e^t; xy - y^2 = k$
13. $x^4 + x^2 + y^2 = k$. The trajectories are closed curves in the xy plane. We cannot find parametric equations for the trajectories since the equation for x requires elliptic functions for its solution.
15. (*a*) $x = t^3 - 2t + 2, y = 3t^2 - 2$
    (*b*) $x = t^3 - 11t + 15, y = 3t^2 - 11$
    More than one trajectory of a nonautonomous system may pass through a particular point. Only one trajectory of an autonomous system can pass through a particular point.

Chapter 14, Sec. 14.3

1. Saddle point, necessarily unstable.
3. Stable node. **5.** Unstable node.

7. No, because the property of stability is not included.

Chapter 14, Sec. 14.4

1. Saddle point, necessarily unstable.
3. Saddle point, necessarily unstable.
5. Unstable spiral point.
7. Unstable node.
9. $x^2y + 3xy - 4x^2 + 2y^2 = k$
11. The critical point $(0, 0)$ is an asymptotically stable node. The critical point $(1, -\frac{1}{2})$ is a saddle point, necessarily unstable.
13. The critical point $(0, 0)$ is an unstable node. The critical point $(-1, 1)$ is a saddle point, necessarily unstable.
15. The critical point $(1, 1)$ is an asymptotically stable spiral point. The critical point $(-1, 1)$ is a saddle point, necessarily unstable.
17. In Example 2, the formula for the slope of a trajectory is

$$m_t = -\frac{r\cos\theta + r^2\sin^2\theta}{r\sin\theta} = -\frac{\cos\theta}{\sin\theta} - r\sin\theta$$
$$= m_c - r\sin\theta$$

Since $r\sin\theta$ can be either positive or negative, it cannot be concluded that $m_t > m_c$, as in Example 3.

Chapter 14, Sec. 14.5

1. Use $4x^2 + y^2$.
3. Use $-x^2 + y^2$.
5. Asymptotically stable
7. Unstable
9. Unstable
11. $\dot{V}$ is negative semidefinite for $V = x^2 + y^2$. $\dot{V}$ is negative definite for $V = 3x^2 + y^2$.

Chapter 14, Sec. 14.6

1. $r = 2,\ \theta = t + t_0$, stable
3. $r = 1,\ \theta = t + t_0$, stable; $r = 2,\ \theta = t + t_0$, unstable
5. $r = 1,\ \theta = -t + t_0$, stable
$r = 2,\ \theta = -t + t_0$, semistable
$r = 3,\ \theta = -t + t_0$, unstable

7. (*a*) $r = \dfrac{2}{\sqrt{1 + ke^{-8t}}},\ \theta = t + t_0$

(*b*) $\ln\left|\dfrac{r}{r-1}\right| - \dfrac{1}{r-1} = t + k,\ \theta = -t + t_0$

(*c*) $\dfrac{r(r-2)}{(r-1)^2} = ke^{2t},\ \theta = t + t_0$

21. If Γ is a closed trajectory defined by the particular periodic solution $x = f(t)$, $y = g(t)$, then $x = f(t + t_0)$, $y = g(t + t_0)$ are other periodic solutions defining the same closed trajectory.

Appendix A.1

1. 1.8414 **3.** 1.1462 **5.** 0.7391

7. $y_2 = 1 + x$, $y_3 = 1 + x + \frac{x^2}{2}$, $y_4 = 1 + x + \frac{x^2}{2} + \frac{x^3}{6}$

The exact solution is $y = e^x = 1 + x + \frac{x^2}{2!} + \frac{x^3}{3!} + \cdots$

9. $y_2 = y_0 + y_0^2 x$, $y_3 = y_0 + y_0^2 x + y_0^3 x^2 + y_0^4 \frac{x^3}{3}$

$y_4 = y_0 + y_0^2 x + y_0^3 x^2 + y_0^4 x^3 + \frac{2}{3} y_0^5 x^4$

The exact solution is

$$y = \frac{y_0}{1 - y_0 x} = y_0 + y_0 x + y_0^3 x^2 + y_0^4 x^3 + \cdots$$

Appendix A.2

1. If b/M is the smaller of the two quantities $(a, b/M)$, so that $h = b/M$ or $M = b/h$.

INDEX

Page numbers in *italic* indicate exercises.